Ultracold Gases and Quantum Information

École d'été de Physique des Houches in Singapore

Session XCI, 29 June–24 July 2009
École thématique du CNRS

Ultracold Gases and Quantum Information

Edited by

C. Miniatura, L-C. Kwek, M. Ducloy, B. Grémaud,
B-G. Englert, L.F. Cugliandolo, A. Ekert, and K.K. Phua

OXFORD
UNIVERSITY PRESS

OXFORD
UNIVERSITY PRESS

Great Clarendon Street, Oxford OX2 6DP

Oxford University Press is a department of the University of Oxford.
It furthers the University's objective of excellence in research, scholarship,
and education by publishing worldwide in

Oxford New York

Auckland Cape Town Dar es Salaam Hong Kong Karachi
Kuala Lumpur Madrid Melbourne Mexico City Nairobi
New Delhi Shanghai Taipei Toronto

With offices in

Argentina Austria Brazil Chile Czech Republic France Greece
Guatemala Hungary Italy Japan Poland Portugal Singapore
South Korea Switzerland Thailand Turkey Ukraine Vietnam

Oxford is a registered trade mark of Oxford University Press
in the UK and in certain other countries

Published in the United States
by Oxford University Press Inc., New York

© Oxford University Press 2011

The moral rights of the author have been asserted
Database right Oxford University Press (maker)

First published 2011

British Library Cataloguing in Publication Data

Data available

Library of Congress Cataloging in Publication Data

Data available

Typeset by SPI Publisher Services, Pondicherry, India
Printed in Great Britain
on acid-free paper by
CPI Antony Rowe, Chippenham, Wiltshire

ISBN 978–0–19–960365–7

1 3 5 7 9 10 8 6 4 2

École de Physique des Houches
Service inter-universitaire commun à l'Université Joseph Fourier de Grenoble et à
l'Institut National Polytechnique de Grenoble

Subventionné par l'Université Joseph Fourier de Grenoble,
le Centre National de la Recherche Scientifique, le
Commissariat à l'Énergie Atomique

Directeur:
Leticia F. Cugliandolo, Université Pierre et Marie Curie – Paris VI, France

Directeurs scientifiques de la session XCI:
Artur Ekert, Centre for Quantum Technologies, National University of Singapore.
Kok Khoo Phua, Institute of Advanced Studies, Nanyang Technological University,
Singapore.

Previous sessions

Publishers

- Session VIII: Dunod, Wiley, Methuen
- Sessions IX and X: Herman, Wiley
- Session XI: Gordon and Breach, Presses Universitaires
- Sessions XII–XXV: Gordon and Breach
- Sessions XXVI–LXVIII: North Holland
- Session LXIX–LXXVIII: EDP Sciences, Springer
- Session LXXIX–LXXXVIII: Elsevier
- Session LXXXIX– : Oxford University Press

Organizers

MINIATURA Christian, CNRS and Université de Nice Sophia, France; and National University of Singapore

KWEK Leong-Chuan, National University of Singapore and Nanyang Technological University, Singapore

DUCLOY Martial, CNRS and Université Paris 13, France

GRÉMAUD Benoît, CNRS and Université Pierre et Marie Curie, Paris VI, France; and National University of Singapore

ENGLERT Berthold-Georg, National University of Singapore

CUGLIANDOLO Leticia, Université Pierre et Marie Curie, Paris VI, France

Preface

This volume contains the lecture notes of the courses that were given on ultracold gases and quantum information during the XCI session of the École de Physique des Houches. This extraordinary session was organized in Singapore from June 29th to July 24th 2009 at the Nanyang Technological University (NTU). The topics covered were degenerate quantum gases (bosons and fermions), weak and strong localization phenomena, quantum phase transitions, quantum Hall effects, Fermi and Lüttinger liquids, quantum information, quantum computing and entanglement, quantum cryptography, and quantum information processing using ions, atoms, and optical devices.

P.1 Why Singapore?

It all began over a cup of coffee. Sometime in April 2007, a gang of four (Christian Miniatura, Leong-Chuan Kwek, Martial Ducloy and Berge Englert) gathered at the Spinelli Coffee outlet at the University Hall of the National University of Singapore (NUS) and came up with the idea of a Les Houches summer school session that would be organized in Singapore.

To many physicists around the world, the "École des Houches" is synonymous with the best advanced physics education that a young physicist could get. Since its establishment in 1951 by Cécile DeWitt-Morette, the École des Houches has maintained a high standard in the organization of physics summer schools, combining in-depth courses in the most advanced fields with appropriate pedagogy accessible to beginners in the field. The school has trained generations of high-level scientists, some of whom have since become Nobel prize-winners after their stint at the Les Houches summer schools, either as students or lecturers. With the fast-growing scientific and economic development of Asian countries, we felt that such a session outside France would significantly increase the visibility of the Les Houches school in East Asia and encourage greater participation from physics students from this part of the world. The hope was also to "export" the excellent Les Houches summer school structure to East Asia and to strengthen more scientific and academic links between France and Europe and East Asian countries. There were several reasons for proposing Singapore for this first Asian school: its central location in Asia; the large investments made by the Singapore Government in higher education as well as in science and technology research; the history of collaboration between Singapore and France, and in particular between the Centre for Quantum Technologies (CQT) at NUS and the Institut National de Physique (INP) du Centre National de la Recherche Scientifique (CNRS). It was in this spirit that we finally submitted our proposal for a Singapore school of Physics to the Les Houches Executive Committee.

After considerable deliberation, the Les Houches Summer School Executive Committee decided to accept this unique project and Singapore was chosen as the site for the first Les Houches summer school session ever organized outside of France since 1951. It bears the number XCI (91) in the long history of the Les Houches summer school sessions. Within Singapore, NTU was chosen as the most suitable location since it could offer good facilities with a sufficiently remote location, almost completely isolated from town. All lecturers and participants stayed at the Nanyang Executive Centre (NEC) as it is divided into two wings, one where the lectures were held and another one with hotel rooms for the lecturers and participants.

P.2 Our sponsors

This Singapore Les Houches session was directed by the Les Houches Summer School, in collaboration with the Institute of Advanced Studies (IAS) at NTU and the CQT at NUS and organized by the gang of four joined by Benoît Grémaud.

The organization of the school would not have been possible without generous financial sponsoring. We are particularly thankful to the Nanyang Technological University, the National University of Singapore, the Institute for Advanced Studies at NTU, The Centre for Quantum Technologies at NUS, the French Embassy in Singapore (Merlion programme), the Institut National de Physique du CNRS (PICS grant 4159) and the University of Nice Sophia (BQR funding).

P.3 Primary objective and scientific themes of the Singapore session

The primary objective of the Les Houches session in Singapore was to provide to the best students within the Asia–Pacific region an opportunity to attend top-level courses as typically provided by the Les Houches school. Indeed, it is a documented fact that Asian students seldom attend the sessions in France due to the distance and the lack of funding. We have the hope that the organization of this summer course will enhance closer scientific and technological cooperation between Asian and European research centers. In particular, we hope that Asian students will be keener to consider European and French universities and laboratories for their future studies.

With a bird's eye perspective, it seemed obvious to us that (at least) two fields in physics had seen an explosive growth in the past ten years, namely quantum gases and quantum information. Indeed, in 1994, Peter Shor proved that a quantum computer could, in principle, factor very large numbers into their prime factors much more efficiently than a conventional silicon-chip computer. An experimental realization of a quantum computer became a Holy Grail, though a formidable challenge. Almost concomitantly, in 1995, the first gaseous Bose–Einstein condensates were produced in labs, and a few years later degenerate Fermi gases, revolutionizing the field of atomic physics by letting it tread the condensed-matter turf. Both fields bubbled over with ideas and realizations: nuclear magnetic resonance based quantum bits (qubits), ion trap architecture, superconducting qubits, Mott–superfluid transition, BEC-BCS crossover and so forth. This quantum realm has reached a stage where the current developments are now addressing technological aspects with the potential to impact deeply on our everyday lives, bringing with it important industrial, economic, as well as societal stakes. Many different scientific communities are involved and we anticipate that the synergy between them will spark many new ideas in science in the very near future. It is therefore not surprising to witness a growth in quantum science activities in the most developed Asian countries, and in Singapore in particular. We felt that Singapore, being a hub of education and research as well as a melting pot where Asian and Western cultures merge harmoniously, would offer an appropriate and fertile place to welcome students from all the surrounding countries and have them benefit from a school devoted to ultracold gases and quantum information.

However, the topics in both quantum degenerate gases and quantum information sciences are so vast that it was simply impossible to provide an exhaustive and comprehensive view of these subjects in one single school. We had to make a selection, sometimes driven by the haphazard acceptance or refusal of our invitations. The set of lectures that we have chosen are detailed in Section P.5 of this preface.

P.4   Demographics

When the applications for the school closed in February 2009, there were more than 110 postdoctoral fellows, graduates, and PhD students who had applied for the Singapore session. More than half of the applicants were from Asian countries. A total of 64 participants were eventually admitted to the school and we were pleased to welcome 19 female students. This ratio of one third is a nice indication that women in Asia do not hesitate to embrace a scientific path. Most of the selected students were graduate students rather than postdoctoral fellows (five only). In terms of demographics, there were 21 different nationalities at the school (see Figure P.1). These participants came from the universities and colleges of 16 different countries (see Figure P.3), with more than half of the participants (35) coming from Asian countries. 18 of the participants came from Singapore (NTU and NUS). Part of the reason for this strong representation from Singapore was due to the intensive research in cold atoms and quantum information in the country. It is important to note that amongst the 18 participants from Singapore, there were about six different nationalities, reflecting the cosmopolitan nature of the Singapore education system. There were also 23 participants from Europe with the majority (eight) from France.

As a net result, the school did indeed successfully increase the number of participants from Asian countries compared to sessions in France. However, what was particularly important was the active integration of postdoctoral fellows and students from French and other European universities with students from Asian countries.

P.5   Lecturers, courses, and content of this book

There were two principal themes for this school: ultracold atomic gases and quantum information. The session was organized around four fundamental courses supplemented by seven topical courses. To ensure a high level of pedagogical lectures, the lecturers

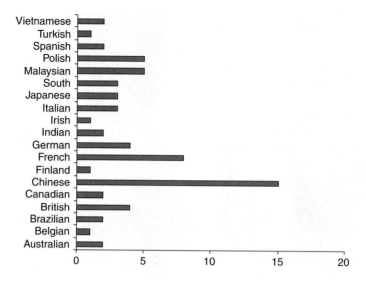

Fig. P.1 Bar chart showing the distribution of nationalities among the participants.

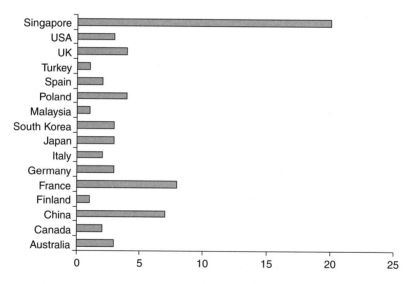

Fig. P.2 The bar chart shows the distribution of participants from different countries. NTU and NUS contributed 18 students to the total number of participants.

for the school were carefully selected for their expertise and their ability to deliver clear and succinct lectures at a graduate level. A total of 86 hours of lectures were delivered over four weeks, in addition to the special talks given by distinguished guests such as Anthony Leggett (2003 physics Nobel Prize winner), Frédéric Chevy (LKB, Paris), and Xing Zhizhong (IHEP, China).

Fig. P.3 Some participants chatting with Tony Leggett during the school.

The volume first starts with fundamental aspects of degenerate Bose and Fermi gases on the one hand and on foundational aspects of quantum theory and quantum information processing on the other hand. David Guéry-Odelin (LCAR, Toulouse, France) details the basic theory behind Bose–Einstein condensation, from the ideal gas to bimodal condensates, through mean-field theory and beyond. This introduction is then followed by an exposition by Patrizia Vignolo (INLN, Valbonne, France) on the fascinating realm of degenerate Fermi gases and the link between two worlds offered by the BEC–BCS crossover. To cover fundamental concepts in quantum theory, Valerio Scarani (CQT, NUS, Singapore) speaks about some of the intriguing aspects of quantum correlations, ranging from quantum cloning to quantum teleportation and ending with the power of Bell. This material is followed by lecture notes by Dagmar Bruss (Institut für Theoretische Physik, Düsseldorf, Germany) and Chiara Macchiavello (Istituto Nazionale di Fisicadella Materia, Pavia, Italy) on quantum networks, quantum algorithms, quantum error correction, and the one-way computing paradigm.

The second part of the book focuses on specific topics in both fields. Jürgen Eschner (ICFO, Barcelona, Spain) reviews the basic experimental techniques with trapped ions and explains how quantum bit encoding, logic gates, and quantum computation processing can be performed. Mark Goerbig (LPS, Orsay, France) presents a comprehensive set of lectures on the quantum Hall effects. In recent years there have been proposals to implement topological codes on solid-state systems exhibiting quantum Hall effects. Mark's lecture notes therefore serve as a gentle introduction to those who would like to delve deeper into the subject. In the next lecture, George Batrouni (INLN, Valbonne, France) explains quantum phase transitions, i.e. the radical change that occurs in the topology of the ground state of a many-body system at zero temperature when the parameters of the system are varied. This is followed by Thierry Giamarchi (DPMC, Geneva, Switzerland) who presents fundamental tools for studying one-dimensional quantum fluids, a world where the Fermi-liquid description is doomed to fail and where collective-excitation physics is the rule. Then, Cord Müller (Physikalisches Institut, Bayreuth, Germany) and Dominique Delande (LKB, Paris, France) introduce the physics of weak and strong localization, i.e. the subtle interplay between disorder and interference that is observed when a wave propagates in a disordered medium and suffers many scattering events. These interference corrections to transport can lead, under suitable circumstances, to a subtle metal–insulator disorder-induced transition. Quantum cryptography has been deciphered by Norbert Lütkenhaus (Institute for Quantum Computing, Waterloo, Canada).[1] Finally, Christian Kurtisefer and Antía Lamas Linares (both at CQT, NUS, Singapore) close the volume by presenting how quantum information processes can be performed using quantum optical devices such as one-photon sources.

P.6   Entertainment and social events

The summer school was not all work and no play. The official opening for the school on Monday, June 29 2009 (first day of the school) was attended by the French Ambassador to Singapore, His Excellency Pierre Buhler. One of the enjoyable moments during the school was the celebration of the French National Day on July 14 with wine and cheese tasting, a primer for most of the Asian students.

Participants at the school had access to many sports facilities, including the swimming pool nearby at NTU, and enjoyed several sports events like basketball and football matches. The Staff Club next to NEC also provided an interesting venue at night where the participants could relax, mingle, and enjoy chats, guitar performances, and good laughs as well as play darts or billiards. Many students loved the place and it was not uncommon for the lecturers and students to exchange ideas over a mug of beer or a glass of wine at the club. During the weekends the students were free to arrange their own activities. However, special activities, such as trips to the Sungei Buloh natural reserve or Pulau Ubin island, were organized every Saturday for all participants. During these weekends, the more adventurous participants visited neighboring countries such as Malaysia, Indonesia, and Thailand. The students, especially those coming from Europe, were deeply impressed by what they saw during these getaways and shared their adventures with other participants at the school. We believe that it was a unique

---

[1] We regret that the quantum cryptography lecture notes are unfortunately missing from the book.

Fig. P.4 Opening ceremony on 29 June 2009. First row, from left to right: Dr Martial Ducloy (co-organizor of the Les Houches session in S'pore), Prof. Leticia Cugliandolo (Director of the Les Houches school), Prof. K-K Phua (Director of IAS), His Excellency Pierre Buhler (French Ambassador to Singapore) and Dr Guaning Su (President of NTU). Second row, from left to right: Marc Piton (French Counsellor for Culture, Science and Education), Walid Benzarti (French Attaché for Science and Higher Education), Tony Mayer, Monique van Donzel.

experience that they will not forget. It was surely an attractive flavor of the Singapore session. There were also other special "outings," including a visit to CQT and tours of the research laboratories at the School of Physical and Mathematical Sciences at NTU.

P.7 Acknowledgments

This is always the most difficult part, as the fear of forgetting someone is always lurking in the back of your mind...

First, we would like to express our warmest gratitude to all the colleagues who believed in us and in this "crazy project." In this respect we would particularly like to thank Antoine Mynard, from the French Embassy in Singapore, for his enthusiastic and effective constant support from the very start. In France, we benefited enormously from the decisive help of Christian Chardonnet and Patricio Leboeuf (INP, CNRS). Making this project fly would have been virtually impossible without them.

We are deeply thankful to the Les Houches Scientific Advisory Board for their bold decision, which allowed this extraordinary session to happen, the first one ever organized outside the small village in the French Alps, and which made this adventure possible. It is our great pleasure to acknowledge the tremendous help received from the staff at the French Embassy in Singapore, particularly from his Excellency Pierre Buhler (French Ambassador to Singapore), Olivier Guyonvarch (French Deputy Chief of Mission), Marc Piton (French Counsellor for Culture, Science and Education), and Walid Benzarti (French Attaché for Science and Higher Education).

Special thanks go to our sponsors whose generous fundings allowed us to organize a session with such high quality standards.

The hidden side of the iceberg is always the largest ... We thank all the staff at CQT and IAS for their kind and immense help. We particularly wish to thank Professor Choy Sin Hew, Dr Jean Yong Wan Hong, Ms Chris Ong, Ms Corinne Lam, and Ms Ang Wuan Suan from IAS for their logistic and secretarial support and advice in the preparation and running of the session, and Ms Evon Tan, Ms Chan Chui Theng, and Mr Darwin Gosal from CQT who helped us organize some of the social activities. At this point we would like to point out that some social events benefited greatly from the Singapore branch of the French company Carrefour who generously donated a large number of bottles of French wine.

Special thanks to Toh-Miang Ng, our wizard webmaster, who crafted an efficient site for the conference. Special thoughts to Nathalie Hamel (INLN) for organizing the travel plans of the French participants and to Jorge Tredicce (INLN), a Singapore-lover, for his friendly and humorous support.

The daily organization of the school would not have been so smooth without the wonderful and dedicated assistance given by many student helpers: Ang Zhongzhi, Hu Jiazhong, Ni Xiaotong, Setiawan, Wu Peisan, Anton Peshkov, and Yenny Widjaya. They have provided relentless support for all the logistics during the school. Also, we would like to thank members of the Center for Excellence in Learning and Teaching (CELT), particularly Professor Daniel Tan, Goh Wee Sen, Andrew Chua, Paul Ang, Mohd Nazli bin Johar, and Ong Day Cheng for their help in various forms of technical support for the school.

Finally, we would like to thank all the lecturers for their dedication, the excellent quality of their performances, and the youthful spirit they showed when mingling with the students. And, above all, thank you to all the participants who created a magic, incredible, and unique atmosphere blended with different, amazing, and cross-fertilizing cultures. Thank you for the wonderful pictures that you have all taken. Thank you Thierry for your magic card tricks that left us all stunned and filled with wonder, thank you Cord and Dominique for setting up a CBS experiment on the spot, thank you Valerio and Han Rui for organizing football and basketball matches, thank you Jean-François for playing rock songs on your guitar – we almost broke chairs – and thank you Bess and Gabriel for your wonderful cello concerts!

Thank you all, we hope we met your expectations.
We'll miss every wonderful moment of this unique adventure!

Singapore, April 2010

Christian Miniatura

Leticia Cugliandolo

Leong Chuan Kwek

Artur Ekert

Martial Ducloy

Kok Khoo Phua

Benoît Grémaud

Berthold-Georg Englert

Contents

List of participants

DIRECTORS

CUGLIANDOLO LETICIA
École de Physique des Houches, Côte des Chavants, 74310 Les Houches, France.

EKERT ARTUR
Centre for Quantum Technologies, National University of Singapore, S15 # 03-18, 3 Science Drive 2, Singapore 117543, Singapore.

PHUA KOK KHOO
Institute of Advanced Studies, Nanyang Technological University, Nanyang Executive Centre # 02-18, 60 Nanyang View, Singapore 639673 Singapore.

ORGANIZERS

MINIATURA CHRISTIAN
Institut Non Linéaire de Nice, UMR 6618, Université de Nice Sophia, CNRS, 1361, route des Lucioles, 06560 Valbonne, France ; Centre for Quantum Technologies and Physics Department, National University of Singapore S15 #03-18, 3 Science Drive 2, Singapore 117543, Singapore.

KWEK LEONG CHUAN
Centre for Quantum Technologies, National University of Singapore S15 #03-18, 3 Science Drive 2, Singapore 117543, Singapore; Institute of Advanced Studies, Nanyang Technological University, #02-18 60 Nanyang View, Singapore 639673, Singapore

DUCLOY MARTIAL
Laboratoire de Physique des Lasers, UMR 7538, Université Paris 13, Institut Galilée, CNRS, 99, avenue Jean-Baptiste Clément, F-93430 Villetaneuse, France.

GRÉMAUD BENOÎT
Laboratoire Kastler Brossel, UMR 8552, Université Paris 6, CNRS, ENS, 4, Place Jussieu, F-75252 Paris Cedex 05, France ; Centre for Quantum Technologies and Physics Department, National University of Singapore S15 #03-18, 3 Science Drive 2, Singapore 117543, Singapore.

ENGLERT BERTHOLD-GEORG
Centre for Quantum Technologies and Physics Department, National University of Singapore S15 #03-18, 3 Science Drive 2, Singapore 117543, Singapore.

LECTURERS

BATROUNI GEORGE
Institut Non Linéaire de Nice, Université of Nice Sophia, CNRS, 1361, route des Lucioles, 06560 Valbonne, France.

BRUSS DAGMAR
Institut für Theoretische Physik III, Heinrich-Heine-Universität Düsseldorf, Universitätsstr. 1, Geb. 25.32, D-40225 Düsseldorf, Germany.

DELANDE DOMINIQUE
Laboratoire Kastler-Brossel, Université Paris 6, CNRS, ENS, 4 Place Jussieu, F-75005 Paris, France

ESCHNER JÜRGEN
ICFO - Institut de Ciències Fotòniques, Parc Mediterrani de la Tecnologia, Av. del Canal Olimpic s/n, 08860 Castelldefels, Barcelona, Spain.

GIAMARCHI THIERRY
DPMC-MaNEP, University of Geneva, 24, quai Ernest-Ansermet, CH1211 Geneva 4, Switzerland.

GOERBIG MARK OLIVER
Laboratoire de Physique des Solides, CNRS, Université Paris Sud, Bât. 510, F-91405 Orsay cedex, France.

GUÉRY-ODELIN DAVID
Université Paul Sabatier - Toulouse 3, Laboratoire de Collisions - Agrégats - Réactivité, 118 Route de Narbonne, Bât. 3R1b4, F-31062 Toulouse Cedex 9, France.

KURTSIEFER CHRISTIAN
Centre for Quantum Technologies and Physics Department, National University of Singapore, 3 Science Drive 2, Singapore 117543, Singapore.

LAMAS LINARES ANTÍA
Centre for Quantum Technologies and Physics Department, National University of Singapore, 3 Science Drive 2, Singapore 117543, Singapore.

LÜTKENHAUS NORBERT
Institute for Quantum Computing and Department of Physics and Astronomy, University of Waterloo, 200 University Avenue West, Waterloo, Ontario, Canada N2L 3G1, Canada.

MACCHIAVELLO CHIARA
Dipartimento di Fisica A. Volta, Via Bassi 6, 27100 Pavia, Italy.

MÜLLER CORD
Physikalisches Institut, Universität Bayreuth, 95440 Bayreuth, Germany.

SCARANI VALERIO
Centre for Quantum Technologies and Physics Department, National University of Singapore, S15 # 03-18, 3 Science Drive 2 Singapore 117543, Singapore.

VIGNOLO PATRIZIA
Institut Non Linéaire de Nice, Université de Nice Sophia, CNRS, 1361 route des Lucioles, 06560 Valbonne, France.

PARTICIPANTS

ARNOLD KYLE (M)
Centre for Quantum Technologies, National University of Singapore, Block S15, 3, Science Drive 2, Singapore 117543, Singapore.

AULBACH MARTIN (M)
School of Physics and Astronomy, E C Stoner Building, University of Leeds, Leeds LS2 9JT, UK.

BALACHANDRAN VINITHA (F)
Physics Department, Blk S12, Faculty of Science National University of Singapore, 2, Science Drive 3, Singapore 117542, Singapore.

BERNARD ALAIN (M)
Laboratoire Charles Fabry, Institut d'Optique Graduate School, Université de Paris-Sud, CNRS, Campus Polytechnique, 2 Avenue Augustin Fresnel, RD 128 F-91127, Palaiseau cedex, France.

BOADA-KERANS OCTAVI (M)
Departament ECM and ICCUB, Departament de Fisica, Universitat de Barcelona, 08028 Barcelona, Spain.

BROWN KATHERINE (F)
School of Physics and Astronomy, E C Stoner Building, University of Leeds, Leeds LS2 9JT, UK.

CAMERER STEPHAN (M)
Ludwig-Maximilians-Universität München, Schellingstraße 4 / III floor, D-80799 München, Germany.

CERRILLO-MORENO JAVIER (M)
The Institute for Mathematical Sciences, Imperial College, 53 Prince's Gate, South Kensington, London SW7 2PG, UK.

CHAN KIN SUNG (M)
Division of Physics and Applied Physics, School of Physical and Mathematical Sciences, Nanyang Technological University, 21 Nanyang Link, Singapore 637371, Singapore.

CORRO IVAN (M)
School of Physics (David Caro Building), The University of Melbourne, VIC 3010, Australia.

DE GREVE KRISTIAAN (M)
Department of applied Physics and Engineering, Edward L. Ginzton Laboratory, Stanford University, Stanford, CA 94305-4088, USA.

DE PASQUALE ANTONELLA (F)
Physics Department of University of Bari, Via Amendola 173, Bari, I-70126, Italy.

DUBOST BRICE (M)
ICFO - Institut de Ciències Fotòniques, Parc Mediterrani de la Tecnologia, Av. del Canal Olimpic s/n, 08860 Castelldefels, Barcelona, Spain.

FABRE CHARLOTTE (F)
Laboratoire Collisions Agrégats Réactivité, Université Paul Sabatier - Bât. 3R1b4, 118 route de Narbonne, 31062 Toulouse Cedex 09, France.

FANG YIYUAN BESS (F)
Centre for Quantum Technologies, National University of Singapore, Block S15, 3 Science Drive 2, Singapore 117543, Singapore.

GAUL CHRISTOPHER (M)
Universität Bayreuth, Physikalisches Institut, 95440 Bayreuth, Germany.

GAWRYLUK KRZYSZTOF (M)
Wydział Fizyki, Uniwersytet w Białymstoku, ul. Lipowa 41, Białystok 15 424, Poland.

GOETSCHY ARTHUR (M)
Laboratoire de Physique et Modélisation des Milieux Condensés, Maison des Magistères, 25 rue des Martyrs, BP 166, 38042 Grenoble Cedex 9, France.

GUDGEON EILIDH (F)
Centre for Quantum Technologies, National University of Singapore, Block S15, 3, Science Drive 2, Singapore 117543, Singapore.

GÜRKAN ZEYNEP NILHAN (F)
Izmir Institute of Technology, Department of Mathematics, Gülbahçe Köyü, Urla 35430, Izmir, Turkey.

HAN RUI (F)
Centre for Quantum Technologies, National University of Singapore, Block S15, 3, Science Drive 2, Singapore 117543, Singapore.

HEIKKINEN MIIKKA (M)
Department of Applied Physics, Helsinki University of Technology, P.O. Box 5100, FI-02015 TKK, Finland.

HUANG WUJIE (M)
Department of Physics, Tsinghua University, Beijing 100084, P.R. China.

HUO MINGXIA (F)
Department of Physics, Nankai University, Tianjin 300071, P.R. China.

JI SE-WAN (M)
Korea Institute for Advanced Study, Hoegiro 87(207-43 Cheongnyangni-dong), Dongdaemun-gu, Seoul 130-722, South Korea.

KUNTZ KATANYA BRIANNE (F)
The University of New South Wales, Australian Defence Force Academy, Canberra ACT 2600, Australia.

LE HUY NGUYEN (M)
Centre for Quantum Technologies, National University of Singapore, Block S15, 3, Science Drive 2, Singapore 117543, Singapore.

LOC LE XUAN (M)
Laboratoire de Photonique Quantique et Moléculaire, CNRS, ENS Cachan, 61 avenue du Président Wilson, France.

LEE JUHUI (F)
Sookmyung Women's University, 52 Hyochangwon-gil, Yongsan-gu, Seoul 140-742, South Korea.

LEE CHANGHYOUP (M)
Hanyang University, Department of Physics, Quantum Information Processing Group, Seoul 133-791, South Korea.

LEE JIANWEI (M)
Centre for Quantum Technologies, National University of Singapore, Block S15, 3, Science Drive 2, Singapore 117543, Singapore.

LEE KEAN LOON (M)
Centre for Quantum Technologies, National University of Singapore, Block S15, 3, Science Drive 2, Singapore 117543, Singapore.

LEMARIÉ GABRIEL (M)
Laboratoire Kastler Brossel, Université Paris 6, CNRS, ENS, 4 Place Jussieu, Paris 75005, France.

LEWTY NICHOLAS (M)
Centre for Quantum Technologies, National University of Singapore, Block S15, 3, Science Drive 2, Singapore 117543, Singapore.

LI YING (M)
Department of Physics, Nankai University, Tianjin 300071, P.R. China.

LIENNARD THOMAS (M)
Laboratoire de physique des lasers, Université Paris 13, CNRS, Institut Galilée, 99, avenue J.-B. Clément, F-93430 Villetaneuse, France.

LU LI-HUA (F)
Zhejiang University, Department of Physics, Zhe Da Road No. 38, Hangzhou city of Zhejiang Province, 310027, P.R. China.

LU YIN (F)
Centre for Quantum Technologies, National University of Singapore, Block S15, 3, Science Drive 2, Singapore 117543, Singapore.

LÜ XIN (M)
Centre for Quantum Technologies, National University of Singapore, Block S15, 3, Science Drive 2, Singapore 117543, Singapore.

MA RUI CHAO (M)
Division of Physics and Applied Physics, School of Physical and Mathematical Sciences, Nanyang Technological University, 21, Nanyang Link, Singapore 637371, Singapore.

MARTIN ANTHONY (M)
Laboratoire de Physique de la Matière Condensèe, Université de Nice Sophia, CNRS, Parc Valrose, F-06108 Nice Cedex 2, France.

MARQUES FURTADO DE MENDONÇA PAULO EDUARDO (M)
The University of Queensland, Queensland 4072, Australia.

MIDGLEY SARAH (F)
The University of Queensland, School of Physical Sciences, ACQAO, Brisbane, QLD 4072, Australia.

MOHAN ANUSHYAM (M)
Division of Physics and Applied Physics, School of Physical and Mathematical Sciences, Nanyang Technological University, 21, Nanyang Link, Singapore 637371, Singapore.

MOTZOI FELIX (M)
University of Waterloo, Department of Physics and Astronomy, 200 University Avenue West, Waterloo, Ontario N2L 3G1, Canada.

PAWŁOWSKI KRZYSZTOF (M)
Centrum Fizyki Teoretycznej, Polska Akademia Nauk, al. Lotników 32/46, Warszawa 02-668, Poland.

PIELAWA SUSANNE (F)
Harvard Physics, 17 Oxford Street, Cambridge MA 02138, USA.

SCHAFF JEAN-FRANÇOIS (M)
Institut Non Linéaire de Nice, Université de Nice Sophia, CNRS, 1361 route des Lucioles, 06560 Valbonne, France.

SHADMAN ZAHRA (F)
Heinrich-Heine-Universität Düsseldorf, Institut für Theoretische Physik III, Universitätstraße 1, Geb. 25.32, D-40225 Düsseldorf, Germany.

SHAARI JESNI SHAMSUL (M)
International Islamic University Malaysia, Faculty of Science, Jalan Istana, Bandar Indera Mahkota Kuantan, Pahang 25200, Malaysia.

SHINSUKE FUJISAWA (M)
The University of Tokyo, School of Science, Department of Physics, 9th floor, Faculty of Science, Building 1, 7-3-1 Hongo, Bunkyo-ku, Tokyo 113-0033, Japan.

SKOWRONEK ŁUKASZ (M)
Uniwersytet Jagielloński, Instytut Fizyki, ul. Reymonta 4, Kraków 30-059, Poland.

SOEDA AKIHITO (M)
University of Tokyo, Graduate School of Science, Department of Physics, 7-3-1 Hongo, Bunkyo-ku, Tokyo 113-0033, Japan.

SUZUKI JUN (M)
National Institute of Informatics, Quantum Information Science Group, 2-1-2 Hitotsubashi, Chiyoda-ku, Tokyo 101-8430, Japan.

TACLA ALEXANDRE BARON (M)
University of New Mexico, Department of Physics and Astronomy, 800 Yale Blvd. NE, MSC07 4220 Albuquerque, NM 87131, USA.

VITELLI CHIARA (F)
Università di Roma, piazzale Aldo Moro 5, Roma, Italy.

WANG GUANGQUAN (M)
Centre for Quantum Technologies, National University of Singapore, Block S15, 3, Science Drive 2, Singapore 117543, Singapore.

WOJCIECHOWSKI ADAM (M)
Uniwersytet Jagielloński, Instytut Fizyki, ul. Reymonta 4, Kraków 30-059, Poland.

WOLAK MARTA (F)
Centre for Quantum Technologies, National University of Singapore, Block S15, 3, Science Drive 2, Singapore 117543, Singapore.

WU XING (M)
Division of Physics and Applied Physics, School of Physical and Mathematical Sciences, Nanyang Technological University, 21 Nanyang Link, Singapore 637371, Singapore.

XU NANYANG (M)
University of Science and Technology of China, Hefei National Laboratory for Physical Sciences at Microscale and Department of Modern Physics, Room 323-627, East Campus Hefei City, Anhui Prov. 230026, P.R. China.

ZHANG JIANG-MIN (F)
Tsinghua University, Center for Advanced Study, Beijing 100084, P.R. China.

ZHANG JIANG-MIN (M)
Chinese academy of science, Institute of physics, Zhong-guan-cun, Beijing 100080, China.

ZHU HUANGJUN (M)
Centre for Quantum Technologies, National University of Singapore, Block S15, 3, Science Drive 2, Singapore 117543, Singapore.

1
Basics on Bose–Einstein condensation

D. GUÉRY-ODELIN and T. LAHAYE

Laboratoire Collisions Agrégats Réactivité, CNRS UMR 5589, IRSAMC, Université
Paul Sabatier, 118 Route de Narbonne, 31062 Toulouse CEDEX 4, France

1.1 Introduction

These lecture notes provide some basic results on Bose–Einstein condensation. They are the written version of the set of lectures given by one of the authors (D. G.-O.) at the Les Houches School of Physics on Ultracold gases and Quantum Information held in Singapore from June, 29 to July, 24 2009.

Degenerate quantum fluids (i.e. fluids in which quantum statistical effects play a key role) are encountered in nature in very different systems, ranging from atomic nuclei, superfluid helium, conduction electrons in metals, to neutron stars, and give rise to spectacular physical properties. The systems just mentioned, although differing by orders of magnitude in density[1] for instance, have a common characteristics: the interactions between particles are strong and cannot be controlled easily. The achievement of Bose–Einstein condensation in dilute gases in 1995 (Anderson *et al.*, 1995; Davis *et al.*, 1995), shortly followed by Fermi degeneracy (DeMarco and Jin, 1999), has allowed physicists to study degenerate quantum *gases*, in which interactions can be weak, and, more importantly, controlled. The possibility to tailor almost at will the external potentials in which the particles evolve, as well as the interactions, has led to fascinating studies that are nowadays at the interface between atomic physics, condensed matter physics, and even high-energy physics.

The purpose of this set of lectures is to introduce basic notions about Bose–Einstein condensates (BECs). These notes are organized as follows. We first study ideal Bose gases, putting an emphasis on the role of the trapping geometry and of the dimensionality. We introduce the correlation functions characterizing coherence. The second chapter deals with weakly interacting BECs, which are described, in a mean-field approach, by the so-called Gross–Pitaevskii equation. We derive this equation and apply it to a variety of experimentally relevant situations. The third chapter deals with beyond-mean-field effects, and introduces in particular the Bogolubov approximation. The last chapter is devoted to the study of BECs in double-well potentials, a situation where beyond-mean-field effects can appear quite easily, and that remains simple enough to be studied theoretically in details.

Even restricted to the above-mentioned topics, the subject is vast and growing, and many interesting aspects of Bose–Einstein condensation will not be described in these notes. We thus give below a few general references dealing with topics not covered here:

- the lecture notes written by Yvan Castin for two previous sessions of the Les Houches School of Physics (Castin, 2001, 2004) are worth reading;
- an introduction to the important subject of ultracold collisions is given in (Dalibard, 1998);
- the two standard textbooks in the field are (Pethick and Smith, 2002) and (Pitaevskii and Stringari, 2003);
- for a recent review dealing with the many-body physics that can be explored with ultracold gases, the reader is referred to (Bloch *et al.*, 2008).

1.2 Ideal Bose–Einstein condensates

1.2.1 Indistinguishable particles and the Bose gas

The theoretical treatment of the Bose gas requires to account properly for the indistinguishability of the atoms of the gas. We therefore start by a short review on the second quantization formalism, which is a powerful framework to deal with identical particles. Its use in combination with statistical physics is summarized afterwards.

1.2.1.1 Second quantization

Let us consider the orthonormal basis of one-particle states $\{|\varphi_\alpha\rangle\}$. Any arbitrary N-particle state can be expanded in a basis that is the tensor product of those one-particle states, namely

$$|\Psi(1,\dots,N)\rangle = \sum_{\{n_\alpha\}} C(n_1,\dots,n_N)|\varphi_{n_1}\rangle \otimes \dots \otimes |\varphi_{n_N}\rangle. \tag{1.1}$$

For bosons, $|\Psi(1,\dots,N)\rangle$ is symmetric under an arbitrary exchange of particles $(j) \longleftrightarrow (k)$, and therefore the coefficients $C(n_1,\dots,n_N)$ must be symmetric.

The second quantization procedure requires to consider an enlarged space of states in which the number of particles is not fixed. If we denote by $\mathcal{H}_0$ the Hilbert space with no particles, $\mathcal{H}_1$ the Hilbert space with only one particle and, in general, $\mathcal{H}_N$ the Hilbert space for N particles, the direct sum of these spaces is called the Fock space: $\mathcal{H} = \mathcal{H}_0 \oplus \mathcal{H}_1 \oplus \dots \oplus \mathcal{H}_N \oplus \dots$. An arbitrary state $|\Psi\rangle$ in Fock space is the sum over all the subspaces $\mathcal{H}_N$: $|\Psi\rangle = |\Psi^{(0)}\rangle + |\Psi^{(1)}\rangle + \dots + |\Psi^{(N)}\rangle + \dots$. The subspace with no particles is denoted $|0\rangle$ and is called the vacuum.

One can define creation and annihilation operators that act in the enlarged Hilbert space $\mathcal{H}$ for bosons where $\hat{a}_\alpha^\dagger$ creates a particle in the state $|\varphi_\alpha\rangle$, $\hat{a}_\alpha^\dagger|0\rangle = |\varphi_\alpha\rangle$, and $\hat{a}_\alpha$ destroys a particle in $|\varphi_\alpha\rangle$, $\hat{a}_\alpha|\varphi_\alpha\rangle = |0\rangle$. It is thus convenient to express the states of $\mathcal{H}$ using the orthonormal basis $\{|\dots n_\alpha \dots n_\beta \dots\rangle\}$, where n_α is the occupation number of state $|\varphi_\alpha\rangle$. In the case of bosons, the n_αs can be any non-negative integer. One can show that the symmetry of the states of systems made of many identical bosons is then simply expressed as commutation relations for the operators a_α and $a_\alpha^\dagger$:

$$\left[\hat{a}_\alpha, \hat{a}_\beta^\dagger\right] = \delta_{\alpha\beta}, \text{ and } [\hat{a}_\alpha, \hat{a}_\beta] = \left[\hat{a}_\alpha^\dagger, \hat{a}_\beta^\dagger\right] = 0. \tag{1.2}$$

The action of the operators a_α and $a_\alpha^\dagger$ over a Fock state $|\dots n_\alpha \dots n_\beta \dots\rangle$ is given by

$$\begin{aligned}
\hat{a}_\alpha^\dagger|\dots n_\alpha \dots n_\beta \dots\rangle &= \sqrt{n_\alpha + 1}\,|\dots n_\alpha + 1 \dots n_\beta \dots\rangle, \\
\hat{a}_\alpha|\dots n_\alpha \dots n_\beta \dots\rangle &= \sqrt{n_\alpha}\,|\dots n_\alpha - 1 \dots n_\beta \dots\rangle.
\end{aligned} \tag{1.3}$$

Let $f(i)$ be a one-particle operator. For instance, $f(i)$ can represent the kinetic energy of particle i. In first quantization, the corresponding operator for a system made of N particles reads $F = \sum_{i=1}^{N} f(i)$. One can show (Landau and Lifshitz, 1958) that this operator becomes in second quantization

$$\hat{F} = \sum_{\alpha} \sum_{\beta} \langle \varphi_{\beta} | f | \varphi_{\alpha} \rangle \, \hat{a}_{\beta}^{\dagger} \hat{a}_{\alpha}. \tag{1.4}$$

Similarly, the two-particle operator acting on a system made of N particles is given, in first quantization, by $G = (1/2) \sum_{i=1}^{N} \sum_{j \neq i} g(i,j)$, where $g(i,j) = g(j,i)$ and, in second quantization by

$$\hat{G} = \frac{1}{2} \sum_{\alpha} \sum_{\beta} \sum_{\gamma} \sum_{\delta} \hat{a}_{\delta}^{\dagger} \hat{a}_{\gamma}^{\dagger} \hat{a}_{\beta} \hat{a}_{\alpha} \, \langle \varphi_{\delta}(1) \varphi_{\gamma}(2) | g(1,2) | \varphi_{\alpha}(1) \varphi_{\beta}(2) \rangle. \tag{1.5}$$

1.2.1.2 Grand-canonical ensemble

The statistical ensemble well adapted to describe indistinguishable particles is the grand-canonical ensemble (Huang, 1963). Within this formalism, the system can exchange two extensive quantities with reservoirs: *energy* and *particles*. The equilibrium state is obtained by determining the density matrix $\hat{\rho}$ of the system. This is achieved by maximizing the missing information, or otherwise stated the statistical entropy $S(\hat{\rho}) = -k_B \text{Tr}[\hat{\rho} \log \hat{\rho}]$ with the two constraints

- of a fixed mean number of particles $\langle \hat{N} \rangle = N$;
- and of a fixed mean energy $\langle \hat{H} \rangle = E$, where $\hat{H}$ is the hamiltonian of the system.

This maximization is readily carried out using Lagrange multipliers. One finds

$$\hat{\rho} = \frac{e^{-\alpha \hat{N} - \beta \hat{H}}}{Z_G}, \quad \text{where} \quad Z_G = \text{Tr}\left(e^{-\alpha \hat{N} - \beta \hat{H}}\right) \tag{1.6}$$

is the grand-canonical partition function. α is the Lagrange multiplier associated with the constraint on the mean number of particles. The Lagrange multiplier β associated with the constraint on the mean energy can be related to the temperature T by $\beta = 1/k_B T$, and α to the chemical potential μ, that is the energy required to add one particle to the system, by $\alpha = \beta \mu$. It is convenient for calculations to introduce the *fugacity*, a dimensionless quantity defined by $z = e^{\beta \mu}$.

Let us consider that the system is in a three-dimensional box of volume V. We fix the mean number of particles N and the mean total energy E, so that all thermodynamical quantities depend on the three extensive parameters (V, N, E). Once the expression for the partition function Z_G is known, the value of the fugacity $z(V, N, E)$ and of $\beta(V, N, E)$ is obtained by the relations

$$N = z \frac{\partial}{\partial z} \ln Z_G(V, z, \beta), \quad \text{and} \quad E = -\frac{\partial}{\partial \beta} \ln Z_G(V, z, \beta). \tag{1.7}$$

1.2.2 The ideal quantum gas

The explicit determination of Z_G is in general impossible for an interacting gas. However, the expression of Z_G can readily be derived for an ideal gas. The symmetrization principle in quantum mechanics, applied to an ideal gas at thermodynamical equilibrium yields astonishing properties that are still essentially valid in the dilute limit. This is the reason why we begin these lecture notes by considering the case of an ideal gas.

For a system made of N particles that do not interact, the total hamiltonian $\hat{H}$ is the sum of the individual one-body hamiltonians: $\hat{H} = \hat{h}(1) + \ldots + \hat{h}(N)$. Let us introduce the eigenbasis $\{|\lambda\rangle\}$ of the one-body hamiltonian $\hat{h}$: $\hat{h}|\lambda\rangle = \varepsilon_\lambda|\lambda\rangle$.

If we denote, as in the previous section, the operator a_λ for destruction and $a_\lambda^\dagger$ for creation of a particle in the individual state $|\lambda\rangle$, the hamiltonian operator and the total number of particles operator can then be recast in the form

$$\hat{H} = \sum_\lambda \varepsilon_\lambda a_\lambda^\dagger a_\lambda \text{ and } \hat{N} = \sum_\lambda a_\lambda^\dagger a_\lambda. \tag{1.8}$$

These operators are obviously diagonal in the Fock basis $\{|N_\lambda, N_{\lambda'}, \ldots\rangle\}$ where the N_λ are the occupation numbers of the individual quantum states. For a given microscopic configuration $|\ell\rangle = |N_\lambda, N_{\lambda'}, \ldots\rangle$ one has:

$$\hat{N}|\ell\rangle = N_\ell|\ell\rangle \text{ with } N_\ell = \sum_\lambda N_\lambda,$$

$$\hat{H}|\ell\rangle = E_\ell|\ell\rangle \text{ with } E_\ell = \sum_\lambda N_\lambda \varepsilon_\lambda.$$

The grand-canonical partition function takes a simple form in the $\{|\ell\rangle\}$ basis:

$$Z_{\mathrm{G}} = \sum_\ell e^{-\alpha N_\ell - \beta E_\ell} = \sum_{N_\lambda, N_{\lambda'}, \ldots} e^{-(\alpha + \beta \varepsilon_\lambda)N_\lambda} \times e^{-(\alpha + \beta \varepsilon_{\lambda'})N_{\lambda'}} \times \cdots$$

$$= \prod_\lambda \zeta_\lambda \text{ with } \zeta_\lambda = \sum_{N_\lambda} e^{-(\alpha + \beta \varepsilon_\lambda)N_\lambda}. \tag{1.9}$$

We therefore obtain a factorization of the partition function Z_{G} as a product of elementary partition functions, each of them related to an individual quantum state $|\lambda\rangle$. This is the major advantage of using the grand-canonical formalism. Let us derive the explicit form of the partition function for fermions and bosons:

- *The fermionic case*

$$\zeta_\lambda = 1 + e^{-(\alpha + \beta \varepsilon_\lambda)} \implies \log Z_{\mathrm{G}} = \sum_\lambda \log\left(1 + e^{-(\alpha + \beta \varepsilon_\lambda)}\right), \tag{1.10}$$

since the occupation number of a given state λ can take only two values $N_\lambda = 0$ or $N_\lambda = 1$ due to the Pauli exclusion principle. We deduce the expression for the mean total number of particles:

$$N = \sum_\lambda N_\lambda \text{ with } N_\lambda = \frac{1}{1 + e^{\beta(\varepsilon_\lambda - \mu)}}. \tag{1.11}$$

The chemical potential μ can take any real value. In the limit $|\mu| \gg k_{\mathrm{B}}T$ and $\mu < 0$, $N_\lambda \simeq ze^{-\beta\varepsilon_\lambda}$ and we recover the Boltzmann result for classical statistics. In the opposite limit, $|\mu| \gg k_{\mathrm{B}}T$ and $\mu > 0$, we find that $N_\lambda = 1$ for $\varepsilon_\lambda < \mu$ and 0 otherwise. Particles fully occupy the so-called Fermi sea. This regime for fermions is referred to as the degenerate regime, i.e. the regime where statistics play a key role at the macroscopic level for the whole gas.

- *The bosonic case*

$$\zeta_\lambda = \sum_{N_\lambda=0}^{\infty} e^{-(\alpha+\beta\varepsilon_\lambda)N_\lambda} = \frac{1}{1 - e^{\beta(\varepsilon_\lambda+\mu)}} \qquad (1.12)$$

so that the mean total number of particles reads:

$$N = \sum_\lambda N_\lambda \quad \text{with} \quad N_\lambda = \frac{1}{e^{\beta(\varepsilon_\lambda-\mu)} - 1}. \qquad (1.13)$$

The chemical potential cannot take any value. Indeed, μ must remain smaller than the minimum energy $\varepsilon_{\min}$ to avoid an unphysical negative occupation number. The limit $|\mu| \gg k_{\mathrm{B}}T$ and $\mu < \varepsilon_{\min}$ yields, as for fermions, the classical statistics result. The quantum degenerate regime, that will be extensively studied in the following, is reached when the chemical potential takes a value smaller than $\varepsilon_{\min}$ but very close to this upper bound. For convenience, we usually set $\varepsilon_{\min} = 0$. Indeed, the value $\varepsilon_{\min}$ can always be absorbed into the definition of the chemical potential.

In the grand-canonical formalism, $\beta = 1/k_{\mathrm{B}}T$ and z are fixed, and the mean number of occupation N_λ of an individual energy state ε_λ is fixed and given by Eq. (1.13). Another related problem, closer to the experimental situation for dilute gases, consists in studying an isolated system made of N bosons at the microcanonical thermal equilibrium. One may wonder how the bosons will share on the individual energy levels. If we ignore the fluctuations of the number of particles about their average value, we can use the grand-canonical result with the constraint $N = \sum_\lambda N_\lambda(z, \beta)$. This equation gives the implicit value of the fugacity z as a function of N and T.

1.2.2.1   *Bose–Einstein condensation in a harmonic trap*

In the presence of an external confinement, the spectrum of the one-body hamiltonian $\hat{h}$ is discrete. The number N' of particles in the excited levels has *a priori* an upper bound:

$$N' = \sideset{}{'}\sum_\lambda \frac{1}{e^{\beta(\varepsilon_\lambda-\mu)} - 1} < N'_{\max} = \sideset{}{'}\sum_\lambda \frac{1}{e^{\beta(\varepsilon_\lambda-\varepsilon_{\min})} - 1}, \qquad (1.14)$$

where $\sum_\lambda'$ denotes the sum over all eigenstates λ except the ground state. This upper bound $N'_{\max}$ is called the saturation number and depends only on the temperature. Its precise value requires the knowledge of the type of confinement and of the dimensionality. Note that $N'_{\max}$ may not be finite; an example of this situation is discussed in Section 1.2.2.4. In the following we assume that $N'_{\max}(T) < \infty$.

If, at a given temperature T, we put in the trap a number N of particles larger than $N'_{\text{max}}(T)$, we are sure that $N - N'_{\text{max}}$ particles are in the ground state. As a direct application of the saturation of the number of atoms in the excited states, let us consider a three-dimensional isotropic harmonic oscillator trap with an angular frequency ω. The eigenenergies of the individual states read:

$$\varepsilon_{n_x,n_y,n_z} = \hbar\omega \left(n_x + n_y + n_z + \frac{3}{2} \right). \tag{1.15}$$

The degeneracy of a level ε_n with $n = n_x + n_y + n_z$ is $g_n = (n+1)(n+2)/2$, and the saturation number is therefore given by

$$N'_{\text{max}} = \sum_{(n_x,n_y,n_z)\neq(0,0,0)} \frac{1}{e^{\xi(n_x+n_y+n_z)} - 1} = \sum_{n=1}^{\infty} \frac{g_n}{e^{n\xi} - 1}, \tag{1.16}$$

where $\xi = \beta\hbar\omega$. In the limit $\xi \ll 1$, this discrete sum can be replaced by an integral[2]:

$$N'_{\text{max}} \simeq \frac{1}{2\xi^3} \int_0^\infty \frac{x^2}{e^x - 1} \, \mathrm{d}x = \frac{1}{2\xi^3} \int_0^\infty x^2 e^{-x} \sum_{n=0}^{\infty} e^{-nx} \, \mathrm{d}x = \frac{g_3(1)}{\xi^3}, \tag{1.17}$$

where the $g_\alpha(z)$ are the Bose functions defined by

$$g_\alpha(z) = \sum_{n=1}^{\infty} \frac{z^n}{n^\alpha}. \tag{1.18}$$

From the expression of the saturation number we deduce the critical temperature below which a macroscopic fraction of particles occupies the ground state: $N'_{\text{max}}(T_c) = N$. For the case of a 3D isotropic harmonic oscillator, we find $k_B T_c \simeq 0.94\hbar\omega N^{1/3}$. The non-trivial feature of Bose–Einstein condensation lies in the fact that the critical thermal energy $k_B T_c$ is very large compared to the energy-level spacing, as exemplified with the 3D harmonic oscillator for which the critical temperature is larger than $\hbar\omega$ by a factor $N^{1/3}$. Below the critical temperature, the number of atoms in the excited states of the 3D isotropic harmonic confinement is therefore $N'(T) = N(T/T_c)^3$ (see Figure 1.1):

$$N = N_0 + N'(T) \implies \frac{N_0}{N} = 1 - \left(\frac{T}{T_c} \right)^3. \tag{1.19}$$

1.2.2.2   *Bose–Einstein condensation in the semi-classical limit*

The semi-classical approximation, used in the previous section to estimate the critical temperature by replacing the discrete sums by continuous integrals, is justified when the energy difference between successive energy levels remains small compared to $k_B T$: $\delta\varepsilon \ll k_B T$. This is precisely the case, as illustrated in the previous section, for the thermodynamical description of the Bose–Einstein condensation phenomena since $k_B T_c \gg \delta\varepsilon$ if the system has a large enough number of particles. Within this

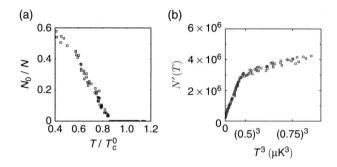

Fig. 1.1 (a) Condensate fraction N_0/N as a function of the temperature for a three-dimensional harmonic potential. (b) Number N' of atoms in the excited states as a function of temperature. One observes that below T_c, $N' \propto T^3$. The decrease in atom number above T_c for decreasing T is due to the evaporation of the most energetic particles. Figure taken from (Guéry-Odelin, 1998).

approximation, the calculation of all thermodynamic quantities requires only the knowledge of the density of states $\rho(\varepsilon)$, which depends on the dimensionality and on the type of confinement.

The general expression for the single-particle density of states is:

$$\rho(\varepsilon) = \int \frac{\mathrm{d}^D r \mathrm{d}^D p}{h^D} \delta\left[h(\vec{r}, \vec{p}) - \varepsilon\right], \qquad (1.20)$$

where $h(\vec{r}, \vec{p})$ is the one-body hamiltonian, and D the dimensionality. The expressions of the density of states for a box or an isotropic harmonic potential in 2D and 3D are given in Table 1.1.

1.2.2.3 Bose–Einstein condensation in a 3D box

For a Bose gas confined in a 3D box and described by the semi-classical formalism, the expression for the fugacity as a function of the temperature T and the total number of particles N is given by

$$N = \int_0^\infty \mathrm{d}\varepsilon \frac{\rho(\varepsilon)}{z^{-1} e^{\beta \varepsilon} - 1} = \frac{2}{\sqrt{\pi}} \frac{V}{\lambda_{\mathrm{dB}}^3} I(z), \qquad (1.21)$$

Table 1.1 Density of states for different types of confinement.

3D box	2D box	3D harmonic trap	2D harmonic trap
$\dfrac{V m^{3/2} \sqrt{2\varepsilon}}{2\pi^2 \hbar^3}$	$\dfrac{2\pi m L_x L_y}{h^2}$	$\dfrac{\varepsilon^2}{2(\hbar\omega)^3}$	$\dfrac{\varepsilon}{(\hbar\omega)^2}$

where V is the volume of the box, $\lambda_{dB} = h/\sqrt{2\pi m k_B T}$ the thermal de Broglie wavelength, and $I(z)$ a dimensionless integral whose expression in terms of Bose functions is:

$$I(z) = \int_0^\infty dx \frac{\sqrt{x}}{z^{-1}e^x - 1} = \sum_{\ell=1}^\infty \frac{z^\ell}{\ell^{3/2}} \int_0^\infty du \sqrt{u} e^{-u} = \frac{\sqrt{\pi}}{2} g_{3/2}(z). \tag{1.22}$$

We infer the expression for the saturation number of particles:

$$N'(T) = \frac{V}{\lambda_{dB}^3} g_{3/2}(1). \tag{1.23}$$

Figure 1.2 shows the variation of the Bose function $g_{3/2}(z)$ in the range $0 < z < 1$. It has an upper bound equal to $g_{3/2}(1) \simeq 2.612\ldots$. As the density of states vanishes for $\varepsilon = 0$, one has to add explicitly the contribution of the number of particles in the ground state to the total number of particles:

$$N = \frac{z}{1-z} + \frac{V}{\lambda_{dB}^3} g_{3/2}(z) \text{ for } T > T_c, \tag{1.24}$$

$$N = \frac{z}{1-z} + \frac{V}{\lambda_{dB}^3} g_{3/2}(1) \text{ for } T < T_c, \tag{1.25}$$

where the critical temperature that enters the expression of λ_{dB} is defined by the equality $(N/V)\lambda_{dB}^3 = g_{3/2}(1)$. This relation shows explicitly that the degenerate regime is reached when the de Broglie wavelength becomes larger than the mean inter-particle distance. For a given density, it requires a sufficiently low temperature. This is the reason why laser cooling turned out to be a required step towards the production of dilute Bose–Einstein condensate with alkali atoms.

It is instructive to evaluate the number of particles in the first excited state of energy ε_1. For the sake of simplicity, let us consider that half of the particles are in the

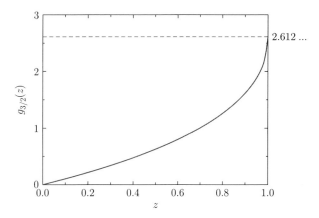

Fig. 1.2 Graph of the polylog function $g_{3/2}(z)$.

ground state: $N/2 = z/(1 - z)$ so that $z \simeq 1 - 2/N$. As we are dealing with the semi-classical approximation ($N \gg 1$), we have $\varepsilon_1 - \varepsilon_0 = \varepsilon_1 \ll k_B T$, and $e^{\beta \varepsilon_1} \simeq 1 - \beta \varepsilon_1$. The fraction of atoms in the first excited state can therefore be estimated:

$$\frac{N_1}{N} = \frac{1}{N} \frac{z e^{-\beta \varepsilon_1}}{1 - z e^{-\beta \varepsilon_1}} \simeq \frac{1}{N} \frac{1}{1 - (1 - 2/N)(1 - \beta \varepsilon_1)} = \frac{1}{2 + N^{1/3}} \simeq N^{-1/3} \ll 1. \quad (1.26)$$

We conclude that only the ground state is macroscopically occupied. We shall come back to this question when interactions are taken into account (see Section 1.4.2.2).

1.2.2.4   The role of dimensionality

In order to investigate the role played by the dimensionality, let us now consider a 2D box. The implicit expression for the fugacity is given by

$$N = \int_0^\infty \frac{\rho(\varepsilon) d\varepsilon}{e^{\beta(\varepsilon - \mu)} - 1} = \frac{L_x L_y}{\lambda_{dB}^2} \ln\left(\frac{e^{-\beta \mu}}{e^{-\beta \mu} - 1}\right), \quad (1.27)$$

where we have used the expression of the density of states given in Table 1.1. The fraction of atoms in the ground state is

$$\frac{N_0}{N} = \frac{1}{N} \frac{z}{1 - z} = \frac{e^{\sigma \lambda_{dB}^2} - 1}{N}, \quad (1.28)$$

where $\sigma = N/L_x L_y$ is the density of atoms per unit surface. Consider the thermodynamical limit for which the size of the system goes to infinity, while keeping constant the value of the intensive parameters such as the density σ and the temperature T. From Eq. (1.28), we find that N_0/N tends to zero when N tends to infinity. We conclude that there is no possible macroscopic occupation, and thus no Bose–Einstein condensation in a two-dimensional box in the thermodynamical limit. Otherwise stated, the saturation number tends to infinity faster than N when one takes the thermodynamical limit.

Let us emphasize that this result is intrinsically connected to the form of the density of states, which determines whether N'_{max} is finite or not. For a 2D box, the density of states does not depend on the energy ε. It is also the case for a one-dimensional harmonic oscillator for which $\rho(\varepsilon) = 1/\hbar \omega$. In this case also there is no possible condensation within the thermodynamical limit.

We conclude that confinement and dimensionality play a key role in the existence or not of Bose-Einstein condensation. Low-dimensional systems in the degenerate regime are particularly interesting in the presence of interactions. This topic is beyond the scope of this set of lectures (see (Pricoupenko *et al.*, 2004; Bloch *et al.*, 2008) for reviews).

1.2.2.5   Bose–Einstein condensation in an arbitrary trap

We consider now an ideal Bose gas confined in a general potential well $U(\vec{r})$. In the semi-classical approximation, one can show (Castin, 2001) using the Wigner distribution, that the density is given by:

$$n(\vec{r}) = \frac{1}{\lambda_{dB}^3} \sum_\ell \frac{z^\ell}{\ell^{3/2}} e^{-\beta \ell U(\vec{r})} = \frac{1}{\lambda_{dB}^3} g_{3/2}\left(z e^{-\beta U(\vec{r})}\right). \tag{1.29}$$

In the classical regime, $z \ll 1$, we recover the Boltzmann form for the atomic density $n(\vec{r}) \propto z e^{-\beta U(\vec{r})}$. If the trapping potential has its minimum of energy at $\vec{r} = \vec{0}$, the BEC transition condition is reached first at the center of the trap and the critical temperature is given by the relation: $n(\vec{0})\lambda_{dB}^3 = g_{3/2}(1)$.

1.2.3 Coherence properties of a Bose-Einstein condensate

The second quantized formulation developed in the first section is a powerful tool for introducing field operators $\hat{\Psi}^\dagger(\vec{r})$ and $\hat{\Psi}(\vec{r})$ that create and destroy an atom at $\vec{r}$. They are in close analogy with the the the electric-field operators $E^{(-)}(\vec{r})$ and $E^{(+)}(\vec{r})$ introduced in quantum optics to account for the coherence properties of light, and are useful for analyzing the coherence properties of Bose–Einstein condensates, as illustrated in the following with the calculation of the first- and second-order correlation functions.

1.2.3.1 Field operators

The field operators are linear combinations of the creation and annihilation operators where the coefficients are the single-particle wave functions:

$$\hat{\Psi}(\vec{r}) = \sum_\alpha \varphi_\alpha(\vec{r})\, \hat{a}_\alpha \quad \text{and} \quad \hat{\Psi}^\dagger(\vec{r}) = \sum_\alpha \varphi_\alpha^*(\vec{r})\, \hat{a}_\alpha^\dagger, \tag{1.30}$$

where the sum runs over the complete set of single-particle quantum numbers. Those operators are by construction defined at each space point $\vec{r}$. Their interpretation becomes clear when one calculates the action of the operator $\Psi^\dagger(\vec{r})$ on the vacuum state $|0\rangle$:

$$\hat{\Psi}^\dagger(\vec{r})\,|0\rangle = \sum_\alpha \varphi_\alpha^*(\vec{r})\, \hat{a}_\alpha^\dagger\,|0\rangle = \sum_\alpha \varphi_\alpha^*(\vec{r})\,|\varphi_\alpha\rangle. \tag{1.31}$$

Using the property $\varphi_\alpha^*(\vec{r}) = \langle \varphi_\alpha | \vec{r} \rangle$ and the fact that $\{|\varphi_\alpha\rangle\}$ forms a complete basis for the single-particle Hilbert space $\sum_\alpha |\varphi_\alpha\rangle \langle \varphi_\alpha| = 1$, one finds

$$\hat{\Psi}^\dagger(\vec{r})\,|0\rangle = \sum_\alpha \varphi_\alpha^*(\vec{r})\,|\varphi_\alpha\rangle = \sum_\alpha |\varphi_\alpha\rangle \langle \varphi_\alpha \mid \vec{r} \rangle = |\vec{r}\rangle. \tag{1.32}$$

The field operator $\hat{\Psi}^\dagger(\vec{r})$ is therefore an operator that creates an atom at $\vec{r}$. Similarly, the operator $\hat{\Psi}(\vec{r})$ annihilates an atom at $\vec{r}$. The field operators for bosons satisfy simple commutation relations

$$\left[\hat{\Psi}(\vec{r}), \hat{\Psi}^\dagger(\vec{r}\,')\right] = \delta(\vec{r} - \vec{r}\,') \quad \text{and} \quad \left[\hat{\Psi}(\vec{r}), \hat{\Psi}(\vec{r}\,')\right] = \left[\hat{\Psi}^\dagger(\vec{r}), \hat{\Psi}^\dagger(\vec{r}\,')\right] = 0. \tag{1.33}$$

The field operators $\hat{\Psi}^\dagger(\vec{r})$ and $\hat{\Psi}(\vec{r})$ play the same role in the basis $\{|\vec{r}\rangle\}$ as the operators $\hat{a}_\alpha^\dagger$ and $\hat{a}_\alpha$ in the basis $\{|\varphi_\alpha\rangle\}$.

1.2.3.2 *First-order correlation function*

The first-order correlation function $\left\langle \hat{\Psi}^\dagger(\vec{r})\hat{\Psi}(\vec{r}')\right\rangle$ is directly proportional to the visibility in an interference experiment. Let us choose the single-particle basis corresponding to a 3D box with periodic boundary conditions to work out the explicit expression of the first-order function correlation. The eigenstates are plane waves

$$\psi_{\vec{k}}(\vec{r}) = \frac{e^{i\vec{k}\cdot\vec{r}}}{L^{3/2}}, \qquad \text{where} \qquad k_i = \frac{2\pi}{L}n_i, \tag{1.34}$$

and n_i is any integer. The field operators are defined by

$$\hat{\Psi}(\vec{r}) = \frac{1}{L^{3/2}}\sum_{\vec{k}}\hat{a}_{\vec{k}}e^{i\vec{k}\cdot\vec{r}} \qquad \text{and} \qquad \hat{\Psi}^\dagger(\vec{r}) = \frac{1}{L^{3/2}}\sum_{\vec{k}}\hat{a}_{\vec{k}}^\dagger e^{-i\vec{k}\cdot\vec{r}}. \tag{1.35}$$

The first-order correlation function $G^{(1)}(\vec{r},\vec{r}')$ is defined by:

$$G^{(1)}(\vec{r},\vec{r}') = \left\langle \hat{\Psi}^\dagger(\vec{r})\hat{\Psi}(\vec{r}')\right\rangle. \tag{1.36}$$

We consider an ideal Bose gas whose hamiltonian is thus

$$\hat{H} = \sum_{\vec{k}}\varepsilon_{\vec{k}}\hat{a}_{\vec{k}}^\dagger\hat{a}_{\vec{k}}, \qquad \text{with} \qquad \varepsilon_{\vec{k}} = \frac{\hbar^2 k^2}{2m}. \tag{1.37}$$

The calculation of $G^{(1)}(\vec{r},\vec{r}')$ is performed using the density matrix whose expression in the grand-canonical formalism is $\hat{\rho}_{\text{eq}} = e^{-\beta(\hat{H}-\mu\hat{N})}/Z_{\text{G}}$:

$$G^{(1)}(\vec{r},\vec{r}') = \frac{1}{L^3}\sum_{\vec{k},\vec{k}'}e^{-i\left(\vec{k}\cdot\vec{r}-\vec{k}'\cdot\vec{r}'\right)}\left\langle a_{\vec{k}}^\dagger a_{\vec{k}'}\right\rangle. \tag{1.38}$$

The invariance by translation implies $\langle a_{\vec{k}}^\dagger a_{\vec{k}'}\rangle = \langle n_{\vec{k}}\rangle\delta_{\vec{k},\vec{k}'}$. The mean value $\langle n_{\vec{k}}\rangle$ obtained for an ideal Bose gas in the grand-canonical ensemble is

$$\langle n_{\vec{k}}\rangle = \text{Tr}(\hat{\rho}_{\text{eq}}a_{\vec{k}}^\dagger a_{\vec{k}}) = \frac{ze^{-\beta\varepsilon_k}}{1 - ze^{-\beta\varepsilon_k}} = \sum_{\ell=1}^{\infty} z^\ell e^{-\ell\beta\varepsilon_k}. \tag{1.39}$$

The first-order correlation function is therefore given by the Fourier transform of the momentum distribution. For a 3D ideal Bose gas, we therefore find

$$\begin{aligned}
G^{(1)}(\vec{r},\vec{r}') &= \frac{N_0}{L^3} + \frac{1}{(2\pi)^3}\int d^3\vec{k}\, e^{i\vec{k}\cdot(\vec{r}'-\vec{r})}\sum_{\ell=1}^{\infty} z^\ell e^{-\ell\beta\hbar^2 k^2/2m} \\
&= \frac{N_0}{L^3} + \frac{1}{\lambda_{\text{dB}}^3}\sum_{\ell=1}^{\infty}\frac{z^\ell}{\ell^{3/2}}\exp\left(-\frac{\pi(\vec{r}'-\vec{r})^2}{\ell\lambda_{\text{dB}}^2}\right).
\end{aligned} \tag{1.40}$$

The fact that $G^{(1)}(\vec{r},\vec{r}')$ depends only on the relative distance $|\vec{r}'-\vec{r}|$ is a direct consequence of the invariance by translation. The first term of the left-hand side of Eq. (1.40) accounts for the contribution of Bose condensed atoms, and the second term to that of thermal atoms.

The first-order correlation for $\vec{r}=\vec{r}'$ is simply the atomic density that is uniform in a box with periodic boundaries:

$$G^{(1)}(\vec{r},\vec{r}) = n(\vec{r}) = \frac{N_0}{L^3} + \frac{1}{\lambda_{\mathrm{dB}}^3}\sum_{\ell=1}^{\infty}\frac{z^\ell}{\ell^{3/2}} = \frac{N}{L^3}. \tag{1.41}$$

It is instructive to work out two limits:

- The limit where classical statistics is valid $n\lambda_{\mathrm{dB}}^3 \ll 1$, $N_0 \ll N$ and $z \ll 1$,

$$G^{(1)}(\vec{r},\vec{r}') \simeq \frac{N}{L^3}\exp^{-\pi(\vec{r}'-\vec{r})^2/\lambda_{\mathrm{dB}}^2}. \tag{1.42}$$

  We find that for a classical gas the coherence length is $\lambda_{\mathrm{dB}}/\sqrt{\pi}$.

- Below the critical temperature, the contribution of the condensed atoms yields an infinite coherence length. It reveals the presence of a long-range spatial order due to the condensate (see Figure 1.3a).

The previous results have been derived for an ideal gas in 3D. More generally, the first-order correlation function of the field operators allows one to identify the one-particle state having a macroscopic population that is at the origin of the long-range order. The wave function of this state appears to be the wave function of the condensate and is called the *order parameter*.

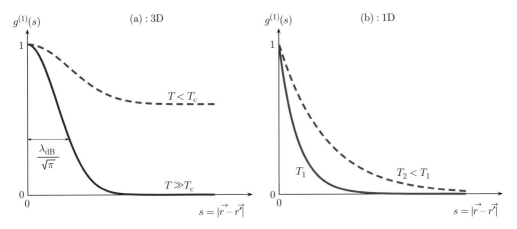

Fig. 1.3 Normalized first-order correlation function $g^{(1)}$ for a homogeneous Bose gas in 3D (a) in 1D (b). In 3D, below T_c, $g^{(1)}(s)$ goes to the constant value N_0/N when $s \to \infty$, clearly displaying off-diagonal long-range order. In contrast, in 1D, the correlation function always decays exponentially.

Coherence: the role of the dimensionality

For a 1D ideal Bose gas in the degenerate regime and within the validity range of the semi-classical description, $k_B T \gg |\mu| \gg h/L$, the first-order correlation function reads

$$G^{(1)}(x,0) = \frac{1}{L} \int \frac{e^{-ipx/\hbar}}{z^{-1}e^{-\beta p^2/2m} - 1} \, dp \propto \frac{1}{L} \int \frac{e^{-ipx/\hbar}}{p^2 + p_c^2} \, dp \propto e^{-|x|/\xi},$$

where $p_c^2 = 2m|\mu|$ and $\xi = h/p_c$. From the expression of the total number of particle, and using the expansion for $n(p) \simeq (p^2 + p_c^2)^{-1}$, one finds $p_c = h/(n_{1D}\lambda_T^2)$. The first-order correlation exhibits an exponential decay with a coherence length $\xi = n_{1D}\lambda_T^2/2\pi$ and therefore there is no long-range order (see Figure 1.3b). We recover the fact that in the thermodynamical limit there is no Bose–Einstein condensation for a 1D Bose gas confined in a box. The first-order correlation function of a very elongated BEC has been investigated experimentally in (Dettmer *et al.*, 2001; Richard *et al.*, 2003).

Let us mention two experimental techniques that have been used to study the first-order coherence of Bose–Einstein condensates by interferometric means. The first one consists in putting a condensate, initially at rest, in a superposition of states with different momenta. This is realized using Bragg pulses. As the two condensates separate, one monitors the contrast of the matterwave interference fringes (Hagley *et al.*, 1999). The second method consists in outcoupling selectively atoms from two different (and adjustable) locations in the condensate; the visibility of the interference pattern observed on the outcoupled matterwaves then gives the value of the first-order correlation function (Bloch *et al.*, 2000).

1.2.3.3   *Higher-order correlation function*

The second-order correlation function is defined as

$$G^{(2)}(\vec{r}, \vec{r}') = \left\langle \hat{\Psi}^\dagger(\vec{r})\hat{\Psi}^\dagger(\vec{r}')\hat{\Psi}(\vec{r}')\hat{\Psi}(\vec{r}) \right\rangle. \tag{1.43}$$

It is related to the conditional probability of presence of a particle at $\vec{r}$ knowing that another one is at $\vec{r}'$. One often uses the normalized second-order correlation function:

$$g^{(2)}(\vec{r}, \vec{r}') = \frac{G^{(2)}(\vec{r}, \vec{r}')}{G^{(1)}(\vec{r}, \vec{r})G^{(1)}(\vec{r}', \vec{r}')}. \tag{1.44}$$

For the ideal Bose gas above the critical temperature, $G^{(2)}$ can be expressed in terms of $G^{(1)}$ in the grand canonical ensemble using Wick's theorem (Cohen-Tannoudji and Robillard, 2001):

$$g^{(2)}(\vec{r}, \vec{r}') = 1 + \frac{|G^{(1)}(\vec{r}, \vec{r}')|^2}{G^{(1)}(\vec{r}, \vec{r})G^{(1)}(\vec{r}', \vec{r}')}. \tag{1.45}$$

One finds $g^{(2)}(\vec{r}, \vec{r}) = 2$, a result that is referred to as the bosonic bunching effect. These quantum correlations are the atomic analog of the Hanbury-Brown–Twiss effect (Hanbury-Brown and Twiss, 1956). This correlation function can also be calculated for a condensate. Using a Fock state to describe the condensate, one finds $g^{(2)}(\vec{r}, \vec{r}) \simeq 1$ if the number of particles is sufficiently large. To measure this correlation

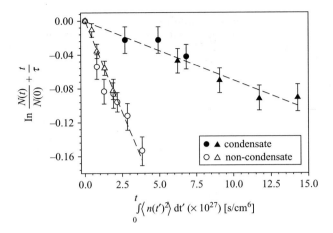

Fig. 1.4 Three-body losses in a cloud of ^{87}Rb. The slopes are proportional to $g^{(3)}(\vec{r},\vec{r},\vec{r})$. Figure taken from (Burt *et al.*, 1997), copyright American Physical Society.

function directly one needs to detect atoms one by one and to correlate their relative distances. This has been realized successfully using multichannel plate detectors for metastable helium atoms (Jeltes *et al.*, 2007), high-finesse optical cavities (Ritter *et al.*, 2007) or single-atom-sensitive fluorescence imaging (Manz *et al.*, 2009) for the alkalis, etc. The inhibition of the bosonic bunching for Bose–Einstein condensates has been observed, in good agreement with theory. Alternatively, $g^{(2)}(\vec{r},\vec{r})$ can be inferred from the energy released during the ballistic expansion of a Bose–Einstein condensate (Ketterle and Miesner, 1997).

Finally, let us emphasize that third-order correlations can be investigated through the study of three-body losses (Burt *et al.*, 1997). The loss rate obeys the equation:

$$\frac{\mathrm{d}N}{\mathrm{d}t} = -\kappa \int n^3(\vec{r},t)\,\mathrm{d}^3r, \tag{1.46}$$

where κ is proportional to $g^{(3)}(\vec{r},\vec{r},\vec{r})$. One expects $g^{(3)}(\vec{r},\vec{r},\vec{r}) = 3! = 6$ for a thermal gas, and 1 for a condensate as a signature of its coherence. The loss rates in both situations are plotted in figure 1.4; the ratio of the slopes, proportional to $g^{(3)}(\vec{r},\vec{r},\vec{r})$, is measured to be 7.4 ± 2.6. This experiment therefore clearly demonstrates the high coherence of a 3D Bose–Einstein condensate (in particular, the reduction of density fluctuations as compared to the thermal gas).

1.3 Mean-field theory

1.3.1 Introduction

In this section we study zero-temperature Bose–Einstein condensates in the presence of weak atom–atom interactions. Section 1.3.2 is devoted to the derivation, by a variational calculation, of the equation fulfilled by the condensate at equilibrium when interactions are taken into account within the mean-field approximation. This equation is referred to as the Gross–Pitaevskii equation. We discuss the various physical quantities characterizing the condensate, and determine their dependence with the number N of condensed atoms (Sections 1.3.3 and 1.3.4). Another important issue is the role played by the sign of the scattering length, as it has dramatic consequences on the stability of the condensate, as discussed in Section 1.3.5.

In several experiments performed with Bose–Einstein condensates, the trapping potential is varied in time. To describe the condensate dynamics in those contexts, we derive the time-dependent Gross–Pitaevskii equation from a least-action principle (Section 1.3.6). We also recast this equation as a set of hydrodynamic equations (Section 1.3.7). This formalism is used to describe the ballistic expansion of the condensate and the low-lying excitations for a trapped condensate in the Thomas–Fermi regime (Section 1.3.8). The study of such elementary excitations provides us with a powerful tool for probing the fundamental properties of quantum many-body systems. They have been, for instance, extensively studied in the context of solid-state physics (Pines, 1999; Leggett, 2006), superfluid helium (Pines and Nozieres, 1966) and nuclear physics (Mottelson, 1976). In the context of trapped Bose–Einstein condensates, the measurement of the excitation frequencies helped in establishing the time-dependent Gross–Pitaevskii equation as an excellent description of condensate dynamics at low temperatures. The collective modes are finally discussed for a harmonically trapped Bose–Einstein condensate in Section 1.3.9.

1.3.2 Mean-field description of the condensate

In this section, we consider N identical bosons trapped in an external potential $V_{\text{trap}}(\vec{r})$ at equilibrium and at temperature $T = 0$. In the absence of interactions, all atoms are in the ground state of the trap, and the N-body wave function of the condensate reads $|\psi\rangle = |\varphi_0(1)\rangle \otimes |\varphi_0(2)\rangle \otimes \ldots \otimes |\varphi_0(N)\rangle$, where $|\varphi_0\rangle$ denotes the wave function of the ground state determined by the confinement.

In the presence of interactions, the ground-state wave function that describes the structure of the condensate is that of the following N-body hamiltonian:

$$H = \sum_{i=1}^{N} \left[\frac{\vec{p}_i^2}{2m} + V_{\text{trap}}(\vec{r}_i) \right] + \frac{1}{2} \sum_i \sum_{j \neq i} V(\vec{r}_i - \vec{r}_j), \qquad (1.47)$$

where the terms $V(\vec{r}_i - \vec{r}_j)$ account for two-body interactions. In most cases, it is impossible to determine from this hamiltonian the exact expression for the ground-state energy and the associated N-body wave function. Alternatively, one can resort to an approximate determination of the ground state using the variational approach.

1.3.2.1 *Variational calculation of the condensate wave function*

Such a calculation is performed within a given family of functions. By extension of the exact N-body wave function in the absence of interactions, we restrict ourselves to the family of tensor products of N single-particle identical states:

$$|\psi\rangle = |\varphi(1)\rangle \otimes |\varphi(2)\rangle \otimes \ldots \otimes |\varphi(N)\rangle. \tag{1.48}$$

Those states are by definition symmetric in particle permutations, as required for identical bosons. Being tensor products of N states, they cannot describe the quantum correlations between the N atoms.

In the subspace generated by the vectors (1.48), the best state to approximate the ground state minimizes the energy functional $E_{\text{tot}}[\varphi]$ defined by:

$$E_{\text{tot}}[\varphi, N] = \langle H \rangle = \frac{\langle \psi | H | \psi \rangle}{\langle \psi | \psi \rangle}, \tag{1.49}$$

with the constraint $\langle \psi | \psi \rangle = 1$, or equivalently $\langle \varphi | \varphi \rangle = 1$. This best state will describe how the state of each atom is modified by the mean-field of the $N - 1$ other atoms.

The method of Lagrange multipliers permits one to recast the problem into the minimization of $\langle \psi | H | \psi \rangle - \mu \langle \psi | \psi \rangle$, where μ is the Lagrange multiplier associated with the conservation of the norm of the wave function. The functional differentiation $\delta(\langle \psi | H | \psi \rangle - \mu \langle \psi | \psi \rangle) = 0$ gives

$$N \int \mathrm{d}^3 r \delta \varphi^*(\vec{r}) \left\{ -\frac{\hbar^2}{2m} \Delta \varphi(\vec{r}) + V_{\text{ext}}(\vec{r}) \varphi(\vec{r}) \right.$$

$$\left. + (N-1) \left[\int \mathrm{d}^3 r' V(\vec{r} - \vec{r}') |\varphi(\vec{r}')|^2 \right] \varphi(\vec{r}) - \mu \varphi(\vec{r}) \right\} + \text{c.c.} = 0. \tag{1.50}$$

Since the variations of $\delta \varphi^*$ and $\delta \varphi$ can be considered as independent, the coefficient of $\delta \varphi^*$ must vanish, yielding:

$$-\frac{\hbar^2}{2m} \Delta \varphi(\vec{r}) + V_{\text{trap}}(\vec{r}) \varphi(\vec{r})$$

$$+ (N-1) \left[\int \mathrm{d}^3 r' \, V(\vec{r} - \vec{r}') \, |\varphi(\vec{r}')|^2 \right] \varphi(\vec{r}) = \mu \varphi(\vec{r}). \tag{1.51}$$

This equation, which resembles the Schrödinger equation, gives the evolution of each atom in the trapping potential and in the mean-field created at its position by the $(N - 1)$ other atoms.[3]

1.3.2.2 *Stationary Gross–Pitaevskii equation*

The variational method neglects the correlations between atoms at short distances. The gas is therefore supposed to be dilute. In this approximation, atoms are essentially far away one from another, and the interactions are governed by the large-distance asymptotic behavior of the wave function. Under this assumption, one can replace the true interacting potential by the corresponding pseudo-potential $V_{\text{pseudo}}(\vec{r} - \vec{r}') = g\, \delta(\vec{r} - \vec{r}') = (4\,\pi \hbar^2 a / m)\, \delta(\vec{r} - \vec{r}')$, where a is the scattering length of the real potential Dalibard (1998); Castin (2001). With such a contact potential, Eq. (1.51) takes the simple form

$$-\frac{\hbar^2}{2m}\Delta\varphi(\vec{r}) + V_{\text{trap}}(\vec{r})\,\varphi(\vec{r}) + (N-1)\,g\,|\varphi(\vec{r})|^2\,\varphi(\vec{r}) = \mu\,\varphi(\vec{r}). \qquad (1.52)$$

Usually, we are dealing with a sufficiently large number of atoms ($N \gg 1$) so that we can replace in the previous equation $N-1$ by N. Equation (1.52), referred to as the stationary Gross–Pitaevskii equation, plays a central role in the study of the static properties of Bose–Einstein condensation in the dilute limit.

In order to relate the Lagrange multiplier μ to a known physical quantity, we substitute into the energy functional (1.49) the real interaction potential by V_{pseudo}, and get:

$$E_{\text{tot}}\,[\varphi, N] = N \int \mathrm{d}^3 r\,\varphi^*\,(\vec{r}) \left[-\frac{\hbar^2}{2m}\Delta + V_{\text{trap}}\,(\vec{r}) + \frac{(N-1)g}{2}\,|\varphi\,(\vec{r})|^2 \right] \varphi\,(\vec{r}). \quad (1.53)$$

$E_{\text{tot}}\,[\varphi, N]$ depends explicitly on the number of atoms N, and also implicitly through the N dependence of φ so that:

$$\frac{\mathrm{d}E_{\text{tot}}\,[\varphi, N]}{\mathrm{d}N} = \frac{\partial E_{\text{tot}}\,[\varphi, N]}{\partial N} + \frac{\delta E_{\text{tot}}\,[\varphi, N]}{\delta\varphi}\frac{\partial\varphi}{\partial N} = \frac{\partial E_{\text{tot}}\,[\varphi, N]}{\partial N} + 0$$

$$= \int \mathrm{d}^3 r\,\varphi^*\,(\vec{r}) \left[-\frac{\hbar^2}{2m}\Delta + V_{\text{trap}}\,(\vec{r}) + \left(N - \frac{1}{2}\right) g\,|\varphi\,(\vec{r})|^2 \right] \varphi\,(\vec{r}), \quad (1.54)$$

where we have used explicitly the fact that the functional derivative $\delta E_{\text{tot}}\,[\varphi, N]\,/\delta\varphi$ vanishes since φ is such that $E_{\text{tot}}\,[\varphi, N]$ is extremal for any variation of φ. The Gross–Pitaevskii equation (1.52) gives an integral expression for the Lagrangian multiplier μ:

$$\mu = \int \mathrm{d}^3 r\,\varphi^*\,(\vec{r}) \left[-\frac{\hbar^2}{2m}\Delta + V_{\text{trap}}\,(\vec{r}) + (N-1)g\,|\varphi\,(\vec{r})|^2 \right] \varphi\,(\vec{r}), \qquad (1.55)$$

where we have used the normalization property $\langle\varphi|\varphi\rangle = 1$. If we compare Eq. (1.54) with Eq. (1.55), in the limit of large N, we deduce that

$$\mu = \frac{\partial E_{\text{tot}}\,[\varphi]}{\partial N} = E_{\text{tot}}\,[\varphi, N] - E_{\text{tot}}\,[\varphi, N - 1].$$

The Lagrange multiplier μ therefore corresponds to the variation of the total mean energy when N varies by one unit, which is simply the definition of the chemical potential.

1.3.2.3   *Expression of the various quantities in terms of the spatial density*

From Eq. (1.53), we can write the total energy E_{tot} as a sum of three terms $E_{\text{tot}} = E_{\text{kin}} + E_{\text{trap}} + E_{\text{int}}$ that can be expressed in terms of the spatial density $n\,(\vec{r}) = N\,|\varphi\,(\vec{r})|^2$:

- the kinetic energy due to the confinement[4]:

$$E_{\text{kin}} = N \frac{\hbar^2}{2m} \int d^3 r \left| \vec{\nabla} \varphi (\vec{r}) \right|^2 = \frac{\hbar^2}{2m} \int d^3 r \left[\vec{\nabla} \sqrt{n (\vec{r})} \right]^2, \tag{1.56}$$

- the trapping energy:

$$E_{\text{trap}} = N \int d^3 r V_{\text{trap}} (\vec{r}) \left| \varphi (\vec{r}) \right|^2 = \int d^3 r \, V_{\text{trap}} (\vec{r}) \, n (\vec{r}), \tag{1.57}$$

- and the interaction energy:

$$E_{\text{int}} = \frac{N (N - 1)}{2} \, g \int d^3 r \left| \varphi (\vec{r}) \right|^4 \simeq \frac{g}{2} \int d^3 r \left[n (\vec{r}) \right]^2. \tag{1.58}$$

It is instructive to rewrite the chemical potential in terms of those three energies. Multiplying the Gross–Pitaevskii equation (1.52) by $\varphi^*(\vec{r})$ and integrating over r gives

$$\mu = \frac{1}{N} \left(E_{\text{kin}} + E_{\text{trap}} + 2 E_{\text{int}} \right). \tag{1.59}$$

We conclude that the chemical potential is not equal to the mean total energy per atom ($\mu \neq E_{\text{tot}}/N$). This is due to the fact that, contrary to E_{trap} and E_{kin}, E_{int} does not increase linearly with N.

Finally, we give an extra relation between the three energies that enter the expression of the total energy and that is valid for a harmonic trapping potential:

$$2 E_{\text{kin}} - 2 E_{\text{trap}} + 3 E_{\text{int}} = 0. \tag{1.60}$$

This equation results from the virial theorem (Dalfovo *et al.*, 1999).

1.3.3 Condensate in a box and healing length

1.3.3.1 Condensate in a 1D box

The atom–atom interactions yield a new characteristic length, the healing length. Its physical meaning appears clearly by considering a 3D condensate in a box of volume L^3 with periodic boundary conditions along two axes and strict boundary conditions along the planes $z = 0$ and $z = L$. In the absence of interactions, all the atoms are in the ground state of the trap. Their wave function is thus given by

$$\Psi (x, y, z) = \frac{1}{L} \varphi_0 (z) \quad \text{with} \quad \varphi_0 (z) = \frac{2}{\sqrt{L}} \sin \left(\frac{\pi z}{L} \right), \tag{1.61}$$

and the corresponding atomic linear density $N |\varphi_0|^2$ is inhomogeneous. In the presence of interactions, the wave function $\varphi(z)$ still has to vanish at $z = 0$ and $z = L$, but tends to be homogeneous far from the walls since this minimizes the interaction energy for repulsive interactions. This behavior can be shown in the following manner (Cohen-Tannoudji, 1998). Let us calculate the interaction energy for a homogeneous density $n_0 = N/L$ over a distance L:

$$E_{\text{int}}^{\text{hom}} = \frac{g}{2} \int dz\, n_0^2 = \frac{g\, n_0^2 L}{2}. \tag{1.62}$$

If we consider another state with the same total number N of atoms but with an inhomogeneous density $n(z)$, it has an interaction energy:

$$E_{\text{int}}^{\text{inh}} = \frac{g}{2} \int dz\, n^2(z). \tag{1.63}$$

The comparison between those two interaction energies is obtained by calculating the difference of energies:

$$E_{\text{int}}^{\text{inh}} - E_{\text{int}}^{\text{hom}} = \frac{g}{2} \int dz\, \left[n^2(z) - n_0^2 \right]. \tag{1.64}$$

Using the normalization relation $\int n(z)dz = \int n_0 dz$, we can recast the previous equation in the following form:

$$E_{\text{int}}^{\text{inh}} - E_{\text{int}}^{\text{hom}} = \frac{g}{2} \int dz\, \left[n(z) - n_0 \right]^2 \geq 0. \tag{1.65}$$

This inequality shows that the homogeneous distribution of atoms is the one giving the smallest interaction energy, it also gives the smallest kinetic (or confinement) energy since $d\varphi/dz = 0$, and it is thus the one privileged by the system.

1.3.3.2   *Healing length*

One may wonder what is the characteristic length scale ξ over which the wave function, in the presence of interactions, varies from 0 at a wall position to its constant value $\tilde{n}_0$ (see Figure 1.5). The total number of particles N being fixed, the removal of atoms near the walls increases the spatial density $\tilde{n}_0$ far from the walls and thus the interaction energy E_{int}. We deduce that when ξ increases, the interaction energy E_{int} increases, and correlatively, the kinetic energy E_{kin} decreases since the gradient of density is weaker. The equilibrium shape of the condensate corresponds to the

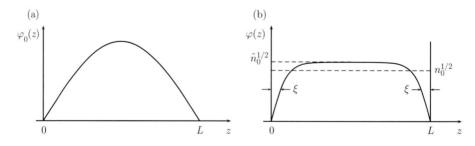

Fig. 1.5   (a) Ground-state wave function in a one-dimensional box of size L with strict boundary conditions. (b) Ground-state wave function with the same confinement but in the presence of interactions that tend to flatten the wave function. The healing length ξ is the distance over which the boundary condition no longer affects the wave function.

value of ξ for which the sum of these two energies is minimum. This characteristic length ξ is called the healing length, and more generally, represents the length after which the condensate recovers from a local perturbation (which, here, is due to the walls).

The order of magnitude of the healing length is readily obtained by the scaling of both the kinetic and interaction energies. The order of magnitude of E_{kin} is given by

$$E_{\mathrm{kin}}(\xi) = \frac{\hbar^2}{2m} \int dz \left(\frac{\partial \sqrt{n}}{\partial z}\right)^2 \simeq \frac{\hbar^2}{2m} 2\xi \frac{n}{\xi^2} \simeq \frac{\hbar^2 n_0}{m\xi}. \tag{1.66}$$

In order to estimate the interaction energy, one needs the expression for the density plateau $\tilde{n}_0$. From Figure 1.5, one finds approximately $\tilde{n}_0 - n_0 \simeq (2\xi/L)\, n_0$, which implies $\tilde{n}_0 \simeq (1 + 2\xi/L)\, n_0$ assuming $\xi \ll L$. The order of magnitude of E_{int} is given by

$$E_{\mathrm{int}}(\xi) = \frac{g}{2} \int dz\, n^2 \simeq \frac{g}{2}(L - 2\xi)\, n_0^2 (1 + 2\xi/L)^2 \simeq \frac{g n_0^2 L}{2} + g\, n_0^2\, \xi. \tag{1.67}$$

As expected intuitively, $E_{\mathrm{kin}}(\xi)$ is a decreasing function of ξ and $E_{\mathrm{int}}(\xi)$ an increasing function of ξ. The order of magnitude of the healing length is obtained by minimizing $E_{\mathrm{kin}}(\xi) + E_{\mathrm{int}}(\xi)$:

$$\frac{\partial}{\partial \xi}\left(\frac{\hbar^2}{m\xi} + g\, n_0 \xi\right) = 0 \quad \Rightarrow \quad \xi \simeq \frac{\hbar}{\sqrt{m\, g\, n_0}} = \frac{1}{\sqrt{4\pi\, a\, n_0}}. \tag{1.68}$$

Alternatively, one can solve the Gross–Pitaevskii equation for a one-dimensional box:

$$-\frac{\hbar^2}{2m}\frac{d^2\varphi}{dz^2} + Ng\varphi^3(z) = \mu\varphi(z), \tag{1.69}$$

with $\varphi(z)$ a real function that obeys the boundary conditions $\varphi(z{=}0) = \varphi(z = L) = 0$. Far from the walls $(z \sim L/2)$, one can neglect $d^2\varphi/dz^2$ and deduce the approximate value for the chemical potential $\mu \simeq Ng\varphi^2(z) = gn(z) \simeq g\tilde{n}_0$. Let us introduce the *standard definition of the healing length,*

$$\xi_0 = \left(\frac{\hbar^2}{2mg\tilde{n}_0}\right)^{1/2} = \frac{1}{(8\pi a\tilde{n}_0)^{1/2}}. \tag{1.70}$$

Using the dimensionless variable $\zeta = z/\xi_0$, the stationary Gross–Pitaevskii equation (1.69) can be rewritten in the form:

$$\frac{d^2\varphi(\zeta)}{d\zeta^2} - \frac{N}{n_0}\varphi^3(\zeta) + \varphi(\zeta) = 0. \tag{1.71}$$

The solution of Eq. (1.71) reads $\varphi(\zeta) = \sqrt{n_0/N}\,\mathrm{th}\left(\zeta/\sqrt{2}\right)$. Starting from 0 at $z = 0$, the wave function reaches a constant value after a few healing lengths ξ_0.

1.3.4   Condensate in a harmonic trap

Experimentally, one uses either a magnetic trap (Pritchard, 1983) or a far-off resonance dipole trap (Grimm *et al.*, 2000) to confine the atoms, and the condensate experiences a harmonic confinement. Let us first consider an isotropic harmonic trap of angular frequency ω_0. In the absence of interactions, the spatial extent of the ground state is given by the oscillator length $a_{\text{ho}} = (\hbar/m\omega_0)^{1/2}$. We work out in the following the scalings of the different contributions to the total energy for a non-ideal Bose–Einstein condensate held in a harmonic trap. These straightforward estimates are useful to classify the different interacting regimes.

1.3.4.1   Scaling

Let us denote by R the typical radius of the condensate. In the absence of interactions, R is on the order of a_{ho}. We estimate in the following how R is modified by interactions. To answer this question it is convenient to express R in units of a_{ho}: $w = R/a_{\text{ho}}$, and to use the following gaussian ansatz for the wave function of atoms in an isotropic harmonic trap:

$$\varphi(\vec{r}) = \frac{1}{\pi^{3/4}\left(w^3 a_{\text{ho}}^3\right)^{1/2}} \exp\left[-\frac{r^2}{2w^2 a_{\text{ho}}^2}\right]. \tag{1.72}$$

The three energies that contribute to the total energy can be readily calculated analytically, and one finds:

$$E_{\text{tot}}[w] = E_{\text{kin}} + E_{\text{trap}} + E_{\text{int}} = N\hbar\omega_0 \left[\frac{3}{4}\frac{1}{w^2} + \frac{3}{4}w^2 + \frac{1}{\sqrt{2\pi}}\frac{aN}{a_{\text{ho}}}\frac{1}{w^3}\right]. \tag{1.73}$$

1.3.4.2   Different interacting regimes

In the absence of interactions ($a = 0$), the last term of Eq. (1.73) vanishes, and the minimum of the sum of kinetic and trapping energy is obtained for $w = 1$. In this limit, we recover the well-known expression for the ground-state wave function of an harmonic oscillator. If the scattering length is non-zero, the order of magnitude of the interaction energy compared to the kinetic and trapping energies for $w = 1$ is determined by the dimensionless parameter $\chi = Na/a_{\text{ho}}$. If $\chi \ll 1$, interactions can be ignored. In the opposite limit ($\chi \gg 1$), the last term of Eq. (1.73) plays a crucial role and one must determine the new value of w that minimizes the total energy. The result depends on the sign of the scattering length: if $a > 0$, the effective interactions are repulsive and w scales as $N^{1/5}$, i.e. the size of the ground-state wave function increases with the number of condensed atoms, if $a < 0$, the effective interactions are attractive and one finds $w < 1$ when a solution exists (see Section 1.3.5).

1.3.4.3   Condensate with a positive scattering length and the Thomas–Fermi limit

For a positive scattering length, both the kinetic and interaction energies are decreasing functions of w, whereas the trapping energy increases with w. There is always

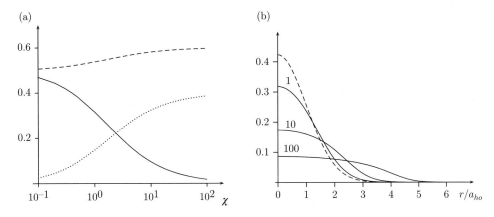

Fig. 1.6 (a) Relative contributions of the different energy terms to the total energy as a function of the dimensionless parameter $\chi = Na/a_{\text{ho}}$: $E_{\text{trap}}/E_{\text{tot}}$ (dashed line), $E_{\text{kin}}/E_{\text{tot}}$ (solid line), $E_{\text{int}}/E_{\text{tot}}$ (dotted line). (b) Condensate wave function, at $T = 0$, obtained by solving numerically the stationary Gross–Pitaevskii equation Eq. (1.52) in a spherical trap and with repulsive interactions. The dashed line corresponds to the ideal gas ($a = 0$); the solid lines to $\chi = 1, 10, 100$. Figure courtesy of S. Giorgini.

a value of w that minimizes the total energy and that corresponds to a stable condensate. The radius of the condensate increases when the strength of the repulsive interactions increases as illustrated in Figure 1.6 where different numerical solutions of the stationary Gross–Pitaevskii equation for increasing values of the χ parameter are represented.

We have plotted in Figure 1.6.a the relative contribution to the total energy of the trapping, kinetic and interaction energies as a function of the interacting parameter χ. In the limit $\chi \gg 1$, referred to as the Thomas–Fermi limit, one can neglect the kinetic energy term, and the Gross–Pitaevskii equation becomes a simple algebraic equation

$$V_{\text{trap}}(\vec{r}) + Ng|\varphi(\vec{r})|^2 = V_{\text{trap}} + gn_0(\vec{r}) = \mu. \tag{1.74}$$

For a harmonic confinement, $V_{\text{trap}}(\vec{r}) = m\omega_0^2 r^2/2$, and the spatial density $n_0(\vec{r}) = N|\varphi(\vec{r})|^2$ has the shape of an inverted parabola: $n(\vec{r}) = [\mu - m\omega_0^2 r^2/2]/g$ that starts from the value μ/g for $r = 0$ and that vanishes for $r \geq r_{\text{max}} = (2\mu/m\omega_0^2)^{1/2}$. The expression for the chemical potential is obtained from the normalization condition. Integrating the density profile over r, we find

$$\mu(N) = \frac{\hbar\omega_0}{2}\left(15\frac{Na}{a_{\text{ho}}}\right)^{2/5}. \tag{1.75}$$

The total energy is obtained by integrating Eq. (1.55):

$$E_{\text{tot}}(N) = \int_0^N \mu(N')\,\mathrm{d}N' = \frac{\hbar\omega_0}{2}\left(15\frac{a}{a_{\text{ho}}}\right)^{2/5}\frac{5\,N^{7/5}}{7}, \tag{1.76}$$

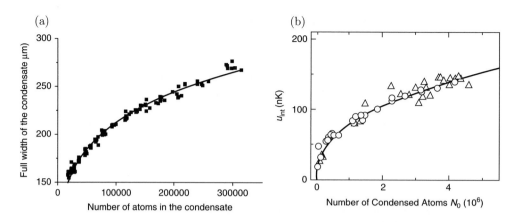

Fig. 1.7 (a) Variations of the full width of the condensate along one axis, as a function of the number of condensed atoms N. The solid line is a fit proportional to $N^{1/5}$, figure taken from (Söding *et al.*, 1999). (b) Mean-field energy per condensed atom versus the number of atoms in the condensate. The solid line is a fit proportional to $N^{2/5}$. Figure from (Mewes *et al.*, 1996). Copyright American Physical Society.

and the total energy per particle is equal to $E_{\text{tot}}\left(N\right)/N = 5\mu/7$. Using the relations (1.59) and (1.60), one deduces the expression for the interaction energy per particle in the Thomas–Fermi limit for which the kinetic energy[5] is negligible:

$$\frac{E_{\text{int}}\left(N\right)}{N} = \frac{2}{7}\mu\left(N\right) = \frac{\hbar\omega_0}{7}\left(15\frac{Na}{a_{\text{ho}}}\right)^{2/5}. \tag{1.77}$$

The size of the condensate depends on the number of atoms through the chemical potential:

$$r_{\max}\left(N\right) = \sqrt{\frac{2\mu}{m\omega_0^2}} = a_{\text{ho}}\left(\frac{15Na}{a_{\text{ho}}}\right)^{1/5}. \tag{1.78}$$

This atom number dependence of the size is illustrated in Figure 1.7.a where the fit proportional to $N^{1/5}$ is in good agreement with the size observed experimentally for different condensed atoms number.

The interaction energy can be measured by removing abruptly the confinement (Mewes *et al.*, 1996). Just after the switch off, the total energy is equal to the interaction energy (which has not changed) plus the kinetic energy, which is negligible in the Thomas–Fermi limit. The interaction energy is converted into kinetic energy during the expansion of the condensate (see Section 1.3.9.2). The measurement of this energy shows that it varies with the number of condensed atoms as $N^{2/5}$ (see Figure 1.7.b), as expected from the Thomas–Fermi limit result (1.77).

Most experiments use cylindrically symmetric harmonic traps:

$$V\left(z, r\right) = \left(m/2\right) \left[\omega_z^2 \, z^2 + \omega_\perp^2 \, r^2\right].$$

The inverted parabola shape of the density profile is limited axially to $\pm z_{\mathrm{max}}$ and radially to $\pm r_{\mathrm{max}}$ defined, respectively, by $m\omega_z^2 \, z_{\mathrm{max}}^2 = 2\mu$ and $m\omega_\perp^2 \, r_{\mathrm{max}}^2 = 2\mu$. The aspect ratio of the condensate is therefore given by $z_{\mathrm{max}}/r_{\mathrm{max}} = \omega_\perp/\omega_z$. In the absence of interactions, this aspect ratio is equal to the ratio of the oscillator lengths of each oscillator:

$$\frac{z_{\mathrm{max}}}{r_{\mathrm{max}}} = \frac{a_{\mathrm{ho}}^z}{a_{\mathrm{ho}}^\perp} = \sqrt{\frac{\omega_\perp}{\omega_z}} < \frac{\omega_\perp}{\omega_z}, \quad \text{if} \quad \frac{\omega_\perp}{\omega_z} > 1. \tag{1.79}$$

We deduce that interactions tend to magnify the aspect ratio of the condensate with respect to the ideal Bose gas.

A remarkable feature of Bose–Einstein condensates in the Thomas–Fermi limit is that interactions can be important whilst the gas is dilute. Consider a condensate contained in a volume R^3. It can be considered as dilute as soon as the mean distance $d = (N/R^3)^{-1/3}$ between atoms is large compared to the scattering length a. Since R is always larger than a_{ho}, one has:

$$\frac{a}{d} < \frac{a}{a_{\mathrm{ho}}} N^{1/3}. \tag{1.80}$$

The ratio a/d increases at most as $N^{1/3}$ only, whereas $\chi = Na/a_{\mathrm{ho}}$, which characterizes the importance of interactions increases as N. One can therefore have a condensate in the Thomas–Fermi regime ($\chi \gg 1$) while remaining in the dilute regime for which $a \ll d$. For example, let us consider the case of ^{87}Rb atoms, whose scattering length is $a = 5\,\mathrm{nm}$, in a harmonic trap of frequency $\omega/2\pi = 250\,\mathrm{Hz}$ yielding an oscillator length $a_{\mathrm{ho}} = 0.68\,\mu\mathrm{m}$. For a condensate of $N = 10^6$ atoms, one finds for the interaction parameter $\chi = Na/a_{\mathrm{ho}} \simeq 7400 \gg 1$. The Thomas–fermi radius is then $r_{\mathrm{max}} \simeq 6.9\,\mu\mathrm{m}$ and the parameter that characterizes the diluteness of the gas $a/d \sim (a/r_{\mathrm{max}}) N^{1/3} \simeq 7.2 \times 10^{-2} \ll 1$.

1.3.5 Condensate with a negative scattering length

1.3.5.1 *Condition of stability*

Theoretical studies predict that a homogeneous Bose–Einstein condensate with attractive interactions is unstable (Stoof, 1994). This can be understood simply from the spectrum of elementary excitations (see Section 1.3.6.4): for $a < 0$, the frequency of small momenta excitations is imaginary, yielding the so-called *phonon instability*. In the presence of a confinement, a condensate may form if the atom number is not too large. Physically, the stability originates from the tradeoff between the attractive interaction energy, that tends to contract the cloud, and the kinetic energy term resulting from the position–momentum uncertainty in the presence of confinement.[6]

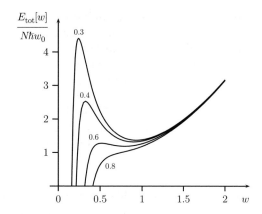

Fig. 1.8   Total energy, normalized to $N\hbar\omega_0$, as a function of the width parameter w for the Gaussian model of Section 1.3.4.1, for different values of the interaction parameter $|\chi| = 0.3, 0.4, 0.6, 0.8$.

This balance is illustrated in the case of an isotropic harmonic confinement in Figure 1.8.(i) where we have plotted the total energy per particle normalized to $\hbar\omega_0$ as a function of the effective width w for the Gaussian model of Eq. (1.73), for several values of the interaction parameter $|\chi| = N|a|/a_{\mathrm{ho}}$. There exists a critical value χ_c such that if $|\chi|$ is larger than χ_c, there is no longer any local minimum of the total energy E_{tot}. In other words, for a given negative value of the scattering length a, the condensate can accommodate only a finite number of atoms $N < N_c = \chi_c a_{\mathrm{ho}}/|a|$ (Bradley *et al.*, 1997). The critical value of the interaction parameter is equal to $\chi_c \simeq 0.671$ within the gaussian ansatz approximation (see Section 1.3.4.1). A more refined theoretical analysis yields $\chi_c = 0.574$ (see (Dalfovo *et al.*, 1999) and references therein), in rather good agreement with experimental observations (Roberts *et al.*, 2001).

1.3.5.2   *Collapse and explosion of a condensate with a negative scattering length*

Beyond the stability limit, self-attraction overwhelms the repulsion due to the quantum pressure and causes the condensate to collapse. During the collapse, the density rises yielding a dramatic increase of the inelastic collision rate such as three-body recombination and therefore induces atom losses. One may wonder what happens after the collapse.

The group of R. Hulet was able to observe a ${}^{7}\mathrm{Li}$ atoms condensate regrowing after the collapse (Gerton *et al.*, 2000). Indeed, just after the collapse the gas is out of equilibrium and a condensate may form filled through elastic collisions between thermal atoms in the gas. The condensate is therefore found to undergo many cycles of growth and collapses before reaching a stationary regime (Sackett *et al.*, 1998).

The JILA group has explored the dynamics of collapse and its subsequent explosion when the balance of forces governing the Bose–Einstein condensate size and shape is suddenly altered. The collapse is induced by tuning abruptly the interactions from repulsive to attractive using an externally applied magnetic field close to a Feshbach resonance. This technique has allowed for the observation of an exploding atomic ejection from the collapsing Bose–Einstein condensate (Donley *et al.*, 2001).

Surprisingly, after this dramatic event, a remnant and highly excited condensate with a number of atoms greater than N_c was sometimes observed. This apparent contradiction with the stability criterium was recently solved. Indeed, it turns out that the remnant is composed of multiple solitons that have relative phases such that they repel each other and oscillate in the trapping potential for a long time without degradation (Cornish *et al.*, 2006; Strecker *et al.*, 2002).

1.3.6 Time-dependent Gross–Pitaevskii equation

In several experiments performed with Bose–Einstein condensates, the trapping potential is varied in time. For instance, time-of-flight experiments where one switches off suddenly the confinement to observe the ballistic expansion of the condensate provide a way to investigate the properties of the many-body ground state and in particular the role played by atom–atom interactions. To describe the dynamics of the condensate in those contexts, one needs to extend (using a least-action principle) to time-dependent phenomena the Gross–Pitaevskii equation introduced in Section 1.3.2 for analyzing static properties of condensates.

1.3.6.1 Derivation of Schrödinger equation from a principle of least action

As a starting point, we recall (Cohen-Tannoudji, 1998) that the Schrödinger equation for a particle in a confining potential $V_{\text{trap}}(\vec{r}, t)$,

$$i\hbar \frac{\partial}{\partial t}\psi\left(\vec{r}, t\right) = -\frac{\hbar^2}{2m}\Delta\psi\left(\vec{r}, t\right) + V_{\text{trap}}\left(\vec{r}, t\right)\psi\left(\vec{r}, t\right), \tag{1.81}$$

can be obtained by minimizing the action $S = \int_{t_1}^{t_2} \mathrm{d}t \int \mathrm{d}^3 r \, \mathcal{L}$ related to the Lagrangian density

$$\mathcal{L}\left(\psi, \psi^*, \vec{\nabla}\psi, \vec{\nabla}\psi^*, \dot{\psi}, \dot{\psi}^*\right) = i\frac{\hbar}{2}\left[\psi^*\dot{\psi} - \dot{\psi}^*\psi\right] - \frac{\hbar^2}{2m}\vec{\nabla}\psi^* \cdot \vec{\nabla}\psi - V_{\text{trap}}(\vec{r}, t)\psi^*\psi, \tag{1.82}$$

where $\dot{\psi}$ and $\dot{\psi}^*$ refer to the time derivatives of ψ and ψ^*, respectively.

Similarly, the Schrödinger equation for N interacting bosons

$$i\hbar \frac{\partial}{\partial t}\psi\left(\vec{r}_1, \dots \vec{r}_N, t\right) = -\frac{\hbar^2}{2m}\sum_{i=1}^{N}\Delta_i\psi\left(\vec{r}_1, \dots \vec{r}_N, t\right) + \sum_{i=1}^{N}V_{\text{trap}}\left(\vec{r}_i, t\right)\psi\left(\vec{r}_1, \dots \vec{r}_N, t\right)$$

$$+ \frac{1}{2}\sum_{i=1}^{N}\sum_{j=1}^{N}V\left(\vec{r}_i - \vec{r}_j\right)\psi\left(\vec{r}_1, \dots \vec{r}_N, t\right) \tag{1.83}$$

can be derived by minimizing the Lagrangian density

$$
\mathcal{L} = i\frac{\hbar}{2}\left[\psi^*(\vec{r}_1,\ldots\vec{r}_N,t)\,\dot{\psi}(\vec{r}_1,\ldots\vec{r}_N,t) - \dot{\psi}^*(\vec{r}_1,\ldots\vec{r}_N,t)\,\psi(\vec{r}_1,\ldots\vec{r}_N,t)\right]
$$

$$
-\frac{\hbar^2}{2m}\sum_{i=1}^{N}\left(\vec{\nabla}_{r_i}\psi^*(\vec{r}_1,\ldots\vec{r}_N,t)\right)\cdot\left(\vec{\nabla}_{r_i}\psi(\vec{r}_1,\ldots\vec{r}_N,t)\right)
$$

$$
-\left[\sum_{i=1}^{N}V_{\text{trap}}(\vec{r}_i,t)+\frac{1}{2}\sum_{i=1}^{N}\sum_{j=1}^{N}V(\vec{r}_i-\vec{r}_j)\right]\psi^*(\vec{r}_1,\ldots\vec{r}_N,t)\,\psi(\vec{r}_1,\ldots\vec{r}_N,t). \quad (1.84)
$$

1.3.6.2 *Determination of the best time-dependent N-particle state*

As for the time-independent case, we look for a N-particle wave function equal to a product of N identical single-particle functions:

$$
\psi(\vec{r}_1,\ldots\vec{r}_N,t) = \varphi(\vec{r}_1,t)\,\varphi(\vec{r}_2,t)\ldots\varphi(\vec{r}_N,t). \quad (1.85)
$$

Such a fully symmetric state describes a situation where all N bosons evolve in the same way, and neglects quantum correlations between the atoms.

Inserting the ansatz (1.85) into the Lagrangian density (1.84) and expressing that the variation δS of the corresponding action S vanishes to first order in $\delta\varphi$ for any variation $\delta\varphi$ of φ leads to the following time-dependent Gross–Pitaevskii equation:

$$
i\hbar\frac{\partial}{\partial t}\varphi(\vec{r},t) = -\frac{\hbar^2}{2m}\Delta\varphi(\vec{r},t) + V_{\text{trap}}(\vec{r},t)\,\varphi(\vec{r},t) + N\,g\,|\varphi(\vec{r},t)|^2\,\varphi(\vec{r},t), \quad (1.86)
$$

where we have replaced the interaction potential V by the contact potential V_{pseudo} (see Section 1.3.2.2).

If $V_{\text{trap}}(\vec{r},t) = V_0(\vec{r})$ does not depend on time, one can look for stationary solutions of Eq. (1.86) of the form $\varphi(\vec{r},t) = \varphi_0(\vec{r})\exp(-i\mu t/\hbar)$. In this way, one exactly recovers the time-independent Gross–Pitaevskii equation (1.52).

1.3.6.3 *Response of a condensate to a time-dependent perturbation*

The time-dependent Gross–Pitaevskii equation allows one to infer the response of the condensate to a small perturbation. To address this problem, we consider a small time-dependent perturbation $\delta V(\vec{r},t)$ that is added to the confining potential $V_0(\vec{r})$: $V_{\text{trap}}(\vec{r},t) = V_0(\vec{r}) + \delta V(\vec{r},t)$. As an example, the perturbation δV can account for a slight modulation of the strength of the trapping potential at a controlled frequency. The calculation of the response of the condensate to this excitation and to first order in δV is obtained by searching a solution of Eq. (1.86) in the form:

$$
\varphi(\vec{r},t) = [\varphi_0(\vec{r}) + \delta\varphi(\vec{r},t)]\,e^{-i\mu t/\hbar}, \quad (1.87)
$$

where $\delta\varphi(\vec{r},t)$ accounts for the time-dependent modifications of the wave function driven by $\delta V(\vec{r},t)$. In this way, one readily obtains the following set of coupled equations:

$$ih\frac{\partial}{\partial t}\begin{pmatrix}\delta\varphi\\\delta\varphi^*\end{pmatrix} = \mathcal{L}_{\mathrm{GP}}\begin{pmatrix}\delta\varphi\\\delta\varphi^*\end{pmatrix} + \begin{pmatrix}\varphi_0\delta V\\-\varphi_0^*\delta V\end{pmatrix}, \tag{1.88}$$

where $\mathcal{L}_{\mathrm{GP}}$ denotes the following time-independent 2×2 matrix:

$$\mathcal{L}_{\mathrm{GP}} = \begin{pmatrix} H_0 - \mu + 2Ng|\varphi_0|^2 & Ng\varphi_0^2 \\ Ng(\varphi_0^*)^2 & -\left(H_0 - \mu + 2Ng|\varphi_0|^2\right) \end{pmatrix} \tag{1.89}$$

and

$$H_0 = -\frac{\hbar^2}{2m}\Delta + V_0(\vec{r}). \tag{1.90}$$

The second term in the r.h.s. of Eq. (1.88) is proportional to δV and acts as a source term. Equation (1.88) is simply the linearization of the time-dependent Gross–Pitaevskii equation (1.86), and is therefore valid for $|\delta V| \ll |V_0|$.

1.3.6.4 *Frequencies of the small-amplitude oscillations*

Let us suppose that one applies a small perturbation δV for a given duration. Just after the application of the perturbation, the condensate is out of equilibrium, and, as a result, oscillates with a small amplitude at its eigenfrequencies ω. To determine these frequencies, one must diagonalize the 2×2 matrix $\mathcal{L}_{\mathrm{GP}}$, and solve the two coupled equations:

$$[H_0 - \mu + 2Ngn_0(\vec{r})]\,u(\vec{r}) + Ngn_0(\vec{r})\,v(\vec{r}) = \hbar\omega u(\vec{r}),$$
$$-Ngn_0(\vec{r})\,u(\vec{r}) - [H_0 - \mu + 2Ngn_0(\vec{r})]\,v(\vec{r}) = \hbar\omega v(\vec{r}), \tag{1.91}$$

where u and v are defined by $\delta\varphi(\vec{r},t) = u(\vec{r})e^{-i\omega t} + v^*(\vec{r})e^{i\omega t}$. This set of equations is commonly referred to as the Bogolubov–de Gennes equations, and is generally solved numerically to infer the frequencies ω for any value of the spatial density n_0.

Consider as an example a condensate in a box with cyclic boundary conditions. The density is constant and equal to $n_0 = N/L^3$. The spectrum $\omega(k)$ for solutions of $u(\vec{r})$ and $v(\vec{r})$ in form of plane waves $\exp(i\vec{k}\cdot\vec{r})$ is given by the determinant deduced from $\mathcal{L}_{\mathrm{GP}}$:

$$\begin{vmatrix} \dfrac{\hbar^2k^2}{2m} + gn_0 - \hbar\omega & gn_0 \\[2ex] -gn_0 & -\dfrac{\hbar^2k^2}{2m} - gn_0 - \hbar\omega \end{vmatrix} = 0. \tag{1.92}$$

One infers from Eq. (1.92) the dispersion law of the elementary excitations of a condensate in a box, known as the Bogolubov spectrum:

$$\hbar\omega = \left[\frac{\hbar^2k^2}{2m}\left(\frac{\hbar^2k^2}{2m} + 2gn_0\right)\right]^{1/2}. \tag{1.93}$$

The dispersion law varies linearly with the wave vector k in the large-wavelength regime: $\omega = kc$, where $c = (gn_0/m)^{1/2}$ is the sound velocity and plays a key role for the superfluidity properties of a Bose–Einstein condensate (see Section 1.4.3.4).

1.3.7 Analogy with hydrodynamic equations

The Gross–Pitaevskii equation can be rewritten in the form of hydrodynamic equations. Their expression is particularly simple in the strong-interaction regime, and turns out to be very useful for the interpretation of several physical effects.

1.3.7.1 *Density field and velocity field*

In order to derive the set of hydrodynamic equations that is equivalent to the time-dependent Gross–Pitaevskii equation, it is convenient to normalize the wave function to the number of particles: $\int \mathrm{d}^3 r \, |\varphi(\vec{r}, t)|^2 = N$. The spatial density is then given by the modulus of the wave function: $n(\vec{r}, t) = |\varphi(\vec{r}, t)|^2$. With such a choice for the normalization of the wave function, there is no longer an explicit dependence of the time-dependent Gross–Pitaevskii with the number of atoms N:

$$i\hbar \frac{\partial}{\partial t} \varphi(\vec{r}, t) = -\frac{\hbar^2}{2m} \Delta\varphi(\vec{r}, t) + V_{\text{trap}}(\vec{r}, t)\, \varphi(\vec{r}, t) + g\, |\varphi(\vec{r}, t)|^2\, \varphi(\vec{r}, t). \tag{1.94}$$

It is instructive to rewrite the wave function $\varphi(\vec{r}, t)$ in terms of its phase $S(\vec{r}, t)$ and its modulus: $\varphi(\vec{r}, t) = \sqrt{n(\vec{r}, t)} \exp[i\, S(\vec{r}, t)]$.

1.3.7.2 *Continuity equation. Evolution equation of the velocity field*

Using Eq. (1.94) to derive the equation fulfilled by the modulus $n(\vec{r}, t)$, one obtains:

$$\frac{\partial}{\partial t} n(\vec{r}, t) + \vec{\nabla} \cdot [n(\vec{r}, t)\, \vec{v}(\vec{r}, t)] = 0, \tag{1.95}$$

where $\vec{v}$ denotes the velocity field and is proportional to the gradient of the phase $S(\vec{r}, t)$:

$$\vec{v}(\vec{r}, t) = \frac{\hbar}{m} \vec{\nabla} S(\vec{r}, t). \tag{1.96}$$

Equation (1.95) is the continuity equation. It shows that the integral of $n(\vec{r}, t)$ over space does not change in time, or otherwise stated, that, if φ is normalized at $t = 0$, it remains normalized for all $t > 0$. The quantity $n(\vec{r}, t)\, \vec{v}(\vec{r}, t)$ is the current density. We emphasize that the velocity field obeys the equation $\vec{\nabla} \times \vec{v}(\vec{r}, t) = \vec{0}$ and is therefore irrotational.[7]

Similarly, the equation for the evolution of the phase $S(\vec{r}, t)$, and thus of the velocity field $\vec{v}(\vec{r}, t)$ is readily inferred from Eq. (1.94):

$$m\frac{\partial \vec{v}}{\partial t} = \vec{\nabla} \left[\frac{\hbar^2}{2m} \frac{1}{\sqrt{n}} \Delta\sqrt{n} - \frac{1}{2}m\vec{v}^2 - V_{\text{trap}} - gn \right]. \tag{1.97}$$

The continuity equation (1.95) for the density and the Euler-like equation (1.97) for the velocity field are often referred to as hydrodynamics equations, and are strictly equivalent to the time-dependent Gross–Pitaevskii equation (1.94) as soon as the solutions do not exhibit singularities.

1.3.7.3 Quantum pressure

The first term of the right-hand side of the Euler equation (1.97) is the only one that contains explicitly $\hbar$, and is called the quantum pressure term. It originates from the kinetic-energy term arising from density gradients, and is a direct consequence of the Heisenberg uncertainty principle.

In order to determine in which circumstances this term plays a role, let us denote by R the characteristic length of the spatial variations of the atomic density $n(\vec{r})$. The quantum pressure term scales as $\hbar^2/2mR^2$ and is negligible compared to the interaction term gn when

$$R \gg \left(\frac{\hbar^2}{2mgn}\right)^{1/2} = \xi, \tag{1.98}$$

where ξ is the healing length. The healing length thus appears as the characteristic length ξ such that the energy of confinement in a volume ξ^3 is equal to the interaction energy.

1.3.7.4 Small-amplitude oscillations of an homogeneous condensate. Bogolubov dispersion law

The hydrodynamic equations can be used to find the frequencies of the small-amplitude oscillations of the condensate around equilibrium in the absence of confining potential $V_{\text{trap}} = 0$ (homogeneous condensate). To perform this calculation, we start by expanding the density and the velocity field with respect to their equilibrium values in the form

$$n(\vec{r}, t) = n_0 + \delta n(\vec{r}, t) \quad \text{and} \quad \vec{v}(\vec{r}, t) = \vec{0} + \delta\vec{v}(\vec{r}, t). \tag{1.99}$$

By inserting these relations into the hydrodynamics equations and restricting the expansion to the first order in δn and δv, one obtains:

$$\frac{\partial^2 \delta n}{\partial t^2} + n_0 \vec{\nabla} \left(\frac{\partial \delta \vec{v}}{\partial t}\right) = 0 \quad \text{and} \quad \frac{\partial \delta \vec{v}}{\partial t} = \vec{\nabla} \left(\frac{\hbar^2}{4m^2 n_0} \Delta \delta n - g \, \delta n\right). \tag{1.100}$$

Combining those two equations, we get the equation obeyed by the density perturbation $\delta n(\vec{r}, t)$:

$$\frac{\partial^2 \delta n}{\partial t^2} + \frac{\hbar^2}{4m} \Delta(\Delta \delta n) - \frac{gn_0}{m} \Delta \delta n = 0. \tag{1.101}$$

By inserting in Eq. (1.101) solutions having the form of a plane wave propagating through the homogeneous condensate, $\delta n(\vec{r}, t) = \delta n_0 \exp[i(kx - \omega t)]$, one recovers the Bogolubov dispersion law:

$$-\omega^2 + \frac{\hbar^2}{4m^2}k^4 + \frac{gn_0}{m}k^2 = 0 \iff \omega = ck\sqrt{1 + \frac{1}{k^2c^2}\left(\frac{\hbar k^2}{2m}\right)^2}. \quad (1.102)$$

1.3.8 Thomas–Fermi approximation for time-dependent problems

In the stationary case, the time-independent Gross–Pitaevskii equation can be simplified in the limit of strong interactions ($\chi \gg 1$) by neglecting the kinetic-energy term.

In problems involving the dynamics of the condensate, it is not correct, in the limit $\chi \gg 1$, to neglect the kinetic-energy term in the time-dependent Gross–Pitaevskii equation (1.94). For example, in the ballistic expansion of a condensate, after switching off suddenly the confinement, the kinetic energy term is small at the beginning of the expansion compared to the interaction energy, but the interaction energy is converted in the course of the expansion into kinetic energy that therefore becomes very large and even dominant for long expansion times.

However, the Thomas–Fermi approximation takes a simple form on the Euler hydrodynamic equation where the contributions of the amplitude gradient and phase gradient to the kinetic energy are clearly separated, being represented, respectively, by the first two terms of the right-hand side of Eq. (1.97). When $\chi \gg 1$, the amplitude gradient (appearing in the quantum pressure term) remains small at all times, whereas the second term, coming from phase gradients, can become very large.

The Thomas–Fermi limit, in the time-dependent case, thus corresponds to a situation where the quantum pressure term can be neglected, so that the set of hydrodynamic equations is, in this limit, equivalent to the following two equations

$$\frac{\partial}{\partial t}n + \vec{\nabla} \cdot [n\vec{v}] = 0,$$

$$m\frac{\partial \vec{v}}{\partial t} = \vec{\nabla}\left[-\frac{1}{2}m\vec{v}^2 - V_{\text{trap}} - gn\right]. \quad (1.103)$$

It is worth noticing that, in the regime $\chi \gg 1$, $\hbar$ no longer appears in the equation for the velocity field, which consequently appears as a classical Euler equation describing the motion of a fluid in the trapping potential and in the pressure field due to the density of other particles. The motion of a condensate, in the Thomas–Fermi limit, and in the mean-field approximation, can thus be described by classical, irrotational hydrodynamics (since by definition the velocity field obeys the relation $\vec{\nabla} \times \vec{v}(\vec{r}, t) = \vec{0}$).

In the static case, for which there is no global motion of the condensate, the velocity field $\vec{v}$ is equal to $\vec{0}$, and Eq. (1.103) gives:

$$\vec{\nabla}\left[V_{\text{trap}}(\vec{r}) + gn(\vec{r})\right] = \vec{0}. \quad (1.104)$$

We recover here the equilibrium shape in the Thomas–Fermi limit derived in Section 1.3.4.3.

The hydrodynamic equations (1.103) also give access to the motion of the small-amplitude oscillations of the condensate in the presence of a confinement ($V_{\text{trap}} \neq 0$). The calculation is similar to the one performed in Section 1.3.7.4 to derive the Bogolubov dispersion law. One linearizes the hydrodynamic equations around the equilibrium state defined by

$$n_0(\vec{r}) = [\mu - V_{\text{trap}}(\vec{r})]/g, \qquad \text{and} \qquad \vec{v}_0(\vec{r}) = \vec{0}. \tag{1.105}$$

One finds that the first-order corrections δn to the density and $\delta \vec{v}$ to the velocity field obey the linear set of equations:

$$\frac{\partial \delta n}{\partial t} = -\vec{\nabla} \cdot (n_0 \delta \vec{v}), \tag{1.106}$$

$$m \frac{\partial \delta \vec{v}}{\partial t} = -\vec{\nabla}(V_{\text{trap}} + g n_0 + g \,\delta n) = -g \vec{\nabla} \delta n. \tag{1.107}$$

The equation governing the density perturbation $\delta n(\vec{r}, t)$ therefore reads:

$$\frac{\partial^2 \delta n(\vec{r}, t)}{\partial t^2} = \vec{\nabla} \cdot \left[c^2(\vec{r}) \vec{\nabla} \delta n(\vec{r}, t)\right], \text{ with } c^2(\vec{r}) = \frac{g}{m} n_0(\vec{r}). \tag{1.108}$$

The quantity $c(\vec{r})$ is a local sound velocity. Sound waves can propagate in a non-uniform medium. For a cylindrical geometry with a transverse harmonic confinement, the sound velocity in the longitudinal direction is $(\mu/2M)^{1/2}$ (Zaremba, 1998; Kavoulakis and Pethick, 1998; Stringari, 1998b). This result differs from the one obtained in a box by a factor of $1/\sqrt{2}$ since it is the average density over the radial direction, and not the peak density, that dictates the value of the sound velocity in such an elongated geometry. The propagation of sound waves in such a geometry has been studied experimentally in (Andrews et al., 1998).

1.3.9   Thomas–Fermi dynamics for harmonic confinement

We suppose in this section that the trapping potential is harmonic but not necessarily isotropic:

$$V_{\text{trap}}(\vec{r}, t) = \frac{1}{2} \sum_{i=x,y,z} m \,\omega_i^2(t) \, r_i^2, \tag{1.109}$$

with $(r_1, r_2, r_3) = (x, y, z)$, and that the condensate is in the Thomas–Fermi regime ($\chi \gg 1$), so that we can use the hydrodynamic equations (1.103) without the quantum pressure term. The time dependence of the trapping frequencies $\omega_i(t)$ will allow us to analyze several problems that can be readily investigated experimentally.

1.3.9.1   Scaling transformation

When the strength of the confining potential is changed as a function of time, the time-dependent Gross–Pitaevskii can be solved using a scaling transformation. In the following, we denote by $\omega_i(0)$ the angular frequency along the $i = x, y, z$ axis for a time

$t \leq 0$, and $\omega_i(t)$ for $t \geq 0$. The time-dependent density resulting from the excitation is searched in the form (Kagan *et al.*, 1997):

$$n(r_i, t) = \frac{1}{b_x(t) b_y(t) b_z(t)} n_0 \left(\frac{r_i}{b_i(t)} \right) = \frac{1}{\Pi_j b_j(t)} n_0 \left(\frac{r_i}{b_i(t)} \right), \qquad (1.110)$$

where n_0 is the initial equilibrium density distribution. The prefactor $\Pi_j b_j^{-1}(t)$ ensures the normalization of the density to the number of atoms. The ansatz (1.110) inserted in the continuity equation gives the expression for the velocity field:

$$v_j(\vec{r}, t) = \frac{\dot{b}_j(t)}{b_j(t)} r_j. \qquad (1.111)$$

The Euler equation (1.103) yields:

$$m \frac{\partial v_j}{\partial t} + \frac{\partial}{\partial r_j} \left(\frac{1}{2} m v^2 + V_{\text{trap}} \right) = \frac{\ddot{b}_j(t)}{b_j(t)} r_j + m \omega_j^2(t) r_j - g \frac{\partial n}{\partial r_j}. \qquad (1.112)$$

The calculation of the last term requires the knowledge of the equilibrium Thomas–Fermi profile (1.74), from which one deduces

$$n(r_i, t) = \frac{1}{\Pi_j b_j(t)} \left[\frac{\mu}{g} - \frac{m}{2g} \right] \sum_i \omega_i^2(0) \frac{r_i^2}{b_i^2(t)}. \qquad (1.113)$$

Combining this expression with Eq. (1.112), we find the set of non-linear coupled equations fulfilled by the dilation factors b_j:

$$\ddot{b}_j(t) + \omega_j^2(t) b_j(t) - \frac{\omega_j^2(0)}{b_j(t)} \frac{1}{\Pi_i b_i(t)} = 0. \qquad (1.114)$$

This set of equations for the scaling factors means that it is possible to account for large-amplitude oscillations, and to investigate non-linear features associated, for example, with the dynamics of the expansion of the gas, by simply solving a set of three non-linear coupled ordinary differential equations. It is worth noticing, however, that such an approach is restricted to quadratic potentials.

1.3.9.2 *Ballistic expansion*

To analyze the properties of the condensate, the standard method consists in monitoring the evolution of the shape of the condensate after having switched off suddenly the trapping potential (Anderson *et al.*, 1995). Such a ballistic expansion is usually unavoidable for an optical detection since the in-trap transverse size of the condensate is on the order of 1 μm, and, in addition its optical density is very large.

For an ideal gas, the *in situ* position dispersion $\Delta x_i(0)$ along a given axis gives rise, through the Heisenberg principle, to a velocity dispersion $\Delta v_i = \hbar/(2m \Delta x_i(0))$. After switching off the trapping, the cloud expands and the position dispersion evolves according to $\Delta x_i(t) = [(\Delta x_i(0))^2 + (\Delta v_i)^2 t^2]^{1/2}$. For long expansion times, the size is dominated by the velocity dispersion term. If the initial harmonic potential is anisotropic, the expansion reflects this anisotropy, since the velocity dispersion along one axis is inversely proportional to the initial position dispersion. The ellipticity of the clouds is reversed in the course of the expansion; this *inversion of ellipticity* is usually considered as the "smoking-gun" evidence for Bose–Einstein condensation.

Let us consider now a condensate in the presence of atom–atom interactions and that is initially well described by the Thomas–Fermi approximation. We can therefore apply the scaling-factors formalism of Section 1.3.9.1 to the analysis of time-of-flight experiments. The angular frequencies are modified at time $t = 0$ from $\omega_j\,(t < 0) = \omega_j \neq 0$ to $\omega_j\,(t \geq 0) = 0$. Let us suppose that the trap has a cylindrical symmetry around Oz, and a cigar shape: $\omega_x = \omega_y = \omega_\perp \gg \omega_z$. For $t \geq 0$, the equations (1.114) can be recast in the form

$$\frac{\mathrm{d}^2 b_\perp\,(\tau)}{\mathrm{d}\tau^2} = \frac{1}{b_\perp^3\,(\tau)\,b_z\,(\tau)}, \quad \text{and} \quad \frac{\mathrm{d}^2 b_z\,(\tau)}{\mathrm{d}\tau^2} = \frac{\lambda^2}{b_\perp^2\,(\tau)\,b_z^2\,(\tau)}, \tag{1.115}$$

where $\tau = \omega_\perp t$ and $\lambda = \omega_z / \omega_\perp$. The solution to the second order in $\lambda \ll 1$ gives the evolution of the ellipticity $E(\tau)$ during the expansion (Castin and Dum, 1996):

$$E\,(\tau) = \frac{R_\perp\,(0)\,b_\perp\,(\tau)}{R_z\,(0)\,b_z\,(\tau)} = E\,(0)\,\frac{\sqrt{1 + \tau^2}}{1 + \lambda^2\left[\tau \arctan\,(\tau) - \ln\sqrt{1 + \tau^2}\right]}. \tag{1.116}$$

It is worth mentioning that this equation means that the information on the initial anisotropy of the wave function is not lost during the expansion. Actually, this anisotropy is transferred to the velocity field, and, during the expansion, which acts as a magnifier for the condensate wave function, one observes an inversion of the ellipticity of the cloud (see Figure 1.9). This is to be contrasted with what one would observe with a Boltzmann gas initially at equilibrium in an anisotropic trap. The density profile of a trapped cloud before switching off the confinement reflects the anisotropy of the trap. However, after a sufficiently long expansion time, the information on the initial sizes are lost and the shape of the cloud is spherical, reflecting the isotropy of the initial velocities Boltzmann distribution.[8]

1.3.9.3   Normal modes

Equation (1.108) allows one to calculate the frequencies of the collective modes of a trapped Bose–Einstein condensate in the Thomas–Fermi regime by searching for solutions with a time dependence $\delta n \propto e^{-i\omega t}$:

$$-\omega^2 \delta n\,(\vec{r}) = \vec{\nabla} \cdot \left[c^2(\vec{r})\vec{\nabla}\delta n\,(\vec{r})\right]. \tag{1.117}$$

Experimentally, the measurement of those frequencies can be carried out with a high precision, and therefore allows for precise quantitative comparisons with theory.

We first consider the case of an isotropic harmonic confinement $V_{\mathrm{trap}}(\vec{r}) = m\omega_0^2 r^2/2$ for which the normal modes can be obtained analytically (Stringari, 1996).

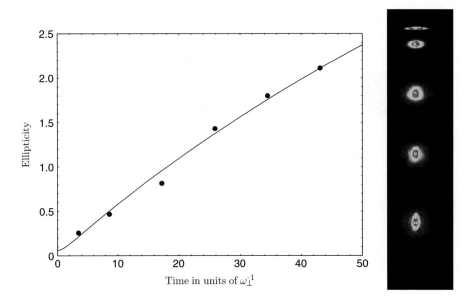

Fig. 1.9   Aspect ratio $E(\tau)$ (see text) of an expanding condensate as a function of time. The experimental data were taken with an ellipticity at time $t = 0$ equal to $E(0) = 0.1$, the solid line is obtained from Eq. (1.116). Figure taken from (Guéry-Odelin, 1998).

According to Eq. (1.74), the density profile at equilibrium is given by $n_0(\vec{r}) = \mu[1 - r^2/R^2]/g$. One finds that the atomic density for the normal modes is of the form

$$\delta n(\vec{r}) = P_\ell^{(2n)}(r/R)\, r^\ell Y_m^\ell(\theta, \varphi), \tag{1.118}$$

where $P_\ell^{(2n)}$ are polynomials of degree $2n$, and $Y_m^\ell(\theta, \varphi)$ are the spherical harmonics. From Eq. (1.117), we deduce the dispersion law of the normal modes[9]:

$$\omega(n, \ell) = \omega_0 \left(2n^2 + 2n\ell + 3n + \ell\right)^{1/2}. \tag{1.119}$$

By contrast to the uniform case, we find that the dispersion law does not depend on the interaction parameter. Actually, for large values of n and ℓ the density modulation occurs on small spatial scales, and it is no longer correct to neglect the quantum pressure term (Dalfovo *et al.*, 1999).

Using Eqs. (1.118) and (1.107), we find that the modes with $n = 0$ have a velocity field of the form $\delta\vec{v}(\vec{r}) \sim \vec{\nabla}\left(r^\ell Y_m^\ell(\theta, \varphi)\right)$. Because of the properties of the spherical harmonics, $\vec{\nabla} \cdot \delta\vec{v} \sim \Delta\left(r^\ell Y_m^\ell(\theta, \varphi)\right) = 0$, and the normal modes with $n = 0$ are divergence free. For this particular case, the condensate behaves as if it was incompressible: the "surface" of the condensate is modified while its volume remains constant. This subset of normal modes is for this reason referred to as surface modes. Their eigenfrequencies are $\omega_0\sqrt{\ell}$.

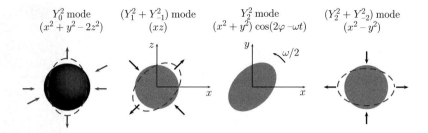

Fig. 1.10 Examples of quadrupole modes for a spherical trap.

Dipole modes. The $n = 0$, $\ell = 1$ modes are the dipole modes, they have the same angular frequency ω_0 as that of the confinement. They describe a global translation of the cloud. This is easily justified by considering for instance the $\ell = 1$ and $m = 0$ mode for which $\delta n(\vec{r}) = (\delta n_0/R)(rY_0^1) = (\delta n_0/R)z$, where δn_0 is a constant and R is the Thomas–Fermi radius. The density can therefore be put in the following form to the lowest order in $(g\delta n_0/\mu)$:

$$n(\vec{r}) = n_0(\vec{r}) + \delta n(\vec{r}, t) = \frac{\mu}{g}\left(1 - \frac{x^2 + y^2}{R^2} - \frac{(z - z_0(t))^2}{R^2}\right), \tag{1.120}$$

where $z_0(t) = R(g\delta n_0/2\mu)\cos(\omega_0 t)$ is the position of the center of mass of the condensate. The dipole mode, also referred to as Kohn's mode, characterizes the center of mass oscillation, and is therefore not affected by two-body interactions,[10] nor by the temperature or the statistics. Experimentally, this mode is often used to calibrate the value of the trap frequencies.

Quadrupole modes. The surface modes with $\ell = 2$ are referred to as quadrupole modes and have a characteristic frequency equal to $\sqrt{2}\omega_0$. Let us detail the kind of density deformation they generate (see also Figure 1.10):

- The mode $\ell = 2, m = 0$ corresponds to a density perturbation

  $$\delta n(\vec{r}, t) \sim \mathrm{Re}(r^2 Y_0^2(\theta, \varphi)e^{-i\omega t}) \sim (2z^2 - x^2 - y^2)\cos(\omega t).$$

  The compression in the transverse plane is accompanied by a dilatation along the longitudinal axis, and conversely.

- The modes $\ell = 2, m = \pm 1$ generate an elliptical deformation along an axis at $45°$ with respect to the Ox-Oz axis and Oy-Oz axis, respectively.

- The mode $\ell = 2, m = 2$ corresponds to a density perturbation

  $$\delta n(\vec{r}, t) \sim \mathrm{Re}(r^2 Y_2^2(\theta, \varphi)e^{-i\omega t}) \sim (x^2 + y^2)\cos(2\varphi - \omega t).$$

  It corresponds to an elliptic deformation in the plane xOy that rotates at a velocity $\omega/2$ (see Figure 1.10). The same conclusion holds for the mode $\ell = 2, m = -2$ but with a reverse rotation velocity. The linear superposition of those two modes with equal real weights gives a density perturbation of the form $\delta n(\vec{r}, t) \sim (x^2 - y^2)\cos\omega t$ that has an opposite sign along the x and y axis, and is therefore commonly referred to as the $x^2 - y^2$ mode. The linear superposition with opposite sign gives rise to the xy mode. These modes do not affect the z axis.

A general feature of surface modes is that they are insensitive to the form of the equation of state $\mu(n)$. This is the reason why they have the same expression for a Bose gas above the critical temperature when it can be described by hydrodynamic equations, i.e. in the regime for which the collision rate is very large in comparison with the trapping angular frequency (Griffin *et al.*, 1997).

Monopole mode. This peculiarity of surface modes is to be contrasted with the monopole mode $n = 1$, $\ell = 0$ that has a frequency $\sqrt{5}\omega_0$ for a Bose–Einstein condensate in the Thomas–Fermi regime in a spherical harmonic trap. The frequency of this mode is intrinsically linked to the compressibility of the condensate (Singh and Rokhsar, 1996). For negative scattering length, the compressibility increases all the more than the number of atoms is close to the critical value N_c (defined in Section 1.3.5.1) (Ueda and Leggett, 1998). As intuitively expected, the larger the compressibility, the lower the monopole mode frequency.[11]

This mode corresponds to a breathing of the radius, and therefore does not conserve the volume. For a Bose gas above the critical temperature, or for a one-species Fermi gas, this mode has a frequency $2\omega_0$ regardless of the elastic collision rate (Guéry-Odelin *et al.*, 1999).

Actually, the monopole mode and the quadrupole modes $(x^2 + y^2 - 2z^2)$ and $(x^2 - y^2)$ can readily be obtained using the scaling factors approach developed in Section 1.3.9.1 by linearizing the set of equations (1.114) around $b_i = 1$. This method provides the modes that have the symmetry adapted to the scaling factors, but can also be used for non-spherical traps.

Surface modes for cylindrically-symmetric traps. Most experiments are performed with a trapping geometry having a cylindrical symmetry. Atoms thus experience a confining potential of the form:

$$V_{\text{trap}}(x, y, z) = \frac{1}{2}m\omega_\perp^2(x^2 + y^2) + \frac{1}{2}m\omega_z^2 z^2. \tag{1.121}$$

For instance, Ioffe–Pritchard traps give rise to cigar-shaped clouds ($\omega_z \ll \omega_\perp$), and TOP traps to pancake-shaped traps with $\omega_z = \sqrt{8}\omega_\perp$ (see e.g. Chapter 4 of (Pethick and Smith, 2002)). The normal modes are strongly affected by the geometry. However, the azimuthal quantum number m remains a good quantum number because of the axial symmetry. This is no longer the case of the quantum number ℓ. One finds that a perturbation density of the form $\delta n(\vec{r}) = r^2 Y_m^2(\theta, \varphi)$ is still a solution of Eq. (1.117) with eigenfrequency $\omega = \omega_\perp\sqrt{2}$ for $m = \pm 2$, and $\omega = (\omega_\perp^2 + \omega_z^2)^{1/2}$ for $m = \pm 1$. A density perturbation of the form $r^2 Y_0^2(\theta, \varphi)$ is no longer an eigenmode. However, two linear superpositions of the $m = 0$ quadrupole $r^2 Y_0^2$ and monopole[12] $r^2 Y_0^2$ gives rise to the eigenfrequencies[13] (Stringari, 1996):

$$\omega_\pm^2 = 2\omega_\perp^2 + \frac{3}{2}\omega_z^2 \mp \frac{1}{2}\sqrt{9\omega_z^2 - 16\omega_z^2\omega_\perp^2 + 16\omega_\perp^4}. \tag{1.122}$$

This prediction has been confronted to experimental results in (Jin *et al.*, 1996). Results are depicted in Figure 1.11 for the mode $m = 0$ with the ω_- frequency and the mode $m = 2$. The measured frequencies deviate clearly from the non-interacting

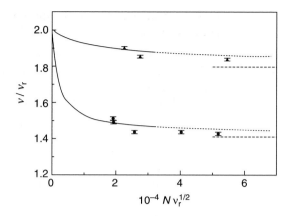

Fig. 1.11 Measurement of two quadrupole mode frequencies for a Bose–Einstein condensate in a TOP trap as a function of the interaction strength parameter $\chi \propto N\nu_r^{1/2}$, where ν_r is the radial frequency. Solid lines show the mean-field calculation by Edwards and co-workers (Edwards *et al.*, 1999), dotted lines show the results of similar calculations by Esry (Esry, 1997). For the strongly interacting limit(dashed line), we have $\omega = \omega_0\sqrt{2}$ for the $m = 2$ quadrupole mode, and $\omega_- = 1.797\omega_0$ (Stringari, 1996). Figure taken from (Jin *et al.*, 1996), copyright American Physical Society.

model that predicts a frequency equal to $2\omega_0$, and tend to the Thomas–Fermi limit given by Eq. (1.122) when the interaction strength parameter χ is sufficiently large. A full mean-field calculation is required if χ is not large enough since the quantum pressure term plays a non-negligible role in this regime (Edwards *et al.*, 1999).

Let us conclude this section devoted to the normal modes by mentioning the origin of other corrections to the hydrodynamics results within the Thomas–Fermi limit:

- the finite-size corrections (Zambelli and Stringari, 2002).
- the beyond-mean-field effects that introduce corrections to the equation of state of the condensate $\mu(n)$, on the order of the gas parameter $(na^3)^{1/2}$ and thus affect the compression modes (for instance the monopole mode) but not the surface modes (Pitaevskii and Stringari, 1998; Braaten and Pearson, 1999).
- the interaction with the thermal cloud that leads to damping and frequency shifts of condensate collective modes, as measured in several experiments (Jin *et al.*, 1997; Stamper-Kurn *et al.*, 1998; Maragò *et al.*, 2001; Chevy *et al.*, 2002).

1.4 Beyond mean-field theory

1.4.1 Introduction

In the previous section, we have investigated the ground state of a weakly interacting N-boson system trapped at zero temperature. This state has been determined using the variational approach that describes the motion of each particle in the mean-field created by the other $N - 1$ particles. In such an approximate treatment, one neglects the correlations between atoms and assumes that they share the same one-particle state. In this limit, the state of the condensate is described by a classical field that obeys the Gross–Pitaevskii equation.

In this section, we provide a more elaborate theoretical framework for the weakly interacting Bose gas. The gas will be described by a quantum field operator, and we will take into account the quantum fluctuations about the state in which all atoms share the same single quantum state. As a consequence of quantum fluctuations, even at $T = 0$, the atoms are not all in the same one-particle state. This effect is known as *quantum depletion*. One may wonder how far does the exact ground state differ from the ideal Bose gas situation for which the ground state is simply given by a product of N one-particle states, i.e. what is the proportion of atoms occupying other states than the ground state.

This is precisely the purpose of the Bogolubov theory that requires the gas to be sufficiently dilute so that only two-body collisions with small momentum transfers play an important role. The small parameter that allows a perturbative treatment is $na^3 \ll 1$, where n is the atomic spatial density and a is the s-wave scattering length. The Bogolubov approach relies on the description of the N-particle state in Fock space. Its formulation is therefore performed within the second quantization formalism.

This microscopic theory allows one to obtain more insight into atom–atom correlations and quantum depletion, and also to describe the first excited states known as the *elementary excitations* to which quasi-particles can be associated.

In the following, the Bogolubov approach is developed for a homogeneous Bose gas where particles are assumed to be in a box. This case is simpler to handle since the density does not vary in space, and yields analytical predictions. In addition, we assume a positive scattering length to ensure that the system is stable against collapse.

In the second quantized theory, one introduces field operators $\hat{\Psi}^\dagger(\vec{r})$ and $\hat{\Psi}(\vec{r})$ that create and destroy an atom at $\vec{r}$. The second-order correlation function of these field operators will be presented in Section 1.4.4. It is useful for analyzing the coherence properties of BECs, as illustrated in Section 1.1 for the ideal gas.

1.4.2   Homogeneous condensates

In 1947, Bogolubov introduced the modern description of a dilute interacting low-temperature Bose gas for an homogeneous Bose gas (Bogolubov, 1947). In this case, the one-particle state that has a macroscopic population is always the single-particle ground state of the box.[14]

The second quantized form of the many-body Hamiltonian $H = H_{\text{kin}} + H_{\text{int}}$ contains (i) a kinetic term H_{kin} that is the sum of products of two creation and annihilation operators ($\hat{a}_{\vec{k}}^\dagger$ and $\hat{a}_{\vec{k}}$, respectively) of an atom in the state $\vec{k}$ of the box, and (ii) the two-body interaction terms H_{int} that involve the sum of products of four creation and annihilation operators.

The first step of the Bogolubov approach relies on the fact that the Bose gas being sufficiently dilute, it should differ only slightly from an ideal gas, so that the ground state with zero momentum remains macroscopically occupied. Mathematically, this assumption is exploited by replacing the operators $\hat{a}_0$ and $\hat{a}_0^\dagger$ by c-numbers equal to $N^{1/2}$, where N is the total number of particles. In a second step, one neglects in the expression of H_{int} the products containing less than two $\hat{a}_0$ or $\hat{a}_0^\dagger$. One is then left for the total hamiltonian H with a quadratic expression in $\hat{a}_k$ and $\hat{a}_k^\dagger$. The Bogolubov theory consists in diagonalizing this quadratic hamiltonian.

1.4.2.1 *System of N bosons in a box*

For a gas confined in a box of volume L^3 with periodic boundary conditions, the one-particle states are plane waves

$$\langle \vec{r} | \varphi_{\vec{k}} \rangle = \varphi_{\vec{k}}(\vec{r}) = \frac{1}{L^{3/2}} \exp(i\,\vec{k} \cdot \vec{r}), \quad \text{with} \quad k_{x,y,z} = \frac{2\pi}{L} n_{x,y,z}, \tag{1.123}$$

where $n_{x,y,z}$ is any integer. The operator $\hat{a}_{\vec{k}}$ (resp. $\hat{a}_{\vec{k}}^\dagger$) destroys (respectively, creates) an atom in state $|\varphi_{\vec{k}}\rangle$. The two contributions to the many-body Hamiltonian H are

- the kinetic-energy term

$$\hat{H}_{\text{kin}} = \sum_{\vec{k}} \frac{\hbar^2 k^2}{2m} \hat{a}_{\vec{k}}^\dagger \hat{a}_{\vec{k}}, \tag{1.124}$$

- and the interaction term

$$\hat{H}_{\text{int}} = \frac{1}{2L^3} \sum_{\vec{k}_1} \sum_{\vec{k}_2} \sum_{\vec{k}} \tilde{V}(\vec{k}) \hat{a}_{\vec{k}_1}^\dagger \hat{a}_{\vec{k}_2}^\dagger \hat{a}_{\vec{k}_2+\vec{k}} \hat{a}_{\vec{k}_1-\vec{k}}, \tag{1.125}$$

where we have explicitly used the conservation of the total linear momentum, and introduced the Fourier transform of the two-body interaction potential

$$\tilde{V}(\vec{k}) = \int \mathrm{d}^3 r\, V(\vec{r}) e^{-i\vec{k}\cdot\vec{r}}. \tag{1.126}$$

In the dilute limit, the inter-particle potential may be approximated by a pseudo-potential[15] $V(\vec{r}) = g\,\delta(\vec{r})$ that has a constant Fourier transform $\tilde{V}(\vec{k}) = g$, and the original Hamiltonian $\hat{H}_{\text{int}}$ takes the simple form

$$\hat{H}_{\text{int}} = \frac{g}{2L^3} \sum_{\vec{k}_1} \sum_{\vec{k}_2} \sum_{\vec{k}} \hat{a}_{\vec{k}_1}^\dagger \hat{a}_{\vec{k}_2}^\dagger \hat{a}_{\vec{k}_2+\vec{k}} \hat{a}_{\vec{k}_1-\vec{k}}. \tag{1.127}$$

We emphasize that the latter expression assumes implicitly that the interactions are repulsive ($g > 0$) since a homogeneous condensate with attractive interactions cannot be stable.

1.4.2.2 *Inhibition of condensate fragmentation in a box with repulsive interactions*

One may wonder if, in the presence of repulsive atom–atom interactions, a macroscopic fraction of the atom could occupy not only one level but two, three, . . . To answer this question, we consider the system of the previous paragraph made of N interacting bosons confined in a box with periodic boundary conditions, and we compare the energy for two different situations: (*i*) all atoms are in the single-particle ground state, and (*ii*) atoms are occupying the first two states.

(*i*) For the first case, all particles share the same ground state with zero momentum $|\vec{k}_0 = \vec{0}\rangle$. The corresponding N-body wave function is the Fock state $|\psi_0\rangle = |N : \vec{k}_0, 0 \dots, 0 \dots\rangle$, and the calculation of the mean energy of this state reads:

$$E_0 = \langle \psi_0 | \hat{H}_{\rm kin} + \hat{H}_{\rm int} | \psi_0 \rangle = \langle \psi_0 | \hat{H}_{\rm int} | \psi_0 \rangle = \frac{g}{2L^3} \langle \hat{a}_0^\dagger \hat{a}_0^\dagger \hat{a}_0 \hat{a}_0 \rangle$$

$$= \frac{g}{2L^3} \langle \hat{a}_0^\dagger (\hat{a}_0 \hat{a}_0^\dagger - 1) \hat{a}_0 \rangle = \frac{g}{2L^3} N(N-1) \simeq \frac{gn}{2} N. \tag{1.128}$$

(*ii*) In the second case, we assume that N_1 atoms occupy the state $|\vec{k}_0 = \vec{0}\rangle$ and $N_2 = N - N_1$ the state $|\vec{k}_1\rangle$ so that the N-body wave function is given by the Fock state $|\psi\rangle = |N_1 : \vec{k}_0, N_2 : \vec{k}_1, 0 \dots, 0 \dots\rangle$. We assume the volume of the box to be sufficiently large to make the contribution of the kinetic energy of the N_2 atoms in the $|\vec{k}_1\rangle$ state negligible in comparison with the interaction energy. The calculation of the mean energy $\tilde{E}_0$ in the state $|\psi\rangle$ is therefore given by the contribution of the interaction energy $\tilde{E}_0 = \langle \psi | H_{\rm int} | \psi \rangle$. According to Eq. (1.127), only six terms have a non-vanishing contribution for the calculation of $\tilde{E}_0$, four with $\vec{k} = \vec{0}$ and two with $\vec{k} = \pm(\vec{k}_0 - \vec{k}_1)$:

$$\langle \psi | H_{\rm int} | \psi \rangle = \frac{g}{2L^3} \langle \hat{a}_0^\dagger \hat{a}_0^\dagger \hat{a}_0 \hat{a}_0 + \hat{a}_1^\dagger \hat{a}_1^\dagger \hat{a}_1 \hat{a}_1 + \hat{a}_0^\dagger \hat{a}_1^\dagger \hat{a}_1 \hat{a}_0 + \hat{a}_1^\dagger \hat{a}_0^\dagger \hat{a}_0 \hat{a}_1 \rangle$$

$$+ \frac{g}{2L^3} \langle \hat{a}_0^\dagger \hat{a}_1^\dagger \hat{a}_0 \hat{a}_1 + \hat{a}_1^\dagger \hat{a}_0^\dagger \hat{a}_1 \hat{a}_0 \rangle$$

$$= \frac{g}{2L^3} [N_1(N_1 - 1) + N_2(N_2 - 1) + 2N_1 N_2] + \frac{g}{2L^3} 2N_1 N_2$$

$$\simeq E_0 + \frac{gN_1 N_2}{L^3}. \tag{1.129}$$

We find that the fragmentation of the condensate over two states is not energetically favorable since $\tilde{E}_0 > E_0$, and it is therefore inhibited. This result is usually referred to as the Nozières's argument against fragmentation (Nozières, 1995).

As it clearly appears from the calculation, this conclusion originates from the two "exchange" terms. Those terms involve the overlapping of the wave functions, and add a mean-field contribution to the energy.[16]

1.4.3 Bogolubov theory

1.4.3.1 *Bogolubov prescription and diagonalization*

The ground state of an ideal Bose gas in a box is characterized by N particles in the single-particle mode:

$$|\Psi_0(1, \dots, N)\rangle = |N, 0, \dots, 0, \dots\rangle = \frac{(a_0^\dagger)^N}{(N!)^{1/2}} |0, 0, \dots, 0, \dots\rangle. \tag{1.130}$$

In a dilute Bose gas near $T = 0$, the population of the ground state is macroscopic and very large compared to the total population in all other states. It is then justified to make a theory limited to the first order in

$$\varepsilon = \frac{1}{N} \sum_{k \neq 0} n_k \ll 1. \tag{1.131}$$

The action of $\hat{a}_0$ on a Fock state can therefore be approximated by

$$\hat{a}_0 \left| \{n_k\} \right\rangle = n_0^{1/2} \left| \{n_k - \delta_{k0}\} \right\rangle = N^{1/2} (1 - \varepsilon)^{1/2} \left| \{n_k - \delta_{k0}\} \right\rangle$$
$$\simeq N^{1/2} \left| \{n_k - \delta_{k0}\} \right\rangle. \tag{1.132}$$

The matrix elements of $\hat{a}_0$ and $\hat{a}_0^\dagger$ are thus on the order of $N^{1/2} \gg 1$. We will neglect[17] the non-commutation of $\hat{a}_0$ and $\hat{a}_0^\dagger$ and use the so-called Bogolubov prescription by replacing them by a c-number $N^{1/2}$: $\hat{a}_0^\dagger \simeq \hat{a}_0 \simeq N^{1/2}$.

In Eq. (1.127), there are terms containing four, two or zero operators a_0 or $a_0^\dagger$. There are no terms with an odd number of a_0 or $a_0^\dagger$ because of the conservation of the total linear momentum. The term with four creation and annihilation operators requires some care since it is very large:

$$\hat{a}_0^\dagger \hat{a}_0^\dagger \hat{a}_0 \hat{a}_0 = \hat{a}_0^\dagger \left(\hat{a}_0 \hat{a}_0^\dagger - 1 \right) \hat{a}_0 = \left(\hat{a}_0^\dagger \hat{a}_0 \right)^2 - \hat{a}_0^\dagger \hat{a}_0$$

$$= \left(\hat{N} - \sum_{k \neq 0} \hat{a}_{\vec{k}}^\dagger \hat{a}_{\vec{k}} \right)^2 - \hat{a}_0^\dagger \hat{a}_0 \simeq \hat{N}^2 - 2\hat{N} \sum_{k \neq 0} \hat{a}_{\vec{k}}^\dagger \hat{a}_{\vec{k}} - \hat{N}. \tag{1.133}$$

Substituting the Bogolubov prescription into the other terms of Eq. (1.127) and keeping only the leading terms of order N^2 and N, one can finally rewrite the hamiltonian $H = H_{\text{kin}} + H_{\text{int}}$ in a quadratic form in the creation and annihilation operators.

$$\hat{H} = \frac{gn}{2} N + \sum_{\vec{k}} \frac{\hbar^2 k^2}{2m} \hat{a}_{\vec{k}}^\dagger \hat{a}_{\vec{k}}$$

$$+ \frac{gn}{2} \sum_{\vec{k} \neq \vec{0}} \left[\hat{a}_{\vec{k}}^\dagger \hat{a}_{\vec{k}} + \hat{a}_{-\vec{k}}^\dagger \hat{a}_{-\vec{k}} + \hat{a}_{\vec{k}}^\dagger \hat{a}_{-\vec{k}}^\dagger + \hat{a}_{\vec{k}} \hat{a}_{-\vec{k}} \right]. \tag{1.134}$$

This truncated form of the hamiltonian neglects interaction of particles out of the condensate. The important feature of Eq. (1.134) is that it can be solved exactly since it is a quadratic form in the operators $\hat{a}_{\vec{k}}$ and $\hat{a}_{\vec{k}}^\dagger$.

If we write the Heisenberg equation for the time evolution of the operator $\hat{a}_{\vec{k}}$, we find that $\hat{a}_{\vec{k}}$ is coupled to $\hat{a}_{-\vec{k}}^\dagger$, and similarly that the time evolution of $\hat{a}_{-\vec{k}}^\dagger$ is coupled to $\hat{a}_{\vec{k}}$:

$$i\hbar \frac{d\hat{a}_{\vec{k}}}{dt} = [\hat{a}_{\vec{k}}, \hat{H}] = \left(\frac{\hbar^2 k^2}{2m} + gn \right) \hat{a}_{\vec{k}} + gn\hat{a}^{\dagger}_{-\vec{k}},$$

$$i\hbar \frac{d\hat{a}^{\dagger}_{-\vec{k}}}{dt} = [\hat{a}^{\dagger}_{-\vec{k}}, \hat{H}] = -\left(\frac{\hbar^2 k^2}{2m} + gn \right) \hat{a}^{\dagger}_{-\vec{k}} - gn\hat{a}_{\vec{k}}. \tag{1.135}$$

It is therefore natural to search for the diagonalization of the Hamiltonian (1.134) a canonical transformation where a linear combination of these two operators is involved:

$$\hat{b}_{\vec{k}} = u_k \hat{a}_{\vec{k}} + v_k \hat{a}^{\dagger}_{-\vec{k}} \quad \text{and} \quad \hat{b}^{\dagger}_{-\vec{k}} = u_k \hat{a}^{\dagger}_{-\vec{k}} + v_k \hat{a}_{\vec{k}}, \tag{1.136}$$

where u_k and v_k are real coefficients depending only on $k = |\vec{k}|$. The corresponding transformation is canonical if the new set of operators $\hat{b}_{\vec{k}}$ and $\hat{b}^{\dagger}_{-\vec{k}}$ obey the bosonic commutation relations

$$\left[\hat{b}_{\vec{k}}, \hat{b}^{\dagger}_{\vec{k}} \right] = u_k^2 \left[\hat{a}_{\vec{k}}, \hat{a}^{\dagger}_{\vec{k}} \right] + v_k^2 \left[\hat{a}^{\dagger}_{-\vec{k}}, \hat{a}_{-\vec{k}} \right] = u_k^2 - v_k^2 = 1, \tag{1.137}$$

for $k \neq 0$. This relation is automatically fulfilled by searching the coefficients u_k and v_k in the form $u_k = \cosh \theta_k$ and $v_k = \sinh \theta_k$. The parameter θ_k is determined by imposing that the expression of the Hamiltonian H in terms of the creation and annihilation operators $\hat{b}_{\vec{k}}$ and $\hat{b}^{\dagger}_{\vec{k}}$ is that of a sum of independent harmonic oscillators. This condition is fulfilled by imposing

$$\tanh 2\theta_k = \frac{gn}{gn + \hbar^2 k^2/2m} = \frac{8\pi na}{8\pi na + k^2} = \frac{k_0^2}{k_0^2 + k^2}, \tag{1.138}$$

where $k_0 = (8\pi na)^{1/2} = 1/\xi$ is the inverse of the healing length ξ introduced in the previous section. The final expression for the diagonalized Hamiltonian is thus

$$\hat{H} = E_0 + \sum_{\vec{k} \neq \vec{0}} \hbar\omega_k^B \, \hat{b}^{\dagger}_{\vec{k}} \hat{b}_{\vec{k}} \quad \text{with} \quad \hbar\omega_k^B = \sqrt{\frac{\hbar^2 k^2}{2m} \left(\frac{\hbar^2 k^2}{2m} + 2gn \right)} \tag{1.139}$$

and

$$E_0 = \frac{gn}{2} \left(N + \sum_{\vec{k} \neq \vec{0}} \left[x\sqrt{x^2 + 2} - 1 - x^2 \right] \right) \quad \text{with} \quad x = \frac{k}{k_0}. \tag{1.140}$$

1.4.3.2   *Quasi-particles and dispersion relation*

The Hamiltonian (1.139) is a sum of Hamiltonians of harmonic oscillators. The ground state $|\Psi_0\rangle$ of the system is therefore obtained when all these oscillators are in their ground state. The wavefunction $|\Psi_0\rangle$ obeys the relations $\hat{b}_{\vec{k}}|\psi_0\rangle = 0$ for all $\vec{k} \neq \vec{0}$, and is therefore of the form:

$$|\psi_0\rangle = \alpha|n_0 = N, 0, \ldots, 0, \ldots\rangle + \sum_{\vec{k} \neq \vec{0}} \beta_{\vec{k}}|N - 2, 0, \ldots, n_{\vec{k}} = 1, n_{-\vec{k}} = 1, 0, \ldots\rangle + \ldots$$

$$(1.141)$$

It contains pairs of atoms with opposite momenta.

The first excited states of the system with momentum $\vec{k}$ and energy $\hbar\omega_{\vec{k}}^B$ are given by $\hat{b}_{\vec{k}}^\dagger|\psi_0\rangle = u_k\,\hat{a}_{\vec{k}}^\dagger|\psi_0\rangle + v_k\,\hat{a}_{-\vec{k}}|\psi_0\rangle$. They describe elementary excitations, or quasi-particles, whose dispersion relation is given by $\hbar\omega_{\vec{k}}^B(k)$ (see Eq. 1.139). A quasi-particle $\{\vec{k}, \omega_{\vec{k}}^B\}$ is therefore obtained by creating from the ground state $|\psi_0\rangle$ a particle of wave vector $\vec{k}$ and by annihilating a particle with an opposite wavevector $-\vec{k}$, i.e. by creating a hole. A quasi-particle thus appears as a linear superposition of a particle and a hole of opposite momenta, with amplitudes u_k and v_k.

1.4.3.3 *Dispersion relation*

In the long-wavelength limit for which $\lambda = 2\pi/k$ is large compared to the healing length $\xi = 1/k_0$, the chemical potential $\mu = gn$ is large compared to the kinetic energy $gn \gg \hbar^2k^2/2m$ and the dispersion relation becomes linear with k: $\hbar\omega^B(k) \simeq ck$, where $c = (gn/m)^{1/2}$. This linear dispersion relation describes phonons that propagate in the condensate with the sound velocity c. Furthermore, we have, in this limit, according to Eq. (1.138), $\tanh 2\theta_k \simeq 1$, so that the coefficients $u_k \simeq v_k$. The elementary excitations are linear superposition with equal amplitudes of a particle and a hole.

In the opposite limit $\hbar^2k^2/2m \gg gn$, the dispersion relation is given by

$$\hbar\omega_k^B \simeq \frac{\hbar^2k^2}{2m} + \mu. \tag{1.142}$$

One simply finds the kinetic energy of a free particle shifted by the chemical potential μ. In this regime, we have, according to Eq. (1.138), $\tanh 2\theta_k \simeq 0$, so that u_k is on the order of 1 and v_k becomes negligible. The elementary excitation is then very close to an ordinary particle propagating in the condensate and having its energy slightly modified by an amount μ due to its interaction with the condensed particles. These results concerning the dispersion law are summarized in Figure 1.12.

The excitation spectrum ω_k^B gives important insight into the superfluidity properties. It has been measured in liquid helium by scattering of a monochromatic beam of neutrons (Henshaw and Woods, 1961). The scattering process can be considered as a collision of a neutron with a quasi-particle. Knowing the initial energy of the neutrons and their energy change after the scattering under different angles, it is possible to infer the spectrum of the quasi-particles. For trapped condensates of dilute alkali gases, the Bogolubov spectrum has been measured through the investigation of the structure factor using Bragg spectroscopy (Steinhauer *et al.*, 2002) which consists in determining the maximum probability of excitation from an absorption-stimulated emission elementary process with a well-controlled angular frequency ω and wavevector $\vec{k}$.

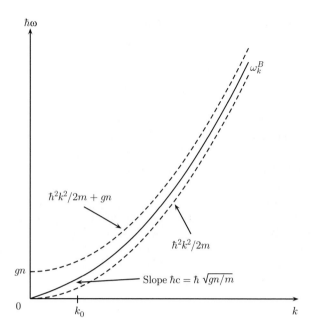

Fig. 1.12 Bogolubov dispersion law for the elementary excitations. The transition between a phonon-like dispersion, $\omega_k^B \simeq ck$, and a free particle, $\omega_k^B \simeq \hbar^2 k^2/2m$, occurs around $k = k_0 = \xi^{-1}$, where ξ is the healing length.

1.4.3.4 *Landau's criterion for superfluidity*

A major consequence of the Bogolubov dispersion relation (1.139) for the elementary excitations is that, according to Landau's criterion, a Bose–Einstein condensate is superfluid, with a critical velocity given by the speed of sound $\sqrt{gn/m}$. In this section, we recall briefly a basic derivation of Landau's criterion for superfluidity.

We consider a point-like[18] test particle of mass M (an impurity) moving in the condensate with a velocity $\vec{v}$. The particle will experience a drag force if and only if it interacts with the condensate, i.e, if it can create Bogolubov elementary excitations having a momentum $\hbar \vec{k}$ and an energy ω_k^B; the impurity acquires in the process a new momentum $M\vec{v}'$. Conservation of energy and momentum reads

$$\frac{1}{2}Mv^2 = \frac{1}{2}Mv'^2 + \hbar\omega_k^B \quad \text{and} \quad M\vec{v} = M\vec{v}' + \hbar\vec{k}.$$

Eliminating $\vec{v}'$ between those two equations, we get

$$\frac{1}{2}Mv^2 - \hbar\omega_k^B = \frac{1}{2}M\left(\vec{v} - \frac{\hbar\vec{k}}{M}\right)^2.$$

Therefore, in the process, one must necessarily have

$$\vec{v} \cdot \vec{k} = \omega_k^B + \frac{\hbar k^2}{2M} \geq \omega_k^B. \tag{1.143}$$

If we assume that the critical velocity v_c defined as

$$v_c = \min_k \frac{\omega_k^B}{k}$$

is finite, then we see that for $v \leq v_c$, the relation (1.143) cannot be fulfilled; therefore the test particle cannot experience any drag in the condensate. For the Bogolubov dispersion relation, one finds that $v_c = \sqrt{gn/m}$. We thus conclude that in the presence of atom–atom interactions, a condensate at $T = 0$ is superfluid, with a critical velocity equal to the speed of sound. For the ideal gas, one has the usual quadratic dispersion relation of free particles, which yields a vanishing critical velocity. Therefore, an ideal BEC is not superfluid.

Experimentally, manifestations of superfluidity in gaseous condensates have been observed in a variety of ways, not only by measuring the critical velocity, but also by studying extensively collective oscillations and rotating condensates, with, e.g. the apparition of quantized vortices above a critical rotation velocity (Madison *et al.*, 2000). A description of these experiments is beyond the scope of these lecture notes and the reader is referred to the general references given at the end of Section 1.1.

1.4.3.5 *Quantum depletion*

As already emphasized, the ground state $|\psi_0\rangle$ does not coincide with the Fock state of Eq. (1.130) where all the atoms are in the ground state $|\varphi_0\rangle$ of the box. The Bogolubov theory enables one to determine the number of particles out of the state $|\varphi_0\rangle$ when the gas of bosons is in its ground state $|\psi_0\rangle$:

$$N - N_0 = \sum_{\vec{k} \neq \vec{0}} \langle \psi_0 | \hat{a}_{\vec{k}}^\dagger \hat{a}_{\vec{k}} | \psi_0 \rangle. \tag{1.144}$$

To calculate explicitly this quantity, one expresses the operators $\hat{a}_{\vec{k}}^\dagger$ and $\hat{a}_{\vec{k}}$ in terms of the quasi-particle operators $\hat{b}_{\vec{k}}^\dagger$ and $\hat{b}_{\vec{k}}$:

$$N - N_0 = \sum_{\vec{k} \neq \vec{0}} \langle \psi_0 | \left(u_k \hat{b}_{\vec{k}}^\dagger - v_k \hat{b}_{-\vec{k}} \right) \left(u_k \hat{b}_{\vec{k}} - v_k \hat{b}_{-\vec{k}}^\dagger \right) | \psi_0 \rangle,$$

$$= \sum_{\vec{k} \neq \vec{0}} v_k^2 \langle \psi_0 | \hat{b}_{-\vec{k}} \hat{b}_{-\vec{k}}^\dagger | \psi_0 \rangle = \sum_{\vec{k} \neq \vec{0}} v_k^2. \tag{1.145}$$

Using Eq. (1.138) and the definition of v_k, one can calculate the depletion by replacing the discrete sum by an integral. One finds

$$\frac{N - N_0}{N} = \frac{8}{3\sqrt{\pi}} \left(n a^3 \right)^{1/2}. \tag{1.146}$$

The quantum depletion is therefore small since our calculation is performed under the diluteness assumption $na^3 \ll 1$. The corrections to the mean-field theory turn out to be relatively small for a dilute Bose gas.

1.4.3.6 *Energy of the ground state and chemical potential*

The energy E_0 of the ground state $|\Psi_0\rangle$ is given by Eq. (1.140) and can be calculated explicitly by replacing the sum over k by an integral. Actually, one gets in this way a divergent result. This is an artefact due to the use of a delta-function for the atom–atom interaction potential. Indeed, the Fourier components of the potential do not tend to zero at large k for a delta potential. A more realistic potential would yield a decrease of these large k components, and therefore a convergence of the integral. Renormalizing the relation between the scattering length and the pseudo-potential leads to a convergent integral for E_0, the result of this calculation is (Lee and Yang, 1957; Lee *et al.*, 1957)

$$E_0 = \frac{Ngn}{2}\left[1 + \frac{128}{15\sqrt{\pi}}\left(na^3\right)^{1/2}\right]. \tag{1.147}$$

Here also, we find that the correction to the mean-field result $(Ngn/2)$ scales as $(na^3)^{1/2}$ and is therefore small if the gas is dilute. The same conclusion holds for the equation of state $\mu(n)$ that is directly related to the energy of the ground state:

$$\mu = \frac{\partial E_0}{\partial N} = gn\left[1 + \frac{32}{3\sqrt{\pi}}\left(na^3\right)^{1/2}\right], \tag{1.148}$$

and is modified with respect to the mean-field results $\mu(n) = gn$. The beyond-mean-field correction to the equation of state yields *a priori* measurable effects on diverse quantities such as the speed of sound c. Combining the expression (1.148) for the chemical potential with the linearized hydrodynamic equations, one finds indeed

$$mc^2 = gn\left(1 + \frac{16}{\sqrt{\pi}}\left(na^3\right)^{1/2}\right). \tag{1.149}$$

1.4.3.7 *Comparison with the mean-field results*

The dispersion relation (1.139) can also be obtained from the time-dependent Gross–Pitaevskii equation (see Section 1.3.6.4 of the previous chapter). However, the Bogolubov formalism gives a clear interpretation in terms of quasi-particles having a well-defined energy and momentum, and obtained from the vacuum $|\psi_0\rangle$ by the creation of a particle–hole pair.

The quantum depletion (1.146) and the energy of the ground state (1.147) are new results that cannot be obtained in the mean-field approach since they originate from the quantum fluctuations of the matterwave field. They are, however, small for a dilute gas since the corresponding corrections scale as $(na^3)^{1/2} \ll 1$.

1.4.4 Second-order correlation function

The second quantized formulation used in this section is convenient for introducing field operators and their correlation functions, as already emphasized in Section 1.1 with the calculation of first- and second-order correlations for the ideal Bose gas.

In the following, we show how the second-order correlation function of the field operators

$$\left\langle \hat{\Psi}^{\dagger}(\vec{r}_0)\hat{\Psi}^{\dagger}(\vec{r}_0')\hat{\Psi}(\vec{r}_0')\hat{\Psi}(\vec{r}_0) \right\rangle, \tag{1.150}$$

when calculated in the Bogolubov theory, predicts correlations between the positions of atoms that cannot be predicted by the mean-field theory. In the following, we thus work out the expression of this quantity within both the mean-field and Bogolubov theories.

Using the complete set of single-particle quantum wave functions (1.123), the quantum matter field can be rewritten in the following manner:

$$\hat{\Psi}(\vec{r}) = \sum_{\vec{k}} \frac{1}{L^{3/2}} e^{i\vec{k}\cdot\vec{r}} \hat{a}_{\vec{k}}. \tag{1.151}$$

Within the mean-field theory, all particle states with $\vec{k} \neq \vec{0}$ are empty, and therefore only the term proportional to $\hat{a}_0^{\dagger}\hat{a}_0^{\dagger}\hat{a}_0\hat{a}_0$ in the expression (1.150) of the second-order correlation function has a non-zero contribution:

$$\left\langle \hat{\Psi}^{\dagger}(\vec{r}_0)\hat{\Psi}^{\dagger}(\vec{r}_0')\hat{\Psi}(\vec{r}_0')\hat{\Psi}(\vec{r}_0) \right\rangle = \frac{1}{L^6} \langle N_0 = N | \hat{a}_0^{\dagger}\hat{a}_0^{\dagger}\hat{a}_0\hat{a}_0 | N_0 = N \rangle = \frac{N(N-1)}{L^6},$$

$$\simeq \frac{N^2}{L^6} = \rho^2 = \left\langle \hat{\Psi}^{\dagger}(\vec{r}_0)\hat{\Psi}(\vec{r}_0) \right\rangle \left\langle \hat{\Psi}^{\dagger}(\vec{r}_0')\hat{\Psi}(\vec{r}_0') \right\rangle, \tag{1.152}$$

since $\left\langle \hat{\Psi}^{\dagger}(\vec{r}_0)\hat{\Psi}(\vec{r}_0) \right\rangle = N/L^3 = \rho$ is the average atomic density, which is independent of $\vec{r}_0$ for an homogeneous condensate. The second-order correlation function between two points is just given by the square of the mean spatial density. Consequently, according to mean-field theory, there are no correlations between the positions of two atoms.

Calculating the average value of the density correlations within the Bogolubov theory requires the calculation of $\langle \psi_0 | \hat{\Psi}^{\dagger}(\vec{r}_0)\hat{\Psi}^{\dagger}(\vec{r}_0')\hat{\Psi}(\vec{r}_0')\hat{\Psi}(\vec{r}_0) | \psi_0 \rangle$. In order to calculate this quantity, we proceed similarly to Section 1.4.3.1 for obtaining the quadratic expression of the Bogolubov Hamiltonian. In the product of four operators $\hat{a}_{\vec{k}}$ and $\hat{a}_{\vec{k}}^{\dagger}$, we replace $\hat{a}_0$ and $\hat{a}_0^{\dagger}$ by $\sqrt{N}$ (except for the term $\hat{a}_0^{\dagger}\hat{a}_0^{\dagger}\hat{a}_0\hat{a}_0$) and we neglect all products containing less than two operators $\hat{a}_0$ or $\hat{a}_0^{\dagger}$. These calculations yield the following result (Cohen-Tannoudji, 1998):

$$\langle \psi_0 | \hat{\Psi}^{\dagger}(\vec{r}_0)\hat{\Psi}^{\dagger}(\vec{r}_0')\hat{\Psi}(\vec{r}_0')\hat{\Psi}(\vec{r}_0) | \psi_0 \rangle = \rho^2 \left[1 - \chi(\vec{r}_0 - \vec{r}_0')\right], \tag{1.153}$$

with

$$\chi(\vec{s}) = \frac{2}{N} \sum_{\vec{k}\neq\vec{0}} e^{i\vec{k}\cdot\vec{s}} \left[v_k^2 - u_k v_k\right] = \frac{1}{(2\pi)^3\rho} \int d^3k\, e^{i\vec{k}\cdot\vec{s}} \left(\frac{k}{(k^2 + 2k_0^2)^{1/2}} - 1\right). \tag{1.154}$$

In Bogolubov theory, the second-order correlation function no longer factorizes, which shows the existence of spatial correlations between the positions of two atoms.

This effect was not predicted by the mean-field theory. Furthermore, we get an analytical expression for the Fourier transform of the difference between the second-order correlation function and ρ^2. The asymptotic behavior of $\chi(\vec{s})$ can be readily worked out:

- In the small-distance regime $s \ll \xi$, or equivalently for $k \gg k_0$, one finds:

$$\chi(\vec{r} - \vec{r}') \simeq -\frac{2a}{|\vec{r} - \vec{r}'|}. \tag{1.155}$$

  Actually this expression is not valid for arbitrarily small values of the relative distance between the particles. Its derivation assumes implicitly that $|\vec{r} - \vec{r}'| \gg a$, where a is the scattering length. Repulsive interactions yield a decrease of the probability to have two atoms very close to each other.

- In the large-distance regime $s \gg \xi$,

$$\chi(\vec{r} - \vec{r}') \propto -\frac{1}{|\vec{r} - \vec{r}'|^4} \tag{1.156}$$

  tends to zero algebraically.

1.4.5   Conclusion

The mean-field theory developed in the previous section is a description in terms of classical fields. A more rigorous treatment of many-particle assemblies requires the use of quantum field theory and Green's functions (Fetter and Walecka, 1971). However, for many systems, it is instructive to work out a direct approach that consists in simplifying the second-quantized hamiltonian and obtaining an approximate problem that can be solved exactly. This approach is at the heart of the Bogolubov theory for the weakly interacting Bose gas. It relies on a canonical transformation for the creation and annihilation operators after having recast the second quantized hamiltonian in a quadratic form using the so-called Bogolubov prescription for which the operators associated to the macroscopically occupied state that described the condensate are replaced by c-numbers. The analytical results obtained for homogeneous condensates within this theoretical framework bring corrections to the mean-field results, which remain small for dilute gases.

In addition, the second quantization formulation of the problem used in the Bogolubov approach allows a simple introduction of the field operators and of their correlation functions. These correlation functions allow fruitful analogies with quantum optics and are used to investigate the coherence properties of Bose–Einstein condensates.

When the small parameter na^3 of the Bogolubov theory is magnified, one deals with strongly interacting Bose–Einstein condensate, and the theoretical framework developed in this section is no longer valid. Experimentally, this regime can be obtained by tuning atom–atom interactions in the condensate with a magnetically induced Feshbach resonance. In the strongly interacting regime quantum depletion and beyond mean-field corrections are significant, and the excitation spectrum is strongly modified, as shown experimentally in (Papp *et al.*, 2008).

1.5   Bimodal condensates

1.5.1   Introduction

We consider in this section that the atoms can be in two states $|\phi_a\rangle = a^\dagger|0\rangle$ and $|\phi_b\rangle = b^\dagger|0\rangle$. The typical physical situation we have in mind is the case of a symmetric double-well potential (see Figure 1.13) where $|\phi_a\rangle$ and $|\phi_b\rangle$ describe the (spatially separated) ground states in the left and right wells, respectively. Due to tunneling through the central barrier, there is a coupling between these two states.[19]

In second quantization, the hamiltonian of the system reads:

$$H = -\frac{\hbar J}{2}\left(a^\dagger b + b^\dagger a\right) + \frac{\hbar U}{2}\left[a^\dagger a\left(a^\dagger a - 1\right) + b^\dagger b\left(b^\dagger b - 1\right)\right], \qquad (1.157)$$

where J is the tunneling rate (which varies exponentially with the barrier height and can thus be tuned easily) and U accounts for the interactions (and can be tuned either by changing the confinement or by means of Feshbach resonances). We will consider, unless otherwise stated, repulsive interactions $U > 0$. The creation and annihilation operators $a^\dagger$ and a create and destroy, respectively, a particle in well a, and the same holds for $b^\dagger$ and b. They satisfy the usual bosonic commutation relations $[a, a^\dagger] = 1$, $[b, b^\dagger] = 1$, and $[a, b] = 0$. Due to symmetry, the interaction coupling constant is the same for the states a and b (this would not necessarily be the case for spinor condensates; in addition, in that case, an interspecies interaction term $U_{ab}a^\dagger ab^\dagger b$ would be present[20]). Note that the hamiltonian (1.157) commutes with the total number of particles $\hat{N} = a^\dagger a + b^\dagger b$, and thus we restrict the analysis to a fixed total number N of atoms.[21]

The situation in which a superfluid can tunnel through a barrier between two regions of space is well known in condensed matter physics, when two superconductors are separated by a thin insulating layer, forming a so-called *Josephson junction*, and gives rise to a wealth of physical effects. Here, we will see that the ultracold atom analog of such a junction gives a unique control on all parameters entering the problem, and allows us to investigate regimes that are not encountered in condensed matter physics. In particular, as we shall see, beyond mean-field effects due to interactions can be much more dramatic in this two-mode problem than in the single-mode one. This section is organized as follows: we first treat the problem within the mean-field approximation (Section 1.5.2), and then switch to a second quantized treatment that allows to treat simply all regimes, from the non-interacting one (Section 1.5.3.1) to the interaction

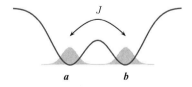

Fig. 1.13   A double-well potential in which tunneling between the two wells a and b occurs at a rate J.

dominated ones (Sections 1.5.3.2, 1.5.3.3 and 1.5.3.4). Finally, we shall see that the in the case of attractive interactions ($U < 0$), interesting many-body ground states with a "Schrödinger cat" character emerge (Section 1.5.4).

1.5.2   Mean-field: Josephson oscillations and non-linear self-trapping

In this section we will study the dynamics of a condensate in a double-well potential, when the initial number of atoms N_a and N_b in the two wells are different. In analogy with Josephson junctions, one expects intuitively to observe oscillations of the superfluid between the two wells as a consequence of tunneling. We shall see that repulsive interactions can lead to a non-intuitive result, namely the non-linear self-trapping of the condensate (absence of oscillations) when the initial population imbalance $\ell \equiv (N_a - N_b)/2$ is too large.

Equations of motion.   The mean-field approximation consists in replacing the operators appearing in (1.157) by c-numbers:

$$a \rightarrow \sqrt{N_a}$$
$$a^\dagger \rightarrow \sqrt{N_a}$$
$$b \rightarrow \sqrt{N_b}\exp(i\phi)$$
$$b^\dagger \rightarrow \sqrt{N_b}\exp(-i\phi),$$

where we have chosen $\langle a \rangle$ to be real since we are free to chose the global phase. We then get a classical Hamiltonian depending on the variables N_a, N_b, and ϕ:

$$H(N_a, N_b, \phi) = -\hbar J \sqrt{N_a\,N_b}\cos\phi + \frac{\hbar U}{2}\left[N_a(N_a - 1) + N_b(N_b - 1)\right].$$

Introducing the population imbalance by

$$\ell = \frac{N_a - N_b}{2} \quad \Longleftrightarrow \quad \begin{cases} N_a = N/2 + \ell \\ N_b = N/2 - \ell \end{cases},$$

we can rewrite the hamiltonian as

$$H(\ell, \phi) = \hbar U \ell^2 - \frac{\hbar J}{2}\sqrt{N^2 - 4\ell^2}\cos\phi, \tag{1.158}$$

where we have omitted the constant term $\hbar U(N^2/4 - N/2)$.

Josephson oscillations and self-trapping.   If we could neglect the ℓ^2 term in the square root in front of the $\cos\phi$ term of Eq. (1.158), we would have to deal with the hamiltonian of a simple pendulum (ϕ being the angle of the pendulum with respect to the vertical and ℓ being proportional to the angular momentum). For such a system, it is well known that two kinds of motion can occur: for small initial angular momentum, one observes oscillations around $\phi = 0$, while if the initial angular momentum is large enough, the pendulum rotates and the phase ϕ increases without bound. Here, in spite of the fact that our "pendulum" has a momentum-dependent "length", those two kinds of motion also exist. If the initial value of ℓ is above a critical value ℓ_c, the phase ϕ winds up (almost linearly, with a small periodic component on top) while the population imbalance stays almost constant around a non-zero value.

The physical reason for this behavior is simple: when the imbalance is so large that the increase in the phase difference between the two wells due to the repulsive interactions cannot be compensated for by tunneling (which tends on the contrary to equalize the phases), one observes self-trapping.

Of course, for $\ell(0) < \ell_c$, one observes oscillations of both the phase and the imbalance. To find the oscillation frequency, one simply expands Eq. (1.158) around $(\ell, \phi) = (0, 0)$:

$$H(\ell, \phi) \simeq \left(\hbar U + \frac{\hbar J}{N}\right) \ell^2 + \frac{N \hbar J}{4} \phi^2.$$

The equations of motion then yield

$$\ddot{\ell} = -NJ\left(U + J/N\right) \ell,$$

which correspond to a harmonic oscillator with angular frequency

$$\omega_{\mathrm{J}} = \sqrt{J(J + NU)}. \tag{1.159}$$

We shall see below that this result can be recovered by the Bogolubov method (see Section 1.5.3.3).

Remark In the literature, there has been some controversy in recent years on the terminology "Josephson oscillations" (Levy *et al.*, 2007). Indeed, in condensed matter physics, the term *Josephson oscillations* describes the existence of an ac current (in the microwave range for typical experimental numbers) flowing between two superconductors separated by an insulating barrier when a (fixed) dc bias voltage is applied to the junction. The corresponding situation in a BEC double-well junction would correspond to a fixed chemical potential difference (i.e. population imbalance) and an ac superfluid current between the two wells. This is in fact what happens in the non-linear self-trapping regime described above (due to the finite total number of atoms, the chemical potential difference actually undergoes small oscillations, but around a non-zero value). To keep the terminology that is the most widely used in the cold-atom community, we shall continue to call Josephson oscillations the oscillation with frequency ω_{J}, the other regime being called *non-linear self-trapping*.

Experimental realization. A pioneering experiment demonstrating the transition from Josephson oscillations to non-linear self-trapping was performed in M. K. Oberthaler's group (Albiez *et al.*, 2005). The double-well potential was realized by superimposing a shallow one-dimensional optical lattice (with a periodicity of a few micrometers) onto a harmonic trap, in such a way that two (and only two) potential minima are created (see the dashed box in Figure 1.14a). The parameters J and U can be tuned by changing the light intensity (in particular, raising the barrier height decreases exponentially

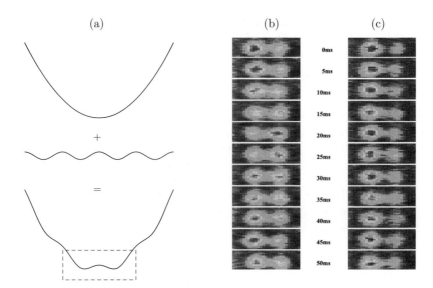

Fig. 1.14 (a) Realization of an optical double-well potential. Experimental observation of Josephson oscillations (b) and non-linear self-trapping (c) with a BEC in a double-well potential. In (b) (respectively, (c)), the initial population imbalance was below (respectively, above) the critical value ℓ_c. Figure adapted from (Albiez *et al.*, 2005), courtesy of M. K. Oberthaler.

the tunneling rate J. By varying the initial atom numbers in each well (which can be achieved by performing evaporative cooling in a double well with different energy offsets in each well, this being obtained with a shift of the lattice with respect to the harmonic trap minimum), the transition from the Josephson oscillation regime to the non-linear self-trapping could be observed (see Figures 1.14b and c). More recently, Josephson oscillations have been observed in other setups (Levy *et al.*, 2007).

1.5.3   Beyond mean-field

In the previous section, we have assumed from the beginning that a condensate is present, i.e. that a single-particle state is macroscopically occupied. However, it is by no means obvious that such a situation is always fulfilled, and we shall see below that for strong repulsions $(U \gg NJ)$ the opposite is true. This section is thus devoted to the study of the hamiltonian (1.157) without any mean-field approximation.

1.5.3.1   *Ground state in the absence of interactions*

In the absence of interactions $(U = 0)$, the particles will condense into the lowest-energy single-particle state. Introducing the operators $a_\pm \equiv (a \pm b)/\sqrt{2}$, one can rewrite the Hamiltonian as

$$-\frac{\hbar J}{2}\left(a^\dagger b + b^\dagger a\right) = -\frac{\hbar J}{2}\left(a_+^\dagger a_+ - a_-^\dagger a_-\right),$$

which clearly shows that the single-particle ground state is the symmetric combination $(|\phi_a\rangle + |\phi_b\rangle)/\sqrt{2}$. The N-body ground state therefore reads

$$|\Psi\rangle_{U=0} = \frac{1}{\sqrt{N!}} \left(\frac{a^\dagger + b^\dagger}{\sqrt{2}} \right)^N |0\rangle.$$

It is instructive to expand this state on the Fock state basis $\{|n_a, N - n_a\rangle\}$. Using the binomial theorem, one finds a distribution of Fock states centered on $N/2$, with variance $\Delta \hat{n}_a = N/4$. This shows that one has Poissonian number fluctuations in the wells, due to the tunneling of particles. It is intuitively clear that the repulsive interactions U will lead to a suppression of such quantum fluctuations that are not energetically favorable. One can also calculate the average value $\langle a^\dagger b \rangle$, which is a measure of coherence (see below). One finds easily $\langle a^\dagger b \rangle = N/2$, i.e. one has full coherence between the wells, as expected.

1.5.3.2 *Bimodal condensate in the absence of tunneling*

We now study the opposite limit, namely the absence of tunneling ($J = 0$). In this case the hamiltonian (1.157) is diagonal in the Fock-state basis:

$$H|n_a, N - n_a\rangle = E(n_a)|n_a, N - n_a\rangle,$$

with $E(n_a) = \hbar U \left[(n_a - N/2)^2 + N/2(N/2 - 1) \right]$. In this case, the repulsive interactions ($U > 0$) clearly favor an equal number of particles in each well and the ground state is obtained for $n_a = N/2$, i.e. it reads

$$|\Psi\rangle_{J=0} = \frac{1}{(N/2)!} \left(a^\dagger \right)^{N/2} \left(b^\dagger \right)^{N/2} |0\rangle = |N/2, N/2\rangle.$$

In this case, the fluctuations of the atom numbers in each well are frozen ($\Delta \hat{n}_a = 0$). It is easy to see that for such a ground state, one has $\langle a^\dagger b \rangle = 0$, meaning that coherence between the two wells is lost. Repulsive interactions lead to a *fragmentation* of the condensate into the two modes a and b.

1.5.3.3 *Bogolubov approach*

This section is devoted to the perturbative analysis of the effect of the interactions when they are small compared to tunneling (the precise condition to be fulfilled by U/J will be clarified later). We assume that the symmetric mode $a_+ \equiv (a + b)/\sqrt{2}$ is still macroscopically occupied, as in the non-interacting case, and we follow the Bogolubov prescription (see Section 1.4.3.1):

$$a_+ \simeq a_+^\dagger \simeq \sqrt{N}. \tag{1.160}$$

We shall now rewrite the various terms of the hamiltonian in terms of the operators $a_\pm = (a \pm b)/\sqrt{2}$ (and their hermitian conjugates), which annihilate (and create) particles in the symmetric and antisymmetric modes, respectively:

- As we have seen above, the tunneling term is rewritten in terms of $a_\pm$ and $a_\pm^\dagger$ as

$$-\frac{\hbar J}{2}\left(a^\dagger b + b^\dagger a\right) = -\frac{\hbar J}{2}\left(a_+^\dagger a_+ - a_-^\dagger a_-\right) = -\frac{\hbar J}{2}\left(N - 2a_-^\dagger a_-\right),$$

  where the last equality is obtained using the exact identity $a_+^\dagger a_+ + a_-^\dagger a_- = N$.

- For the interaction term, which involves terms such as $a^\dagger a\, a^\dagger a$ that are large (of order N^2), one must use the Bogolubov prescription (1.160) with care (see the corresponding discussion in Section 1.4.3.1). From the exact identity

$$a^\dagger a = \frac{1}{2}\left(N + a_-^\dagger a_+ + a_+^\dagger a_-\right)$$

we get, using the Bogolubov prescription,

$$a^\dagger a = \frac{1}{2}\left[N + \sqrt{N}\left(a_-^\dagger + a_-\right)\right],$$

which we square to get:

$$a^\dagger a\, a^\dagger a = \frac{1}{4}\left[N^2 + 2N\sqrt{N}\left(a_-^\dagger + a_-\right) + N\left(a_-^\dagger + a_-\right)^2\right].$$

In a similar way, one obtains

$$b^\dagger b\, b^\dagger b = \frac{1}{4}\left[N^2 - 2N\sqrt{N}\left(a_-^\dagger + a_-\right) + N\left(a_-^\dagger + a_-\right)^2\right].$$

Finally, the interaction term can be rewritten as:

$$\frac{\hbar N U}{4}\left(a_-^\dagger + a_-\right)^2,$$

where we have omitted some constant terms, which are a function of N only.

The hamiltonian (1.157) now reads:

$$H = \hbar\left(J + \frac{NU}{2}\right)a_-^\dagger a_- + \frac{\hbar N U}{4}\left(a_-^\dagger a_-^\dagger + a_- a_-\right). \tag{1.161}$$

It is a quadratic form in $a_-^\dagger$ and a_-, for which we shall now seek a diagonal form. For that purpose, we introduce new operators β and $\beta^\dagger$ defined by the following relation:

$$a_- = u\beta + v\beta^\dagger, \quad \text{with } (u, v) \in \mathbb{R}^2. \tag{1.162}$$

We impose that those operators obey the usual bosonic commutation rules $[\beta, \beta^\dagger] = 1$, which implies, using $[a_-, a_-^\dagger] = 1$, that

$$u^2 - v^2 = 1. \tag{1.163}$$

Rewriting Eq. (1.161) in terms of the operators β and $\beta^\dagger$, we obtain, after a straight-forward calculation:

$$H = \left[\hbar NU \, uv + \hbar \left(J + \frac{NU}{2}\right)(u^2 + v^2)\right]\beta^\dagger \beta$$

$$+ \left[\frac{\hbar NU}{4}(u^2 + v^2) + \hbar\left(J + \frac{NU}{2}\right)uv\right]\left(\beta^2 + \beta^{\dagger 2}\right),$$

where we have again omitted constant terms. This expression has the required form of the hamiltonian of independent quasi-particles if the coefficient of the $\beta^2 + \beta^{\dagger 2}$ term vanishes. We thus get another equation fulfilled by u and v. Together with Eq. (1.163), this allows us to find the values of u and v. Indeed, one gets a biquadratic equation for v, which can be solved to give

$$v^2 = \frac{1}{2}\left[-1 + \sqrt{1 + \frac{N^2U^2}{4J(NU + J)}}\right] \text{ and } u^2 = \frac{1}{2}\left[1 + \sqrt{1 + \frac{N^2U^2}{4J(NU + J)}}\right].$$

The hamiltonian can then be recast in the very simple form:

$$H = \hbar\sqrt{J(J + NU)}\,\beta^\dagger \beta.$$

The excitations of the system thus have an energy $\hbar\omega_J$ given by the Josephson oscillation frequency (1.159). For very small interactions $U/J \ll 1/N$, one recovers oscillations at the tunneling frequency J, as expected for the single-particle regime. On the contrary, if U/J is not so small (and we shall see in the paragraph below that the Bogolubov approach is valid as long as $U/J \ll N$), the oscillation angular frequency is $\sqrt{NJU}$.

We now turn to the conditions that N, U and J must fulfill for the Bogolubov approach to be applicable. The basic hypothesis is that the occupation of the symmetric mode a_+ is macroscopic, or, in other words, that the *quantum depletion* $\left\langle a_-^\dagger a_-\right\rangle$ is small compared to N (the average is taken here in the ground state $|\psi_0\rangle$ of the system, which corresponds to the vacuum of quasi-particles and therefore fulfills $\beta|\psi_0\rangle = 0$, as well as the hermitian conjugate relation $\langle\psi_0|\beta^\dagger = 0$). We thus calculate

$$\left\langle a_-^\dagger a_-\right\rangle = \langle\psi_0|\,u^2\beta^\dagger\beta + v^2\beta\beta^\dagger + uv\left(\beta^2 + \beta^{\dagger 2}\right)|\psi_0\rangle = v^2. \tag{1.164}$$

From the expression of v^2, one sees that the condition $\left\langle a_-^\dagger a_-\right\rangle \ll N$ is then equivalent to

$$U \ll NJ.$$

It is interesting to study, as a function of the ratio U/J, (i) the average $\left\langle a^\dagger b\right\rangle$ that characterizes coherence;[22] (ii) the fluctuations ΔN_a of the number of atoms in, say, the well a.

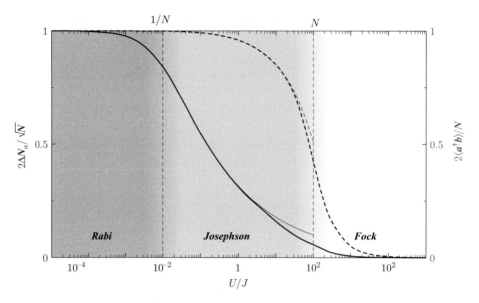

Fig. 1.15 Evolution of the variance of the number of atoms in site a (solid line) and of the coherence (dashed line) as a function of U/J. The thin lines correspond to the results obtained with the Bogolubov approach (valid only as long as $U/J \ll N$), while the thick lines are obtained numerically. One distinguishes three regimes: for $U \ll J/N$, the effect of interactions is negligible, and one observes the single-particle behavior, with strong fluctuations of the particle number and full coherence between the wells. This is called the *Rabi regime*. For $1/N \ll U/J \ll N$, repulsive interactions tend to reduce number fluctuations (one has some number squeezing), but the coherence is still high. In this *Josephson regime*, the mean-field description of the system is thus still valid. When $U \gg NJ$, interactions suppress both fluctuations and coherence: one observes fragmentation of the condensate towards the state $|N/2, N/2\rangle$. This is the so-called *Fock regime*.

The calculation of these average quantities is done in a way very similar to that of $\langle a_-^\dagger a_- \rangle$ in (1.164). One easily finds $\langle a^\dagger b \rangle = N/2 - v^2$. Concerning the fluctuations of the atom number in well a, one finds $\Delta N_a = \sqrt{N} \left(u^2 + v^2 + 2uv \right)/2$. The variations of the coherence factor, as well as of the number fluctuations, as a function of U/J, are shown as thin lines in Figure 1.15 for $N = 100$. We shall now compare those predictions with exact numerical results, and see that the Bogolubov approach gives an extremely good approximation.

1.5.3.4 *Numerical diagonalization of the hamiltonian*

In fact, it is quite simple to diagonalize numerically the hamiltonian (1.157): it is indeed a $(N+1) \times (N+1)$ tridiagonal matrix when expressed in the Fock basis. It is then easy to calculate any quantity of interest, such as the coherence factor or the number fluctuations. Figure 1.15 shows the result of such a numerical calculation. One distinguishes easily three regimes when U/J is varied:

- For $U \ll J/N$, the effect of the interactions is always negligible. One thus has the same ground state as in the non-interacting case, with full coherence between the wells and Poissonian fluctuations in the number of atoms in each well. This is called the Rabi (or single-particle) regime.

- For $1/N \ll U/J \ll N$, the coherence is still very close to its maximum value; however, the atom-number fluctuations on a site start to decrease significantly (one observes number squeezing). In this regime, one still has a condensate. This is the regime where Josephson oscillations and non-linear self-trapping can be observed, and is called for this reason the *Jospehson* regime (one also encounters the term *plasmon* regime originating from condensed matter physics). Experimentally, number squeezing was observed recently in such a system (Estève *et al.*, 2008).

- Finally, when $U \gg NJ$, the system loses its coherence and number fluctuations are almost completely frozen. One observes the fragmentation of the condensate into two distinct macroscopically populated modes (the ground state is close to $|N/2, N/2\rangle$). This is called, for obvious reasons, the *Fock* regime. In this regime, the Bogolubov approach breaks down. We come back to the physical reasons for this fragmentation in the next section.

1.5.3.5   *Fragmentation of the condensate*

We have shown from energy considerations in Section 1.4.2.2 that repulsive interactions prevent the fragmentation of the condensate in the case of a homogeneous gas. Here, in the case of a double-well potential, we have just come to the opposite conclusion: for very strong repulsive interactions, one observes fragmentation into two macroscopically occupied states. The physical reason for this apparent paradox lies in the fact that Nozières' argument applies to states with overlapping wavefunctions, for which one needs to take exchange effects into account. Here, the wavefunctions $|\phi_a\rangle$ and $|\phi_b\rangle$ are spatially separated, and this implies that the average interaction energy in the fragmented state $|N/2, N/2\rangle$ is lower than in the non-interacting ground state.[23]

1.5.4   Small condensates with attractive interactions

In this section we mention briefly what happens when the scattering length is negative, i.e. when the contact interaction is attractive: $U < 0$ (Cirac *et al.*, 1998; Ho and Ciobanu, 2004). In the absence of tunneling, the Hamiltonian is again diagonal in the Fock basis, but the energy is now minimized when all the particles are in the same well (see Figure 1.16 for the case $N = 6$). The two states $|N, 0\rangle$ and $|0, N\rangle$ are thus degenerate.

The inclusion of tunneling lifts the degeneracy between the states $|N, 0\rangle$ and $|0, N\rangle$. Although they are not coupled directly by the tunneling term, they are coupled indirectly, to Nth order, via all the intermediate states $|N - k, k\rangle$. As a result, at $|U| \gg J$, the ground state is (very close to) the symmetric combination

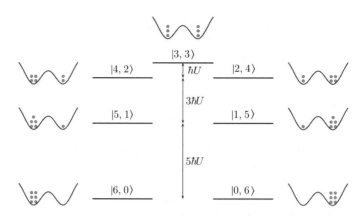

Fig. 1.16 Energy levels for $N = 6$ attractively interacting particles in a double well (in the absence of tunneling).

$\left(|N, 0\rangle + |0, N\rangle\right)/\sqrt{2}$. This is a quantum superposition of two macroscopically different states (all the particles in well a and all the particles in well b), and can thus be called a "Schrödinger cat" (Cirac *et al.*, 1998; Ho and Ciobanu, 2004).[24] Obviously, such macroscopic (or even mesoscopic) superpositions are very sensitive to decoherence (the loss of a single atom transforms the cat into a statistical mixture). At large U/J, the gap between this symmetric cat state and the antisymmetric one $\left(|N, 0\rangle - |0, N\rangle\right)/\sqrt{2}$ scales as $J^N/|U|^{N-1}$, i.e. it vanishes exponentially fast when the number N of atoms increases.[25]

Experimentally, a similar system with $N = 2$, but with repulsive interactions,[26] has been studied recently (Fölling *et al.*, 2007) using a superlattice, in which many copies of a double well containing two atoms are created simultaneously. The co-tunneling of atoms as pairs has been observed, and it was measured to occur with a rate J^2/U, as expected.

Notes

1. The density of liquid helium is 125 kg/m^3, while that of a neutron star is typically $5 \times 10^{17} \text{ kg/m}^3$.
2. $g_3(1) \simeq 1.202\ldots$.
3. The method used here is analogous to the Hartree–Fock approach used in atomic physics to describe poly-electronic atoms, and that consists in describing the evolution of each electron by an effective potential that takes into account the attraction of the nucleus and the average repulsive effect of the other electrons (except for the fact that in that case, one must obviously use antisymmetric functions).
4. We have used the relation $\vec{\nabla}(\varphi^*\vec{\nabla}\varphi) = \varphi^*\Delta\varphi + |\vec{\nabla}\varphi|^2$ and the divergence theorem.
5. Actually, the calculation of the kinetic energy using Eq. (1.56) and the Thomas–Fermi density profile gives rise to a logarithmic divergence since, close to the

boundary, the Thomas–Fermi approximation breaks down and the kinetic energy term can no longer be ignored. A more refined analysis can be carried out to circumvent this artefact (Dalfovo *et al.*, 1996).

6. Or, in an equivalent way, of a lower limit, equal to the inverse of the trap size, for the possible wavevector of phonons.

7. In the presence of vortices, the phase is not defined at the location of the vortices, and the use of the modulus-phase formalism requires some care to cope with singularities.

8. Actually, this is only true if elastic collisions play a minor role during the expansion. If the classical gas enters the hydrodynamic regime, i.e. if the collision rate is not negligible with respect to the trapping frequencies, the asymptotic expansion does reflect the initial anisotropy of the trap (Pedri *et al.*, 2003).

9. For an ideal Bose gas, one finds $\omega(n, \ell) = \omega_0(2n + \ell)$.

10. This is a general property of the center of mass motion that can easily be shown using the Ehrenfest theorem. Actually, the same is true for the classical counterpart and this mode is therefore unaffected by the temperature of the gas: it remains the same above and below the critical temperature T_c.

11. A sum rule approach can be usefully employed to work out the link with compressibility (Zambelli and Stringari, 2002).

12. The polynomial $P_0^2 \propto r^2$.

13. For a spherical trap, one recovers the frequency $\omega_- = \sqrt{2}\omega_0$ of the quadrupole mode $n = 0, \ell = 2, m = 0$ and the frequency $\omega_+ = \sqrt{5}\omega_0$ of the monopole mode $n = 1, \ell = 0, m = 0$.

14. This is to be contrasted with the case of a trapped condensate, where the ground state is modified when the number of atoms increases.

15. This expression for the pseudo-potential may lead to artificial divergences. They are solved by using the model potential introduced by Enrico Fermi: $V(r) = g\delta(r)\partial(r\cdot)/\partial r$ (Fermi, 1936).

16. The situation is radically different if the two states do not overlap in position space as for a condensate in a double well potential (see Section 1.5); in this case repulsive interactions do favor the fragmentation of the condensate.

17. This rather crude, but frequently used, treatment of the operators $\hat{a}_0$ and $\hat{a}_0^\dagger$ can be improved in a more refined presentation described for example in (Castin and Dum, 1998).

18. An impurity with a finite size can generate other types of excitations, different from the Bogolubov quasi-particles (for instance vortex rings), which changes the conclusions of the present paragraph.

19. Other physical situations also correspond to such a two-mode problem: for instance, spinor condensates where two different internal states can be coherently coupled, e.g. by RF or microwave fields. In that case, however, the two wavefunctions overlap in space.

20. Long-range interactions, such as the dipolar interaction (Lahaye *et al.*, 2009), would imply the presence of this term even in the case of two spatially separated modes.

21. That we will assume to be even to simplify the discussions. This assumption is of course not fundamental.
22. That the quantity $\langle a^\dagger b \rangle$ is a measure of the degree of coherence can be understood in the two following (and related) ways:

    — The momentum distribution of the system has a sinusoidal modulation at wavevector $2\pi/\ell$ (ℓ being the distance between the two wells) with a coefficient proportional to $\langle a^\dagger b \rangle$ (see e.g. (Pitaevskii and Stringari, 2003), Chapter 13). Upon time of flight this gives rise to interference fringes whose visibility (averaged over many realizations) is proportional to $\langle a^\dagger b \rangle$.

    — If one forms the one-body density matrix $\rho(\vec{r}, \vec{r'}) = \left\langle \hat{\Psi}^\dagger(\vec{r})\hat{\Psi}(\vec{r'}) \right\rangle$, the "off-diagonal" terms $\rho(\vec{r}_a, \vec{r}_b)$ are proportional to $\langle a^\dagger b \rangle$. Since Bose–Einstein condensation is precisely defined as the existence of off-diagonal long-range order, a vanishing $\langle a^\dagger b \rangle$ is a signature of the absence of Bose–Einstein condensation for the whole system (one still does have two independent BECs, one in each well; in other words, the BEC is fragmented).

23. A straightforward calculation shows that these energies read $\hbar U N(N-2)/4$ and $\hbar U N(N-1)/4$. Fragmentation is thus energetically favored by a macroscopic quantity $\hbar U N/4$.
24. In quantum optics, such states for photons have been termed "NOON"-states.
25. A reader familiar with basic atomic physics will notice the analogy between the coupling $J^2/|U|$ for $N=2$ between the degenerate ground states in the absence of coupling, and the effective Rabi frequency for the *stimulated Raman effect* at large detuning Δ from the excited state, which reads $\Omega_1\Omega_2/\Delta$, with the single-laser Rabi frequencies $\Omega_{1,2}$ replacing J as the couplings to the excited state.
26. This changes obviously the nature of the ground state, but not the fact that $|2,0\rangle$ and $|0,2\rangle$ are coupled, via $|1,1\rangle$, by an effective matrix element J^2/U.

Acknowledgments

We would like to thank C. Cohen-Tannoudji for his illuminating lectures on Bose–Einstein condensation at Collège de France (1997–2002) and at City University of Hong Kong in 2008. We thank C. Miniatura for inviting one of us (D. G.-O.) to give a set of lectures on BECs at the Les Houches School of Physics on Ultracold Gases and Quantum Information held in Singapore in July 2009. Finally, we wish to thank the students that attended the lectures for their useful questions and comments.

References

Albiez, M., Gati, R., Fölling, J., Hunsmann, S., Cristiani, M., and Oberthaler, M. K. (2005). *Phys. Rev. Lett.*, **95**, 010402.

Anderson, M. H., Ensher, J. R., Matthews, M. R., Wieman, C. E., and Cornell, E. A. (1995). *Science*, **269**, 198.

Andrews, M. R., Kurn, D. M., Miesner, H.-J., Durfee, D. S., Townsend, C. G., Inouye, S., and Ketterle, W. (1998). *Phys. Rev. Lett.*, **79**, 553.

Bloch, I., Dalibard, J., and Zwerger, W. (2008). *Rev. Mod. Phys.*, **80**, 885.

Bloch, I., Hänsch, T. W., and Esslinger, T. (2000). *Nature*, **403**, 166.

Bogolubov, N. N. (1947). *J. Phys. (USSR)*, **11**, 23.

Braaten, E. and Pearson, J. (1999). *Phys. Rev. Lett.*, **82**, 252.

Bradley, C. C., Sackett, C. A., and Hulet, R. G. (1997). *Phys. Rev. Lett.*, **78**, 985.

Burt, E. A., Ghrist, R. W., Myatt, C. J., Holland, M. J., Cornell, E. A., and Wieman, C. E. (1997). *Phys. Rev. Lett.*, **79**, 337.

Castin, Y. (2001). In *Coherent atomic matter waves* (ed. R. Kaiser, C. Westbrook, and F. David), EDP Sciences and Springer-Verlag, Berlin, p. 1.

Castin, Y. (2004). *J. Phys. IV France*, **116**, 89.

Castin, Y. and Dum, R. (1996). *Phys. Rev. Lett.*, **77**, 5315.

Castin, Y. and Dum, R. (1998). *Phys. Rev. A*, **57**, 3008.

Chevy, F., Bretin, V., Rosenbusch, P., Madison, K. W., and Dalibard, J. (2002). *Phys. Rev. Lett.*, **88**, 250402.

Cirac, J. I., Lewenstein, M., Mølmer, K., and Zoller, P. (1998). *Phys. Rev. A*, **57**, 1208.

Cohen-Tannoudji, C. (1998). Cours de physique atomique et moléculaire 1998-1999 du Collège de France (in French), available at www.phys.ens.fr/cours/college-de-france/1998-99/.

Cohen-Tannoudji, C. and Robillard, C. (2001). *C. R. Acad. Sci. Paris*, **2**, 445.

Cornish, S. L., Thompson, S. T., and Wieman, C. E. (2006). *Phys. Rev. Lett.*, **96**, 170401.

Dalfovo, F., Giorgini, S., Pitaevskii, L. P., and Stringari, S. (1999). *Rev. Mod. Phys.*, **71**, 463.

Dalfovo, F., Pitaevskii, L., and Stringari, S. (1996). *Phys. Rev. A*, **54**, 4213.

Dalibard, J. (1998). In *Bose-Einstein condensation in atomic gases, proceedings of the international school of physics Enrico Fermi* (ed. M. Inguscio, S. Stringari, and C. E. Wieman), IOS Press, Amsterdam.

Davis, K. B., Mewes, M.-O., Andrews, M. R., van Druten, N. J., Durfee, D. S., Kurn, D. M., and Ketterle, W. (1995). *Phys. Rev. Lett.*, **75**, 3969.

DeMarco, B. and Jin, D. S. (1999). *Science*, **285**, 1703.

Dettmer, S., Hellweg, D., Ryytty, P., Arlt, J. J., Ertmer, W., Sengstock, K., Petrov, D. S., Shlyapnikov, G. V., Kreutzmann, H., Santos, L., and Lewenstein, M. (2001). *Phys. Rev. Lett.*, **87**, 160406.

Donley, E. A., Claussen, N. R., Cornish, S. L., Roberts, J. L., Cornell, E. A., and Wieman, C. E. (2001). *Nature*, **412**, 295.

Edwards, M., Ruprecht, P. A., Burnett, K., Dodd, R. J., and Clark, C. W. (1999). *Phys. Rev. Lett.*, **77**, 1671.

Esry, B. D. (1997). *Phys. Rev. A*, **55**, 1147.

Estève, J., Gross, C., Weller, A., Giovanazzi, S., and Oberthaler, M. K. (2008). *Nature*, **455**, 1216.

Fermi, E. (1936). *Ricerca Sci.*, **7**, 13.

Fetter, A. L. and Walecka, J. D. (1971). *Quantum theory of many-particle systems*. McGraw-Hill, San Francisco.

Fölling, S., Trotzky, S., Cheinet, P., Feld, M., Saers, R., Widera, A., Muller, T., and Bloch, I. (2007). *Nature*, **448**, 1029.

Gerton, J. M., Strekalov, D., Prodan, I., and Hulet, R. D. (2000). *Nature*, **408**, 692.

Griffin, A., Wu, W.-C., and Stringari, S. (1997). *Phys. Rev. Lett.*, **78**, 1838.

Grimm, R., Weidemüller, M., and Ovchinnikov, Yu. B. (2000). *Adv. At. Mol. Opt. Phys.*, **42**, 95.

Guéry-Odelin, D. (1998). Ph.D. thesis, University Paris IV.

Guéry-Odelin, D., Zambelli, F., Dalibard, J., and Stringari, S. (1999). *Phys. Rev. A*, **60**, 4851.

Hagley, E. W., Deng, L., Kozuma, M., Wen, J., Helmerson, K., Rolston, S. L., and Phillips, W. D. (1999). *Science*, **283**, 1706.

Hanbury-Brown, R. and Twiss, R. Q. (1956). *Nature*, **177**, 27.

Henshaw, D. G. and Woods, A. D. (1961). *Phys. Rev.*, **121**, 1266.

Ho, T.-L. and Ciobanu, C. V. (2004). *J. Low Temp. Phys.*, **135**, 257.

Huang, K. (1963). *Statistical mechanics*. Wiley, New York.

Jeltes, T., McNamara, J. M., Hogervorst, W., Vassen, W., Krachmalnicoff, V., Schellekens, M., Perrin, A., Chang, H., Boiron, D., Aspect, A., and Westbrook, C. I. (2007). *Nature*, **445**, 402.

Jin, D. S., Ensher, J. R., Matthews, M. R., Wieman, C. E., and Cornell, E. A. (1996). *Phys. Rev. Lett.*, **77**, 420.

Jin, D. S., Matthews, M. R., Ensher, J. R., Wieman, C. E., and Cornell, E. A. (1997). *Phys. Rev. Lett.*, **78**, 764.

Kagan, Yu., Surkov, E. L., and Shlyapnikov, G. V. (1997). *Phys. Rev. A*, **55**, 18(R).

Kavoulakis, G. M. and Pethick, C. J. (1998). *Phys. Rev. A*, **58**, 1563.

Ketterle, W. and Miesner, H.-J. (1997). *Phys. Rev. A*, **56**, 3291.

Lahaye, T., Menotti, C., Santos, L., Lewenstein, M., and Pfau, T. (2009). *Rep. Prog. Phys.*, **72**, 126401.

Landau, L. D. and Lifshitz, E. M. (1958). *Quantum mechanics*. Butterworth Heinemann, Oxford.

Lee, T. D., Huang, K., and Yang, C. N. (1957). *Phys. Rev.*, **106**, 1135.

Lee, T. D. and Yang, C. N. (1957). *Phys. Rev.*, **105**, 1119.

Leggett, A. J. (2006). *Quantum liquids: Bose condensation and Cooper pairing in condensed-matter systems*. Oxford Graduate Texts, Oxford.

Levy, S., Lahoud, E., Shomroni, I., and Steinhauer, J. (2007). *Nature*, **449**, 579.

Madison, K. W., Chevy, F., Wohlleben, W., and Dalibard, J. (2000). *Phys. Rev. Lett.*, **84**, 806.

Manz, S., Bucker, R., Betz, T., Koller, Ch, Hofferberth, S., Mazets, I. E., Imambekov, A., Demler, E., Perrin, A., Schmiedmayer, J., and Schumm, T. (2010). *Phys. Rev. A*, **81**, 031610(R)

Maragò, O., Hechenblaikner, G., Hodby, E., and Foot, C. (2001). *Phys. Rev. Lett.*, **86**, 3938.

Mewes, M. O., Andrews, M. R., van Druten, N. J., Kurn, D. M., Durfee, D. S., and Ketterle, W. (1996). *Phys. Rev. Lett.*, **77**, 416.

Mottelson, B. (1976). *Rev. Mod. Phys.*, **48**, 375.

Nozières, P. (1995). In *Bose-Einstein condensation* (ed. A. Griffin, D. Snoke, and S. Stringari), Cambridge University Press, Cambridge.

Papp, S. B., Pino, J. M., Wild, R. J., Ronen, S., Wieman, C. E., Jin, D. S., and Cornell, E. A. (2008). *Phys. Rev. Lett.*, **101**, 135301.

Pedri, P., Guéry-Odelin, D., and Stringari, S. (2003). *Phys. Rev. A*, **68**, 043608.

Pethick, C. J. and Smith, H. (2002). *Bose-Einstein condensation in dilute gases.* Cambridge University Press, Cambridge.

Pines, D. (1999). *Elementary excitations in solids : Lectures on phonons, electrons, and plasmons.* Advanced Book Classics, New York.

Pines, D. and Nozieres, P. (1966). *The theory of quantum liquids.* W. A. Benjamin, New York.

Pitaevskii, L. and Stringari, S. (1998). *Phys. Rev. Lett.*, **81**, 4541.

Pitaevskii, L. and Stringari, S. (2003). *Bose-Einstein condensation.* Oxford University Press, Oxford.

Pricoupenko, L., Perrin, H., and Olshanii, M. (2004). *J. Phys. IV (France)*, **116**, 1.

Pritchard, D. E. (1983). *Phys. Rev. Lett.*, **51**, 1336.

Richard, S., Gerbier, F., Thywissen, J. H., Hugbart, M., Bouyer, P., and Aspect, A. (2003). *Phys. Rev. Lett.*, **91**, 010405.

Ritter, S., Öttl, A., Donner, T., Bourdel, T., Köhl, M., and Esslinger, T. (2007). *Phys. Rev. Lett.*, **98**, 090402.

Roberts, J. L., Claussen, N. R., Cornish, S. L., Donley, E. A., Cornell, E. A., and Wieman, C. E. (2001). *Phys. Rev. Lett.*, **86**, 4211.

Sackett, C. A., Stoof, H. T. C., and Hulet, R. D. (1998). *Phys. Rev. Lett.*, **80**, 2031.

Singh, K. G. and Rokhsar, D. S. (1996). *Phys. Rev. Lett.*, **77**, 1667.

Söding, J., Guéry-Odelin, D., Desbiolles, P., Chevy, F., Inamori, H., and Dalibard, J. (1999). *Appl. Phys. B*, **69**, 257.

Stamper-Kurn, D. M., Miesner, H.-J., Inouye, S., Andrews, M. R., and Ketterle, W. (1998). *Phys. Rev. Lett.*, **81**, 500.

Steinhauer, J., Ozeri, R., Katz, N., and Davidson, N. (2002). *Phys. Rev. Lett.*, **88**, 120407.

Stoof, H. T. C. (1994). *Phys. Rev. A*, **49**, 3824.

Strecker, K. E., Partridge, G. B., Truscott, A. G., and Hulet, R. G. (2002). *Nature*, **417**, 150.

Stringari, S. (1996). *Phys. Rev. Lett.*, **77**, 2360.

Stringari, S. (1998). *Phys. Rev. A*, **58**, 2385.

Ueda, M. and Leggett, A. J. (1998). *Phys. Rev. Lett.*, **80**, 1576.

Zambelli, F. and Stringari, S. (2002). *Las. Phys.*, **12**, 240.

Zaremba, E. (1998). *Phys. Rev. A*, **57**, 518.

2

Degenerate Fermi gases

Patrizia VIGNOLO

Institut Non Linéaire de Nice, Université de Nice Sophia, CNRS;
1361 route des Lucioles, 06560 Valbonne, France

2.1   Introduction

The techniques that have led to the achievement of Bose–Einstein condensation in vapors of bosonic atoms (Anderson *et al.*, 1995; Davis *et al.*, 1995; Bradley *et al.*, 1995) have allowed in the last ten years to trap and cool dilute gases of fermionic alkali atoms down to the degenerate regime (DeMarco and Jin, 1999) and to achieve fermionic superfluids (Greiner *et al.*, 2003; Bourdel *et al.*, 2004; Kinast *et al.*, 2004; Bartenstein *et al.*, 2004).

Whereas the standard evaporative cooling scheme is based on *s*-wave collisions as this is the only effective channel at ultralow temperatures, identical fermions in a magnetic trap cannot collide in this channel owing to the antisymmetry of the fermionic wave function. This problem has been tackled by trapping either a fermion gas of two components in different internal states or two different atomic species. Indeed, fermionic atomic gases were brought to the degenerate regime in two spin-component Fermi mixtures such as ^{6}Li–^{6}Li (DeMarco and Jin, 1999; Granade *et al.*, 2002; Jochim *et al.*, 2002; Dieckmann *et al.*, 2002) and ^{40}K–^{40}K (Gensemer and Jin, 2001), or together with bosonic atoms in several alkali atom mixtures, such as ^{7}Li–^{6}Li (Truscott *et al.*, 2001; Schreck *et al.*, 2001), ^{23}Na-^{6}Li (Hadzibabic *et al.*, 2002), ^{87}Rb-^{40}K (Goldwin *et al.*, 2002; Roati *et al.*, 2002; Modugno *et al.*, 2002), and very recently in a mixed gas of ytterbium (Yb) isotopes, ^{174}Yb–^{173}Yb (Fukuhara *et al.*, 2009).

The geometry of the confinement where atomic samples are cooled down can be isotropic or very anisotropic. Atoms can be confined in pancake-or cigar-shaped traps, in planes or wires by means of one-dimensional or bidimensional optical lattices. The interplay between dimensionality, quantum statistics and interactions gives rise to a variety of physical effects that are amenable to observation.

Low temperatures enhance the effect of quantum statistics and, with respect to Bose gases, Fermi systems exhibit important differences. A first consequence is that, below the Fermi temperature T_F, which is the temperature characterizing the Fermi quantum degeneracy, a spin-polarized Fermi gas can be considered as non-interacting since *s*-wave scattering events are suppressed by the Pauli principle. This provides the opportunity to have access to a quasi-ideal Fermi gas and to observe single-orbital occupation effects in the presence of a very anisotropic confinement. Even in a two-spin-component Fermi gas, the filling of the Fermi sphere strongly suppresses collisions and the gas can be driven from a collisional to a collisionless regime, as shown in a ^{40}K experiment at JILA (Gensemer and Jin, 2001).

By further lowering the temperature, a two-component Fermi gas, with attractive interactions between the spin-up and -down components, becomes stable at forming pairs, at T_{pair}, and forms a condensate of pairs below T_{c}, the critical temperature (see Figure 2.1). This is a consequence of the interplay between Fermi statistics and interactions: below T_{c} a Fermi liquid with attractive interactions is unstable and phase transition towards superfluid allows the system to stabilize. The microscopic feature of the superfluid depends on the attraction strength: at weak interaction the many-body wave function corresponds to largely overlapping fermion pairs (BCS), while at strong interaction it corresponds to a condensate of composed molecules (BEC).

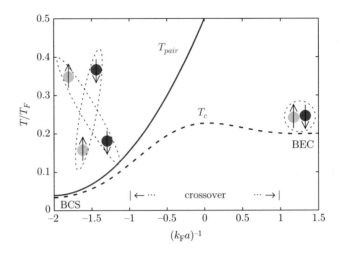

Fig. 2.1   Illustrative phase diagram for a two-component Fermi gas.

Interactions can be tuned by applying a suitable elecromagnetic field (Feshbach resonances). This allows the study of the collisional properties of Fermi gas above T_{pair}, to explore, below T_{c}, the evolution from a BCS superfluid to a molecular BEC (Zwierlein *et al.*, 2003; Greiner *et al.*, 2003; Bourdel *et al.*, 2004; Kinast *et al.*, 2004; Bartenstein *et al.*, 2004; Partridge *et al.*, 2005), and to prove theories describing this crossover.

This chapter focuses on the theorical description of degenerate Fermi gases, from the ideal Fermi gas case to the BCS–BEC crossover, and on the main observables that are accessible in actual experiments. Here is a brief outline of the chapter. We first remind ourselves about the Fermi–Dirac distribution (Section 2.2) and introduce the concept of Fermi energy and Fermi sphere for an homogeneous gas in one, two and three dimensions. The case of a harmonically confined ideal Fermi gas is discussed within the framework of the local density approximation (LDA) in Section 2.2.2, and that of an exact Green's function method in Section 2.2.3. The latter allows, without knowing the many-body wave function, to calculate all moments of the one-body density matrix and two-body correlations (Section 2.2.3.3). The discussion on the dynamical properties of a harmonically confined spin-polarized Fermi gas (Section 2.2.4) concludes the first part of the course devoted to the ideal Fermi gas.

The second part focuses on the two-component Fermi gas (Section 2.3). We first mention the scattering theory in the low-energy limit (Section 2.3.1) and discuss a simple model to describe the mechanism of Feshbach resonances, which allow the tuning of the interactions (Section 2.3.1.1).

Above T_{pair} the two-component Fermi gas is described by Vlasov–Boltzmann trasport equations for the phase-space density distribution (see Section 2.3.2). We show that this framework allows us to describe the collisional–collisionless transition due to Pauli blocking in agreement with experimental findings (Section 2.3.2.1).

Below T_{c} (Section 2.3.3) the description of the gas depends on the interaction strength. Section 2.3.3.1 recalls the BCS theory, while for the BEC limit

(Section 2.3.3.2) students should use the chapter on Degenerate Bose Gases of this school as reference. The crossover can be described by a BCS mean-field model, or by more sophisticated approaches that take into account higher order correlations. The equation of states (EOS) predicted by the different theories is discussed in Section 2.3.3.4. For a trapped system the EOS can be inferred by mesuring the equilibrium density profile and the spectrum of collective excitations. The evaluation of some collective modes and their comparison with the experimental findings conclude the course (Section 2.3.3.5).

2.2 Ideal Fermi gas

2.2.1 Fermi–Dirac distribution

The Fermi–Dirac distribution gives the distribution of N identical fermions over single-particle energy states, where no more than one fermion can occupy a state. Let us consider a group of levels labelled by an index j, containing g_j single-particle states at energy ε_j and occupied by N_j fermions. The statistical weight of the macrostate of the gas is $w = \prod_j w_j$,

$$w_j = \frac{g_j!}{N_j!(g_j - N_j)!} \tag{2.1}$$

being the number of ways in which N_j identical objects may be distributed in g_j boxes with the condition that each box cannot contain more that one object. Using the Stirling formula ($\ln N! \simeq N \ln N - N$ at large N), the entropy of the Fermi gas can be written as

$$S = -k_B \ln w = -k_B \sum_j g_j[(1 - n_j)\ln(1 - n_j) + n_j \ln(n_j)], \tag{2.2}$$

where $n_j = N_j/g_j$ is the average occupation number for the jth group of states.

A state at thermal equilibrium at fixed volume obeys the thermodynamic identity $T dS + \mu dN - dU = 0$, with $N = \sum_j g_j n_j$ and $U = \sum_j g_j n_j \varepsilon_j$ and μ being the chemical potential. Thus, we find the Fermi–Dirac distribution

$$\langle n_j \rangle = \mathcal{F}(\varepsilon_j) = \frac{1}{e^{(\varepsilon_j - \mu)/k_B T} + 1}, \tag{2.3}$$

where $\mathcal{F}(\varepsilon)$ is known as the Fermi function. The average number of fermions with energy ε_j can be found by multiplying the Fermi–Dirac distribution (2.3) by the degeneracy g_j, namely

$$\langle n(\varepsilon_j) \rangle = \frac{g_j}{e^{(\varepsilon_j - \mu)/k_B T} + 1}. \tag{2.4}$$

For the case of a quasi-continuum stectrum, the number of fermions per energy unity is

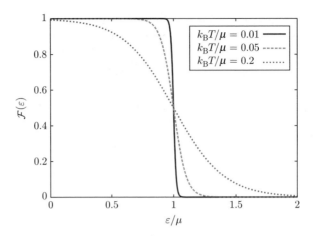

Fig. 2.2 The Fermi distribution for different values of the temperature.

$$\langle n(\varepsilon) \rangle = \rho_{\rm s}(\varepsilon) \mathcal{F}(\varepsilon) = \frac{\rho_{\rm s}(\varepsilon)}{e^{(\varepsilon - \mu)/k_{\rm B}T} + 1}, \tag{2.5}$$

where $\rho_{\rm s}(\varepsilon)$ is the density of states. The total number of fermions is given by integrating Eq. (2.5) over energy. The behavior of the function $\mathcal{F}(\varepsilon)$ for different values of temperature is shown in Figure 2.2. In the limit of $T \to 0$, $\mathcal{F}(\varepsilon)$ coincides with the step function $\theta(\varepsilon_F - \varepsilon)$, the Fermi energy ε_F being the chemical potential at zero temperature.

The density of states for a D-dimensional Fermi gas with spin degeneracy g, trapped in a box of volume L^D is given by

$$\rho_{\rm s}(\varepsilon) = \frac{g}{(2\pi\hbar)^D} \int {\rm d}^D x \int {\rm d}^D p \, \delta \left(\frac{p^2}{2m} - \varepsilon \right) \tag{2.6}$$

$$= c_D \frac{g L^D (2m\varepsilon)^{D/2}}{\varepsilon \, \hbar^D},$$

with $c_1 = 1/2\pi$, $c_2 = 1/4\pi$ and $c_3 = 1/4\pi^2$. The momentum distribution can be deduced by imposing $\langle n(p) \rangle {\rm d}^D p = \langle n(\varepsilon) \rangle {\rm d}\varepsilon$, with $\varepsilon = p^2/2m$ being the relation between the energy ε and the momentum p. One finds that

$$\langle n(p) \rangle = \frac{g L^D}{(2\pi\hbar)^D} \frac{1}{e^{(p^2/2m - \mu)/k_{\rm B}T} + 1}, \tag{2.7}$$

which reduces to

$$\langle n(p) \rangle = \frac{g L^D}{(2\pi\hbar)^D} \theta(p_F - |p|), \tag{2.8}$$

at zero temperature, with $p_F = \hbar k_F = \sqrt{2m\varepsilon_F}$. The density of the homogeneous Fermi gas $n_D = \frac{1}{L^D} \int \langle n(p) \rangle \mathrm{d}^D p$ at zero temperature reduces to

$$n_D = \mathcal{C}_D \, g \, k_F^D = \mathcal{C}_D \frac{g}{\hbar^D} (2m\varepsilon_F)^{D/2}, \tag{2.9}$$

with $\mathcal{C}_1 = 1/\pi$, $\mathcal{C}_2 = 1/4\pi$ and $\mathcal{C}_3 = 1/6\pi^2$. Equation (2.9) provides the explicit relation between the Fermi energy and the density

$$\varepsilon_F = \frac{\hbar^2}{2m} \left(\frac{n_D}{\mathcal{C}_D g} \right)^{2/D}, \tag{2.10}$$

and the Fermi wavevector and the density

$$k_F = \left(\frac{n_D}{\mathcal{C}_D g} \right)^{1/D}. \tag{2.11}$$

2.2.2   Ideal Fermi gas in a harmonic confinement: the LDA approach

We consider now a system of N non-interacting Fermi particles subjected to an external confinement given by a real potential $V_{ext}(\mathbf{r})$.

In the LDA we replace the actual equation of state $\mu(n)$ by the function $\mu_{loc} = \bar{\mu} - V_{ext}$, $\bar{\mu}(N)$ being the chemical potential of the unhomogeneous system. The spatial density distribution can be written as

$$n_D(\mathbf{r}) = \mathcal{C}_D \frac{g}{\hbar^D} [2m(\varepsilon_F - V_{ext}(\mathbf{r}))]^{D/2}. \tag{2.12}$$

In order to evaluate the Fermi energy of the trapped gas, let us first calculate the density of states at energy ε. In the semi-classical approximation, one has

$$\rho_s(\varepsilon) = \frac{g}{(2\pi\hbar)^D} \int \mathrm{d}^D r \int \mathrm{d}^D p \, \delta \left(\frac{p^2}{2m} + V_{ext}(\mathbf{r}) - \varepsilon \right) \tag{2.13}$$

$$= c_D \frac{g(2m)^{D/2}}{\hbar^D} \int \mathrm{d}^D r \, (\varepsilon - V_{ext}(\mathbf{r}))^{D/2-1}.$$

For a harmonic potential, $V_{ext} = \frac{1}{2} m\omega^2 r^2$, one obtains

$$\rho_s(\varepsilon) = \frac{1}{(D-1)!} \frac{g \varepsilon^{D-1}}{(\hbar\omega)^D}. \tag{2.14}$$

The Fermi energy ε_F is found by the normalization condition $\int \langle n(\varepsilon) \rangle \mathrm{d}\varepsilon = N$. At zero temperature,

$$\int_0^{\varepsilon_F} \rho_s(\varepsilon) \mathrm{d}\varepsilon = \frac{g}{D!} \frac{\varepsilon_F^D}{(\hbar\omega)^D} = N, \tag{2.15}$$

giving

$$\varepsilon_F = \left(D! \frac{N}{g} \right)^{1/D} \hbar\omega. \tag{2.16}$$

We remark that the density profile (2.12) can be written in the form

$$n_D(\mathbf{r}) = C_D \frac{g}{\hbar^D} (2m\varepsilon_F)^{D/2} \left(1 - \frac{r^2}{R_{TF}^2} \right)^{D/2}, \tag{2.17}$$

with

$$R_{TF} = \sqrt{2\varepsilon_F/m\omega^2} = k_F a_{ho}^2 = \sqrt{2} \left(D! \frac{N}{g} \right)^{1/(2D)} a_{ho} \tag{2.18}$$

being the Thomas–Fermi radius of the fermionic cloud in the semi-classical approximation, and $a_{ho} = \sqrt{\hbar/m\omega}$ being the harmonic oscillator length. In contrast, the radius of an ideal Bose gas, trapped in a harmonic confinement, is simply a_{ho}. Thus, an observable that distinguishes a trapped ideal Fermi gas from a trapped ideal Bose gas at zero temperature is the size of the cloud.

2.2.3   1D spin-polarized Fermi gas: exact density profile

A spin-polarized Fermi gas (spin degeneracy $g = 1$) at very low temperature can be considered as non-interacting since s-wave scattering events are forbidden by the Pauli principle and higher-order interaction terms are negligible. The many-body wave function, with the suitable antisymmetric properties under exchange of two fermions, can be written in terms of the Slater determinant

$$\Phi(x_1, x_2, \ldots, x_N) = \frac{1}{\sqrt{N!}} \begin{vmatrix} \psi_1(x_1) & \psi_1(x_2) & \psi_1(x_3) & \ldots & \psi_1(x_N) \\ \psi_2(x_1) & \psi_2(x_2) & \psi_2(x_3) & \ldots & \psi_2(x_N) \\ \psi_3(x_1) & \psi_3(x_2) & \psi_3(x_3) & \ldots & \psi_3(x_N) \\ \ldots & \ldots & \ldots & \ddots & \ldots \\ \psi_N(x_1) & \psi_N(x_2) & \psi_N(x_3) & \ldots & \psi_N(x_N) \end{vmatrix}, \tag{2.19}$$

where $\psi_i(x)$ is the single-particle orbital with eigenvalue ε_i. For a one-dimensional (1D) gas we may take the eigenfunctions $\psi_i(x)$ as real.[1] The one-body density matrix at zero temperature reads

$$\rho(x, x') = N \int dx_2 \int dx_3 \ldots \int dx_N \, \Phi^*(x, x_2, \ldots, x_N) \Phi(x', x_2, \ldots, x_N)$$

$$= \sum_{i=1}^{N} \psi_i(x)\psi_i(x') = \sum_{i=1}^{N} \psi_i(x') e^{i\hat{p}(x-x')} \psi_i(x'), \tag{2.20}$$

showing how distant points are correlated through the momentum operator $\hat{p}$. Expansion in powers of the relative coordinate $r = x - x'$ yields physical observables, such as the zero-moment that gives the particle density profile $n(x)$,

$$n(x) = \rho(x, x')|_{x'=x} = \sum_{i=1}^{N} \langle \psi_i \,|\, \delta(x - x_i) \,|\, \psi_i \rangle = \sum_{i=1}^{N} |\psi_i(x)|^2. \qquad (2.21)$$

For the even moments of higher-order several definitions have been considered in the literature (for a review see (Ziff *et al.*, 1977)), among which

$$P_n(x) = (-i)^n \left[\frac{\partial^n}{\partial x_1^n} \rho(x, x_1) \right]_{x_1 = x} \qquad (2.22)$$

gives twice the kinetic energy density for $n = 2$ (setting $\hbar = 1$ and $m = 1$).

2.2.3.1   *The Green's function method*

This method allows calculation of all moments of the one-body density matrix without evaluating the many-body wavefunction. The main idea is to rewrite Eqs. (2.21) and (2.22) as the imaginary part of the ground-state average of the Green's function in coordinate space $\hat{G}(x) = (x - \hat{x} + i\epsilon)^{-1}$. We have

$$n(x) = -\frac{1}{\pi} \lim_{\epsilon \to 0^+} \mathrm{Im} \sum_{i=1}^{N} \langle \psi_i \,|\, \hat{G}(x) \,|\, \psi_i \rangle \qquad (2.23)$$

for the density, and

$$P_n(x) = -\frac{1}{\pi} \lim_{\epsilon \to 0^+} \mathrm{Im} \sum_{i=1}^{N} \langle \psi_i \,|\, \hat{G}(x) \hat{p}^n \,|\, \psi_i \rangle \qquad (2.24)$$

for higher moments (Eq. (2.22)). The equivalence between expressions (2.21) and (2.23) is easily proved in the coordinate representation, where the density profile in Eq. (2.23) reads

$$n(x) = -\frac{1}{\pi} \lim_{\epsilon \to 0^+} \mathrm{Im} \sum_{i=1}^{N} \int dx_i |\psi_i(x_i)|^2 \frac{1}{x - x_i + i\epsilon}, \qquad (2.25)$$

yielding Eq. (2.21) when one takes the limit $\epsilon \to 0^+$. A similar demostration applies to Eqs. (2.22) and (2.24).

   To evaluate the Green's functions in the specific case of harmonic confinement, we make use of the representation for the position and the momentum operators in the basis of the eigenstates of the harmonic oscillator, i.e. $\hat{x} = (a + a^\dagger)/\sqrt{2}$ and $\hat{p} = i(a^\dagger - a)/\sqrt{2}$ with $a \,|\, \psi_i \rangle = \sqrt{i-1} \,|\, \psi_{i-1} \rangle$ and $a^\dagger \,|\, \psi_i \rangle = \sqrt{i} \,|\, \psi_{i+1} \rangle$ (we have set the harmonic oscillator frequency $\omega = 1$). Explicitly, the representation of $\hat{x}$ in matrix form is given by

$$\hat{x} = \begin{pmatrix} 0 & 1/\sqrt{2} & & & & \\ 1/\sqrt{2} & 0 & 1 & & & \\ & 1 & 0 & \sqrt{3/2} & & \\ & & \sqrt{3/2} & 0 & \sqrt{2} & \\ & & & \sqrt{2} & 0 & \sqrt{5/2} \\ & & & & \ddots & \ddots & \ddots \end{pmatrix}, \tag{2.26}$$

and similarly for $\hat{p}$.

The matrix (2.26) is semi-infinite, thus the calculation of $\hat{G}(x)$ by inverting the matrix $x - \hat{x} + i\epsilon$ is not suitable. Since one needs to know explicitly only the first N diagonal terms of the Green's operator, we write an effective position operator $\hat{\xi}(x)$ of dimension $N \times N$ exploiting a decimation/renormalization procedure (see Appendix A), so that

$$\left[(x - \hat{x} + i\epsilon)^{-1}\right]_{i,j} = \left[(x - \hat{\xi}(x) + i\epsilon)^{-1}\right]_{i,j} \tag{2.27}$$

in the subspace $\{1, 2, \ldots, N\}$. This is defined by setting $[\hat{\xi}(x)]_{i,j} = [\hat{x}]_{i,j}$ if $(i, j) \neq (N, N)$ and $[\hat{\xi}(x)]_{N,N} = \tilde{x}_{N,N}(x)$, with

$$\tilde{x}_{N,N}(x) = \cfrac{N/2}{x + i\epsilon - \cfrac{(N+1)/2}{x + i\epsilon - \ldots}}. \tag{2.28}$$

The single term $\tilde{x}_{N,N}(x)$ contains the contribution of all the states that are not occupied by the fermions. Its asymptotic value is $\tilde{x}_{N,N}(x) = i\sqrt{2N - x^2}$ for $N \to \infty$.

Since in the subspace $\{1, 2, \ldots, N\}$, $\hat{G}(x)$ can be written as the inverse of a finite tridiagonal matrix, for the calculation of the trace we can: (i) invert the matrix $x - \hat{\xi}(x) + i\varepsilon$ (not convenient for large N), or (ii) exploit an extension of the Kirkman–Pendry relation (Kirkman and Pendry, 1984; Vignolo *et al.*, 1999; Farchioni *et al.*, 2000), that connects the trace to an off-diagonal element, allowing a further reduction of the dimensionality of the problem.

The Kirkman–Pendry relation. We detail here the Kirkman–Pendry relation, which is more suitable for the calculation of the density.

We define the Green's function

$$\hat{\mathcal{G}}(\delta, x) = \frac{1}{x + \delta - \hat{\xi}(x) + i\epsilon}, \tag{2.29}$$

where δ is an auxiliary continuous variable. The operator $\hat{\mathcal{G}}(\delta, x)$ evaluated at $\delta = 0$ corresponds to the Green's function $\hat{G}(x)$ projected on the subspace $\{1, 2, \ldots, N\}$.

The expression for $\mathrm{Tr}_N \hat{G}(x)$ is obtained from the Green's function element $\hat{\mathcal{G}}_{1,N}(\delta, x)$ between the first and the last occupied state by using the expression

$$\frac{\partial}{\partial \delta} \ln \hat{\mathcal{G}}_{1,N}(\delta, x) = \frac{\partial}{\partial \delta} \ln \frac{\prod_{i=1}^{N-1}[\hat{\xi}(x)]_{i,i+1}}{\det[x + \delta - \hat{\xi}(x) + i\epsilon]} = \frac{\partial}{\partial \delta} \ln \frac{\prod_{i=1}^{N-1}[\hat{\xi}(x)]_{i,i+1}}{\prod_{i=1}^{N}[x + \delta - \xi_i(x) + i\epsilon]}$$

$$= \frac{\partial}{\partial \delta} \ln \prod_{i=1}^{N} \frac{1}{x + \delta - \xi_i(x) + i\epsilon} = \sum_{i=1}^{N} \frac{-1}{x + \delta - \xi_i(x) + i\epsilon}, \quad (2.30)$$

where $\xi_i(x)$ are the eigenvalues of the operator $\hat{\xi}(x)$. Therefore, the particle density is given by

$$n(x) = -\frac{1}{\pi} \lim_{\epsilon \to 0^+} \mathrm{Im} \mathrm{Tr}_N \hat{G}(x) = \frac{1}{\pi} \lim_{\epsilon \to 0^+} \mathrm{Im} \left[\frac{\partial}{\partial \delta} \ln \hat{\mathcal{G}}_{1,N}(\delta, x) \right]_{\delta=0}. \quad (2.31)$$

The matrix element $\hat{\mathcal{G}}_{1,N}(\delta, x)$ can be calculated by a further renormalization (Vignolo et al., 1999) of $\hat{\xi}(x)$, obtaining an effective operator $\hat{\xi}'(x)$ of dimension 2×2, acting on the space $\{1, N\}$.

The density profile of a spin-polarized Fermi gas, trapped in a 1D harmonic confinement, is shown in Figure 2.3 for different numbers of particles. The profiles exhibit a number of peaks equal to the number of trapped fermions and the width of this oscillating structure decreases as the number of particles increases. A mapping exists in 1D between spin-polarized fermions and impenetrable bosons, that takes the form of an identity between the many-body wave function of the bosons and the modulus of the wave function of the fermions (Girardeau, 1960, 1965; Lenard, 1966). Thus, a 1D gas of impenetrable bosons subjected to the same confinement, exhibits an identical density profile to that shown in Figure 2.3.

The momentum distribution $n(p)$ can be analogously written in terms of the trace of the Green's function $\hat{G}(p) = (p - \hat{p} + i\epsilon)^{-1}$ in momentum space, taken on the eigenvectors $|\phi_i\rangle$ of the Hamiltonian in the momentum representation:

$$n(p) = \sum_{i=1}^{N} \int dp_i |\phi_i(p_i)|^2 \delta(p - p_i), \quad (2.32)$$

or

$$n(p) = -\frac{1}{\pi} \lim_{\epsilon \to 0^+} \mathrm{Im} \sum_{i=1}^{N} \langle \phi_i | \hat{G}(p) | \phi_i \rangle, \quad (2.33)$$

where p_i are the eigenvalues of the momentum operator.

Another expression for the evaluation of the trace. The partial trace Tr_N of a generic matrix Q is related to the determinant of the inverse matrix Q^{-1} by

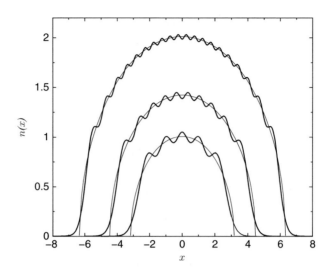

Fig. 2.3 From Ref. (Vignolo *et al.*, 2000). Exact particle density profile (bold lines) for $N = 5$, 10 and 20 harmonically confined fermions, compared with the corresponding profiles evaluated in the local density approximation. Positions are in units of the characteristic length of the harmonic oscillator $a_{ho} = \sqrt{\hbar/(m\omega)}$ and the particle density in units of a_{ho}^{-1}.

$$\text{Tr}_N Q = \partial \left[\ln \det(Q^{-1} + \lambda \mathbb{I}_N)\right] / \partial\lambda|_{\lambda=0}. \tag{2.34}$$

This relation is demonstrated as the following,

$$\partial \left[\ln \det(Q^{-1} + \lambda \mathbb{I}_N)\right] / \partial\lambda|_{\lambda=0} = \partial \left[\text{Tr}\ln(Q^{-1} + \lambda \mathbb{I}_N)\right] / \partial\lambda|_{\lambda=0}$$

$$= \text{Tr}\left[(Q^{-1} + \lambda \mathbb{I}_N)^{-1}\mathbb{I}_N\right]_{\lambda=0} = \text{Tr}(Q\mathbb{I}_N) = \text{Tr}_N Q. \tag{2.35}$$

Here, we take $Q = \hat{G}(x)\hat{p}^n$. In the case of harmonic confinement this matrix is the product of a $(2n + 1)$-diagonal matrix and of the inverse of a tridiagonal matrix. Using Eq. (2.34) we can write

$$P_n(x) = -\frac{1}{\pi} \lim_{\epsilon \to 0^+} \text{Im} \frac{\partial}{\partial\lambda} \left[\ln \det(x + i\epsilon - \hat{K}^n)\right]_{\lambda=0}, \tag{2.36}$$

where $\hat{K}^n = \hat{x} - \hat{p}^n \lambda \mathbb{I}_N$. We have thus reduced the problem to the evaluation of the determinant of the matrix $(x + i\epsilon - \hat{K}^n)$, which is tridiagonal in the tail and whose first N rows have only a few non-vanishing elements for low values of n.

If we make a partition in blocks of dimension $n \times n$ for the first part of the matrix, the whole operator can be written as

$$\hat{K}^n = \lim_{M \to \infty} \hat{K}^n_M = \lim_{M \to \infty} \begin{pmatrix} \mathcal{A}_1 & \mathcal{B}_{1,2} & & & & \\ \mathcal{B}_{2,1} & \mathcal{A}_2 & \mathcal{B}_{2,3} & & & \\ & \mathcal{B}_{3,2} & \ddots & \mathcal{B}_{3,4} & & \\ & & \ddots & \ddots & \ddots & \\ & & & \ddots & \ddots & \mathcal{B}_{M-1,M} \\ & & & & \mathcal{B}_{M,M-1} & \mathcal{A}_M \end{pmatrix}. \tag{2.37}$$

The determinant of the matrix $(x + i\epsilon - \hat{K}^n)$ with a tridiagonal representation such as that in Eq. (2.37) can be factorized into a product of determinants of matrices having the same dimension as that of the blocks of the partition (Farchioni *et al.*, 2000).

As a first step, it is easy to show that if we separate the matrix $\hat{K}^n_M$ into four blocks, thereby defining $\tilde{\mathcal{B}}_{i,j}$ and $\hat{K}^n_i$ from

$$\hat{K}^n_M = \begin{pmatrix} \hat{K}^n_{M-1} & \tilde{\mathcal{B}}_{M-1,M} \\ \tilde{\mathcal{B}}_{M,M-1} & \mathcal{A}_M \end{pmatrix}, \tag{2.38}$$

we can write

$$\det(x + i\epsilon - \hat{K}^n_M) = \det(x + i\epsilon - \hat{K}^n_{M-1})$$
$$\cdot \det(x + i\epsilon - \mathcal{A}_M - \tilde{\mathcal{B}}_{M,M-1}(x + i\epsilon - \hat{K}^n_{M-1})^{-1}\tilde{\mathcal{B}}_{M-1,M}). \tag{2.39}$$

Applying recursively this procedure and taking the limit of $M \to \infty$ we obtain

$$\det(x + i\epsilon - \hat{K}^n) = \det(x + i\epsilon - \mathcal{A}_1) \tag{2.40}$$

$$\cdot \lim_{M \to \infty} \prod_{j=2}^{M} \det(x + i\epsilon - \mathcal{A}_j - \tilde{\mathcal{B}}_{j,j-1}(x + i\epsilon - \hat{K}^n_{j-1})^{-1}\tilde{\mathcal{B}}_{j-1,j}).$$

It is now important to notice that, owing to the particular form of the matrices $\tilde{\mathcal{B}}_{j-1,j}$ and $\tilde{\mathcal{B}}_{j,j-1}$, it is not necessary to explicitly invert the matrices $(x + i\epsilon - \hat{K}^n_{j-1})$. Rather, we may use a renormalization procedure for the operator $\hat{K}^n_{j-1}$ to further simplify the expression (2.40) and to calculate the inverse of matrices with dimensions at most equal to $n \times n$. Renormalization allows us to write

$$\det(x + i\epsilon - \hat{K}^n) = \prod_{j=1}^{\infty} \det(x + i\epsilon - \tilde{\mathcal{A}}_j), \tag{2.41}$$

where $\tilde{\mathcal{A}}_1 = \mathcal{A}_1$ and

$$\tilde{\mathcal{A}}_j = \mathcal{A}_j + \mathcal{B}_{j,j-1}(x + i\epsilon - \tilde{\mathcal{A}}_{j-1})^{-1}\mathcal{B}_{j-1,j} \tag{2.42}$$

for $j > 1$.

Both the Kirkman–Pendry relation and the determinant procedure are very convenient from a computational point of view. The advantage of the determinant trick is that it can be easily applied for the evaluation of higher-order moments, for the case of a 2D, 3D anisotropic gas (Vignolo and Minguzzi, 2003), or the case of a finite-temperature Fermi gas (Akdeniz *et al.*, 2002).

2.2.3.2   Finite-temperature effect

At finite temperature, the one-body coherence properties of a quantum system are described by the generalized grand-canonical density matrix. For fermions (March and Murray, 1960) it can be written as

$$D(x, x'; \beta, \mu) = \sum_{i=1}^{\infty} \frac{1}{1 + \exp\left[\beta(\varepsilon_i - \mu)\right]} \psi_i^*(x') e^{i\hat{p}(x-x')} \psi_i(x'). \qquad (2.43)$$

Here, $\beta = 1/k_B T$ and μ is the chemical potential, while ψ_i and ε_i are the single-particle orbitals and the corresponding energy eigenvalues. The zero-temperature limit of Eq. (2.43) leads to the Dirac density matrix given in Eq. (2.20).

The particle-density profile $n(x)$ of the gas at temperature T and chemical potential μ is the zero-order moment of the matrix $D(x, x_1)$,

$$n(x) = D(x, x'; \beta, \mu)|_{x'=x}$$

$$= \sum_{i=1}^{\infty} \frac{1}{1 + \exp\left[\beta(\varepsilon_i - \mu)\right]} \langle \psi_i | \delta(x - \hat{x}) | \psi_i \rangle. \qquad (2.44)$$

Equation (2.44) can be rewritten in terms of $\hat{G}(x)$,

$$n(x) = -\frac{1}{\pi} \lim_{\epsilon \to 0^+} \text{Im} \sum_{i=1}^{\infty} \frac{1}{1 + \exp\left[\beta(\varepsilon_i - \mu)\right]} \langle \psi_i | \hat{G}(x) | \psi_i \rangle$$

$$= -\frac{1}{\pi} \lim_{\epsilon \to 0^+} \text{Im} \text{Tr} \left(\mathcal{T} \cdot \hat{G}(x)\right). \qquad (2.45)$$

Here, we have used a matrix formalism, by introducing the *temperature matrix* $\mathcal{T}$ whose diagonal elements $[\mathcal{T}]_{i,i} = 1/\{1 + \exp\left[\beta(\varepsilon_i - \mu)\right]\}^{-1}$ contain the statistical Fermi factors, while the off-diagonal elements $[\mathcal{T}]_{i,j}$ with $i \neq j$ are null.

It follows from Eq. (2.34) that Eq. (2.45) can be written as

$$n(x) = -\frac{1}{\pi} \lim_{\epsilon \to 0^+} \text{Im} \frac{\partial}{\partial \lambda} \left[\ln \det(x - \hat{x} + \lambda \mathcal{T} + i\epsilon)\right]|_{\lambda=0}. \qquad (2.46)$$

In the specific case of harmonic confinement, the tridiagonal form of the matrix representing $\hat{x}$ allows us to evaluate the determinants in Eq. (2.46) by making use of the recursive algorithm outlined in the previous section. In Figure 2.4 we compute Eq. (2.45) for 4 fermions at various temperatures. The effect of the temperature is to smooth out the density peaks, which disappear at $T \simeq 0.5\hbar\omega/k_B$. For $T > 0.5\hbar\omega/k_B$ the density profiles evaluated within the semi-classical approximation are superposed to the exact ones (right panel).

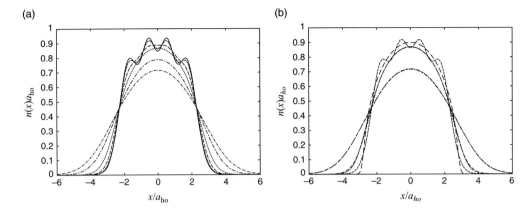

Fig. 2.4 From (Akdeniz *et al.*, 2002). Particle density profile for $N = 4$ harmonically confined fermions at various values of the temperature. Left panel: *exact* profiles at $T = 0$ (bold continuous line) and $T = 0.2\hbar\omega/k_B$ (dashed curve); the other curves refer to $k_B T/\hbar\omega = 0.5$, 1.0, 2.0 and 3.0, in order of decreasing peak height. Right panel: *exact* density profiles at $k_B T/\hbar\omega = 0.2$, 1.0 and 3.0, together with those calculated in the semiclassical approximation (light lines). Positions are in units of the harmonic oscillator length $a_{ho} = \sqrt{\hbar/(m\omega)}$ and the particle density is in units of a_{ho}^{-1}.

2.2.3.3 *Pair distribution function*

Higher-order correlations are described in the equal-times pair distribution function. The pair distribution function of a spin-polarized 1D Fermi gas is defined as

$$\rho_2(x_1, x_2) = n(x_1)n(x_2) - F(x_1, x_2), \tag{2.47}$$

where the function $F(x_1, x_2)$ is given by

$$F(x_1, x_2) = \sum_{i,j=1}^{N} \phi_i^*(x_1)\phi_j(x_1)\phi_j^*(x_2)\phi_i(x_2). \tag{2.48}$$

This function can be calculated by an extension of the Green's function method used in the previous sections. We first rewrite Eq. (2.48) in terms of the Green's function $\hat{G}(x)$,

$$F(x_1, x_2) = \frac{1}{\pi^2} \lim_{\varepsilon \to 0^+} \sum_{i,j=1}^{N} \mathrm{Im}\langle\psi_i|\hat{G}(x_1)|\psi_j\rangle \, \mathrm{Im}\langle\psi_j|\hat{G}(x_2)|\psi_i\rangle; \tag{2.49}$$

we then use the property $\mathrm{Im}A \cdot \mathrm{Im}B = (1/2)[\mathrm{Re}(A \cdot B^*) - \mathrm{Re}(A \cdot B)]$ to obtain the final expression

$$F(x_1, x_2) = \frac{1}{2\pi^2} \lim_{\varepsilon \to 0^+} \left\{ \mathrm{ReTr}\left[\hat{G}_N(x_1) \cdot \hat{G}_N^*(x_2)\right] - \mathrm{ReTr}\left[\hat{G}_N(x_1) \cdot \hat{G}_N(x_2)\right] \right\}, \tag{2.50}$$

where $\hat{G}_N(x)$ is the first $N \times N$ block of the matrix $\hat{G}(x)$. In the case of harmonic confinement this can be evaluated again by making use of renormalization techniques and recursive methods: the evaluation of $F(x_1, x_2)$ is reduced to the calculation of the determinant of pentadiagonal $N \times N$ matrices, which can be factorized into the product of 2×2 matrices.

2.2.4 Dynamical properties

2.2.4.1 *The dynamic structure factor*

The number of states per energy unit that can be excited to a state of momentum $\hbar\mathbf{k}$ and energy $\hbar\Omega$ is given by the dynamic structure factor

$$S(\mathbf{k}, \Omega) = \int d^3r_1 \int d^3r_2\, e^{-i\mathbf{k}\cdot(\mathbf{r}_1 - \mathbf{r}_2)} \int dt e^{-i\Omega t}\, S(1, 2)|_{t=t_2-t_1}, \qquad (2.51)$$

which is the Fourier transform of the two-point correlation function,

$$S(1, 2) = \langle \hat{\Phi}(\mathbf{r}_1, t_1)\hat{\Phi}^\dagger(\mathbf{r}_1, t_1)\hat{\Phi}(\mathbf{r}_2, t_2)\hat{\Phi}^\dagger(\mathbf{r}_2, t_2)\rangle - n(\mathbf{r}_1, t_1)n(\mathbf{r}_2, t_2). \qquad (2.52)$$

For a non-interacting Fermi gas in the collisionless regime, we can expand the field operator in energy modes as $\hat{\Phi}(\mathbf{r}_1, t_1) = \sum_i \psi_i(\mathbf{r}_1)\exp(-i\varepsilon_i t_1)\hat{a}_i$ and we can write the dynamic structure factor as

$$S(\mathbf{k}, \Omega) = \sum_{i,j} \left| \int d^3r\, e^{-i\mathbf{k}\cdot\mathbf{r}}\psi_i^*(\mathbf{r})\psi_j(\mathbf{r}) \right|^2 \mathcal{F}(\varepsilon_i)[1 - \mathcal{F}(\varepsilon_j)]\, 2\pi\delta(\Omega - (\varepsilon_j - \varepsilon_i)/\hbar).$$

$$(2.53)$$

For the dynamic factor structure of a 3D non-interacting Fermi gas see Ref. (Nozières and Pines, 1966). To have access to a 1D dynamics, one has to excite the longitudinal modes without involving the transverse excited states. This occurs in the experiments where *(i)* the transverse component $k_\perp$ of the momentum transfer vanishes, due to orthogonality of harmonic oscillator wave functions, or *(ii)* the energy transfer Ω is smaller than the gap between the chemical potential and the first transverse excited state.

In this limit, Eq. (2.53) reduces to a one-dimensional problem and the transverse-state wave function factorizes out.

By neglecting the effect of the longitudinal confinement one finds the dynamic structure factor for the homogeneous ideal Fermi gas in a 1D box of length L

$$S_{hom}(k_x, \Omega) = N^2 \int \frac{dk_i}{2k_F} \int \frac{dk_j}{2k_F} \frac{\sin^2[(k_x + k_i - k_j)L/2]}{[(k_x + k_i - k_j)L/2]^2} \qquad (2.54)$$

$$2\pi\delta(\Omega - (\varepsilon_j - \varepsilon_i)/\hbar)\theta(k_F - |k_i|)\theta(|k_j| - k_F),$$

namely

$$
S_{hom}(k_x, \Omega) = \begin{cases} \dfrac{2\pi N}{k_x k_F} \dfrac{m}{\hbar} & \text{if} \qquad |\Omega_2(k_x)| < \Omega < \Omega_1(k_x) \\[2ex] 0 & \text{otherwise,} \end{cases} \tag{2.55}
$$

with $\Omega_{1,2} = \hbar k_x^2/2m \pm \hbar k_x k_F/m$ (see Figure 2.5), implying that the system can be excited with a frequency Ω and at a wavelength k_x only if the Fermi wavevector satisfies the condition $k_{F,-} < k_F < k_{F,+}$ with

$$
k_{F,\pm} = \frac{m}{\hbar} \frac{\Omega}{k_x} \pm \frac{k_x}{2}. \tag{2.56}
$$

A good description for the spectrum by taking into account the axial potential $V_{ext} = \frac{1}{2} m\omega^2 r^2$ is given by the LDA, defined as

$$
S_{LDA}(k_x, \Omega) = \frac{1}{N} \int dx \, n(x) S_{hom}(k_x, \omega; \mu(x)) . \tag{2.57}
$$

The integral in Eq. (2.57) can be evaluated analytically, giving

$$
S_{LDA}(k_x, \Omega) = \frac{2}{k_x} \frac{m}{\hbar} x(k_{F,loc})\big|_{k_{F,+}}^{k_{F,-}} \tag{2.58}
$$

$$
= \frac{2}{k_x \omega} \frac{m}{\hbar} \left[\sqrt{\frac{2N\hbar\omega}{m} - \left(\frac{\Omega}{k_x} - \frac{\hbar k_x}{2m} \right)^2} - \sqrt{\frac{2N\hbar\omega}{m} - \left(\frac{\Omega}{k_x} + \frac{\hbar k_x}{2m} \right)^2} \right].
$$

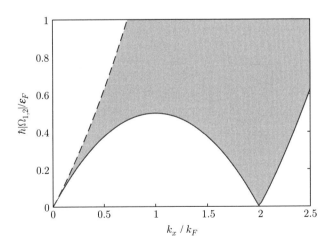

Fig. 2.5 The frequencies Ω_1 (dashed line) and $|\Omega_2|$ (continuous line) in units of $\varepsilon_F/\hbar$ as functions of k_x/k_F. The dynamic structure factor $S_{hom}(k_x, \Omega)$ defined in Eq. (2.54) is non-zero in the gray region.

To obtain Eq. (2.58), we have made use of Eq. (2.12) and the fact that in 1D $k_{F,loc} = \pi n$.

The exact spectrum can be evaluated by exploiting the properties of the Hermite polynomials (Gradshteyn and Ryzhik, 2000). One obtains (Vignolo *et al.*, 2001)

$$S(\mathbf{k}, h) = 2\pi e^{-k_\perp^2/2\lambda} e^{-k_x^2/2} \sum_{i=\max\{N-h,0\}}^{N-1} \frac{i!}{(i+h)!} \left(\frac{k_x^2}{2} \right)^h \left[L_i^h \left(k_x^2/2 \right) \right]^2. \qquad (2.59)$$

Here, h is an integer corresponding to a single-atom excitation of h quanta of the harmonic oscillator, $L_i^h(x)$ is the ith generalized Laguerre polynomial of parameter h, λ is the anisotropy parameter for the 3D harmonic oscillator and we have set $\hbar = 1$, $m = 1$ and $\omega = 1$.

Figure 2.6 show the dynamic structure factor as a function of exciting frequency Ω for two different values of k_x. We can see that the LDA description captures the main features of the spectrum.

2.2.4.2 Collective excitations

Now we turn to the long-wavelength limit and investigate the analog of sound-wave propagation. Due to the presence of the external confinement, the collective excitations are quantized. We evaluate here the spectrum and the expression for the density fluctuations in the linear regime.

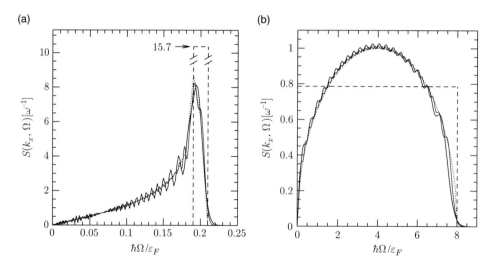

Fig. 2.6 From (Vignolo *et al.*, 2001). Dynamic structure factor $S(k_x, \Omega)$ of a 1D harmonically confined Fermi gas (in units of ω^{-1}) as a function of $\hbar\Omega/\varepsilon_F$, where $\varepsilon_F = N\hbar\omega$. Left panel: at $k_x = 0.1\, k_F$ and numbers of fermions $N=500$ (solid line) and $N=1000$ (solid bold line). Right panel: at $k_x = 2\, k_F$ and $N=20$ (solid line) and $N=40$ (solid bold line). In both panels the LDA spectrum (dotted line) and the dynamic structure factor of a homogeneous 1D Fermi gas (dashed line) are plotted in the same units.

Assuming that the gas reaches the hydrodynamic regime because of the presence of few impurities (Amoruso *et al.*, 1999), we use the the continuity equation,

$$\partial_t n + \nabla \cdot (n\mathbf{v}) = 0 \tag{2.60}$$

and the Euler equation

$$m\partial_t \mathbf{v} + \nabla \left[\mu(n) + V_{ext}(x) + \frac{1}{2}mv^2 \right] = \mathbf{0} \tag{2.61}$$

to derive the equation of motion for the density profile $n = n(x,t)$. We define the hydrodynamic regime by requiring that the chemical potential $\mu(n)$ depends locally on the particle density

$$\mu(n) = \frac{\pi\hbar^2 n^2}{2m} \tag{2.62}$$

and we include the effect of the confinement in the frame of the local density approximation, setting that $\mu(n) = \bar{\mu} - V_{ext}$.

In the linear regime Eqs. (2.60) and (2.61) lead to a closed equation for the density fluctuation $\delta n(x,t) = n(x,t) - n_0(x)$, which reads

$$-m\partial_t^2 \delta n + \partial_z \left[n_0 \partial_z \left(\frac{\partial\mu}{\partial n}\bigg|_{n=n_0} \delta n \right) \right] = 0. \tag{2.63}$$

By setting $\delta n(x,t) = \delta n(x)e^{-i\Omega t}$ and $z = x/R_{TF}$, one finds the differential equation

$$(1 - z^2)\partial_z^2 \delta n - 3z\partial_z \delta n + \left(\frac{\Omega^2}{\omega^2} - 1 \right) \delta n = 0. \tag{2.64}$$

The solutions of Eq. (2.64) are the Chebyshev polynomials. With the aim to obtain the dispersion relation, we assume that Eq. (2.64) also holds outside the classical radius $R_{TF} = \sqrt{2m\bar{\mu}/\omega^2}$. This hypothesis is necessary since the Chebyshev polynomials diverge at the classical boundary as $|x^2 - R_{TF}^2|^{-1/2}$. At higher dimensionality one imposes that the solution should vanish at $x = R_{TF}$, as in the 3D case (Amoruso *et al.*, 1999). By matching the solutions $\delta n(x)\sqrt{|x^2 - R_{TF}^2|}$ inside and outside the classical radius we obtain the spectrum (Minguzzi *et al.*, 2001)

$$\Omega = n\omega , \tag{2.65}$$

with corresponding solutions inside the classical radius

$$\delta n_{in}(z) = \frac{\cos(n\arccos(x))}{\sqrt{1 - z^2}} \tag{2.66}$$

and outside the classical radius

$$\delta n_{out}(z) = \frac{\left(|z| - \sqrt{z^2 - 1}\right)^n}{\sqrt{z^2 - 1}} . \tag{2.67}$$

The divergence of the solutions at $z = 1$, i.e. at $x = R_{TF}$, is unphysical, and is a consequence of our approximations: the linearized LDA solutions are not valid around the classical turning points.

2.3   Two-component Fermi gas

Let us consider a Fermi gas polarized in two spin states, that we call up and down, and let us consider the case of balanced populations, namely $N_\uparrow = N_\downarrow = N/2$. At low temperature $(T < T_F)$, interactions can be characterized by the physics of two-body s-wave collisions between fermions with different spin states. In fact, for the case of short-range potential, at low energy (i.e. low temperature), s-wave scattering is the dominant contribution. Thus, due to the Pauli exclusion, one can neglect interactions between identical fermions. In the next section, we recall the behavior of the s-wave scattering as a function of the potential range R_0 and the potential depth V_0. For a most exhaustive treatment of the scattering theory, the reader can refer, for instance, to (Messiah, 1999).

2.3.1   Scattering length and effective potential in the low-energy limit

Let us describe the interparticle potential $V_{1,2}(r)$ as a spherical well

$$V_{1,2}(r) = -V_0 \qquad 0 \le r \le R_0$$
$$V_{1,2}(r) = 0 \qquad r > R_0. \tag{2.68}$$

The radial Schrödinger equation as a function of the reduced mass m_r

$$\left(-\frac{d^2}{dr^2} + \frac{2m_r}{\hbar^2} V_{1,2}(r) \right) u(r) = \frac{2m_r}{\hbar^2} E u(r) \tag{2.69}$$

takes the explicit form of

$$\left(-\frac{d^2}{dr^2} + k^2 \right) u(r) = 0 \qquad r > R_0,$$
$$\left(-\frac{d^2}{dr^2} + k'^2 \right) u(r) = 0 \qquad 0 \le r \le R_0, \tag{2.70}$$

with $k^2 = 2mE/\hbar^2$ and $k'^2 = 2m_r(E + V_0)/\hbar^2$. Here, $u(r) = r\psi_{1,2}(r)$ is the reduced radial wave function for the relative motion of particles 1 and 2. The solution of Eq. (2.70) taking into account the condition $u(0) = 0$ is

$$u(r) = A\sin(kr + \delta) \qquad r > R_0,$$
$$u(r) = B\sin(k'r) \qquad 0 \le r \le R_0, \tag{2.71}$$

if $E > 0$. If $E < 0$, the solution is a bound state given by

$$u(r) = A e^{-\kappa r} \qquad r > R_0,$$
$$u(r) = B \sin(k'r) \qquad 0 \leq r \leq R_0, \qquad (2.72)$$

with $\kappa = -ik$.

The scattering length is defined as

$$a = \lim_{k \to 0} -\frac{\delta}{k}. \qquad (2.73)$$

By imposing the continuity of the derivative of $\ln(u(r))$ at $r = R_0$, from Eq. (2.70) one obtains

$$k \cot(kR_0 + \delta) = k' \cot k' R_0. \qquad (2.74)$$

Thus, for $k \to 0$ and $k' \to k_0 = (2m_r V_0/\hbar^2)^{1/2}$, in the limit of small δ

$$\delta \simeq -k \left(R_0 - \frac{\tan k_0 R_0}{k_0} \right) \qquad (2.75)$$

giving

$$a = R_0 \left(1 - \frac{\tan k_0 R_0}{k_0 R_0} \right). \qquad (2.76)$$

If the potential is attractive ($V_{1,2}(r) < 0$), as always happens at short distances because of the van der Waals interaction, the scattering length can be positive or negative. However, if the potential is repulsive ($V_{1,2}(r) > 0$) the scattering length is always positive. Bound states occur for attractive inter-particle potentials, and from the continuity of the logarithmic derivative of Eq. (2.72), one notices that bound states are only possible for negative values of $k' \cot g k' R_0$, namely for positive values of the scattering length (see Eq. (2.76)). In the strong-coupling limit $k_0 R_0 \gg 1$, if the condition for the existence of bound states is satisfied ($k' \cot k' R_0 \simeq k_0 \cot k_0 R_0 \simeq -1/a < 0$), the reduced wave function takes the form of $u(r) \propto e^{-r/a}$. In this regime, larger values of the potential V_0 correspond to lower (positive) values of the scattering length and tighter molecules.

A diverging scattering length corresponds to $k_0 R_0 = (1 + n)\pi/2$, thus to a dephasing $\delta \simeq \pi/2$ (see Eq. (2.74)). Nothing dramatic happens when $a \to \infty$ but the scattering length is no longer a length scale characterizing the system. In this regime, the so-called *unitary regime*, the only relevant length for the two-body problem is $1/k$.

In the absence of bound states, one can replace the *real* potential $V_{1,2}(r)$ by an effective potential $V_{eff}(r)$ that gives rise to the same scattering length.

For cold atoms as well as low-energy neutrons, one uses

$$V_{eff} u(r) = g \delta(\mathbf{r}) \frac{\mathrm{d}}{\mathrm{d}r} (r\, u(r)), \qquad (2.77)$$

with

$$g = \frac{2\pi\hbar^2}{m_r}a = \frac{4\pi\hbar^2}{m}a \tag{2.78}$$

for particles with equal masses. The effective potential behaves like a repulsive (attractive) potential for positive (negative) values of the scattering length.

2.3.1.1 *Feshbach resonances*

We consider now an atom–atom scattering problem where atoms have two internal states, the spin states of the valence electron. The effective inter-particle potential depends on if the two spins are in the singlet or triplet state. These two interaction "channels" $V_{S,T}$ are not independent, since they are coupled by the hyperfine interaction W between electron spin and nucleus spin. Moreover, the presence of a magnetic field $\mathscr{B}$ induces a different Zeeman shift $\mathscr{M}\mathscr{B}$. The Schödinger equation for the above model reads

$$\begin{pmatrix} -\dfrac{\hbar^2\nabla^2}{2m_r} + V_T(r) - E & W \\[2ex] W & -\dfrac{\hbar^2\nabla^2}{2m_r} + \mathscr{M}\mathscr{B} + V_S(r) - E \end{pmatrix} \begin{pmatrix} \Phi_T \\ \Phi_S \end{pmatrix} = \begin{pmatrix} 0 \\ 0 \end{pmatrix}. \tag{2.79}$$

As a specific example we use for both interaction potentials (Duine and Stoof, 2004) a spherical model with range R_0, namely

$$V_{S,T}(r) = -V_{S,T} \qquad 0 \le r \le R_0,$$
$$V_{S,T}(r) = 0 \qquad r > R_0, \tag{2.80}$$

as shown in Figure 2.7. To apply the scattering theory, one has to diagonalize the Hamiltonian (2.79) for $r > R_0$. Since the kinetic energy operator does not depend on the internal state, one diagonalizes the Hamiltonian,

$$\mathcal{H} = \begin{pmatrix} 0 & W \\ W & \mathscr{M}\mathscr{B} \end{pmatrix}, \tag{2.81}$$

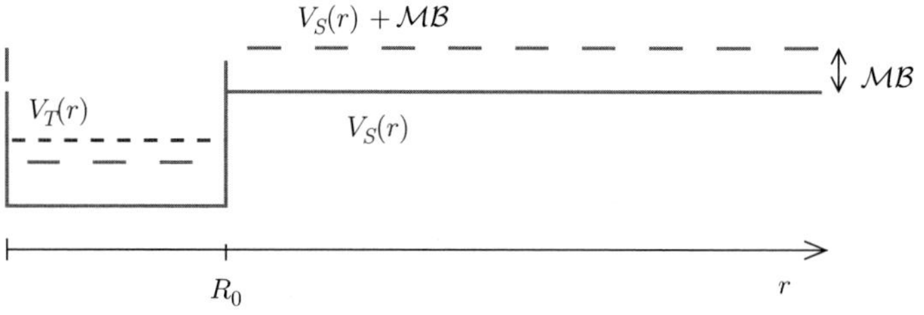

Fig. 2.7 Illustration of the two-channel model.

whose eigenvalues are

$$E_{\pm} = \frac{\mathscr{M}\mathscr{B}}{2} \pm \frac{1}{2}\sqrt{(\mathscr{M}\mathscr{B})^2 + 4W^2}, \tag{2.82}$$

and can be approximated to $E_- \simeq 0$, $E_+ \simeq \mathscr{M}\mathscr{B}$ in the limit of small hyperfine coupling. The eigenstate relative to the eigenvalue E_- is called the "open" channel, while that with eigenvalue E_+ is the so-called "closed" channel.

In cold-atom experiments, usually $\mathscr{M}\mathscr{B} \gg k_B T$ so that no atom scatters in the closed channel. The scattering length for the open channel can be evaluated by imposing the continuity in R_0 of the logaritmic derivative of the solutions

$$\begin{cases} u_-(r) = A_- \sin(kr + \delta) \\ u_+(r) = A_+ e^{-\kappa r} \end{cases} \quad r > R_0,$$

$$\begin{cases} u_-(r) = B_- \sin(k'_- r) \\ u_+(r) = B_+ \sin(k'_+ r) \end{cases} \quad 0 \leq r \leq R_0, \tag{2.83}$$

where $\kappa^2 = 2m_r(\mathscr{M}\mathscr{B} - E)/\hbar^2$, and $k'^2_{\pm} = 2m_r(E + E'_{\pm})/\hbar^2$, where

$$E'_{\pm} = \frac{\mathscr{M}\mathscr{B} - V_T - V_S}{2} \pm \frac{1}{2}\sqrt{(V_S - V_T - \mathscr{M}\mathscr{B})^2 + 4W^2} \tag{2.84}$$

are the eigenvalues of the matrix

$$\mathcal{H}' = \begin{pmatrix} -V_T & W \\ W & \mathscr{M}\mathscr{B} - V_S \end{pmatrix}. \tag{2.85}$$

The scattering length is then given by Eq. (2.76), namely $a = R_0(1 - \tan k_0 R_0 / k_0 R_0)$, with $k_0 = k'_+(E = 0)$.

Thus, the scattering in the presence of the singlet and triplet channels, at low energy and in the limit of weak coupling, is equivalent to a single-channel scattering where the depth of the potential well can be controlled by the external magnetic field $\mathscr{B}$. This allows us to vary the magnitude and the sign of the effective scattering length *via* the function $\tan(k_0 R_0)$. Obviously, the resonance is located at $k_0 R_0 = \pi/2$.

Feshbach resonances allow us to modify the effective potential and to create bound states (pairs). The threshold temperature $T_{pair} = E_{pair}/k_B$ for the formation of pairs is generally in the experiments lower than T_F. In the next section we discuss the equilibrium and the dynamical properties of a two-component Fermi gas at a finite temperature T with $T_{pair} < T \ll T_F$.

2.3.2 Above T_{pair}: Vlasov–Boltzmann approach versus hydrodynamics

In this temperature regime, one can assume that the eigenstates of the interacting gas can be connected to those of the ideal gas by switching on the interactions adiabatically. The Fermi surface is deformed and the real ground state follows adiabatically from the excited state of the non-interacting system. This is the assumption that defines the normal Fermi gas.

Each spin component j (with $j = 1$ or 2) experiences a trapping potential $V_{\text{ext}}^{(j)}(\mathbf{r}) = \frac{1}{2}m_j\omega_j^2 r^2$ and a mean-field interaction potential $gn^{(\bar{j})}(\mathbf{r}, t)$, with $\bar{j}$ denoting the species different from j, $g = 2\pi\hbar^2 a/m_r$, a the s-wave scattering length between two atoms of different species, and m_r the reduced mass. Thus, the total effective potential for each species reads $U^{(j)}(\mathbf{r}, t) = V_{\text{ext}}^{(j)}(\mathbf{r}) + gn^{(\bar{j})}(\mathbf{r}, t)$.

The equilibrium density profiles, at given temperature $T = 1/k_B\beta$, can be written in a semi-classical approach as

$$n^{(j)}(\mathbf{r}) = \int \frac{d^3p}{(2\pi\hbar)^3} f_0^{(j)}(\mathbf{r}, \mathbf{p}) = \int \frac{d^3p}{(2\pi\hbar)^3} \left(e^{\beta(p^2/2m_j + U^{(j)}(\mathbf{r}) - \mu_j)} + 1 \right)^{-1}, \quad (2.86)$$

where μ_j is the chemical potential for the component j ensuring the normalization condition of the local Fermi–Dirac distributions, namely $\int f_0^{(j)}(\mathbf{r}, \mathbf{p}) d^3r\, d^3p/h^3 = N_j$.

Dynamics can be studied by means of quantum kinetic transport equations for the distribution functions $f^{(j)}(\mathbf{r}, \mathbf{p}, t)$,

$$\partial_t f^{(j)} + \frac{\mathbf{p}}{m} \cdot \nabla_{\mathbf{r}} f^{(j)} - \nabla_{\mathbf{r}} U^{(j)} \cdot \nabla_{\mathbf{p}} f^{(j)} = C_{12}[f^{(j)}], \quad (2.87)$$

where we have set $\hbar = 1$, and $n^{(j)}(\mathbf{r}, t)$ is the density obtained by integrating $f^{(j)}(\mathbf{r}, \mathbf{p}, t)$ over momentum.

The term $\nabla_{\mathbf{r}} U^{(j)} \cdot \nabla_{\mathbf{p}} f^{(j)}$ in Eq. (2.87) may be interpreted as a flow of particles dragged by the external force $-\nabla_{\mathbf{r}} V_{\text{ext}}^{(j)}$ and by the inhomogeneity of the other compenent.

The collision term C_{12} in Eq. (2.87) involves only collisions between particles that are polarized in two different Zeeman states. We have:

$$C_{12}[f^{(j)}] \equiv g^2 \frac{2(2\pi)^4}{V^3} \sum_{\mathbf{p}_2,\mathbf{p}_3,\mathbf{p}_4} \Delta_{\mathbf{p}}\Delta_{\varepsilon} [\bar{f}^{(j)} \bar{f}_2^{(\bar{j})} f_3^{(j)} f_4^{(\bar{j})} - f^{(j)} f_2^{(\bar{j})} \bar{f}_3^{(j)} \bar{f}_4^{(\bar{j})}], \quad (2.88)$$

with $f^{(j)} \equiv f^{(j)}(\mathbf{r}, \mathbf{p}, t)$, $\bar{f}^{(j)} \equiv 1 - f^{(j)}$, $f_i^{(j)} \equiv f^{(j)}(\mathbf{r}, \mathbf{p}_i, t)$, $\bar{f}_i^{(j)} \equiv 1 - f_i^{(j)}$. V is the volume occupied by the gas and the factors $\Delta_{\mathbf{p}}$ and Δ_{ε} are the usual delta functions accounting for conservation of momentum and energy, with the energies given by $p_i^2/2m_j + U^{(j)}(\mathbf{r}, t)$.

At very low temperature, near equilibrium, collisions are inhibited by the exclusion principle: the filling of the Fermi sea makes collisions unlikely and the collisional frequency ω_c becomes smaller and smaller. If the collisional time $\tau_c \propto 1/\omega_c$ becomes longer than the dynamics timescale, collisions are negligible and the system behaves as a non-interacting gas. Thus, a signature that a two-spin component Fermi gas has reached the degenerate regime could be the observation of the transition from a collisional to a collisionless dymanics at lowering the temperature (DeMarco and Jin, 2002).

2.3.2.1 *Dipolar oscillation: the collisionless–collisional transition*

The first cold-atom experiments in which fermions were cooled down below the Fermi temperature were realized at Jila (Gensemer and Jin, 2001; DeMarco and Jin, 2002; Loftus *et al.*, 2002; Regal and Jin, 2003).

The system was excited with a sudden displacing of the center of the trap via a bias magnetic field (Gensemer and Jin, 2001) and the gas started oscillating in the trap (dipolar modes). In such experiment, the time-evolution of the center of mass of the two species depends on the ratio between the collisional frequency ω_c and the trap frequencies ω_j (see Figure 2.8). For $\omega_c/\omega_j \ll 1$ the two clouds oscillate independently without appreciable damping, while as ω_c/ω_j approaches unity the two centers of mass start moving together but incoherently and therefore their oscillations are damped. At higher collisionality ($\omega_c/\omega_j \gg 1$) the two species are locked and oscillate at the same frequency without damping.

The transition was observed in the experiment by varying the density. Otherwise one could drive ω_c by exploiting Feshbach resonances or by lowering the temperature in order to fill the Fermi sphere.

2.3.2.2 *The zero-first-sound transition*

The fact that collective modes are "well" defined in the collisional/hydrodynamic limit and in the collisionless limit too, for a Fermi gas, has deep roots in the existence of zero sound. In a normal gas, sound propagation is possible only in the presence of collisions, but for a collisionless Fermi gas, a pressure/density perturbation propagates at a

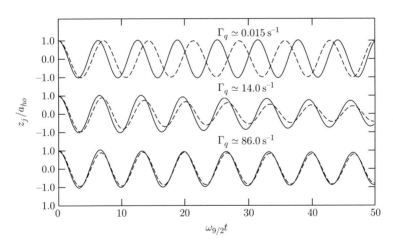

Fig. 2.8 Reprinted figure with permission from F. Toschi, P. Capuzzi, S. Succi, P. Vignolo and M. P. Tosi, J. Phys. B: At. Mol. Opt. Phys. 37, S91 (2004). Copyright (2004) by the Institue of Physics Publishing. Center-of-mass position $z_j(t)$ (in units of the oscillator length $a_{ho} = \sqrt{\hbar/m\omega_{9/2}}$) as functions of $\omega_{9/2}t$ for a mixture of ^{40}K atoms at $T = 0.3T_F$. From top to bottom, the indicated values of the collisional rate $\Gamma_q = \omega_c/2\pi$ correspond to the collisionless, the intermediate and the hydrodynamic regime.

well-defined speed, the zero-sound velocity c_0. Obviously, in the absence of interactions (mean-field), c_0 coincides with v_F. In the presence of interactions, for the homogeneous gas, one solves the transport equation (2.87) with $C_{1,2} = 0$ and for a distribution function

$$f^{(j)}(\mathbf{r}, \mathbf{p}) = f_0^{(j)}(\mathbf{p}) + f_1^{(j)}(\mathbf{p})e^{i(\mathbf{q}\cdot\mathbf{r} - \Omega t)}, \tag{2.89}$$

where $f_0^{(j)} = \theta[\mu^{(j)} - \frac{p^2}{2m} - gn_0^{(\bar{j})}]$. Equation (2.87) then becomes

$$\left(\frac{\mathbf{q}\cdot\mathbf{p}}{m} - \Omega\right) f_1^{(j)}(\mathbf{p}) - \frac{\mathbf{q}\cdot\mathbf{p}}{m}\frac{\partial f_0^{(j)}(\mathbf{p})}{\partial \varepsilon} \int \frac{d\mathbf{p}'}{(2\pi\hbar)^3} g f_1^{(\bar{j})}(\mathbf{p}') = 0, \tag{2.90}$$

where $\varepsilon = p^2/(2m)$. The form of Eq. (2.90) suggests to write $f_1^{(j)}(\mathbf{p})$ as proportional to $\nu \partial f_0^{(j)}/\partial \varepsilon$, with ν being defined as

$$\nu = \frac{\cos\theta}{\cos\theta - \eta}, \tag{2.91}$$

where $\eta = \omega/(qv_F^{(j)})$, $v_F^{(j)}$ being the Fermi velocity for the component j, and $\cos\theta = \mathbf{p}\cdot\mathbf{q}/pq$. For the case of equal spin-populations the Fermi velocities coincide, $v_F^{(j)} = v_F^{(\bar{j})} = v_F$, and one obtains (Nozières and Pines, 1966; Akdeniz *et al.*, 2003)

$$\frac{\eta}{2}\log\frac{\eta+1}{\eta-1} - 1 = \frac{\pi\hbar}{2mv_F a}. \tag{2.92}$$

In a cigar-shaped geometry, as would be suitable for a sound-propagation experiment (Andrews *et al.*, 1997; Joseph *et al.*, 2007), one finds (Capuzzi *et al.*, 2006*a*):

$$\eta = \sqrt{1 + \frac{2\hbar a}{\pi\tilde{a}_\perp^2 m v_F^{(j)}}}, \tag{2.93}$$

with $a_\perp = (\hbar/m\tilde{\omega}_\perp)$ being the radial harmonic oscillator length related to the effective radial harmonic oscillator frequency $\tilde{\omega}_\perp = \omega_\perp[1 - 3gn_0^{(\bar{s})}(0)/(2\mu)]^{1/2}$. We remark that Eq. (2.92) does not have real solutions for a negative scattering length (the mixture collapses), while the cylindrical geometry stabilizes the mixture in the presence of (small) attractive interactions.

When the gas is collisional, the Fermi gas behaves as a liquid and one can use hydrodynamics to deduce the usual (first) sound velocity c_1. In the homogeneous 3D gas (in the absence of mean-field interactions),

$$c_1 = \sqrt{\frac{n_0}{m}\frac{\partial\mu(n)}{\partial n}\bigg|_{n=n_0}} = \sqrt{\frac{2}{3}\frac{v_F^2}{2}} = \frac{v_F}{\sqrt{3}}. \tag{2.94}$$

In a cigar-shaped geometry, the effect of the radial confinement can be taken into account by solving the hydrodynamics equations (2.60) and (2.61) neglecting the axial confinement, and looking for pertubations propagating in the axial confinement, i.e. setting $n(r_\perp, z) = n_0(r_\perp) + \delta n_q(r_\perp) e^{i\Omega t - qz}$. In the linear regime one obtains

$$m\Omega^2 \delta n_q = q^2 \left(n_0 \left. \frac{\partial \mu}{\partial n} \right|_{n=n_0} \delta n_q \right) - \nabla_\perp \cdot \left[n_0 \nabla_\perp \left(\left. \frac{\partial \mu}{\partial n} \right|_{n=n_0} \delta n_q \right) \right]. \qquad (2.95)$$

The calculation of the sound velocity requires only the expression of δn_q calculated at $q = 0$. By using $\delta n_{q=0} = (\partial \mu / \partial n|_{n=n_0})^{-1}$ we obtain

$$c_1 = \left[\frac{1}{m} \int n_0 \, d^2 r_\perp \bigg/ \int \left(\left. \frac{\partial \mu}{\partial n} \right|_{n=n_0} \right)^{-1} d^2 r_\perp \right]^{1/2}. \qquad (2.96)$$

This expression provides the exact velocity of sound propagation in cylindrically confined hydrodynamic gases with equation of states $\mu(n)$. For $\mu(n) = \frac{\hbar^2}{2m}(6\pi n)^{2/3}$, Eq. (2.96) leads to

$$c_1 = \frac{v_F}{\sqrt{5}}. \qquad (2.97)$$

2.3.3 Below T_c: the BCS–BEC crossover

Below T_{pair} a two-component Fermi gas minimizes the energy by forming pairs between atoms polarized in different spin states. At T_c the system undergoes a phase transition and the pairs condense. For weak interactions ($V_0 \to 0$ and thus $a \to 0^-$, as can be deduced from Eq. (2.76)), fermions with different polarizations form Cooper pairs (the BCS regime). For strong attractive interactions ($a \to 0^+$), fermions form bosonic molecules and the molecules form a Bose–Einstein condensate (the BEC regime). In this section we discuss the different mechanisms of pairing in the crossover from weak to strong interactions (see, for instance, (Giorgini et al., 2008)), as well as some physical observables that can identify the different regimes.

2.3.3.1 The BCS limit: Cooper pairs

For negative scattering lengths, one should not expect bound states from the two-body scattering theory. However, Cooper showed that two interacting fermions above a filled Fermi sea ($T \ll T_F$) have a bound state for an arbitrarily weak attractive interaction, and that the quasi-particle picture is unsuited to describe the ground state.

The guess is that the ground state can be constructed by pairs of fermions (see, for instance, (Ketterson and Song, 1999)), as

$$\Phi_N = \hat{A}\{\phi(1,2)\phi(3,4)\ldots\phi(N-1,N)\}, \qquad (2.98)$$

with

$$\phi(1,2) = \psi(\mathbf{r}_1, \mathbf{r}_2)\chi(\sigma_1, \sigma_2), \qquad (2.99)$$

$\hat{A}$ being the antisysymmetrizer, and χ the singlet state for spins σ_1 and σ_2. Performing the Fourier expansion $\psi(\mathbf{r}) = \sum_\mathbf{k} g_\mathbf{k} e^{i\mathbf{k}\cdot\mathbf{r}}$, one obtains

$$\Phi_N = \sum_{\mathbf{k}_1} \cdots \sum_{\mathbf{k}_{N/2}} g_{\mathbf{k}_1} \cdots g_{\mathbf{k}_{N/2}} \hat{A}\left(e^{i\mathbf{k}_1\cdot(\mathbf{r}_1-\mathbf{r}_2)} \ldots e^{i\mathbf{k}_{N/2}\cdot(\mathbf{r}_{N-1}-\mathbf{r}_N)}\right),$$

$$\times [(1\uparrow)(2\downarrow)\ldots(N-1\uparrow)(N\downarrow)]. \tag{2.100}$$

In second quantization, Eq. (2.100) reads

$$\Phi_N = \sum_{\mathbf{k}_1} \cdots \sum_{\mathbf{k}_{N/2}} g_{\mathbf{k}_1} \cdots g_{\mathbf{k}_{N/2}} \hat{c}^\dagger_{\mathbf{k}_1\uparrow} \hat{c}^\dagger_{-\mathbf{k}_1\downarrow} \cdots \hat{c}^\dagger_{\mathbf{k}_{N/2}\uparrow} \hat{c}^\dagger_{-\mathbf{k}_{N/2}\downarrow} |0\rangle, \tag{2.101}$$

where $|0\rangle$ is the vacuum state and $\hat{c}^\dagger_{\mathbf{q}\sigma}$ the creation operator for a fermion with momentum $\mathbf{q}$ and spin σ. Bardeen, Cooper and Schrieffer proposed the alternative wave function (Bardeen *et al.*, 1957)

$$\Phi_{BCS} = \prod_\mathbf{k}(u_\mathbf{k} + v_\mathbf{k}\hat{c}^\dagger_{\mathbf{k}\uparrow}\hat{c}^\dagger_{-\mathbf{k}\downarrow})|0\rangle, \tag{2.102}$$

where the product extends over all plane-wave states and the complex variation parameters $u_\mathbf{k}$ and $v_\mathbf{k}$ satisfy the condition $|u_\mathbf{k}|^2 + |v_\mathbf{k}|^2 = 1$. Equation (2.102) is a coherent state and represents a condensate of pairs.

The average number of particles $\bar{N}$, associated to the wave function (2.102) is given by

$$\bar{N} = \langle\Phi_{BCS}|\hat{N}|\Phi_{BCS}\rangle = \sum_\mathbf{k} 2|v_\mathbf{k}|^2, \tag{2.103}$$

with $\hat{N} = \sum_{\mathbf{k},\sigma} \hat{c}^\dagger_{\mathbf{k},\sigma}\hat{c}_{\mathbf{k},\sigma}$ being the number operator. The $u_\mathbf{k}$ parameter is instead related to the mean square fluctuation of the number of particles

$$\delta\bar{N}^2 \equiv \bar{N^2} - \bar{N}^2 = \sum_\mathbf{k} 4|v_\mathbf{k}|^2|u_\mathbf{k}|^2. \tag{2.104}$$

Thus, neglecting $u_\mathbf{k}$ means neglecting number fluctuations.

The BCS ground state and its chemical potential is obtained by using Eq. (2.102) as a trial wave function, and minimizing the expectation value of the fermionic Hamiltonian $\hat{H}' = \hat{H} - \mu\hat{N}$ where the constraint condition on the average number of particles has been included. The Hamiltonian $\hat{H}'$ reads

$$\hat{H}' = \sum_{\mathbf{k},\sigma}(\varepsilon_\mathbf{k} - \mu)\hat{c}^\dagger_{\mathbf{k}\sigma}\hat{c}_{\mathbf{k}\sigma} + \sum_{\mathbf{k},\mathbf{k}'} V_{\mathbf{k},\mathbf{k}'}\hat{c}^\dagger_{\mathbf{k}'\uparrow}\hat{c}^\dagger_{-\mathbf{k}'\downarrow}\hat{c}_{-\mathbf{k}\downarrow}\hat{c}_{\mathbf{k}\uparrow}, \tag{2.105}$$

where

$$V_{\mathbf{k},\mathbf{k}'} = \frac{1}{L^3}\int V_{1,2}(r)e^{i(\mathbf{k}-\mathbf{k}')\cdot\mathbf{r}}d^3r = g, \tag{2.106}$$

if the contact potential is of the form $g\delta(\mathbf{r})$. The ground-state energy can thus be written as

$$E = \langle \Phi_{BCS} | \hat{H}' | \Phi_{BCS} \rangle = 2 \sum_{\mathbf{k}} |v_{\mathbf{k}}|^2 (\varepsilon_{\mathbf{k}} - \mu) + g \sum_{\mathbf{k},\mathbf{k}'} v_{\mathbf{k}}^* u_{\mathbf{k}} v_{\mathbf{k}'} u_{\mathbf{k}'}^*. \tag{2.107}$$

Mimizing with respect to $v_{\mathbf{k}}$, one finds

$$u_{\mathbf{k}}^2 = \frac{1}{2} \left(1 + \frac{\varepsilon_{\mathbf{k}} - \mu}{E_{\mathbf{k}}} \right), \tag{2.108}$$

and

$$v_{\mathbf{k}}^2 = \frac{1}{2} \left(1 - \frac{(\varepsilon_{\mathbf{k}} - \mu)}{E_{\mathbf{k}}} \right), \tag{2.109}$$

with $E_{\mathbf{k}} = \sqrt{((\varepsilon_{\mathbf{k}} - \mu)^2 + \Delta^2}$ being the quasi-particle energy spectrum. The gap energy Δ satisfies the relation $1 = -g \sum_{\mathbf{k}} 1/(2E_{\mathbf{k}})$. By using the renormalized value for the coupling (Randeira, 1995), one obtains the so-called *gap equation*

$$\frac{m}{4\pi\hbar^2 a} = \int \frac{d\mathbf{k}}{(2\pi\hbar)^3} \left(\frac{1}{2\varepsilon_{\mathbf{k}}} - \frac{1}{2E_{\mathbf{k}}} \right), \tag{2.110}$$

where ultraviolet divergence have been regulated. Equation (2.110) has to be solved together with the number equation

$$\bar{n} = \frac{N}{V} = \int \frac{d\mathbf{k}}{(2\pi\hbar)^3} n(\mathbf{k}) = \int \frac{d\mathbf{k}}{(2\pi\hbar)^3} \left(1 - \frac{(\varepsilon_{\mathbf{k}} - \mu)}{E_{\mathbf{k}}} \right), \tag{2.111}$$

obtained from Eqs. (2.103) and (2.109).

In the weak-coupling limit the solution of Eqs. (2.110) and (2.111) leads to

$$\mu(n) \simeq \varepsilon_F(n), \tag{2.112}$$

$$\Delta \simeq \frac{4k_F^2}{m} e^{-2\pi/|mk_F g|}, \tag{2.113}$$

and to a momentum distribution

$$n(\mathbf{k}) \simeq \theta(\mu - \varepsilon_{\mathbf{k}}). \tag{2.114}$$

The effect of the interactions on the momentum distribution is a broadening of width Δ around $k = k_F$. In this regime the critical temperature is determined by the energy needed to break a pair, thus $T_c \propto \Delta/k_B$.

2.3.3.2   *The BEC limit*

For positive and small values of the scattering length ($k_F a \ll 1$), the system can be described by the theory of Bose–Einstein condensation (the so-called BEC limit). The atoms are bonded in dimers that can be described by the wave function

$$\psi_b = e^{-r/a}/\sqrt{2\pi a} r, \tag{2.115}$$

whose binding energy is

$$\varepsilon_b = -\frac{\hbar^2}{2m_r a^2}, \tag{2.116}$$

as confirmed by the experiment of (Regal *et al.*, 2003) using radio-frequency spectroscopy. The uniform Fermi gas undergoes transition into a superfluid when the dimers condense at the critical temperature

$$T_c = T_{\mathrm{BEC}} = 0.218 T_F, \tag{2.117}$$

with $T_{\mathrm{BEC}} = (2\pi\hbar^2)/(k_B m)(n_d/\zeta(3/2))^{3/2}$ being the condensation temperature of bosonic molecules of mass $2m$ and density n_d (equal to the density of each spin component), and $\zeta(3/2) = 2.612$ (see in this book the chapter on Degenerate Bose Gases).

At $T = 0$ the atomic chemical potential $\mu(n) = [\epsilon_b + \mu_d(n_d)]/2$ is half the sum of the binding energy and the chemical potential for dimers

$$\mu_d(n_d) = g_{dd} n_d = \frac{2\pi\hbar^2 a_{dd} n_d}{m}, \tag{2.118}$$

with g_{dd} being the dimer–dimer coupling and a_{dd} the dimer–dimer scattering length, which is a function of the fermion–fermion scattering length a.[2]

The gap energy is given by half the energy to break a pair, namely

$$\Delta = \frac{|\varepsilon_b|}{2}. \tag{2.119}$$

The momentum distribution in this limit is proportional to the square of the Fourier transform of the molecular wave function (2.115) and reads

$$n(p) = \frac{4}{3(2\pi\hbar L)^3} \frac{(k_F a)^3}{(1 + p^2 a^2/\hbar^2)^2}. \tag{2.120}$$

2.3.3.3   *The unitary limit*

In this regime the scattering length is no longer a relevant scale length of the many-body problem. The scattering dephasing δ and the scattering cross-section σ take the universal values $\pm\frac{\pi}{2}$ and $4\pi/k^2$, apart from the feature of the interatomic potential. Thus, one formulates the universal hypothesis that the only relevant scale length is the mean distance between particles, namely $n^{-1/3} \propto 1/k_F$. Accordingly, one expects all thermodynamic quantities to be functions of ε_F. For instance, at $T = 0$ the interaction energy has to be proportional to the Fermi energy, $E_{int} = \beta\varepsilon_F$, β being a universal constant. Following this hypothesis, the chemical potential must read

$$\mu(n) = (1 + \beta)\varepsilon_F(n). \tag{2.121}$$

The universal hypothesis has been confirmed for a two-component Fermi gas at unitary by observing that this strongly interacting many-body system obeys the virial theorem for an ideal gas over a wide range of temperatures (Thomas *et al.*, 2005). The value of β in Eq. (2.121) has been fixed to -0.58 ± 0.01 using quantum Monte-Carlo (QMC) techniques (Carlson *et al.*, 2003; Astrakharchik *et al.*, 2004; Carlson and Reddy, 2005). Generally, in the regime $k_F|a| \gg 1$, an exact solution does not exist (yet), and approximated theory or QMC simulations are employed.

The first evidence for superfluidity at unitarity was shown by J. Kinast and collaborators (Kinast *et al.*, 2004) studying dipolar oscillations as functions of the temperature. In this experiment they observed that both components oscillate at the same (hydrodynamic) frequency and that damping becomes negligible by lowering the temperature. The same group measured the heat capacity as a function of the temperature, deducing the critical temperature in this regime (Kinast *et al.*, 2005).

2.3.3.4 *Mean-field model for the crossover*

Regardless of the value of the scattering length, at $T = 0$, the ground state for a two-component attractive Fermi gas is a superfluid and nothing dramatic happens by varying the interaction strength. Thus, the passage from a BCS ($V_0/\varepsilon_F \to 0$, $a \to 0^-$) to a BEC ($V_0/\varepsilon_F \gg 1$, $a \to 0^+$) for s-wave interactions is a crossover and not a phase transition. The simplest model for the crossover is the BCS mean-field theory (Eagles (1969); Leggett (1980)) where one supposes that Eqs. (2.110) and (2.111) are valid for any coupling. This is a good approach in the BCS side of the crossover, but it is not accurate in the BEC side where molecule–molecule interactions give rise to correlations involving four fermions (rather than just two). Figure 2.9 shows the prediction of the BCS mean-field theory for the chemical potential in comparison with the results of

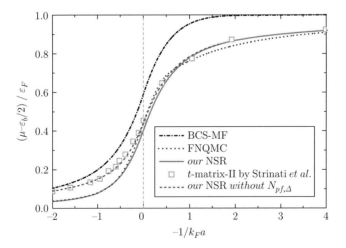

Fig. 2.9 Reprinted figure with permission from H. Hu, X.-J Liu and P.D. Drummond, Europhys. Lett. **74**, 574 (2006). Copyright (2006) by the EDP Sciences. Chemical potential vs. the dimensionless parameter $-1/k_F a$. Comparison between different theories.

other methods that take into account higher-order correlations: a QMC calculation (Astrakharchik *et al.*, 2004; Giorgini *et al.*, 2008), a finite-temperature Green function approach (Perali *et al.*, 2004), a Nozière and Schmitt–Rink-like approach to take into account pair fluctuations in the superfluid phase (Hu *et al.*, 2006).

The different theories for the BCS–BES crossover can be tested in the experiments by measuring

- the radius of a superfluid Fermi gas trapped in a harmonic confinement, since $R_{TF} = \sqrt{2m\bar{\mu}/\omega^2}$;
- the response of the cloud after being perturbed, namely measuring the spectrum of the collective excitations.

2.3.3.5   *Collective excitations*

The hydrodynamic approach. The dynamics of a superfluid gas can be described by the equations of hydrodynamics (2.60) and (2.61) provided that $\mu(n)$ is the correct EOS for the superfluid. For the case of a homogeneous gas, by setting $\delta n(\mathbf{r}, t) = \delta n_q e^{i\Omega t - \mathbf{q} \cdot \mathbf{r}}$, one leads to the eigenvalue equation for small-amplitude density modes δn_q

$$m\Omega^2 \delta n_q = q^2 \left(n_0 \left. \frac{\partial \mu}{\partial n} \right|_{n=n_0} \delta n_q \right), \tag{2.122}$$

which provides the sound velocity

$$c_1 = \sqrt{\frac{n_0}{m} \left. \frac{\partial \mu(n)}{\partial n} \right|_{n=n_0}}. \tag{2.123}$$

For a power-law EOS $\mu(n) = Cn^\gamma$, Eq. (2.123) reduces to

$$c_1 = \sqrt{\frac{\gamma \bar{\mu}}{m}}, \tag{2.124}$$

namely $c_1 = v_F/\sqrt{3}$ in the BCS limit, $c_1 = v_F\sqrt{(1+\beta)/3}$ in the unitary limit and $c_1 = v_B$ in the BEC limit, with v_B being the Bogolubov sound velocity.

In a trapped system there is a discretized spectrum of collectives modes, and their frequencies depend on both the EOS and the geometry of the confinement. We recall that for a cigar-shaped trap, the eigenvalue equation for small-amplitude density modes $\delta n_q(\mathbf{r}_\perp)e^{iqz}$ reads

$$m\Omega^2 \delta n_q = q^2 \left(n_0 \left. \frac{\partial \mu}{\partial n} \right|_{n=n_0} \delta n_q \right) - \nabla_\perp \cdot \left[n_0 \nabla_\perp \left(\left. \frac{\partial \mu}{\partial n} \right|_{n=n_0} \delta n_q \right) \right]. \tag{2.125}$$

The sound velocity can be obtained as outlined in Section 2.3.2.2. For a power-law EOS one obtains

$$c_1 = \sqrt{\frac{\gamma}{2\gamma + 2}} \sqrt{\frac{2\bar{\mu}}{m}}, \tag{2.126}$$

namely $c_1 = v_F/\sqrt{5}$ in the BCS limit, $c_1 = v_F\sqrt{(1+\beta)/5}$ in the unitary limit and $c_1 = v_B\sqrt{2}$ in the BEC limit. In the whole crossover one can use the relation (2.96), namely

$$c_1 = \left[\frac{1}{m} \int n_0 \, d^2r_\perp \Big/ \int \left(\frac{\partial\mu}{\partial n}\Big|_{n=n_0}\right)^{-1} d^2r_\perp\right]^{1/2}. \tag{2.127}$$

The sound propagation in a cigar-shaped superfluid Fermi gas through the BCS–BEC crossover was mesured at Duke University by the group of Thomas (Joseph *et al.*, 2007). By using the EOS evaluated by (Hu *et al.*, 2006) in Eq. (2.127), one obtains a very good agreement with the experimental data, as shown in Figure 2.10.

   This proves the validity of the theory developed by (Hu *et al.*, 2006) for the crossover and that hydrodynamics equations describe correctly the collective excitations of a superfluid.

The scaling Ansatz. Higher-energy modes at $q = 0$, i.e. breathing modes, can excited by varying the trap frequencies at $t > 0$. They can be evaluated by adopting the following scaling form of the time-dependent density profile,

$$n(\mathbf{r}, t) = \left[\prod_\alpha b_\alpha(t)\right]^{-1} n_0(x/b_x(t), y/b_y(t), z/b_z(t)), \tag{2.128}$$

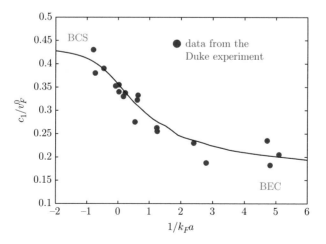

Fig. 2.10 First-sound velocity as a function of the dimensionless parameter $1/k_F a$. Data courtesy of the group of Thomas (Joseph *et al.*, 2007).

where the dependence on time is entirely contained in the scaling parameters $b_\alpha(t)$ (Kagan *et al.*, 1996). The continuity equation fixes the velocity field as $v_\alpha(\mathbf{r}, t) = \dot{b}_\alpha(t) r_\alpha / b_\alpha(t)$. Equation (2.128) is an exact solution of the equation of motion if the EOS is a power-law, $\mu(n) = Cn^\gamma$. In this case, the scaling parameters obey the coupled differential equations

$$\ddot{b}_\alpha + \omega_\alpha(t)^2 b_\alpha - \frac{\omega_\alpha^2(0)}{b_\alpha \left(\prod_\beta b_\beta\right)^\gamma} = 0. \tag{2.129}$$

By taking $\omega_\alpha(t) = \omega_\alpha(t=0)(1 + \varepsilon \sin(\Omega)t)$ with $\varepsilon \ll 1$, one finds three frequencies. The first corresponds to a surface mode with projected angular momentum $m = \pm 2$ and lies at $\Omega_0 = \sqrt{2}\omega_\perp$ independently of the coupling. The other two modes are given by

$$\Omega_\pm^2 / \omega_\perp^2 = 1 + \gamma + (2 + \gamma)\lambda^2/2$$
$$\pm \sqrt{[1 + \gamma + (2 + \gamma)\lambda^2/2]^2 - 2(2 + 3\gamma)\lambda^2}, \tag{2.130}$$

where $\lambda = \omega_z/\omega_\perp$ and $\pm$ signs refer to the radial and axial mode, respectively. In a highly elongated trap ($\lambda \ll 1$), the breathing modes reduce to

$$\Omega_+ = \sqrt{2 + 2\gamma}\,\omega_\perp, \tag{2.131}$$

$$\Omega_- = \sqrt{\frac{2 + 3\gamma}{1 + \gamma}}\,\omega_z. \tag{2.132}$$

In the crossover, when the EOS is not a power law, one can introduces the effective exponent (Hu *et al.*, 2004)

$$\gamma(n) = 1 + n[(d^2\mu/dn^2)/(d\mu/dn)], \tag{2.133}$$

and, for an inhomogeneous gas, the averaged effective exponent

$$\bar{\gamma} = \left(\int d^3\mathbf{r}\, n_0 r_\alpha^2 \gamma(n_0)\right) \Big/ \left(\int d^3\mathbf{r}\, n_0 r_\alpha^2\right). \tag{2.134}$$

Thus, Eq. (2.130) holds in the whole crossover by replacing γ with $\bar{\gamma}$. Note that the choice of the weight function $r_\alpha^2 n_0$ leads to the same breathing-mode frequencies as predicted by a sum-rule approach for various systems (Liu and Hu (2003)).

Figure 2.11 shows the comparison between the breathing modes evaluated via Eqs. (2.130) and (2.134) and the experimental data from two experiments (Kinast *et al.*, 2004; Bartenstein *et al.*, 2004).

To obtain the dispersion relation $\Omega(q)$ for any value of q one must resort to the numerical solution of the eigenvalue equation (2.125). A possible method to solve such an equation consists of expanding the eigenmodes δn_q in a complete set of basis

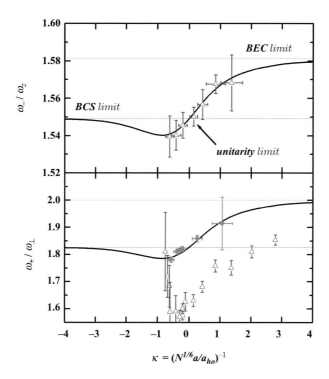

Fig. 2.11 Reprinted figure with permission from H. Hu, A. Minguzzi, X.-J Liu, M. P. Tosi, Phys. Rev. Lett. 93, 190403 (2004). Copyright (2004) by the American Physical Society. The frequency of the axial and transverse breathing modes in an elongated trapped Fermi gas as functions of $\kappa = (N^{1/6}a/a_{ho})^{-1}$. The solid circles and empty functions of triangles with error bars are the experimental results given by (Kinast *et al.*, 2004) and by (Bartenstein *et al.*, 2004), respectively.

functions (see (Zaremba, 1998)) as

$$\delta n(\mathbf{r}_\perp) = \sum_\alpha b_\alpha \, \delta n_\alpha(\mathbf{r}_\perp), \qquad (2.135)$$

where $\alpha = (n_r, m)$ labels the basis functions, n_r is the radial number and m the number for the azimuthal angular momentum. By inserting this expression into Eq. (2.125) and projecting the result onto an element of the basis, a matrix representation of the eigenvalue equation is found, which is suitable for a numerical solution. The results for the two lowest-frequency modes as functions of q are shown in Figures 2.12 and 2.13. The lowest mode, shown in Figure 2.12, is sound-like and has the phononic dispersion relation at long wavelengths given by $\Omega = c_1 q$.

The second mode is a monopolar compressional mode that for $q = 0$ is purely radial and at coincides with the frequency of the breathing mode Ω_+.

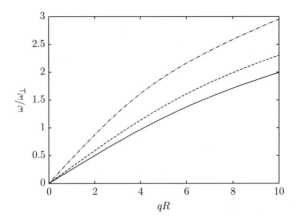

Fig. 2.12 From (Capuzzi *et al.*, 2006*b*). Dispersion relation $\Omega(q)$ (in units of $\omega_\perp$) for the lowest-frequency (sound) mode as a function of qR, with R the radius of the fermion density profile in the BCS limit. The dash-dotted, dashed, and solid lines correspond to fermions in the BCS, unitary, and BEC limits, respectively.

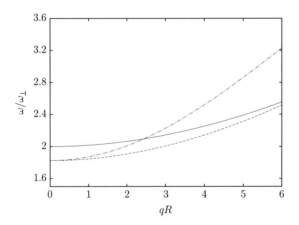

Fig. 2.13 From (Capuzzi *et al.*, 2006*b*). The same as in Figure 2.12 for the monopole mode.

Appendix A: The renormalization procedure

This method allows us to evaluate the exact Green's function relative to an operator $\hat{x}$ in a chosen subspace $\mathcal{A}$ of the Hilbert space. Let us write the operator $\hat{x}$ as

$$\hat{x} = \hat{x}_0 + \delta\hat{x}. \tag{A.1}$$

The Green's function $\hat{G}(x) = (x - \hat{x})^{-1}$ can then be written using the Dyson identity

$$\hat{G} = \hat{G}_0 + \hat{G}_0 \delta\hat{x} \hat{G}, \tag{A.2}$$

where $\hat{G}_0 = (x - \hat{x}_0)^{-1}$. Let P_A and P_B be the projectors relative to the subspaces $\mathcal{A}$ and $\mathcal{B}$, respectively, $\mathcal{B}$ being the complementary subspace of $\mathcal{A}$. We define

$$\hat{x}_0 = \hat{x}^{AA} + \hat{x}^{BB}, \qquad \delta\hat{x} = \hat{x}^{AB} + \hat{x}^{BA}, \tag{A.3}$$

where we set $\hat{x}^{ij} = P_i \hat{x} P_j$. Equation (A.2) takes the form of

$$\hat{G} = \hat{G}_0 + \hat{G}_0 \hat{x}^{AB} \hat{G} + \hat{G}_0 \hat{x}^{BA} \hat{G}, \tag{A.4}$$

where $\hat{G}_0 = \hat{G}_0^{AA} + \hat{G}_0^{BB}$. It follows that

$$\hat{G}^{AA} = \hat{G}_0^{AA} + \hat{G}_0^{AA} \hat{x}^{AB} \hat{G}_0^{BB} \hat{x}^{BA} \hat{G}^{AA}. \tag{A.5}$$

Multiplying by $[\hat{G}^{AA}]^{-1}$ (right side)

$$1 = \hat{G}_0^{AA} [\hat{G}^{AA}]^{-1} + \hat{G}_0^{AA} \hat{x}^{AB} \hat{G}_0^{BB} \hat{x}^{BA}, \tag{A.6}$$

or equivalently

$$1 - \hat{G}_0^{AA} \hat{x}^{AB} \hat{G}_0^{BB} \hat{x}^{BA} = \hat{G}_0^{AA} [\hat{G}^{AA}]^{-1}, \tag{A.7}$$

and multiplying by $[\hat{G}_0^{AA}]^{-1}$ (left side)

$$x - \hat{x}^{AA} - \hat{x}^{AB} \hat{G}_0^{BB} \hat{x}^{BA} = [\hat{G}^{AA}]^{-1}, \tag{A.8}$$

one gets the result

$$\hat{G}^{AA} = \frac{1}{x - \hat{\xi}(x)}, \tag{A.9}$$

where $\hat{\xi}(x)$ is an effective operator acting on the subspace $\mathcal{A}$ given by

$$\hat{\xi}(x) = \hat{x}^{AA} + \hat{x}^{AB} \hat{G}_0^{BB} \hat{x}^{BA}. \tag{A.10}$$

Notes

1. This condition is not restrictive since in 1D we can use the non-degeneration theorem (Landau and Lifshitz, 1959) to show that all eigenfunctions vanishing at infinity are real.
2. Petrov and collaborators have calculated that $a_{dd} = 0.6a$ if $k_F a \gg 1$ (Petrov *et al.*, 2004).

Acknowledgments

I warmly thank the colleagues involved in the work presented here. I am indebted to Bess Fang for having read the manuscript and to Mario Gattobigio for many useful discussions and encouragement. I am very grateful to Christian Miniatura, Benoît

Grémaud, Martial Ducloy, Leong Chuan Kwek and Berthold-Georg Englert for having organized such a stimulating summer school.

References

Akdeniz, Z., Vignolo, P., Minguzzi, A., and Tosi, M. P. (2002). *Phys. Rev. A*, **66**, 055601.

Akdeniz, Z., Vignolo, P., and Tosi, M. P. (2003). *Phys. Lett. A*, **311**, 246.

Amoruso, M., Meccoli, I., Minguzzi, A., and Tosi, M. P. (1999). *Eur. Phys. J. D*, **7**, 441.

Anderson, M. H., Ensher, J. R., Matthews, M. R., Wieman, C. E., and Cornell, E. A. (1995, July). *Science*, **269**, 198.

Andrews, M. R., Kurn, D. M., Miesner, H. J., Durfee, D. S., Townsend, C. G., Inouye, S., and Ketterle, W. (1997, July). *Phys. Rev. Lett.*, **79**, 553. *ibid.* 80 (1998) 2967.

Astrakharchik, G. E., Boronat, G., Casulleras, J., and Giorgini, S. (2004). *Phys. Rev. Lett.*, **93**, 200404.

Bardeen, J., Cooper, L. N., and Schrieffer, J. R. (1957). *Phys. Rev.*, **106**, 162.

Bartenstein, M., Altmeyer, A., Riedl, S., Jochim, S., Chin, C., Denschlag, J. Hecker, and Grimm, R. (2004). *Phys. Rev. Lett*, **92**, 203201.

Bourdel, T., Khaykovich, L., J. Cubizolles, Zhang, J., Chevy, F., Teichmann, M., Tarruell, L., Kokkelmans, S.J.J.M.F., and Salomon, C. (2004). *Phys. Rev. Lett.*, **93**, 05040.

Bradley, C. C., Sackett, C. A., Tollett, J. J., and Hulet, R. G. (1995, August). *Phys. Rev. Lett.*, **75**, 1687. *ibid.* 79 (1997) 1170.

Capuzzi, P., Vignolo, P., Federici, F., and Tosi, M. P. (2006*a*). *J. Phys. B: At. Mol. Opt. Phys.*, **39**, S25.

Capuzzi, P, Vignolo, P, Federici, F, and Tosi, M P (2006*b*). *Phys. Rev. A*, **74**, 057601.

Carlson, J., Chang, S.-Y., Pandharipande, V. R., and Schmidt, K. E. (2003). *Phys. Rev. Lett.*, **91**, 050-401.

Carlson, J. and Reddy, S. (2005). *Phys. Rev. Lett.*, **95**, 060401.

Davis, K. B., Mewes, M. O., Andrews, M. R., van Druten, N. J., Durfee, D. S., Kurn, D. M., and Ketterle, W. (1995, November). *Phys. Rev. Lett.*, **75**, 3969.

DeMarco, B. and Jin, D. S. (1999, September). *Science*, **285**, 1703.

DeMarco, B. and Jin, D. S. (2002). *Phys. Rev. Lett.*, **88**, 040405.

Dieckmann, K., Stan, C. A., Gupta, S., Hadzibabic, Z., Schunck, C. H., and Ketterle, W. (2002). *Phys. Rev. Lett.*, **89**, 203201.

Duine, R. A. and Stoof, H. T. C. (2004). *Phys. Rep.*, **396**, 115.

Eagles, D. M. (1969). *Phys. Rev.*, **186**, 456.

Farchioni, R., Grosso, G., and Vignolo, P. (2000). *Phys. Rev. B*, **62**, 12565.

Fukuhara, T., Sugawa, S., Takasu, Y., and Takahashi, Y. (2009). *Phys. Rev. A*, **79**, R021601.

Gensemer, S. D. and Jin, D. S. (2001). *Phys. Rev. Lett.*, **87**, 173201.

Giorgini, S., Pitaevskii, L. P., and Stringari, S. (2008). *Theory of ultracold atomic Fermi gases*, **80**, 1215.

Girardeau, M. D. (1960). *J. Math. Phys.*, **1**, 516.

Girardeau, M. D. (1965). *Phys. Rev.*, **139**, B500.

Goldwin, J., Papp, S. B., DeMarco, B., and Jin, D. S. (2002). *Phys. Rev. A*, **65**, 021402.

Gradshteyn, I. S. and Ryzhik, I. M. (2000). *Table of integrals, series and productes* (6th). Academic Press, London 16

Granade, S. R., Gehm, M. E., O'Hara, K. M., and Thomas, J. E. (2002). *Phys. Rev. Lett.*, **88**, 120405.

Greiner, M., Regal, C. A., and Jin, D. S. (2003). *Nature*, **426**, 537.

Hadzibabic, Z., Stan, C. A., Dieckmann, K., Gupta, S., Zwierlein, M. W., Görlitz, A., and Ketterle, W. (2002). *Phys. Rev. Lett.*, **88**, 160401.

Hu, H., Liu, X.-J, and Drummond, P.D. (2006). *Europhys. Lett.*, **74**, 574.

Hu, H., Minguzzi, A., Liu, X.-J, and Tosi, M. P. (2004). *Phys. Rev. Lett.*, **93**, 190403.

Jochim, S., Bartenstein, M., Hendl, G., Hecker-Denschlag, J., Grimm, R., Mosk, A., and Weidemüller, M. (2002). *Phys. Rev. Lett.*, **89**, 273202.

Joseph, J., Clancy, B., Luo, L., Kinast, J., Turlapov, A., and Thomas, J. E. (2007). *Phys. Rev. Lett.*, **98**, 170401.

Kagan, Yu., Surkov, E. L., and Shlyapnikov, G. V. (1996). *Phys. Rev. A*, **54**, R1753.

Ketterson, J. B. and Song, S. N. (1999). *Superconductivity*. Cambridge University Press, Cambridge.

Kinast, J., Hemmer, S. L., Gehm, M. E., Turlapov, A., and Thomas, J. E. (2004). *Phys. Rev. Lett.*, **92**, 150402.

Kinast, J., Turlapov, A., Thomas, J. E., Chen, Q., Stajic, J., and Levin, K. (2005). *Science*, **307**, 1296.

Kirkman, P. D. and Pendry, J. B. (1984). *J. Phys. C*, **17**, 4327.

Landau, L. D. and Lifshitz, E. M. (1959). *Quantum mechanics: non-relativistic theory*. Pergamon, Oxford.

Leggett, A. J. (1980). In *Modern trends in the theory of condensed matter* (ed. A. Pekalski and J. Przystawa). Springer, Berlin.

Lenard, A. (1966). *J. Math. Phys.*, **7**, 1268.

Liu, Xia-Ji and Hu, Hui (2003). *Phys. Rev. A*, **67**, 023613.

Loftus, T., Regal, C. A., Ticknor, C., Bohn, J. L., and Jin, D. S. (2002). *Phys. Rev. Lett.*, **88**, 173201.

March, N. H. and Murray, A. M. (1960). *Phys. Rev.*, **120**, 830.

Messiah, A. (1999). *Quantum mechanics*. Dover Dover Publishing Company, New York.

Minguzzi, A., Vignolo, P., Chiofalo, M. L., and Tosi, M. P. (2001). *Phys. Rev. A*, **64**, 033605.

Modugno, G., Roati, G., Riboli, F., Ferlaino, F., Brecha, R. J., and Inguscio, M. (2002). *Science*, **297**, 2240.

Nozières, P. and Pines, D. (1966). *Theory of quantum liquids*. Benjamin, New York.

Partridge, G. B., Strecker, K. E., Kamar, R. I., Jack, M. W., and Hulet, R. G. (2005). *Phys. Rev. Lett.*, **95**, 020404.

Perali, A., Pieri, P., Pisani, L., and Strinati, G. C. (2004). *Phys. Rev. Lett.*, **92**, 220404.

Petrov, D. S., Salomon, C., and Shlyapnikov, G. V. (2004). *Phys. Rev. Lett.*, **93**, 090404.

Randeira, M. (1995). In *Bose-Einstein condensation* (ed. A. Griffin, D. W. Snoke, and S. Stringari), p. 355. Cambridge University Press, Cambridge.

Regal, C. A. and Jin, D. S. (2003). *Phys. Rev. Lett.*, **90**, 230404.

Regal, C. A., Ticknor, C., Bohn, J. L., and Jin, D. S. (2003). *Nature*, **424**, 47.

Roati, G., Riboli, F., Modugno, G., and Inguscio, M. (2002). *Phys. Rev. Lett.*, **89**, 150403.

Schreck, F., Khaykovich, L., Corwin, K. L., Ferrari, G., Bourdel, T., Cubizolles, J., and Salomon, C. (2001, August). *Phys. Rev. Lett.*, **87**, 080403.

Thomas, J. E., Kinast, J., and Turlapov, A. (2005). *Phys. Rev. Lett.*, **95**, 120402.

Truscott, Andrew G., Strecker, Kevin E., McAlexander, William I., Partridge, Guthrie B., and Hulet, Randall G. (2001, March). *Science*, **291**, 2570.

Vignolo, P., Farchioni, R., and Grosso, G. (1999). *Phys. Rev. B*, **59**, 16065.

Vignolo, P. and Minguzzi, A. (2003). *Phys. Rev. A*, **67**, 053601.

Vignolo, P., Minguzzi, A., and Tosi, M. P. (2000). *Phys. Rev. Lett.*, **85**, 2850.

Vignolo, P., Minguzzi, A., and Tosi, M. P. (2001). *Phys. Rev. A*, **64**, 023421.

Zaremba, E. (1998). *Phys. Rev. A*, **57**, 518.

Ziff, R. M., Uhlenbeck, G. E., and Kag, M. (1977). *Phys. Rep*, **32**, 169.

Zwierlein, M. W., Stan, C. A., Schunck, C. H., Raupach, S. M. F., Gupta, S., Hadzibabic, Z., and Ketterle, W. (2003). *Phys. Rev. Lett.*, **91**, 250401.

3

Quantum information: primitive notions and quantum correlations

Valerio SCARANI

Centre for Quantum Technologies and Department of Physics
National University of Singapore

3.1 Quantum theory

This chapter covers basic material, that can be found in many sources. If I'd have to suggest some general bibliography, here are my preferences of the moment: for quantum physics in general, the book by Sakurai (Sakurai, 1993); for a selection of important topics that does not follow the usual treatises, the book by Peres (Peres, 1995); and for quantum information, the books of Nielsen and Chuang (Nielsen and Chuang, 2000) or the short treatise by Le Bellac (Le Bellac, 2006).

3.1.1 Axioms, assumption and theorems

3.1.1.1 *Classical versus quantum physics*

It is convenient to start this chapter by recalling the meaning of some frequently used expressions:

Physical system: the degrees of freedom under study. For instance, the system "point-like particle" is defined by the degrees of freedom $(\vec{x}, \vec{p})$. A system is called *composite* if several degrees of freedom can be identified: for instance, "Earth and Moon".

State at a given time: the collection of all the properties of the system; a *pure state* is a state of maximal knowledge, while a *mixed state* is a state of partial knowledge.

An extremely important distinction has to be made now: the distinction between the *description* (or kinematics) of a physical system and its *evolution* (or dynamics). The distinction is crucial because *the specificity of quantum physics lies in the way physical systems are described, not in the way they evolve!*

Classical physics. In classical physics, the issue of the description of a physical system is not given a strong emphasis because it is, in a sense, trivial. Of course, even in classical physics, before studying "how it evolves", we have to specify "what it is", i.e. to identify the degrees of freedom under study. But once the system is identified, its properties combine according to the usual rules of set theory. Indeed, let P represent properties and s pure states. In classical physics, s is a point in the configuration space and P a subset of the same space. Classical logic translates into the usual rules of set theory:

- the fact that a system in state s possesses property P translates as $s \in P$;
- for any two properties P_1 and P_2, one can construct the set $P_1 \cap P_2$ of the states that possess both properties with certainty;
- if a property P_1 implies another property P_2, then $P_2 \subset P_1$;

and so on. In particular, by construction $s = \bigcap_k P_k$, where P_k are all the properties that the system possesses at that time. As is well known, only part of this structure can be found in quantum physics: in particular, a pure state is specified by giving the list of all *compatible* physical properties that it possesses with certainty.

Another interesting remark can be made about composite systems. Let us consider for instance "Earth and Moon": the two subsystems are obviously interacting, therefore the dynamics of the Earth will be influenced by the Moon. However, at each time one can consider the state of the Earth, i.e. give its position and its momentum. If the initial state was assumed to be pure, the state of each subsystem at any given time will remain pure (i.e. the values are sharply defined). This is not the case in quantum physics: when two subsystems interact, generically they become "entangled" in a way that only the global state remains pure.

Quantum physics. In this chapter, I shall focus on *non-relativistic quantum physics* and use the following definition of quantum physics:

(i) Physical systems (degrees of freedom) are described by a Hilbert space.

(ii) The dynamics of a closed system is reversible.

Admittedly, (i) is not a very physical definition: it is rather a description of the formalism. Several attempts have been made to found quantum physics on more "physical" or "axiomatic" grounds, with some success; but I consider none of them conclusive enough to be worthwhile adopting in a school. We shall come back to this issue in Section 3.5. Requirement (ii) is not specific to quantum physics: in all of modern physics, irreversibility is practical but not fundamental (the result of the system interacting with a large number of degrees of freedom, which one does not manage to keep track of).

3.1.1.2 Kinematics: Hilbert space

Note: the students of this school are supposed to be already familiar with the basic linear algebra used in quantum physics, as well as with the Dirac notation. Therefore, I skip the mathematical definitions and focus only on the correspondence between physical notions and mathematical objects.

Pure states, Born's rule. Basically, the only statement that should be taken as an axiom is the description of states in quantum physics: the space of states is not a set, like the classical configuration space, but a *vector space with scalar product*, defined on the complex field, called *Hilbert space*[1] and written $\mathcal{H}$. In this space, *pure states are described by one-dimensional subspaces*. A one-dimensional subspace is identified by the corresponding projector P; or alternatively, one can choose any vector $|v\rangle$ in the subspace as representative, with the convention, however, that every vector $c|v\rangle$ differing by a constant represents the same state.

With this construction, as is well known, after a measurement a system may be found to possess a property that it did not possess with certainty before the measurement. In other words, given a state $|v_1\rangle$, there is a non-zero probability that a measurement finds the system in a *different* state $|v_2\rangle$. The probability is given by

$$P(v_2|v_1) = \text{Tr}\,(P_1\,P_2) = |\langle v_1|v_2\rangle|^2 \qquad (3.1)$$

where we assume, as we shall always do from now onwards, that the vectors are normalized as $\langle v|v\rangle = 1$. This is called *Born's rule for probabilities*.

What we have discussed in this paragraph is the essence of quantum physics. It is generally called the *superposition principle*: states are treated as vectors, i.e. the vector sum of states defines another state that is not unrelated to its components.[2] All the usual rules of quantum kinematics follow from the vector-space assumption. It is useful to sketch this derivation.

Ideal measurements. Consider a measurement, in which different positions of the pointer can be associated with d different states $|\phi_1\rangle, ..., |\phi_d\rangle$. Among the features of an ideal measurement, one tends to request that the device correctly identifies the state. In quantum physics, this cannot be enforced for all states because of Born's rule; but at least one can request the following: if the input system is in state $|\phi_j\rangle$, the measurement should produce the outcome associated to that state with certainty (for non-destructive measurements, this implies that the measurement outcome is reproducible). Therefore, given Born's rule, *an ideal measurement is defined by an orthonormal basis, i.e. a set of orthogonal vectors.*

A very compact way of defining an ideal measurement is provided by *self-adjoint operators*, because any self-ajoint operator A can be diagonalized on an orthonormal basis: $A = \sum_k a_k |\phi_k\rangle\langle\phi_k|$, with a_k real numbers. In this case, one typically considers A to be the "physical quantity" that is measured and the a_k to be the "outcomes" of the measurement. In turn, one is able to compute derived quantities, in particular average values over repeated measurements. Indeed, let $|\psi\rangle$ be the state produced by the source, $a^{(i)}$ be the outcome registered in the ith measurement and n_k the number of times the outcome a_k was registered: then one has

$$\langle A \rangle_\psi = \lim_{N\to\infty} \frac{1}{N} \sum_{i=1}^{N} a^{(i)} = \sum_{k=1}^{d} \left(\lim_{N\to\infty} \frac{n_k}{N} \right) a_k = \sum_{k=1}^{d} P(\phi_k|\psi)\, a_k = \langle\psi|A|\psi\rangle.$$

It is, however, important to keep in mind that, strictly speaking, *only the orthonormal basis defines an ideal measurement, while the labelling of the outcomes of a measurement is arbitrary*: even if real numbers are a very convenient choice in many cases, one can use complex numbers, vectors, colors or any other symbol. In other words, *what is directly measured in an ideal measurement are the probabilities of each outcome*; average values are derived quantities.

Mixed states. The simplest way of introducing mixed states is to think of a source that fluctuates, so that it produces the pure state $|\psi_1\rangle$ with probability p_1, $|\psi_2\rangle$ with probability p_2 and so on. If nothing is known about these fluctuations, the statistics of any ideal measurement will look like

$$P(\phi_j|\rho) = \sum_k p_k\, P(\phi_j|\psi_k) = \mathrm{Tr}\left(P_{\phi_j}\rho\right) \text{ with } \rho = \sum_k p_k P_{\psi_k}.$$

This is the usual definition of the *density matrix*. It is again important to stress that the notion of mixed state is not proper to quantum physics: in fact, in the formula above, only the $P(\phi_j|\psi_k)$ are typically quantum, but the probabilities p_k are classical (at least formally; their ultimate origin may be entanglement, see below, but this is another matter).

A remark on Gleason's theorem.   At this stage, it is worthwhile mentioning Gleason's theorem (Gleason, 1957). Suppose that properties P are defined by subspaces $\mathcal{E}_P$ of a Hilbert space $\mathcal{H}$; and suppose in addition that orthogonal subspaces are associated to distinguishable properties. Let ω be an assignment of probabilities, i.e. $\omega(\mathcal{E}_P) \in [0, 1]$, $\omega(\mathcal{H}) = 1$ and $\omega(\mathcal{E}_P \oplus \mathcal{E}_{P'}) = \omega(\mathcal{E}_P) + \omega(\mathcal{E}_{P'})$ if $\mathcal{E}_P \perp \mathcal{E}_{P'}$. Then, if the dimension of the Hilbert space is $d \geq 3$, to each such ω one can associate a non-negative Hermitian operator ρ_ω such that $\omega(\mathcal{E}_P) = \mathrm{Tr}(\rho_\omega \Pi_P)$ with Π_P the projector on the subspace $\mathcal{E}_P$.

This complicated statement basically says that Born's rule can be *derived* if one decides to associate properties to subspaces of a Hilbert space and to identify orthogonality with distinguishability. Curiously enough, if one allows for generalized measurements (see below), the theorem becomes valid for $d = 2$ and the proof is considerably simplified (Caves *et al.*, 2004).

3.1.1.3   *Dynamics of closed systems: reversibility*

Why reversible evolution must be unitary.   The requirement of reversible dynamics for closed systems implies that *the evolution operator must be unitary*. As a simple way of understanding this, let us take an analogy with classical computation, that can be seen as an evolution of an initial string of bits. A computation is obviously reversible if and only if it consists in a permutation: indeed, a permutation is reversible; any other operation, that would map two strings onto the same one, is clearly not reversible. A reversible evolution in a vector space is slightly more general: it is *a change of basis*. A change of basis is obviously reversible; the brute force proof of the converse is hard, but we can provide a proof "by consistency of the theory" that is quite instructive.

We are first going to argue that any operation that describes *reversible* evolution in time must preserve the modulus of the scalar product, $\chi = |\langle \psi_1 | \psi_2 \rangle|$, between any two vectors. The point is that, as a direct consequence of Born's rule for probabilities, χ has a crucial operational interpretation in the theory: two states can be perfectly discriminated if and only if $\chi = 0$, and are identical if and only if $\chi = 1$; this indicates (and it can be proved rigorously, see Section 3.3.1) that χ quantifies the "distinguishability" between the two states in the best possible measurement. At this stage, we have only spoken of preparation and measurement and invoked Born's rule; so we have made no assumption on time evolution. So now suppose that an evolution in time can change χ: if χ decreases, then the states become more distinguishable, contradicting the fact that χ quantifies the maximal distinguishability (in other words, in this case the best measurement would consist in waiting for some time before performing the measurement itself). So, for our measurement theory to be meaningful, χ can only increase. But if the evolution is reversible, by reverting it we would have a valid evolution in which χ decreases. The only remaining option is that χ does not change during reversible evolution. Unitarity follows from the fact that only unitary and anti-unitary operations preserve χ, but anti-unitary operations cannot be used for a symmetry with a continuous parameter as translation in time is.

The status of the Schrödinger equation.   If the generator of the evolution is supposed to be independent of time, standard group theoretical arguments and correspondence with classical dynamics allow writing $U = e^{-iHt/\hbar}$, where H is the Hamilton operator; from this expression, the corresponding differential equation $i\hbar \frac{d}{dt}|\psi\rangle = H|\psi\rangle$ (Schrödinger equation) is readily derived. For the case where the dynamics itself varies with time, there is no such derivation: rather, one *assumes* the same Schrödinger equation to hold with $H = H(t)$. Needless to say, the corresponding dynamics remains unitary.

3.1.2   Composite systems

In the classical case, as we have argued above, a pure state of a composite system is always a *product state*, i.e. a state of the form $s_A \times s_B$, where s_j is a pure state of system j; in the space of states, this translates by the Cartesian product of the sets of properties of each system.

In quantum theory, product states obviously exist as well, as they correspond to possible physical situations. But the space of states of the whole system cannot be simply $\mathcal{H}_A \times \mathcal{H}_B$, because this is not a vector space. The vector space that contains all product states and their linear combinations is the *tensor product*

$$\mathcal{H} = \mathcal{H}_A \otimes \mathcal{H}_B.$$

We start with a rapid survey of the algebraic structure of the tensor product – it really behaves like a product.

3.1.2.1   *Tensor product algebra*

The space $\mathcal{H} = \mathcal{H}_A \otimes \mathcal{H}_B$ is constructively defined out of its components by requiring that: for any basis $\{|\alpha_j\rangle\}_{j=1...d_A}$ of $\mathcal{H}_A$ and for any basis $\{|\beta_k\rangle\}_{j=1...d_B}$ of $\mathcal{H}_B$, the set of states $\{|\alpha_j\rangle \otimes |\beta_k\rangle\}_{j,k=...}$ forms a basis of $\mathcal{H}$. It follows immediately that the dimension of $\mathcal{H}$ is $d_A d_B$, as it should.[3]

The scalar product is defined on product states as

$$((\langle\psi| \otimes \langle\phi|)(|\psi'\rangle \otimes |\phi'\rangle)) = \langle\psi|\psi'\rangle \langle\phi|\phi'\rangle$$

and extended to the whole space by linearity.

The space of linear operators on $\mathcal{H}$ corresponds with the tensor product of the spaces of linear operators on $\mathcal{H}_A$ and on $\mathcal{H}_B$. An operator of the product form, $A \otimes B$, acts on product states as

$$(A \otimes B)(|\psi\rangle \otimes |\phi\rangle) = A|\psi\rangle \otimes B|\phi\rangle \, ;$$

by linearity, this defines uniquely the action of the most general operator on the most general state.

3.1.2.2   *Entanglement*

Any linear combination of product states defines a possible pure state of the composite system. However, a state like

$$|\psi(\theta)\rangle = \cos\theta|\alpha_1\rangle \otimes |\beta_1\rangle + \sin\theta|\alpha_2\rangle \otimes |\beta_2\rangle \qquad (3.2)$$

cannot be written as a product state, as it is easily checked by writing down the most general product state $\left(\sum_j a_j |\alpha_j\rangle\right) \otimes \left(\sum_k b_k |\beta_k\rangle\right)$ and equating the coefficients. In fact, it is easy to convince oneself that product states form a set of measure zero (according to any reasonable measure) in the whole set of states. *A pure state that cannot be written as a product state, is called "entangled".*

The most astonishing feature of entanglement is that we have a pure state of the composite system that does not arise from pure states of the components: in other words, the properties of the whole are sharply defined, while the properties of the subsystems are not.

Case study: the singlet state. We are going to study a specific example that plays an important role in what follows. Like most of the examples in this chapter, this one involves *qubits*, i.e. two-level systems. The algebra of two-level systems is supposed to be known from basic quantum physics; for convenience, it is recalled as an Appendix to this section (Section 3.1.3).

Consider two qubits. The state

$$|\Psi^-\rangle = \frac{1}{\sqrt{2}} (|0\rangle \otimes |1\rangle - |1\rangle \otimes |0\rangle) \tag{3.3}$$

is called *singlet* because of its particular status in the theory of addition of angular momenta. The projector on this state reads

$$P_{\Psi^-} = \frac{1}{4} (\mathbb{1} \otimes \mathbb{1} - \sigma_x \otimes \sigma_x - \sigma_y \otimes \sigma_y - \sigma_z \otimes \sigma_z). \tag{3.4}$$

It appears clearly from the projector[4] that this state is *invariant by bilateral rotation*: $u \otimes u |\Psi^-\rangle = |\Psi^-\rangle$ (possibly up to a global phase). In particular, consider a measurement in which the first qubit is measured along direction $\vec{a}$ and the second along direction $\vec{b}$. The statistics of the outcomes $r_A, r_B \in \{-1, +1\}$ are given by $P_{\vec{a},\vec{b}}(r_A, r_B) = \left| \left(\langle r_A \vec{a}| \otimes \langle r_B \vec{b}|\right) |\Psi^-\rangle \right|^2$. The calculation leads to

$$P_{\vec{a},\vec{b}}(++) = P_{\vec{a},\vec{b}}(--) = \frac{1}{4}(1 - \vec{a} \otimes \vec{b}), \tag{3.5}$$

$$P_{\vec{a},\vec{b}}(+-) = P_{\vec{a},\vec{b}}(-+) = \frac{1}{4}(1 + \vec{a} \otimes \vec{b}). \tag{3.6}$$

Much information can be extracted from those statistics, we shall come back to them in Section 3.4 For the moment, let us just stress the following features:

- On the one hand, $P_{\vec{a},\vec{b}}(r_A = +) = P_{\vec{a},\vec{b}}(r_A = -) = \frac{1}{2}$ and $P_{\vec{a},\vec{b}}(r_B = +) = P_{\vec{a},\vec{b}}(r_B = -) = \frac{1}{2}$: the outcomes of the measurements of each qubit appear completely random, independent of the directions of the measurements. This implies the operational meaning of the fact that the state of each qubit cannot be pure (and here it is actually maximally mixed, see below).

- On the other hand, there are very sharp correlations: in particular, whenever $\vec{a} = \vec{b}$, the two outcomes are rigorously opposite.

This can be seen as defining the relation "being opposite" for two arrows in itself, without specifying in which direction each arrow points. While this possibility is logically compelling, it is worthwhile recalling that this is impossible in our everyday life.

A warning: we stressed that the singlet has a special role to play in the addition of two spins 1/2: indeed, it defines a one-dimensional subspace of total spin 0, as opposite to the three states that form the spin-1 triplet. However, in quantum information theory this feature is *not* crucial. In other words, in quantum information the state $|\Psi^-\rangle$ is as good as any other state $u_1 \otimes u_2 |\Psi^-\rangle = \frac{1}{\sqrt{2}} \left(|+\hat{m}\rangle \otimes |+\hat{n}\rangle + |-\hat{m}\rangle \otimes |-\hat{n}\rangle \right)$ that can obtained from it with local unitaries. In particular, it is customary to define the states

$$|\Psi^+\rangle = \frac{1}{\sqrt{2}} \left(|0\rangle \otimes |1\rangle + |1\rangle \otimes |0\rangle \right), \tag{3.7}$$

$$|\Phi^+\rangle = \frac{1}{\sqrt{2}} \left(|0\rangle \otimes |0\rangle + |1\rangle \otimes |1\rangle \right), \tag{3.8}$$

$$|\Phi^-\rangle = \frac{1}{\sqrt{2}} \left(|0\rangle \otimes |0\rangle - |1\rangle \otimes |1\rangle \right) \tag{3.9}$$

that are orthogonal to $|\Psi^-\rangle$ and with which they form the so-called *Bell basis*.

Entanglement for mixed states. It is in principle not difficult to decide whether a pure state is entangled or not: just check if it is a product state; if not, then it is entangled. For mixed states, the definition is more subtle, because a mixed state may exhibit classical correlations. As an example, consider the mixture "half of the times I prepare $|0\rangle \otimes |0\rangle$, and half of the times I prepare $|1\rangle \otimes |1\rangle$", that is, $\rho = \frac{1}{2}|00\rangle\langle 00| + \frac{1}{2}|11\rangle\langle 11|$ (with obvious shortcut notations). This state exhibits correlations: whenever both particles are measured in the basis defined by $|0\rangle$ and $|1\rangle$, the outcomes of both measurements are the same. However, by the very way the state was prepared, it is clear that no entanglement is involved.

We shall then call *separable* a mixed state for which a decomposition as a convex sum of product state exists:

$$\rho = \sum_k p_k \left(|\psi_k\rangle\langle\psi_k| \right)_A \otimes \left(|\phi_k\rangle\langle\phi_k| \right)_B. \tag{3.10}$$

It is enough that a single mixture[5] of product states exists, for the state to be separable. A mixed state is called *entangled* if it is not separable, i.e. if some entangled state must be used in order to prepare it (Werner, 1989). Apart from some special cases, to date no general criterion is known to decide whether a given mixed state is separable or entangled.

3.1.2.3 *Partial states, no-signalling and purification*

We have just seen that, in the presence of entanglement, one cannot assign a separate pure state to each of the subsystems. Still, suppose two entangled particles are sent apart from one another to two physicists, Alice and Bob. Alice on her location holds particle A and can make measurements on it, without possibly even knowing that a guy named Bob holds a particle that is correlated with hers. Quantum physics should provide Alice with rules to compute the probabilities for the outcome she observes: there must be a "state" (positive, unit trace hermitian operator) that describes the information available on Alice's side. Such a state is called a "partial state" or a "local state". The local state cannot be pure if the state of the global system is entangled.

Suppose that the full state is ρ_{AB}. The partial state ρ_A is defined as follows: for any observable $\mathcal{A}$ on Alice's particle, it must hold

$$\mathrm{Tr}_A(\rho_A \mathcal{A}) = \mathrm{Tr}_{AB}(\rho_{AB}(\mathcal{A} \otimes \mathbb{1}_B)) \tag{3.11}$$

$$= \sum_{j=1}^{d_A} \sum_{k=1}^{d_B} \langle a_j, b_k | \rho_{AB}(\mathcal{A} \otimes \mathbb{1}_B) | a_j, b_k \rangle$$

$$= \sum_{j=1}^{d_A} \langle a_j | (\mathrm{Tr}_B(\rho_{AB})) \mathcal{A} | a_j \rangle = \mathrm{Tr}_A(\mathrm{Tr}_B(\rho_{AB}) \mathcal{A}),$$

so by identification

$$\rho_A = \mathrm{Tr}_B(\rho_{AB}) = \sum_{k=1}^{d_B} \langle b_k | \rho_{AB} | b_k \rangle. \tag{3.12}$$

This is the general definition of the partial state: the partial state on A is obtained by *partial trace* over the other system B. The result (3.12) is not ambiguous since the trace is a unitary invariant, that is, gives the same result for any choice of basis. This implies that *whatever Bob does, the partial state of Alice will remain unchanged.* This fact has an important consequence, namely that Bob cannot use entanglement to send any message to Alice. This is known as the principle of *no-signalling through entanglement.*[6]

When it comes to computing partial states, one can always come back to the general definition (3.12), but there are often more direct ways. For instance, no-signalling itself can be used: to obtain (say) ρ_A, one may consider that Bob performs a given measurement and compute the mixture that would be correspondingly prepared on Alice's side. This mixture must be ρ_A itself, since the partial state is the same for any of Bob's actions.

The notion of "purification". The notion of *purification* is, in a sense, the reverse of the notion of partial trace. It amounts to the following: any mixed state can be seen as the partial state of a pure state in a larger Hilbert space. To see that this is the case, note that any state ρ can be diagonalized, i.e. there exist a set of orthogonal vectors $\{|\varphi_k\rangle\}_{k=1...r}$ (r being the rank of ρ) such that $\rho = \sum_{k=1}^r p_k |\varphi_k\rangle\langle\varphi_k|$. Then, one can always take r orthogonal vectors $\{|e_k\rangle\}_{k=1...r}$ of an ancilla and construct the pure state $|\Psi\rangle = \sum_{k=1}^r \sqrt{p_k}|\varphi_k\rangle \otimes |e_k\rangle$, of which ρ is the partial state for the first subsystem.

This construction shows that a purification always exists; but because the decomposition of a mixed state onto pure states is not unique, the purification is not unique as well. In particular, if $|\Psi\rangle$ is a purification of ρ and U is a unitary on the ancilla, then $\mathbb{1} \otimes U|\Psi\rangle$ is also a purification of ρ, because it just amounts to choosing another set $\{|e'_k\rangle\}_{k=1\dots r}$ of orthogonal vectors. Remarkably though, this is the only freedom: it can indeed be proved that all purification are equivalent up to local operations on the ancilla. In this sense, the purification can be said to be unique.

3.1.2.4 *Measurement and evolution revisited*

To conclude this first Section, we have to mention one of the most important extensions of the notion of tensor product, namely the definitions of *generalized measurements and evolution*. I shall not go into details of the formalism; whenever these notions appear later in this chapter, it will always be in a rather natural way.

Evolution: CP maps, decoherence. We have stressed above that the evolution of a closed system is reversible, hence unitary. In general, however, it is impossible to have a perfectly isolated system: even if some degrees of freedom ("the system") are carefully selected and prepared in a given state, the evolution may imply some interaction with other degrees of freedom ("the environment"). In principle, it is obvious what has to be done to describe such a situation: consider both the system and the environment, study the unitary evolution, then trace the environment out and study the resulting state of the system. By doing this study for any given time t, one defines an effective map T_t such that $\rho_S(t) = T_t[\rho_S(0)]$. It turns out that the maps defined in this way coincide with the family of *trace-preserving, completely positive (CP) maps*. Let us comment on each of these important properties:

- A map is positive if it transforms positive operators into positive operators; in our case, this means that the density matrix will never develop any negative eigenvalue;

- The trace-preserving property is self-explained: $\text{Tr}[\rho_S(t)] = 1$ for all t. Together with the previous, it implies that a density matrix remains a density matrix – an obvious necessity.

- Not all positive maps, however, define possible evolutions of a subsystem. For instance, time-reversal is a positive map, but if one applies time-reversal only to a subsystem, it seems clear that one may get into trouble. A map T is called "completely positive" if $T \otimes \mathbb{1}$ is also a positive map for any possible enlarged system. Interestingly, non-CP maps play an important role in quantum information: since the non-positivity can appear only if the system is entangled with the environment, these maps act as entanglement witnesses (see chapter on entanglement theory).

Trace-preserving CP maps are the most general evolution that a quantum system can undergo. Whenever the map on the system is not unitary, the state of the system becomes mixed by entanglement with the environment: this is called *decoherence*. An obvious property of these maps is that states can only become "less distinguishable".

A last remark: it would of course be desirable to derive a differential equation for the state of the system, whose solution implements the CP-map, without having to study the environment fully. In full generality, this task has been elusive; a general form has been derived by Lindblad for the case where the environment has no memory (Lindblad, 1976). For a detailed introduction to the theory of open quantum systems, we refer to the book by Breuer and Petruccione (Breuer and Petruccione, 2002).

Generalized measurements. For measurements, a similar discussion can be made. The most general measurement on a quantum system consists in appending other degrees of freedom, then performing a measurement on the enlarged system. These additional degrees of freedom are called *ancillae* (latin for "servant maids") rather than environment, but play exactly the same role: the final measurement may project on states in which the system and the ancillae are entangled. Note that the ancillae start in a state that is independent of the state of the system: they are part of the measuring apparatus, and the whole measurement can truly be said to be a measurement on the system alone.

The effective result on the system is captured by a family of positive operators: for all possible generalized measurement with D outcomes, there exist a family $\{A_k\}_{k=1,\dots,D}$ of positive operators, such that $\sum_k A_k^\dagger A_k = \mathbb{1}$. If the system has been prepared in the state ρ, the probability of obtaining outcome k is given by $p_k = \mathrm{Tr}(A_k \rho A_k^\dagger)$ and the state is subsequently prepared as $\rho_k = \frac{1}{p_k} A_k \rho A_k^\dagger$. Note that D must be at most the dimension of the total Hilbert space "system+ancillae", but can of course be much larger than the dimension of the Hilbert space of the system alone.

For some reason, the name that has been retained for such generalized measurements is *positive-operator-valued measurements, or POVMs*. Some specific examples will be presented in the coming section.

3.1.3 Appendix: one-qubit algebra

Since most of the examples in this chapter will be done on spin $1/2$ systems (*qubits*), I recall here for convenience some basic elements of the Hilbert space $\mathcal{H} = \mathbb{C}^2$ and of the linear operators on that space.

We start by recalling the Pauli matrices:

$$\sigma_x = \begin{pmatrix} 0 & 1 \\ 1 & 0 \end{pmatrix}, \ \sigma_y = \begin{pmatrix} 0 & -i \\ i & 0 \end{pmatrix}, \ \sigma_z = \begin{pmatrix} 1 & 0 \\ 0 & -1 \end{pmatrix}. \tag{3.13}$$

The computational basis $\{|0\rangle, |1\rangle\}$ is universally assumed to be the eigenbasis of σ_z, so that:

$$\sigma_z|0\rangle = |0\rangle, \ \sigma_z|1\rangle = -|1\rangle, \tag{3.14}$$

$$\sigma_x|0\rangle = |1\rangle, \ \sigma_x|1\rangle = |0\rangle, \tag{3.15}$$

$$\sigma_y|0\rangle = i|1\rangle, \ \sigma_y|1\rangle = -i|0\rangle. \tag{3.16}$$

It is useful sometimes to write the Pauli matrices as

$$\sigma_x = |0\rangle\langle 1| + |1\rangle\langle 0|, \ \sigma_y = -i|0\rangle\langle 1| + i|1\rangle\langle 0|, \ \sigma_z = |0\rangle\langle 0| - |1\rangle\langle 1|. \tag{3.17}$$

One has $\text{Tr}(\sigma_k) = 0$ and $\sigma_k^2 = \mathbb{1}$ for $k = x, y, z$; moreover,

$$\sigma_x \sigma_y = -\sigma_y \sigma_x = i\sigma_z \quad + \text{cyclic permutations.} \tag{3.18}$$

The generic *pure state* of a qubit is written $|\psi\rangle = \alpha|0\rangle + \beta|1\rangle$. The associated projector is therefore

$$|\psi\rangle\langle\psi| = |\alpha|^2|0\rangle\langle0| + |\beta|^2|1\rangle\langle1| + \alpha\beta^*|0\rangle\langle1| + \alpha^*\beta|1\rangle\langle0| \tag{3.19}$$

$$= \frac{1}{2}\left(\mathbb{1} + \left(|\alpha|^2 - |\beta|^2\right)\sigma_z + 2\text{Re}(\alpha\beta^*)\sigma_x + 2\text{Im}(\alpha\beta^*)\sigma_y\right). \tag{3.20}$$

It follows immediately from $\sigma_k^2 = \mathbb{1}$ that $|\psi\rangle\langle\psi| = \frac{1}{2}\left(\mathbb{1} + \sum_k \langle\sigma_k\rangle_\psi \sigma_k\right)$. One writes

$$|\psi\rangle\langle\psi| = \frac{1}{2}\left(\mathbb{1} + \hat{n}\cdot\vec{\sigma}\right), \quad \text{with} \quad \hat{n} = \begin{pmatrix} \langle\sigma_x\rangle_\psi \\ \langle\sigma_y\rangle_\psi \\ \langle\sigma_z\rangle_\psi \end{pmatrix} = \begin{pmatrix} 2\text{Re}(\alpha\beta^*) \\ 2\text{Im}(\alpha\beta^*) \\ |\alpha|^2 - |\beta|^2 \end{pmatrix}. \tag{3.21}$$

The vector $\hat{n}$ is called the Bloch vector, it corresponds to the expectation value of the "magnetic moment" $\vec{\sigma}$ in the given state. For pure states (the case we are considering here), its norm is one. It is actually well known that such vectors cover the unit sphere (called the *Bloch sphere*, or the *Poincaré sphere* if the two-level system is the polarization of light). In fact, there is a one-to-one correspondence between unit vectors and pure states of a two-level system given by the following parametrization in spherical coordinates:

$$|\psi\rangle \equiv |+\hat{n}\rangle = \cos\frac{\theta}{2}|0\rangle + e^{i\varphi}\sin\frac{\theta}{2}|1\rangle \leftrightarrow \hat{n} \equiv \hat{n}(\theta, \varphi) = \begin{pmatrix} \sin\theta\cos\varphi \\ \sin\theta\sin\varphi \\ \cos\theta \end{pmatrix}. \tag{3.22}$$

In turn, $|+\hat{n}\rangle$ is the eigenstate of $\hat{n}\cdot\vec{\sigma}$ for the eigenvalue $+1$; or alternatively,

$$\hat{n}\cdot\vec{\sigma} \equiv \sigma_n = |+\hat{n}\rangle\langle+\hat{n}| - |-\hat{n}\rangle\langle-\hat{n}|. \tag{3.23}$$

One has $\text{Tr}(\sigma_n) = 0$ and $\sigma_n^2 = \mathbb{1}$. For any basis $\{|+\hat{n}\rangle, |-\hat{n}\rangle\}$, one has the closure (completeness) relation $|+\hat{n}\rangle\langle+\hat{n}| + |-\hat{n}\rangle\langle-\hat{n}| = \mathbb{1}$. The eigenstates of σ_x and σ_y are frequently used. They read, up to a global phase:

$$|\pm x\rangle = \frac{1}{\sqrt{2}}\left(|0\rangle \pm |1\rangle\right), \quad |\pm y\rangle = \frac{1}{\sqrt{2}}\left(|0\rangle \pm i|1\rangle\right). \tag{3.24}$$

Let's move to the study of *mixed states*. Given that any projector can be written as Eq. (3.21), obviously any mixed state takes exactly the same form. In fact, consider just the mixture of two pure states:

$$\rho = p_1|\psi_1\rangle\langle\psi_1| + p_2|\psi_2\rangle\langle\psi_2| = \frac{1}{2}\left(\mathbb{1} + \left(p_1\hat{n}_1 + p_2\hat{n}_2\right)\cdot\vec{\sigma}\right).$$

It is easy to verify that the resulting Bloch vector $\vec{m} = p_1\hat{n}_1 + p_2\hat{n}_2$ lies *inside* the unit sphere. In fact, the points in the volume of the Bloch sphere are in a one-to-one correspondence with all possible states of a single qubit, be they pure (in which case their Bloch vector lies on the surface) or mixed.

We can summarize all that should be known on the states of a single qubit as follows: generically, a state of a single qubits reads

$$\rho = \frac{1}{2}\left(\mathbb{1} + \vec{m} \cdot \vec{\sigma}\right) = \frac{1}{2}\left(\mathbb{1} + |\vec{m}|\sigma_m\right), \text{ with } \vec{m} = \begin{pmatrix} \text{Tr}(\sigma_x \rho) \\ \text{Tr}(\sigma_y \rho) \\ \text{Tr}(\sigma_z \rho) \end{pmatrix}. \quad (3.25)$$

The norm of the Bloch vector is $|\vec{m}| \leq 1$, with equality if and only if the state is pure. Since ρ is hermitian, there is one and only one decomposition as the sum of two orthogonal projectors. Clearly, the eigenstates of this decomposition are the eigenstates $|+\hat{m}\rangle$ and $|-\hat{m}\rangle$ of $\hat{m} \cdot \vec{\sigma}$, where $\hat{m} = \vec{m}/|\vec{m}|$. The orthogonal decomposition reads

$$\rho = \left(\frac{1 + |\vec{m}|}{2}\right)|+\hat{m}\rangle\langle+\hat{m}| + \left(\frac{1 - |\vec{m}|}{2}\right)|-\hat{m}\rangle\langle-\hat{m}|. \quad (3.26)$$

Recall that any density matrix ρ that is not a projector admits an infinity of decompositions as the sum of projectors. All these decompositions are equivalent, because the density matrix carries all the information on the state, that is, on the actual properties of the system.

Finally, a useful result: the probability of finding $|+\hat{n}\rangle$ given the state $\rho = \frac{1}{2}(\mathbb{1} + \vec{m} \cdot \vec{\sigma})$ is

$$\text{Prob}(+\hat{n}\,|+\vec{m}\,) = \frac{1}{4}\,\text{Tr}\left[(\mathbb{1} + \hat{n} \cdot \vec{\sigma})(\mathbb{1} + \vec{m} \cdot \vec{\sigma})\right] = \frac{1 + \hat{n} \cdot \vec{m}}{2}. \quad (3.27)$$

3.1.4 Tutorials

3.1.4.1 Problems

Exercise 1.1
Which of the following states are entangled?

1. $|\Psi_1\rangle = \cos\theta|0\rangle|0\rangle + \sin\theta|1\rangle|1\rangle$.
2. $|\Psi_2\rangle = \cos\theta|0\rangle|0\rangle + \sin\theta|1\rangle|0\rangle$.
3. $|\Psi_3\rangle = \frac{1}{2}(|0\rangle|0\rangle + |0\rangle|1\rangle - |1\rangle|0\rangle - |1\rangle|1\rangle)$.
4. $|\Psi_4\rangle = \frac{1}{2}(|0\rangle|0\rangle + |0\rangle|1\rangle + |1\rangle|0\rangle - |1\rangle|1\rangle)$.

For those that are not entangled, give the decomposition as a product state. For simplicity of notation, we write $|\psi_1\rangle|\psi_2\rangle$ instead of $|\psi_1\rangle \otimes |\psi_2\rangle$.

Exercise 1.2
Compute the partial states ρ_A and ρ_B for the following states of two qubits:

1. The pure state $|\Psi\rangle = \sqrt{\frac{2}{3}}|0\rangle|+\rangle + \sqrt{\frac{1}{3}}|+\rangle|-\rangle$; where $|\pm\rangle = \frac{1}{\sqrt{2}}(|0\rangle \pm |1\rangle)$.
2. The mixed state $W(\lambda) = \lambda|\Psi^-\rangle\langle\Psi^-| + (1-\lambda)\frac{\mathbb{1}}{4}$, called a Werner state [R.F. Werner, Phys. Rev. A **40**, 4277 (1989)].

Verify in both cases that ρ_A and ρ_B are mixed by computing the norm of their Bloch vectors.

Exercise 1.3
We consider the following decoherent channel (Scarani *et al.*, 2002). A qubit, initially prepared in a state ρ, undergoes sequential "collisions" with qubits coming from a reservoir. All the qubits of the reservoir are supposed to be in state $\xi = p|0\rangle\langle0| + (1-p)|1\rangle\langle1|$. Each collision implements the evolution

$$U: \begin{cases} |0\rangle|0\rangle \longrightarrow |0\rangle|0\rangle \\ |0\rangle|1\rangle \longrightarrow \cos\phi|0\rangle|1\rangle + i\sin\phi|1\rangle|0\rangle \\ |1\rangle|0\rangle \longrightarrow \cos\phi|1\rangle|0\rangle + i\sin\phi|0\rangle|1\rangle \\ |1\rangle|1\rangle \longrightarrow |1\rangle|1\rangle \end{cases} \tag{3.28}$$

with $\sin\phi \neq 0$. We assume that each qubit of the reservoir interacts only once with the system qubit. Therefore, the state of the system after collision with $n+1$ qubits of the bath is defined recursively as

$$\rho^{(n+1)} \equiv T_\xi^{n+1}[\rho] = \mathrm{Tr}_B\left(U\rho^{(n)} \otimes \xi U^\dagger\right). \tag{3.29}$$

1. Let $\rho^{(n)} = d^{(n)}|0\rangle\langle0| + (1-d^{(n)})|1\rangle\langle1| + k^{(n)}|0\rangle\langle1| + k^{(n)*}|1\rangle\langle0|$. Prove that the CP-map (3.29) induces the recursive relations

$$d^{(n+1)} = c^2 d^{(n)} + s^2 p, \quad k^{(n+1)} = ck^{(n)}, \tag{3.30}$$

with $c = \cos\phi$ and $s = \sin\phi$.
2. By iteration, provide $d^{(n+1)}$ and $k^{(n+1)}$ as a function of the parameters of the initial state $d^{(0)}$ and $k^{(0)}$. Conclude that $T_\xi^n[\rho] \longrightarrow \xi$ when $n \to \infty$, whatever the initial state ρ (pure or mixed).
3. We have just studied an example of "thermalization": a system, put in contact with a large reservoir, ultimately assumes the same state as the particles in the reservoir. Naively, one would have described this process as $\rho \otimes \xi^{\otimes N} \to \xi^{\otimes N+1}$ for all ρ. Why is such a process not allowed by quantum physics?
4. The condition $\sin\phi \neq 0$ is necessary to have a non-trivial evolution during each collision; however, to have a meaningful model of thermalization one has to enforce $\cos\phi \gg |\sin\phi|$. What is the meaning of this condition? *Hint*: as a counter-example, consider the extreme case $\sin\phi = 1$: what is then U? What does the process look like in this case?

3.1.4.2 Solutions

Exercise 1.1
$|\Psi_1\rangle$ is entangled. $|\Psi_2\rangle = (\cos\theta|0\rangle + \sin\theta|1\rangle)|0\rangle$ is not entangled. $|\Psi_3\rangle = \frac{1}{2}(|0\rangle - |1\rangle)(|0\rangle + |1\rangle) = |-\rangle|+\rangle$ is not entangled. $|\Psi_4\rangle$ is entangled: this can be verified by direct calculation, or also by noticing that $|\Psi_4\rangle = \frac{1}{\sqrt{2}}(|0\rangle|+\rangle + |1\rangle|-\rangle)$; by just relabelling the basis of the second system, one sees that this state has the same form as $|\Psi_1\rangle$ with $\cos\theta = \sin\theta = \frac{1}{\sqrt{2}}$.

Exercise 1.2

1. For the pure state under study, $\rho_A = \frac{2}{3}|0\rangle\langle 0| + \frac{1}{3}|+\rangle\langle +| = \frac{1}{2}\left(\mathbb{1} + \frac{2}{3}\sigma_z + \frac{1}{3}\sigma_x\right)$. In order to compute ρ_B, here is a possibility (maybe not the fastest one): first, rewrite the state as $|\Psi\rangle = |0\rangle\left(\sqrt{\frac{2}{3}}|+\rangle + \sqrt{\frac{1}{6}}|-\rangle\right) + \sqrt{\frac{1}{6}}|1\rangle|-\rangle$. Then

$$
\begin{aligned}
\rho_B &= \left(\sqrt{\frac{2}{3}}|+\rangle + \sqrt{\frac{1}{6}}|-\rangle\right)\left(\sqrt{\frac{2}{3}}\langle +| + \sqrt{\frac{1}{6}}\langle -|\right) + \frac{1}{6}|-\rangle\langle -| \\
&= \frac{2}{3}|+\rangle\langle +| + \frac{1}{3}|-\rangle\langle -| + \frac{1}{3}|+\rangle\langle -| + \frac{1}{3}|-\rangle\langle +| = \frac{2}{3}|0\rangle\langle 0| + \frac{1}{3}|+\rangle\langle +| = \rho_A
\end{aligned}
$$

since $|0\rangle = \frac{1}{\sqrt{2}}(|+\rangle + |-\rangle)$. Note how, in this last calculation, the normalization is taken care of automatically.

The Bloch vector of $\rho_A = \rho_B$ has norm $|\vec{m}| = \frac{\sqrt{5}}{3} < 1$, therefore the states are mixed.

2. For the Werner state: $\rho_A = \rho_B = \frac{\mathbb{1}}{2}$. These states are maximally mixed, indeed $|\vec{m}| = 0$.

Exercise 1.3

1. The first point is a matter of patience in writing down explicitly $U\rho^{(n)} \otimes \xi U^\dagger$, then noticing that $\mathrm{Tr}(|0\rangle\langle 0|) = \mathrm{Tr}(|1\rangle\langle 1|) = 1$ and $\mathrm{Tr}(|0\rangle\langle 1|) = \mathrm{Tr}(|1\rangle\langle 0|) = 0$.

2. For the off-diagonal term, the recursion is obviously

$$
k^{(n+1)} = c^{n+1}k^{(0)}.
$$

For the diagonal term, one has

$$
d^{(n+1)} = c^2\left[c^2 d^{(n-1)} + s^2 p\right] + s^2 p = c^4 d^{(n-1)} + s^2(1 + c^2)p = \ldots
$$

$$
= c^{2(n+1)}d^{(0)} + s^2 \sum_{k=0}^{n} c^{2k}\, p = c^{2(n+1)}d^{(0)} + [1 - c^{2(n+1)}]p
$$

because $\sum_{k=0}^{n} c^{2k} = \frac{1 - c^{2(n+1)}}{1 - c^2} = \frac{1 - c^{2(n+1)}}{s^2}$. Therefore, $d^{(n+1)} \to p$ and $k^{(n+1)} \to 0$ for $n \to \infty$.

3. The evolution $\rho \otimes \xi^{\otimes N} \to \xi^{\otimes N+1}$ is not unitary, since two initially different states would end up being the same.

4. For $\sin\phi = 1$, U is the swap operation. In this case, the "thermalization" would consist in dumping the initial system in the reservoir and replacing it with one of the qubits of the reservoir. Such a process would introduce a very large fluctuation in the reservoir. By setting $\cos\phi \approx 1$, on the contrary, one has $\mathrm{Tr}(\rho_j A) \approx \mathrm{Tr}(\xi A)$ for any qubit j, for any single-particle physical quantity A. In other words, the system *appears* to be completely thermalized and one has to measure some multi-particle physical quantities to see some differences. This view is perfectly consistent with the idea that irreversibility is only apparent.

3.2   Primitives of quantum information (I)

3.2.1   A tentative list of primitives

The main tasks of quantum information, at the present stage of its development, are quantum computing and quantum cryptography. These tasks are complex: they rely on simpler notions, most of which are of interest in themselves. These notions that subtend the whole field are those that I call "primitives". Here is my tentative list of primitives, listed in chronological order of their appearance in the development of quantum physics:

1. violation of Bell's inequalities;
2. quantum cloning;
3. state discrimination;
4. quantum coding;
5. teleportation;
6. error correction;
7. entanglement distillation.

The common feature of all these primitives is that they have been studied in great detail. This does not mean that there are no open issues left; however, with a few remarkable exceptions, those are generally difficult points of rather technical nature. This is why you may not hear many talks dedicated to these topics in research conferences – but the notions are there and will appear over and over again, as something anyone should know. This is why it is important to review those basic notions in a school like this one.

In this chapter section, I shall deal with quantum cloning, teleportation and entanglement distillation (this section), state discrimination, quantum coding (section 3.3) and the violation of Bell's inequalities in greater detail (Section 3.4–3.6). Error correction is presented in this school by other lecturers.

3.2.2   Quantum cloning

The first primitive that should be considered is *quantum cloning*. The famous no-go theorem was formulated in 1982–83 (Wootters and Zurek, 1982; Dieks, 1982; Milonni and Hardies, 1982; Mandel, 1983); much later, in 1996, came the idea of studying optimal cloning (Bužek and Hillery, 1996). Since then, the subject has been the object of rather thorough investigations; two very comprehensive review articles are available (Scarani *et al.*, 2005; Cerf and Fiurášek, 2006).

3.2.2.1   The no-go theorem

It is well known that one cannot measure the state $|\psi\rangle$ of a single quantum system: the result of any single measurement of an observable A is one of its eigenstates, unrelated to the input state $|\psi\rangle$. To reconstruct $|\psi\rangle$ (or more generally ρ) one has to measure the average values of several observables; this implies a statistic over a large number of identically prepared systems (see Section 3.3).

One can imagine to circumvent the problem in the following way: take the system in the unknown state $|\psi\rangle$ and let it interact with N other systems previously prepared in a blank reference state $|R\rangle$, in order to obtain $N + 1$ copies of the initial state:

$$|\psi\rangle \otimes |R\rangle \otimes |R\rangle ... \otimes |R\rangle \xrightarrow{?} |\psi\rangle \otimes |\psi\rangle \otimes |\psi\rangle ... \otimes |\psi\rangle. \tag{3.31}$$

Such a procedure would allow one to determine the quantum state of a single system, without even measuring it because one could measure the N new copies and leave the original untouched. The no-cloning theorem of quantum information formalizes the suspicion that such a procedure is impossible:

No-cloning theorem: no quantum operation exists that can duplicate perfectly an unknown quantum state.

The theorem can be proved by considering the $1 \to 2$ cloning. Suppose first that perfect cloning is possible without any ancilla: this means that there exists a unitary operation such that

$$|\text{in}(\psi)\rangle \equiv |\psi\rangle \otimes |R\rangle \xrightarrow{?} |\psi\rangle \otimes |\psi\rangle \equiv |\text{out}(\psi)\rangle. \tag{3.32}$$

But such an operation cannot be unitary, because it does not preserve the scalar product:

$$\langle \text{in}(\psi)|\text{in}(\varphi)\rangle = \langle \psi|\varphi\rangle \neq \langle \text{out}(\psi)|\text{out}(\varphi)\rangle = \langle \psi|\varphi\rangle^2.$$

Now we have to prove that perfect cloning is impossible also for CP maps, the most general evolution. So let's add an ancilla (the "machine") and suppose that

$$|\psi\rangle \otimes |R\rangle \otimes |M\rangle \xrightarrow{?} |\psi\rangle \otimes |\psi\rangle \otimes |M(\psi)\rangle \tag{3.33}$$

is unitary. The same type of proof as before can be done as in Sections 9–4 of Peres' book (Peres, 1995); here we give a different one, closer to the Wootters–Zurek proof (Wootters and Zurek, 1982). We suppose that Eq. (3.33) holds for two orthogonal states, labelled $|0\rangle$ and $|1\rangle$:

$$|0\rangle \otimes |R\rangle \otimes |M\rangle \longrightarrow |0\rangle \otimes |0\rangle \otimes |M(0)\rangle$$
$$|1\rangle \otimes |R\rangle \otimes |M\rangle \longrightarrow |1\rangle \otimes |1\rangle \otimes |M(1)\rangle.$$

Because of linearity (we omit tensor products) then:

$$\big(|0\rangle + |1\rangle\big)|R\rangle\,|M\rangle \longrightarrow |00\rangle\,|M(0)\rangle + |11\rangle\,|M(1)\rangle$$

that cannot be equal to $\big(|0\rangle + |1\rangle\big)\big(|0\rangle + |1\rangle\big)|M(0+1)\rangle = \big(|00\rangle + |10\rangle + |01\rangle + |11\rangle\big)|M(0+1)\rangle$. So Eq. (3.33) may hold for states of a basis, but cannot hold for all states. Since a unitary evolution with an ancilla is the most general evolution allowed for quantum systems, the proof of the theorem is concluded.

3.2.2.2   *How no-cloning was actually discovered*

Sometimes one learns more from mistakes than from perfect thought. This is, in my opinion, the case with the paper that triggered the discovery of no-cloning (Herbert, 1982). The author, Herbert, reasoned as follows. Consider two particles in the singlet state: one particle goes to Alice, the other to Bob. If Alice measures σ_z, she prepares effectively on Bob's side either $|+z\rangle$ or $|-z\rangle$, with equal probability: therefore Bob's local state is $\rho_z = \frac{1}{2}|+z\rangle\langle+z| + \frac{1}{2}|-z\rangle\langle-z| = \frac{1}{2}\mathbb{1}$. If Alice measures σ_x, she prepares effectively on Bob's side either $|+x\rangle$ or $|-x\rangle$, with equal probability: Bob's local state is now $\rho_x = \frac{1}{2}|+x\rangle\langle+x| + \frac{1}{2}|-x\rangle\langle-x| = \frac{1}{2}\mathbb{1}$, equal to ρ_z, as it should because of no-signaling.

But suppose now that Bob can make a perfect copy of his qubit. Now, if Alice measures σ_z, Bob ends up with either $|+z\rangle|+z\rangle$ or $|-z\rangle|-z\rangle$, with equal probability: therefore Bob's local state is $\rho_z = \frac{1}{2}|+z+z\rangle\langle+z+z| + \frac{1}{2}|-z-z\rangle\langle-z-z|$. A similar reasoning leads to the conclusion that, if Alice measures σ_x, Bob's local state is $\rho_x = \frac{1}{2}|+x+x\rangle\langle+x+x| + \frac{1}{2}|-x-x\rangle\langle-x-x|$. But now, $\rho_x \neq \rho_z$! Herbert suggested that he had discovered a method to send signals faster than light – had he been a bit more careful, he would have discovered the no-cloning theorem.

Astonishingly enough, Herbert's paper was published. Both referees are known: the late Asher Peres explained he knew the paper was wrong, but guessed it was going to trigger interesting developments (Peres, 2002); GianCarlo Ghirardi pointed out the mistake in his report, which may therefore be the first "proof" of the no-cloning theorem. Later, Gisin reconsidered Herbert's scheme and studied how the two-copy state must be modified in order for ρ_z and ρ_x to be equal after imperfect duplication; he obtained an upper bound for the fidelity of the copies, that can actually be reached (Gisin, 1998).

A final point is worth noting. The whole reasoning of Herbert implicitly assumes that something does change instantaneously on a particle upon measuring an entangled particle at a distance. This view is not shared by most physicists. The whole debate might have gone astray on discussions about "collapse", just as the vague reply of Bohr to the Einstein–Podolski–Rosen paper prevented people from defining local variables in a precise way. It is fortunate that people seemed to have learned the lesson, and the problem was immediately cast in an *operational* way. This is the "spirit" that later led to the rise of quantum information science.

3.2.2.3   *The notion of quantum cloning machines*

Perfect cloning of an unknown quantum state is impossible; conversely, cloning of orthogonal states belonging to a known basis is trivially possible: simply measure in the basis, and produce as many copies as you like of the state you obtained. What is also possible, is to *swap* the state from one system to the other: $|\psi\rangle|R\rangle \to |R\rangle|\psi\rangle$ is unitary; one has then created a perfect image of the input state on the second system, at the price of destroying the initial one.

The notion of *quantum cloning machines* (QCM) is a wide notion encompassing all possible intermediate cases. One needs a figure of merit. Here, we focus on the *single-copy fidelity*, called "fidelity" for short. For each copied system j, this is defined as $F_j = \langle \psi | \rho_j | \psi \rangle$ for the initial state $|\psi\rangle$.

Here are some intuitive statements:

- If *perfect* cloning of an unknown quantum state is impossible, *imperfect* cloning should be possible. In particular, there should be an operation that allows us to copy equally well, with a fidelity $F < 1$, any unknown state. Any such operation will imply a "degradation" of the state of the original system.

- The fidelity of the "original" and of the "copy" after the cloning need not be the same; the better the copy, the more the original is perturbed.

- One can also consider cloning $N \to M = N + k$; if $N \to \infty$, the fidelity of the final M copies can be arbitrarily close to 1.

- Also, one may be willing to copy only a subset of all states.

Along with the variety of possible approaches, some terminology has been created:

Universal QCM: copies equally well all the states; non-universal QCM, called state dependent, have been studied only in some cases, mostly related to the attacks of the spy on some cryptography protocols.

Symmetric QCM: the original(s) and the copy(ies) have the same fidelity.

Optimal QCM: for a given fidelity of the original(s) after interaction, the fidelities of the copy(ies) is the maximal one.

3.2.2.4   Case study: universal symmetric QCM 1 → 2 for qubits

We study in detail the first example of a quantum cloning machine, the universal symmetric QCM $1 \to 2$ for qubits found by Bužek and Hillery (Bužek and Hillery, 1996). It is, however, instructive to start by analyzing first some trivial cloning strategies.

Trivial cloning.   Consider the following strategy: *Let the prepared qubit fly unperturbed, and produce a new qubit in a randomly chosen state, say* $|0\rangle$. *Don't keep track of which qubit is which.*

Let us compute the single-copy fidelity. We detect one particle: the original one with probability $\frac{1}{2}$, the new one with the same probability. Thus, the average single-copy fidelity is

$$F_{triv} = \frac{1}{2} \times 1 + \frac{1}{2} \times \left(\frac{1}{4\pi} \int_0^{2\pi} d\varphi \int_{-1}^1 d(\cos\theta) \langle \psi | P_0 | \psi \rangle \right)$$

$$= \frac{1}{2} + \frac{1}{2} \left(\frac{1}{2} \int_{-1}^1 d(\cos\theta) \frac{1 + \cos\theta}{2} \right) = \frac{3}{4}. \tag{3.34}$$

It is interesting to see that a fidelity of 75% can be reached by such an uninteresting strategy. In particular, this implies that one must show $F > \frac{3}{4}$ in order to demonstrate non-trivial cloning.

Note that one can consider another trivial cloning strategy, namely: *measure the state in an arbitrary basis and produce two copies of the outcome.* A similar calculation to the one above shows that this strategy leads to $F = \frac{2}{3}$ for the average single-copy fidelity and is therefore worse than the previous one.

The Bužek–Hillery (B–H) QCM for qubits. The Bužek–Hillery (B–H) QCM is a universal symmetric QCM for $1 \rightarrow 2$ qubits, that was soon afterwards proved to be the optimal one. We give no derivation, but rather start from the definition and verify all the properties *a posteriori*.

The B–H cloner uses *three qubits*: the original (A), the copy (B) and an ancilla (C). For convention, B and C are initially set in the state $|0\rangle$. Here is the action in the computational basis of A:

$$
\begin{aligned}
|0\rangle|0\rangle|0\rangle &\rightarrow \sqrt{\tfrac{2}{3}}\,|0\rangle|0\rangle|0\rangle + \sqrt{\tfrac{1}{6}}\,\big[|0\rangle|1\rangle + |1\rangle|0\rangle\big]|1\rangle \\
|1\rangle|0\rangle|0\rangle &\rightarrow \sqrt{\tfrac{2}{3}}\,|1\rangle|1\rangle|1\rangle + \sqrt{\tfrac{1}{6}}\,\big[|1\rangle|0\rangle + |0\rangle|1\rangle\big]|0\rangle
\end{aligned}
\tag{3.35}
$$

These two relations induce the following action on the most general input state $|\psi\rangle = \alpha|0\rangle + \beta|1\rangle$:

$$
|\psi\rangle|0\rangle|0\rangle \rightarrow \sqrt{\tfrac{2}{3}}\,|\psi\rangle|\psi\rangle|\psi^*\rangle + \sqrt{\tfrac{1}{6}}\,\big[|\psi\rangle|\psi^\perp\rangle + |\psi^\perp\rangle|\psi\rangle\big]|\psi^{*\perp}\rangle.
\tag{3.36}
$$

We have written $|\psi^\perp\rangle = \beta^*|0\rangle - \alpha^*|1\rangle$, $|\psi^*\rangle = \alpha^*|0\rangle + \beta^*|1\rangle$; combining the two definitions one finds that $|\psi^{*\perp}\rangle = |\psi^{\perp *}\rangle$. Equation (3.36) is the starting point for the subsequent analysis.

The verification that Eq. (3.36) follows from Eq. (3.35) is made as follows: from Eq. (3.35), because of linearity,

$$
|\psi\rangle|0\rangle|0\rangle \rightarrow \alpha\sqrt{\tfrac{2}{3}}|000\rangle + \alpha\sqrt{\tfrac{1}{6}}[|011\rangle + |101\rangle] + \beta\sqrt{\tfrac{2}{3}}|111\rangle + \beta\sqrt{\tfrac{1}{6}}[|100\rangle + |010\rangle].
$$

Then one writes explicitly the r.h.s. of Eq. (3.36) and finds the same state.

B–H: State of A and B. From Eq. (3.36), one sees immediately that A and B can be exchanged, and in addition, that the transformation has the same coefficients for all input state $|\psi\rangle$. Thus, the B–H QCM is symmetric and universal. Explicitly, the partial states are

$$
\rho_A = \rho_B = \frac{2}{3}|\psi\rangle\langle\psi| + \frac{1}{3}\frac{\mathbb{1}}{2} = \frac{5}{6}|\psi\rangle\langle\psi| + \frac{1}{6}|\psi^\perp\rangle\langle\psi^\perp| = \frac{1}{2}\left(\mathbb{1} + \frac{2}{3}\hat{m}\cdot\vec{\sigma}\right).
\tag{3.37}
$$

From the standpoint of A then, the B–H cloner "shrinks" the Bloch vector by a factor $\frac{2}{3}$ without changing its direction.

For both the original and the copy, the B–H cloner gives the fidelity

$$
F_A = F_B = \langle\psi|\rho_A|\psi\rangle = \frac{5}{6}.
\tag{3.38}
$$

This is the optimal fidelity for a symmetric universal $1 \longrightarrow 2$ cloner of qubits, a statement that is not evident and was proved in later papers (Gisin, 1998; Bruss *et al.*, 1998; Gisin and Massar, 1997).

B–H: State of C. Although it is a departure from the main theme, I find it interesting to spend some words about the state of the ancilla C after cloning. No condition has been imposed on this, but it turns out to have a quite interesting meaning. We have

$$\rho_C = \frac{2}{3}|\psi^*\rangle\langle\psi^*| + \frac{1}{3}|\psi^{*\perp}\rangle\langle\psi^{*\perp}| = \frac{1}{2}\left(\mathbb{1} + \frac{1}{3}\hat{m}_* \cdot \vec{\sigma}\right), \tag{3.39}$$

with $\hat{m}_* = (m_x, -m_y, m_z)$. This state is related to another operation that, like cloning, is impossible to achieve perfectly, namely the NOT operation that transforms $|\psi\rangle = \alpha|0\rangle + \beta|1\rangle$ into $|\psi^\perp\rangle = \beta^*|0\rangle - \alpha^*|1\rangle$. Because of the need for complex conjugation of the coefficients, the NOT transformation is anti-unitary and cannot be performed.[7] Just as for the cloning theorem, one can choose to achieve the NOT on some states while leaving other states unchanged; or one can find the operation that approximates at best the NOT on all states, called the universal NOT (Bužek *et al.*, 1999). This operation needs some ancilla, and reads $|\psi\rangle \rightarrow \rho_{NOT} = \frac{2}{3}|\psi^\perp\rangle\langle\psi^\perp| + \frac{1}{3}|\psi\rangle\langle\psi|$. Now, it is easy to verify that $\rho_{NOT} = \sigma_y \rho_C \sigma_y$ (just see the definition of $\hat{m}_*$). Thus, the ancilla qubit of the B–H QCM carries (up to a rotation of π around the y axis of the Bloch sphere) the universal NOT of the input state.

B–H as the coherent version of trivial cloning. There is an intriguing link between the B–H QCM and the trivial cloning presented above. Recall that in the trivial cloning one has to "forget" which qubit is which in order to pick up one of the two qubits at random. In other words, the process involves summing over the classical permutations. One might ask what happens if this classical permutation is replaced by the *quantum (coherent) permutation*. This operation is defined as the following CP-map:

$$T[\rho] = \frac{2}{3} S_2 \left(\rho \otimes \mathbb{1}\right) S_2, \tag{3.40}$$

where S_2 is the projector on the symmetric space of two qubits, i.e. the 3-dimensional subspace spanned by $\{|00\rangle, |11\rangle, |\Psi^+\rangle\}$; the factor $\frac{2}{3}$ guarantees that the map is trace-preserving. Remarkably, for $\rho = |\psi\rangle\langle\psi|$, $T[\rho] = \rho_{AB}(\psi)$ as obtained by applying the B–H QCM to $|\psi\rangle$. The map T is not unitary, which justifies the need for the ancilla. This elegant construction was noted by Werner, who used it to find universal symmetric $N \rightarrow M$ cloning (Werner, 1998).

3.2.3 Teleportation

The second primitive we are considering is *teleportation*. Discovered over a black-board discussion, the teleportation protocol (Bennett *et al.*, 1993) is a fascinating physical phenomenon (what is more, it has a catchy name). Of course, contrary to quantum cloning, teleportation in itself admits few and rather obvious generalizations; in other words, "teleportation" is not and has never been a subfield of quantum information.

However, it is a seed for many other ideas and plays an important role in entanglement theory.

3.2.3.1 The protocol

As usual in this chapter, we present the protocol with qubits. Consider three qubits: qubit A is prepared in an arbitrary state $|\psi\rangle$; qubits B and C are prepared in a maximally entangled state, say $|\Phi^+\rangle$.

By writing $|\psi\rangle = \alpha|0\rangle + \beta|1\rangle$ and expanding the terms, one can readily verify

$$
|\psi\rangle_A|\Phi^+\rangle_{BC} = \frac{1}{2}\Big[|\Phi^+\rangle_{AB}|\psi\rangle_C + |\Phi^-\rangle_{AB}\left(\sigma_z|\psi\rangle\right)_C
$$
$$
+ |\Psi^+\rangle_{AB}\left(\sigma_x|\psi\rangle\right)_C + |\Psi^-\rangle_{AB}\left(\tilde{\sigma}_y|\psi\rangle\right)_C\Big], \tag{3.41}
$$

where $\tilde{\sigma}_y = -i\sigma_y$. This identity is the basis of the *teleportation protocol*:

1. Prepare the three qubits as described above; bring qubits A and B together.
2. Perform a *Bell-state measurement* on qubits A and B, and send the result of the measurement (2 bits) to the location of qubit C.
3. Upon reception of this information, apply the suitable unitary operation to C in order to recover $|\psi\rangle$.

That is all for the protocol. Still, some remarks are worth making:

- It is customary to emphasize that *information, not matter, is teleported:* qubit C had to exist in order to receive the state of qubit A. That being clarified, the name "teleportation" is well chosen: the information has been transferred from A to C without ever being available in the region when particle B has propagated.

- Another point that is usually stressed is the fact that, of course, this task respects *no-signalling*: indeed, the teleportation can be achieved only when the two bits of classical communication are sent, and these cannot travel faster than light. The phenomenon is nonetheless remarkable, because the two classical bits are definitely not sufficient to carry information about a state of a qubit (a vector in the Poincare sphere is defined by three continuous parameters).

- Even though some of the particles may be "propagating", the qubit degree of freedom that is going to be teleported does not evolve in the protocol (there is no hamiltonian anywhere). Teleportation is due to the *purely kinematical* identity (3.41).

- Since the protocol can teleport every pure state with perfect fidelity, it can *teleport any mixed state* as well.

3.2.3.2 *Entanglement swapping*

If the qubit to be teleported is itself part of an entangled pair, the identity (3.41) applied to B–CD leads to

$$|\Phi^+\rangle_{AB}|\Phi^+\rangle_{CD} = \frac{1}{2}\Big[|\Phi^+\rangle_{AD}|\Phi^+\rangle_{BC} + |\Phi^-\rangle_{AD}|\Phi^-\rangle_{BC}$$

$$+ |\Psi^+\rangle_{AD}|\Psi^+\rangle_{BC} + |\Psi^-\rangle_{AD}|\Psi^-\rangle_{BC}\Big] \qquad (3.42)$$

because $\mathbb{1} \otimes \sigma_z|\Phi^+\rangle = |\Phi^-\rangle$, $\mathbb{1} \otimes \sigma_x|\Phi^+\rangle = |\Psi^+\rangle$ and $\mathbb{1} \otimes \tilde{\sigma}_y|\Phi^+\rangle = |\Psi^-\rangle$. Therefore, by performing the Bell-state measurement on B and C and sending the result to D, one can prepare the particles A and D in the state $|\Phi^+\rangle$, even if they have never interacted. This protocol is called *entanglement swapping* (Yurke and Stoler, 1992; Zukowski *et al.*, 1993).

Entanglement swapping, in itself, is simply a special case of teleportation, in which the particle to be teleported is itself part of an entangled pair. However, it shows that *direct interaction is not needed to create entanglement.*

3.2.3.3 *Teleportation and entanglement swapping as primitives for other tasks*

Two-qubit maximally entangled states as universal resources. The possibility of teleportation has a fundamental consequence in entanglement theory, namely that bipartite maximally entangled states are universal resources to distribute entanglement. Indeed, suppose that N partners want to share a fully N-partite entangled state. It is enough for *one* of the partners, Paul, to act as a provider. Indeed, if Paul shares a bipartite maximally entangled state with each of the others, he can prepare locally the N-partite state, then teleport the state of each particle to the suitable person.

Now, maximally entangled states of any dimension can be created from enough copies of two-qubit maximally entangled states (at least on paper), as the following example makes clear:

$$|\Phi^+\rangle_{AB}|\Phi^+\rangle_{A'B'} = \frac{1}{2}\Big[|00\rangle_{AA'}|00\rangle_{BB'} + |01\rangle_{AA'}|01\rangle_{BB'}$$

$$+ |10\rangle_{AA'}|10\rangle_{BB'} + |11\rangle_{AA'}|11\rangle_{BB'}\Big]$$

$$= \frac{1}{2}\Big[|0\rangle_A|0\rangle_B + |1\rangle_A|1\rangle_B + |2\rangle_A|2\rangle_B + |3\rangle_A|3\rangle_B\Big]. \qquad (3.43)$$

In conclusion, as soon as the partners can share pairwise maximally entangled states of two-qubits with one provider, they can share any entangled state of arbitrary many parties and dimensions. Let me stress again that this is a theoretical statement about the behavior of entanglement as a resource, not necessarily a practical or even feasible scheme to realize an experiment.

Quantum repeaters. At the other end of the spectrum, the idea of quantum repeaters is triggered by a very practical problem. In quantum communication, one normally uses photons because they propagate well and interact little. In this case, decoherence is by far not the most problematic issue: *losses* are. In other words, quantum communication schemes reach their limits when the photons just don't arrive often enough for the signal to overcome the noise of the detectors and other local apparatuses. Since you cannot amplify your signal because of the no-cloning theorem, losses seem unbeatable. Quantum repeaters are a clever solution to the problem (Briegel *et al.*, 1998; Duan *et al.*, 2001). For exhaustive information, a recent review article on the topic is available (Sangouard *et al.*, 2009).

Suppose you want to share an entangled pair between locations A and B, at a distance ℓ. The transmission typically scales exponentially $t = 10^{-\alpha \ell}$. What happens if A and B can both send one photon to an intermediate location C, where someone performs the Bell-state measurement? This does not seem to help, because the transmission from A to C is $\sqrt{t}$, and so is the transmission from B to C, so the probability that both photons arrive in C is still t. However, if C has *quantum memories*, the picture changes significantly: now, C can establish a link with A and *independently* a link with B.

This effect is best understood in terms of the time needed to establish an entangled pair between A and B. If photons have to travel from A to B, this time is (in suitable units) $\tau_{AB}^{(1)} = \frac{1}{t}$. In the same units, $\tau_{AC}^{(1)} = \tau_{BC}^{(1)} = \frac{1}{\sqrt{t}}$. If the two links can be established independently, C needs on average $\frac{1}{2} \frac{1}{\sqrt{t}}$ to establish the first of the links, because the first can be either; conditioned now on the fact that one link was established, C has to wait in average $\frac{1}{\sqrt{t}}$ to establish the second link. After this, entanglement swapping can be performed, thus establishing the entangled pair between A and B in a time $\tau_{AB}^{(1)} \approx \frac{3}{2} \frac{1}{\sqrt{t}}$. For more details on this calculation, see Appendix B in (Scarani *et al.*, 2009).

3.2.4 Entanglement distillation

This section is scandalously short. The reason is that I thought other lecturers would introduce the notion of distillation, but it turned out they had planned their lectures differently and there was no time for it! When we realized our lack of coordination, we decided that I would at least mention rapidly the idea, for the sake of completeness. So here it is. All the meaningful notions, extensions and references can be found for instance in the comprehensive review paper devoted to entanglement theory written by the Horodecki family (Horodecki *et al.*, 2009).

3.2.4.1 The notion of distillation

We have seen above that two-qubit maximally entangled states are a universal resource for distributing quantum states. Suppose now two parties Alice and Bob (it can of course be generalized) share many copies of some other, less entangled state ρ_{AB}: can they somehow "concentrate" or "distill" the entanglement they have, in order to end up with fewer copies of those very useful maximally entangled state? Of course, for the task to make sense, the distillation must be performed only using operations that themselves do not increase entanglement on a single-copy level: these are *local operations and classical communication (LOCC)*. In other words, Alice and Bob are in different locations and they cannot send quantum systems to each other (otherwise trivially Alice would prepare a maximally entangled state and give or send one particle to Bob).

So, is it possible to achieve

$$\rho_{AB}{}^{\otimes N} \xrightarrow{LOCC} \Phi^{\otimes m} \otimes (\text{garbage}), \tag{3.44}$$

with $\Phi = |\Phi\rangle\langle\Phi|$ a maximally entangled state of two qubits? The answer is: often yes but sometimes no! For instance, entanglement is distillable for all entangled pure states; also for all entangled states of two qubits, or of a qubit and a qutrit. Some entangled mixed state of higher dimensions and/or of more parties, however, are such that their entanglement is not distillable! In other words, you need entanglement to create them, but this entanglement cannot be recovered. Such states are called *bound-entangled*.

At the moment of writing, the question of assessing whether or not an entangled state is distillable is still open in general. Those who want to have an idea of how many notions I am skipping here – partial transpose (a positive, but non-completely positive map), different measures of entanglement, etc. – may browse the review paper mentioned above. But, in the context of a school, I cannot resist spelling out one of the most fascinating examples of bound-entangled states.

3.2.4.2   *A nice example of bound-entangled state*

The example of bound-entangled state that we are going to consider (Bennett *et al.*, 1999) is a three-qubit state. Suppose that, for any reason, one starts out writing a basis of the Hilbert space as

$$|\varphi_1\rangle = |0\rangle|1\rangle|+\rangle, \ |\varphi_2\rangle = |1\rangle|+\rangle|0\rangle, \ |\varphi_3\rangle = |+\rangle|0\rangle|1\rangle, \ |\varphi_4\rangle = |-\rangle|-\rangle|-\rangle:$$

these states are obviously orthogonal, but four more states are needed to have a basis. Now, it turns out that the four remaining states cannot be product states: they must contain some entanglement! The four product states we started with form a so-called *unextendible product basis*.

The next ingredient is the fact that the remaining four states can all be taken to be entangled only between the first two qubits; i.e. one can find four orthogonal vectors $|\varphi_k(AB|C)\rangle = |\Psi_k\rangle_{AB}|\psi_k\rangle_C$ for $k = 5, 6, 7, 8$ (I leave the interested reader to find out the explicit expressions.) But of course, since the initial four states are symmetric under permutations, one might just as well choose the remaining $|\varphi_k\rangle$ to be of the form $|\varphi_k(CA|B)\rangle = |\Psi_k\rangle_{CA}|\psi_k\rangle_B$ or $|\varphi_k(BC|A)\rangle = |\Psi_k\rangle_{BC}|\psi_k\rangle_A$.

Consider now the state

$$\rho = \frac{1}{4}\left(\mathbb{1} - \sum_{k=1}^{4} |\varphi_k\rangle\langle\varphi_k|\right). \tag{3.45}$$

This is the maximally mixed state defined on the subspace that is complementary to the unextendible product basis. As such, it is obviously entangled, because there cannot be any product state in its support. But *where* is the entanglement? Notice that

$$\rho = \frac{1}{4} \sum_{k=5}^{8} |\varphi_k(AB|C)\rangle\langle\varphi_k(AB|C)|$$

$$= \frac{1}{4} \sum_{k=5}^{8} |\varphi_k(CA|B)\rangle\langle\varphi_k(CA|B)|$$

$$= \frac{1}{4} \sum_{k=5}^{8} |\varphi_k(BC|A)\rangle\langle\varphi_k(BC|A)| :$$

therefore, according to *each* possible bipartition, the state is separable. It is therefore impossible that separate parties can distill entanglement: ρ is bound entangled.

Is bound entanglement "useful"? For a long time, the answer was supposed to be negative. In 2005, a theoretical breakthrough showed that some bound-entangled states contain secrecy, i.e. can be used for cryptography. In order to explore this topic, I suggest to read first (Scarani *et al.*, 2009, II.B.2), then the original references given there. However, no explicit protocol to distribute such states has ever been devised, nor will probably ever be: other protocols are much simpler and efficient for the task.

3.2.5   Tutorials

3.2.5.1   *Problems*

Exercise 2.1

Amplification of light is of course compatible with the no-cloning theorem, because spontaneous emission prevents amplification to be perfect (Mandel, 1983). Actually, if the amplifier is independent of the polarization, universal symmetric cloning of that degree of freedom is implemented (Simon *et al.*, 2000; Kempe *et al.*, 2000). In this problem, we explore the basics of this correspondence.

Consider a single spatial mode of the electromagnetic field and focus on the polarization states; we denote by $|n, m\rangle$ the state in which n photons are polarized H and m photons are polarized V. Suppose one photon in mode H is initially present in the amplifier, and that after amplification 2 photons have been produced.

1. Compute the single-copy fidelity of this cloning process. *Hint:* if you don't remember the physics of amplification, you can reach the result by comparing $a_H^\dagger|1,0\rangle$ with $a_V^\dagger|1,0\rangle$.
2. How would you describe the state of the system (field + amplifier medium) in this process? *Hint:* compare with the B–H QCM.

3.2.5.2   *Solutions*

Exercise 2.1

1. The theory of spontaneous and stimulated emission implies that, starting with $|1,0\rangle$, the probability of creating $|2,0\rangle$ is twice as large as the probability of creating $|1,1\rangle$. The single-copy fidelity is defined as the probability of finding one

of the photons in the initial state, whence obviously $F = \frac{2}{3} \times 1 + \frac{1}{3} \times \frac{1}{2} = \frac{5}{6}$. This is identical to the fidelity for optimal universal symmetric cloning.

2. The analogy with cloning is actually exact: indeed, by conservation of angular momentum, the emission of an H photon and of a V photon cannot be due to the same process. Therefore, after amplification and post-selection of the emission of two photons, the state of the system "field + amplifying medium" reads $\sqrt{\frac{2}{3}}|2,0\rangle \otimes |e_H\rangle + \sqrt{\frac{1}{3}}|1,1\rangle \otimes |e_V\rangle$, i.e. in first-quantized notation

$$\sqrt{\frac{2}{3}}|H\rangle|H\rangle \otimes |e_H\rangle + \sqrt{\frac{1}{3}}|\Psi^+\rangle \otimes |e_V\rangle$$

and this exactly the state produced by the B–H QCM.

3.3   Primitives of quantum information (II)

3.3.1   State discrimination

Under the heading of state discrimination, a large variety of tasks can be accommodated. For a school, rather than reviewing each of them exhaustively, I find it more useful to present concrete examples of each. A review article was written by Chelfes in 2000 (Chefles, 2000): it contains most of the basic ideas; some recent developments will be mentioned below.

3.3.1.1   *Overview*

As the name indicates, state discrimination refers to obtaining information about the quantum state produced by a source that is not fully characterized. We can broadly divide the possible tasks into two categories:

- *Single-shot tasks:* the goal is to obtain information on each signal emitted by the source. Without further information, as is well known, the task is almost hopeless: basically, after measuring one system, the only thing one can be sure of is that the state was not orthogonal to the one that has been detected. However, the task becomes much more appealing if some additional knowledge is present: for instance, if one is guaranteed that each system can be either in state ρ_1 or in state ρ_2, with the ρ_j two well-specified states. In such situations, one can consider *probabilistic state discrimination* and try to minimize the probability of a wrong guess; or even *unambiguous state discrimination*, a POVM that either identifies the state perfectly or informs that the discrimination was inconclusive.

- *Multi-copy tasks:* if the source is guaranteed to always produce the same state, the state can be exactly reconstructed asymptotically; this process is called *state reconstruction*, or *state estimation*, or *state tomography*. If, in addition, one knows that the state is either ρ_1 or ρ_2, one can ask how fast the probability of a wrong guess decreases with the number of copies and obtain a quantum version of the *Chernoff bound*. If the source is not guaranteed to always produce the same state, the task seems hopeless, and in full generality it is; but if the observed statistics are symmetric under permutation, a quantum version of the *de Finetti theorem* exists.

3.3.1.2   *Single-shot tasks (I): Probabilistic discrimination*

Probabilistic discrimination of two states.   Let us consider the simplest case: two states ρ_1 and ρ_2 are given each with probability η_1 and $\eta_2 = 1 - \eta_1$. At each run, one performs a measurement, whose outcome is used to guess which state was given; we want to *minimize the probability P_{error} that the guess is wrong*.

Since ultimately we want two outcomes, without loss of generality the measurement can be described by two projectors Π_1 and $\Pi_2 = \mathbb{1} - \Pi_1$, where $\mathbb{1}$ is the identity over the subspace spanned by ρ_1 and ρ_2. Therefore, the probability of guessing ρ_j correctly, i.e. the probability of guessing j given ρ_j, is given by $\text{Tr}(\Pi_j \rho_j)$; whence the average probability of error for this measurement is

$$P_{\text{error}}(\Pi_1) = 1 - \sum_{j=1,2} \eta_j \text{Tr}(\Pi_j \rho_j) = \eta_1 - \text{Tr}\big(\Pi_1 (\eta_1 \rho_1 - \eta_2 \rho_2)\big). \qquad (3.46)$$

In order to minimize this, we have to find the projector Π_1 that maximizes the second term on the right-hand side. The result is (Helstrom, 1976; Herzog and Bergou, 2004)

$$P_{\text{error}} = \frac{1}{2}\big[1 - \text{Tr}\,|\eta_1 \rho_1 - \eta_2 \rho_2|\big]. \qquad (3.47)$$

A constructive measurement strategy that would lead to this optimal result is the following: measure the Hermitian operator $M = \eta_1 \rho_1 - \eta_2 \rho_2$; if the outcome is a positive eigenvalue, guess ρ_1, if it's a negative eigenvalue, guess ρ_2.

Let us prove Eq. (3.47) for the special case of *equal a priori probabilities* $\eta_1 = \eta_2 = \frac{1}{2}$. In this case, Π_1 is the projector on the subspace of positive eigenvalues of $\rho_1 - \rho_2$; but since $\text{Tr}(\rho_1 - \rho_2) = 0$, the sum of the positive eigenvalues and of the negative ones must be the same in absolute value. Therefore, we find $\max_{\Pi_1} \text{Tr}\big(\Pi_1(\rho_1 - \rho_2)\big) = \frac{1}{2}\text{Tr}\,|\rho_1 - \rho_2|$ and finally

$$P_{\text{error}} = \frac{1}{2}\left[1 - \frac{1}{2}\text{Tr}\,|\rho_1 - \rho_2|\right] \quad (\eta_1 = \eta_2 = \tfrac{1}{2}). \qquad (3.48)$$

Intermezzo: trace distance.   The mathematical object

$$D(\rho, \sigma) = \frac{1}{2}\,\text{Tr}|\rho - \sigma| \qquad (3.49)$$

that appeared in the previous proof is called the *trace-distance* between two states ρ and σ. This quantity appears often in quantum information, so it is worth while spending some time on it (for all proofs and more information, refer to Chapter 9 of (Nielsen and Chuang, 2000). Some simple properties of the trace distance are: $D(\rho, \sigma) = 0$ if and only if $\rho = \sigma$; the maximal value $D(\rho, \sigma) = 1$ is reached for orthogonal states; $D(\rho, \sigma) = D(\sigma, \rho)$. It can, moreover, be proved that the triangle inequality $D(\rho, \tau) \leq D(\rho, \sigma) + D(\sigma, \tau)$; therefore it has the mathematical properties of a "metric", i.e. it defines a valid distance between states.

As we have seen, two states can be distinguished with probability at most $\frac{1}{2}[1 + D]$. But $\frac{1}{2}$ is sheer random guessing: rewriting $\frac{1}{2}[1 + D] = D \times 1 + (1 - D) \times \frac{1}{2}$, we see that, *in any task*, the two states behave differently with probability at most D. Indeed, suppose there is a task for which the two behaviors are more distinguishable: we would use that task as measurement for discrimination, thus violating the bound (3.48).

The usefulness of this remark becomes even more apparent when phrased in a slightly different context. Instead of having to discriminate two states, suppose one has a state ρ and wants to compare it to an "ideal" state ρ_{ideal}. Then $D(\rho, \rho_{\text{ideal}})$ is the *maximal probability of failure*, i.e. the maximal probability that the real state will produce a result different than the one the ideal state would have produced. This plays a central role, for instance, in the definition of security in quantum cryptography; see II.C.2 in (Scarani *et al.*, 2009) for a discussion and original references. We shall find this idea in a different context in Section 3.6.2.2.

PSD of more than two states.   In the most general case, as expected, the optimal PSD strategy is not known. Suboptimal strategies are trivially found: just invent a measurement strategy and compute the probabilities of failure. A particular strategy performs often quite well, so much so that it has been called *pretty good measurement (PGM)* (Hausladen and Wootters, 1994). It is defined as follows: let $\{\rho_k, p_k\}$ be the set of states to be distinguished, with the corresponding *a priori* probabilities; and define $M = \sum_k p_k \rho_k$. Then the PGM is the POVM whose elements are defined by $E_k = p_k M^{-1/2} \rho_k M^{-1/2}$. The special case, in which all the states to be distinguished are pure ($\rho_k = |\psi_k\rangle\langle\psi_k|$) and equally probable, is known as *square-root measurement*; in this case, the elements of the POVM are the suitably normalized projectors on the states $|\chi_k\rangle \propto M^{-1/2}|\psi_k\rangle$ (Hausladen *et al.*, 1996). There is a significant amount of literature on these measurements, that in some cases define the optimal measurement strategy. Its review goes beyond our scope.

Let us finally mention that an *upper bound* on the guessing probabilities can be found by studying the dual problem[8] (Koenig *et al.*, 2008).

PSD and cloning.   There is an interesting, somehow intuitive, link between optimal cloning and PSD, namely: for any ensemble of *pure states* $\{(|\psi_k\rangle, \eta_k)\}_{k=1...n}$ to be distinguished, optimal PSD is equivalent to optimal symmetric $1 \to N$ cloning in the limit $N \to \infty$ (Bae and Acín, 2006). Note that this fact does not help to find the explicit strategy, since optimal state-dependent cloners are not known in general either.

The argument goes as follows. For convenience, define F_C as the single-copy fidelity of the optimal $1 \to \infty$ cloner and F_M as the fidelity of the state reconstructed after the optimal PSD measurement. It is obvious that $F_M \leq F_C$: after measurement, we have a guess for the state, so we can just create as many copies as we want of that state, so this defines a possible cloner. The proof of the converse is more tricky. Basically, one applies the cloner C to a half of a maximally entangled state: $(\mathbb{1} \otimes C)|\Phi^+\rangle = \rho_{AB_1...B_N}$. By assumption, all the ρ_{AB_j} are equal. In the limit $N \to \infty$ the information of B_j is "infinitely shareable" and a theorem then guarantees that ρ_{AB_j} is separable. But then, the restriction $\tilde{C}$ defined as $(\mathbb{1} \otimes \tilde{C})|\Phi^+\rangle = \rho_{AB_1}$ describes an entanglement-breaking channel, and it is known that any such channel is equivalent to performing a measurement and forwarding the collapsed state. In conclusion, there exists a measurement strategy that achieves the single-copy fidelity of the optimal cloner.

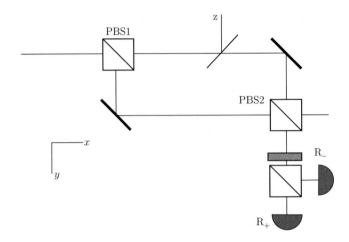

Fig. 3.1 Optical setup for unambiguous discrimination of non-orthogonal polarization states.

3.3.1.3 Single-shot tasks (II): Unambiguous state discrimination

We consider now a different situation: now we want discrimination to be unambiguous; the price to pay is that sometimes the procedure will output an inconclusive outcome.

Unambiguous discrimination of two pure states. The case where the two states have equal *a priori* probabilities was solved independently by Ivanovics, Dieks and Peres (Ivanovic, 1987; Dieks, 1988; Peres, 1988); Jaeger and Shimony later solved the problem for arbitrary probabilities (Jaeger and Shimony, 1995). See also (Peres, 1995), section 9-5. For two *pure* states, the discrimination succeeds at most with probability $1 - |\langle\psi_1|\psi_2\rangle|$; of course, with probability $|\langle\psi_1|\psi_2\rangle|$ one obtains the inconclusive outcome.

Let us begin by describing a simple setup that achieves USD of two pure states (Huttner *et al.*, 1996); an even simpler one is given as a tutorial. The setup is shown in Figure 3.1. The polarizing beamsplitters PBS1 and PBS2 are oriented such that $|H\rangle$ is transmitted and $|V\rangle$ is reflected. The goal is to distinguish between

$$|\psi_\pm\rangle = \cos\alpha|H\rangle \pm \sin\alpha|V\rangle \qquad (3.50)$$

whose overlap is $\chi = \cos 2\alpha$. Indeed, we have:

$$
\begin{aligned}
\big(\cos\alpha|H\rangle \pm \sin\alpha|V\rangle\big)|x\rangle \;\; &\overset{PBS1}{\to} \;\; \cos\alpha|H\rangle|x\rangle \pm \sin\alpha|V\rangle|y\rangle \\
&\overset{BS\,x-z}{\to} \;\; \cos\alpha\sqrt{t}\,|H\rangle|x\rangle \pm \sin\alpha|V\rangle|y\rangle + \cos\alpha\sqrt{1-t}\,|H\rangle|z\rangle \\
&\overset{mirrors}{\to} \;\; i\cos\alpha\sqrt{t}\,|H\rangle|y\rangle \pm i\sin\alpha|V\rangle|x\rangle + \cos\alpha\sqrt{1-t}\,|H\rangle|z\rangle \\
&\overset{PBS2}{\to} \;\; i\big(\cos\alpha\sqrt{t}\,|H\rangle \pm \sin\alpha|V\rangle\big)|y\rangle + \cos\alpha\sqrt{1-t}\,|H\rangle|z\rangle.
\end{aligned}
$$

If $\cos \alpha \sqrt{t} = \sin \alpha$, that is for transmission of the BS $t = \tan^2 \alpha$, then the two states that can appear in mode y are orthogonal and can be discriminated by a usual projective measurement (output modes $|R_\pm\rangle$ of a PBS). In summary, we have implemented the transformation

$$\left(\cos \alpha |H\rangle \pm \sin \alpha |V\rangle \right)|x\rangle \longrightarrow i\sqrt{2} \sin \alpha \,|\pm\rangle|R_\pm\rangle + \cos \alpha \sqrt{1 - \tan \alpha} \,|H\rangle|z\rangle, \quad (3.51)$$

where the probability of obtaining a conclusive result is the optimal one, since $2 \sin^2 \alpha = 1 - \cos 2\alpha = 1 - |\langle \psi_+ | \psi_- \rangle|$.

Taking a more formal view, this setup realizes a POVM: there are three outcomes for a qubit. The two conclusive outcomes are described by

$$A_+ = \eta |+\rangle\langle|\psi_-^\perp\rangle| \, , \ A_- = \eta |-\rangle\langle|\psi_+^\perp\rangle|, \quad (3.52)$$

with $|\psi_\pm^\perp\rangle = \sin \alpha |H\rangle \mp \cos \alpha |V\rangle$. The factor η is determined by the constraint that the largest eigenvalue of $A_+^\dagger A_+ + A_-^\dagger A_-$ should be 1; direct inspection leads to $\eta = \frac{1}{\sqrt{2}\cos\alpha}$. The probability of guessing correctly $|\psi_\omega\rangle$ is given by $\mathrm{Tr}\left(A_\omega |\psi_\omega\rangle\langle\psi_\omega| A_\omega^\dagger \right) = 2 \sin^2 \alpha = 1 - \cos 2\alpha$, as it should.

Generalizations: more pure states, mixed states. For any number *pure states*, unambiguous state discrimination is possible if and only if the states are linearly independent. This implies in particular that no set consisting of $N > d$ states, where d is the dimension of the Hilbert space, can be discriminated unambiguously.

Though not optimal in general, the following strategy can always be implemented. Consider N linearly independent states $\{|\psi_k\rangle\}_{k=1,...,N}$ generating the N-dimensional subspace $\mathcal{E}$. Let $|\varphi_k\rangle$ be the vector in $\mathcal{E}$ that is orthogonal to all the vectors except $|\psi_k\rangle$: of course, if one performs a measurement and finds the result $|\varphi_k\rangle$, one can unambiguously identify $|\psi_k\rangle$ as being the measured state. This looks very much like a usual projective measurement defined by the $P_k = |\varphi_k\rangle\langle\varphi_k|$; however, the $|\varphi_k\rangle$s are in general not orthogonal (if they are all orthogonal, they coincide with $|\psi_k\rangle$ that are therefore already orthogonal), so $\sum_k P_k$ does not need to be proportional to $\mathbb{1}$ and its largest eigenvalue may be $\lambda > 1$. A possible way of obtaining a valid POVM consists in defining $A_k = P_k/\sqrt{\lambda}$ and associate the inconclusive outcome to $A_0^\dagger A_0 = \mathbb{1} - \sum_k P_k/\lambda$.

When it comes to unambiguous discrimination of *mixed states*, the situation is far more complex. For instance, even two different mixed states cannot be discriminated unambiguously if their supports are identical (obviously, two pure states have identical support if and only if they are the same state). Some non-trivial examples have been worked out (Raynal and Lütkenhaus, 2005).

3.3.1.4   *Multi-copy tasks (I): tomography, Chernoff bound*

Tomography of an a priori completely unknown state. The idea of tomography is inherent to the notion of state itself. A state is a description of our statistical knowledge. In classical statistics, a probability distribution can be reconstructed by collecting many independent samples that follow the distribution. Similarly, if one performs suitable measurements on N copies of an unknown quantum state ρ, one can obtain a faithful description of the state itself.

Some examples will suffice to clarify the principle. Any one-qubit state can be written as $\rho = \frac{1}{2}\left(\mathbb{1} + \vec{m}\cdot\vec{\sigma}\right)$, where $m_k = \langle\sigma_k\rangle$. Therefore, if one can estimate the average values $\langle\sigma_x\rangle$, $\langle\sigma_y\rangle$ and $\langle\sigma_z\rangle$, one can reconstruct the state. Similarly, for two qubits, one has to estimate the fifteen average values $\langle\sigma_k \otimes \mathbb{1}\rangle$ (three), $\langle\mathbb{1} \otimes \sigma_k\rangle$ (three) and $\langle\sigma_j \otimes \sigma_k\rangle$ (nine). It's exactly like reconstructing a classical distribution, but for the fact that one has to perform several different samplings (measurements) because of the existence of incompatible physical quantities.

While the principle is simple and necessary for the notion of state to have any meaning, the topic is not closed. For instance, one can look for "optimal" tomography according to different figure of merit (smallest size of the POVM, faster convergence with the number of samples N, etc.).

Among the rigorous results that have been obtained only recently, let us mention the computation of the *quantum Chernoff bound*. The task is somehow half-way between state estimation and tomography: given the promise that the state is either ρ_1 or ρ_2, one wants to estimate *how fast* the probability of distinguishing increases for increasing N. The result is most easily formulated in terms of the probability of error, which has been proved to decrease exponentially: $\frac{1}{2}\left[1 - \frac{1}{2}\left|\eta_1\rho_1^{\otimes N} - \eta_2\rho_2^{\otimes N}\right|\right] \sim e^{-\xi N}$ with $\xi = -\log\left(\min_{0\leq s\leq 1} \mathrm{Tr}(\rho_0^s\rho_1^{1-s})\right)$ (Audenaert *et al.*, 2007).

3.3.1.5 *Multi-copy tasks (II): de Finetti theorem and extensions*

The tomography discussed above works only under the assumption that the source produces always the same state, i.e. that the state of N emitted particles is simply $\rho^{\otimes N}$. Operationally, this cannot be guaranteed: all that one can guarantee is that the experimental procedure is the same for each realization. This implies that the N-particle state ξ_N, whatever it is, is such that all possible statistics are invariant by permutation of the particles (for we have no information that allows us to distinguish some realizations from others).

Now, the statement we want to make is the following: *in the limit of large N, any state ξ_N that is invariant by permutation is "close" to a product state $\rho^{\otimes N}$ or to a classical mixture of such states.*[9] It is not superfluous to notice that we use this result, implicitly, in every laboratory experiment, be it about classical or quantum physics.

There are several proofs of this statement, differing in important details (what "to be close" exactly means) and consequences. Most of them are called "de Finetti theorems", from the name of the Italian mathematician who first proved a similar statement in the context of classical probability theory; they consist of an estimate of the trace distance between the real state and the set of product states. For a very clear presentation, including a review of previous works, we refer the reader to (Renner, 2007). The most recent extension, that obtains a much better bound under relaxed assumptions, is based on the idea of post-selection (Christandl *et al.*, 2009).

3.3.2 Quantum coding

By *quantum coding* I denote the generalization to quantum physics of the main results of classical information theory. A very thorough presentation of the basic material is already available in Chapters 11 and 12 of Nielsen and Chuang's book (Nielsen and Chuang, 2000): unless other references are given, this section is a distillation of that source, to which the reader should also refer to obtain the references to the original works. The studies of security of quantum cryptography are of course an offspring of this field, so several notions can be learned in that context.

3.3.2.1 *Shannon entropy and derived notions*

It should be well known that "entropy" is associated first with "uncertainty" in information theory. The elementary entropic quantity in quantum physics is the *von Neumann entropy*, which is simply the Shannon entropy of the eigenvalues $\{\lambda_k\}$ of a given state ρ:

$$S(\rho) = -\text{Tr}(\rho \log \rho) = -\sum_k \lambda_k \log \lambda_k. \tag{3.53}$$

Many of the properties of Shannon entropy translate directly to von Neumann entropy. For instance, it's a concave function: the entropy of a mixture is larger than the average of the entropies, formally

$$S\left(\sum_k p_k \rho_k\right) \geq \sum_k p_k S(\rho_k). \tag{3.54}$$

One can also define *relative entropy* $S(\rho||\sigma) = -S(\rho) - \text{Tr}(\rho \log \sigma)$, with similar properties as the classical analog. Of course, the quantum *joint entropy* is just defined as $S(A,B) = S(\rho_{AB})$. Again, it behaves like the classical analog; in particular, the property called "strong subadditivity" holds: $S(A,B,C) + S(B) \leq S(A,B) + S(B,C)$.

Not exactly everything is a copy of classical information theory, though: a most remarkable exception is the possibility for *conditional entropy* to be negative. Classical conditional entropy for a probability distribution $P(a,b)$ is the uncertainty on the distribution on A knowing B: it's the entropy of the conditional probabilities, averaged over all possible conditions, i.e. $H(A|B) = \sum_b P(b) \left[-\sum_a P(a|b) \log P(a|b) \right]$. This definition cannot be generalized as such in quantum physics, because there is no obvious analog of $P(a|b)$; however, it is simple to show that $H(A|B) = H(A,B) - H(B)$, and this expression can be generalized:

$$S(A|B) = S(\rho_{AB}) - S(\rho_B). \tag{3.55}$$

Consider now, as an example, a maximally entangled state of two qubits $\rho_{AB} = |\Phi^+\rangle\langle\Phi^+|$: this state is pure, so $S(\rho_{AB}) = 0$; however, ρ_B is maximally mixed, whence $S(\rho_B) = 1$; so in all $S(A|B) = -1$. This is a manifestation of one of the unexpected features of entanglement: the fact that one has sharp properties for the composite system that do not derive from sharp properties of the subsystems.

Interestingly, conditional entropy has an operational interpretation as the amount of quantum information needed to transfer a state ρ_{AB}, initially shared between two parties, to one of the parties, while keeping the coherence with possible purifying systems that are not available to either party (Horodecki *et al.*, 2005). When the quantity is negative, it basically means that, after the transfer, the parties still keep some quantum correlations[10] that can be used to transfer additional quantum informations via teleportation.

3.3.2.2   *From Holevo to Schumacher, and beyond*

We start by presenting the two best-known early results in quantum coding: the Holevo bound and Schumacher's compression.

Holevo bound: classical information coded in quantum states.   Suppose Alice prepares any of the states $\{\rho_x\}_{x=1...N}$, each with some probability p_x and sends them to Bob on a noiseless channel. Bob performs a measurement, generically a POVM, with outcomes $y \in \{1,...,m\}$. Since ultimately both Alice's and Bob's variables are classical, the amount of information that Bob has obtained through the measurement is given by the Shannon mutual information $I(X:Y) = H(X) + H(Y) - H(X,Y)$. The *Holevo bound* is an upper bound on this amount of information:

$$I(X:Y) \leq \chi_{\{p_x,\rho_x\}} \equiv S\left(\sum_x p_x\rho_x\right) - \sum_x p_x S(\rho_x). \tag{3.56}$$

The left-hand side is a purely classical quantity, because X and Y are classical; the right-hand side is quantum, because classical information has been coded in quantum states. The bound can be saturated only if the states ρ_x are mutually commutative. We shall meet below an extension of this bound.

Maybe it's convenient to stress something at this point. The configuration envisaged in this section looks at first like a mathematical exercise with little or no usefulness in practice. Indeed, the task is "Alice wants to send a non-secret message to Bob". To achieve this, Alice most probably would not use quantum states at all in the first place; or more precisely, she'd use orthogonal states (this is classical communication), so that Bob would have no trouble discriminating them. So, why should one bother about such an artificial task as sending messages with non-orthogonal states? The answer is the following: in itself, the situation is artificial indeed; but this situation may appear as natural in the context of another task. For instance, consider the effective channel linking Alice to Eve in quantum cryptography. Here, Alice does *not* want Eve to learn the message: it's Eve that sneaks in and gets whatever she can. It is therefore normal that the channel Alice–Eve is not optimized for direct communication. It turns out that this channel is precisely of the Holevo type: Alice's bits are encoded in quantum states that Eve keeps (see also Tutorial).

Schumacher compression of quantum information.   Consider a source producing classical symbols x, independent and identically distributed (i.i.d.) according to a distribution $p(x)$. One wants to know how much a message can be compressed and later decompressed without introducing errors, i.e. $m(n)$ in $(x_1,...,x_n) \xrightarrow{\mathcal{C}} (y_1,...,y_m) \xrightarrow{\mathcal{D}}$

$(x_1, ..., x_n)$. A well-known theorem by Shannon says that, asymptotically, $m(n) = nH(X)$.

Schumacher's coding is the quantum generalization of this theorem. The source is now represented by the mixed state $\rho = \sum_x p(x)|x\rangle\langle x|$ living in a d-dimensional Hilbert space $\mathcal{H}$ (i.e. $\log d$ qubits). The compression – decompression procedure is now $\rho^{\otimes n} \xrightarrow{\mathcal{C}} \sigma \xrightarrow{\mathcal{D}} \rho^{\otimes n}$. The result is that σ must live in a Hilbert space of dimension $2^{nS(\rho)}$.

The richness of quantum information theory. With classical signals, one can basically send classical information. When one realizes the huge body of knowledge that this has generated (Cover and Thomas, 2006), the complexity of opening the box to quantum physics becomes striking. Indeed, once quantum systems are brought into the game, the number of possible situations that one can envisage explodes:

- The *nature of the coding*: classical information theory has classical bits (*c-bits*). In quantum information theory, one has of course to add quantum bits (*q-bits or qubits*), but also more complex units like bits of entanglement (*e-bits*) because states of two qubits are not the same of two states of qubits.

- The *available resources*: in classical information theory, one has classical channels for communication and shared randomness for possible pre-established correlations. Now we have to add quantum channels (those that allow us to send qubits), pre-established entanglement ... For instance, one can study what can be done with a classical channel assisted with shared entangled pairs; and all possible combinations.

We have reviewed many results, including Schumacher's, which may give the impression that ultimately one will always find the analog of known results in classical information theory. However, this is *not* the case, and the reason is *entanglement*. Indeed, a crucial assumption in Schumacher's result is that the source is i.i.d. We are going to gain more insight on this point by studying *channel capacities*, a field in which several ground-breaking results have been found only recently.

3.3.2.3 *Channel capacities: a rapid overview*

Definition of quantum channel capacity. Alice wants to send m x-bits to Bob (x can be c, q, e...). She encodes her information in a state of n qubits (possibly entangled) and sends these qubits over a quantum channel T to Bob, who decodes the information correctly with probability $\varepsilon(m, n)$. The *capacity* of this channel is given by

$$C_x(T) = \sup \lim_{n \to \infty} \frac{m}{n} \text{ such that } \lim_{n \to \infty} \varepsilon(m, n) = 0. \tag{3.57}$$

The supremum is taken over all possible choice of coding and decoding. In general therefore, for a given quantum channel T, several different capacities can be defined, depending on the nature of the information to be transmitted (x-bits), but also on possible additional resources, on the requested speed of convergence of the error rate, etc.

Case study: classical capacity of a quantum channel. For definiteness, let us focus on the *classical capacity* of a quantum channel, written $\mathcal{C}(T)$. The scenario is the one considered by Holevo: Alice codes classical information X in quantum states, sends them to Bob who performs a POVM and extracts classical information Y. The only difference is that now the quantum systems are sent over a non-trivial channel T: i.e. if Alice sends ρ_k, Bob receives $T[\rho_x]$.

Under the assumption of an i.i.d. source, we can easily understand that

$$\mathcal{C}_{i.i.d.}(T) = \chi(T) = \max_{p_x, \rho_x} \chi_{\{p_x, T[\rho_x]\}}. \tag{3.58}$$

But is it possible to achieve a larger capacity with a non-i.i.d. source? In classical information, This is known not to help: one can always maximize the rate of a channel with i.i.d. sources. In this case, one says that *capacity is additive*. But quantum physics allows for entanglement: Alice may associate the sequence (x_1, x_2) to an entangled state $\rho_{x_1 x_2}$. Can this help? For a long time, the answer was conjectured to be negative, though explicit proofs were available only for some particular channels. In September 2008, Hastings (2009) proved that the conjecture is in general wrong: there exist channels, whose full classical capacity cannot be reached by i.i.d. sources – in other words, for some channels, entanglement does help even if you are using the channel to share classical information!

Because of this, the general expression of the classical capacity of a quantum channel is

$$\mathcal{C}(T) = \lim_{n \to \infty} \frac{1}{n} \chi(T^{\otimes n}) \text{ with } \chi(T^{\otimes n}) = \max_{p_x, \rho_x} \chi_{\{p_x, T^{\otimes n}[\rho_x]\}} : \tag{3.59}$$

now the maximum must be taken over all possible choice of n-qubit states, each coding for the classical information $\mathbf{x} = (x_1, ..., x_n)$. There is manifestly no hope of computing such a maximum by brute force. This is the reason why even the "simple" classical capacity of quantum channels in not known for arbitrary channels.

Other capacities. Among the other possible capacities of a quantum channel, two are worth at least mentioning:

- The *classical private capacity* is associated with the task of sending classical information while keeping it secret from the environment. Its expression is (Devetak, 2005):

$$\mathcal{P}(T) = \lim_{n \to \infty} \frac{1}{n} \max_{p_x, \rho_x} \left[\chi_{\{p_x, T^{\otimes n}[\rho_x]\}} - \chi_{\{p_x, \tilde{T}^{\otimes n}[\rho_x]\}} \right]. \tag{3.60}$$

  where $\tilde{T}$ is the "complementary channel", i.e. the information that leaks into the environment – as such, the formula is somewhat intuitive.

- The *quantum capacity* is the capacity of the channel when Alice wants to send quantum information. It is generally written $\mathcal{Q}$; its expression would require introducing additional notions that are beyond the scope of our survey.

In general, it holds

$$C(T) \geq P(T) \geq Q(T). \tag{3.61}$$

The fact that the first inequality can be strict is almost obvious; less obvious is the fact that the second inequality can also be strict (Horodecki, Horodecki *et al.*, 2005). The proofs of non-additivity are all very recent: Smith and Yard had proved the non-additivity of the quantum capacity in Summer 2008 (Smith and Yard, 2008). Later, the conjecture that the private capacity may also be non-additive was formulated (Smith and Smolin, 2009) and proved (Li *et al.*, 2009). At the moment of writing, the field is still very active.

3.3.3 Tutorials

3.3.3.1 *Problems*

Exercise 3.1
    Prove that the trace distance between two pure states $|\psi_1\rangle$ and $|\psi_2\rangle$ is given by

$$D(\rho_1, \rho_2) = \sqrt{1 - |\langle\psi_1|\psi_2\rangle|^2}. \tag{3.62}$$

Hint: Note that you can always find a basis in which $|\psi_1\rangle = c|0\rangle + s|1\rangle$ and $|\psi_2\rangle = e^{i\varphi}(c|0\rangle - s|1\rangle)$ with $c = \cos\theta$ and $s = \sin\theta$.

Exercise 3.2
    A short laser pulse can be sent either at time t_1 or at time t_2. By detecting the time of arrival, one can obviously discriminate between these two cases. This rather trivial process is actually an example of *unambiguous state discrimination* of the two two-mode coherent states $|\psi_1\rangle = |0\rangle|\alpha\rangle$ and $|\psi_2\rangle = |\alpha\rangle|0\rangle$; it is used to create the raw key in the quantum cryptography protocol called COW (Stucki *et al.*, 2005). We recall the decomposition of the coherent state $|\alpha\rangle$, $\alpha \in \mathbb{C}$, on the number basis:

$$|\alpha\rangle = e^{-|\alpha|^2/2} \sum_{n=0}^{\infty} \frac{\alpha^n}{\sqrt{n!}} |n\rangle. \tag{3.63}$$

1. What is the probability of success for optimal USD?
2. Prove that the POVM for optimal USD can simply be realized by detecting the time of arrival (with a perfect detector). *Hint:* What are the "inconclusive" events?
3. Discuss what happens if the detector is not perfect, in particular how the discussion is modified by (i) efficiency $\eta < 1$; (ii) dark counts.

Exercise 3.3
    Let $\{|e_k\rangle\}_{k=1...4}$ be an orthonormal set of four vectors. We define $|\psi_1^{\pm}\rangle = \sqrt{1-\varepsilon}|e_1\rangle \pm \sqrt{\varepsilon}|e_2\rangle$ and $|\psi_2^{\pm}\rangle = \sqrt{1-\varepsilon}|e_3\rangle \pm \sqrt{\varepsilon}|e_4\rangle$; and we construct in turn the mixtures $\rho_0 = (1-\varepsilon)|\psi_1^+\rangle\langle\psi_1^+| + \varepsilon|\psi_2^+\rangle\langle\psi_2^+|$ and $\rho_1 = (1-\varepsilon)|\psi_1^-\rangle\langle\psi_1^-| + \varepsilon|\psi_2^-\rangle\langle\psi_2^-|$.

1. Compute the Holevo bound $\chi(\rho_0, \rho_1)$, assuming $p_0 = p_1 = \frac{1}{2}$.
2. The states given above, of course, have a meaning: they describe Eve's states in the optimal eavesdropping on the BB84 protocol of quantum cryptography, when an error rate ε is measured by Alice and Bob – see Section III.B.2 and Appendix A of (Scarani *et al.*, 2009). In this scenario, what does the Holevo bound mean? *Hint:* the index a or the matrices ρ_a is Alice's bit.

3.3.3.2   Solutions

Exercise 3.1

By writing $|\psi_1\rangle = c|0\rangle + s|1\rangle$ and $|\psi_2\rangle = e^{i\varphi}(c|0\rangle - s|1\rangle)$, we have $\langle \psi_1 | \psi_2 \rangle = e^{i\varphi}(c^2 - s^2) = e^{i\varphi} \cos 2\theta$ and

$$\rho_1 - \rho_2 = 2cs\sigma_x,$$

whence $D(\rho_1, \rho_2) = \frac{1}{2}(|+2cs| + |-2cs|) = 2cs = \sin 2\theta.$ The   result   follows immediately.

Exercise 3.2

1. The probability of optimal USD is $p_{USD} = 1 - |\langle \psi_1 | \psi_2 \rangle|$; here $|\langle \psi_1 | \psi_2 \rangle| = |\langle 0 | \alpha \rangle|^2 = e^{-|\alpha|^2}$.
2. As soon as the detector fires, the two states can be distinguished; so the "inconclusive" events are the events in which the detector did not fire; if the detector has perfect efficiency, this can only happen because of the vacuum component of the state. In both $|\psi_1\rangle$ and $|\psi_2\rangle$, the amplitude of the vacuum component is $e^{-|\alpha|^2/2}$; therefore the probability that the detector fires is $1 - e^{-|\alpha|^2} = p_{USD}$.
3. A detector with efficiency $\eta < 1$ is equivalent to losses $\sqrt{\eta}$; the discrimination is still unambiguous but succeeds only with probability $p = 1 - e^{-\eta|\alpha|^2} < p_{USD}$ (note that this is still the optimal procedure, under the constraint that one has to use such imperfect detectors). If dark counts are present, the detector may fire even if there was no photon; therefore the discrimination is no longer unambiguous.

Exercise 3.3

1. We have to compute $\chi(\rho_0, \rho_1) = S(\rho) - \frac{1}{2}[S(\rho_0) + S(\rho_1)]$ with $\rho = \frac{1}{2}(\rho_0 + \rho_1)$. Now, ρ_0 and ρ_1 are both incoherent mixtures of two orthogonal states with the same weights; therefore $S(\rho_0) = S(\rho_1) = -(1 - \varepsilon) \log(1 - \varepsilon) - \varepsilon \log \varepsilon \equiv h(\varepsilon)$. Moreover, $\rho = (1 - \varepsilon)^2 |e_1\rangle\langle e_1| + \varepsilon(1 - \varepsilon)|e_2\rangle\langle e_2| + \varepsilon(1 - \varepsilon)|e_3\rangle\langle e_3| + \varepsilon^2 |e_4\rangle\langle e_4|$, whence $S(\rho) = 2h(\varepsilon)$. All in all, $\chi(\rho_0, \rho_1) = h(\varepsilon)$.
2. One can see the relation between Alice and Eve as a channel, in which Alice's bit value a has been encoded in a state ρ_a. Therefore, the Holevo bound represents the maximal information that Eve might extract about Alice's bit.

3.4 Quantum correlations (I): the failure of alternative descriptions

This and the following two sections are devoted to the violation of Bell's inequalities and related topics. This means that we shall focus on a restricted family of quantum phenomena: the establishment of *correlations between distant partners through separated measurement of entangled particles*. This kind of phenomenon by no means exhausts the possibilities of quantum physics. However, it is there that the discrepancy between classical and quantum physics manifests itself in the most straightforward way (this section). As such, one might expect this feature to be useful for some quantum information tasks: this is indeed the case, but curiously enough, this awareness is only rather recent (Section 3.6). Section 3.5 will be an excursion into an extended theoretical framework that may be very promising or may just be a wrong track, but is funny and worth at least having heard about.

3.4.1 Correlations at a distance

3.4.1.1 Generic setup under study

The kind of experiment we are considering is sketched in Figure. 3.2. A source sends out two particle to two distant locations. In each location, a user chooses a possible measurement (A, B) and registers the outcome (a, b). The procedure is repeated a large number of times. Later, the two users come together, compare their results and derive the *probability distribution*

$$P_{AB}(a, b). \tag{3.64}$$

This probability distribution is often written as a conditional probability $P(a, b|A, B)$. Ultimately, it's a choice of notation that matters little; I use the notation (3.64) to stress that the origin of the statistics on (a, b) is the quantum randomness we want to query, while the statistics on (A, B) are of a very different nature (just the choice of the users on how often to perform each measurement).

A few crucial remarks:

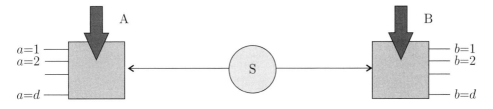

Fig. 3.2 The setup to be kept in mind for Bell-type experiments: a source S distributed two quantum systems to separated locations. On each location, the physicist is free to choose which measurement to perform (A, B); as a result, they obtain outcomes a, b. Of course, this setup can be generalized to more parties, or to the case where the number of outcomes is different between the parties.

(i) There is a conceptual distinction between the *users* and the physicists who have constructed the whole setup. To perform the task, the users do not even have to know what state the source has prepared, what physical systems are sent, which measurements are being performed. For the users, the measurement is just a knob they can freely set at any position; A and B refer to the label of the scale in their instruments. In what follows, the names of Alice and Bob will be always referring to the users.

(ii) The number of possible outcomes a and b plays an important role, but the labelling of the outcomes does not matter: the probability distribution $P(a, b)$ fully characterizes the process of measuring A and B, irrespective of whether the outcomes are real numbers, multiples of $\hbar$, complex numbers, colors... Because of this, we are free to choose the most convenient labelling without loss of generality. For instance, if the number of outcomes is two on both sides, then the correlation coefficient is defined as

$$E_{AB} = P_{AB}(a = b) - P_{AB}(a \neq b). \qquad (3.65)$$

This definition is unambiguous and always valid. If in addition, one chooses the labelling $a, b \in \{-1, +1\}$, then one obtains the handy relation $E_{AB} = \langle ab \rangle_{AB}$ the average of the product of the outcomes.

3.4.1.2 *Classical mechanisms for correlations*

We are dealing with a *family* of probability distributions: Alice picks up measurement A, Bob picks up measurement B, and the outcomes are guaranteed to be distributed according to the probability distribution P_{AB}. Now, in general, $P_{AB}(a, b) \neq P_A(a)P_B(b)$, where $P_A(a) = \sum_b P_{AB}(a, b)$ and $P_B(b) = \sum_a P_{AB}(a, b)$ are the *marginal distributions*. In other words, random variables distributed according to P_{AB} are correlated. The question that we are going to study is: can these correlations be ascribed to a classical mechanism? The question seems vague, but it becomes less vague once one realizes that there are only *two* classical mechanisms for distributing correlations.

The first and most obvious is *communication*: for instance, the information about Alice's choice of measurement A is available to the particle measured by Bob. This mechanism can be checked by arranging *space-like separation*: loosely speaking, if the two particles reach the measurement devices at the same time and the choice of measurement is done at the very last moment, a signal informing a particle about what is happening in the other location should travel faster than light. If the correlations persist in this configuration, the mechanism of communication becomes highly problematic.

The second mechanism consists in using *pre-established strategies*: each particle might have left the source with a set of instructions, specifying how it should behave in any measurement. Interestingly, this mechanism can explain the behavior of the singlet state in the cases where Alice and Bob choose to perform the same measurement (P_{MM}). Indeed, assume that the particles are carrying lists of pre-determined results

$\lambda_A = \{a_M\}_M$ and $\lambda_B = \{b_M\}_M$ such that $a_M = -b_M$ for each measurement M: one always gets $P_{MM}(a \neq b) = 1$, and the local randomness can be easily accounted for by varying the lists at each run. However, it is the milestone result by John Bell in 1964 to prove that the *whole* family of probabilities predicted by quantum physics cannot be reproduced with pre-established strategies (Bell, 1964).

We embark now on a more detailed study of each of the two classical mechanisms and their failure to reproduce the results observed in experiments.

3.4.2   Pre-established strategies ("Local variables")

3.4.2.1   The model

By definition, a pre-established strategy is some hypothetical information λ that the particles carry out with themselves from the source. Each particle is supposed to produce its outcome taking into account only this λ and the measurement to which it is submitted (plus some possible information encountered along the path, that we neglect for simplicity here). In other words, for given λ, the two random processes are supposed to be independent: $P_{AB}(a, b|\lambda) = P_A(a|\lambda)P_B(b|\lambda)$. The only freedom left is the possibility of changing the information λ at each run. If λ is drawn from a distribution $\rho(\lambda)$, the observed probability distribution will be

$$P_{AB}(a, b) = \int d\lambda \, \rho(\lambda) \, P_A(a|\lambda)P_B(b|\lambda). \qquad (3.66)$$

Here comes an important mathematical characterization:

Theorem. $P_{AB}(a, b)$ can be obtained by pre-established strategies if and only if it can be written as a convex sum of local deterministic strategies.

A *deterministic strategy* is a strategy in which, for each possible measurement, the result is determined. A *local* deterministic strategy is defined by $P_A(a|\lambda) = \delta_{a=f(A,\lambda)}$ and $P_B(b|\lambda) = \delta_{b=g(B,\lambda)}$. The "if" implication is therefore trivial. The "only if" implication stems from the fact that any classical random process can be mathematically decomposed as a convex sum of deterministic processes. In other words, for each λ, there exist an additional random variable $\mu = \mu(\lambda)$ with distribution $\rho'(\mu)$ such that $P_A(a|\lambda) = \int d\mu \rho'(\mu)\delta_{a=f(A,\lambda,\mu)}$. One can therefore just "enlarge" the definition of the local variable to $\lambda' = (\lambda, \mu(\lambda))$.

Note that no restriction of the family of pre-established strategies under study is being made. The theorem is a purely mathematical result. It is useful because it allows deriving results by arguing with deterministic strategies that are very easy to handle; if the result is stable under convex combination, it is automatically guaranteed to hold for *all* possible pre-established strategies, deterministic or not.

3.4.2.2   Two remarks

Before continuing, it is worthwhile making two remarks.

About terminology.   Historically, λ was called a "local hidden variable". While the expression *"local variable"* is ultimately convenient, the adjective "hidden" is definitely superfluous and even misleading: quantum physics is at odds with local variables, irrespective of whether they are supposed to be hidden or not. For instance, the local variable may be a description of the total quantum state (Gisin, 2009). Much more recently, in the interaction between physicists and computer scientists, the name of *"shared randomness"* has also become fashionable to denote pre-established strategies.

Regarding the fact itself, that quantum correlations cannot be attributed to pre-established strategies and local parameters, the linguistic debate is even more involved. Maybe the most precise expression would be "falsification of crypto-determinism", as proposed, e.g., by Asher Peres (Peres, 1995); but it is hardly used. The two most common expressions found in the literature are "quantum non-locality" and "violation of local realism". Both have their shortcomings, some people in the field have rather strong feeling against one or the other.[11] You must above all keep in mind that there are plenty of unfortunate expressions in science – the essence of "quantum" physics is the superposition of states, certainly not the discreteness! – but there is little danger in using them, as long as one knows what their meaning is. I shall use *"non-locality"*, without the "quantum", and refer to *"non-local correlations"* as a shortcut for "probability distributions that cannot be reproduced by pre-established agreement".

Pre-established values for single systems. If quantum systems are considered as a whole, i.e. not consisting of separate subsystems, one can always find a local variable model that reproduces the correlations. For dimensions larger than two, the Kochen–Specker theorem proves that the local variable model must take into account the whole measurement (it must be "contextual"); but contextuality is not a problem for local-variable theories.

Though its interest may be rather limited, for the sake of illustration, let me introduce a well-known explicit local variable model that reproduces the quantum predictions for one qubit. The goal is to reproduce the statistics of the quantum state $\rho = \frac{1}{2}(\mathbb{1} + \vec{m} \cdot \vec{\sigma})$ under all possible von Neumann measurements, i.e. $P(\pm\vec{a}) = \frac{1}{2}(\mathbb{1} \pm \vec{m} \cdot \vec{a})$. The local variable is a vector $\vec{\lambda} = [\sin\theta\cos\phi, \sin\theta\sin\phi, \cos\theta]$ on the surface of the unit sphere; for each realization, $\vec{\lambda}$ is drawn randomly with uniform measure. The rule for the outcome of a measurement is deterministic: the outcome is $r(\vec{a}, \vec{\lambda}) = \text{sign}[(\vec{m} - \vec{\lambda}) \cdot \vec{a}]$. Then one can prove that $\langle r \rangle(\vec{a}) = \int_{S^2} r(\vec{a}, \vec{\lambda}) \sin\theta d\theta d\phi = \vec{m} \cdot \vec{a}$, which is exactly the quantum expectation value (the proof is simple using a geometrical visualization: the integral is the difference between the surface of two spherical hulls).

3.4.2.3   CHSH inequality: derivation

Let us now present the derivation of the most famous Bell inequality, the one derived by Clauser, Horne, Shimony and Holt and therefore known as CHSH (Clauser *et al.*, 1969). This inequality is defined by the fact that both Alice and Bob can make only two possible measurements (we label them A, A' for Alice, B, B' for Bob) and the outcomes are binary (we choose the labelling $a, b \in \{+1, -1\}$).

Let us first consider a local deterministic strategy λ_D: here, it is just any list $\lambda_D = (a, a', b, b')$ specifying the two outcomes of Alice and the two outcomes of Bob. If this list is defined, then the number $S(\lambda_D) = (a + a')b + (a - a')b'$ is also defined. By inspection, it is obvious that $S(\lambda_D)$ can only take the values $+2$ or -2. If we now take a convex combination of such strategies with distribution ρ, it is obvious that $\langle S \rangle = \int d\lambda_D \rho(\lambda_D) S(\lambda_D)$ must lie between -2 and 2. Moreover, since the average of a sum is the sum of the averages, and since with our labelling $\langle ab \rangle$ is the correlation coefficient, we have found

$$|\langle S \rangle| = |E_{AB} + E_{A'B} + E_{AB'} - E_{A'B'}| \leq 2. \tag{3.67}$$

This is the *CHSH inequality*: it holds for all convex combinations of local deterministic strategies and therefore, by virtue of the theorem above, it holds for all pre-established strategies. An important remark here: the derivation of the inequality is particularly simple when the outcomes are labelled $+1$ and -1; this is how we did it, and we shall keep this convention in this whole lecture (for the topic of the next lecture, another labelling will prove more convenient). However, the inequality itself is independent of this choice: if one chooses another labelling, the inequality still holds with the general definition (3.65).

3.4.2.4 *CHSH inequality: violation in quantum physics*

In quantum physics, a measurement that can give two outcomes, $+1$ and -1, is described by an Hermitian operator with those eigenvalues (possibly degenerate). Therefore,

$$E_{AB} \rightarrow \langle A \otimes B \rangle, \tag{3.68}$$

where A and B are two such operators. In other words, in quantum physics $\langle S \rangle$ becomes the expectation value of the *CHSH operator*

$$\mathcal{S} = A \otimes B + A' \otimes B + A \otimes B' - A' \otimes B'. \tag{3.69}$$

The largest possible value of $\langle S \rangle$ is therefore[12] the largest eigenvalue of $\mathcal{S}$. As noted by Tsirelson,[13] in the case of the CHSH inequality it is easy to find a bound for the maximal eigenvalue (Cirel'son, 1980). Indeed, using the fact that $A^2 = A'^2 = B^2 = B'^2 = \mathbb{1}$, one can easily compute $\mathcal{S}^2 = 4\mathbb{1} \otimes \mathbb{1} + [A, A'] \otimes [B, B']$. The maximal eigenvalue of $[A, A']$ cannot exceed 2, because $|\langle [A, A'] \rangle| \leq |\langle AA' \rangle| + |\langle A'A \rangle|$ and the spectrum of both A and A' contains only $+1$ and -1. So the maximal eigenvalue of $\mathcal{S}^2$ cannot exceed 8, which implies that in quantum physics

$$|\langle S \rangle| \leq 2\sqrt{2}. \tag{3.70}$$

This bound can be saturated already with two-qubit states. Indeed, consider the correlations of the singlet state $E_{AB} = -\vec{a} \cdot \vec{b}$: by choosing $\vec{a} = \hat{z}$, $\vec{a}' = \hat{x}$, $\vec{b} = \frac{1}{\sqrt{2}}(\hat{z} + \hat{x})$, $\vec{b}' = \frac{1}{\sqrt{2}}(\hat{z} - \hat{x})$, one obtains $E_{AB} = E_{AB'} = E_{A'B} = -E_{A'B'} = -\frac{1}{\sqrt{2}}$, whence $|\langle S \rangle| = 2\sqrt{2}$.

Two important remarks:

- It is remarkable that local variables can be ruled out already in the *simplest* scenario: with the smallest Hilbert space that allows entanglement and with the least possible number of measurements and of outcomes. Indeed, if a party performs only one measurement, one can always find a local variable model that explains the statistics; and a measurement with only one outcome would be obviously trivial.

- As we have just seen, there is not a unique Bell operator, but a family parametrized by the measurement settings. It is trivial to find "bad settings", that don't give any violation even if the state is maximally entangled: for instance, remember that the correlations of the singlet cannot violate any Bell inequality if one requests Alice and Bob to perform the same measurements, i.e. if $A = B$, $A' = B'$.

3.4.2.5   *Ask Nature: experiments and loopholes*

Experiments are reviewed in other series of lectures in this school; comprehensive review articles are also available (Tittel and Weihs, 2001; Pan *et al.*, 2008). Here, I just address the following question: to which extent alternative models have been falsified.

The non-locality of quantum correlations is so striking an effect that it has been scrutinized very closely. Basically two possible *loopholes* have been identified:

- If one does not arrange the timing properly, the detections may be attributed to a subluminal signal: this is called *locality loophole*. In order to close this loophole, the events must be ordered in space-time as in Figure. 3.3. Note in particular that the choice of setting on Alice's side must be space-like separated from the end of the measurement on Bob's side, and vice versa. Now, the "end of a measurement" is one of the most fuzzy notions in quantum theory! Consider a photon impinging on a detector: when does quantum coherence leaves place to classical results? Already when the photon generates the first photo-electron? Or when an avalanche of photo-electrons is produced? Or when the result is registered in a computer? There is even an interpretation (Everett's, also called many-worlds) in which no measurement ever happens, the whole evolution of the universe being just a developing of quantum entanglements. All these options are compatible with our current understanding and practice of quantum theory (a fact expressed sometimes by saying that there is a "quantum measurement problem"). As long as this is the situation, strictly speaking it is impossible to close the locality loophole. However, many physicists adopt the reasonable assumption that the measurement is finished "not too long a time" after the particle impinges on the detector[14] and I shall adopt such a position below. In this view, τ_M is of the order of the microsecond, in which case the distance between Alice and Bob should at least be 300 m.

- The other possible problem is called a *detection loophole*. In all experiments, the violation of Bell's inequalities is measured on the events in which *both* particles

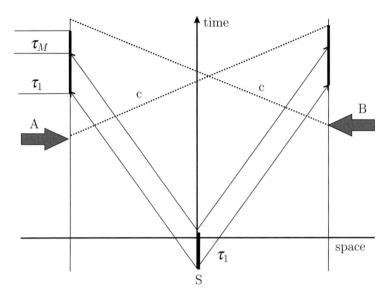

Fig. 3.3 Space-time diagram to close the locality loophole: the measurement A must be chosen at a time, such that this information reaches the location of B after the measurement on particle B is completed; and symmetrically. The dotted lines denoted by c represent the light cone in vacuum; the particles, even when they are photons, may propagate at a slower speed (e.g. if they are sent through optical fibers). The emission of a pair happens with some uncertainty τ_1 called the single-particle coherence time. The parameter τ_M indicates the time it takes to perform a measurement, i.e. to "collapse" the state: a very problematic notion in this discussion, see text.

have been detected. Since detectors don't have perfect efficiency, there is also a large number of events in which only one particle is detected, while the other has been missed. The detection loophole assumes a form of conspiracy, in which the undetected particle "chose" not to be detected after learning to which measurement it was being submitted. In this scenario, if the detection efficiency is not too high, it is pretty simple to produce an apparent violation of Bell's inequalities with local variables: when the local variable does not have the desired value, the particle opts not to reply altogether! For CHSH and maximally entangled states, the threshold value of the detection efficiency required to close this loophole is around 80% (see Tutorial).

At the moment of writing, no experiment has closed both loopholes simultaneously;[15] some groups are aiming at it, but the requirements are really demanding. Is it really worthwhile? After all, all experimental data are perfectly described by our current physical theory and are in agreement with our understanding of how devices (e.g. detectors) work. By the *usual* standards in physics, the violation of Bell's inequalities by separated entangled pairs is experimentally established beyond doubt. Given the implications of this statement for our world-view, though, some people think that *much higher standards* should be applied in this case before making a final claim.

I leave it to the readers to choose their camp. I just note here that the detection loophole, so artificial in this context, has recently acquired a much more serious status in a different kind of problem: we shall have to study it in detail in Section 3.6.

3.4.2.6 *Entanglement and non-locality*

The literature devoted to Bell's inequalities is huge. All possible generalizations of the CHSH inequality have been presented: more than two measurements, more than two outcomes, more than two particles, and all possible combinations. There are still several open questions (Gisin, 2007; Scarani and Méthot, 2007). Just to give an idea of this complexity, let me review rapidly the development of one of the main questions: *do all entangled states violate a Bell inequality?* This seems an obvious question, but as we shall see, the answers are rather intricate.

The easy part is the following: *all pure entangled states violate some Bell-type inequality.* This statement is known under the name of *Gisin's theorem*: indeed Gisin seems to be the first to have addressed the question, and he proved the statement for bipartite states (Gisin, 1991); shortly afterwards, Popescu and Rohrlich provided the extension to multipartite states[16] (Popescu and Rohrlich, 1992).

Problems begin when one considers *mixed states*. Even before Gisin proved his theorem, Werner had already shown that there exist mixed states such that (i) they are definitely entangled, in the usual sense that any decomposition as a mixture of pure states must contain some entangled state; (ii) the statistics obtained for all possible von Neumann measurements can be reproduced with pre-established strategies (Werner, 1998). Several years later, Barrett showed that the result holds true for the same states even when POVMs are taken into account (Barrett, 2002). So *there exist mixed entangled states that do not violate any Bell-type inequality.*

For two qubits, the one-parameter family of Werner states is

$$\rho_W = W|\Psi^-\rangle\langle\Psi^-| + (1-W)\frac{\mathbb{1}}{4}, \tag{3.71}$$

with $0 \le W \le 1$. These states are entangled for $W > \frac{1}{3}$. The best extension of Werner's result shows that the statistics of von Neumann measurements are reproducible with local variables for $W \lesssim 0.6595$ (Acín *et al.*, 2006). On the other hand, Werner states provably violate some Bell inequality as soon as $W \gtrsim 0.7056$ (Vértesi, 2008). Nobody has been able to close the gap at the moment of writing: this is an open problem, that one may legitimately consider of moderate interest, but that is frustrating in its apparent simplicity.[17]

Now, having learned about *distillation of entanglement*, you may legitimately be surprised by Werner's result. Indeed, Werner's states are distillable: why can't one just distill enough entanglement to violate a Bell's inequality? Of course, one can! The previous discussion has been made under the tacit assumption that Alice and Bob do not perform collective measurements, but measure each of their particles individually. This is the way experiments are usually done, but is not the most general scenario allowed by quantum physics. Distillation involves collective measurements, so there is no contradiction between the distillability of Werner's state and Werner's result on local variable models. Actually, it is worthwhile stressing that the first distillation protocol was invented by Popescu *precisely* in order to extract a violation of Bell's inequalities out of "local" Werner states (Popescu, 1995).

It has recently been shown that *all entangled mixed states (including bound-entangled ones), if submitted to suitable multi-copy processing, can produce statistics that violate some Bell-type inequality* (Masanes *et al.*, 2008). This is somehow a comforting result: in this very general scenario, entanglement and the impossibility of pre-established strategies coincide. Still, in my opinion, Werner's result and all its extensions keep their astonishing character: each pair being entangled and produced independently of the others, why does it take complex collective measurements in order to reveal the non-classicality of the source?

3.4.3 Superluminal communication

3.4.3.1 General considerations

The issue of superluminal communication as a possible explanation for quantum correlations is delicate. Several physicists think that it is just not worthwhile addressing in the first place. While I am definitely convinced that no superluminal communication is indeed going on, instead of shunning this topic completely, I prefer to adopt a more pragmatic view based on the following considerations: while local variables have been directly and conclusively disproved as a possible mechanism, superluminal communication seems to be excluded "only" on the belief that nothing, really nothing, should propagate faster than light. Now, strictly speaking, relativity forbids faster-than-light propagation of signals that can be used by us (because this would open causality loops and allow signalling in the past). And, as far as I know, nobody has really proved that all possible models based on communication are intrinsically inconsistent – some serious authors have actually argued the opposite (Reuse, 1984; Caban and Rembielinski, 1999). Therefore, I think it is worthwhile trying to invent such models and falsify them in experiments.

Still, the family of such models seems to be much more complex than the clear-cut definition of pre-established strategies. Here are a few elements to be taken into account:

- First, when dealing with a hypothetical signal travelling faster than light, one has to decide *in which frame this communication is defined*. There are basically two alternatives: either one considers a *global preferred frame*; or one envisages *separate preferred frames for each particle*. Both alternatives have been explored.

- By admission of its very founders, Bohmian mechanics can be seen as a theory in which information (the deformation of the quantum potential due to a measurement) propagates at infinite speed in a global preferred frame (Bohm and Hiley, 1993). Now, however problematic the full Bohmian program may be, Bohmian mechanics is mathematically equivalent to quantum mechanics. Therefore, we know that there is at least one model with superluminal communication that cannot be checked against quantum predictions, because its predictions are exactly the same! In other words, there cannot be an analog of Bell's theorem ruling out *all* possible models with communication.

- When constructing a model, it is customary to enforce the fact that the users should not be able to signal faster than light (i.e. the fact that superluminal communication must be "hidden"). The models we shall review here are meant to be of this kind; note, however, that, as soon as one is willing to introduce a preferred frame in physics, there is no compelling reason to enforce no-signalling (Eberhard, 1989).

3.4.3.2 *Bounds on the speed in a preferred frame*

Let us first suppose that the hypothetical superluminal communication is defined in a *preferred frame*. One has to arrange the detection events to be as simultaneous as possible in this frame; upon observing that the correlations persist, one obtains a lower bound on the speed of this hypothetical communication. Now, we don't know any preferred frame, so which one should one choose? It can be anything, from the local rest frame of the town to the frame in which the cosmic background radiation has no Doppler shift.

It turns out that one does not really have to choose! Indeed, suppose one arranges simultaneity in the rest frame of the town: by virtue of Lorentz transformation,[18] simultaneity is automatically guaranteed in all those frames whose relative speed is orthogonal to the direction A–B. If in addition one arranges the line A–B in the East–West orientation, the rotation of the Earth will scan *all* possible frames in twelve hours!

When experimental data are analyzed, the bounds are striking: the hypothetical communication should travel at speeds that exceed $10\,000c$ (Scarani *et al.*, 2000; Salart *et al.*, 2008)! This is a very strong suggestion that the mechanism of communication in a preferred frame is not a valid explanation for quantum correlations.

3.4.3.3 *Before-before arrangement in the local frames*

It seems that the previous discussion settles the problem. However, we have another alternative thanks to the inventiveness of Antoine Suarez. This is sometimes known as the Suarez–Scarani model, because I happened to help Antoine in formalizing his intuition.[19] The idea is that there may not be a unique preferred frame: rather, each particle, upon detection, would send out superluminal signals in the rest frame of the measurement device (Suarez and Scarani, 1997).

Again, one has to define which is the meaningful measurement device: the choice of the setting, the detector... Whichever one chooses though, how does one falsify such a model? By arranging a *before-before* timing. Consider the space-time diagram sketched in Figure 3.4, representing the situation in which the two measurement devices move away from one another: in that arrangement, each particle arrives before the other one at its respective device! If the model is correct, in this situation one should cease to violate Bell's inequalities. Experiments have been performed and the violation does not disappear (Zbinden *et al.*, 2001; Stefanov *et al.*, 2002, 2003): the Suarez–Scarani model adds to the long list of falsified alternative descriptions of quantum phenomena.

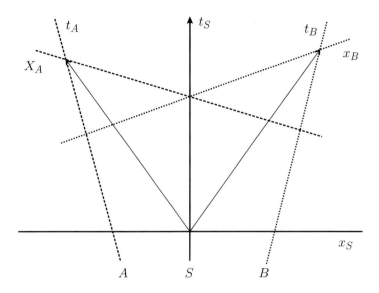

Fig. 3.4 Space-time diagram for the before-before experiment (coherence times have been omitted for simplicity, of course they must be taken into account in a detailed analysis). When particle A is detected, in the reference frame of its device particle B had not been detected yet; and symmetrically.

3.4.4 Leggett's model

Let's leave aside the details and admit the failure of the two classical explanations for correlations: something "non-classical" is present. Still, one may try to *save as much as possible of a classical world-view*. This is a quite general idea but, as a matter of fact, only one specific model inspired by this line of thought has been proposed so far. This first such proposal is due to Leggett (Leggett, 2003), whose initial intuition was rediscovered by the Vienna group (Gröblacher *et al.*, 2007) and has undergone further clarification (Branciard *et al.*, 2008).

Recall that one of the astonishing aspects of entanglement is the fact that a composite system can be in an overall pure state, while none of its components is. One can try to check this statement directly. The problem can be formulated as follows: whether it is possible to find a decomposition

$$P_{AB}(a,b) = \int d\xi\, P_{AB}(a,b|\xi),\qquad(3.72)$$

such that $P_{AB}(a,b|\xi)$ is no-signalling but describes correlations that violate Bell's inequalities (the necessary "non-classical" element) but the marginals are like those of a pure, single-particle state. If this is possible, one could have sharp properties for both the composite and the individual systems.

Without entering the details of the derivation, a decomposition like Eq. (3.72) fails to recover the quantum probabilities. Experiments have been performed, whose results are in excellent agreement with the latter and falsify therefore the alternative model. In addition, one can prove that, to reproduce the correlations of the singlet state, the marginal of *all* the $P_{AB}(a, b|\xi)$ cannot be anything else than completely random.

From Leggett's model we have learned that not only the full classical mechanisms are ruled out, but also attempts at sneaking back some elements of classicality in an otherwise non-classical theory seem destined to fail. I dare say that this is another remarkable proof of the "structural accuracy" of quantum theory: not only, as is widely known, this theory is most powerful in its numerical predictions: its structure itself has uncovered properties of reality that we could not have imagined otherwise.

3.4.5   Balance

Let us summarize the extent of the failure of alternative descriptions to quantum physics, inspired by the phenomenon of quantum correlations at a distance:

Pre-established agreement: All possible models based on pre-established agreement (local variables) are incompatible with the quantum predictions. All the experiments are in excellent agreement with the latter; there is a universal consensus that the remaining loopholes are technical issues whose closure is only a matter of time and skill.

Superluminal communication: No direct check of all models seems to be possible, and at least one of them is known to be compatible with all quantum predictions. However, all the models that have been tested, together with the implausibility of the assumption itself, clearly uphold the view that no communication is involved in the establishment of quantum correlations (in other words, that there is no *time-ordering* between the correlated events).

There is no third classical way: if one considers (as I do) that we have enough evidence to rule out both pre-established agreement and communication, the conclusion is that *no mechanism in space and time* can reproduce quantum correlations. These correlations "happen", we can predict them, but they are a fundamental, irreducible phenomenon. We don't have an explanation in terms of more basic phenomena, and, most remarkably, *we know we shall never find one.*

Note added in proof: after writing this text, I was reminded of another presentation of Bell's result, which relates to contextuality. It is due to Fine (Fine, 1982) and should be mentioned here. In order to present it in the simplest way, I have to change the notation a little: let me now write $P(a_A, b_B)$ instead of $P_{AB}(a, b)$; and let's suppose for definiteness that both A and B take two values, say 1 and 2, as in the CHSH inequality. Then one can ask whether there exists a global probability distribution $P(a_1, a_2, b_1, b_2)$, of which the $P(a_A, b_B)$ would be the marginal distributions [i.e. for instance, $P(a_1, b_1) = \sum_{a_2} \sum_{b_2} P(a_1, a_2, b_1, b_2)$]. The requirement that the global P exists is equivalent to the requirement of the existence of local hidden variables.

3.4.6   Tutorials

3.4.6.1   Problems

Exercise 4.1
This calculations shows that all pure entangled states of two qubits violate the CHSH inequality (Gisin, 1998). Using the Schmidt decomposition, any two-qubit pure

state can be written, in a suitable basis, as $|\psi(\theta)\rangle = \cos\theta|00\rangle + \sin\theta|11\rangle$ with $\cos\theta \geq \sin\theta \geq 0$.

1. Write down the Bell-CHSH operator S for $A = \sigma_z$, $A' = \sigma_x$, $B = \cos\beta\sigma_z + \sin\beta\sigma_x$ and $B = \cos\beta\sigma_z - \sin\beta\sigma_x$.
2. Compute $S(\theta|\beta) = \langle\psi(\theta)|S|\psi(\theta)\rangle$, then choose the settings of Bob to achieve the optimal violation $S(\theta) = \max_\beta S(\theta|\beta) = 2\sqrt{1 + \sin^2 2\theta}$.

Note: it can be proved that the family of settings above is optimal for the family of states under study. Therefore, $S(\theta)$ is the maximal violation of CHSH achievable with the state $|\psi(\theta)\rangle$. For CHSH and two qubits, the optimal violation is also known for mixed states (Horodecki *et al.*, 1995).

Exercise 4.21
In this exercise, we study the Greenberger–Horne–Zeilinger (GHZ) argument for non-locality (Greenberger *et al.*, 1989; Mermin, 1990). Consider the three-qubit state

$$|GHZ\rangle = \frac{1}{\sqrt{2}}(|000\rangle + |111\rangle). \tag{3.73}$$

Compute the four expectation values $\langle\sigma_x \otimes \sigma_x \otimes \sigma_x\rangle$, $\langle\sigma_x \otimes \sigma_y \otimes \sigma_y\rangle$, $\langle\sigma_y \otimes \sigma_x \otimes \sigma_y\rangle$ and $\langle\sigma_y \otimes \sigma_y \otimes \sigma_x\rangle$ on this state. Try to find a local deterministic strategy that would have such correlations and obtain a contradiction.

3.4.6.2 Solutions

Exercise 4.1
Inserting the given settings in the definition of the Bell-CHSH operator, one finds

$$S = 2\cos\beta\,\sigma_z \otimes \sigma_z + 2\sin\beta\,\sigma_x \otimes \sigma_x, \tag{3.74}$$

whence $\langle\psi(\theta)|S|\psi(\theta)\rangle = \cos\beta + \sin\beta\sin 2\theta$. As is well known, $\max_x(a\cos x + b\sin x) = \sqrt{a^2 + b^2}$ obtained for $\cos x = \frac{a}{\sqrt{a^2+b^2}}$; therefore

$$S(\theta) = 2\sqrt{1 + \sin^2 2\theta} \quad \text{for} \quad \cos\beta = \frac{1}{\sqrt{1 + \sin^2 2\theta}}. \tag{3.75}$$

Note that $S(\theta)$ is always larger than 2, apart from the case $\theta = 0$ of a product state.

Exercise 4.2
For the GHZ state, one has

$$\begin{aligned}
\langle\sigma_x \otimes \sigma_x \otimes \sigma_x\rangle &= 1 \\
\langle\sigma_x \otimes \sigma_y \otimes \sigma_y\rangle &= -1 \\
\langle\sigma_y \otimes \sigma_x \otimes \sigma_y\rangle &= -1 \\
\langle\sigma_y \otimes \sigma_y \otimes \sigma_x\rangle &= -1,
\end{aligned} \tag{3.76}$$

i.e. the state is an eigenvector of those operators. Explicitly, the first line means that only four possible triples of results are possible when all particles are measured in the

x direction: $(+,+,+)$, $(+,-,-)$, $(-,+,-)$ and $(-,-,+)$; in the other three cases, the other four triples are possible.

In a local deterministic strategy, all the outcomes should be known in advance, so we want to find six numbers $\{a_X, a_Y; b_X, b_Y; c_X, c_Y\} \in \{+1, -1\}^6$ such that

$$
\begin{aligned}
a_X b_X c_X &= 1 \\
a_X b_Y c_Y &= -1 \\
a_Y b_X c_Y &= -1 \\
a_Y b_Y c_X &= -1
\end{aligned}
\tag{3.77}
$$

These relations are, however, obviously impossible to satisfy (notice for instance that the product of the four lines gives $+1 = -1$). So there is no deterministic point satisfying the correlations of the GHZ state. Since these are perfect correlations, no additional element of randomness can create them: in other words, no convex combination of deterministic points can satisfy the correlations either.

3.5 Quantum correlations (II): the mathematics of no-signalling

3.5.1 The question of Popescu and Rohrlich

We saw in the first Section that the present definition of quantum physics is rather a description of its formalism. Still, there is no shortage of "typically quantum features": intrinsic randomness, incompatible measurements and uncertainty relations, no-cloning, teleportation, non-local correlations... Can't one find the physical definition of quantum physics among those?

Popsecu and Rohrlich asked this question for *non-locality without signaling* (Popescu and Rohrlich, 1994): does quantum physics define the set of all probability distributions that are possibly non-local but are compatible with the no-signalling condition? The answer is: NO. Consider the following probability distribution for binary input $A, B \in \{0, 1\}$ and binary output $a, b \in \{0, 1\}$ (compared to the previous lecture, we use again our freedom of choosing the labels in the most convenient way):

$$
P_{AB}(a \oplus b = AB) = 1 \,, \ P_{AB}(a) = P_{AB}(b) = \frac{1}{2}.
\tag{3.78}
$$

The distribution is obviously no-signalling: the marginal of Alice does not depend on Bob's measurement (nor on Alice's, in this specific case) and the same for Bob. The correlation says that for $(A, B) \in \{(0,0), (0,1), (1,0)\}$ one has $a = b$ (perfect correlation) while for $(A, B) = (1, 1)$ one has $a = b \oplus 1$ (perfect anti-correlation). Therefore, if one evaluates the CHSH expression on this distribution, one finds $\langle S \rangle = 4$! In other words, this innocent-looking probability distribution reaches the largest possible value of CHSH, while we have seen that quantum physics cannot go beyond the Tsirelson bound $\langle S \rangle = 2\sqrt{2}$.

The hypothetical resource that would produce the distribution (3.78) has been called *PR-box*. After the Poescu–Rohrlich paper, the PR-box remained rather in the shadows for a few years. Interest in it was revived mainly by two studies. Computer scientists were astonished by the result of van Dam, who noticed that this box would make "communication complexity" tasks trivial (van Dam, 2005). For physicists, the result of Cerf, Gisin, Massar and Popescu is maybe more appealing (Cerf *et al.*, 2005; Degorre *et al.*, 2005): they proved that the correlations of the singlet can be simulated by local variables plus a single use of the PR-box (thus improving on a previous result by Toner and Bacon (Toner and Bacon, 2003), who showed the same but using one bit of communication as a non-local resource). This latter work raised the hope that the PR-box would play the same role of elementary building-block for non-local distributions, as the singlet plays for quantum states. This hope was later shattered: there are multi-partite non-local distributions that cannot be obtained even if the partners, pairwise, share arbitrarily many PR-boxes (Barrett and Pironio, 2005), and even for bipartite states it seems that full simulation will be impossible beyond the two-qubit case (Bacciagaluppi, 2008).

However, the fact that initial hopes are shattered is not new in physics (nor in life, for that matter): creativity is not stopped by that, and indeed, several further interesting results have been obtained along the line of thought started by Popescu and Rohrlich. In order to appreciate them, we have to go a step further. Given the gap between 4 and $2\sqrt{2}$, one can easily guess that the PR-box is not the only no-signalling resource outside quantum physics: there is actually a continuous family of such objects. The next section is devoted to the formal framework in which no-signalling distributions can be studied.

3.5.2 Formal framework to study no-signalling distributions

3.5.2.1 *A playful interlude*

As a warm up, we consider a family of games in which the two players, Alice and Bob, after learning the rules, are sent to two different locations. There, each of them is submitted to some external input and has to react. The game is won if the reactions are in agreement with the rules that have been fixed in advance.

Specifically, consider the following[20]

Game 1: *each of the players receives a paper to referee. The game is won if, whenever the two players receive the same paper, they produce the same answer (i.e. either both accept or both reject it).*

This game is easy to win: Alice and Bob can just agree in advance that, whatever paper they receive, they will accept it. Admittedly, if many runs of the game are played, this strategy may be a bit boring; but more elaborated ones are possible, for instance: they accept in the first run, reject in the second and third, then accept again in the fourth... Also, the strategy in each round may be much more subtle than a simple "accept all" or "reject all": for instance, in one round they can decide to "accept when ([first author name starts with A–M] OR [has been submitted to IJQI]) AND [does not come from CQT]; reject otherwise". This family of strategies are called *pre-established strategies*. Obviously, there are uncountably many pre-established strategies. Remarkably, though, their set can be bounded, as we already know: it is the convex body, whose extremal points are local deterministic strategies.

Let us now change the rules of the game:

Game 2: *each of the players receives a paper to referee. The game is won if the two players produce the same answer only when they receive the same paper, and produce different answers otherwise.*

If the set of possible inputs consists only of two papers, the game can still be won by pre-established strategy: the players have just to agree on which paper to accept and which to reject. As soon as there are more than two possible inputs, however, pre-established strategies cannot win the game with certainty. For instance, consider the case where there are three possible papers and write down the conditions for winning: $a_1 = b_1$, $a_2 = b_2$, $a_3 = b_3$, $a_1 \neq \{b_2, b_3\}$, etc. It's easy to convince oneself that no set of six numbers $\{a_1, a_2, a_3\} \times \{b_1, b_2, b_3\}$ can fulfill all the conditions. Since no local deterministic strategy can win the game, no pre-established strategy can.

In order to win such a game, in the classical world one has to use the other resource: *communication*. In view of what we know about quantum physics, let us stress here that communication is, in a sense, an exaggerated resource for Game 2. Indeed, with communication, one can win all possible games, for instance

Game 3: *each of the players receives a paper to referee. The game is won under the condition: Alice accepts her paper if and only if the paper received by Bob has been authored in CQT.*

Why does Game 3 seem more extreme than Game 2? Because Game 3 requires Alice to learn *specific information* about the input received by Bob; while in Game 2, the criteria for winning include only *relations* between Alice's and Bob's inputs and outputs. In other words, communication is intrinsically required to win Game 3; while one might hope to win Game 2 without communication – with no-signalling resources.

3.5.2.2   *Local and no-signalling polytopes: a case study*

Turning now to the formalism, I think it more useful, for the purpose of this book to do a fully developed case study, rather than giving the general formulas, that those who are going to work in the field will easily find in the literature. Therefore, we focus on the simplest non-trivial case, the same as for the CHSH inequality and the PR-box: two partners, each with binary input $A, B \in \{0, 1\}$ and binary output $a, b \in \{0, 1\}$.

Probability space.   The first element to be studied is the *dimensionality of the probability space*: how many numbers are required to specify the four no-signalling probability distributions $P_{AB}(a, b)$ completely? In all, there are sixteen probabilities; but since each of the four P_{AB} must be normalized, we can already reduce to twelve. Actually, a cleverer reduction to only *eight* parameters can be achieved by exploiting the no-signalling condition. Indeed, note first that the three numbers needed to specify P_{AB} can be chosen as being $P_{AB}(a = 0)$, $P_{AB}(b = 0)$ and $P_{AB}(a = 0, b = 0)$, since $P_{AB}(a = 0, b = 1) = P_{AB}(a = 0) - P_{AB}(a = 0, b = 0)$, $P_{AB}(a = 1, b = 0) = P_{AB}(b = 0) - P_{AB}(a = 0, b = 0)$ and $P_{AB}(a = 1, b = 1)$ follows by normalization. Furthermore, because of no-signalling, $P_{AB}(a = 0) = P_A(a = 0)$ for all A and $P_{AB}(b = 0) = P_B(b = 0)$ for all B. Therefore, we are left with eight probabilities that can be conveniently arranged as a table:

$$\mathbf{P} = \begin{array}{c|cc} & P_{B=0}(b=0) & P_{B=1}(b=0) \\ \hline P_{A=0}(a=0) & P_{00}(0,0) & P_{01}(0,0) \\ P_{A=1}(a=0) & P_{10}(0,0) & P_{11}(0,0) \end{array} . \tag{3.79}$$

For instance, the probability distribution of a PR-box (3.78) and the one associated to the best measurements on a maximally entangled state read, respectively,

$$\mathbf{P}_{PR} = \begin{array}{c|cc} & 1/2 & 1/2 \\ \hline 1/2 & 1/2 & 1/2 \\ 1/2 & 1/2 & 0 \end{array} , \mathbf{P}_{ME} = \begin{array}{c|cc} & 1/2 & 1/2 \\ \hline 1/2 & \frac{1+1/\sqrt{2}}{4} & \frac{1+1/\sqrt{2}}{4} \\ 1/2 & \frac{1+1/\sqrt{2}}{4} & \frac{1-1/\sqrt{2}}{4} \end{array} . \tag{3.80}$$

A priori, Bell's inequalities will appear naturally later in the construction; however, since we have already derived the CHSH inequality in a different way, we can legitimately study here what the inequality looks like in this notation. Noting that $E_{AB} = 1 - 2[P_{AB}(0,1) - P_{AB}(1,0)] = 4P_{AB}(0,0) - 2P_A(a=0) - 2P_B(b=0) + 1$, we find

$$\langle S \rangle = 4\left[P_{00}(0,0) + P_{01}(0,0) + P_{10}(0,0) - P_{11}(0,0) - P_{A=0}(a=0) - P_{B=0}(b=0)\right] + 2.$$

Remembering that the inequality reads $-2 \leq \langle S \rangle \leq 2$, by simply rearranging the terms we find

$$-1 \leq P_{00}(0,0) + P_{01}(0,0) + P_{10}(0,0) - P_{11}(0,0) - P_{A=0}(a=0)$$
$$-P_{B=0}(b=0) \leq 0 \tag{3.81}$$

i.e. the inequality known as Clauser–Horne (CH)–which is therefore *strictly equivalent* to CHSH, under the assumption of no-signalling. Written as a table, we have

$$\mathbf{T}_{CH} = \begin{array}{c|cc} & -1 & 0 \\ \hline -1 & 1 & 1 \\ 0 & 1 & -1 \end{array} \tag{3.82}$$

and the inequality reads

$$-1 \leq \mathbf{T}_{CH} \cdot \mathbf{P} \leq 0, \tag{3.83}$$

where $\cdot$ represents term-by-term multiplication. For instance, $\mathbf{T}_{CH} \cdot \mathbf{P}_{PR} = \frac{1}{2}$ and $\mathbf{T}_{CH} \cdot \mathbf{P}_{ME} = \frac{1}{\sqrt{2}} - \frac{1}{2}$.

Local deterministic points (vertices of the local polytope). The next step consists in identifying all the *local deterministic strategies*. There are only four deterministic functions $a = f(A)$ from one bit to one bit: $f_1(A) = 0$, $f_2(A) = 1$, $f_3(A) = A$ and $f_4(A) = A \oplus 1$. Therefore, there are sixteen local deterministic strategies $D_{AB}^{ij}(a,b) = \delta_{a=f_i(A)}\delta_{b=f_j(B)}$. Let us write down explicitly:

$$D_{11} = \begin{array}{c|cc} & 1 & 1 \\ \hline 1 & 1 & 1 \\ 1 & 1 & 1 \end{array}, \quad D_{12} = \begin{array}{c|cc} & 0 & 0 \\ \hline 1 & 0 & 0 \\ 1 & 0 & 0 \end{array}, \quad D_{13} = \begin{array}{c|cc} & 1 & 0 \\ \hline 1 & 1 & 0 \\ 1 & 1 & 0 \end{array}, \quad D_{14} = \begin{array}{c|cc} & 0 & 1 \\ \hline 1 & 0 & 1 \\ 1 & 0 & 1 \end{array},$$

$$D_{21} = \begin{array}{c|cc} & 1 & 1 \\ \hline 0 & 0 & 0 \\ 0 & 0 & 0 \end{array}, \quad D_{22} = \begin{array}{c|cc} & 0 & 0 \\ \hline 0 & 0 & 0 \\ 0 & 0 & 0 \end{array}, \quad D_{23} = \begin{array}{c|cc} & 1 & 0 \\ \hline 0 & 0 & 0 \\ 0 & 0 & 0 \end{array}, \quad D_{24} = \begin{array}{c|cc} & 0 & 1 \\ \hline 0 & 0 & 0 \\ 0 & 0 & 0 \end{array},$$

$$D_{31} = \begin{array}{c|cc} & 1 & 1 \\ \hline 1 & 1 & 1 \\ 0 & 0 & 0 \end{array}, \quad D_{32} = \begin{array}{c|cc} & 0 & 0 \\ \hline 1 & 0 & 0 \\ 0 & 0 & 0 \end{array}, \quad D_{33} = \begin{array}{c|cc} & 1 & 0 \\ \hline 1 & 1 & 0 \\ 0 & 0 & 0 \end{array}, \quad D_{34} = \begin{array}{c|cc} & 0 & 1 \\ \hline 1 & 0 & 1 \\ 0 & 0 & 0 \end{array},$$

$$D_{41} = \begin{array}{c|cc} & 1 & 1 \\ \hline 0 & 0 & 0 \\ 1 & 1 & 1 \end{array}, \quad D_{42} = \begin{array}{c|cc} & 0 & 0 \\ \hline 0 & 0 & 0 \\ 1 & 0 & 0 \end{array}, \quad D_{43} = \begin{array}{c|cc} & 1 & 0 \\ \hline 0 & 0 & 0 \\ 1 & 1 & 0 \end{array}, \quad D_{44} = \begin{array}{c|cc} & 0 & 1 \\ \hline 0 & 0 & 0 \\ 1 & 0 & 1 \end{array}.$$

According to the theorem shown in Section 3.4, we know that the set of local distributions (distributions that can be obtained with pre-established strategies) is the convex set whose extremal points are the deterministic strategies. A convex set with a finite number of extremal point is called a "polytope", therefore we shall call this set a *local polytope*. The extremal points of a polytope are called *vertices*.

Facets of the local polytope: Bell's inequalities. The vertices of the local polytope define its *facets*, i.e. the planes that bound the set. If a point is below the facet, the corresponding probability distribution can be reproduced with local variables; if a point is above the facet, it cannot. Therefore, facets are the geometric representation of Bell's inequalities! Actually, in addition to Bell's inequalities, there are many *trivial facets* that can never be violated: conditions like $P_{AB}(a, b) = 0$ or $P_{AB}(a, b) = 1$ obviously define boundaries within which every local distribution must be found... because *any* distribution must be found therein! We forget about these trivial facets in what follows.

In our simple case, the local polytope is a polygone in an 8-dimensional space, so its facets are 7-dimensional planes. One has to identify the sets of eight points that define one of these planes, then write the equation that defines each plane. This can be done by brute force, but we just use some intuition and then quote the known result. The CH inequality indeed defines facets: we have $T_{CH} \cdot D = 0$ for D_{11}, D_{13}, D_{22}, D_{24}, D_{31}, D_{34}, D_{42}, D_{43}; and one can check that these eight points indeed define a plane of dimension 7. Above this facet, the most non-local point is the PR-box (3.78).

Also, $T_{CH} \cdot D = -1$ for the other eight deterministic points: this facet is "opposite" to the previous one, with the local polytope between the two. The most non-local point above this facet is also a PR-box, the one defined by the rule $a \oplus b = AB \oplus 1$. Note that this PR-box is obtained from the "original" one by trivial local processing: e.g. Alice flips her outcome.

A simple argument of symmetry gives us immediately six other equivalent facets. Indeed, given a Bell inequality, a relabelling of the inputs and/or the outputs provides another Bell inequality. Their number is most easily counted by studying how many different rules for PR-boxes one can find, and these are obviously $a \oplus b = (A \oplus 1)B$, $a \oplus b = A(B \oplus 1)$ and $a \oplus b = (A \oplus 1)(B \oplus 1)$, with the corresponding opposite facets obtained as before by adding 1 on either side.

Now, it can be proved that there are no other Bell's inequalities for this case: only eight versions of CHSH, each with eight extremal points on the corresponding facet[21] and one single PR-box on top. All these facets being equivalent to the others up to trivial relabelling of the inputs and/or the outputs, it is customary to say that *there is only one Bell inequality for the case of two users, each with binary inputs and outputs; namely, CHSH (or CH)*.

The no-signalling polytope and the quantum set. We have studied at length the set of local distributions. Now we have to say a few words about the two other meaningful sets, namely the no-signalling distributions and the distributions that can be obtained with quantum physics. The image of the probability space is usually drawn as in Figuer 3.5. This drawing is of course a rather poor representation of an 8-dimensional object.

The no-signalling conditions obviously define a convex set (if two distributions are no-signalling, any convex combination will also be no-signalling). It turns out that this set is also a polytope, i.e. it has a finite number of extremal points; obviously, it is called a *no-signalling polytope*. All the local deterministic points are also extremal for the no-signalling polytope; but in addition to those, of course, there are some non-local ones, that can in principle be found.[22] For the simple example we studied, the only additional extremal points are the eight PR-boxes defined above (Barrett *et al.*, 2005).

The *quantum set* is also convex, but is not a polytope: it has an uncountable number of extremal points. Interestingly, at the time of writing, the shape of this convex body has not been characterized in full generality yet, not even for the simple case under study here. A necessary condition is the following (Tsirelson, 1987; Landau, 1988; Masanes, 2003): for any probability distribution coming from quantum physics, the correlation coefficients must satisfy an inequality that reminds of CHSH, namely

$$|\arcsin(E_{AB}) + \arcsin(E_{AB'}) + \arcsin(E_{A'B}) - \arcsin(E_{A'B'})| \le \pi. \quad (3.84)$$

However, this condition is provably not sufficient in general: there are probability distributions that satisfy this inequality but cannot be produced with quantum states (Navascues *et al.*, 2008).

Final remarks. Let us conclude this section by saying that most of the simple features of this case study are *not* maintained as soon as one considers more complex situations, i.e. more than two inputs, or more than two outputs, or more than two parties.

For a general polytope, the task of finding the facets given the vertices is increasingly complex (in fact, it's provably computationally hard). Many of Bell's inequalities have by now been found with computers finding facets – and possibly sorting the equivalent ones, otherwise the number is just overwhelming.[23]

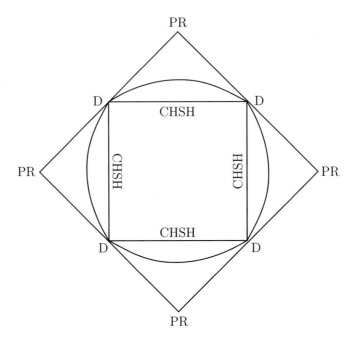

Fig. 3.5 Representation of the local polytope, the set of quantum correlations and the no-signalling polytope, for the case under study (2 parties, 2 inputs and 2 outputs per party). The local polytope (inner square) is delimited by versions of the CHSH inequality; the quantum set (round body) exceeds the local polytope, and is contained in the no-signalling polytope (external square), whose extremal points are PR-boxes. Local deterministic points (D) are extremal points of all three sets (not obvious in the drawing, which is just a projection on a two-dimensional slice: the deterministic points do not lie on this slice).

Also, the characterization of the no-signalling polytope becomes cumbersome. Once moving away from the simplest case, it is no longer the case that all the extremal no-signalling points are equivalent up to symmetries; nor that only one extremal no-signalling point can be found above each facet. The number of objects increases rapidly and the impossibility of visualizing these high-dimensional structures is a powerful deterrent not to embark on such studies unless driven by a very good reason.

3.5.3 The question revisited: why $2\sqrt{2}$?

3.5.3.1 What no-signalling non-local points share with quantum physics

Most of the features that are usually highlighted as "typically quantum" are actually shared by all possible theories that allow non-locality without signalling. This vindicates, in a sense, the intuition of Popescu and Rohrlich: non-locality without signalling is a deep physical principle, to which many observed facts can be attributed.

Here is a rapid list of features that seem to be the share of no-signalling non-local theories, rather than of quantum physics alone:

- *Intrinsic randomness.* Consider first the PR-box: if one enforces the rule $a \oplus b = xy$ (i.e. maximal algebraic violation of CHSH), it is easy to verify that any prescription for the local statistics other than $P(a = 0) = P(b = 0) = \frac{1}{2}$ leads to signalling. In other words, the local outcomes of the PR-box *must* be random in order to satisfy no-signalling. Now, this holds in fact for any non-local probability distribution: as soon as $\mathbf{P}$ violates a Bell inequality (be it compatible or not with quantum physics), the local outcomes cannot be deterministic. We'll see in the next section that this can be quantified to provide guaranteed randomness.

- *No-cloning theorem.* A form of no-cloning theorem can be defined for all non-local no-signalling probability distributions (Masanes *et al.*, 2006; Barnum *et al.*, 2007). The formulation goes as follows: suppose there exist $\mathbf{P}_{ABB'} = P(a, b, b'|x, y, y')$ such that $\mathbf{P}_{AB}$ and $\mathbf{P}_{AB'}$ are the same distribution; then $\mathbf{P}_{AB}$ is local. In other words, if $\mathbf{P}_{AB}$ is non-local, one cannot find an extension $\mathbf{P}_{ABB'}$ satisfying no-signalling and such that $\mathbf{P}_{AB'}$ is equal to $\mathbf{P}_{AB}$: the box of B cannot be cloned. The proof is particularly simple in the case of the PR box, see Tutorial.

- *Possibility of secure cryptography.* It seems that the security of cryptography can be proved only on the basis of no-signalling, without invoking at all the formalism of quantum physics (Barrett, *et al.*, 2005; Acín, *et al.*, 2006). The exact statement is slightly more involved, since a few technical problems in the security proofs have not been sorted out yet: see the most recent developments for all details (Masanes, 2008; Hänggi *et al.*, 2009).

- *Uncertainty relations, i.e. information-disturbance tradeoff.* Such a tradeoff has been discussed in the context of cryptographic protocols (Scarani *et al.*, 2006).

- *Teleportation and swapping of correlations.* After a first negative attempt (Short *et al.*, 2006), it seems that they can actually be defined as well within the general no-signalling framework (Barnum *et al.*, 2008; Skrzypczyk *et al.*, 2009). Still work is in progress.

3.5.3.2 And what they do not

Still, there is overwhelming evidence that our world is well described by quantum physics, while there seems to be no evidence of more-than-quantum correlations. Where does the difference lie? This is an open question. Again, I present a rapid list of what is known.

- *Poor dynamics.* Recall that we are playing with purely kinematical concepts: the $\mathbf{P}$s are, at least at first sight, the analog of the measurement of an entangled state, not of the state itself! There are suggestions that the allowed dynamics of objects like PR-boxes would be seriously restricted (Barrett, 2007; Short and Barrett, 2009; Gross *et al.*, 2009).

- *Communication tasks becoming trivial.* A few communication tasks have been found, for which quantum physics does not lead to any significant advantage over classical physics, while some more-than-quantum correlations would start helping (the task becoming ultimately trivial if PR-boxes would be available). As mentioned above, the first such example was *communication complexity* (van

Dam, 2005; Brassard *et al.*, 2006; Brunner and Skrzypczyk, 2009). Two further tasks were proved to become more efficient than allowed in the quantum world as soon as $S > 2\sqrt{2}$: *non-local distributed computing* (Linden *et al.*, 2007) and a form of *random access codes* (Pawłowski *et al.*, 2009; Allcock *et al.*, 2009). These last works formulated the principle of "information causality" as a possible criterion to rule more-than-quantum correlations out.

- *Classical limit.* Finally, it was noticed that most stronger-than-quantum correlations would not recover the classical world in the limit of many copies (Navascues and Wunderlich, 2009). This criterion of "macroscopic locality" is related to a family of experiments rather than to an information-theoretical task.

3.5.4 Tutorials

3.5.4.1 *Problems*

Exercise 5.1
Which of the following games can be won by pre-established strategies?

1. Each of the players receives a paper to referee. The game is won if both players produce always the same answer, unless both papers come from CQT, in which case they must produce different answers.
2. Generalization of Game 2 of the lectures: the two players must produce the same output if and only if they received the same input. *Hint:* Consider the cases where the number of possible inputs is smaller, equal to or larger than the number of possible outputs.

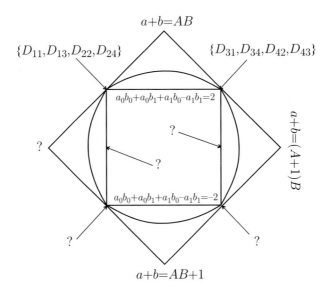

Fig. 3.6 Slice of the no-signalling polytope for 2 parties, 2 inputs and 2 outputs per party.

Exercise 5.2

Figure 3.6 represents a two-dimensional slice of the no-signalling polytope for 2 parties, 2 inputs and 2 outputs per party. Add the missing information. *Hint:* in order to assign the deterministic points, rewrite the inequality in the form of a table, as is done in the text.

Exercise 5.3

Prove that a tripartite distribution $\mathbf{P}_{ABB'} = P(a, b, b'|x, y, y')$, such that $a \oplus b = xy$ and $a \oplus b' = xy'$, violates the no-signalling constraint. This is the proof of the no-cloning theorem for the specific case of the PR-box.

3.5.4.2 Solutions

Exercise 5.1

The first game is obviously non-local: it looks like a generalized PR-box with a large number N of inputs (the possible affiliations) and two outcomes. Interestingly, in the limit $N \to \infty$, these correlations can be achieved with quantum states (Barrett *et al.*, 2006).

The second game can be won with local strategies if and only if the number of outputs is larger Than or equal to the number of inputs. Indeed, if this is the case, the players agree to output their input, or a function thereof on which they had agreed upon in advance. If this is not the case, a generalization of the argument given in the lectures leads to the conclusion that the game cannot be won.

Exercise 5.2

The inequality below the PR-box $a \oplus b = (x+1)y$ is obtained from Eq. (3.82) by flipping Alice's input, i.e. by flipping the lines:

$$\mathbf{T}_{CH} = \begin{array}{c|cc} & -1 & 0 \\ \hline 0 & 1 & -1 \\ -1 & 1 & 1 \end{array}. \tag{3.85}$$

The rest follows quite immediately; the full result is given in Figure 3.7.

Exercise 5.3

The proof is very simple: the two conditions $a \oplus b = xy$ and $a \oplus b' = xy'$ imply $b \oplus b' = x(y \oplus y')$. So, if both B and B' are given to Bob, he knows b, b', y and y' and can therefore reconstruct Alice's input x. In other words, signalling is possible from A to (B, B').

3.6 Quantum correlations (III): the power of Bell

3.6.1 The model, again

In Section 3.4.2.1, the "local variable" model was presented in its original flavor (though clarified by some more modern terminology than the one that was used in the 1960s to the 1980s): namely, a possible alternative description of nature, supposedly

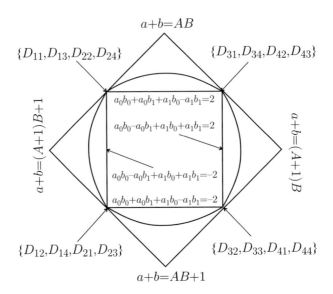

$a+b=AB$

$\{D_{11},D_{13},D_{22},D_{24}\}$

$\{D_{31},D_{34},D_{42},D_{43}\}$

$a+b=(A+1)B+1$

$a+b=(A+1)B$

$a_0b_0+a_0b_1+a_1b_0-a_1b_1=2$

$a_0b_0-a_0b_1+a_1b_0+a_1b_1=2$

$a_0b_0-a_0b_1+a_1b_0+a_1b_1=-2$

$a_0b_0+a_0b_1+a_1b_0-a_1b_1=-2$

$\{D_{12},D_{14},D_{21},D_{23}\}$

$\{D_{32},D_{33},D_{41},D_{44}\}$

$a+b=AB+1$

Fig. 3.7 Slice of the no-signalling polytope for 2 parties, 2 inputs and 2 outputs per party.

along the line Einstein had in mind. Through the check of Bell's inequalities, experiment is called to rule out this particular model. I find it useful to open this last lecture by re-presenting the model under a different light (Gisin, 2007), inspired by the present discussions in quantum information science.

The scenario is about two players, Alice and Bob, who are asked to reproduce some quantum statistics without actually measuring the state. Specifically, Alice and Bob are brought together and told that the shall have to reproduce the statistics of a two-particle state ρ under a family of measurements $\{A_j\}$ for Alice and $\{B_k\}$ for Bob. Let's stress that Alice and Bob are given a *complete and exact description on paper* of both the state and the measurements: therefore, they know the statistics $P_{A_j,B_k}(a,b)$ they should obtain for all j and k. While still together, they are also allowed to agree on a *common strategy* denoted s. After this, Alice and Bob are brought far apart from one another, without possibility of communication, and the game starts: run after run, an external party gives Alice a value of j and Bob a value of k, upon which Alice and Bob have to produce outcomes a and b that will satisfy the expected statistics.

Now, Alice and Bob can succeed if and only if the set of $P_{A_j,B_k}(a,b)$ does not violate a Bell-type inequality. While this conclusion should not surprise the reader at this stage, it may be useful to stress that Alice knows quite a lot about Bob's part: she knows the whole state they are supposed to share, she knows the set of measurements Bob will be asked to choose from, and she may have shared a strategy with him. The only thing Alice does not know about Bob is which k he will receive in each specific run. The same holds for Bob's knowledge about Alice's part. In other words, the "local variable" can be as large as $\lambda = \{ ``\rho", ``\{A_j\}", ``\{B_k\}", s\}$.

The idea can be presented in yet another way: *if Alice and Bob observe a violation of Bell's inequalities and the correlations cannot be attributed to a signal (e.g. because the emission of the signal and the choice of the measurement are space-like separated),* *then Alice and Bob must be measuring entangled states.* In other words, it makes a difference whether Alice and Bob actually share entangled particles and measure them, or whether they try to simulate the same statistics with purely classical means.

3.6.2 Device-independent: quantum information in a black box

The fact that violation of Bell's inequalities implies entanglement was recognized very early in quantum information: in particular, this was precisely Ekert's intuition when he rediscovered quantum cryptography (Ekert, 1991). The new development, which I personally find very exciting, is the awareness that *the amount of violation of Bell's inequalities makes it possible to derive quantitative statements on the "quantumness" of a black box.* Here follow the specific examples that have been studied to date. Since each case refers to a very limited number of papers, I keep this text quite short, hoping to encourage the reader to read the original papers for all details.

3.6.2.1 *Device-independent quantum cryptography*

The first task, to which the idea of device-independent assessment has been applied, is quantum cryptography. This research project amounts precisely at making Ekert's intuition quantitative: if the observed value of the CHSH parameter is S, how much information could the eavesdropper Eve have got? Here is *cryptography at its most paranoid*: the authorized partners Alice and Bob could have bought even the source and the measurement devices from the eavesdropper, provided they control the choices of the measurements!

I direct the reader to the original references for a thorough discussion of the scenario, the results and the open issues (Acín *et al.*, 2007; Ling *et al.*, 2008; Pironio, *et al.*, 2009). See also (Scarani and Kurtsiefer, 2009) for a more personal assessment of the importance of these results for quantum cryptography in general.

3.6.2.2 *Device-independent source characterization*

Once the idea of device-independent assessment is understood, a rather obvious task is the direct assessment of the quality of a source: the observed violation of a Bell inequality should be related to the amount of entanglement of the produced state. This question can be given a slightly different turn, inspired by the idea of device testing (Mayers and Yao, 2004; Magniez *et al.*, 2006): if I use the source that yields a given violation for a specific task, how probable it is that the result differs from the one I would have obtained with an ideal source? As the reader recalls from Section 3.3.1.2, this amounts to finding a bound for the trace distance as a function of the violation of a Bell's inequality. The derivation of such bounds seems very complex even in the simplest case, only partial results are known (Bardyn *et al.*, 2009).

3.6.2.3 *Guaranteed randomness*

Yet another intuitive statement that can be turned quantitative: if one violates Bell's inequalities, the individual outcomes must have some intrinsic randomness – because pseudorandomness is ultimately deterministic and therefore cannot violate

Bell's inequalities! Phrased in a different way: if one observes a violation of Bell's inequalities, one can be sure that the underlying process is random.

This can be an extremely exciting development. As of today, in order to certify a random number generator, one has to look in detail into the complicated process and somehow become convinced that something random (or, at the very least, very hard to predict) is happening. On the contrary, Bell's inequalities provide a very straightforward way of certifying randomness, based on the observed statistics only.

The first such study lay deeply buried in the Ph.D. thesis of Roger Colbeck (Cambridge University). For some reason, it was never turned into a publication; a preprint (Colbeck, 2009) was made available only after the appearance of the work by Pironio, Acín and Massar. This is a much more thorough investigation, complemented with an experimental realization by the group of Monroe (Pironio, Acín, Massar *et al.*, 2009).

3.6.2.4   *Dimension witnesses*

Finally, Bell's inequalities can be used to estimate bounds on the dimension of the Hilbert space of the measured particles. More specifically, one can have the following statement: if an observed violation of some inequality exceeds some threshold, then the system that is being measured must be at least of dimension D. The derivation of such statements is technically demanding, in particular because one wants to rule out also POVMs: in order, for instance, to exclude qubits, it is not enough to observe that three outcomes are possible!

As it turns out, the CHSH inequality cannot be used as a dimension witness, because one can already obtain all possible violations with two-qubit states. The first results were obtained independently for inequalities with ternary outcomes (Brunner *et al.*, 2008) and for inequalities with binary outcomes and more than two measurements (Vértesi and Pál, 2009). Note that, in itself, the problem of bounding the dimension of the Hilbert space does not require the use of Bell's inequalities (Wehner *et al.*, 2008).

3.6.3   Detection loophole: a warning

In all the previous section, I have written "violation of Bell's inequality" several times. This expression refers to a *conclusive* violation, i.e. one that is not obtained by post-selection; in other words, *the detection loophole must be closed*. We have already encountered this loophole in Section 3.4.2.5. There, in the context of experiments, it looked as a very minor point, an absurd conspiracy theory invented by the die-hards of classical mechanisms. Indeed, the vast majority of physicists is certain that the detection loophole will be closed one day: as John Bell himself used to note, it's hard to believe that quantum physics will suddenly be falsified just by increasing the efficiency of our detectors.

However, in device-independent assessment, the situation is quite different. In this scenario, we have not built devices in order to ask questions to a benevolent (or at least, not malevolent) Nature: rather, we are testing devices built by someone else – and this someone else may be adversarial. Eve in cryptography, the vendor of an allegedly good source or trusted random number generator: all may have very good reasons to try and cheat us! In other words, if we are not careful, they may engineer a purely classical device that exhibits "violations of Bell's inequalities" using the detection loophole. This situation has triggered a revival of interest in the detection loophole. For those who want to know more, there is no review paper available, but the latest work (Vértesi, *et al.*, 2009) summarizes pretty well the situation.

3.6.4 Challenges ahead

When I entered the field of quantum information in the year 2000, the general atmosphere was rather cold about Bell's inequalities. Indeed, the general argument one could hear was: "we know by now that local hidden variables are not there, it's time to study entanglement theory". Some of us tried to counter such an argument by vague statements like "Bell's inequalities uncover one of the most astonishing quantum features, they must be useful for something" – not very convincing... but true! It took several years and several detours, but at the time of writing, the role of Bell's inequalities in quantum information has been vindicated: they are the only tool for device-independent quantum information.

One can scarcely imagine quantum information without the notion of "qubit" and still, the assessment on the dimensionality of a quantum system requires a very careful characterization. I would not go as far as suggesting that quantum information will ultimately be formulated without qubits; were it only because there are scenarios in which the physical system is well characterized and there is no reason to be paranoid. But the possibility of device-independent assessment is fascinating, useful... and with plenty of challenges ahead for both theorists and experimentalists!

3.6.5 Tutorials

3.6.5.1 Problems

Exercise 6.1
In this exercise, we consider the detection loophole for a maximally entangled state and the CHSH inequality; but the construction can be easily generalized to study other situations, e.g. non-maximally entangled states (Eberhard, 1993) or asymmetric detection efficiencies (Cabello and Larsson, 2007; Brunner *et al.*, 2007).

Suppose for simplicity that Alice and Bob have a heralded pair source, so that they know when a pair of particles is expected to arrive. They are not allowed to discard any event, so they agree on the following: if Alice's detector fires, she keeps the result of the detection ($+1$ or -1); if not, she arbitrarily decides that the result is $+1$. Bob follows the same strategy. Let now η be the efficiency of the detectors (supposed identical for all detectors): then the observed average of CHSH will be

$$\langle S \rangle = \eta^2 S_2 + 2\eta(1-\eta)S_1 + (1-\eta^2)S_0, \tag{3.86}$$

where S_2 is the expected value of CHSH when both Alice's and Bob's detectors have fired, S_1 when only either Alice's or Bob's have fired, and S_0 where none has fired.

1. Supposing everything else is perfect (ideal measurements, no dark counts, etc.), what are the values of S_2, S_1 and S_0?
2. Prove that $\langle S \rangle > 2$ can be achieved only if $\eta > \frac{2}{\sqrt{2}+1} \approx 82\%$.

3.6.5.2   Solutions

Exercise 6.1

If everything is perfect, when both photons are detected we have $S_2 = 2\sqrt{2}$. However, $S_1 = 0$: when one party detects and the other does not, there are no correlations between the results; and since the detection comes from half of a maximally entangled state, that bit is completely random. Finally, $S_0 = 2$: by pre-established agreement, both Alice and Bob output $+1$, thus achieving the local bound. The threshold for η follows immediately.

At the address of experimentalists: the fact that, in case of no detection, the local bound is achieved, allows to get rid of the no-detection events altogether (these events are not even defined for most real sources that are not heralded).

Notes

1. Strictly speaking, the Hilbert space is the infinite-dimensional vector space used in quantum physics to describe the point-like particle. But it has been customary to extend the name to all the vector spaces of quantum theory, including the finite-dimensional ones on which we mostly focus here.

2. The classical space of states has the structure of a set, not of a vector space, in spite of being identified with $\mathbb{R}^n$. Indeed, the vector sum is not defined in the configuration space: one can formally describe a point as the vector sum of two others, but there is no relation between the physical states (the "sum" of a car going East at 80 km/h and a car going West at 80 km/h is a parked car located at the mid-point). Yet another example of a set of numbers, on which operations are not defined, is the set of telephone numbers.

3. Indeed, one can do the counting on product states: two product states of the composite system are perfectly distinguishable if either of the two states of the first system are perfectly distinguishable, or the two states of the second system are perfectly distinguishable, or both.

4. For those who have never done it, it is useful to check that indeed the state has the same form in any basis. To do so, write a formal singlet in a different basis $\frac{1}{\sqrt{2}}(|+\hat{n}\rangle \otimes |-\hat{n}\rangle + |-\hat{n}\rangle \otimes |+\hat{n}\rangle)$, open the expression up using Eq. (3.22) and verify that this state is exactly the same state as Eq. (3.3).

5. Recall that any mixed state that is not pure can be decomposed in an infinite number of ways as a mixture of pure states (although generally there is only one mixture of pure *mutually orthogonal* states, because of hermiticity).

6. This principle is sometimes stated by saying that entanglement does not allow us to signal "faster-than-light". Indeed, if entanglement would allow signalling, this signal would travel faster than light; and this is why we feel relieved when we notice that quantum physics does not allow such a signalling. However, entanglement does not allow us to signal *tout court*, be it faster or slower than light.

7. Here is an intuitive version of this impossibility result: any unitary operation on a qubit acts as a rotation around an axis in the Bloch sphere, while the NOT is achieved as the point symmetry of the Bloch sphere through its center. Obviously,

no rotation around an axis can implement a point symmetry. A rotation of π around the axis z achieves the NOT only for the states in the (x, y) plane, while leaving the eigenstates of σ_z invariant.

8. The dual problem is related to the definition of min-entropies. Let $\{(\rho_k, \eta_k)\}_{k=1...n}$ be the states to be distinguished with the respective *a priori* probabilities. Then one forms the *classical-quantum* state $\rho_{AB} = \sum_k \eta_k |k\rangle\langle k| \otimes \rho_k$. Choose now a state σ_B and compute the minimum λ such that $M = \lambda \mathbb{1}_A \otimes \sigma_B - \rho_{AB}$ is a non-negative operator: then $\lambda \geq 1 - P_{\text{error}}$, with the guarantee of equality for the minimum over all possible σ_B.

9. The careful reader may immediately notice that the statement, as loosely stated, cannot be strictly true: for instance, the GHZ state $|000\ldots0\rangle + |111\ldots1\rangle$ is invariant under permutation but is far, under any measure, from a product state. Indeed, the exact statement says that: if ξ_N is invariant under permutation, there exist $k < N$ such that $\text{Tr}_k\xi_N$ is close to a mixture of product states (see how this is the case for the GHZ state, already with $k = 1$).

10. In the extreme case of pure states, this is pretty clear. Alice and Bob know which state they share. If the state is pure, there is no coherence with a purifying system to be preserved! Therefore, Bob can just generate the state in his own location: no resources are used, and the shared states are still available for teleportation.

11. For instance, in the language of quantum field theory (which historically pre-dates Bell's inequalities), "non-locality" had been used to mean what we call here "signalling": if you stick to this meaning, quantum theory and quantum field theory are "local", because they are no-signalling.

    As another example, I tend not to like "violation of local realism", because in philosophy "realism" is the school of thought that accepts that there is something outside our brain and we can have some convenient knowledge about it. In particular then, a "philosophical realist" is someone who accepts that "local realism is violated".

12. We are using the well-known fact that the maximal eigenvalue is equal to the largest average value over the set of possible states.

13. This author used to spell his name as Cirel'son until the mid-1980s.

14. Some theorists having speculated on a link between gravitation and "collapse", a recent experiment has bothered connecting the detectors to a piezo-electric device: each time a photon impinges is detected, in addition to the signal being registered by a computer, a *"large" mass* (a few grams) is set into motion (Salart *et al.*, 2008). If those theoretical speculations have a foundation in nature, this experiment may be extremely meaningful. If not ... it had at least the merit of raising the awareness.

15. The locality loophole can be closed (and has been closed) in experiments with photons; but typical detection efficiencies, including losses, are far from reaching the threshold that would close the detection loophole. In turn, this loophole has been closed in experiments with trapped ions and atoms; but the micrometric separation and long detection times definitely prevent exclusion of some subluminal communication.

16. The scheme of Popescu and Rohrlich goes as follows: if the system consists of N particles, measurements are performed on $N - 2$ of them and the results are

communicated to the last two measuring stations: conditioned on this knowledge, the two particles are now in a well-defined bipartite state, so Gisin's original theorem applies. Though it uses communication, the scheme is perfectly valid because the last two parties do *not* communicate with each other: for them, the information communicated by the others acts as pre-established information, which cannot create a violation of Bell's inequalities. For a reason unknown to me, a quite large set of people working in the field are convinced that Gisin's theorem has not been proved in general. Admittedly, for many cases we may not know any compact and elegant inequality that is violated by all pure states; but the scenario of Popescu and Rohrlich does prove the statement in general.

17. For the anecdote, let us stress that for many years the known bound for violation was $W > \frac{1}{\sqrt{2}} \sim 0.7071$; this is the bound directly obtained using CHSH. In order to obtain the minor improvement reported above, namely 0.7056, one needs inequalities that use at least 465 settings per party!

18. We assume that standard Lorentz transformation applies to the description of classical events, such as detection; we refer to the discussion above for the problem of defining classical events at all.

19. Some years later, I realized that this formalized version was not very brilliant after all: if one tries to extend it to three particles, it leads to signalling (Scarani and Gisin, 2002)! But by then, experiments had already falsified the model anyway (Zbinden *et al.*, 2001).

20. In all the following examples, we assume that there is no correlation between the content of a paper and its acceptance or rejection by a referee. This is not a very strong assumption, especially if the referee has some other interests, like here: winning a game.

21. Since there are exactly eight points on each facet, the CHSH polytope is a generalized *tetrahedron* (in 3-dimensional space, the tetrahedron is the simplex that has exactly 3 points on each facet). Like in a tetrahedron, one must take convex combinations of all the eight extremal points to define a point "inside" the facet; any combination of fewer points defines only points on the "edges".

22. The facets of the no-signalling polytope are rather easy to characterize: they are the "trivial" facets that satisfy no-signalling. By intersecting them, a computer program finds the extremal points. So the procedure is somewhat the reverse of the one used to find Bell's inequalities from local deterministic strategies.

23. Note that "lower" Bell inequalities remain facets even in more general cases (Pironio, 2005): for instance, in any local polytope there are CHSH-like facets that involve only two of the users, two of the inputs and two of the outputs. However, even removing those obvious embeds from lower-dimensional polytopes, the number of inequivalent Bell's inequalities grows extremely fast. As an anecdote: Pitowsky and Svozil listed 684 inequalities for the case where the users have each ternary input and binary output (Pitowski and Svozil, 2001). A later inspection by Collins and Gisin (Collins and Gisin, 2004) proved that there are actually only *two* non-equivalent ones: CHSH for sure, and a new one that they baptized I_{3322} but that in fact had been already proposed several years earlier in a forgotten paper (Froissard, 1981).

References

Acín, A., N. Gisin, L. Masanes, *Phys. Rev. Lett.* 97, 120405 (2006)

Acín, A., B. Toner, N. Gisin, *Phys. Rev. A* **73**, 062105 (2006)

Acín, A., N. Brunner, N. Gisin, S. Massar, S. Pironio, V. Scarani, *Phys. Rev. Lett.* **98**, 230501 (2007)

Allcock, J., N. Brunner, M. Pawłowski, V. Scarani, arXiv:0906.3464 (2009)

Audenaert, K.M.R., J. Calsamiglia, L. Masanes, R. Muñoz-Tapia, A. Acín, E. Bagan, F. Verstraete, *Phys. Rev. Lett.* **98**, 160501 (2007)

Bacciagaluppi, G., arXiv:0811.3444 (2008)

Bae, J., A. Acín, *Phys. Rev. Lett.* **97**, 030402 (2006)

Bardyn, C.-E., T.C.H. Liew, S. Massar, M. McKague, V. Scarani, arXiv:0907.2170 (2009)

Barnum, H., J. Barrett, M. Leifer, A. Wilce, *Phys. Rev. Lett.* **99**, 240501 (2007)

Barnum, H., J. Barrett, M. Leifer, A. Wilce, arXiv:0805.3553 (2008)

Barrett, J., *Phys. Rev. A* **65**, 042302 (2002)

Barrett, J., S. Pironio, *Phys. Rev. Lett.* **95**, 140401 (2005)

Barrett, J., N. Linden, S. Massar, S. Pironio, S. Popescu, D. Roberts, *Phys. Rev. A* **71**, 022101 (2005)

Barrett, J., L. Hardy, A. Kent, *Phys. Rev. Lett.* **95**, 010503 (2005)

Barrett, J., A. Kent, S. Pironio, *Phys. Rev. Lett.* **97**, 170409 (2006)

Barrett, J., *Phys. Rev. A* **75**, 032304 (2007)

Bell, J.S., *Physics* **1**, 195 (1964)

Bennett, C.H., G. Brassard, C. Crépeau, R.Jozsa, A. Peres, W.K. Wootters, *Phys. Rev. Lett.* **70**, 1895(1993)

Bennett, C.H., D. DiVincenzo, T. Mor, P.W. Shor, J.A. Smolin, B.M. Terhal, *Phys. Rev. Lett.* **82**, 5385 (1999)

Bohm, D., B.J. Hiley, *The undivided universe* (Routledge, New York, 1993)

Brassard, G., H. Buhrman, N. Linden, A. A. Méthot, A. Tapp, F. Unger, *Phys. Rev. Lett.* **96**, 250401 (2006)

Branciard, C., N. Brunner, N. Gisin, C. Kurtsiefer, A. Lamas-Linares, A. Ling, V. Scarani, *Nature Phys.* **4**, 681 (2008)

Breuer, H.-P., F. Petruccione, *The theory of open quantum systems* (Oxford University Press, Oxford, 2002)

Briegel, H.-J., W. Dür, J.I. Cirac, P. Zoller, *Phys. Rev. Lett.* **81**, 5932 (1998)

Brunner, N., N. Gisin, V. Scarani, C. Simon, *Phys. Rev. Lett.* **98**, 220403 (2007)

Brunner, N., S. Pironio, A. Acín, N. Gisin, A. A. Méthot, V. Scarani, *Phys. Rev. Lett.* **100**, 210503 (2008)

Brunner, N., P. Skrzypczyk, *Phys. Rev. Lett.* **102**, 160403 (2009)

Bruss, D., D.P. DiVincenzo, A. Ekert, C.A. Fuchs, C. Macchiavello, J.A. Smolin, *Phys. Rev. A* **57**, 2368 (1998)

Bužek, V., M. Hillery, *Phys. Rev. A* **54**, 1844 (1996)

Bužek, V., M. Hillery, R.F. Werner, *Phys. Rev. A***60**, R2626 (1999)

Caban, P., J. Rembielinski, *Phys. Rev. A* **59**, 4187 (1999)

Cabello, A., J.-A. Larsson, *Phys. Rev. Lett.* **98**, 220402 (2007)

Caves, C.M., C.A. Fuchs, K. Manne, J.M. Renes, *Found. Phys.* **34**, 193 (2004)

Cerf, N.J., N. Gisin, S. Massar, S. Popescu, *Phys. Rev. Lett.* **94**, 220403 (2005)

Cerf, N.J., J. Fiurášek, *Progress in optics,* vol. 49, ed. E. Wolf (Elsevier, Amsterdam, 2006), p. 455

Chefles, A., *Contemp. Phys.* **41**, 401 (2000)

Christandl, M., R. Koenig, R. Renner, *Phys. Rev. Lett.* **102**, 020504 (2009)

Cirel'son, B.S., Lett. *Math. Phys.* **4**, 93 (1980)

Clauser, J.F., M.A. Horne, A. Shimony, R.A. Holt, Phys. Rev. Lett. **23**, 880 (1969)

Colbeck, R., arXiv:0911.3814

Collins, D., N. Gisin, *J. Phys. A: Math. Gen.* **37**, 1175 (2004)

Cover, T.M., J.A. Thomas, *Elements of information theory* (Wiley-Interscience, Hoboken, 2nd edn 2006)

Degorre, J., S. Laplante, J. Roland, *Phys. Rev. A* **72**, 062314 (2005)

Devetak, I., *IEEE Trans. Inf. Theory* **51**, 44 (2005)

Dieks, D., *Phys. Lett. A* **92**, 271 (1982)

Dieks, D., *Phys. Lett. A* **126**, 303 (1988)

Duan, L.M., M.D. Lukin, J.I. Cirac, P. Zoller, *Nature* **414**, 413 (2001)

Eberhard, P.H., in: W. Schommers (ed.), *Quantum theory and pictures of reality* (Springer, Berlin, 1989)

Eberhard, P.H., *Phys. Rev. A* **47**, R747 (1993)

Ekert, A.K., *Phys. Rev. Lett.* **67**, 661 (1991)

Fine, A., *Phys. Rev. Lett.* **48**, 291 (1982)

Froissard, M., *Nuovo Cim. B* **64**, 241 (1981)

Gisin, N., *Phys. Lett. A* **154**, 201 (1991)

Gisin, N., S. Massar, *Phys. Rev. Lett.* **79**, 2153 (1997)

Gisin, N., *Phys. Lett. A* **242**, 1 (1998)

Gisin, N., quant-ph/0702021 (2007)

Gisin, N., arXiv:0901.4255 (2009)

Gleason, A.M., *J. Math. Mech.* **6**, 885 (1957)

Greenberger, D.M., M. Horne, A. Zeilinger, in: E. Kafatos (ed.), *Bells theorem, quantum theory, and conceptions of the universe* (Kluwer, Dordrecht, 1989), p. 69

Gröblacher, S., T. Paterek, R. Kaltenbaek, Č. Brukner, M. Żukowski, M. Aspelmeyer, A. Zeilinger, *Nature* **446**, 871 (2007)

Gross, D., M. Müller, R. Colberck, O.C.O. Dahlsten, arXiv:0910.1840 (2009)

Hänggi, E., R. Renner, S. Wolf, arXiv:0906.4760 (2009)

Hastings, M.B., *Nature Phys* **5**, 255 (2009)

Hausladen, P., W. K. Wootters, *J. Mod. Opt.* **41**, 2385 (1994)

Hausladen, P., R. Jozsa, B. Schumacher, M. Westmoreland, W.K. Wootters, *Phys. Rev. A* **54**, 1869 (1996)

Helstrom, C.W., *Quantum detection and estimation theory* (Academic Press, New York, 1976)

Herbert, N., *Found. Phys.* **12**, 1171 (1982)

Herzog, U., J.A. Bergou, *Phys. Rev. A* **70**, 022302 (2004)

Horodecki, R., P. Horodecki, M. Horodecki, *Phys. Lett. A* **200**, 340 (1995)

Horodecki, K., M. Horodecki, P. Horodecki, J. Oppenheim, *Phys. Rev. Lett.* **94**, 160502 (2005)

Horodecki, M., J. Oppenheim, A. Winter, *Nature* **436**, 673 (2005)

Horodecki, R., P. Horodecki, M. Horodecki, K. Horodecki, *Rev. Mod. Phys.* **81**, 865 (2009)

Huttner, B., A. Muller, J.D. Gautier, H. Zbinden, N. Gisin, *Phys. Rev. A* **54**, 3783 (1996)

Ivanovic, I.D., *Phys. Lett. A* **123**, 257 (1987)

Jaeger, G.A. Shimony, *Phys. Lett. A* **197**, 83 (1995)

Kempe, J., C. Simon, G. Weihs, *Phys. Rev. A* **62**, 032302 (2000)

Koenig, R., R. Renner, C. Schaffner, arXiv:0807.1338 (2008)

Landau, L., *Found. Phys.* **18**, 449 (1988)

Le Bellac, M., A short introduction to quantum information and quantum computation (Cambridge University Press, Cambridge, 2006)

Leggett, A.J., *Found. Phys.* **33**, 1469 (2003)

Li, K., A. Winter, X. Zou, G. Guo, arXiv:0903.4308 (2009)

Lindblad, G., *Comm. Math. Phys.* **48**, 119 (1976)

Linden, N., S. Popescu, A.J. Short, A. Winter, *Phys. Rev. Lett.* **99**, 180502 (2007)

Ling, A., M.P. Peloso, I. Marcikic, V. Scarani, A. Lamas-Linares, C. Kurtsiefer, *Phys. Rev. A* **78**, 020301(R) (2008)

Magniez, F., D. Mayers, M. Mosca, H. Ollivier, *Self-testing of quantum circuits.* In Proceedings of ICALP2006, Part I, M. Bugliesi *et al.* (ed.), Lecture Notes in Computer Science **4051**, pp. 72–83 (2006) [quant-ph/0512111]

Mandel, L., *Nature* **304**, 188 (1983)

Masanes, L., quant-ph/0309137 (2003)

Masanes, L., A. Acín, N. Gisin, *Phys. Rev. A.* **73**, 012112 (2006)

Masanes, L., arXiv:0807.2158 (2008)

Masanes, L., Y.-C. Liang, A. Doherty, *Phys. Rev. Lett.* **100**, 090403 (2008)

Mayers, D., A. Yao, *Quant. Inf. Comput.*, **4**, 273 (2004)

Mermin, N.D., *Am. J. Phys.* **58**, 731 (1990)

Milonni, P.W., M.L. Hardies, *Phys. Lett A.* **92**, 321 (1982)

Navascues, M., S. Pironio, A. Acín, New J. Phys. **10**, 073013 (2008)

Navascues, M., H. Wunderlich, arXiv:0907.0372 (2009)

Nielsen, M., I. Chuang, *Quantum computation and quantum information* (Cambridge University Press, Cambridge, 2000)

Pan, J.-W., Z.-B. Chen, M. Żukowski, H. Weinfurter, A. Zeilinger, arXiv:0805.2853 (2008)

Pawłowski, M., T. Paterek, D. Kaszlikowski, V. Scarani, A. Winter, M. Żukowski, *Nature* **461**, 1101 (2009)

Peres, A., *Phys. Lett. A* **128**, 19 (1988)

Peres, A., *Quantum theory: concepts and methods* (Kluwer, Dordrecht, 1995)

Peres, A., quant-ph/0205076 (2002)

Pironio, S., *J. Math. Phys.* **46**, 062112 (2005)

Pironio, S., A. Acín, N. Brunner, N. Gisin, S. Massar, V. Scarani, *New J. Phys.* **11**, 045021 (2009)

Pironio, S., A. Acín, S. Massar, A. Boyer de la Giroday, D.N. Matsukevich, P. Maunz, S. Olmschenk, D. Hayes, L. Luo, T. A. Manning, C. Monroe, arXiv:0911.3427

Pitowski, I., K. Svozil, *Phys. Rev. A* **64**, 014102 (2001)

Popescu, S., D. Rohrlich, *Phys. Lett. A* **166**, 293 (1992)

Popescu, S., D. Rohrlich, *Found. Phys.* **24**, 379 (1994)

Popescu, S., *Phys. Rev. Lett.* **74**, 2619 (1995)

Raynal, P., N. Lütkenhaus, *Phys. Rev. A* **72**, 022342 (2005)

Renner, R., *Nature Phys* **3**, 645 (2007)

Reuse, F., Ann. Phys. **154**, 161 (1984)

Sakurai, J.J., *Modern quantum mechanics* (Addison Wesley, Reading, revised edition 1993)

Salart, D., A. Baas, J. van Houwelingen, N. Gisin, H. Zbinden, *Phys. Rev. Lett.* **100**, 220404 (2008)

Salart, D., A. Baas, C. Branciard, N. Gisin, H. Zbinden, *Nature* **454**, 861 (2008)

Sangouard, N., C. Simon, H. De Riedmatten, N. Gisin, arXiv:0906.2699 (2009)

Scarani, V., W. Tittel, H. Zbinden, N. Gisin, *Phys. Lett. A* **276**, 1 (2000)

Scarani, V., M. Ziman, P. Štelmachovič, N. Gisin, V. Bužek, *Phys. Rev. Lett.* **88**, 090705 (2002)

Scarani, V., N. Gisin, *Phys. Lett. A* **295**, 167 (2002)

Scarani, V., S. Iblisdir, N. Gisin, A. Acín, *Rev. Mod. Phys.* 77, 1225–1256 (2005)

Scarani, V., N. Gisin, N. Brunner, Ll. Masanes, S. Pino, A. Acín, *Phys. Rev. A* **74**, 042339 (2006)

Scarani, V., A.A. Méthot, *Quantum Inf. Comput.* **7**, 157 (2007)

Scarani, V., H. Bechmann-Pasquinucci, N.J. Cerf, M. Dušek, N. Lütkenhaus, M. Peev, *Rev. Mod. Phys.* **81**, 1301 (2009)

Scarani, V., C. Kurtsiefer, arXiv:0909.2601 (2009)

Short, A.J., J. Barrett, arXiv:0909.2601 (2009)

Short, A.J., N. Gisin, S. Popescu, *Phys. Rev. A* **73**, 012101 (2006)

Simon, C., G. Weihs, A. Zeilinger, Phys. Rev. Lett **84**, 2993 (2000)

Skrzypczyk, P., N. Brunner, S. Popescu, *Phys. Rev. Lett.* **102**, 110402 (2009)

Smith, G., J. Yard, *Science* **321**, 1812 (2008)

Smith, G., J.A. Smolin, *Phys. Rev. Lett.* **102**, 010501 (2009)

Stefanov, A., H. Zbinden, N. Gisin, A. Suarez, *Phys. Rev. Lett.* **88**, 120404 (2002)

Stefanov, A., H. Zbinden, N. Gisin, A. Suarez, *Phys. Rev. A* **67**, 042115 (2003)

Stucki, D., N. Brunner, N. Gisin, V. Scarani, H. Zbinden, *Appl. Phys. Lett.* **87**, 194108 (2005)

Suarez, A., V. Scarani, *Phys. Lett. A* **232**, 9 (1997)

Tittel, W., G. Weihs, *Quantum Inf. Comput.* **1**, 3 (2001)

Toner, B.F., D. Bacon, *Phys. Rev. Lett.* **91**, 187904 (2003)

Tsirelson, B., *J. Sov. Math.* **36**, 557 (1987)

van Dam, W., quant-ph/0501159 (2005)

Vértesi, T., *Phys. Rev. A* **78**, 032112 (2008)

Vértesi, T., K.F. Pál, *Phys. Rev. A* **79**, 042106 (2009)

Vértesi, T., S. Pironio, N. Brunner, arXiv:0909.3171 (2009)

Wehner, S., M. Christandl, A.C. Doherty, *Phys. Rev. A* **78**, 062111 (2008)

Werner, R.F., *Phys. Rev. A* **40**, 4277 (1989)
Werner, R.F., *Phys. Rev. A* **58**, 1827 (1998)
Wootters, W.K., W.H. Żurek, *Nature* **299**, 802(1982)
Yurke, B., D. Stoler, *Phys. Rev. A* **46**, 2229 (1992)
Zbinden, H., J. Brendel, N. Gisin, W. Tittel, *Phys. Rev. A* **63**, 022111 (2001)
Zukowski, M., A. Zeilinger, M.A. Horne, A.K. Ekert,*Phys. Rev. Lett.* **71**, 4287 (1993)

4

Quantum computing and entanglement

Les Houches School Singapore,
June 29th–July 24th 2009

Dagmar Bruß[1] and Chiara Macchiavello[2]

[1]Institut für Theoretische Physik III, Heinrich-Heine-Universität Düsseldorf,
Universitätsstr. 1, Geb. 25.32, D-40225 Düsseldorf, Germany
[2]Dipartimento di Fisica A. Volta, Universita di Pavia, Via Bassi 6, 27100 Pavia, Italy

4.1 Introduction

Since there are excellent and complete textbooks on quantum computing (see, for example, (Nielsen and Chuang, 2000; Bruß and Leuchs, 2007)), this chapter is just meant to be a report of the basic concepts that were presented during the Les Houches School in Singapore. The chapter is organized as follows. We will first briefly review the basic concepts of computational complexity theory, then, in Section 4.2, we will introduce the concepts of quantum gates and quantum networks, and review the issue of universality in quantum computation. In Section 4.3 we will review the main quantum algorithms, while in Section 4.4 we will describe the basic concepts of quantum error correction. In Section 4.5 an introduction to the theory of entanglement will be given. In Section 4.6 the role of entanglement in quantum algorithms is discussed. In the final Section, 4.7, the basic ideas of one-way quantum computation are presented.

4.1.1 Computational complexity

In order to solve a particular problem computers follow a precise set of instructions that can be mechanically applied to yield the solution to any given instance of the problem. A specification of this set of instructions is called an algorithm. Examples of algorithms are the procedures taught in elementary schools for adding and multiplying whole numbers; when these procedures are mechanically applied, they always yield the correct result for any pair of whole numbers. Some algorithms are fast (e.g. multiplication), others are very slow (e.g. factorization, playing chess). An algorithm is said to be fast or efficient if the time taken to execute it increases no faster than a polynomial function of the size of the input. We generally take the input size to be the total number of bits needed to specify the input (for example, a number n requires $\log_2 n$ bits of binary storage in a computer) and measure the execution time as the number of computational steps. Thus, an efficient algorithm on a general input n runs in $poly(\log n)$ time. The two crucial cases roughly corresponding to what we would loosely call "good" or "bad" algorithms are polynomial and exponential time algorithms, respectively.

Computational complexity theory classifies problems according to the efficiency of algorithms required to solve them. The theory looks at the minimum time and space (memory) required to solve the hardest instance of the problem. Figure 4.1 shows some of the complexity classes and their presumed relationships (not much about this classification has been proved mathematically). The class **P** consists of all problems that can be solved in polynomial time (such as addition and multiplication). A presumably more general class called **NP** contains problems that cannot be solved (or we do not know how to solve them) in polynomial time but verifying that an attempted solution is indeed a solution can be performed in polynomial time (e.g. factoring belongs to **NP** because we do not know how to factor in polynomial time but we can easily check any proposed solution by multiplication). The next important class in the diagram, known in the literature as **BPP**, is somewhat different. It refers to randomized computation and concerns decision problems, i.e. problems for which the output is just "yes" or "no" (problems in other complexity classes can also be

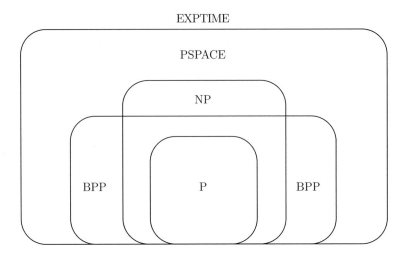

Fig. 4.1 Simplified diagram showing some of the complexity classes and their presumed relationships. The diagram refers to classical computation–quantum complexity classes could be quite different.

phrased this way, e.g. given n and $m < n$, is there a factor of n less than m?). A **BPP** algorithm is an efficient algorithm providing an answer that for any input is correct with a probability greater than some constant $\delta > 1/2$ (e.g. greater than $2/3$). We cannot check easily if the answer is correct or not but we may repeat the algorithm some fixed number k times and then take a majority vote of all the k answers. For sufficiently large k the majority answer will be correct with probability as close to 1 as desired (in fact the probability even converges to 1 *exponentially* fast in k). In computational complexity theory it is customary to view problems in **BPP** as being "tractable" or "solvable in practice" and problems not in **BPP** as "intractable" or "unsolvable in practice on a computer". Further out in the complexity hierarchy is **PSPACE**. Problems in this class can be solved using only polynomial memory space, but not necessarily polynomial time. **PSPACE** includes **NP** but there are problems in **PSPACE** that are thought to be harder than **NP** (needless to say, this isn't proven either). Finally, there is the class of problems called **EXPTIME** that contains problems solvable in exponential time, some of them can actually be proven not to be solvable in deterministic polynomial time. A rich source of **PSPACE** and **EXPTIME** problems has been the area of combinatorial games, e.g. deciding whether or not the first player in a game such as Go or Chess on an $l \times l$ board has a winning strategy (for a more complete treatment of computational complexity theory see, for example, (Papadimitriou, 1994)).

We point out that the definitions of efficient and inefficient algorithms have been carefully constructed to avoid any reference to a physical hardware. Thus, it may happen that the computing power of a quantum device can exceed that of any classical device and new quantum complexity classes are needed to assess the "difficulty" of some mathematical operations such as factoring. We will show how the factoring problem can be solved in polynomial time with the Shor algorithm.

4.2   Quantum networks

In classical computation any Boolean function can be constructed as a Boolean network composed of elementary logic gates. Extending this idea to the quantum domain we will now express quantum algorithms in terms of quantum computational networks. A quantum network is a quantum computing device consisting of quantum logic gates whose computational steps are synchronized in time. The outputs of some of the gates are connected by wires to the inputs of others. Quantum gates can be viewed as the active components of quantum networks—in an n-bit quantum gate, n qubits undergo a coherent interaction. Wires are the passive components that allow quantum states to be carried from one computational step to another.

4.2.1   Single-qubit gates

Let us denote by $\{|0\rangle, |1\rangle\}$ a basis in a Hilbert space describing the state of a single qubit (it can represent, for example, the eigenstates of the σ_z operator). We can conveniently express the state of the qubit in the Bloch vector representation as follows

$$|\psi\rangle = \cos\frac{\theta}{2}|0\rangle + e^{i\phi}\sin\frac{\theta}{2}|1\rangle, \tag{4.1}$$

where $0 \leq \theta \leq \pi$ and $0 \leq \phi \leq 2\pi$. We can easily see that the pure state of a single qubit can then be represented in the Bloch sphere as a point on a unitary spherical surface, located by the azimuthal and longitudinal angles θ and ϕ. The three components of the unit-length vector representing the state in the Bloch picture are called Bloch vector components (denoted by s_x, s_y, s_z) and are related to the expectation values of the Pauli operators as follows

$$\langle\psi|\sigma_x|\psi\rangle = \sin\theta\cos\phi = s_x,$$
$$\langle\psi|\sigma_y|\psi\rangle = \sin\theta\sin\phi = s_y,$$
$$\langle\psi|\sigma_z|\psi\rangle = \cos\theta = s_z. \tag{4.2}$$

Let us introduce some elementary quantum gates. One of the main gates in quantum computing is the Hadamard gate, which implements the Hadamard transform. It is the single qubit gate performing the unitary transformation

$$\mathrm{H} = \frac{1}{\sqrt{2}}\begin{pmatrix} 1 & 1 \\ 1 & -1 \end{pmatrix} \qquad |x\rangle \;-\boxed{\mathrm{H}}-\; (-1)^x|x\rangle + |1 - x\rangle, \tag{4.3}$$

where the diagram on the right provides a schematic representation of the gate H acting on a qubit in state $|x\rangle$, with $x = 0, 1$. The Hadamard gate performs the following basis change

$$H|0\rangle = |+\rangle = \frac{1}{\sqrt{2}}(|0\rangle + |1\rangle),$$

$$H|1\rangle = |-\rangle = \frac{1}{\sqrt{2}}(|0\rangle - |1\rangle). \tag{4.4}$$

Another single-qubit operation that is central in quantum computing is the phase-shift gate, denoted by R(δ), which is given by

$$R(\delta) = \begin{pmatrix} 1 & 0 \\ 0 & e^{i\delta} \end{pmatrix}, \tag{4.5}$$

namely it just introduces a phase factor $e^{i\delta}$ in front of the state $|1\rangle$. The above transformation corresponds to a rotation of the angle δ around the z-axis in the Bloch vector representation.

Any single qubit unitary operation can be written as

$$R_{\mathbf{n}}(\delta) = e^{i\frac{\delta}{2}\mathbf{n}\cdot\sigma}, \tag{4.6}$$

which corresponds to a rotation of the angle δ around the direction identified by the unit-vector $\mathbf{n}$.

4.2.2 Two-qubit gates

An important two-qubit gate is the quantum controlled-NOT (C-NOT or XOR) operation defined as

$$C\text{-}NOT = \begin{pmatrix} 1 & 0 & 0 & 0 \\ 0 & 1 & 0 & 0 \\ 0 & 0 & 0 & 1 \\ 0 & 0 & 1 & 0 \end{pmatrix} \tag{4.7}$$

where $x, y = 0$ or 1 and $\oplus$ denotes addition modulo 2. The matrix is written in the basis $\{|0\rangle|0\rangle, |0\rangle|1\rangle, |1\rangle|0\rangle, |1\rangle|1\rangle\}$ (the diagram on the right shows the structure of the gate). The action of this gate is to flip the value of the second qubit (also called target qubit), when the first qubit (also called the control qubit) is in state $|1\rangle$.

Another important two-qubit gate is the conditional phase shift $C - R(\delta)$ defined as

$$C\text{-}R(\delta) = \begin{pmatrix} 1 & 0 & 0 & 0 \\ 0 & 1 & 0 & 0 \\ 0 & 0 & 1 & 0 \\ 0 & 0 & 0 & e^{i\delta} \end{pmatrix} \tag{4.8}$$

Notice that the above transformation corresponds to the application of a phase-shift gate R(δ) to the target qubit when the control qubit is in state $|1\rangle$.

The above two examples are particular cases of a class of two-qubit gates called controlled unitary C-U, defined as

$$C\text{-}U(|c\rangle|t\rangle) = |c\rangle U^c|t\rangle, \tag{4.9}$$

where U represent an arbitrary single qubit unitary operation that is applied to the target qubit $|t\rangle$ when the control qubit is in state $|1\rangle$.

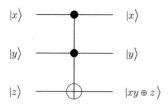

Fig. 4.2 Diagrammatic representation of the controlled-controlled-NOT (Toffoli) gate.

4.2.3 Universality

A central result in quantum computing is that we do not need to worry about constructing complicated multi-qubit gates in order to perform quantum computations because any n-qubit quantum gate can be decomposed in a sequence of single-qubit and two-qubit gates. It is possible to identify sets of universal quantum gates, namely a set of gates out of which any multi-qubit gate can be constructed. An example of a universal set of gates is the one composed of $\{C\text{-}NOT, H, R(\delta)\}$.

This result highlights an important difference of quantum computing with respect to its classical counterpart. In reversible classical computation it is not possible to realize a universal set of classical gates that involves less than three bits (a typical universal gate for reversible classical computation is the Toffoli gate T, which acts on three bits as: $T(x, y, z) = (x, y, z \oplus x \cdot y)$, shown in Figure 4.2), while in quantum computation this is possible with quantum gates that involve at most two qubits. For a complete proof of universality see, for example, (Nielsen and Chuang, 2000).

4.3 Quantum algorithms

In this Section we will describe the main quantum algorithms, namely the Deutsch, Deutsch–Jozsa, Grover and Shor algorithms. In general, the solution of the algorithm is based on the evaluation of a $\{0, 1\}^n \to \{0, 1\}^m$ Boolean function f. The algorithm is run on two registers, composed of n and m qubits, respectively, that are needed to compute f in a reversible way. The initial step of the algorithm consists in preparing an equally weighted superposition of all the 2^n computational basis states. This is typically done, for example, by starting from preparing all qubits in state $|0\rangle$, and then applying a Hadamard transform on all of them, leading to the desired state

$$|\psi_0\rangle \equiv \frac{1}{\sqrt{2^n}} \sum_{x=0}^{2^n-1} |x\rangle = H^{\otimes n}|0\rangle, \tag{4.10}$$

where in the above expression $|x\rangle$ represent the computational basis states of the n qubits. The main body of the quantum algorithm then consists in performing a unitary transformation U_f on the two registers, defined as

$$U_f|x\rangle|y\rangle = |x\rangle|f(x) \oplus y\rangle, \tag{4.11}$$

where x refers to the state of the first register (also called the control register) expressed in the computational basis, y refers to the state of the second register (target register) and the addition is performed bit by bit modulo two. The fact that by applying the above transformation we process in parallel within the superposition all the possible input values for the function f is usually referred to as quantum parallelism. In some instances, such as the problems we will describe in this chapter, it is possible to exploit this feature to solve problems faster than classically. The final step in a quantum algorithm is a measurement, which can be designed in order to efficiently retrieve the desired information about f. In the following we will see some examples, which are interesting from the pedagogical and historical point of view, and allow us to understand the essential features of quantum algorithms. For a recent and complete review on quantum algorithms, see for example (Mosca, 2008).

4.3.1 Deutsch's algorithm

Deutsch's algorithm, proposed by David Deutsch in 1985 (Deutsch, 1985), is the first example of a quantum algorithm that outperforms classical computational means. The problem can be formulated as follows (Cleve *et al.*, 1998): given a Boolean function $f : \{0,1\} \rightarrow \{0,1\}$, find out whether it is constant (i.e. $f(0) = f(1)$) or balanced (i.e. $f(0) \neq f(1)$) by computing f once. Notice that the question is about a property of the function, independently of the specific values it takes for different inputs.

The quantum algorithm then uses two registers, each composed of a single qubit. The target qubit is initially prepared in the state $|-\rangle \equiv (|0\rangle - |1\rangle)/\sqrt{2}$. In this way the action of U_f leads to the state

$$U_f|x\rangle|-\rangle = (-1)^{f(x)}|x\rangle|-\rangle, \tag{4.12}$$

where $(-1)^{f(x)} = \pm 1$ is just a real phase factor. The global state of the two registers after the function evaluation step is then given by

$$\frac{1}{\sqrt{2}}[(-1)^{f(0)}|0\rangle + (-1)^{f(1)}|1\rangle]|-\rangle. \tag{4.13}$$

Notice that the second qubit is always disentangled from the first and is left unchanged by the transformation U_f. The information we are looking for about the function f is contained in the state of the first qubit. The first qubit then undergoes a Hadamard transform and takes the form

$$(-1)^{f(0)}|f(0) \oplus f(1)\rangle. \tag{4.14}$$

The desired information about f is then retrieved by performing a measurement of the first qubit in the computational basis $\{|0\rangle, |1\rangle\}$. If the result is 0 the function is constant, while it is balanced otherwise. The network for the Deutsch algorithm is given in Figure 4.3.

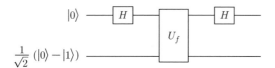

Fig. 4.3   Quantum network for the Deutsch algorithm.

4.3.2   Deutsch–Jozsa's algorithm

An interesting generalization of the above problem is the Deutsch–Jozsa's algorithm (Deutsch and Jozsa, 1992). Given a $\{0,1\}^n \to \{0,1\}$ function, which is promised to be either constant or balanced (balanced means that for half of the inputs it takes output value 0, while for the other half it takes output value 1), the question now is to decide which one by computing it only once (notice that the worst case requires $2^{n-1} + 1$ evaluations of f by classical means).

The algorithm is a generalization of Deutsch's algorithm, with a control register now composed of n qubits and initially prepared in state $|\psi_0\rangle$. As in the previous case, the qubit in the second register is prepared in state $|-\rangle$. The structure of the algorithm is the same as above: the two registers undergo the transformation (4.11), which leads to the state

$$\frac{1}{\sqrt{2^n}} \sum_{x \in \{0,1\}^n} (-1)^{f(x)} |x\rangle |-\rangle. \tag{4.15}$$

As we can see, the target qubit is always not entangled with the first register and it carries no information about the function f. We can therefore restrict our attention to the n-qubit state of the first register and omit the second register in the following analysis. The action of the Hadamard transform on each qubit can be compactly written as

$$H^{\otimes n} |x\rangle = \frac{1}{\sqrt{2^n}} \sum_{z \in \{0,1\}^n} (-1)^{x \cdot z} |z\rangle, \tag{4.16}$$

where $x \cdot z = x_1 z_1 \oplus x_2 z_2 \oplus \ldots x_n z_n$ (x_i represent the binary digits for the number x) and $\oplus$ denotes addition modulo two. The first register is then transformed by $H^{\otimes n}$, leading to the state

$$\frac{1}{2^n} \sum_{x,z} (-1)^{f(x) + x \cdot z} |z\rangle. \tag{4.17}$$

As before, the final step is a measurement in the computational basis, which can be realized as a measurement in the computational basis $\{|0\rangle, |1\rangle\}$ independently for each qubit. The probability $P(0)$ that all the qubits outcome 0 as a result of the measurement is given by

$$P(0) = \frac{1}{2^{2n}} |\sum_{x,z} (-1)^{f(x) + x \cdot z} \delta_{z0}|^2 = \frac{1}{2^{2n}} |\sum_x (-1)^{f(x)}|^2. \tag{4.18}$$

As we can easily see from eq. (4.18), $P(0) = 1$ for a constant function f, while $P(0) = 0$ for a balanced function. Therefore, if the result of the measurement is 0 for all qubits then we know for sure that the function is constant, while it is balanced in all the other cases. The problem can then be solved exactly by performing the function evaluation step U_f just once.

An interesting variation of this problem was proposed by Bernstein and Vazirani (Bernstein and Vazirani, 1993). The $\{0,1\}^n \rightarrow \{0,1\}$ function in this case is promised to be of the form $f(x) = a \cdot x \oplus b$, and the question is to find the unknown n-bit string a (b is a bit). This problem can be solved by the same algorithm used for the Deutsch–Jozsa problem. The state of the first register after the application of U_f has now the form

$$\frac{(-1)^b}{\sqrt{2^n}} \sum_{x=0}^{2^n-1} (-1)^{a \cdot x} |x\rangle. \tag{4.19}$$

After the application of $H^{\otimes n}$ it is then given by

$$\frac{(-1)^b}{2^n} \sum_{x,z} (-1)^{x \cdot (a+z)} |z\rangle = (-1)^b |a\rangle, \tag{4.20}$$

where we have used the relation $\frac{1}{2^n} \sum_z (-1)^{z \cdot (x+y)} = \delta_{xy}$. Therefore, the outcomes of the measurement performed in the computational basis correspond to the value of a.

4.3.3 Grover's algorithm

Grover's algorithm (Grover, 1996) gives an efficient way to solve the search problem with a quadratic speed up with respect to classical computational power (the search problem belongs to the class NP). Let us assume that we have an unsorted database containing N elements and we want to find one or more solutions in it. Let us assume that $N \leq 2^n$ and that we have an oracle (or a black box) that computes the function f: $\{0,1\}^n \rightarrow \{0,1\}$ such that $f(x) = 1$ if x is a solution of the search problem, and $f(x) = 0$ otherwise. The quantum algorithm uses a register of n qubits, initially prepared in state $|\psi_0\rangle$. The algorithm consists in the iteration of a block of operations, that is repeated $O(\sqrt{N/M})$ times before the computation stops and a measurement is performed. This block of operations, usually referred to as the Grover iteration G, is composed of four steps.

1) Oracle call O, where the transformation

$$O|x\rangle = (-1)^{f(x)} |x\rangle \tag{4.21}$$

is performed on the n-qubit register. Notice that the above transformation can be realised, as in the Deutsch–Jozsa's algorithm, by adding a second qubit in state $|-\rangle$ and performing (4.11).

2) $H^{\otimes n}$.

3) A conditional phase-shift transformation P_0, defined as

$$P_0|x\rangle = -(-1)^{\delta_{x0}} |x\rangle, \tag{4.22}$$

which introduces a minus sign in front of all computational basis states apart from the one where all qubits are in state 0.

4) $H^{\otimes n}$.

Notice that the conditional phase shift can be also written as $P_0 = 2|0\rangle\langle 0| - I$, where I denotes the identity operator on the Hilbert space of the n-qubit register. The Grover iteration can then be expressed in a compact way as

$$G = H^{\otimes n}(2|0\rangle\langle 0| - I)H^{\otimes n}O. \tag{4.23}$$

By using Eq. (4.10), the above expression can be more conveniently written as

$$G = (2|\psi_0\rangle\langle\psi_0| - I)O. \tag{4.24}$$

It is now easy to see what happens along the algorithm by taking a geometrical point of view. The operation G actually corresponds to a rotation in the two-dimensional space that is generated by the state vector $|\psi_0\rangle$ and the state vector that is a uniform superposition of the states that are solutions of the search problem. Let us denote this state by $|X_1\rangle$, namely $|X_1\rangle = \frac{1}{\sqrt{N-M}}\sum_{x,f(x)=1}|x\rangle$, and let us denote by $|X_0\rangle = \frac{1}{\sqrt{M}}\sum_{x,f(x)=0}|x\rangle$ the superposition of all the states that are not solutions. We can then express the initial state $|\psi_0\rangle$ as

$$|\psi_0\rangle = \sqrt{\frac{N-M}{N}}|X_0\rangle + \sqrt{\frac{M}{N}}|X_1\rangle. \tag{4.25}$$

The initial state then lies in the subspace generated by $|X_0\rangle$ and $|X_1\rangle$. The action of O on a state belonging to such a subspace is given by

$$O(c_0|X_0\rangle + c_1|X_1\rangle) = c_0|X_0\rangle - c_1|X_1\rangle, \tag{4.26}$$

namely if we represent the state on a two-dimensional plane with axes given by $|X_0\rangle$ and $|X_1\rangle$, it corresponds to a reflection around the axis related to $|X_1\rangle$. Analogously, the operator $(2|0\rangle\langle 0| - I)$ corresponds to a reflection in the same plane around the axis in the direction of the state $|\psi_0\rangle$.

We can express more conveniently the initial state $|\psi_0\rangle$ as

$$|\psi_0\rangle = \cos\frac{\theta}{2}|X_0\rangle + \sin\frac{\theta}{2}|X_1\rangle. \tag{4.27}$$

We can then see that the action of G, which is the sequence of the two reflections discussed above, is a rotation in the plane by an angle θ. Therefore, the state $G^k|\psi_0\rangle$ remains in the plane for any value of k (which represents the number of iterations of the Grover operation), and is rotated by an angle $k\theta$ with respect to the initial state $|\psi_0\rangle$.

We can then write

$$G^k|\psi_0\rangle = \cos\theta_k|X_0\rangle + \sin\theta_k|X_1\rangle, \tag{4.28}$$

where $\theta_k = (k+1/2)\theta$. The operation G is then repeated until the state $G^k|\psi_0\rangle$ overlaps as much as possible with $|X_1\rangle$ that contains a superposition of the solutions to the search problem. This occurs for $k_{opt} = CI[(\pi/\theta - 1)/2]$, where $CI[x]$ denotes the closest integer to x. By using $\sin\theta/2 = \sqrt{M/N}$ and $M \ll N$ we get $k_{opt} = CI[\frac{\pi}{4}\sqrt{N/M} - \frac{1}{2}]$, namely the Grover iterations go as the square root of N. When the iterations stop the n-qubit register is measured in the computational basis. The probability p_w of having a wrong answer, i.e. the probability that the measurement gives as a result one of the states that are not solutions to the search problem, is bounded by

$$p_w = |\langle X_0|G^{k_{opt}}|\psi_0\rangle|^2 \le \cos^2\left(\frac{\pi}{2} - \frac{\theta}{2}\right) \simeq \frac{M}{N}, \qquad (4.29)$$

namely it decreases exponentially fast with the number of qubits n.

4.3.4   Shor's algorithm

Shor's quantum factoring algorithm (Shor, 1994) of an integer N is based on calculating the period r of the function $f(x) = a^x \mathrm{mod} N$, for a randomly selected integer a between 1 and N and coprime with N. For any positive integer y, we define $y \mathrm{mod} N$ to be the unique positive integer $\bar{y}$ between 0 and $N-1$ such that N evenly divides $y - \bar{y}$. The period r of the function f is also called the order of a, namely it is the smallest integer such that $a^r = 1 \mathrm{mod} N$. Once r is known, factors of N are obtained by calculating the greatest common divisor of N and $a^{r/2} = \pm 1 \mathrm{mod} N$ (for details about the number-theoretic basis of the factoring method in Shor's algorithm see, for example, (Ekert and Jozsa, 1996; Nielsen and Chuang, 2000)).

Classically, calculating r is at least as difficult as trying to factor N; the execution time of the best currently known algorithms grows exponentially with the number of digits in N. Shor's algorithm allows us to find the order in a polynomial time. We will now review in a simplified way the main steps of the quantum algorithm.

As mentioned above, the goal of Shor's algorithm is to find the period of the function $f(x) = a^x \mathrm{mod} N$. It is clear that $f(x)$ has period r because for any positive integer j we have $f(x+jr) = a^{x+jr}\mathrm{mod} N = a^x \mathrm{mod} N \cdot a^{jr}\mathrm{mod} N = a^x \mathrm{mod} N = f(x)$. The quantum algorithm is run on two registers composed of n qubits each (with $N^2 \le 2^n \le 2N^2$), which are initially prepared in state $|0\rangle$. A main ingredient in the Shor algorithm is the quantum Fourier transform QFT, which acts as follows on a basis $|j\rangle(j = 0, 2^n - 1)$ for the n-qubit register

$$QFT|j\rangle = \frac{1}{2^{n/2}}\sum_{k=0}^{2^n-1} e^{2\pi ijk/2^n}|k\rangle. \qquad (4.30)$$

The first step of the algorithm consists in applying QFT to the first register in order to create an equally weighted superposition of all basis states (notice that on the initial state $|0\rangle$ it gives the same state as $H^{\otimes n}$, namely $|\psi_0\rangle$). Next, the function evaluation step U_f (4.11) is applied, leading to the global state of the two registers

$$\frac{1}{2^{n/2}} \sum_{x=0}^{2^n-1} |x\rangle |a^x \bmod N\rangle = \frac{1}{2^{n/2}} \sum_{x=0}^{r-1} (|x\rangle + |x+r\rangle + |x+2r\rangle \cdots) |a^x \bmod N\rangle. \quad (4.31)$$

Let us assume for clarity of exposition that 2^n is a multiple of r. The above state can then be written more compactly as

$$\sqrt{\frac{r}{2^n}} \sum_{x=0}^{r-1} \left(\sum_{j=0}^{2^n/r-1} |x+jr\rangle \right) |a^x \bmod N\rangle. \quad (4.32)$$

The second register is then measured in the computational basis. The outcome of the measurement is a number $\bar{y} = a^{\bar{x}} \bmod N$ (the probability distribution of the outcomes is uniform), and the state of the first register is then given by

$$|\phi_{\bar{x}}\rangle = \sqrt{\frac{r}{2^n}} \sum_{j=0}^{2^n/r-1} |\bar{x}+jr\rangle, \quad (4.33)$$

namely it is a uniform superposition of the computational states that corresponds to the same output for f. The quantum Fourier transform is again applied to the first register, leading to

$$QFT|\phi_{\bar{x}}\rangle = \frac{\sqrt{r}}{2^n} \sum_{z=0}^{2^n-1} e^{2\pi i z \bar{x}/2^n} \sum_{j=0}^{2^n/r-1} e^{2\pi i z j r/2^n} |z\rangle. \quad (4.34)$$

By recalling that $\sum_{j=0}^{k-1} e^{2\pi i j z/k}$ equals k when z is a multiple of k and vanishes otherwise, we can rewrite more compactly

$$QFT|\phi_{\bar{x}}\rangle = \frac{1}{\sqrt{r}} \sum_{q=0}^{r-1} e^{2\pi i \bar{x} q/r} |q\frac{2^n}{r}\rangle. \quad (4.35)$$

Notice that the final state $QFT|\phi_{\bar{x}}\rangle$ of the first register contains only the computational basis states that correspond to multiples of $2^n/r$. The first register is then measured in the computational basis, leading as a result to a multiple of $2^n/r$, namely $w = \lambda 2^n/r$, with $0 < \lambda < r$. Since w and 2^n are known, it is possible to learn the ratio $\lambda/r = w/2^n$. If λ and r are coprime, r can then be determined by simplifying $w/2^n$ to an irreducible fraction.

It can be shown that the probability of having λ and r coprime is lower bounded by

$$\mathrm{prob}(\gcd(\lambda, r) = 1) \le \frac{1}{\log r}, \quad (4.36)$$

and therefore the probability of having a bad result after k repetitions of the algorithm decreases exponentially with k.

The more general case where 2^n is not a multiple of r can be treated in a similar way, by using methods based on continuous fraction expansions (see for example (Ekert and Jozsa, 1996)).

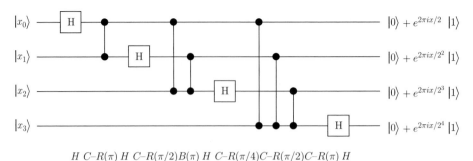

$$H\ C\text{--}R(\pi)\ H\ C\text{--}R(\pi/2)B(\pi)\ H\ C\text{--}R(\pi/4)C\text{--}R(\pi/2)C\text{--}R(\pi)\ H$$

Fig. 4.4 The quantum Fourier transform (QFT) network operating on four qubits. If the input state represents number $x = \sum_k 2^k x_k$ the output state of each qubit is of the form $|0\rangle + e^{i\phi_k}|1\rangle$, where $\phi_k = 2\pi/2^k$ and $k = 0, 1, 3 \ldots$. Note there are three different types of the $C\text{--}R(\phi)$ gate in the network above: $C\text{--}R(\pi)$, $C\text{--}R(\pi/2)$ and $C\text{--}R(\pi/4)$.

Let us finally mention that the above algorithm can be realized with a number of elementary gates that is polynomial in n. First, we notice that quantum factorization contains two major operations: quantum exponentiation (computing $a^x \bmod n$) and the quantum Fourier transform. Quantum exponentiation can be decomposed into a sequence of squaring (linear in n) as follows

$$a^x = a^{2^0 x_0} \cdot a^{2^1 x_1} \cdot \ldots a^{2^{l-1} x_{l-1}}, \tag{4.37}$$

where $x_0, x_1 \ldots$ are the binary digits of x. Squaring is achieved by multiplication and multiplication by a sequence of additions. Following this reduction procedure we end up with a quantum adder as a basic unit for the whole network.

Moreover, the quantum Fourier transform (Eq. 4.30) can be implemented with a network consisting of only the Hadamard gate H and the conditional phase shift $C\text{--}R(\phi)$. In Figure 4.4 we show the network which performs the quantum Fourier transform for $n = 4$. A general case of n qubits requires a trivial extension of the network following the same sequence pattern of gates H and $C\text{--}R$ and containing n Hadamard gates and $n(n-1)/2$ conditional phase shifts, in total $n(n+1)/2$ elementary gates. Thus, the quantum Fourier transform can be performed in an efficient way, the network size grows only as a quadratic function of the size of the input.

4.4 Quantum error correction

4.4.1 Classical repetition codes

The idea of protecting information via encoding and decoding lies at the foundations of classical information theory. It is based on a clever use of redundancy during the data storage or transmission. For example, let us consider the case where the probability of error (bit flip) during a single-bit transmission via a noisy channel is p and each time we want to send bit value 0 or 1 we can encode it by a triple repetition, i.e. by sending 000 or 111. At the receiving end each triplet is decoded as either zero or one following the majority rule–more zeros means 0, more ones means 1. This is the simplest error-correcting protocol that allows to correct up to one error.

In the triple repetition code the signalled bit value is recovered correctly both when there was no error during the transmission of the three bits, which happens with probability $(1 - p)^3$, and when there was one error at any of the three locations, which happens with probability $3p(1 - p)^2$. Thus, the probability of the correct transmission (up to the second order in p) is $1 - 3p^2$, i.e. the probability of error is now $3p^2$, which is much smaller when compared with the probability of error without encoding and decoding p ($p \ll 1$). In this way we can trade the probability of error in the signalled message for a number of transmissions via the channel. In the above example, the reduction of the error rate from p to $3p^2$ requires sending three times more bits. The triple repetition code encodes one bit into three bits and protects against one error, in general we can construct codes that encode l bits into n bits and protect against t errors. The best codes, of course, are those that for a fixed value l minimize n and maximize t.

4.4.2   Quantum noise

Suppose we want to transmit or store a block of l qubits (i.e. two-state quantum systems) in a noisy environment. Here, "noisy" means that each qubit may become entangled with the environment. We will assume that in the block of l qubits each qubit is coupled to a different environment. Basically it allows us to view any noise process of l qubits as a set of independent noise processes of l single qubits.

The qubit–environment interaction leads to the qubit–environment entanglement, which in its most general form can be described as

$$|0\rangle|E\rangle \longrightarrow |0\rangle|E_{00}\rangle + |1\rangle|E_{01}\rangle, \tag{4.38}$$

$$|1\rangle|E\rangle \longrightarrow |0\rangle|E_{10}\rangle + |1\rangle|E_{11}\rangle, \tag{4.39}$$

where states of the environment $|E\rangle$ and $|E_{ij}\rangle$ are in general neither normalized nor orthogonal to each other. The r.h.s. of the formulae above can also be written in a matrix form as

$$\begin{pmatrix} |E_{00}\rangle & |E_{01}\rangle \\ |E_{10}\rangle & |E_{11}\rangle \end{pmatrix} \begin{pmatrix} |0\rangle \\ |1\rangle \end{pmatrix}, \tag{4.40}$$

and the 2×2 matrix can can be subsequently decomposed into some basis matrices, e.g. into the identity and the Pauli matrices

$$|E_0\rangle 1 + |E_1\rangle \sigma_x + i|E_2\rangle \sigma_y + |E_3\rangle \sigma_z, \tag{4.41}$$

where $|E_0\rangle = (|E_{00}\rangle + |E_{11}\rangle)/2$, $|E_3\rangle = (|E_{00}\rangle - |E_{11}\rangle)/2$, $|E_1\rangle = (|E_{01}\rangle + |E_{10}\rangle)/2$, and $|E_2\rangle = (|E_{01}\rangle - |E_{10}\rangle)/2$. Thus, the qubit initially in state $|\Psi\rangle$ will evolve as

$$|\Psi\rangle|E\rangle \longrightarrow \sum_{i=0}^{3} \sigma_i|\Psi\rangle|E_i\rangle \tag{4.42}$$

becoming entangled with the environment (we have relabelled the identity operator and the Pauli matrices $\{1, \sigma_x, \sigma_y, \sigma_z\}$, respectively, as $\{\sigma_0, \sigma_1, \sigma_2, \sigma_3\}$).

The formula (4.42) describes how the environment affects any quantum state of a qubit and shows that a general qubit–environment interaction can be expressed as a superposition of identity and Pauli operators acting on the qubit. As we will see in the following, in the language of error-correcting codes this means that the qubit state is evolved into a superposition of an error-free component and three erroneous components, with errors of the σ_x, σ_y and σ_z type. Notice that errors affecting classical bits can only change their binary values ($0 \leftrightarrow 1$), in contrast quantum errors operators σ_i acting on qubits can change their binary values (σ_x), their phases (σ_z) or both (σ_y).

4.4.3 Three-qubit code

Quantum error correction that protects quantum states is a little more sophisticated simply because the bit flip is not the only "quantum error" that may occur, as we have seen in the previous sections. Moreover, the decoding via the majority rule does not usually work because it may involve measurements that destroy quantum superpositions. Still, the triple repetition code is a good starting point to investigate quantum codes and even to construct the simplest ones.

Let us assume that we want to protect now the unknown state of a single qubit $\alpha|0\rangle + \beta|1\rangle$ and we know that any single qubit that is stored in a register may, with a small probability p, undergo amplitude noise described by Eq. (4.39) with $|E_{00}\rangle = |E_{11}\rangle$ and $|E_{01}\rangle = |E_{10}\rangle$ (this type of noise, which resembles the classical bit-flip error, is called amplitude noise).

Let us now show how to reduce the probability of error to be of the order p^2. We first encode the single qubit by adding two qubits, initially both in state $|0\rangle$, and then perform an encoding unitary transformation

$$|000\rangle \longrightarrow |C_0\rangle = |000\rangle, \qquad |100\rangle \longrightarrow |C_1\rangle = |111\rangle, \qquad (4.43)$$

generating state $\alpha|C_0\rangle + \beta|C_1\rangle$. Now, suppose that only the second qubit was affected by amplitude noise and became entangled with the environment, giving the state

$$(\alpha|C_0\rangle + \beta|C_1\rangle)|E_{00}\rangle + \sigma_x^{(2)}(\alpha|C_0\rangle + \beta|C_1\rangle)|E_{01}\rangle. \qquad (4.44)$$

If vectors $|C_0\rangle$, $|C_1\rangle$, $\sigma_x^{(k)}|C_0\rangle$, and $\sigma_x^{(k)}|C_1\rangle$ are orthogonal to each other we can try to perform a measurement on the qubits and project their state either on the state $\alpha|C_0\rangle + \beta|C_1\rangle$ or on the orthogonal one $\sigma_x^{(2)}(\alpha|C_0\rangle + \beta|C_1\rangle)$. The first case yields the proper state right away, the second one requires one application of σ_x to compensate for the error. The decoding unitary transformation can be constructed using a couple of quantum controlled-NOT gates and the Toffoli gate as shown in Figure 4.5. More careful inspection of the network in Figure 4.5 shows that any single phase-flip $|\bar{0}\rangle \leftrightarrow |\bar{1}\rangle$ will be corrected and the environment will be effectively disentangled from the qubits. In our particular case we obtain

$$(\alpha|0\rangle + \beta|1\rangle)\,[|00\rangle(|E_{00}\rangle + |E_{01}\rangle) + |10\rangle(|E_{00}\rangle - |E_{01}\rangle)]. \qquad (4.45)$$

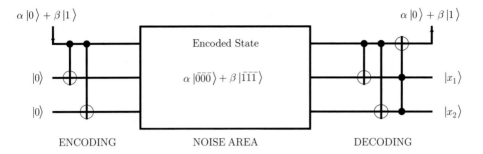

Fig. 4.5 An encoding and decoding (and correcting) network for the three-qubit code.

The two auxiliary outputs carry information about the error syndrome– 00 means no error, 01 means the bit-flip occurred in the third qubit, 10 means the bit-flip in the second qubit and 11 signals the bit flip in the first qubit (c.f. Figure 4.5).

The above quantum error-correcting code can be applied also in the case of dephasing noise

$$|0\rangle|E\rangle \longrightarrow |0\rangle|E_{00}\rangle, \tag{4.46}$$

$$|1\rangle|E\rangle \longrightarrow |1\rangle|E_{11}\rangle, \tag{4.47}$$

namely noise process described by Eq. (4.39) with vanishing $|E_{10}\rangle$ and $|E_{01}\rangle$. Actually we can see that in the Hadamard transformed basis $\{|+\rangle, |-\rangle\}$, phase noise acts as amplitude noise, i.e.

$$|+\rangle|E\rangle \longrightarrow |+\rangle|E_{+}\rangle + |-\rangle|E_{-}\rangle, \tag{4.48}$$

$$|-\rangle|E\rangle \longrightarrow |+\rangle|E_{-}\rangle + |-\rangle|E_{+}\rangle, \tag{4.49}$$

where $|E_{+}\rangle = (|E_{00}\rangle + |E_{11}\rangle)/2$ and $|E_{-}\rangle = (|E_{00}\rangle - |E_{11}\rangle)/2$. The above code then works in the same way as before if the qubits are Hadamard transformed before and after the action of noise.

The first quantum error-correcting code to protect against a noise process of any type acting on a single qubit is the nine-qubit code (Shor, 1995), where the phase and the amplitude correction schemes are combined into one, giving the following encoding transformation

$$|0\rangle \rightarrow \frac{1}{2\sqrt{2}}(|000\rangle + |111\rangle)(|000\rangle + |111\rangle)(|000\rangle + |111\rangle), \tag{4.50}$$

$$|1\rangle \rightarrow \frac{1}{2\sqrt{2}}(|000\rangle - |111\rangle)(|000\rangle - |111\rangle)(|000\rangle - |111\rangle). \tag{4.51}$$

This code involves double encoding, first in basis $|0\rangle$ and $|1\rangle$ and then in basis $|+\rangle$ and $|-\rangle$, and it allows to correct up to one either bit or phase flip. It turns out that the ability to correct both amplitude and phase errors in this combined way suffices to correct any error due to entanglement with the environment.

4.4.4   Conditions and bounds for quantum error correction

Let us now specify the conditions for the existence of quantum error-correcting codes.

We say we can correct a single error $\sigma_i^{(k)}$ (where $i = 0 \ldots 3$ refers to the type of error) if we can find a transformation such that it maps all states with a single error $\sigma_i^{(k)}|\tilde{\phi}\rangle$ into the proper error-free state $|\tilde{\phi}\rangle$:

$$\sigma_i^{(k)}|\tilde{\phi}\rangle \longrightarrow |\tilde{\phi}\rangle. \tag{4.52}$$

To make it unitary we may need an ancilla

$$\sigma_i^{(k)}|\tilde{\phi}\rangle|0\rangle \longrightarrow |\tilde{\phi}\rangle|a_i^k\rangle. \tag{4.53}$$

For encoded basis states of a single qubit $|C_0\rangle$ and $|C_1\rangle$ this implies (Bennett *et al.*, 1996; Knill and Laflamme, 1997)

$$A_k|C_0\rangle|0\rangle \longrightarrow |C_0\rangle|a_k\rangle, \tag{4.54}$$

$$A_k|C_1\rangle|0\rangle \longrightarrow |C_1\rangle|a_k\rangle, \tag{4.55}$$

where A_k denotes all the possible types of independent errors affecting at most one of the qubits. The above requirement leads to the following unitarity conditions

$$\langle C_0|A_k^\dagger A_l|C_0\rangle = \langle C_1|A_k^\dagger A_l|C_1\rangle = \gamma_{kl}, \tag{4.56}$$

$$\langle C_0|A_k^\dagger A_l|C_1\rangle = 0, \tag{4.57}$$

where γ_{kl} represent the entries of an Hermitian matrix. The above conditions can be easily generalized to an arbitrary t error correcting code, which corrects any kind of transformations affecting up to t qubits in the encoded state. In this case the operators A_k are all the possible independent errors affecting up to t qubits, namely operators of the form $\Pi_{i=1}^t \sigma_i$ acting on t different qubits. In the case of the so-called "non-degenerate codes" Eq. (4.56) takes the simple form (Ekert and Macchiavello, 1996),

$$\langle C_0|A_k^\dagger A_l|C_0\rangle = \langle C_1|A_k^\dagger A_l|C_1\rangle = d_k\delta_{kl}. \tag{4.58}$$

This condition requires that all states that are obtained by affecting up to t qubits in the encoded states are all orthogonal to each other, and therefore distinguishable. This ensures that by performing suitable projections of the encoded state we are able to detect the kind of error that occurred and "undo" it to recover the desired error free state. Condition (4.58), even though more restrictive than Eq. (4.56), is quite useful because it allows us to establish bounds on the resources needed in constructing efficient non-degenerate codes. Let us assume that the initial state of l qubits is encoded in a redundant Hilbert space of n qubits. If we want to encode 2^l input basis states and correct up to t errors we must choose the dimension of the encoding Hilbert space 2^n such that all the necessary orthogonal states can be accommodated. According to Eq. (4.58), the total number of orthogonal states that we need in order to be able to correct i errors of the three types σ_x, σ_y and σ_z in an n-qubit state is $3^i\binom{n}{i}$ (this is the number of different ways in which the errors can occur). The argument based on counting orthogonal states then leads to the following condition

$$2^l \sum_{i=0}^{t} 3^i \binom{n}{i} \leq 2^n. \tag{4.59}$$

Equation (4.59) is the quantum version of the Hamming bound for classical error-correcting codes (MacWilliams and Sloane, 1977); given l and t it provides a lower bound on the dimension of the encoding Hilbert space for non-degenerate codes. It follows from Eq. (4.59) that a one-bit quantum error-correcting code to protect a single qubit ($l = 1$, $t = 1$) requires at least 5 encoding qubits (explicit constructions for an optimal five qubit code were first given in (Laflamme *et al.*, 1996; Bennett *et al.*, 1996)).

4.5 Entanglement: introduction and basic definitions

Entanglement is a property that can arise in composite quantum systems, i.e. quantum systems that consist of more than one subsystem (see, e.g., (Bruß and Leuchs, 2007; Horodecki *et al.*, 2009)). The state $|\psi\rangle$ of a composite quantum system is an element of a tensored Hilbert space:

$$|\psi\rangle \in \mathcal{H}_{\text{total}} = \mathcal{H}_A \otimes \mathcal{H}_B \otimes \mathcal{H}_C \otimes \dots. \tag{4.60}$$

It is important to note that the definition of entanglement, as given below, refers to a given tensor structure of the total Hilbert space, as in Eq. (4.60). We will denote the dimensions as $d(\mathcal{H}_A) = d_A, d(\mathcal{H}_B) = d_B, \dots$, and $d(\mathcal{H}_{\text{total}}) = d = d_A \cdot d_B \cdot \dots$. In this Section, we will only consider distinguishable particles. For entanglement of indistinguishable particles, see e.g. (Eckert *et al.*, 2002).

Let us denote the number of subsystems as n. For $n = 2$ we have so-called "bipartite" quantum states, for $n = 3$ we have "tripartite" quantum states, and the case of $n \geq 3$ is also referred to as "multi-partite" quantum states.

4.5.1 Pure bipartite states

Definition 4.1 *A pure state $|\psi\rangle \in \mathcal{H} = \mathcal{H}_A \otimes \mathcal{H}_B$ is separable $\Leftrightarrow |\psi\rangle = |a\rangle \otimes |b\rangle$, otherwise it is entangled.*

Regarding the notation, we will sometimes drop the symbol for the tensor product and denote in an equivalent way: $|0\rangle \otimes |1\rangle \equiv |0\rangle |1\rangle \equiv |01\rangle$.

Example 4.2 A trivial example of a separable pure state of two qubits is given by $|\psi\rangle = |0\rangle \otimes |1\rangle$. A general superposition of separable pure states of two qubits can be written as $|\psi\rangle = a_{00} |00\rangle + a_{01} |01\rangle + a_{10} |10\rangle + a_{11} |11\rangle$ with complex coefficients a_{ij} and $\sum_{i,j} |a_{ij}|^2 = 1$. This state can be separable or entangled, depending on the coefficients c_i. An important example of entangled pure states of two qubits are the *Bell states*:

$$|\psi^{\pm}\rangle = \frac{1}{\sqrt{2}}(|01\rangle \pm |10\rangle), \tag{4.61}$$

$$|\phi^{\pm}\rangle = \frac{1}{\sqrt{2}}(|00\rangle \pm |11\rangle). \tag{4.62}$$

The Bell states form a basis in the four-dimensional Hilbert space.

Here is another example for two similar-looking states of two qubits: $|\psi_1\rangle = \frac{1}{2}(|00\rangle + |01\rangle + |10\rangle + |11\rangle)$ and $|\psi_2\rangle = \frac{1}{2}(|00\rangle + |01\rangle - |10\rangle + |11\rangle)$. Are these states separable or entangled? This question can be easily answered by using the *Schmidt decomposition*.

Lemma 4.3. (Schmidt decomposition) *Every pure bipartite state* $|\psi\rangle \in \mathcal{H}_A \otimes \mathcal{H}_B$, *with dimension* $d = d_A \cdot d_B$, *can be written as a bi-orthogonal sum:*

$$|\psi\rangle = \sum_{k=1}^{r} c_k |e_k\rangle_A \otimes |f_k\rangle_B, \tag{4.63}$$

with real Schmidt coefficients $c_k > 0$ *and* $\sum_k c_k^2 = 1$. *The* Schmidt bases $\{|e_k\rangle\}, \{|f_k\rangle\}$ *are orthonormal, i.e.* $\langle e_k | e_l \rangle = \delta_{kl} = \langle f_k | f_l \rangle$. *The* Schmidt rank r *fulfils* $1 \leq r \leq min(d_A, d_B)$, *where without loss of generality* $d_A \leq d_B$.

Proof (Singular value decomposition) Every complex $d_A \times d_B$ matrix A can be represented as $A = U A_d V^{\dagger}$, where U, V are unitary, i.e. $\sum_i U_{ik}^* U_{il} = \delta_{kl}$ and analogously for V. Here, A_d is the diagonal matrix $A_d = \mathrm{diag}(c_1, \ldots, c_{d_A})$ with c_k real. We denote the elements of A as a_{ij}, and find for every bipartite state $|\psi\rangle$

$$|\psi\rangle = \sum_{i,j=1}^{d_A, d_B} a_{ij} |ij\rangle$$

$$= \sum_{ijk} U_{ik} c_k V_{jk}^* |ij\rangle$$

$$= \sum_k c_k |e_k f_k\rangle, \tag{4.64}$$

where in the last line we have defined $|e_k\rangle = \sum_i U_{ik} |i\rangle$ and $|f_k\rangle = \sum_j V_{jk}^* |j\rangle$. A possible minus sign of c_k can be absorbed into the phase of $|e_k\rangle$ or $|f_k\rangle$. We still have to check orthogonality: $\langle e_k | e_l \rangle = \sum_{ij} \langle j| U_{jk}^* U_{il} |i\rangle = \sum_{ij} U_{jk}^* U_{il} \delta_{ij} = \sum_i U_{ik}^* U_{il} = \delta_{kl}$ (unitarity), and analogously for $|f_k\rangle$. □

Lemma 4.4 *A pure bipartite vector* $|\psi\rangle$ *is separable* $\Leftrightarrow$ *Schmidt rank r=1, otherwise it is entangled.*

One finds the Schmidt decomposition easily by determining the *reduced density matrix*, namely $\rho_A = \mathrm{Tr}_B \rho_{AB} = \sum_k \langle f_k|_B \rho_{AB} |f_k\rangle_B = \sum_k \langle f_k|_B (|\psi\rangle_{AB} \langle\psi|_{AB}) |f_k\rangle_B = \sum_k c_k^2 |e_k\rangle_A \langle e_k|_A$. Thus, the Schmidt coefficients are the square roots of the eigenvalues of the reduced density matrix, i.e. $c_k = \sqrt{\lambda_k}$, where

$\{\lambda_k\} \equiv$ eigenvalues of $\rho_A \equiv$ eigenvalues of ρ_B. Therefore, rank $(\rho_A) =$ Schmidt rank $(\rho_{AB}) = r$. The local Schmidt bases are the eigenbases of the reduced density matrices.

Example 4.5 Let us return to our example: $|\psi_1\rangle = \frac{1}{2}(|00\rangle + |01\rangle + |10\rangle + |11\rangle)$ and $|\psi_1\rangle = \frac{1}{2}(|00\rangle + |01\rangle - |10\rangle + |11\rangle)$. In the following we will use the notation $|\pm\rangle = \frac{1}{\sqrt{2}}(|0\rangle \pm |1\rangle)$. By taking the partial trace we find the reduced density matrix: $\rho_1^A = \frac{1}{4} \cdot 2(|0\rangle + |1\rangle)(\langle 0| + \langle 1|) = |+\rangle\langle +| = \frac{1}{2}\left(\begin{smallmatrix} 1 & 1 \\ 1 & 1 \end{smallmatrix}\right)$. This reduced density matrix has rank $r = 1$, and thus $|\psi_1\rangle$ is separable. For the second state: $\rho_2^A = \frac{1}{4}(|0\rangle - |1\rangle)(\langle 0| - \langle 1|) + \frac{1}{4}(|0\rangle + |1\rangle)(\langle 0| + \langle 1|) = \frac{1}{4}\left(\begin{smallmatrix} 1 & -1 \\ -1 & 1 \end{smallmatrix}\right) + \frac{1}{4}\left(\begin{smallmatrix} 1 & 1 \\ 1 & 1 \end{smallmatrix}\right) = \frac{1}{2}\left(\begin{smallmatrix} 1 & 0 \\ 0 & 1 \end{smallmatrix}\right)$. Here, the reduced density matrix has rank $r = 2$, and thus $|\psi_2\rangle$ is entangled.

In this example one finds that $|\psi_1\rangle$ is separable and $|\psi_2\rangle$ is entangled, although they just differ in one phase. Thus, the minus sign in the second state does matter–it will be important in the context of quantum algorithms that the number and position of minus signs in a superposition of product states with equal (and real) amplitudes determines the entanglement properties of the state.

Note: A pure state $|\psi\rangle$ is separable $\Leftrightarrow$ its reduced density matrices are pure.

Note: The Schmidt decomposition can (in general) *not* be generalized to multipartite systems. An example of a tripartite state that can be written in Schmidt decomposition is $|GHZ\rangle = \frac{1}{\sqrt{2}}(|000\rangle + |111\rangle)$. An example of a tripartite state that cannot be brought into the tri-orthogonal Schmidt form is $|W\rangle = \frac{1}{\sqrt{3}}(|001\rangle + |010\rangle + |100\rangle)$.

4.5.2 Mixed bipartite states

A mixed bipartite state $\rho_{AB} \in \mathcal{B}(\mathcal{H}_A \otimes \mathcal{H}_B)$ is a convex combination of projectors onto pure bipartite states, i.e. $\rho_{AB} = \sum_i p_i |\psi_i\rangle_{AB} \langle\psi_i|_{AB}$. One can also view a mixed bipartite state as the reduced density matrix of a certain tripartite pure state (the *purification* of ρ_{AB}), i.e. $\rho_{AB} = \mathrm{Tr}_E(|\psi\rangle_{ABE} \langle\psi|_{ABE})$ with the purification $|\psi_{ABE}\rangle = \sum_i \sqrt{p_i} |\psi_i\rangle_{AB} |i\rangle_E$.

Definition 4.6. (Entanglement) *A mixed state ρ_{AB}, acting on $\mathcal{H}_A \otimes \mathcal{H}_B$, is separable $\Leftrightarrow$ there exists a decomposition*

$$\rho_{AB} = \sum_i p_i |a_i\rangle \langle a_i| \otimes |b_i\rangle \langle b_i|, \tag{4.65}$$

with $\sum_i p_i = 1$ and $p_i \geq 0$. Otherwise ρ_{AB} is entangled (Werner, 1989).

Note:
- This decomposition is *not* necessarily the spectral decomposition, i.e. in general $\langle a_i|a_j\rangle \neq \delta_{ij}$ and $\langle b_i|b_j\rangle \neq \delta_{ij}$ may hold.
- There exist infinitely many decompositions (no uniqueness).
- Physical interpretation of this definition: a separable state can be prepared *locally*, with the help of classical communication. Entanglement implies *non-locality*.

Example 4.7 (2 qubits)

Example of a separable mixed state–remember that $|\phi^\pm\rangle = \frac{1}{\sqrt{2}}(|00\rangle \pm |11\rangle)$:

$$\rho = \frac{1}{2}|\phi^+\rangle\langle\phi^+| + \frac{1}{2}|\phi^-\rangle\langle\phi^-|$$

$$= \frac{1}{2}|00\rangle\langle00| + \frac{1}{2}|11\rangle\langle11|. \tag{4.66}$$

In this example the "cheapest decomposition" in terms of entanglement is the second line, which proves that ρ is separable.

Example 4.8 (Werner state)

A mixture of a projector onto $|\phi^+\rangle$ and the identity is the so-called Werner state,

$$\rho_W = p|\phi^+\rangle\langle\phi^+| + (1-p)\frac{1}{4}\mathbb{1}. \tag{4.67}$$

It can be easily shown that the Werner state is separable for $0 \le p \le \frac{1}{3}$, and entangled for $\frac{1}{3} < p \le 1$. The proof uses the Peres–Horodecki criterion, see below.

So far we have only met an entanglement criterion for pure bipartite states, namely the Schmidt decomposition. The Peres–Horodecki criterion (or *positive partial transpose* $\equiv$ PPT) is an important entanglement criterion for mixed states.

Definition 4.9 *The partial transpose of ρ with respect to subsystem A is*

$$(\rho^{T_A})_{m\mu,n\nu} = (\rho)_{n\mu,m\nu}, \tag{4.68}$$

where the indices m, n refer to A and μ, ν to B. (Analogous definition for ρ^{T_B}.)

Theorem 4.10. (Peres–Horodecki criterion) *(Peres, 1996; Horodecki et al., 1996) If ρ is a separable state $\Rightarrow$ its partial transpose is positive semi-definite, i.e. $\rho^{T_A} \ge 0$. (The state is said to be PPT.)*
For bipartite systems of dimension 2×2 and 2×3 the reverse holds, i.e. $\rho^{T_A} \ge 0 \Rightarrow \rho$ is separable. (Analogously for $A \leftrightarrow B$.)

Remark 4.11 *A Hermitean matrix χ is positive semi-definite, i.e. $\chi \ge 0 \Leftrightarrow \langle\psi|\chi|\psi\rangle \ge 0 \, \forall\, |\psi\rangle \Leftrightarrow \lambda_i \ge 0 \, \forall i$, where λ_i are the eigenvalues of χ.*

Proof (Proof of Theorem 4.10)

"$\Rightarrow$" obvious, as $\rho^{T_A} = \sum_i p_i(|a_i\rangle\langle a_i|)^T \otimes |b_i\rangle\langle b_i|$, due to linearity, and $(|a_i\rangle\langle a_i|)^T = (|a_i^*\rangle\langle a_i^*|)$. Thus, ρ^{T_A} is a mixture of projectors, i.e. a valid density matrix.
"$\Leftarrow$" The proof uses positive maps (see, e.g. (Horodecki et al., 2009)).   □

Note: • The partial transpose can be interpreted as time reversal in one subsystem. This is an unphysical operation–thus, if the system is entangled, an unphysical (non-positive) matrix may arise.
   • The partial transpose is basis dependent, but its eigenvalues are basis independent.
   • The same arguments as for ρ^{T_A} hold for ρ^{T_B}, as $\rho^{T_B} = (\rho^{T_A})^T$.

- Sometimes one can find a funny notation in the literature: $\rho^{T_A} = \rho^\urcorner, \rho^{T_B} = \rho^\ulcorner$, where $\urcorner$ and $\ulcorner$ stand for "half T".
- In higher dimensions there exist states that are entangled but PPT. They are called *bound entangled* states, see e.g. (Horodecki *et al.*, 2009).

Example 4.12. (Partial transpose of Werner state) Given the Werner state $\rho_W = p|\phi^+\rangle\langle\phi^+| + (1-p)\frac{1}{4}\mathbb{1}$, using the basis $\{|00\rangle, |01\rangle, |10\rangle, |11\rangle\}$ and the general property

$$\rho = \begin{pmatrix} X & Z \\ Z^\dagger & Y \end{pmatrix} \Rightarrow \rho^{T_A} = \begin{pmatrix} X & Z^\dagger \\ Z & Y \end{pmatrix} \text{ and } \rho^{T_B} = \begin{pmatrix} X^T & Z^T \\ Z^* & Y^T \end{pmatrix}, \tag{4.69}$$

one finds that

$$\rho_W^{T_B} = \begin{pmatrix} \frac{1}{4}(1-p) + \frac{p}{2} & 0 & 0 & 0 \\ 0 & \frac{1}{4}(1-p) & \frac{p}{2} & 0 \\ 0 & \frac{p}{2} & \frac{1}{4}(1-p) & 0 \\ 0 & 0 & 0 & \frac{1}{4}(1-p) + \frac{p}{2} \end{pmatrix}. \tag{4.70}$$

The eigenvalues of $\rho_W^{T_B}$ are $\lambda_{1,2,3} = \frac{1}{4}(1+p) \geq 0$ and $\lambda_4 = \frac{1}{4}(1-3p)$, which is negative for $p > \frac{1}{3}$ (i.e. the state is entangled), and positive or zero for $p \leq \frac{1}{3}$ (i.e. the state is separable).

Here, we observe that the Werner state in some surrounding of $\frac{1}{4}\mathbb{1}$ is separable. As we will mention later in the context of NMR quantum computing, there exists a separable ball around the (normalized) identity.

In the following we will consider basic *entanglement measures*, i.e. functions that can quantify entanglement. Starting with the case of pure bipartite states, remember that the reduced density matrix is pure for separable states. Thus, the "degree of mixedness" of the reduced density matrix is related to the degree of entanglement of the total density matrix. A function that quantifies the degree of mixedness of a density matrix ρ is the *von-Neumann entropy*, which is defined as

$$S(\rho) = -\text{Tr}(\rho \log \rho) = -\sum_i \lambda_i \log \lambda_i, \tag{4.71}$$

where λ_i are the eigenvalues of ρ. For $d = 2$ the base of the log is usually chosen to be 2.

Example 4.13 If ρ is a pure state, then $\lambda_1 = 1$ and all other $\lambda_i = 0$. Thus, $S(\rho) = -1\log 1 - \lim_{\epsilon \to 0}(\epsilon \log \epsilon) = 0$. If ρ is the maximally mixed state of a qubit, i.e. $\rho = \frac{1}{2}\begin{pmatrix} 1 & 1 \\ 1 & 1 \end{pmatrix}$, then $S(\rho) = -2 \cdot \frac{1}{2}\log_2 \frac{1}{2} = 1$.

Definition 4.14 *For a pure bipartite state* $|\psi\rangle_{AB}$ *its entropy of entanglement is equal to the von Neumann entropy of the reduced density matrix* $\rho_A = Tr_B(|\psi\rangle\langle\psi|_{AB})$, *i.e.*

$$E(|\psi\rangle) = S(\rho_A) = S(\rho_B) = -\sum_{k=1}^r c_k^2 \log_2 c_k^2, \tag{4.72}$$

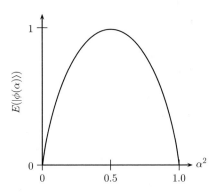

Fig. 4.6 Entropy of entanglement for a pure state of two qubits, see Example 4.15.

where c_k are the Schmidt coefficients and r is the Schmidt rank of $|\psi\rangle_{AB}$.

Example 4.15 For the Bell state $|\phi^+\rangle = \frac{1}{\sqrt{2}}(|00\rangle + |11\rangle)$ we have $\rho_A = \frac{1}{2}\begin{pmatrix} 1 & 0 \\ 0 & 1 \end{pmatrix}$ and thus $E(|\phi^+\rangle) = 1$. For a pure state of two qubits with Schmidt decomposition $|\phi(\alpha)\rangle = \alpha|00\rangle + \sqrt{1-\alpha^2}|11\rangle$, where α is real and positive, one finds $E(|\phi(\alpha)\rangle) = -\alpha^2 \log \alpha^2 - (1-\alpha^2)\log(1-\alpha^2) = h(\alpha^2)$, which is the *binary Shannon entropy*, see Figure 4.6.

Note: For *mixed* bipartite states the von Neumann entropy of ρ_A is not a suitable entanglement measure. To see this, look at the state $\rho_{sep} = \frac{1}{2}(|00\rangle\langle 00| + |11\rangle\langle 11|)$, which has the reduced density matrix $\rho_A = \frac{1}{2}\begin{pmatrix} 1 & 0 \\ 0 & 1 \end{pmatrix}$ and thus $S(\rho_A) = 1$, although ρ_{sep} is clearly separable.

An entanglement measure for pure states can be generalized to mixed states via the *convex roof extension* (the weighted sum of the given measure for the pure states in the decomposition, minimized over all possible decompositions). This construction becomes clear in a famous example for a mixed-state entanglement measure, the *entanglement of formation*.

Definition 4.16. (Entanglement of formation) *The entanglement of formation is the convex-roof extension of the entropy of entanglement, defined as*

$$E_F(\rho_{AB}) = \inf_{\{dec\}} \sum_i p_i S(\rho_{i_A}), \qquad (4.73)$$

where the infimum is taken over all decompositions $\rho_{AB} = \sum_i p_i |\psi_i\rangle_{AB}\langle\psi_i|_{AB}$.

Without a constructive algorithm, a minimization over all decompositions is, in general, "hard" to calculate (also in the sense of complexity theory). The entanglement of formation for two qubits is an entanglement measure for mixed states that can be easily calculated, due to its connection to the *concurrence*, for which Wootters derived an analytic formula (Wootters, 1998).

Definition 4.17. (Concurrence) *Let us denote the spin-flipped density matrix as $\tilde\rho_{AB} = (\sigma_y \otimes \sigma_y)\rho^*_{AB}(\sigma_y \otimes \sigma_y)$. The spectral decomposition of the scalar product of* the original density matrix and its spin-flipped version is $\rho_{AB} \cdot \tilde{\rho}_{AB} = \sum_i \lambda_i^2 \, |\psi_i\rangle \, \langle\psi_i|$. Defining the maximal eigenvalue to be $\lambda_1 = \lambda_{max}$, the concurrence is given as

$$C(\rho_{AB}) = \max\{\lambda_1 - \lambda_2 - \lambda_3 - \lambda_4, 0\}. \tag{4.74}$$

The entanglement of formation is a function of the concurrence (Wootters, 1998),

$$E_F(\rho) = h\left(\frac{1}{2} + \frac{1}{2}\sqrt{1 - C(\rho)^2}\right), \tag{4.75}$$

where $h(x) = -x \log_2 x - (1-x) \log_2(1-x)$ is the binary Shannon entropy. For other entanglement measures, see (Horodecki *et al.*, 2009).

4.5.3   Pure-multipartite states

For more than two subsystems there are several entanglement classes.

Definition 4.18 *A pure state* $|\psi\rangle \in \mathcal{H} = \underbrace{\mathcal{H}_A \otimes \mathcal{H}_B \otimes \ldots \otimes \mathcal{H}_Z}_{n \text{ parties}}$ *is* fully separable *iff*

$$|\psi\rangle = \underbrace{|a\rangle \otimes |b\rangle \otimes \ldots \otimes |z\rangle}_{n}.$$

It is k-separable *with respect to a specific partition, iff*

$$|\psi\rangle = \underbrace{|\alpha\rangle \otimes |\beta\rangle \otimes \ldots \otimes |\omega\rangle}_{k \text{ subsystems with dim } d_\alpha, d_\beta, \ldots}.$$

It is bi-separable *with respect to specific partition, iff k=2, i.e.*

$$|\psi\rangle = \underbrace{|A\rangle}_{m \text{ parties}} \otimes \underbrace{|B\rangle}_{n\text{-}m \text{ parties}}.$$
$$\underbrace{}_{2 \text{ subsystems}}$$

The state $|\psi\rangle$ *is called* fully entangled *(or n-party entangled) iff it is not bi-separable with respect to any bipartition.*

Example 4.19. (GHZ state) A famous example for an n-party entangled state is the GHZ state of n qubits,

$$|\psi_{\text{GHZ}}\rangle = \frac{1}{\sqrt{2}}(|000\ldots00\rangle + |111\ldots11\rangle). \tag{4.76}$$

It has the property that any reduced density matrix of m subsystems is not entangled, as $\rho_{\text{GHZ}}^{\text{red, } m} = \frac{1}{2}(|000\ldots0\rangle \langle 000\ldots0| + |111\ldots1\rangle \langle 111\ldots1|$, where here each state contains m qubits. Remember that a GHZ-like state plays an important role in quantum error correction (see Section 4.4.3). In the case of amplitude noise the encoding is $(\alpha\,|0\rangle + \beta\,|1\rangle) \otimes |00\rangle \xrightarrow{\text{encoding}} \alpha\,|000\rangle + \beta\,|111\rangle$, which is a tripartite entangled state.

For pure states of three qubits there exist inequivalent entanglement classes (Dür et al., 2000):

Definition 4.20 *Two states $|\phi_1\rangle$ and $|\phi_2\rangle$ belong to inequivalent entanglement classes if they cannot be obtained from each other by local operations and classical communication (LOCC) with non-zero probability, i.e. $\nexists$ operators A, B, C, ... (invertible) s.t.*

$$A \otimes B \otimes C \otimes \ldots |\phi_1\rangle = |\phi_2\rangle.$$

Example 4.21. (Three qubits: W versus GHZ states) There exist two inequivalent classes of tripartite entanglement classes for three qubits (Dür et al., 2000), with typical representatives

$$|\psi_{\text{GHZ}}\rangle = \frac{1}{\sqrt{2}}(|000\rangle + |111\rangle), \tag{4.77}$$

$$|\psi_{\text{W}}\rangle = \frac{1}{\sqrt{3}}(|100\rangle + |010\rangle + |001\rangle). \tag{4.78}$$

The general form of the GHZ and W states are given in (Dür et al., 2000). A simple observation shows an important difference between the two types of states: the two-party reduced density matrix of the 3-qubit W state is entangled, as it is given by $\rho_{\text{W}}^{\text{red}} = \frac{1}{3}\{|00\rangle\langle00| + (|01\rangle + |10\rangle)(\langle01| + \langle10|)\}$.

Note that there exists a tripartite entanglement measure, the so-called tangle τ for pure states of three qubits, with $\tau = 0$ for the W state, and $\tau \neq 0$ for the GHZ state.

4.5.4   Mixed multi-partite states

The definition for entanglement of pure multi-partite states can be extended to mixed multi-partite states:

Definition 4.22 *A mixed state ρ, acting on $\mathcal{H} = \underbrace{\mathcal{H}_A \otimes \mathcal{H}_B \otimes \ldots \otimes \mathcal{H}_Z}_{n}$, is fully separable, iff*

$$\rho = \sum_i p_i |a_i\rangle\langle a_i| \otimes |b_i\rangle\langle b_i| \otimes \ldots \otimes |z_i\rangle\langle z_i|,$$

where $p_i \geq 0$ and $\sum_i p_i = 1$.
The state ρ is k-separable with respect to a specific partition, iff

$$\rho = \sum_i p_i \underbrace{|\alpha_i\rangle\langle\alpha_i| \otimes |\beta_i\rangle\langle\beta_i| \otimes \ldots \otimes |\omega_i\rangle\langle\omega_i|}_{k \; subsystems}.$$

The state ρ is bi-separable with respect to a specific partition, iff

$$\rho = \sum_i p_i |A_i\rangle\langle A_i| \otimes |B_i\rangle\langle B_i|.$$

The state ρ is bi-separable, iff it can be written as convex combination of states that are bi-separable w.r.t. a specific partition.

The state ρ is fully entangled, iff it cannot be written as a mixture of biseparable state. There exist inequivalent entanglement classes for fully entangled mixed multi-partite states (Acin et al., 2001).

4.6 Entanglement and quantum algorithms

4.6.1 Role of entanglement in the Deutsch-Jozsa algorithm

Recall the Deutsch algorithm from Section 4.3.1 with the network as in Figure 4.3.

Note that $U_f : |x\rangle |y\rangle \rightarrow |x\rangle |y \oplus f(x)\rangle$ for x,y $= 0,1$ is a gate that may be entangling, depending on the properties of the function $f(x)$:

$$|0\rangle |0\rangle \xrightarrow{H} \frac{1}{\sqrt{2}}(|0\rangle + |1\rangle) |0\rangle \xrightarrow{U_f} \frac{1}{\sqrt{2}}(|0\rangle |0 \oplus f(0)\rangle + |1\rangle |0 \oplus f(1)\rangle)$$

$$= \frac{1}{\sqrt{2}}(|0\rangle |f(0)\rangle + |1\rangle |f(1)\rangle) \begin{cases} \text{for } f(0) = f(1) \rightarrow \text{ separable} \\ \text{for } f(0) \neq f(1) \rightarrow \text{ entangled} \end{cases}. \quad (4.79)$$

In spite of the fact that the gate U_f is in principle an entangling gate, no entanglement occurs during the Deutsch algorithm between the 1st register and the 2nd register, because–as shown in Section 4.3.1:

$$|0\rangle \otimes \frac{1}{\sqrt{2}}(|0\rangle - |1\rangle) \xrightarrow{H} \frac{1}{2}(|0\rangle + |1\rangle) \otimes (|0\rangle - |1\rangle)$$

$$\xrightarrow{U_f} \frac{1}{2}((-1)^{f(0)} |0\rangle + (-1)^{f(1)} |1\rangle) \otimes (|0\rangle - |1\rangle)$$

$$\xrightarrow{H} (-1)^{f(0)} |f(0) \oplus f(1)\rangle \otimes \frac{1}{\sqrt{2}}(|0\rangle - |1\rangle), \quad (4.80)$$

i.e. the total state is separable at all times. For the generalization to the Deutsch–Jozsa algorithm with n inputs, see Chapter 4.3.2, there is also no entanglement between the 1st register (consisting of n qubits) and the 2nd register, because

$$|0\rangle^{\otimes n} \otimes \frac{1}{\sqrt{2}}(|0\rangle - |1\rangle) \xrightarrow{H^{\otimes n} \; U_f \; H^{\otimes n}} \underbrace{\frac{1}{\sqrt{2^n}} \sum_{x=0}^{2^n-1} (-1)^{f(x)} |x\rangle}_{=|\eta_f\rangle} \otimes \frac{1}{\sqrt{2}}(|0\rangle - |1\rangle). \quad (4.81)$$

Now, entanglement in the 1st register may exist: the state $|\eta_f\rangle$ is separable if f is constant, and it is either separable or entangled if f is balanced. A special case is $n = 2$, when the 1st register is always separable. For $n > 2$ one can calculate the number of separable states $|\eta_f\rangle$. It turns out that for $n \rightarrow \infty$ this number is exponentially small, compared with the number of entangled states. For details, see (Bruß and Macchiavello, 2010).

To summarize the role of entanglement in the Deutsch–Jozsa algorithm, there is never entanglement between the 1st and the 2nd register. In any instance of the function f, the 1st register is either separable or entangled, and for large n it is typically entangled. Entanglement is needed in the Deutsch–Jozsa algorithm to accommodate all possible functions.

4.6.2   Role of entanglement in Grover's algorithm

Remember the ingredients of the Grover algorithm (see Section 4.3.3):

i.) The sign-changing operator (oracle) U_f:

$$U_f : |x\rangle \rightarrow (-1)^{f(x)} |x\rangle \quad \text{with} \quad \begin{cases} f(x) = 1 \text{ searched item} \\ f(i \neq x) = 0 \text{ others} \end{cases}. \tag{4.82}$$

This gate may or may not be entangling, depending on the function f. Similarly to the situation in the Deutsch–Jozsa algorithm (see Section 4.6.1), one can study the entanglement properties of the resulting state. As a very simple example, consider a list of four items, represented by two qubits. In the case that only the item $|10\rangle$ is sought, we find

$$|0\rangle \otimes |0\rangle \xrightarrow{H^{\otimes 2}} \frac{1}{2}(|0\rangle \otimes |1\rangle) \otimes (|0\rangle \otimes |1\rangle) = \frac{1}{2}(|00\rangle + |01\rangle + |10\rangle + |11\rangle)$$

$$\xrightarrow{\text{oracle}} \frac{1}{2}(|00\rangle + |01\rangle - |10\rangle + |11\rangle), \tag{4.83}$$

which is entangled (see Schmidt decomposition, Example 4.5). However, if items $|10\rangle$ and $|01\rangle$ are searched, then

$$\cdots \xrightarrow{\text{oracle}} \frac{1}{2}(|00\rangle - |01\rangle - |10\rangle + |11\rangle) = \frac{1}{2}(|0\rangle - |1\rangle)(|0\rangle - |1\rangle), \tag{4.84}$$

which is separable. A more detailed study of the entanglement properties can be found elsewhere (Bruß and Macchiavello, 2010).

ii.) The Grover operator G:

$$G = -H^{\otimes n} U_0 H^{\otimes n} U_f \quad \text{with } U_0 : |0\rangle^{\otimes n} \rightarrow -|0\rangle^{\otimes n}. \tag{4.85}$$

The operator U_0 is entangling, whereas the local operations $H^{\otimes n}$ cannot create entanglement. Application of G generally changes the entanglement properties of the given state.

To summarize the role of entanglement in Grover's algorithm, entanglement may or may not be present in the first step of the algorithm, depending on the number and position of searched items. (For large lists, typically there will be entanglement present, apart from specific search instances.) Entanglement is needed to accommodate all possible searches. One can also study the dynamics of entanglement during the iterated application of the Grover operator (Fang *et al.*, 2005). Note that when interference without entanglement is claimed to be the origin of the speed-up in the

Grover algorithm, see (Lloyd, 1999), exponential resources are needed (i.e. one 2^n-dimensional system instead of n 2-dimensional systems).

4.6.3   Role of entanglement in Shor's algorithm

As we will explain in the following, there is an important difference of the role of entanglement in Shor's algorithm, compared with the Deutsch–Jozsa and Grover algorithm: in Shor's algorithm, entanglement occurs in *any instant*, i.e. for all possible numbers that are factorized. Recall the steps of Shor's algorithm (see Section 4.3.4):

$$|0\rangle^{\otimes n} |0\rangle^{\otimes n} \xrightarrow{H^{\otimes n} \otimes \mathbb{1}^{\otimes n}} \frac{1}{\sqrt{2^n}} \sum_x |x\rangle \otimes |0\rangle^{\otimes n} \xrightarrow{U_f} \frac{1}{\sqrt{2^n}} \sum_x |x\rangle \otimes |a^x \quad \mathrm{mod}\ N\rangle. \quad (4.86)$$

Obviously, in general there exists entanglement between the first and second register (this is another difference to the role of entanglement in Deutsch–Jozsa and Grover). After the next step (measurement of 2nd register), the first register is in the periodic state

$$|\phi_{\bar{x}}\rangle \sim \sum_j |\bar{x} + jr\rangle, \quad (4.87)$$

where r is the period ("order" of $a \ \mathrm{mod}\ N$) and $\bar{x}$ is the offset, e.g. $|\phi_2\rangle \sim |2\rangle + |6\rangle + |10\rangle + \ldots$ for r=4.

Is the periodic state in Eq. (4.87) entangled? Which type of entanglement does it contain–i.e. is there multipartite entanglement? If we restrict our example to three states in the superposition, we have

$$\phi_2 \sim |2\rangle + |6\rangle + |10\rangle \qquad \text{decimal notation}$$
$$= |0_3 0_2 1_1 0_0\rangle + |0_3 1_2 1_1 0_0\rangle + |1_3 0_2 1_1 0_0\rangle \qquad \text{binary notation}$$
$$= (|00\rangle + |01\rangle + |10\rangle)_{32} \otimes |10\rangle_{10}\ .$$

The first factor in this tensor product is an entangled block of two qubits. The total state is bi-separable across the cut of qubits 0 and 1 versus 2 and 3. As shown by Jozsa and Linden (Jozsa and Linden, 2003), for large N periodic states are fully multi-partite entangled.

Lemma 4.23 *(Jozsa and Linden, 2003) With high probability, periodic states for large N have entanglement of an unbounded number of particles.*

Proof (Sketch of proof) Assume the contrary, i.e. there exist blocks of qubits that are in a product state with the rest, i.e.

$$(\overbrace{|a_1\rangle + |a_2\rangle + \ldots |a_{p_a}\rangle}^{\text{increasing order}}) \otimes (\overbrace{|b_1\rangle + |b_2\rangle + \ldots |b_{p_b}\rangle}^{\text{increasing order}}) \otimes \ldots \otimes (\overbrace{|z_1\rangle + |z_2\rangle + \ldots |z_{p_z}\rangle}) .$$

Note that the blocks may arise by reordering the qubits. Thus, the lowest binary string is $|a_1\rangle |b_1\rangle \ldots |z_1\rangle$. Therefore, the second-lowest binary string must be $|a_1\rangle |b_1\rangle \ldots |z_1\rangle + |r\rangle$. Hence, in one of the brackets the difference between the two smallest strings is r, and we take w.l.o.g. $a_2 - a_1 = r$. When N increases, the typical value of r increases. A typical r contains the two-bit string "10" at adjacent places $\frac{1}{4}$ times the number of qubits. For large r, the pair "10" will occur inevitably at positions not included among the qubits of a_i, for a finite block size of a_i. Thus, we arrive at a contradiction to the assumption. □

An important result by Jozsa and Linden (Jozsa and Linden, 2003) is a connection between genuine multipartite entanglement and exponential speed-up.

Theorem 4.24. (Jozsa–Linden) *Genuine multi-partite entanglement (i.e. entanglement of an unbounded number of particles) is necessary for an exponential computational speed-up.*

Proof (Idea of proof) The proof consists of two steps:

- i.) Show that for polynomial time quantum computation with a finite set of *rational* 2-qubit gates, with p being a *fixed positive integer*, where at all time steps $j = 1, \ldots, \mathrm{poly}(m)$ only blocks of maximum size p are entangled in $|\psi_j\rangle$, the final probability distribution can be classically computed in poly(m) steps.
- ii.) Step from rational gates (i.e. rational entries in real and imaginary parts of matrix elements) to all gates (i.e. real numbers).

For details, we refer to the original literature (Jozsa and Linden, 2003). □

Note: This argument does not hold for quantum algorithms employing *mixed* states, because there the parameter counting is different. It is an open question whether an exponential speed-up can be gained via mixed states without multi-partite entanglement.

4.6.4 Quantum computing with NMR

In nuclear magnetic resonance (NMR) a qubit is represented by the spin of an atomic nucleus within a molecule. For example, in Figure 4.7 two qubits correspond to the nuclear spins of two hydrogen atoms. In NMR one applies a magnetic field and studies transitions between Zeeman levels.

Let us denote the transition frequency for qubit i (with $i = 1, 2$) as ω_i, and the two energy levels for qubit i as $E_{i,\pm} = \pm\hbar\omega_i/2$. The population of the levels is determined by the temperature T, and the highly mixed density matrix of qubit i at temperature T reads

$$\rho_i = \begin{pmatrix} p_i(\uparrow) & 0 \\ 0 & p_i(\downarrow) \end{pmatrix} \approx \frac{1}{2}\begin{pmatrix} 1 + \alpha_i & 0 \\ 0 & 1 - \alpha_i \end{pmatrix}, \qquad \text{with } \alpha_i = \frac{\hbar\omega_i}{2kT} \text{ and } i = 1, 2, \quad (4.88)$$

where we have used $p_i \sim \frac{1}{2}e^{-\frac{E_i}{kT}} \approx \frac{1}{2}\left(1 - \frac{E_i}{kT}\right)$, which holds for high temperatures. Here k is the Boltzmann constant.

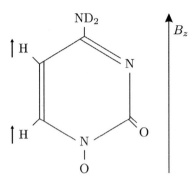

Fig. 4.7 Deuterated cytosine: the nuclear spins of the hydrogen atoms correspond to two qubits.

Then, the tensor product of the density matrices of the two qubits reads

$$\rho = \rho_1 \otimes \rho_2 = \frac{1}{4}\mathbb{1} + \frac{1}{4}\begin{pmatrix} \alpha_1 + \alpha_2 & 0 & 0 & 0 \\ 0 & \alpha_1 - \alpha_2 & 0 & 0 \\ 0 & 0 & -\alpha_1 + \alpha_2 & 0 \\ 0 & 0 & 0 & -\alpha_1 - \alpha_2 \end{pmatrix}. \tag{4.89}$$

One defines the deviation from the normalized identity

$$\rho_\Delta = \rho - \frac{1}{4}\mathbb{1}. \tag{4.90}$$

Here, ρ_Δ is a so-called "pseudo-pure state". The evolution of this state (and therefore the application of quantum gates) can be deduced from the evolution of the original density matrix ρ, as

$$U\rho U^\dagger = \frac{1}{4}\mathbb{1} + U\rho_\Delta U^\dagger \tag{4.91}$$

holds, i.e. the identity is not influenced by a unitary operation.

After the application of unitary operations the "pseudo-state" (i.e. the deviation from the identity) can be entangled or separable, depending on the type of operations. However, it was shown that the original state ρ is never entangled in NMR, as it is close to the identity operator ("close" in the sense of a distance), and as all states in the vicinity of the identity operator are separable.

Theorem 4.25 *(Braunstein et al., 1999) All mixed states of n qubits in a sufficiently small neighborhood of $\frac{1}{2^n}\mathbb{1}$ are separable.*

For the proof we refer to (Braunstein *et al.*, 1999). The separable ball around the identity operator is illustrated in Figure 4.8.

Thus, no "true" entanglement occurs in quantum computing with NMR. (A debate about this topic can be found in the literature.) However, the power of mixed-state

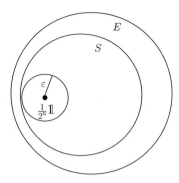

Fig. 4.8 The set of separable states S is a convex compact subset of all states. Entangled states E are not in the set S. The normalized identity operator lies within S, and is surrounded by a separable ball with radius ε. Explicit bounds on ε can be found in the literature.

quantum computing is not yet fully understood, and we will see an example for mixed-state quantum computation without entanglement in the next subsection.

4.6.5 DQC1

The so-called "DQC1" ($\equiv$ Deterministic Quantum Computation with 1 quantum bit) (Knill and Laflamme, 1998) is a computational model in which one has access to one pure qubit and n qubits in the completely mixed state. The DQC1 network, which can estimate the trace of a unitary matrix efficiently, is illustrated in Figure 4.9.

The action of the network in Figure 4.9, acting on $n+1$ qubits, is as follows:

$$
\begin{aligned}
\rho_{n+1} \;&=\; |0\rangle\langle 0| \otimes \frac{1}{2^n}\mathbb{1}_n \\[4pt]
&\xrightarrow{H}\; \frac{1}{2}(|0\rangle + |1\rangle)(\langle 0| + \langle 1|) \otimes \frac{1}{2^n}\mathbb{1}_n \\[4pt]
&\xrightarrow{CU_n}\; \frac{1}{2^{n+1}}(|0\rangle\langle 0| \otimes \mathbb{1}_n + |1\rangle\langle 1| \otimes \mathbb{1}_n + |0\rangle\langle 1| \otimes U_n^\dagger + |1\rangle\langle 0| \otimes U_n).
\end{aligned}
\tag{4.92}
$$

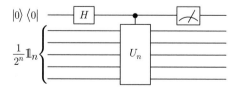

Fig. 4.9 Network for the DQC1, where H is the Hadamard gate, and the action of the control-U_n gate is given by $CU_n:\ |0\rangle \otimes |\psi\rangle \to |0\rangle \otimes |\psi\rangle$ and $|1\rangle \otimes |\psi\rangle \to |1\rangle \otimes U_n|\psi\rangle$.

A measurement of σ_x and σ_y on the *first* qubit reveals the trace of U_n, because

$$\mathrm{tr}(\rho\sigma_x) = \langle\sigma_x\rangle = \frac{1}{2^n}\mathrm{Re}\,[\mathrm{tr}(U_n)],$$

$$\mathrm{tr}(\rho\sigma_y) = \langle\sigma_y\rangle = -\frac{1}{2^n}\mathrm{Im}\,[\mathrm{tr}(U_n)]. \tag{4.93}$$

Here, we have used $\sigma_x = |0\rangle\langle1| + |1\rangle\langle0|$ and $\sigma_y = -i|0\rangle\langle1| + i|1\rangle\langle0|$. The best classical algorithm to determine $\mathrm{tr}(U_n)$ needs $\mathcal{O}(2^n)$ operations. In the DQC1 the number of runs to achieve a fixed accuracy does not depend on n, as the accuracy of the expectation value is $\varepsilon \sim \frac{1}{\sqrt{L}}$, where L is the number of runs. Therefore, one needs $L = \frac{1}{\varepsilon^2}$ runs to achieve a fixed accuracy ε. Thus, the DQC1 model provides an exponential speed-up over the best-known classical algorithm. It is surprising that this speed-up is reached with a state where the first qubit is not entangled with the other n qubits (Poulin *et al.*, 2004), and also the reduced state of the n qubits remains completely mixed during the computation. The full power of mixed-state quantum computing still remains to be explored.

4.7 One-way quantum computing

The concept of "one-way quantum computing" (or "measurement-based quantum computing") was introduced in (Raussendorf and Briegel, 2001). The idea is to implement a computation by performing a sequence of measurements of specific qubits in specific bases (later measurement bases depend on earlier measurement outcomes) on a certain type of entangled state of many qubits (a so-called "cluster state"). The entanglement is consumed during this process. The final result is given by the classical data of all measurement outcomes. It can be shown (Raussendorf and Briegel, 2001; Raussendorf *et al.*, 2003) that the one-way quantum computer is computationally equivalent to quantum circuit model (see Section 4.2).

4.7.1 Cluster states and graph states

Let us first introduce the states that are the basic ingredient of measurement-based quantum computation. We start with a reminder about graphs:

Definition 4.26 *An (undirected, finite) graph $G = (V, E)$ consists of a finite set of vertices V and a finite set of edges E.*

The *adjacency matrix* Γ_G for a graph G with $V = \{a_1, \ldots, a_N\}$ is a symmetric $N \times N$-matrix:

$$(\Gamma_G)_{ij} = \begin{cases} 1 & \text{if } \{a_i, a_j\} \in E \\ 0 & \text{otherwise} \end{cases}. \tag{4.94}$$

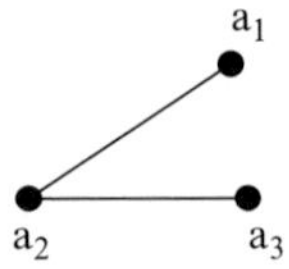

Fig. 4.10 Example of a graph with vertices $V = \{a_1, a_2, a_3\}$.

For the example in Figure 4.10 the adjacency matrix reads

$$\Gamma_G = \begin{pmatrix} 0 & 1 & 0 \\ 1 & 0 & 1 \\ 0 & 1 & 0 \end{pmatrix}. \tag{4.95}$$

The *neighborhood* $N_a \subset V$ of a vertex $a \in V$ is the set of vertices $\{b\}$ that are neighbors of a, i.e. $\{a, b\} \in E$.

Definition 4.27 *A graph state* $|G\rangle$, *assigned to a graph* $G = (V, E)$, *is defined via*

$$K_G^{(a)}|G\rangle = |G\rangle \qquad \forall \, a \in V, \tag{4.96}$$

where $K_G^{(a)}$ is the following correlation operator:

$$K_G^{(a)} = \sigma_x^{(a)} \bigotimes_{b \in N_a} \sigma_z^{(b)} = \sigma_x^{(a)} \bigotimes_{b \in V} (\sigma_z^{(b)})^{\Gamma_{ab}}. \tag{4.97}$$

The set of eigen equations in Eq. (4.96) uniquely (up to a global phase) defines $|G\rangle$. Note that the operators $K_G^{(a)}$ are the generators of a finite Abelian group S_G which is called the *stabilizer* of the graph state $|G\rangle$.

Let us give a very simple example, for the most simple graph of two connected vertices, named 1 and 2: Here, the adjacency matrix is $\Gamma = \begin{pmatrix} 0 & 1 \\ 1 & 0 \end{pmatrix}$, and the stabilizer operators are $K^{(1)} = \sigma_x^{(1)} \otimes \sigma_z^{(2)}$ and $K^{(2)} = \sigma_z^{(1)} \otimes \sigma_x^{(2)}$. Thus,

$$\left. \begin{aligned} \sigma_x^{(1)} \otimes \sigma_z^{(2)}|G_2\rangle &= |G_2\rangle \\ \sigma_z^{(1)} \otimes \sigma_x^{(2)}|G_2\rangle &= |G_2\rangle \end{aligned} \right\} \quad \Rightarrow |G_2\rangle = \frac{1}{\sqrt{2}}(|0\rangle_z|0\rangle_x + |1\rangle_z|1\rangle_x), \tag{4.98}$$

where $|0\rangle_x = \frac{1}{\sqrt{2}}(|0\rangle_z + |1\rangle_z)$ and $|1\rangle_x = \frac{1}{\sqrt{2}}(|0\rangle_z - |1\rangle_z)$. Here, we have used the action of the Pauli operators: $\sigma_z|0\rangle_x = |1\rangle_x$, $\sigma_z|1\rangle_x = |0\rangle_x$, and $\sigma_x|0\rangle_z = |1\rangle_z$, $\sigma_x|1\rangle_z = |0\rangle_z$.

Another example is the "spider" graph, as given in Figure 4.11. It is straightforward to show that the graph state corresponding to this spider graph is a GHZ state for n particles:

$$|G_{\text{spider}}\rangle = \frac{1}{\sqrt{2}}(|0\rangle_z|0\rangle_x|0\rangle_x \cdots + |1\rangle_z|1\rangle_x|1\rangle_x \cdots). \tag{4.99}$$

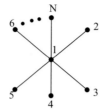

Fig. 4.11 The "spider" graph with $n-1$ "legs".

A *cluster state* is a graph state associated with a particular type of graph, namely a cluster:

Definition 4.28 *A cluster state* $|C\rangle$ *is a graph state on a connected subset (cluster C) of a rectangular lattice in* $d \geq 1$ *dimensions.*

Here, "connected" means that every vertex is reachable from every other vertex via edges. A simple example for a cluster in $d = 2$ is given in Figure 4.12. The most simple examples for cluster states in $d = 1$ are as follows: For $n = 2$ we have already given the cluster state above, see Eq. (4.98).

For $n = 3$ the cluster is given in Figure 4.13. It is a "spider graph" with two legs, and thus the corresponding cluster state is

$$|C_3\rangle = \frac{1}{\sqrt{2}}(|0\rangle_x|0\rangle_z|0\rangle_x + |1\rangle_x|1\rangle_z|1\rangle_x). \tag{4.100}$$

For $n = 4$ the cluster is depicted in Figure 4.14. Note that this is *not* a "spider graph", and therefore the corresponding cluster state is *not* a 4-party GHZ state, but

$$|C_4\rangle = \frac{1}{2}(|+0+0\rangle + |+0-1\rangle + |-1-0\rangle + |-1+1\rangle), \tag{4.101}$$

where we have used the notation $|+\rangle = |0\rangle_x = \frac{1}{\sqrt{2}}(|0\rangle_z + |1\rangle_z)$, and $|-\rangle = |1\rangle_x = \frac{1}{\sqrt{2}}(|0\rangle_z - |1\rangle_z)$.

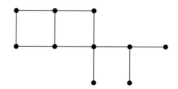

Fig. 4.12 Example for a cluster in $d = 2$.

Fig. 4.13 Cluster with $d = 1$ and $n = 3$.

Fig. 4.14 Cluster with $d = 1$ and $n = 4$.

One can *prepare* a cluster state (or graph state) by performing pairwise Hadamard operations and then turning on an Ising interaction between all neighbors:

i.) Prepare a product state, $|+\rangle_C = \bigotimes_{a \in C} |+\rangle_a$.

ii.) Apply a control phase gate S^{ab} between neighboring qubits:

$$S^{(C)} = \prod_{a,b \text{ neighbours } \in C} S^{ab} .$$

(4.102)

The action of the control phase gate is given by

$$S^{ab} : |00\rangle_{ab} \rightarrow |00\rangle_{ab},$$
$$|01\rangle_{ab} \rightarrow |01\rangle_{ab},$$
$$|10\rangle_{ab} \rightarrow |10\rangle_{ab},$$
$$|11\rangle_{ab} \rightarrow -|11\rangle_{ab}.$$

(4.103)

One can also express it in tensor products of Pauli operators, or in matrix form:

$$S^{ab} = \frac{1}{2} \left(\mathbb{1}^a \otimes \mathbb{1}^b + \sigma_z^{(a)} \otimes \mathbb{1}^{(b)} + \mathbb{1}^{(a)} \otimes \sigma_z^{(b)} - \sigma_z^{(a)} \otimes \sigma_z^{(b)} \right) = \begin{pmatrix} 1 & 0 & 0 & 0 \\ 0 & 1 & 0 & 0 \\ 0 & 0 & 1 & 0 \\ 0 & 0 & 0 & -1 \end{pmatrix}.$$

(4.104)

The claim is that $|C\rangle = S^{(C)}|+\rangle_C$ is a cluster state, i.e. it fulfils the eigenvalue equations for a cluster state, see Eq. (4.96). In order to prove this claim, we first show the action of $S^{(C)}$ on the Pauli operator σ_x, i.e. we calculate $S^{(C)} \sigma_x^{(a)} S^{(C)\dagger}$:

$$S^{ab}(\sigma_x^{(a)} \otimes \mathbb{1}^{(b)})S^{ab\dagger} = \sigma_x^{(a)} \otimes \sigma_z^{(b)},$$
$$S^{ab}(\mathbb{1}^{(a)} \otimes \sigma_x^{(b)})S^{ab\dagger} = \sigma_z^{(a)} \otimes \sigma_x^{(b)},$$
$$S^{ab}(\sigma_x^{(c)} \otimes \mathbb{1})S^{ab\dagger} = \sigma_x^{(c)} \otimes \mathbb{1} \qquad \forall c \in C \setminus \{a, b\},$$

(4.105)

where we have used the commutation relations $[\sigma_x, \sigma_z] = -2i\sigma_y$ etc. We conclude that

$$S^{(C)}(\sigma_x^{(a)} \otimes \mathbb{1})S^{(C)\dagger} = \sigma_x^{(a)} \bigotimes_{b \in N_a} \sigma_z^{(b)}.$$

(4.106)

From here the claim almost immediately follows, because we have

$$S^{(C)}|+\rangle_C = S^{(C)}(\sigma_x^{(a)} \otimes \mathbb{1})|+\rangle_C = S^{(C)}(\sigma_x^{(a)} \otimes \mathbb{1})S^{(C)\dagger}S^{(C)}|+\rangle_C \qquad \forall a \in C,$$

(4.107)

where we have used unitarity of the phase gate. Using Eq. (4.106) concludes the proof.

Thus, we have shown that the gate S^{ab} creates an edge between vertices a and b in the graph. Note that the interaction term of the Hamiltonian that corresponds to this operation is an *Ising* Hamiltonian:

$$H_{\text{int}} = \hbar g \sum_{a,b \text{ neighbours}} \sigma_z^{(a)} \otimes \sigma_z^{(b)}, \qquad (4.108)$$

where g is the coupling constant.

The essential building blocks to implement an algorithm on the one-way quantum computer are *local* measurements:

Lemma 4.29 (Local Pauli measurements for graph states)
Let $G = (V, E)$ be a graph, with associated graph state $|G\rangle$. A measurement of $\sigma_x^{(a)}$, $\sigma_y^{(a)}$ or $\sigma_z^{(a)}$ of the qubit at the vertex $a \in V$ leads to the following state (which depends on the measurement result ± 1):

$$P_{\tau,\pm}^{(a)} |G\rangle = |\tau, \pm\rangle^{(a)} \otimes U_{\tau,\pm}^{(a)} |G'\rangle; \qquad \tau = x, y, z, \qquad (4.109)$$

where $P_{\tau,\pm}^{(a)} = \frac{1 \pm \sigma_\tau^{(a)}}{2}$ is the projector onto $|\tau, \pm\rangle$, e.g. $|z, +\rangle = |0\rangle$, $|x, -\rangle = \frac{1}{\sqrt{2}}(|0\rangle - |1\rangle)$ etc. Here, for the case $\tau = z$ we have

$$U_{z,+}^{(a)} = \mathbb{1}; \quad U_{z,-}^{(a)} = \prod_{b \in N_a} \sigma_z^{(b)}, \qquad (4.110)$$

and the remaining graph G' is $G' = G - \{a\}$ (i.e. all edges that contain a are removed).

The cases $\tau = x, y$ are more involved, i.e. the unitary operations $U_{\tau,\pm}$ and the remaining graph G' are much more complicated. We therefore refer for more details and the proof to the original literature (Raussendorf and Briegel, 2001).

4.7.2 Basic concepts of measurement-based quantum computing

In (Raussendorf and Briegel, 2001; 2002) the computational model underlying the one-way quantum computer was discussed, and it was shown that any quantum network can be implemented on a suitable cluster state.

Theorem 4.30 (*Raussendorf and Briegel, 2001*) *Any quantum logic circuit can be implemented on a sufficiently large cluster state.*

Proof *(Sketch of proof:)*

1. *Remove the superfluous qubits by a measurement in σ_z-basis. Thus, one keeps a suitable subcluster C'; see Lemma 4.29.*

Fig. 4.15 Implementation of a general rotation in measurement-based quantum computing.

2. *Realization of a general rotation $U_R \in SU(2)$:*
   The rotation U_R in Euler representation reads $U_R(\xi, \eta, \varphi) = U_x(\xi), U_z(\eta), U_x(\varphi)$, where $U_x(\alpha) = e^{-i\alpha\frac{\sigma_x}{2}}$ and $U_z(\beta) = e^{-i\beta\frac{\sigma_z}{2}}$.
   On a chain of 5 qubits (see Figure 4.15) one implements the following steps:

   Step1 : Prepare $|\psi\rangle = |\psi_{\text{in}}\rangle_1 \otimes |+\rangle_2 \otimes |+\rangle_3 \otimes |+\rangle_4 \otimes |+\rangle_5$.

   Step2 : Entangle the 5 qubits via the Ising interaction, see eq. (4.102):
   $S^{(C_5)}|\psi\rangle = \frac{1}{2}|\psi_{\text{in}}\rangle|0 - 0-\rangle - \frac{1}{2}|\psi_{\text{in}}\rangle|0 + 1+\rangle - \frac{1}{2}(\sigma_z|\psi_{\text{in}}\rangle)|1 + 0-\rangle + \frac{1}{2}(\sigma_z|\psi_{\text{in}}\rangle)|1 - 1+\rangle$.

   Step3 : Measure qubits 1–4 in the following order and bases:
   3.1 qubit 1 in $\mathcal{B}_1(0)$
   3.2 qubit 2 in $\mathcal{B}_2((-1)^{s_1+1}\xi)$
   3.3 qubit 3 in $\mathcal{B}_3((-1)^{s_2+1}\eta)$
   3.4 qubit 4 in $\mathcal{B}_4((-1)^{s_1+s_3+1}\varphi)$
   with $\mathcal{B}_j(\varphi_j) = \left\{\frac{1}{\sqrt{2}}\left(|0\rangle_j + e^{i\varphi_j}|1\rangle_j\right), \frac{1}{\sqrt{2}}\left(|0\rangle_j - e^{i\varphi_j}|1\rangle_j\right)\right\}$ $s_j \in \{0, 1\}$.
   Here, $s_j = 0(1)$ corresponds to projection onto the first (second) state in $\mathcal{B}_j$, respectively.

   Step4: The resulting state is given by:

   $$|\psi_{\text{out}}\rangle = |s_1\rangle_0 \otimes |s_2\rangle_\xi \otimes |s_3\rangle_\eta \otimes |s_4\rangle_\varphi \otimes U_\Sigma \cdot U_R(\xi, \eta, \varphi)|\psi_{\text{in}}\rangle ,$$

   where $U_\Sigma = \sigma_x^{s_2+s_4}\sigma_z^{s_1+s_3}$ is the so-called "byproduct operator", which has to be corrected at the end of the calculation.

   Again, for details and proofs we refer to the original literature. Note that this scheme has success probability 1, in spite of the probabilistic character of the measurement results.

3. *Realisation of a CNOT-gate:*
   Here, we only sketch the cluster and the operations in Figure 4.16.

4. *Realization of other simple gates:* Here, we only sketch the cluster and the operations in Figure 4.17.

5. *Concatenation of gates:*
   The cluster $C(g)$, realizing the gate g, is composed as follows:

   $$C(g) = \underset{\substack{\downarrow \\ \text{input}}}{C_I(g)} \cup \underset{\substack{\downarrow \\ \text{measurement}}}{C_M(g)} \cup \underset{\substack{\downarrow \\ \text{output}}}{C_O(g)},$$

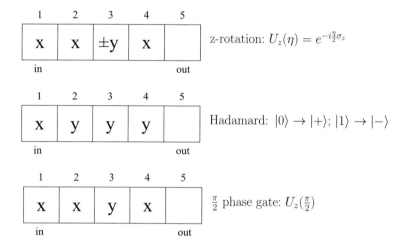

	1	2	3	4	5	6	7
control	x	y	y	y	y	y	out
					y	8	

target	x	x	x	y	x	x	out
	9	10	11	12	13	14	15

Fig. 4.16 Implementation of a CNOT-gate in measurement-based quantum computing.

1	2	3	4	5
x	x	±y	x	
in				out

z-rotation: $U_z(\eta) = e^{-i\frac{\eta}{2}\sigma_z}$

1	2	3	4	5
x	y	y	y	
in				out

Hadamard: $|0\rangle \to |+\rangle; |1\rangle \to |-\rangle$

1	2	3	4	5
x	x	y	x	
in				out

$\frac{\pi}{2}$ phase gate: $U_z(\frac{\pi}{2})$

Fig. 4.17 Implementation of other simple gates in measurement-based quantum computing.

with

$$C_I(g) \cap C_M(g) = \emptyset,$$
$$C_I(g) \cap C_O(g) = \emptyset,$$
$$C_M(g) \cap C_O(g) = \emptyset.$$

The final state results from first entangling the whole cluster and then performing suitable measurements, i.e.

$$|\psi_{\text{out}}\rangle_C = P^{C_I(g) \cup C_M(g)} \underset{\underset{\text{measure}}{\downarrow}}{} S^{(C)} \underset{\underset{\text{entangle}}{\downarrow}}{} |\psi_{\text{in}}\rangle_C.$$

Gates can be either concatenated in a sequential method, or in an equivalent non-sequential method.

Thus, a universal set of gates can be implemented, and the gates can be connected. This ends the sketch of the proof. □

Let us mention that some experiments on one-way quantum computing have already been performed. In (Chen *et al.*, 2007) the Grover search in a list of 4 items has been demonstrated, and in (Gao *et al.*, 2009) the CNOT with 6 qubits has been implemented.

In summary, we have shown in this Section that quantum computing and entanglement are closely interconnected, and that there are many facets of their interplay.

Acknowledgments

We thank Sylvia Bratzik for helping to edit this manuscript.

References

Acin, A. Bruß, D. Lewenstein, M. and Sanpera, A. (2001). *Phys. Rev. Lett.*, **87**, 040401.

Bennett, C.H. et al. (1996). *Phys. Rev. A*, **54**, 3824.

Bernstein, E. and Vazirani, U. (1993). *Proc. 25th Annual ACM Symposium on the Theory of Computation*. ACM Press, New York.

Braunstein, S.L. Caves, C.M. Jozsa, R. Linden, N. Popescu, S. and Schack, R. (1999). *Phys. Rev. Lett.*, **83**, 1054.

Bruß, D. and Leuchs, G. (2007). *Lectures on quantum information*. Wiley-VCH, Weinheim, Germany.

Bruß, D. and Macchiavello, C. (2010). *arXiv: quant-ph/1007.4179*.

Chen, K. Li, C.-M. Zhang, Q. Chen, Y.-A. Goebel, A. Chen, S. Mair, A. and Pan, J.-W. (2007). *Phys. Rev. Lett.*, **99**, 120503.

Cleve, R. Ekert, A. Macchiavello, C. and Mosca, M. (1998). *Proc. R. Soc. Lond. A*, **454**, 339.

Deutsch, D. (1985). *Proc. R. Soc. Lond. A*, **400**, 97.

Deutsch, D. and Jozsa, R. (1992). *Proc. R. Soc. Lond. A*, **439**, 553.

Dür, W. Vidal, G. and Cirac, J.I. (2000). *Phys. Rev. A*, **62**, 062314.

Eckert, K. Schliemann, J. Bruß, D. and Lewenstein, M. (2002). *Ann. Phys.*, **299**, 88.

Ekert, A. and Jozsa, R. (1996). *Rev. Mod. Phys.*, **68**, 733.

Ekert, A. and Macchiavello, C. (1996). *Phys. Rev. Lett.*, **77**, 2585.

Fang, Y. Kaszlikowski, D. Chin, C. Tay, K. Kwek, L.C. and Oh, C.H. (2005). *Phys. Lett. A*, **345**, 265.

Gao, W.-B. et al. (2009). *arXiv: quant-ph/0905.2103*.

Grover, L. (1996). *Proc. 28th Annual ACM Symposium on the Theory of Computation*. ACM Press, New York.

Horodecki, M. Horodecki, R. and Horodecki, P. (1996). *Phys. Rev. A*, **223**, 1.

Horodecki, R. Horodecki, P. Horodecki, M. and Horodecki, K. (2009). *Rev. Mod. Phys.*, **81**, 865.

Jozsa, R. and Linden, N. (2003). *Proc. Roy. Soc. A*, **459**, 2011.

Knill, E. and Laflamme, R. (1997). *Phys. Rev. A*, **55**, 900.

Knill, E. and Laflamme, R. (1998). *Phys. Rev. Lett.*, **81**, 5672.

Laflamme, R. et al. (1996). *Phys. Rev. Lett.*, **77**, 198.

Lloyd, S. (1999). *Phys. Rev. A*, **61**, 010301.

MacWilliams, E.J. and Sloane, N.J.A. (1977). *The theory of error correcting codes.* North Holland.

Mosca, M. (2008). *arXiv: quant-ph/0808.0369*.

Nielsen, M.A. and Chuang, I.L. (2000). *Quantum computation and quantum information.* Cambridge University Press, Cambridge, UK.

Papadimitriou, C.H. (1994). *Computational complexity.* Addison-Wesley, Reading, MA.

Peres, A. (1996). *Phys. Rev. Lett.*, **77**, 1413.

Poulin, D. Blume-Kohout, R. Laflamme, R. and Ollivier, H. (2004). *Phys. Rev. Lett.*, **92**, 177906.

Raussendorf, R. and Briegel, H.J. (2001). *Phys. Rev. Lett.*, **86**, 5188.

Raussendorf, R. and Briegel, H.J. (2002). *Quantum Inf. Comp.*, **6**, 433.

Raussendorf, R. Browne, D. and Briegel, H.J. (2003). *Phys. Rev. A*, **68**, 022312.

Shor, P.W. (1994). *Proc. 35th Annual Symposium on the Foundations of Computer Science.* IEEE Computer Society Press.

Shor, P. (1995). *Phys. Rev. A*, **52**, R2493.

Werner, R.F. (1989). *Phys. Rev. A*, **40**, 4277.

Wootters, W.K. (1998). *Phys. Rev. Lett.*, **80**, 2245.

5
Quantum computation with trapped ions and atoms

F. ROHDE[1], J. ESCHNER[1,2]

[1] ICFO–Institut de Ciències Fotòniques, Mediterranean Technology Park, E-08860 Castelldefels (Barcelona), Spain

[2] Experimentalphysik, Universität des Saarlandes, D-66123 Saarbrücken, Germany

Preface

Since the first preparation of a single trapped, laser-cooled ion by Neuhauser *et al.*, in 1980, a continuously increasing degree of control over the state of single ions has been achieved, such that what used to be spectroscopy has turned into coherent manipulation of the internal (electronic) state, while laser cooling has evolved into the control of the external degree of freedom, i.e. of the motional quantum state of the ion in the trap. Based on these developments, Cirac and Zoller proposed in 1995 to use a trapped ion string for processing quantum information, and they described the operations required to realize a universal two-ion quantum logical gate. This seminal proposal sparked intense experimental activities in many groups and has led to spectacular results.

In these lecture notes, we will review the basic experimental techniques that enable quantum information processing with trapped ions and, much more briefly, with atoms. In particular, we will explain how the fundamental concepts of quantum computing, such as quantum bits (qubits), qubit rotations, and quantum gates, translate into experimental procedures in a quantum optics laboratory. Furthermore, the recent progress of quantum computing with ions and atoms will be summarized, and some of the new approaches to meet future challenges will be mentioned. It is intended to provide an intuitive understanding of the matter that should enable the non-specialist student to appreciate the paradigmatic role and the potential of trapped single ions and atoms in the field of quantum computation. More advanced topics are covered in the original literature and in other recent reviews, e.g. (Häffner *et al.*, 2008*a*; Benhelm *et al.*, 2009; Häffner *et al.*, 2008*b*).

These notes are based on, and extend, an earlier compilation on the same subject (Eschner, 2006). The main addition is a more elaborate discussion of atom-photon qubit interfaces that have shown significant progress in the last few years.

5.1   Ion (and atom) quantum logic

In this section we describe the basics and the state-of-the-art of quantum logic with single-atomic systems, using trapped-ion implementations as the main reference, and addressing briefly some recent advances with neutral atoms.

5.1.1   Single ion and atom storage

5.1.1.1   *Paul trap*

To confine an ion, a focusing force in three dimensions is required. In an electric field a singly charged calcium ion (Ca^+) experiences very strong forces, corresponding to an acceleration on the order of 2×10^8 m/s^2 per V/cm. Although ions cannot be trapped in a static electric field alone, a rapidly oscillating field with quadrupole geometry,

$$\Phi(\mathbf{r}) = (U_0 + V_0 \cos(\xi t)) \frac{x^2 + y^2 - 2z^2}{r_0^2} \tag{5.1}$$

generates an effective (quasi-) potential that under suitably chosen experimental parameters provides 3-dimensional trapping (Paul *et al.*, 1958). The potentials U_0 and V_0 generate a field that alternates too fast for the ion to reach the electrodes of the trap before it reverses its direction. The driven oscillatory motion at the frequency ξ of the applied field (micro-motion) has a kinetic energy that increases quadratically with the distance from the trap center, i.e. from the zero of the quadrupole field amplitude. By averaging over the micro-motion, an effective harmonic potential is obtained, in which the ion oscillates with a smaller frequency, the secular or macro-motion frequency ν; this is the principle of the Paul trap. The three frequencies $\nu_{x,y,z}$ along the three principal axes of the trap may be different. The full trajectory of a trapped ion, consisting of the superimposed secular and micro-motion, is mathematically described by a stable solution of a Mathieu-type differential equation (Ghosh, 1995). Typical macroscopic traps for single ions are roughly 1 mm in size, with a voltage V_0 of several hundred V and frequency ξ of a few tens of MHz, leading to secular motion with frequencies ν in the low-MHz range.

The linear variant of the Paul trap uses an oscillating field with linear quadrupole geometry in two dimensions, providing dynamic confinement along those radial directions, while trapping in the third direction, i.e. along the axis of the 2-D quadrupole, is obtained with a static field. This type of trap is advantageous for trapping several ions simultaneously: under laser cooling and with radial confinement stronger than the axial one, the ions crystallize on the trap axis, thus minimizing their micro-motion that otherwise leads to inefficient laser excitation and unwanted modulation-induced resonances (micro-motion sidebands). A schematic image of a linear Paul trap setup is shown in Figure 5.1.

5.1.1.2   *Quantized motion*

The secular motion of a trapped ion can be regarded as a free oscillation in a 3-dimensional harmonic trap potential, characterized by the three frequencies $\nu_{x,y,z}$. As described in (Stenholm, 1986; Wineland and Itano, 1979; Eschner *et al.*, 2003), laser cooling reduces the thermal energy of that motion to values very close to zero. Therefore, the motion must be accounted for as a quantum-mechanical degree of freedom. For a single ion it is described by the Hamiltonians

$$H_k = \hbar\nu_k \left(a_k^\dagger a_k + \frac{1}{2} \right), \quad (k = x, z, y), \tag{5.2}$$

with energy eigenstates $|n_k\rangle$ at the energy values $\hbar\nu_k(n_k + 1/2)$. The quantum regime is characterized by n being on the order of one or below.

For a string of N ions, there are $3N$ normal modes of vibration, $2N$ radial ones and N axial ones. The two lowest-frequency axial modes are the center-of-mass mode, where all ions oscillate like a rigid body, and the stretch mode, where each ion's oscillation amplitude is proportional to its distance from the center (Steane, 1997; James, 1998). Their respective frequencies are ν_z and $\sqrt{3}\nu_z$, independently of the number of ions, where ν_z is the axial frequency of a single ion. Movies of ion strings with these modes excited can be found in (Nägerl *et al.*, 1998). As will be explained

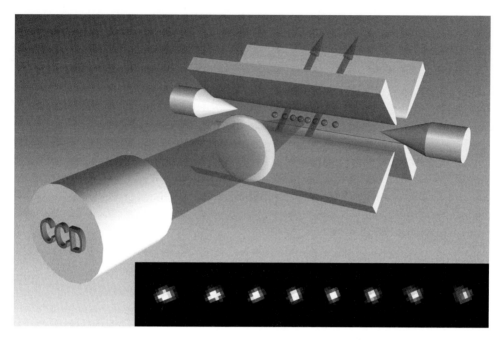

Fig. 5.1 Schematic illustration of a linear ion trap setup with a trapped ion string. The four blades are on high voltage (neighboring blades on opposite potential), oscillating at radio frequency, thus providing Paul-type confinement in the radial directions. The tip electrodes are on positive high voltage and trap the ions axially. A laser addresses the ions individually and manipulates their quantum state. The resonance fluorescence of the ions is imaged onto a CCD camera. (Drawing courtesy of Rainer Blatt, Innsbruck.) In the lower part the CCD image of a string of eight cold, laser-excited ions is shown (Ca^+ ions at ICFO). The distance between the outer ions is about 70 μm.

below, coupling between different ions in a string for QIP purposes is achieved by coherent laser excitation of transitions between the lowest quantum states $|n = 0\rangle$ and $|n = 1\rangle$ of one of the axial vibrational modes, usually the center-of-mass or the stretch mode.

5.1.1.3 Laser cooling

Like trapping, laser cooling is a pivotal ingredient for preparing cold ions or atoms for quantum computing. Laser cooling relies on the photon recoil or, more generally, the mechanical effect of light in a photon-scattering process (absorption – emission cycle), which is caused by the spatial variation of the electric field e^{ikx}. Since several comprehensive reviews of laser cooling exist (Stenholm, 1986; Wineland and Itano, 1979; Eschner *et al.*, 2003), the description shall be limited to a brief summary of Doppler cooling in section 5.1.3.1, an initial cooling technique that is essential for ion storage and cooling to the Lamb–Dicke regime, and sideband cooling, which will be discussed in section 5.1.3.2.

5.1.1.4   *Dipole trap*

Individual neutral atoms are trapped in dipole potentials (Grimm *et al.*, 2000; Adams and Riis, 1997), created by strongly focused laser beams (Schlosser *et al.*, 2001; Frese *et al.*, 2000). In order to trap arrays of atoms while keeping them optically addressable, one may combine several spatially separated laser beams (Beugnon *et al.*, 2006; Yavuz *et al.*, 2006); more sophisticated optical methods have also been demonstrated, such as lens arrays (Dumke *et al.*, 2002) or spatial light modulators (Bergamini *et al.*, 2004).

5.1.2   Qubits

The atomic qubit, in which quantum information is encoded, requires two stable levels that allow coherent laser excitation at Rabi frequencies much higher than all decay rates. Adequate states may be either a ground state and a metastable excited state connected by a forbidden optical transition ("optical qubit"), or two hyperfine sub-levels of the ground state of an ion with non-zero nuclear spin ("hyperfine qubit"). Both cases are treated as atomic two-level systems $\{|g\rangle, |e\rangle\}$ or $\{|\downarrow\rangle, |\uparrow\rangle\}$, see section 5.1.3. Depending on the magnetic properties of the ion species, particular Zeeman substates of the levels are selected, like $S_{1/2}, m = 1/2$ and $D_{5/2}, m = 5/2$ in the optical qubit of Ca^+ (Figure 5.2). An optical qubit transition is driven with a single strong laser with Rabi frequencies $\Omega \gg \Gamma$, while in a hyperfine qubit two lasers, far detuned from an intermediate level, drive a Raman transition with Ω being the Raman Rabi frequency (Wineland *et al.*, 2003).

The ion species that are suitable for QIP can thus be divided in two main classes depending on the implementation of the qubits. For an optical qubit one uses ions with a forbidden, direct optical transition, such as $^{40}Ca^+$, $^{88}Sr^+$, $^{138}Ba^+$, $^{172}Yb^+$, $^{198}Hg^+$, or $^{199}Hg^+$. In this case, the lack of hyperfine structure in the electronic level scheme makes the qubit levels sensitive to ambient magnetic fields and to laser phase fluctuations, which are sources of decoherence. Nevertheless, effective qubits can be constructed from multi-ion quantum states, which form a decoherence-free subspace (Häffner *et al.*, 2005; Kielpinski *et al.*, 2001; Langer *et al.*, 2005).

For a hyperfine qubit one employs odd isotopes such as $^{9}Be^+$, $^{25}Mg^+$, $^{43}Ca^+$, $^{87}Sr^+$, $^{137}Ba^+$, $^{111}Cd^+$, $^{171}Yb^+$. Here, the effect of laser phase noise is reduced because usually the two fields for the Raman transition are derived from the same source, and their difference frequency is stabilized to a microwave oscillator. The magnetic-field sensitivity can also be significantly decreased by using $m = 0$ magnetic sublevels (Haljan *et al.*, 2005), or by applying a static magnetic field at which the qubit levels have the same differential Zeeman shift (Langer *et al.*, 2005). The use of the hyperfine structure, however, comprises a limitation by spontaneous scattering during a Raman excitation pulse. This has been assessed in (Wineland *et al.*, 2003), pointing at heavy ions like Cd^+ or Hg^+ as good qubit candidates.

Both methods are currently pursued in different groups. For example, the NIST group works with $^{9}Be^+$, the Michigan group with $^{111}Cd^+$ and $^{171}Yb^+$, and both the Innsbruck group and the Oxford group use $^{40}Ca^+$. Important progress, such as quantum teleportation, has been achieved with both types of qubits (Riebe *et al.*, 2004; Barrett *et al.*, 2004).

Ca$^+$ ions are attractive for comparing optical qubits (^{40}Ca$^+$) and hyperfine qubits (^{43}Ca$^+$) while using the same trapping and laser setup. Work with ^{43}Ca$^+$ is carried out in Innsbruck (Kirchmair *et al.*, 2009) and Oxford (Lucas *et al.*, 2004). Direct excitation of hyperfine qubits by microwave radiation has also been considered (Mintert and Wunderlich, 2001). It requires the presence of high magnetic field gradients to obtain sufficient sideband coupling (through an effective Lamb–Dicke parameter, see section 5.1.3), which at the same time splits the frequencies of neighboring ions, thus allowing for their addressing.

5.1.3 Laser interaction

Resonant interaction with laser light is used at all stages of QIP with trapped ions: for cooling, for initial-state preparation, qubit manipulation, quantum gates, and for state detection. Various types of optical transitions are employed, and the laser interacts with both the electronic and motional quantum state. Photon scattering is either desired (for cooling and state detection) or unwanted (for qubit manipulations and quantum gates). The relevant processes are summarized in this section, approximately in the order in which they are applied in an experimental cycle: preparation, manipulation, i.e. logic operations, and measurement of the outcome. To give a specific example, Figure 5.2 shows the relevant levels, transitions, and wavelengths of ^{40}Ca$^+$ and explains how these processes are implemented in a real atom.

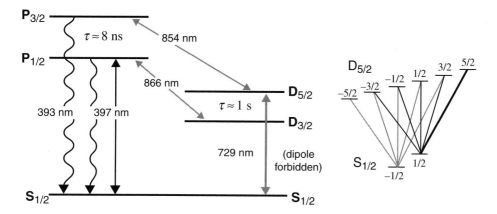

Fig. 5.2 Levels and transitions of ^{40}Ca$^+$. They are employed in the following ways for the processes described in this section: Doppler cooling happens by driving the $S_{1/2}$ to $P_{1/2}$ dipole transition at 397 nm; a laser at 866 nm prevents optical pumping into the $D_{3/2}$ level. Sideband cooling is performed on the dipole-forbidden $S_{1/2}$ to $D_{5/2}$ line at 729 nm, with a laser at 854 nm pumping out the $D_{5/2}$ level to enhance the cycling rate. A Zeeman sublevel of $S_{1/2}$ and one of $D_{5/2}$ form the qubit (for example $m_S = 1/2$ and $m_D = 5/2$, see right-hand diagram), on which coherent operations are driven by the 729-nm laser. The final state is detected by the 397-nm laser that excites resonance fluorescence when the ion is in $S_{1/2}$, but does not couple to $D_{5/2}$.

The atom – laser interaction processes are described by a simple two-level atom with resonance frequency ω_A and states $|g\rangle$ (ground state) and $|e\rangle$ (excited state), and a single vibrational mode with frequency ν, states $|n\rangle$ and creation (annihilation) operator $a^\dagger$ (a). The Hamiltonian of the interaction is

$$H_{int} = -\hbar\Delta|e\rangle\langle e| + \hbar\nu a^\dagger a + \hbar\frac{\Omega}{2}\left(|e\rangle\langle g|e^{i\eta(a^\dagger + a)} + \text{h.c.}\right), \tag{5.3}$$

where $\Delta = \omega_L - \omega_A$ is the detuning between laser frequency ω_L and atomic resonance and Ω is the Rabi frequency characterizing the strength of the atom – laser coupling. The exponential in the bracket stands for the travelling wave of the laser, $\eta(a^\dagger + a) = kz$, with the Lamb–Dicke parameter $\eta = kz_0$ that relates the spatial extension of the motional ground state $z_0 = \langle 0|z^2|0\rangle^{1/2}$ to the laser wavelength $\lambda = 2\pi/k$. Although not strictly necessary, QIP experiments are typically performed in the Lamb–Dicke regime, characterized by $\eta\sqrt{\langle n\rangle} \ll 1$, which means that the ion's motional wave packet is much smaller than the laser wavelength. In this limit transitions $|g, n\rangle \leftrightarrow |e, n'\rangle$ between motional states of different quantum numbers $n \neq n'$ are suppressed as $n - n'$ increases, such that only changes $n - n' = 0, \pm 1$ need to be accounted for. The corresponding laser-driven transitions are called carrier ($|g, n\rangle \leftrightarrow |e, n\rangle$), red sideband ($|g, n\rangle \leftrightarrow |e, n - 1\rangle$), and blue sideband ($|g, n\rangle \leftrightarrow |e, n + 1\rangle$). With respect to Eq. (5.3) this corresponds to expanding the exponential, $e^{i\eta(a^\dagger + a)} \to 1 + i\eta(a^\dagger + a)$, which will be employed in section 5.1.4.

5.1.3.1   *Doppler cooling*

Doppler cooling is performed by exciting the ion(s) on a strong (usually a dipole) transition of natural linewidth $\Gamma > \nu$, with the laser detuned by $\Delta \simeq -\Gamma/2$. It reduces the thermal energy $\hbar\nu\langle n\rangle$ to about $\hbar\Gamma$ in all vibrational modes that have a non-zero projection on the laser beam direction. With a typical dipole transition of 20 MHz width and a trap frequency of 1 MHz, the average residual excitation of the vibrational quantum states is $\langle n\rangle \sim 20$. Much higher trap frequencies allow for Doppler cooling to $\langle n\rangle \sim 1$ (Monroe *et al.*, 1995b). Figure 5.3 shows an illustration of Doppler cooling.

5.1.3.2   *Sideband cooling*

After the ion(s) is(are) cooled to the Lamb–Dicke regime by an initial Doppler cooling stage, sideband cooling is used to prepare the vibrational mode in the motional ground state $|n = 0\rangle$, which is a pure quantum state. Figure 5.3 also illustrates this cooling technique. It is accomplished by exciting an ion on a narrow transition, $\Gamma \ll \nu$, which in practical cases is either a forbidden optical line, such as the $S_{1/2} \to D_{5/2}$ transition in Hg^+ (Diedrich *et al.*, 1989), Ca^+ (Roos *et al.*, 1999), or Sr^+ (Sinclair *et al.*, 2001), or a Raman transition between hyperfine-split levels of the ground state, as in Be^+ (Monroe *et al.*, 1995b; King *et al.*, 1998) or Cd^+ (Deslauriers *et al.*, 2004). The laser is tuned into resonance with the red sideband transition at $\Delta = -\nu$. To increase the cycling rate, the upper level $|e\rangle$ may be pumped out by driving an allowed transition to a higher-lying state that then decays back to $|g\rangle$. Pulsed schemes involving coherent Rabi flopping (π-pulses) are also used (Monroe *et al.*, 1995b; King *et al.*, 1998). Usually only one mode

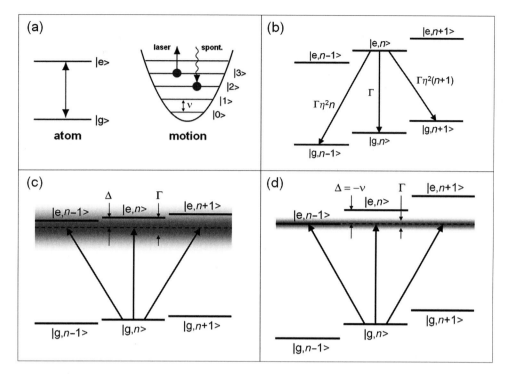

Fig. 5.3 Laser cooling in the Lamb–Dicke regime, $\eta \ll 1$. Only $\delta n = 0, \pm 1$ transitions are relevant. (a) Energy levels and schematic representation of an absorption – emission cycle that cools. (b) Spontaneous emission results in average heating because $|n\rangle \rightarrow |n+1\rangle$ transitions are slightly more probable than $|n\rangle \rightarrow |n-1\rangle$ transitions. (c) In Doppler cooling, when $\Gamma > \nu$, the laser excites simultaneously $|n\rangle \rightarrow |n\rangle$ and $|n\rangle \rightarrow |n \pm 1\rangle$ transitions. By red-detuning the laser to $\Delta \simeq -\Gamma/2$, excitation from $|g,n\rangle$ to $|e,n-1\rangle$ is favored, which thus provides cooling. The final state of Doppler cooling is an equilibrium between these cooling and heating processes. (d) In sideband cooling, when $\Gamma \ll \nu$, the laser is tuned to the $|g,n\rangle$ to $|e,n-1\rangle$ sideband transition at $\Delta = -\nu$, such that all other transitions are off-resonant. This provides a much stronger cooling effect than Doppler cooling, while the heating by spontaneous emission is the same. Thus, the final state is very close to the motional ground state.

is cooled to the ground state, which then serves to establish coherent coupling between ions in a string, see sections 5.1.2 and 5.1.5. Preparation of the motional ground state of the center-of-mass mode with $> 99.9\%$ probability is routinely achieved (Roos *et al.*, 1999; Monroe *et al.*, 1995*b*).

5.1.4 Motional qubits

When ground-state cooling has been achieved, the two lowest levels $|0\rangle$ and $|1\rangle$ of the vibrational mode form a motional qubit, in analogy to the atomic qubit discussed in section 5.1.2. Considering first a single trapped ion, then the basis for quantum logical operations, the "computational subspace" (CS), is formed by a combination

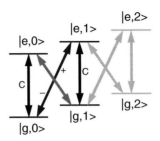

Fig. 5.4 Computational subspace of a single ion with carrier (C) and sideband ($\pm$) transitions.

of motional and atomic qubits, i.e. by the four states $\{|g, 0\rangle, |g, 1\rangle, |e, 0\rangle, |e, 1\rangle\}$. For coherent manipulation of the quantum state within the CS, the red sideband, carrier, or blue sideband transition may be excited by tuning the laser to $\Delta = -\nu$, 0, or $+\nu$, respectively.

The interaction Hamiltonians for these three excitation processes are derived from Eq. (5.3) in the Lamb–Dicke regime and after dropping non-resonant terms,

$$(\text{carrier}) \qquad H_C = \hbar \frac{\Omega}{2} \left(|e\rangle\langle g| + \text{h.c.} \right) \tag{5.4}$$

$$(\text{red sideband}) \qquad H_- = \hbar \frac{\eta\Omega}{2} \left(|e\rangle\langle g|a + \text{h.c.} \right) \tag{5.5}$$

$$(\text{blue sideband}) \qquad H_+ = \hbar \frac{\eta\Omega}{2} \left(|e\rangle\langle g|a^\dagger + \text{h.c.} \right). \tag{5.6}$$

These Hamiltonians induce unitary dynamics in the CS, which are described by the operators

$$(\text{carrier}) \qquad R_C(\theta, \phi) = \exp\left[i\frac{\theta}{2} \left(e^{i\phi}|e\rangle\langle g| + e^{-i\phi}|g\rangle\langle e| \right) \right] \tag{5.7}$$

$$(\text{red sideband}) \qquad R_-(\theta, \phi) = \exp\left[i\frac{\theta}{2} \left(e^{i\phi}|e\rangle\langle g|a + e^{-i\phi}|g\rangle\langle e|a^\dagger \right) \right] \tag{5.8}$$

$$(\text{blue sideband}) \qquad R_+(\theta, \phi) = \exp\left[i\frac{\theta}{2} \left(e^{i\phi}|e\rangle\langle g|a^\dagger + e^{-i\phi}|g\rangle\langle e|a \right) \right], \tag{5.9}$$

where $\theta = \Omega t$ ($\theta = \eta\Omega t$ on the sidebands) is the rotation angle of a pulse of duration t, and ϕ is the phase of the laser. The laser phase is arbitrary when the first of a series of pulses is applied on one transition, but it has to be kept track of in all subsequent operations until the final state measurement. Figure 5.5 shows an example of how the quantum state evolves in the CS under the action of subsequent $R_+(\frac{\pi}{2}, 0)$ pulses.

An example of coherent state manipulation, or unitary qubit rotations, within the CS of a single ion is shown in Figure 5.6. This result was obtained with an optical qubit in a single ^{40}Ca$^+$ ion. Analogous operations with a hyperfine qubit in ^{9}Be$^+$

$$|S,0\rangle \xrightarrow{R^+(\frac{\pi}{2},0)} \frac{1}{\sqrt{2}}(|S,0\rangle + i|D,1\rangle) \xrightarrow{R^+(\frac{\pi}{2},0)} i|D,1\rangle \quad \cdots$$

$$\cdots \xrightarrow{R^+(\frac{\pi}{2},0)} \frac{1}{\sqrt{2}}(-|S,0\rangle + i|D,1\rangle) \xrightarrow{R^+(\frac{\pi}{2},0)} -|S,0\rangle$$

Fig. 5.5 Example of the evolution of a quantum state under application of $\pi/2$-pulses. $|S\rangle$ and $|D\rangle$ stand for $|g\rangle$ and $|e\rangle$ in the text, indication the implementation with Ca$^+$. Note the sign change of the wave function after a 2π-pulse.

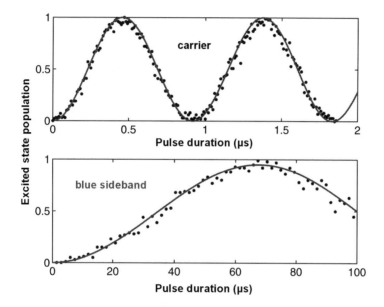

Fig. 5.6 Coherent time evolution of quantum state in the CS of a single ion. After preparation of a single Ca$^+$ ion in the ground state $|g,0\rangle$, a laser pulse of certain duration is applied, tuned to the carrier (top) or blue sideband (bottom) transition. At the end of the pulse, the population of the excited state is measured (see section 5.1.5.1). Points are measured data, the line is a fit. Note the different time-scales of the two plots; the Rabi frequency on the sideband is smaller by a factor of η. From Ref. (Gulde *et al.*, 2002).

were employed to realize the first single-ion quantum gate between an internal and a motional qubit (Monroe *et al.*, 1995a).

5.1.5   Quantum gates

The central concept around which quantum gates with trapped ions are constructed are the common modes of vibration. Sideband excitation of one ion modifies the motional state of the whole string and has an effect on subsequent sideband interaction with all other ions, thus providing the required ion – ion coupling. Due to the strong Coulomb coupling between the ions, the vibrational modes are non-degenerate, therefore they can be spectrally addressed, by tuning the laser to the sideband of one particular mode. In particular, the center-of-mass mode at the lowest frequency ν is separated from the next adjacent mode, the stretch mode, by $(\sqrt{3} - 1)\nu$. Example spectra can be found in Refs. (Schmidt-Kaler *et al.*, 2000; Rohde *et al.*, 2001). Spectral resolution of the sidebands imposes a speed limit on gate operations, which will be discussed in section 5.2.3.

The generalization of the computational subspace from the single-ion case is straightforward: the CS of two ions (these can be any two in a longer string) is the product space of their internal qubits and the selected vibrational mode. The resulting states and transitions are displayed in Figure 5.7.

The role of the vibrational mode is illustrated by considering a basic entangling operation that can be realized with two laser pulses that address the ions individually (see also section 5.1.6.3). The pulses are defined analogously to Eqs. (5.8 and 5.9), with the additional superscript (1) or (2) indicating the addressed ion.

$$|g, g, 0\rangle \xrightarrow{R_+^{(1)}(\frac{\pi}{2}, 0)} \frac{1}{\sqrt{2}} \left(|g, g, 0\rangle + i|e, g, 1\rangle \right) \xrightarrow{R_-^{(2)}(\pi, 0)} \frac{1}{\sqrt{2}} \left(|g, g, 0\rangle - |e, e, 0\rangle \right).$$

The pulse sequence creates an entangled state $(|g, g\rangle - |e, e\rangle)/\sqrt{2}$ of the two internal qubits, while leaving the motional qubit in the ground state.

Several specific gate protocols within the two-ion CS have been proposed and implemented. The classic Cirac–Zoller gate works in the following way (Cirac and Zoller, 1995): first the state of ion 1 is transferred into the motional qubit by a SWAP operation (described below). Then, a single-ion gate is carried out between

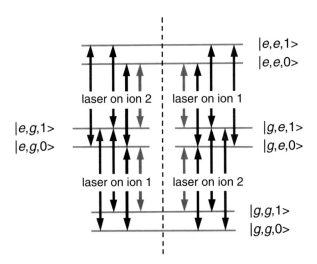

Fig. 5.7 Computational subspace of two ions and one vibrational mode, with all possible transitions. Short, medium, and long arrows are red sideband, carrier, and blue sideband transitions, respectively.

the motional qubit and ion 2. Finally, the motional qubit state is swapped back into ion 1, which completes the gate between the two ions. The experimental realizations of quantum gates are discussed in section 5.1.7.

Note that the motion, although it could be regarded as providing one qubit per vibrational mode, always serves only as a "bus" connecting the internal qubits of the ions. In particular, the motional qubits cannot be measured independently. Thus, at the end of a logic operation, the motional state has to factorize from the state of the internal qubits, which itself can be entangled.

The SWAP operation exchanges the state of two qubits and is employed, e.g., to write the state of the internal into the motional qubit, like in

$$(a|g\rangle + b|e\rangle)(c|1\rangle + d|0\rangle) \rightarrow (c|g\rangle + d|e\rangle)(a|1\rangle + b|0\rangle) .$$

It can be obtained by a π-pulse on the blue sideband, but only if the initial state is not $|g, 1\rangle$. Sideband pulses couple to states outside the computational subspace defined in section 5.1.4, namely $|g, 1\rangle$ to $|e, 2\rangle$ on the blue and $|e, 1\rangle$ to $|g, 2\rangle$ on the red sideband. Moreover, the Rabi frequency on the $|1\rangle \leftrightarrow |2\rangle$ transitions is larger than on $|0\rangle \leftrightarrow |1\rangle$ by a factor of $\sqrt{2}$. This problem can be solved by composite pulses. This technique is well known in NMR spectroscopy (Levitt, 1986) and its application for ion traps has been proposed in Ref. (Childs and Chuang, 2001). A composite pulse sequence,

$$R_{SWAP} = R_+(\pi/\sqrt{2}, 0) \; R_+(\pi\sqrt{2}, \phi_{SWAP}) \; R_+(\pi/\sqrt{2}, 0) ,$$

$$\text{where} \quad \phi_{SWAP} = \cos^{-1}(\cot^2(\pi/\sqrt{2})) \approx 0.303\pi ,$$

performs the SWAP operation for all states of the CS, by acting as a π-pulse on the $|g, 0\rangle \leftrightarrow |e, 1\rangle$ transition and simultaneously as a 4π rotation on $|g, 1\rangle \leftrightarrow |e, 2\rangle$. Composite pulses have been used in the implementation of the Cirac–Zoller quantum gate with Ca$^+$ ions (Schmidt-Kaler *et al.*, 2003*b*).

The single-ion gate that is performed between the motional qubit and ion 2 is a controlled-NOT gate (CNOT). In a CNOT gate the control qubit of the input state, in this case the motional qubit, decides if the target qubit, the atomic qubit of ion 2, is flipped or not. This operation can be realized by a pair of Ramsey pulses enclosing a phase gate. The details of the phase gate used in the Cirac–Zoller proposal are shown in Figure 5.8. In the first experimental realization of this proposal a composite pulse sequence in analogy to the one used for the SWAP operation was applied (Schmidt-Kaler *et al.*, 2003*b*).

5.1.5.1 State detection

At the end of a quantum logical process the state of the qubits has to be determined. This is accomplished by exciting the ions and recording the scattered fluorescence photons on a transition that couples to only one of the qubit levels; see Figure 5.2 for the example of Ca$^+$. More technical details are given in the next section. State detection is a projective measurement that destroys all quantum correlations. The outcome of one individual measurement is a binary signal for each ion, full fluorescence or no photons, corresponding to either $|g\rangle$ or $|e\rangle$. To determine the probabilities that

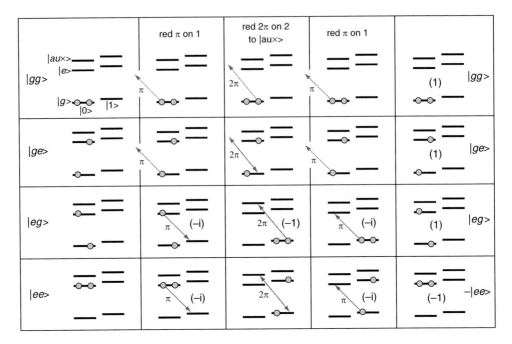

Fig. 5.8 Phase gate for Cirac–Zoller gate operation. Of the four input states $|g, g\rangle$, $|g, e\rangle$, $|e, g\rangle$ and $|e, e\rangle$ only the last one is affected by the pulse sequence that consists of a π-pulse on the red sideband on ion 1, a 2π-pulse on the red sideband resonant with an auxiliary state on ion 2 and a second π-pulse on the red sideband on ion 1. A dashed arrow indicates a pulse that has no effect, either because the ion is in the wrong initial state or because there is no resonant final state. States $|g, g\rangle$ and $|g, e\rangle$ do not couple to any of the three pulses and hence stay unchanged. The phases acquired by $|e, g\rangle$ with each pulse compensate each other, leaving this state also unchanged. Only $|e, e\rangle$ acquires an overall phase of -1 and gets transferred to $-|e, e\rangle$, thus completing the phase gate.

the ion is in either of the two states after the logic operation, one has to repeat the measurement many times. Typically 100 or several 100 runs are used. In the example of Figure 5.6, each data point is an average of 200 single measurements.

In a teleportation experiment, where part of the operations is conditional on the outcome of an intermediate measurement, it is necessary to measure the state of one ion while the other ions are in superposition states. Therefore, these ions have to be protected from the resonant light. This is achieved either by separating the ions spatially, which has been shown to be possible without affecting internal-qubit superpositions (Rowe *et al.*, 2002; Barrett *et al.*, 2004), or by a hiding pulse, as has been demonstrated with Ca$^+$ (Riebe *et al.*, 2004; Häffner *et al.*, 2005*b*). Such a π-pulse transfers the population from the qubit level in S$_{1/2}$, which might couple to the resonant light on the neighbouring ion, to a Zeeman sublevel of D$_{5/2}$, which is different from the qubit level. After the measurement pulse on the adjacent ion, a $-\pi$-pulse reverses the hiding process and restores the superposition of the qubit.

Further examples of techniques that deal with experimental imperfections are spin echo pulses to counteract the effect of magnetic-field fluctuations and gradients (Barrett *et al.*, 2004), and the compensation of Stark shifts that occur during sideband excitation due to non-resonant interaction with the carrier transition (Häffner *et al.*, 2003).

5.1.6 Experimental techniques

5.1.6.1 Initial-state preparation

After the motional state of the ion has been prepared by the laser-cooling techniques explained in section 5.1.1.3 and 5.1.3.2, the atomic state may have to be initialized, e.g. to a specific Zeeman sublevel of the ground state. This is achieved by applying a short optical pumping pulse of appropriate polarization and beam direction, e.g. in Ca^+ by a σ^+-polarized pulse at 397 nm that propagates along the quantization axis and pumps all atomic population into $S_{1/2}, m = +1/2$, see Figure 5.2. Such optical pumping may heat up the ion and is therefore alternated with sideband cooling.

5.1.6.2 Laser pulses

The experimental conditions for qubit manipulations are stringent. Even for a basic operation like a two-ion quantum gate, a sequence of several pulses is required, during which phase-perturbing or decay processes must be negligible. Apart from stable atomic states, this requires the control of environmental magnetic fields, and in particular it puts challenging requirements on the stability of the laser sources. In the example of the optical qubit, where spontaneous decay of the excited state happens on the 1 s timescale, the laser phase fluctuations are critical. A residual laser linewidth in the 100 Hz range or better is desired in order to maintain phase coherence over a significant number of pulses. To achieve this, sophisticated stabilization techniques like those developed in the context of optical clocks have to be applied.

With hyperfine qubits, laser stability is still very important but less critical, because there the difference frequency between two light fields is relevant, which is in the GHz range and can be made very stable using microwave oscillators.

Once the phase-stable light field for driving the qubits is generated, the creation of laser pulses of desired amplitude, detuning, phase shift, and duration is comparatively simple. This is illustrated in Figure 5.9.

5.1.6.3 Addressing individual ions in a string

The conventional way to manipulate individual qubits in a string of ions, as envisioned in the original Cirac–Zoller proposal, is to use a well-focused laser beam that interacts with only one ion at a time, and that can be directed at any ion in the string. This kind of addressing was first demonstrated in (Nägerl *et al.*, 1999). It is only possible if the distance between the ions is significantly larger than the diffraction limit of the focusing lens. This limits the range of possible trap frequencies and thus the speed of gate operations (Steane *et al.*, 2000). With the 729-nm light exciting the optical qubit in Ca^+, focusing to ~ 2.5 μm can typically be achieved, leading to an ion spacing

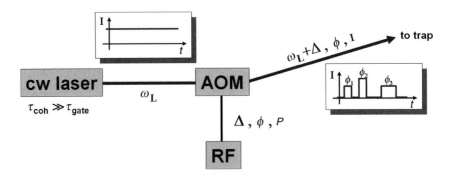

Fig. 5.9 Creation of laser pulses. Continuous light from a stable laser (coherence time ≫ gate time) is sent through an acousto-optical modulator (AOM), which receives a radio-frequency signal from a stable oscillator (RF). The RF frequency Δ is added to the light frequency, thus defining the laser detuning, the RF phase ϕ translates directly into the phase of the light, and the RF power P controls the light intensity I, including switching it on and off. Intensity stability may be improved by monitoring the beam at the trap and compensating unwanted changes by a feedback to the RF power. The AOM process is based on Bragg diffraction from a sound wave in a crystal.

around 5 μm (Nägerl *et al.*, 1999). The Rabi frequency on a neighboring ion may still reach about 5% of that on the addressed ion, introducing addressing errors in the quantum logic operations. Such systematic errors can, however, be compensated for by adjustments to pulse lengths and phases (Schmidt-Kaler *et al.*, 2003*a*).

Addressing in gate operations is not indispensable, since protocols have been developed (Sørensen and Mølmer, 1999) and demonstrated (Sackett *et al.*, 2000; Kielpinski *et al.*, 2001) that employ simultaneous laser interaction with both participating ions. The gate used by the NIST group in their recent experiments relies on the action of spin-dependent dipole forces on both ions when they are simultaneously excited (Leibfried *et al.*, 2003). Still, in general a process is required that allows for shielding one ion from the light that excites the adjacent one. The strategy that appears to be most promising is to separate and combine ions in segmented traps, see section 5.2.1.

5.1.6.4   *State discrimination*

The final state read-out on a string of ions, by a measurement as described in section 5.1.5.1, requires recording the resonance fluorescence of each ion in the string. At the same time, state read-out should happen fast to allow for a high experimental cycle frequency and to avoid state changes by decay processes. Thus, high photon detection efficiency at low dark count rate is desired. A photon-counting camera resolving the individual ions is the appropriate device. Suitable parameters are found by acquiring a histogram of photon counts corresponding to the positions of the ions on the camera image and to the two possible signal levels. The histogram will show Poisson distributions around the mean numbers of fluorescence and dark counts, and efficient state discrimination is achieved when the distributions do not overlap. As

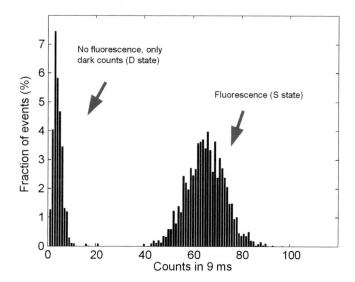

Fig. 5.10 Example histogram for state discrimination with a single Ca$^+$ ion (c.f. (Riebe, 2005)). The ion is prepared with $\sim 70\%$ probability in the S$_{1/2}$ state, which fluoresces, and with $\sim 30\%$ probability in D$_{5/2}$, which is dark. Resonance fluorescence photons are collected for 9 ms. The experiment is repeated several thousand times. The separation between the two Poisson distributions allows one to define a clear threshold that discriminates between the possible outcomes.

shown in the example of Figure 5.10, a few tens of photons are sufficient. In the example of Ca$^+$, in 10 ms a state-discrimination efficiency above 99% can be reached (Riebe, 2005).

In experiments without individual addressing, the ions have to be separated using segmented traps for measuring their final states individually (Barrett *et al.*, 2004). Recently, it has been demonstrated that speed and fidelity of the state readout may be optimized by using time-resolved photon counting and adaptive measurement techniques (Myerson *et al.*, 2008).

5.1.7 Milestone experiments

5.1.7.1 *Trapped ions*

After preliminary work that already implemented the relevant entangling operations (Sackett *et al.*, 2000; Kielpinski *et al.*, 2001), the first two-ion gates were realised in 2003, simultaneously in Boulder with a hyperfine (Leibfried *et al.*, 2003) and in Innsbruck with an optical qubit (Schmidt-Kaler *et al.*, 2003*b*,*a*). The optical implementation with Ca$^+$ ions was very close to the original Cirac–Zoller proposal, while the hyperfine version with Be$^+$ used a protocol based on state-dependent forces, similar to the scheme proposed by Sørensen and Mølmer (Sørensen and Mølmer, 1999).

A spectacular result in 2004 was the first deterministic quantum teleportation of the state of an atom, which again was achieved by the same two groups simultaneously

(Barrett *et al.*, 2004; Riebe *et al.*, 2004). A diagram that illustrates the procedure nicely is found in (Kimble and van Enk, 2004). The NIST experiment involved shuttling ions between zones of a segmented trap, in order to allow for both individual and simultaneous laser manipulation of the ions. This can be considered as a first step towards scalable trap architecture.

The first realisation of a decoherence-free subspace (DFS) was reported in Ref. (Kielpinski *et al.*, 2001). Very long coherence times, of many seconds, of entangled states and states in DFSs have been demonstrated in Refs. (Langer *et al.*, 2005) and (Häffner *et al.*, 2005), and recently a quantum gate within such a DSF was realized (Monz *et al.*, 2009*b*).

Several basic quantum logical operations were also demonstrated, such as quantum dense coding (Schaetz *et al.*, 2004), quantum error correction (Chiaverini *et al.*, 2004), the quantum Fourier transform (Chiaverini *et al.*, 2005*b*), the Toffoli gate (Monz *et al.*, 2009*a*), and entanglement purification (Reichle *et al.*, 2006). Specific entangled 2-ion (Roos *et al.*, 2004*a*) and 3-ion (Roos *et al.*, 2004*b*) quantum states have been deterministically created and characterized by quantum tomography. Also, multi-ion entangled states, like an 8-ion entangled W state forming a "quantum byte" (Häffner *et al.*, 2005*a*) or 6-ion GHZ states (Leibfried *et al.*, 2005), have been created and characterized. Just recently, the first quantum simulations with cold trapped ions were realized (Friedenauer *et al.*, 2008; Gerritsma *et al.*, 2010). Moreover, a programmable two-qubit quantum processor capable of performing arbitrary unitary transformations on a certain set of input states was realized (Hanneke *et al.*, 2010).

5.1.7.2   *Trapped atoms*

For implementing neutral atom quantum gates, various proposals have been put forward including short-range dipolar interactions (Brennen *et al.*, 1999), ground-state collisions (Jaksch *et al.*, 1999), cavity QED coupling (Pellizzari *et al.*, 1995), magnetic dipole – dipole interactions (You and Chapman, 2000), gates with delocalized qubits (Mompart *et al.*, 2003), and Rydberg-state-mediated dipolar interactions (Jaksch *et al.*, 2000). Promising progress has recently been reported with the latter type of approach. Rydberg atoms show long-range dipolar interaction that may be used to couple atoms in independent dipole traps by the so-called Rydberg blockade (Urban *et al.*, 2009; Gaëtan *et al.*, 2009). Two-atom gates (Wilk *et al.*, 2010; Isenhower *et al.*, 2010) have been demonstrated using this mechanism.

5.2   Scalability

After basic few-ion operations have been demonstrated and simple quantum algorithms have been implemented, the question of scalability has to be addressed seriously. One very important aspect of it are quantum error-correction codes, which where discovered simultaneously by Shor (Shor, 1995) and Steane (Steane, 1996). Building an ion trap quantum computer that implements full error correction appears to be an overwhelming task. In this section we summarize ideas and results that are expected to contribute to meeting this challenge. Three main directions are considered, ion shuttling in segmented traps, sympathetic cooling, and fast gates.

5.2.1 Trap architecture

Addressing ions in a static string by steering a laser beam becomes technically difficult as the number of ions increases. Moreover, the conventional two-ion gate protocol employing a vibrational mode of a string becomes difficult to implement as the size of the string grows, because the increasing mass of the ion crystal reduces, through the Lamb–Dicke parameter η, the coupling on the sideband transitions. Therefore, a trap architecture seems favorable that allows for moving the ions. In this scenario, ions are picked from a register and shuttled by time-dependent trap voltages into an interaction zone where two-ion gates are carried out of the kind used in (Barrett *et al.*, 2004). A detailed scheme has been proposed by the NIST group (Kielpinski *et al.*, 2002), and several other groups have started activities to simulate, build, and test various implementations (Metodiev *et al.*, 2005; Steane, 2007; Home and Steane, 2006; Chiaverini *et al.*, 2005*a*; Microtrap, 2006). A notable experimental achievement has been reported by the Michigan group who shuttled an ion around a corner in a T-junction (Hensinger *et al.*, 2006), and just recently the internal-state coherence of an ion could be preserved during transport through a X-junction at NIST (Blakestad *et al.*, 2009). Some important issues of the transport and shuttling through junctions of ions in multidimensional trap arrays are covered in (Hucul *et al.*, 2008). At the moment, different strategies to build scalable micro-traps are being followed in parallel. Some experiments use fabrication techniques from semiconductor technology to build microfabricated traps integrated into silicon (Brownnutt *et al.*, 2006), gallium arsenide (Stick *et al.*, 2006) or gold-coated aluminium oxide (Schulz *et al.*, 2008) based microchips. Another way is to keep fabrication procedure and trap design as simple as possible with regard to enabling large-scale quantum information processing. This can be achieved by employing 2D surface electrode structures that are realized by vapor-deposited gold on glass substrates (Seidelin *et al.*, 2006; Allcock *et al.*, 2009) or vacuum-compatible printed circuit boards (Brown *et al.*, 2007; Splatt *et al.*, 2009). The problem that all micro-traps encounter is that, due to the vicinity of ion and electrodes, the heating rates are relatively high with respect to macroscopic traps. One attempt to overcome this problem is to go to cryogenic temperatures (Deslauriers *et al.*, 2006; Labaziewicz *et al.*, 2008). Recent progress in the fast-moving field of trap technology includes the study of fast-ion transport (Huber *et al.*, 2008), which is important for the speed of quantum algorithms, and an experiment that shows coherent manipulation and sideband cooling close to the motional ground state of a single ion in a micro-trap (Poschinger *et al.*, 2009).

5.2.2 Sympathetic cooling

Part of the future trap technology will be to apply new means of cooling the ions without perturbing their quantum state, i.e. without scattering photons from them. This can be achieved by sympathetic cooling, when the collective modes of a string are cooled by scattering photons only from ions that do not participate in the quantum logic process. Sympathetic Doppler cooling was demonstrated some time ago with $Hg^+ - Be^+$ mixtures (Larson *et al.*, 1986) and with Mg^+ (Bowe *et al.*, 1999). Ground-state cooling of a two-ion crystal through one of the ions was reported in (Rohde *et al.*, 2001). Further relevant results were obtained with mixtures of Cd^+ isotopes (Blinov *et al.*, 2002) and with a $Be^+ - Mg^+$ crystal that was cooled to the motional ground state (Barrett *et al.*, 2003). More recently, sympathetic cooling of a $Be^+ - Al^+$ crystal was employed to perform spectroscopy on Al^+ using quantum logic (Schmidt *et al.*, 2005; Rosenband *et al.*, 2007). In the future this may lead to an accurate frequency standard based on Al^+. Combination of sympathetic cooling with other scalable methods has already been demonstrated (Home *et al.*, 2009).

5.2.3 Fast gates

Another major contribution to scaling up ion-trap QIP is expected from faster gate operations. The current limitation is the use of spectrally resolved sideband resonances, which requires individual laser pulses to be longer than the inverse trap frequency. Smaller traps with larger frequencies can improve the situation, but theoretical and numerical studies indicate that one could in fact use significantly shorter, suitably designed pulses or pulse trains (García-Ripoll *et al.*, 2003; Duan, 2004; Zhu *et al.*, 2006). Their action is, rather than to excite one mode, that they impart photon momentum kicks and thus interact with a superposition of many modes or even with the local modes of one ion, i.e. its local vibration as if all other ions were fixed (Zhu *et al.*, 2006). It will be interesting to see how ideas of quantum coherent control can help advance into this direction (García-Ripoll *et al.*, 2005; Nebendahl *et al.*, 2009). An interesting novel approach is to employ an optical frequency comb for fast coherent qubit control (Hayes *et al.*, 2010).

5.3 Qubit interfacing

An important field of activity related to QIP with trapped ions is qubit interfacing, i.e. linking static quantum information stored in ionic qubits to propagating quantum information stored in photonic quantum states. Such quantum interaction between single trapped ions and single photons is a central objective for many QIP schemes like quantum networking (Kimble, 2008), distributed quantum computing or quantum communication. At present, the main limitation of photon-based quantum communication is the loss of photons in the quantum channel. This limits the distance of a single quantum link to about 250 km with present technology (Stucki *et al.*, 2009). A solution to this problem is to subdivide larger distances into smaller sections over which photons can be faithfully transmitted to create remote atomic entanglement. A quantum repeater situated at each of the nodes of the quantum network would then extend the entanglement to longer distances (Briegel *et al.*, 1998). To accomplish this, a fully coherent and efficient state transfer between flying and stationary qubits, as well as a faithful transmission between the specific locations is necessary.[1] First progress was achieved using atomic ensembles as quantum memory (Julsgaard *et al.*,

2001, 2004; Chou *et al.*, 2005; Chanelière *et al.*, 2005; Eisaman *et al.*, 2005). Compared to single-qubit implementations in ions these systems have the disadvantage that there is no scheme for (scalable) local processing of the stored information.

Atom – photon qubit interfacing can be achieved through a variety of different techniques. The first proposal of coherent coupling of ionic qubits to photonic modes involved a high-finesse resonator (Cirac *et al.*, 1997). Direct photon exchange between remote ions or deterministic gate operations can be realized by such a strongly coupled atom – cavity system. A potentially less complex alternative for efficient atom – light coupling are high numerical aperture optics in free space. There are a number of interesting approaches towards this technique (Tey *et al.*, 2009, 2008; Maiwald *et al.*, 2009; Wrigge *et al.*, 2008; Zumofen *et al.*, 2008; Stobinska *et al.*, 2009).

As mentioned, one application of this interfacing is a quantum network of small ion-trap processing units, coupled with each other by photons. The currently most promising method to realize a quantum network is to create entangled ion – photon states by a projective measurement of spontaneously emitted photons. Subsequently, the interference of such single photons can generate heralded entanglement of remote ions (Cabrillo *et al.*, 1999; Bose *et al.*, 1999; Feng *et al.*, 2003; Duan and Kimble, 2003; Simon and Irvine, 2003). Deterministic quantum gate operations between two remote logic ions would then be realized by probabilistically entangling two distant ancilla ions and performing deterministic local gates between neighboring logic and ancilla ions (Duan *et al.*, 2004).

5.3.1 Experiments

In this section we will briefly review some existing approaches and results in qubit interfacing and quantum networking with single atoms, including both trapped ions and trapped neutral atoms.

The direct coupling of a trapped ion to a high-finesse optical cavity was realized in two groups. The MPQ group demonstrated coupling on a dipole transition (Guthöhrlein *et al.*, 2001; Keller *et al.*, 2003), which was then applied to the creation of single photons with tailored wave packets (Keller *et al.*, 2004). In Innsbruck, Rabi oscillations on a qubit transition induced by a coherent cavity field were observed, and cavity sideband excitation was demonstrated (Mundt *et al.*, 2002, 2003). In the same group, cavity-assisted Raman spectroscopy on the D to P dipole transition was realized (Russo *et al.*, 2009). Here, the vacuum field of the cavity together with a driving laser stimulate a Raman transition from the $D_{3/2}$ to the $S_{1/2}$ state in $^{40}Ca^+$. Moreover, with the same system a single-photon source from a single ion was realized recently (Barros *et al.*, 2009).

Earlier, a single-photon source was implemented with neutral atoms falling through a high-finesse cavity (Kuhn *et al.*, 2002; McKeever *et al.*, 2004). Due to the difficulty of integrating ion-trap electrodes and a resonator setup, neutral-atom experiments with cavities have been historically more accessible. In state-of-the-art experiments, neutral atoms are trapped within the cavity mode by far off-resonant dipole traps (FORT), which are either formed by modes of the cavity itself (Boozer *et al.*, 2007) or by laser beams transverse to the cavity axis (Nussmann *et al.*, 2005). With such a system a coherent state of light was reversibly mapped onto the hyperfine states of an atom trapped within the mode of a high-finesse cavity (Boozer *et al.*, 2007). In an ideal qubit interface the coherent state of light would have to be replaced by a single-photon state.

Neutral atoms have also been used in attempts to reach efficient atom – photon coupling by the use of high numerical aperture optics (Schlosser *et al.*, 2001; Weber *et al.*, 2006; Tey *et al.*, 2008). An example of what can be achieved by this method is the observation of the attenuation and the phase shift of a weak coherent beam induced by a single atom (Tey *et al.*, 2008, 2009). The central challenge of this approach is to mode-match the incoming light to the emission pattern of a single atom (Quabis *et al.*, 2000; van Enk, 2004; Sondermann *et al.*, 2009). One interesting proposal is to use a single trapped ion inside a parabolic mirror (Sondermann *et al.*, 2007). The inward-moving wave is transformed into a time-reversed, radially polarized dipole wave that in principle could excite a single ion, trapped in the focus of the mirror with a new type of Paul trap (Maiwald *et al.*, 2009), with an efficiency close to 100%. There are a number of promising ideas of how to maximize the numerical aperture of lenses for ion – photon coupling, e.g. it has been proposed to use parabolic integrated mirror – ion-trap structures (Luo *et al.*, 2009) or spherical mirrors in combination with aspheric correction optics (Shu *et al.*, 2009*a,b*). A technique applied with quantum dots and single molecules, but less convenient to combine with atom or ion traps is the use of index-matched GaAs immersion lenses (Vamivakas *et al.*, 2007; Wrigge *et al.*, 2008).

An important step towards quantum networking was the demonstration of the entanglement between an ion and the polarization of an emitted photon (Blinov *et al.*, 2004). The ion decays under spontaneous emission of a photon from an exited state into one of two different hyperfine ground states. By selecting a certain emission direction with an aperture, the photon polarization is entangled with the final state of the de-excited ion. This result was reproduced with neutral atoms in a single-atom dipole trap (Volz *et al.*, 2006) and in a high-finesse cavity (Wilk *et al.*, 2007). Using two ions entangled with their emitted photons, it was then possible to post-selectively create remote entanglement of two $^{171}\text{Yb}^+$ ions (Moehring *et al.*, 2007). The quantum interference of indistinguishable photons was used in this experiment to perform a Bell measurement that projects the ions into an entangled state. This seems to be the most promising way to establish distant entanglement as a resource for quantum information processing. A generic scheme is explained in figure 5.11. Another method (Cabrillo *et al.*, 1999), which can be more efficient for small photon detection efficiencies (Zippilli *et al.*, 2008), is based on the indistinguishability of photon scattering from the two ions, as explained in Figure 5.12. The drawback of this method is that it requires interferometric stability, which is experimentally challenging and most likely the reason why this scheme has not yet been realized. The latest highlight is the demonstration of a specific teleportation protocol based on the heralded entanglement of the atoms through two-photon interference, as described above (Olmschenk *et al.*, 2009).

The combination of local quantum logic operations in ion or atom traps with atom photon interfaces that establish connections between distant traps is certainly

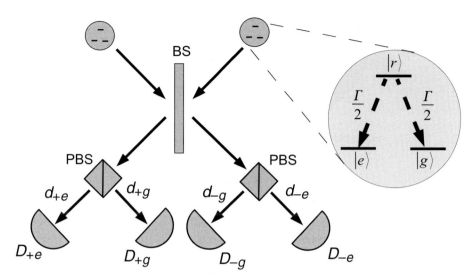

Fig. 5.11 Setup for entangling the internal states of two distant atoms by measurement of two photons (Feng *et al.*, 2003; Duan and Kimble, 2003; Simon and Irvine, 2003). The atoms are prepared in the state $|r_1, r_2\rangle$ and then decay spontaneously. The photon wave packets overlap at a 50:50 beamsplitter (BS). The detection apparatus at the output ports of the BS involves two polarizing beamsplitters (PBS) and four detectors. Coincident clicks at $D_{+,e}$ and $D_{+,g}$ or $D_{-,e}$ and $D_{-,g}$ project the atoms into the state $|\Psi^+\rangle = \frac{1}{\sqrt{2}}(|e_1, g_2\rangle + |e_2, g_1\rangle)$. The state $|\Psi^-\rangle = \frac{1}{\sqrt{2}}(|e_1, g_2\rangle - |e_2, g_1\rangle)$ is found by coincident clicks at $D_{+,e}$ and $D_{-,g}$ or $D_{+,g}$ and $D_{-,e}$. Two indistinguishable photons will always leave the beamsplitter through the same output port (Hong *et al.*, 1987) and thus do not cause coincidence detection.

an exciting aspect of the future perspectives of quantum computing with single ions or atoms.

5.3.2 Entanglement transfer

The currently most efficient and practical resource of remote entanglement is pair-photon generation by spontaneous parametric down-conversion. Using atom – photon qubit interfacing offers thus a promising alternative to create atomic remote entanglement by entanglement transfer. The first proposals for such entanglement distribution involved atom – light coupling via resonators (Lloyd *et al.*, 2001; Kraus and Cirac, 2004). In contrast to the entangling protocol discussed in section 5.3.1, which relies on spontaneous emission of single photons by single ions and a subsequent projective measurement, entanglement transfer requires efficient *absorption* of single photons by single ions. Both processes are complementary and need to be integrated in the same system for efficient bi-directional qubit interfacing.

The scheme in figure 5.13 describes in detail one possible implementation of entanglement transfer from photons to $^{40}\text{Ca}^+$ ions. Two independently trapped ions each absorb one of the two entangled partner photons from a SPDC source, which

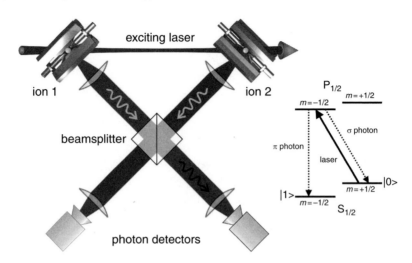

Fig. 5.12 Schematic experimental setup for post-selective entanglement creation between distant trapped ions (Cabrillo *et al.*, 1999). After preparation of both ions in a state $|0\rangle$, a laser pulse excites a state-changing transition $|0\rangle \rightarrow |1\rangle$ with small probability ϵ, thus producing the state $|0,0\rangle + \epsilon(|0,1\rangle + |1,0\rangle) + \epsilon^2|1,1\rangle$. Part of the scattered light from both ions is coherently superimposed on a beamsplitter. A photon-detection event behind the beamsplitter (of π polarization in the example) does not allow us to determine which ion changed its state, and therefore signals creation of the entangled state $(|0,1\rangle + |1,0\rangle)/\sqrt{2}$.

transfers their population to an excited state. Both ions have been prepared such that coincidence detection of the subsequently spontaneously emitted photons projects the ions into an entangled state.

A number of groups are currently developing non-classical light sources tailored to interact with atoms for quantum information applications (Bao *et al.*, 2008; Wolfgramm *et al.*, 2008; Neergaard-Nielsen *et al.*, 2007). In our group we have implemented a tunable narrowband entangled photon pair source that is designed to interact with $^{40}\mathrm{Ca}^+$ and that is intended to be used to realize the scheme explained in figure 5.13 (Haase *et al.*, 2009; Piro *et al.*, 2009).

As a first step towards entanglement transfer it was possible to show resonant interaction of a single ion with the single photons from this down-conversion source (Schuck *et al.*, 2010). Employing a quantum jump scheme and using the temperature dependence of the down-conversion spectrum as well as the tunability of the narrow source, absorption of the down-conversion photons was quantitatively characterized.

In an extension of this work the efficiency of the atom – photon interaction was increased using a pulsed scheme on the $D_{5/2}$ to $P_{3/2}$ transition. The observation of a time correlation between absorption events and the detection of the filtered partner photons (Piro *et al.*, 2011) indicate that entanglement transfer is within the realm of possibility.

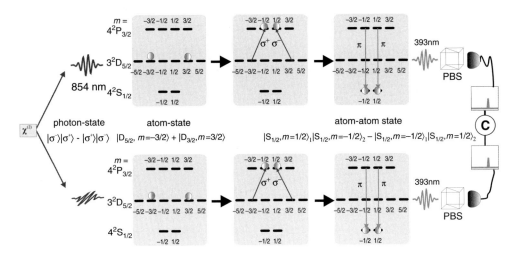

Fig. 5.13 Scheme for transferring photon – photon to atom – atom entanglement. A spontaneous parametric down-conversion source creates the polarization-entangled state $\Psi^- = \frac{1}{\sqrt{2}}(|\sigma^-\rangle|\sigma^+\rangle - |\sigma^+\rangle|\sigma^-\rangle)$ of 854-nm photons, resonant with the $D_{5/2}$ to $P_{3/2}$ transition in ^{40}Ca$^+$. Each partner photon is sent to one of two independently trapped ^{40}Ca$^+$ ions, both of which have previously been prepared in a coherent superposition of the $m = -3/2$ and $m = 3/2$ Zeeman sublevels of the metastable $D_{5/2}$ state. If both ions absorb the photons, their population is transferred to either the $P_{3/2}$, $m = -1/2$ or the $P_{3/2}$, $m = +1/2$ state, depending on the polarization of the absorbed photon. The $P_{3/2}$ state decays rapidly into the $S_{1/2}$ ground state. From the corresponding spontaneous photons only those from π transitions are filtered, such that a coincidence detection of two 393-nm photons projects the system into the entangled atom – atom state $\Psi'^- = \frac{1}{\sqrt{2}}(|+1/2\rangle_1|-1/2\rangle_2 - |-1/2\rangle_1|+1/2\rangle_2)$ composed of the Zeeman sublevels of the $S_{1/2}$ ground state.

Note

1. The QIPC Strategic Report (Zoller et al., 2005) highlights the development of quantum interfaces and repeaters as an integral part of full-scale quantum information systems and identifies their realization as one of the key tasks for the future.

References

Adams, C. S. and Riis, E. (1997). Laser cooling and trapping of neutral atoms. *Prog. Quant. Electr.*, **21**(11), 1–79.

Allcock, D. T. C., Sherman, J. A., Stacey, D. N., Burrell, A. H., Curtis, M. J., Imreh, G., Linke, N. M., Szwer, D. J., Webster, S. C., Steane, A. M., and Lucas, D. M. (2009). Implementation of a symmetric surface electrode ion trap with field compensation using a modulated raman effect. *arXiv:0909.3272v2 [quant-ph]*.

Bao, Xiao-Hui, Qian, Yong, Yang, Jian, Zhang, Han, Chen, Zeng-Bing, Yang, Tao, and Pan, Jian-Wei (2008, Nov). Generation of narrow-band polarization-entangled photon pairs for atomic quantum memories. *Phys. Rev. Lett.*, **101**(19), 190501.

Barrett, M. D., Chiaverini, J., Schaetz, T., Britton, J., Itano, W. M., Jost, J. D., Knill, E., Langer, C., Leibfried, D., Ozeri, R., and Wineland, D. J. (2004). Deterministic quantum teleportation of atomic qubits. *Nature*, **429**, 737–739.

Barrett, M. D., DeMarco, B., Schaetz, T., Meyer, V., Leibfried, D., Britton, J., Chiaverini, J., Itano, W. M., Jelenković, B., Jost, J. D., Langer, C., Rosenband, T., and Wineland, D. J. (2003, Oct). Sympathetic cooling of ^{9}Be$^+$ and ^{24}Mg$^+$ for quantum logic. *Phys. Rev. A*, **68**(4), 042302.

Barros, H. G., Stute, A., Northup, T. E., Russo, C., Schmidt, P. O., and Blatt, R. (2009). Deterministic single-photon source from a single ion. *New J. Phys.*, **11**, 103004.

Benhelm, J., Kirchmair, G., Gerritsma, R., Zähringer, F., Monz, T., Schindler, P., Chwalla, M., Hänsel, W., Hennrich, M., Roos, C. F., and Blatt, R. (2009). Ca+ quantum bits for quantum information processing. In *141st Nobel Symposium on Qubits for Future Quantum Information, Phys. Scr.*, Volume T137, p. 014008.

Bergamini, S., Darquié, B., Jones, M., Jacubowiez, L., Browaeys, A., and Grangier, P. (2004). Holographic generation of microtrap arrays for single atoms by use of a programmable phase modulator. *J. Opt. Soc. Am. B*, **21**(11), 1889–1894.

Beugnon, J., Jones, M. P. A., Dingjan, J., Darquiere, B., Messin, G., Browaeys, A., and Grangier, P. (2006). Quantum interference between two single photons emitted by independently trapped atoms. *Nature*, **440**, 779.

Blakestad, R. B., Ospelkaus, C., VanDevender, A. P., Amini, J. M., Britton, J., Leibfried, D., and Wineland, D. J. (2009, Apr). High-fidelity transport of trapped-ion qubits through an x-junction trap array. *Phys. Rev. Lett.*, **102**(15), 153002.

Blinov, B. B., Deslauriers, L., Lee, P., Madsen, M. J., Miller, R., and Monroe, C. (2002, Apr). Sympathetic cooling of trapped Cd$^+$ isotopes. *Phys. Rev. A*, **65**(4), 040304.

Blinov, B. B., Moehring, D. L., Duan, L. M., and Monroe, C. (2004). Observation of entanglement between a single trapped atom and a single photon. *Nature*, **428**, 153.

Boozer, A. D., Boca, A., Miller, R., Northup, T. E., and Kimble, H. J. (2007). Reversible state transfer between light and a single trapped atom. *Phys. Rev. Lett.*, **98**(19), 193601.

Bose, S., Knight, P. L., Plenio, M. B., and Vedral, V. (1999, Dec). Proposal for teleportation of an atomic state via cavity decay. *Phys. Rev. Lett.*, **83**(24), 5158–5161.

Bowe, P., Hornekær, L., Brodersen, C., Drewsen, M., Hangst, J. S., and Schiffer, J. P. (1999, Mar). Sympathetic crystallization of trapped ions. *Phys. Rev. Lett.*, **82**(10), 2071–2074.

Brennen, G. K., Caves, C. M., Jessen, P. S., and Deutsch, I. H. (1999, Feb). Quantum logic gates in optical lattices. *Phys. Rev. Lett.*, **82**(5), 1060–1063.

Briegel, H.-J., Dür, W., Cirac, J. I., and Zoller, P. (1998, Dec). Quantum repeaters: The role of imperfect local operations in quantum communication. *Phys. Rev. Lett.*, **81**(26), 5932–5935.

Brown, K. R., Clark, R. J., Labaziewicz, J., Richerme, P., Leibrandt, D. R., and Chuang, I. L. (2007, Jan). Loading and characterization of a printed-circuit-board atomic ion trap. *Phys. Rev. A*, **75**(1), 015401.

Brownnutt, M., Wilpers, G., Gill, P., Thompson, R. C., and Sinclair, A. G. (2006). Monolithic microfabricated ion trap chip design for scaleable quantum processors. *New J. Phy.*, **8**(10), 232.

Cabrillo, C., Cirac, J. I., García-Fernández, P., and Zoller, P. (1999, Feb). Creation of entangled states of distant atoms by interference. *Phys. Rev. A*, **59**(2), 1025–1033.

Chanelière, T., Matsukevitch, D. N., Jenkins, S. D., Lan, S. Y., Kennedy, T. A. B., and Kuzmich, A. (2005). Storage and retrieval of single photons transmitted between remote quantum memories. *Nature*, **438**, 833.

Chiaverini, J., Blakestad, R. B., Britton, J., Jost, J. D., Langer, C., Leibfried, D., Ozeri, R., and Wineland, D. J. (2005a). Surface-electrode architecture for ion-trap quantum information processing. *Quant. Inf. Comput.*, **6**, 419–439.

Chiaverini, J., Britton, J., Leibfried, D., Knill, E., Barrett, M. D., Blakestad, R. B., Itano, W. M., Jost, J. D., Langer, C., Ozeri, R., Schaetz, T., and Wineland, D. J. (2005b). Implementation of the Semiclassical Quantum Fourier Transform in a Scalable System. *Science*, **308**(5724), 997–1000.

Chiaverini, J., Leibfried, D., Schaetz, T., Barrett, M. D., Blakestad, R. B., Britton, J., Itano, W. M., Jost, J. D., Knill, E., Langer, C., Ozeri, R., and Wineland, D. J. (2004). Realization of quantum error correction. *Nature*, **432**, 602–605.

Childs, A. M. and Chuang, I. M. (2001). Universal quantum computation with two-level trapped ions. *Phys. Rev. A*, **63**, 012306.

Chou, C. W., de Riedmatten, H., Felinto, D., Polyakov, S. V., van Enk, S. J., and Kimble, H. J. (2005). Measurement-induced entanglement for excitation stored in remote atomic ensembles. *Nature*, **438**, 828–832.

Cirac, J. I. and Zoller, P. (1995, May). Quantum computations with cold trapped ions. *Phys. Rev. Lett.*, **74**(20), 4091–4094.

Cirac, J. I., Zoller, P., Kimble, H. J., and Mabuchi, H. (1997, Apr). Quantum state transfer and entanglement distribution among distant nodes in a quantum network. *Phys. Rev. Lett.*, **78**(16), 3221–3224.

Deslauriers, L., Haljan, P. C., Lee, P. J., Brickman, K.-A., Blinov, B. B., Madsen, M. J., and Monroe, C. (2004). Zero-point cooling and low heating of trapped Cd ions. *Phys. Rev. A*, **70**, 043408.

Deslauriers, L., Olmschenk, S., Stick, D., Hensinger, W. K., Sterk, J., and Monroe, C. (2006). Scaling and suppression of anomalous quantum decoherence in ion traps. *Phys. Rev. Lett.*, **97**, 103007.

Diedrich, F., Bergquist, J. C., Itano, W. M., and Wineland, D. J. (1989). Laser cooling to the zero-point energy of motion. *Phys. Rev. Lett.*, **62**, 403–406.

Duan, L.-M. (2004, Sep). Scaling ion trap quantum computation through fast quantum gates. *Phys. Rev. Lett.*, **93**(10), 100502.

Duan, L.-M., Blinov, B. B., Moehring, D. L., and Monroe, C. (2004). Scalable trapped ion quantum computation with a probabilistic ion-photon mapping. *Quant. Inf. Comp.*, **4**, 165.

Duan, L.-M. and Kimble, H. J. (2003, Jun). Efficient engineering of multiatom entanglement through single-photon detections. *Phys. Rev. Lett.*, **90**(25), 253601.

Dumke, R., Volk, M., Müther, T., Buchkremer, F. B. J., Birkl, G., and Ertmer, W. (2002, Aug). Micro-optical realization of arrays of selectively addressable dipole traps: A scalable configuration for quantum computation with atomic qubits. *Phys. Rev. Lett.*, **89**(9), 097903.

Eisaman, M. D., Andra, A., Massou, F., Fleischhauer, M., Zibrov, A. S., and Lukin, M. D. (2005). Electromagnetically induced transparency with tunable single-photon pulses. *Nature*, **438**, 837.

Eschner, J. (2006). Quantum computation with trapped ions. In *Proceedings of the International School of Physics 'Enrico Fermi' Course CLXII, "Quantum Computers, Algorithms and Chaos"*. IOS Press.

Eschner, J., Morigi, G., Schmidt-Kaler, F., and Blatt, R. (2003). Laser cooling of trapped ions. *J. Opt. Soc. Am. B*, **20**, 1003.

Feng, X.-L., Zhang, Z.-M., Li, X.-D., Gong, S.-Q., and Xu, Z.-Z. (2003, May). Entangling distant atoms by interference of polarized photons. *Phys. Rev. Lett.*, **90**(21), 217902.

Frese, D., Ueberholz, B., Kuhr, S., Alt, W., Schrader, D., Gomer, V., and Meschede, D. (2000, Oct). Single atoms in an optical dipole trap: Towards a deterministic source of cold atoms. *Phys. Rev. Lett.*, **85**(18), 3777–3780.

Friedenauer, A., Schmitz, H., Glueckert, J., Porras, D., and Schaetz, T. (2008). Simulating a quantum magnet with trapped ions. *Nature Phys.*, **4**, 757–761.

Gaëtan, A., Miroshnychenko, Y., Wilk, T., Chotia, A., Viteau, M., Comparat, D., Pillet, P., Browaeys, A., and Grangier, P. (2009). Observation of collective excitation of two individual atoms in the rydberg blockade regime. *Nature Phys.*, **5**, 115–118.

García-Ripoll, J. J., Zoller, P., and Cirac, J. I. (2003, Oct). Speed optimized two-qubit gates with laser coherent control techniques for ion trap quantum computing. *Phys. Rev. Lett.*, **91**(15), 157901.

García-Ripoll, J. J., Zoller, P., and Cirac, J. I. (2005, Jun). Coherent control of trapped ions using off-resonant lasers. *Phys. Rev. A*, **71**(6), 062309.

Gerritsma, R., Kirchmair, G., Zähringer, F., Solano, E., Blatt, R., and Roos, C. F. (2010). Quantum simulation of the dirac equation. *Nature*, **463**, 68–71.

Ghosh, Pradip K. (1995). *Ion Traps*. Clarendon Press, Oxford.

Grimm, R., Weidemüller, M., and Ovchinikov, Y. B. (2000). Optical dipole traps for neutral atoms. *Adv. At. Mol. Opt. Phys.*, **42**, 95.

Gulde, S., Häffner, H., Riebe, M., Lancaster, G., Mundt, A., Kreuter, A., Russo, C., Becher, C., Eschner, J., Schmidt-Kaler, F., Chuang, I. L., and Blatt, R. (2002). *Quantum information processing and cavity QED experiments with trapped Ca+ ions*, Atomic Physics 18. Proceedings of the ICAP.

Guthöhrlein, G. R., Keller, M., Hayasaka, K., Lange, W., and Walther, H. (2001). A single ion as a nanoscopic probe of an optical field. *Nature*, **414**, 49–51.

Haase, A., Piro, N., Eschner, J., and Mitchell, M. W. (2009, January). Tunable narrowband entangled photon pair source for resonant single-photon single-atom interaction. *Opt. Lett.*, **34**(1), 55–57.

Häffner, H., Gulde, S., Riebe, M., Lancaster, G., Becher, C., Eschner, J., Schmidt-Kaler, F., and Blatt, R. (2003, Apr). Precision measurement and compensation of optical stark shifts for an ion-trap quantum processor. *Phys. Rev. Lett.*, **90**(14), 143602.

Häffner, H., Hänsel, W., Roos, C. F., Schmidt, P. O., Riebe, M., Chwalla, M., Chek-al kar, D., Benhelm, J., Rapol, U. D., Körber, T., Becher, C., Gühne, O., Dür, W., and Blatt, R. (2008*a*). Quantum computing with trapped ions. In *Controllable Quantum States* (ed. H. Takayanagi, J. Nitta, and H. Nakano). World Scientific.

Häffner, H., Hänsel, W., Roos, C. F., Benhelm, J., Chek-al-kar, D., Chwalla, M., Körber, T., Rapol, U. D., Riebe, M., Schmidt, P. O., Becher, C., Gühne, O., Dür, W., and Blatt, R. (2005*a*). Scalable multiparticle entanglement of trapped ions. *Nature*, **438**, 643–646.

Häffner, H., Riebe, M., Schmidt-Kaler, F., Hänsel, W., Roos, C., Chwalla, M., Benhelm, J., Körber, T., Lancaster, G., Becher, C., James, D. F. V., and Blatt, R. (2005*b*). Teleportation with atoms. *ATOMIC PHYSICS 19: XIX International Conference on Atomic Physics; ICAP 2004*, **770**(1), 341–349.

Häffner, H., Roos, C. F., and Blatt, R. (2008*b*). Quantum computing with trapped ions. *Phys. Rep.*, **469**, 155.

Häffner, H., Schmidt-Kaler, F., Hänsel, W., Roos, C.F., Körber, T., Chwalla, M., Riebe, M., Benhelm, J., Rapol, U. D., Becher, C., and Blatt, R. (2005). Robust entanglement. *Appl. Phys. B*, **81**, 151–153.

Haljan, P. C., Lee, P. J., Brickman, K.-A., Acton, M., Deslauriers, L., and Monroe, C. (2005). Entanglement of trapped-ion clock states. *Phys. Rev. A*, **72**, 062316.

Hanneke, D., Home, J. P., Jost, J. D., Amini, J. M., Leibfried, D., and Wineland, D. J. (2010). Realization of a programmable two-qubit quantum processor. *Nature Phys.*, **6**(1), 13–16.

Hayes, D., Matsukevich, D. N., Maunz, P., Hucul, D., Quraishi, Q., Olmschenk, S., Campbell, W., Mizrahi, V., Senko, C., and Monroe, C. (2010). Entanglement of atomic qubits using an optical frequency comb. *Phys. Rev. Lett.*, **104**, 140501.

Hensinger, W. K., Olmschenk, S., Stick, D., Hucul, D., Yeo, M., Acton, M., Deslauriers, L., Monroe, C., and Rabchuk, J. (2006). T -junction ion trap array for two-dimensional ion shuttling, storage and manipulation. *App. Phys. Lett.*, **88**, 034101.

Home, J. P., Hanneke, D., Jost, J. D., Amini, J. M., Leibfried, D., and Wineland, D. J. (2009). Complete methods set for scalable ion trap quantum information processing. *Science*, **325**, 1227.

Home, J. P. and Steane, A. M. (2006). Electrode configurations for fast separation of trapped ions. *Quant. Inf. Comput.*, **6**(4&5), 289–325.

Hong, C. K., Ou, Z. Y., and Mandel, L. (1987, Nov). Measurement of subpicosecond time intervals between two photons by interference. *Phys. Rev. Lett.*, **59**(18), 2044–2046.

Huber, G., Deuschle, T., Schnitzler, W., Reichle, R., Singer, K., and Schmidt-Kaler, F. (2008). Transport of ions in a segmented linear paul trap in printed-circuit-board technology. *New J. Phys.*, **10**(1), 013004.

Hucul, D., Yeo, M., Olmschenk, S. and Monroe, C., Hensinger, W. K., and Rabchuk, J. (2008). On the transport of atomic ions in linear and multidimensional ion trap arrays. *Quantum Inf. Comput.*, **8**, 050178.

Isenhower, L., Urban, E., Zhang, X. L., Gill, A. T., Henage, T., Johnson, T. A., Walker, T. G., and Saffman, M. (2010, Jan). Demonstration of a neutral atom controlled-not quantum gate. *Phys. Rev. Lett.*, **104**(1), 010503.

Jaksch, D., Briegel, H.-J., Cirac, J. I., Gardiner, C. W., and Zoller, P. (1999, Mar). Entanglement of atoms via cold controlled collisions. *Phys. Rev. Lett.*, **82**(9), 1975–1978.

Jaksch, D., Cirac, J. I., Zoller, P., Rolston, S. L., Côté, R., and Lukin, M. D. (2000, Sep). Fast quantum gates for neutral atoms. *Phys. Rev. Lett.*, **85**(10), 2208–2211.

James, D. V. F. (1998). Quantum dynamics of cold trapped ions, with application to quantum computation. *Appl. Phys. B*, **66**, 181–190.

Julsgaard, B., Kozhekin, A., and Polzik, E. S. (2001). Experimental long-lived entanglement of two macroscopic objects. *Nature*, **413**, 400–403.

Julsgaard, B., Sherson, J., Cirac, J. I., Fiursek, J., and Polzik, E. S. (2004). Experimental demonstration of quantum memory for light. *Nature*, **432**, 482–486.

Keller, M., Lange, B., Hayasaka, K., Lange, W., and Walther, H. (2003). Deterministic coupling of single ions to an optical cavity. *Appl. Phys. B*, **76**, 125–128.

Keller, M., Lange, B., Hayasaka, K., Lange, W., and Walther, H. (2004). Continuous generation of single photons with controlled waveform in an ion-trap cavity system. *Nature*, **431**, 1075.

Kielpinski, D., Meyer, V., Rowe, M. A., Sackett, C. A., Itano, W. M., Monroe, C., and Wineland, D. J. (2001). A Decoherence-Free Quantum Memory Using Trapped Ions. *Science*, **291**(5506), 1013–1015.

Kielpinski, D., Monroe, C., and Wineland, D. J. (2002). Architecture for a large-scale ion-trap quantum computer. *Nature*, **417**, 709–711.

Kimble, H. J. (2008). The quantum internet. *Nature*, **453**, 1023–1030.

Kimble, H. J. and van Enk, S. J. (2004). Push-button teleportation. *Nature*, **429**, 712.

King, B. E., Myatt, C. J., Turchette, Q. A., Leibfried, D., Itano, W. M., Monroe, C., and Wineland, D. J. (1998). Cooling the collective motion of trapped ions to initialize a quantum register. *Phys. Rev. Lett.*, **81**, 15251528.

Kirchmair, G., Benhelm, J., Zhringer, F., Gerritsma, R., Roos, C. F., and Blatt, R. (2009). High-fidelity entanglement of ions of ^{43}Ca$^+$ hyperfine clock states. *Phys. Rev. A*, **79**, 020304.

Kraus, B. and Cirac, J. I. (2004, Jan). Discrete entanglement distribution with squeezed light. *Phys. Rev. Lett.*, **92**(1), 013602.

Kuhn, A., Hennrich, M., and Rempe, G. (2002). Deterministic single-photon source for distributed quantum networking. *Phys. Rev. Lett.*, **89**, 067901.

Labaziewicz, J., Ge, Y., Leibrandt, D. R., Wang, S. X., Shewmon, R., and Chuang, I. L. (2008, Oct). Temperature dependence of electric field noise above gold surfaces. *Phys. Rev. Lett.*, **101**(18), 180602.

Langer, C., Ozeri, R., Jost, J. D., Chiaverini, J., DeMarco, B., Ben-Kish, A., Blakestad, R. B., Britton, J., Hume, D. B., Itano, W. M., Leibfried, D., Reichle, R., Rosenband, T., Schaetz, T., Schmidt, P. O., and Wineland, D. J. (2005, Aug). Long-lived qubit memory using atomic ions. *Phys. Rev. Lett.*, **95**(6), 060502.

Larson, D. J., Bergquist, J. C., Bollinger, J. J., Itano, Wayne M., and Wineland, D. J. (1986, Jul). Sympathetic cooling of trapped ions: A laser-cooled two-species non-neutral ion plasma. *Phys. Rev. Lett.*, **57**(1), 70–73.

Leibfried, D., DeMarco, B., Meyer, V., Lucas, D., Barrett, M., Britton, J., Itano, W. M., Jelenkovic, B., Langer, C., Rosenband, T., and Wineland, D. J. (2003). Experimental demonstration of a robust, high-fidelity geometric two ion-qubit phase gate. *Nature*, **422**, 412–415.

Leibfried, D., Knill, E., Seidelin, S., Britton, J., Blakestad, R. B., Chiaverini, J., Hume, D. B., Itano, W. M., Jost, J. D., Langer, C., Ozeri, R., Reichle, R., and Wineland, D. J. (2005). Creation of a six-atom 'schrödinger cat' state. *Nature*, **438**, 639–642.

Levitt, M. H. (1986). Composite pulses (NMR spectroscopy). *Prog. Nucl. Magn. Reson. Spectrosc.*, **18**, 61122.

Lloyd, S., Shahriar, M. S., Shapiro, J. H., and Hemmer, P. R. (2001, Sep). Long distance, unconditional teleportation of atomic states via complete Bell state measurements. *Phys. Rev. Lett.*, **87**(16), 167903.

Lucas, D. M., Ramos, A., Home, J. P., McDonnell, M. J., Nakayama, S., Stacey, J.-P., Webster, S. C., Stacey, D. N., and Steane, A. M. (2004). Isotope-selective photoionization for calcium ion trapping. *Phys. Rev. A*, **69**, 012711.

Luo, L., Hayes, D., Manning, T. A., Matsukevich, D. N., Maunz, P., Olmschenk, S., Sterk, J. D., and Monroe, C. (2009). Protocols and techniques for a scalable atomphoton quantum network. *Fortschritte der Physik*, **57**, 1133–1152.

Maiwald, R., Leibfried, D., Britton, J., Bergquist, J. C., Leuchs, G., and Wineland D. J. (2009). Stylus ion trap for enhanced access and sensing. *Nature Phys.*, **5**, 551–554.

McKeever, J., Boca, A., Boozer, A. D., Miller, R., Buck, J. R., Kuzmich, A., and Kimble, H. J. (2004). Deterministic Generation of Single Photons from One Atom Trapped in a Cavity. *Science*, **303**(5666), 1992–1994.

Metodiev, T. S., Thaker, D. D., Cross, A. W., Chong, F. T., and Chuang, I. L. (2005). A quantum logic array microarchitecture: Scalable quantum data movement and computation. *quant-ph/0509051*.

Microtrap (2006). http://www.microtrap.eu.

Mintert, F. and Wunderlich, C. (2001). Ion-trap quantum logic using long-wavelength radiation. *Phys. Rev. Lett.*, **87**, 257904.

Moehring, D. L., Maunz, P., Olmschenk, S., Younge, K. C., Matsukevich, D. N., Duan L.-M., and Monroe, C. (2007). Entanglement of single-atom quantum bits at a distance. *Nature*, **449**, 68.

Mompart, J., Eckert, K., Ertmer, W., Birkl, G., and Lewenstein, M. (2003, Apr). Quantum computing with spatially delocalized qubits. *Phys. Rev. Lett.*, **90**(14), 147901.

Monroe, C., Meekhof, D. M., King, B. E., Itano, W. M., and Wineland, D. J. (1995*a*, Dec). Demonstration of a fundamental quantum logic gate. *Phys. Rev. Lett.*, **75**(25), 4714–4717.

Monroe, C., Meekhof, D. M., King, B. E., Jefferts, S. R., Itano, W. M., Wineland, D. J., and Gould, P. (1995*b*). Resolved-sideband raman cooling of a bound atom to the 3d zero-point energy. *Phys. Rev. Lett.*, **75**, 4011.

Monz, T., Kim, K., Hänsel, W., Riebe, M., Villar, A. S., Schindler, P., Chwalla, M., Hennrich, M., and Blatt, R. (2009a). Realization of the quantum toffoli gate with trapped ions. *Phys. Rev. Lett.*, **102**(4), 040501.

Monz, T., Kim, K., Villar, A. S., Schindler, P., Chwalla, M., Riebe, M., Roos, C. F., Häffner, H., Hänsel, W., Hennrich, M., and Blatt, R. (2009b, Nov). Realization of universal ion-trap quantum computation with decoherence-free qubits. *Phys. Rev. Lett.*, **103**(20), 200503.

Müller, C. A. and Miniatura, C. (2002). Multiple scattering of light by atoms with internal degeneracy, J. Phys. A: Math. Gen. 35, 10163.

Mundt, A. B., Kreuter, A., Becher, C., Leibfried, D., Eschner, J., Schmidt-Kaler, F., and Blatt, R. (2002, Aug). Coupling a single atomic quantum bit to a high finesse optical cavity. *Phys. Rev. Lett.*, **89**(10), 103001.

Mundt, A. B., Kreuter, A., Russo, C., Becher, C., Leibfried, D., Eschner, J., Schmidt-Kaler, F., and Blatt, R. (2003). Coherent coupling of a single Ca+ ion to a high-finesse optical cavity. *Appl. Phys. B*, **76**, 117–124.

Myerson, A. H., Szwer, D. J., Webster, S. C., Allcock, D. T. C., Curtis, M. J., Imreh, G., Sherman, J. A., Stacey, D. N., Steane, A. M., and Lucas, D. M. (2008, May). High-fidelity readout of trapped-ion qubits. *Phys. Rev. Lett.*, **100**(20), 200502.

Nägerl, H. C., Leibfried, D., Rohde, H., Thalhammer, G., Eschner, J., Schmidt-Kaler, F., and Blatt, R. (1999, Jul). Laser addressing of individual ions in a linear ion trap. *Phys. Rev. A*, **60**(1), 145–148.

Nägerl, H. C., Leibfried, D., Schmidt-Kaler, F., Eschner, J., and Blatt, R. (1998). Coherent excitation of normal modes in a string of Ca+ ions. *Opt. Exp.*, **3**, 89–96.

Nebendahl, V., Häffner, H., and Roos, C. F. (2009, Jan). Optimal control of entangling operations for trapped-ion quantum computing. *Phys. Rev. A*, **79**(1), 012312.

Neergaard-Nielsen, J. S., Nielsen, B. M., Takahashi, H., Vistnes, A. I., and Polzik, E. S. (2007). High purity bright single photon source. *Opt. Exp.*, **15**(13), 7940–7949.

Nussmann, S., Hijlkema, M., Weber, B., Rohde, F., Rempe, G., and Kuhn, A. (2005). Submicron positioning of single atoms in a microcavity. *Phys. Rev. Lett.*, **95**(17), 173602.

Olmschenk, S., Matsukevich, D. N., Maunz, P., Hayes, D., Duan, L.-M., and Monroe, C. (2009). Quantum teleportation between distant matter qubits. *Science*, **323**(5913), 486–489.

Paul, W., Osberghaus, O., and Fischer, E. (1958). Ein Ionenkäfig. *Forschungsberichte des Wirtschafts- und Verkehrsministeriums Nordrhein-Westfalen 415, Westfälischer Verlag*, **4**, 15.

Pellizzari, T., Gardiner, S. A., Cirac, J. I., and Zoller, P. (1995, Nov). Decoherence, continuous observation, and quantum computing: A cavity qed model. *Phys. Rev. Lett.*, **75**(21), 3788–3791.

Pierre, F. and Birge, N.O. (2002). Dephasing by extremely dilute magnetic impurities revealed by Aharonov-Bohm oscillations, Phys. Rev. Lett. 89, 206804.

Piro, N., Haase, A., Mitchell, M. W., and Eschner, J. (2009). An entangled photon source for resonant single-photon-single-atom interaction. *J. Phys. B: At. Mol. Opt. Phys.*, **42**, 114002.

Piro, N., Rohde, F., Schuck, C., Almendros, M., Huwer, J., Ghosh, J., Haase, A., Hennrich, M., Dubin, F., and Eschner, J. (2011). Heralded single-photon absorption by a single atom. *Nature Phys.*, **7**, 17.

Poschinger, U. G., Huber, G., Ziesel, F., Deiss, M., Hettrich, M., Schulz, S. A., Singer, K., Poulsen, G., Drewsen, M., Hendricks, R. J., and Schmidt-Kaler, F. (2009). Coherent manipulation of a $^{40}Ca^+$ spin qubit in a micro ion trap. *J. Phys. B: At. Mol. Opt. Phys.*, **42**(15), 154013.

Quabis, S., Dorn, R., Eberler, M., Glöckl, O., and Leuchs, G. (2000). Focusing light to a tighter spot. *Opt. Commun.*, **179**(1-6), 1–7.

Reichle, R., Leibfried, D., Knill, E., Britton, J., Blakestad, R. B., Jost, J. D., Langer, C., Ozeri, R., Seidelin, S., and Wineland, D. J. (2006). Experimental purification of two-atom entanglement. *Nature*, **443**, 838–841.

Riebe, M. (2005). *Preparation of entangled states and quantum teleportation with atomic qubits.* Ph.D. thesis, Leopold-Franzens-Universität Innsbruck.

Riebe, M., Häffner, H., Roos, C. F., Hänsel, W., Benhelm, J., Lancaster, G. P. T., Körber, T. W., Becher, C., Schmidt-Kaler, F., James, D. F. V., and Blatt, R. (2004). Deterministic quantum teleportation with atoms. *Nature*, **429**, 734–737.

Rohde, H., Gulde, S. T., Roos, C. F., Barton, P. A., Leibfried, D., Eschner, J., Schmidt-Kaler, F., and Blatt, R. (2001). Sympathetic ground state cooling and coherent manipulation with two-ion-crystals. *J. Opt. B*, **3**, 34.

Roos, Ch., Zeiger, Th., Rohde, H., Nägerl, H. C., Eschner, J., Leibfried, D., Schmidt-Kaler, F., and Blatt, R. (1999). Quantum state engineering on an optical transition and decoherence in a paul trap. *Phys. Rev. Lett.*, **83**, 4713–4716.

Roos, C. F., Lancaster, G. P. T., Riebe, M., Häffner, H., Hänsel, W., Gulde, S., Becher, C., Eschner, J., Schmidt-Kaler, F., and Blatt, R. (2004a, Jun). Bell states of atoms with ultralong lifetimes and their tomographic state analysis. *Phys. Rev. Lett.*, **92**(22), 220402.

Roos, C. F., Riebe, M., Häffner, H., Hänsel, W., Benhelm, J., Lancaster, G. P. T., Becher, C., Schmidt-Kaler, F., and Blatt, R. (2004b). Control and Measurement of Three-Qubit Entangled States. *Science*, **304**(5676), 1478–1480.

Rosenband, T., Schmidt, P. O., Hume, D. B., Itano, W. M., Fortier, T. M., Stalnaker, J. E., Kim, K., Diddams, S. A., Koelemeij, J. C. J., Bergquist, J. C., and Wineland, D. J. (2007, May). Observation of the $^1S_0 \rightarrow ^3P_0$ clock transition in $^{27}Al^+$. *Phys. Rev. Lett.*, **98**(22), 220801.

Rowe, M. A., Ben-Kish, A., DeMarco, B., Leibfried, D., Meyer, V., Beall, J., Britton, J., Hughes, J., Itano, W. M., Jelenkovic, B., Langer, C., Rosenband, T., and Wineland, D. J. (2002). Transport of quantum states and separation of ions in a dual rf ion trap. *Quant. Informat. Comput.*, **2**, 257–271.

Russo, C., Barros, H. G., Stute, A., Dubin, F., Phillips, E. S., Monz, T., Northup, T. E., Becher, C., Salzburger, T., Ritsch, H., Schmidt, P. O., and Blatt, R. (2009). Raman spectroscopy of a single ion coupled to a high-finesse cavity. *Appl. Phys. B*, **95**, 205–212.

Sackett, C. A., Kielpinski, D., King, B. E., Langer, C., Meyer, V., Myatt, C. J., Rowe, M., Turchette, Q. A., Itano, W. M., Wineland, D. J., and C., Monroe (2000). Experimental entanglement of four particles. *Nature*, **404**, 256–259.

Schaetz, T., Barrett, M. D., Leibfried, D., Chiaverini, J., Britton, J., Itano, W. M., Jost, J. D., Langer, C., and Wineland, D. J. (2004, Jul). Quantum dense coding with atomic qubits. *Phys. Rev. Lett.*, **93**(4), 040505.

Schlosser, N., Reymond, G., Protsenko, I., and Grangier, P. (2001, Jun 28). Subpoissonian loading of single atoms in a microscopic dipole trap. *Nature*, **411**(6841), 1024–1027.

Schmidt, P. O., Rosenband, T., Langer, C., Itano, W. M., Bergquist, J. C., and Wineland, D. J. (2005). Spectroscopy Using Quantum Logic. *Science*, **309**(5735), 749–752.

Schmidt-Kaler, F., Häffner, H., Gulde, S., Riebe, M., Lancaster, G. P. T., Deuschle, T., Becher, C., Hänsel, W., Eschner, J., Roos, C. F., and Blatt, R. (2003a). How to realize a universal quantum gate with trapped ions. *Appl. Phys. B*, **77**, 789.

Schmidt-Kaler, F., Häffner, H., Riebe, M., Gulde, S., Lancaster, G. P. T., Deuschle, T., Becher, C., Roos, C. F., Eschner, J., and Blatt, R. (2003b). Realization of the Cirac-Zoller controlled-not quantum gate. *Nature*, **422**, 408–411.

Schmidt-Kaler, F., Roos, Ch., Nägerl, H. C., Rohde, H., Gulde, S., Mundt, A., Lederbauer, M., Thalhammer, G., Zeiger, Th., Barton, P., Hornekaer, L., Reymond, G., Leibfried, D., Eschner, J., and Blatt, R. (2000). Ground state cooling, quantum state engineering and study of decoherence of ions in Paul traps. *J. Mod. Opt.*, **47**, 2573.

Schuck, C., Rohde, F., Piro, N., Almendros, M., Huwer, J., Mitchell, M. W., Hennrich, M., Haase, A., Dubin, F., and Eschner, J. (2010, Jan). Resonant interaction of a single atom with single photons from a down-conversion source. *Phys. Rev. A*, **81**(1), 011802.

Schulz, S. A., Poschinger, U., Ziesel, F., and Schmidt-Kaler, F. (2008). Sideband cooling and coherent dynamics in a microchip multi-segmented ion trap. *New J. Phys.*, **10**(4), 045007.

Seidelin, S., Chiaverini, J., Reichle, R., Bollinger, J. J., Leibfried, D., Britton, J., Wesenberg, J. H., Blakestad, R. B., Epstein, R. J., Hume, D. B., Itano, W. M., Jost, J. D., Langer, C., Ozeri, R., Shiga, N., and Wineland, D. J. (2006, Jun). Microfabricated surface-electrode ion trap for scalable quantum information processing. *Phys. Rev. Lett.*, **96**(25), 253003.

Shor, P. W. (1995, Oct). Scheme for reducing decoherence in quantum computer memory. *Phys. Rev. A*, **52**(4), R2493–R2496.

Shu, G., Dietrich, M. R., Kurz, N., and Blinov, B. B. (2009a, Aug 14). Trapped ion imaging with a high numerical aperture spherical mirror. *J. Phys. B.: Atom. Molec. Opt. Phys.*, **42**(15), 154005.

Shu, G., Kurz, N., Dietrich, M. R., and Blinov, B. B. (2009b). Efficient fluorescence collection from trapped ion qubits with an integrated spherical mirror. *Phys. Rev. A*, **81**, 042321 (2010).

Simon, C. and Irvine, W. T. M. (2003, Sep). Robust long-distance entanglement and a loophole-free Bell test with ions and photons. *Phys. Rev. Lett.*, **91**(11), 110405.

Sinclair, A. G., Wilson, M. A., Letchumanan, V., and Gill, P. (2001). *Proceedings of the 6th Symposium on Frequency Standards and Metrology, St Andrews, Scotland, UK*, 498–500.

Sondermann, M., Lindlein, N., and Leuchs, G. (2009). Maximizing the electric field strength in the foci of high numerical aperture optics. *arXiv:0811.2098v2 [physics.optics]*.

Sondermann, M., Maiwald, R., Konermann, H., Lindlein, N., Peschel, U., and Leuchs, G. (2007). Design of a mode converter for efficient light-atom coupling in free space. *Appl. Phys. B*, **89**, 489–492.

Sørensen, A. and Mølmer, K. (1999, Mar). Quantum computation with ions in thermal motion. *Phys. Rev. Lett.*, **82**(9), 1971–1974.

Splatt, F., Harlander, M., Brownnutt, M., Zähringer, F., Blatt, R., and Hänsel, W. (2009). Deterministic reordering of ^{40}Ca$^+$ ions in a linear segmented Paul trap. *New J. Phys.*, **11**(10), 103008.

Steane, A. (1997). The ion trap quantum information processor. *Appl. Phys. B*, **64**, 623–642.

Steane, A., Roos, C. F., Stevens, D., Mundt, A., Leibfried, D., Schmidt-Kaler, F., and Blatt, R. (2000, Sep). Speed of ion-trap quantum-information processors. *Phys. Rev. A*, **62**(4), 042305.

Steane, A. M. (1996, Jul). Error correcting codes in quantum theory. *Phys. Rev. Lett.*, **77**(5), 793–797.

Steane, A. M. (2007). How to build a 300 bit, 1 Giga-operation quantum computer. *Quant. Inform. Comput.*, **7**(3), 171–183.

Stenholm, S. (1986). Semiclassical theory of laser cooling. *Rev. Mod. Phys.*, **58**, 699–739.

Stick, D., Hensinger, W. K., Olmschenk, S., Madsen, M. J., Schwab, K., and Monroe, C. (2006). Ion trap in a semiconductor chip. *Nature Phys.*, **2**, 36–39.

Stobinska, M., Alber, G., and Leuchs, G. (2009). Perfect excitation of a matter qubit by a single photon in free space. *EPL*, **86**, 14007.

Stucki, D., Walenta, N., Vannel, F., Thew, R. T., Gisin, N., Zbinden, H., Gray, S., Towery, C. R., and Ten, S. (2009). High rate, long-distance quantum key distribution over 250 km of ultra low loss fibres. *New J. Phys.*, **11**(7), 075003.

Tey, M. K., Maslennikov, G., Liew, T. C. H., Aljunid, S. A., Huber, F., Chng, B., Chen, Z, Scarani, V., and Kurtsiefer, C. (2009). Phase shift of a weak coherent beam induced by a single atom. *Phys. Rev. Lett.*, **103**(15), 153601.

Tey, M. K., Chen, Z., Aljunid, S. A., Chng, B., Huber, F., Maslennikov, G., and Kurtsiefer, C. (2008). Strong interaction between light and a single trapped atom without the need for a cavity. *Nature Phys.*, **4**, 924–927.

Urban, E., Johnson, T. A., Henage, T., Isenhower, L., Yavuz, D. D., Walker, T. G., and Saffman, M. (2009). Observation of Rydberg blockade between two atoms. *Nature Phys.*, **5**, 110–114.

Vamivakas, A. N., Atatüre, M., Dreiser, J., Yilmaz, S. T., Badolato, A., Swan, A. K., Goldberg, B. B., Imamolu, A., and Ünlü, M. S. (2007). Strong extinction of a far-field laser beam by a single quantum dot. *Nano Lett.*, **7 (9)**, 28922896.

van Enk, S. J. (2004, Apr). Atoms, dipole waves, and strongly focused light beams. *Phys. Rev. A*, **69**(4).

Volz, J., Weber, M., Schlenk, D., Rosenfeld, W., Vrana, J., Saucke, K., Kurtsiefer, C., and Weinfurter, H. (2006). Observation of entanglement of a single photon with a trapped atom. *Phys. Rev. Lett.*, **96**(3), 030404.

Weber, M., Volz, J., Saucke, K., Kurtsiefer, C., and Weinfurter, H. (2006, Apr). Analysis of a single-atom dipole trap. *Phys. Rev. A*, **73**(4), 043406.

Wilk, T., Gaëtan, A., Evellin, C., Wolters, J., Miroshnychenko, Y., Grangier, P., and Browaeys, A. (2010, Jan). Entanglement of two individual neutral atoms using Rydberg blockade. *Phys. Rev. Lett.*, **104**(1), 010502.

Wilk, T., Webster, S. C., Kuhn, A., and Rempe, G. (2007). Single-atom single-photon quantum interface. *Science*, **317**(5837), 488–490.

Wineland, D. J., Barrett, M., Britton, J., Chiaverini, J., DeMarco, B. L., Itano, W. M., Jelenkovic, B. M., Langer, C., Leibfried, D., Meyer, V., Rosenband, T., and Schaetz, T. (2003). Quantum information processing with trapped ions. *Phil. Trans. Royal Soc. London A*, **361**, 1349–1361.

Wineland, D. J. and Itano, W. M. (1979). Laser cooling of atoms. *Phys. Rev. A*, **20**, 1521–1540.

Wolfgramm, F., Xing, X., Cerè, A., Predojević, A., Steinberg, A. M., and Mitchell, M. W. (2008). Bright filter-free source of indistinguishable photon pairs. *Opt. Exp.*, **16**(22), 18145–18151.

Wrigge, G., Gerhardt, I., Hwang, J., Zumofen, G., and Sandoghdar, V. (2008). Efficient coupling of photons to a single molecule and the observation of its resonance fluorescence. *Nature Phys.*, **4**, 60–66.

Yavuz, D. D., Kulatunga, P. B., Urban, E., Johnson, T. A., Proite, N., Henage, T., Walker, T. G., and Saffman, M. (2006, Feb). Fast ground state manipulation of neutral atoms in microscopic optical traps. *Phys. Rev. Lett.*, **96**(6), 063001.

You, L. and Chapman, M. S. (2000, Oct). Quantum entanglement using trapped atomic spins. *Phys. Rev. A*, **62**(5), 052302.

Zhu, S.-L., Monroe, C., and Duan, L.-M. (2006). Arbitrary-speed quantum gates within large ion crystals through minimum control of laser beams. *Europhys. Lett.*, **73**(4), 485.

Zippilli, S., Olivares-Renteria, G. A., Morigi, G., Schuck, C., Rohde, F., and Eschner, J. (2008). Entanglement of distant atoms by projective measurement: The role of detection efficiency. *New J. Phys*, **10**, 103003.

Zoller P. et al., (2005). Quantum information processing and communication. *Eur. Phys. J. D*, **36**, 203.

Zumofen, G., Mojarad, N. M., Sandoghdar, V., and Agio, M. (2008). Perfect reflection of light by an oscillating dipole. *Phys. Rev. Lett.*, **101**(18), 180404.

6
Quantum Hall effects

Mark O. Goerbig

Laboratoire de Physique des Solides, CNRS UMR 8502, Université Paris-Sud, France

Preface

This chapter covers a series of three lectures on the quantum Hall effect given at the Singapore session "Ultracold Gases and Quantum Information" at *Les Houches Summer School* 2009. Almost 30 years after the discovery of the quantum Hall effect, the research subject of quantum Hall physics has definitely acquired a high degree of maturity that is reflected by a certain number of excellent reviews and books, of which we can cite only a few (Prange and Girvin, 1990; Yoshioka, 2002; Ezawa, 2000) for possible further or complementary reading. Also, the different sessions of *Les Houches Summer School* have covered in several aspects quantum Hall physics, and S. M. Girvin's series of lectures in 1998 (Girvin, 1999) has certainly become a reference in the field.[1] Girvin's lecture notes were indeed extremely useful for myself when I started to study the quantum Hall effect at the beginning of my Master and PhD studies.

The present chapter is complementary to the existing literature in several aspects. One should first mention its introductory character to the field, which is in no way exhaustive. As a consequence, the presentation of one-particle physics and a detailed discussion of the integer quantum Hall effect occupy the major part of this chapter, whereas the – certainly more interesting – fractional quantum Hall effect, with its relation to strongly correlated electrons, its fractionally charged quasi-particles and fractional statistics, is only briefly introduced.

Furthermore, we have tried to avoid as much as possible the formal aspects of the fractional quantum Hall effect, which is discussed only in the framework of trial wave functions *à la Laughlin*. We have thus omitted, e.g., a presentation of Chern–Simons theories and related quantum-field theoretical approaches, such as the Hamiltonian theory of the fractional quantum Hall effect (Murthy and Shankar, 2003), as much as the relation between the quantum Hall effect and conformal field theories. Although these theories are extremely fruitful and still promising for a deeper understanding of quantum Hall physics, a detailed discussion of them would require more space than this chapter with its introductory character can provide.

Another complementary aspect of the present chapter as compared to existing textbooks consists of an introduction to Landau-level quantization that treats in a parallel manner the usual non-relativistic electrons in semiconductor heterostructures and relativistic electrons in graphene (two-dimensional graphite). Indeed, the 2005 discovery of a quantum Hall effect in this amazing material (Novoselov, *et al.* 2005; Zhang, *et al.* 2005) has given a novel and unexpected boost to research in quantum Hall physics.

As compared to the (oral) lectures, this chapter contains slightly more information. An example is Laughlin's plasma analogy, which is described in Section 6.4.2.5, although it was not discussed in the oral lectures. Furthermore, I have decided to add a section on multi-component quantum Hall systems, which, for completeness, needed to be at least briefly discussed.

Before the Singapore session of *Les Houches Summer School*, this series of lectures had been presented in a similar format at the (French) Summer School of the Research Grouping "Physique Mésoscopique" at the Institute of Scientific Research, Cargèse,

Corsica, in 2008. Furthermore, a longer series of lectures on the quantum Hall effect was prepared in collaboration with my colleague and former PhD advisor Pascal Lederer (Orsay, 2006). Its aim was somewhat different, with an introduction to the Hamiltonian theories of the fractional quantum Hall effect and correlation effects in multi-component systems. As already mentioned above, the latter aspect is only briefly introduced within the present chapter and a discussion of Hamiltonian theories is completely absent. The Orsay series of lectures was repeated by Pascal Lederer at the *Ecole Polytechnique Fédérale* in Lausanne Switzerland, in 2006, and at the University of Recife, Brazil, in 2007. The finalization of these longer and more detailed lecture notes (in French) is currently in progress. The graphene-related aspects of the quantum Hall effect have furthermore been presented in a series of lectures on graphene (Orsay, 2008) prepared in collaboration with Jean-Noël Fuchs, whom I would like to thank for a careful reading of the present Chapter.

6.1 Introduction

Quantum Hall physics – the study of two-dimensional (2D) electrons in a strong perpendicular magnetic field [see Figure 6.1(a)] – has become an extremely important research subject during the last two and a half decades. The interest in quantum Hall physics stems from its position at the borderline between low-dimensional quantum systems and systems with strong electronic correlations, probably the major issues of modern condensed-matter physics. From a theoretical point of view, the study of quantum Hall systems required the elaboration of novel concepts, some of which were better known in quantum-field theories used in high-energy rather than in condensed-matter physics, such as e.g. charge fractionalization, non-commutative geometries and topological field theories.

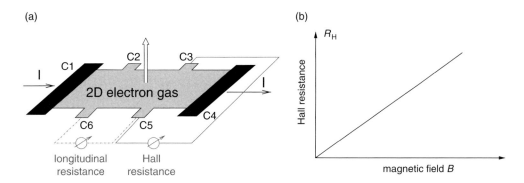

Fig. 6.1 (a) 2D electrons in a perpendicular magnetic field (quantum Hall system). In a typical transport measurement, a current I is driven through the system via the contacts C1 and C4. The longitudinal resistance may be measured between the contacts C5 and C6 (or alternatively between C2 and C3). The transverse (or Hall) resistance is measured, e.g., between the contacts C3 and C5. (b) Classical Hall resistance as a function of the magnetic field.

The motivation of the present Chapter is to provide in an accessible manner the basic knowledge of quantum Hall physics and to enable thus interested graduate students to pursue on her or his own further studies in this subject. I have therefore tried, wherever I feel that a more detailed discussion of some aspects in this large field of physics would go beyond the introductory character of this chapter, to provide references to detailed and pedagogical articles or complementary textbooks.

6.1.1 History of the (Quantum) Hall effect

6.1.1.1 The physical system

Our basic knowledge of quantum Hall systems, i.e. a system of 2D electrons in a per-pendicular magnetic field, stems from electronic transport measurements, where one drives a current I through the sample and where one measures both the *longitudinal* and the *transverse* resistance (also called Hall resistance). The difference between these two resistances is essential and may be defined topologically: consider a current that is driven through the sample via two arbitrary contacts [C1 and C4 in Figure 6.1(a)] and draw (in your mind) a line between these two contacts. A longitudinal resistance is a resistance measured between two (other) contacts that may be connected by a line that does not cross the line connecting C1 and C4. In Figure 6.1(a), we have chosen the contacts C5 and C6 for a possible longitudinal resistance measurement. The transverse resistance is measured between two contacts that are connected by an imaginary line that necessarily crosses the line connecting C1 and C4 [e.g. C3 and C5 in Figure 6.1(b)].

6.1.1.2 Classical Hall effect

Evidently, if there is a *quantum* Hall effect, it is most natural to expect that there exists also a *classical* Hall effect. This is indeed the case, and its history goes back to 1879 when Hall showed that the transverse resistance R_H of a thin metallic plate varies linearly with the strength B of the perpendicular magnetic field [Figure 6.1(b)],

$$R_H = \frac{B}{qn_{el}}, \tag{6.1}$$

where q is the carrier charge ($q = -e$ for electrons in terms of the elementary charge e that we define positive in the remainder of these lectures) and n_{el} is the 2D carrier density. Intuitively, one may understand the effect as due to the Lorentz force, which bends the trajectory of a charged particle such that a density gradient is built up between the two opposite sample sides that are separated by the contacts C1 and C4. Notice that the classical Hall resistance is still used today to determine, in material science, the carrier charge and density of a conducting material.

More quantitatively, the classical Hall effect may be understood within the Drude model for diffusive transport in a metal. Within this model, one considers independent charge carriers of momentum $\mathbf{p}$ described by the equation of motion

$$\frac{d\mathbf{p}}{dt} = -e\left(\mathbf{E} + \frac{\mathbf{p}}{m_b} \times \mathbf{B}\right) - \frac{\mathbf{p}}{\tau},$$

where **E** and **B** are the electric and magnetic fields, respectively. Here, we consider transport of negatively charged particles (i.e. electrons with $q = -e$) with band mass m_b. The last term takes into account relaxation processes due to the diffusion of electrons by generic impurities, with a characteristic relaxation time τ. The macroscopic transport characteristics, i.e. the resistivity or conductivity of the system, are obtained from the stationary solution of the equation of motion, $d\mathbf{p}/dt = 0$, and one finds for 2D electrons with $\mathbf{p} = (p_x, p_y)$

$$eE_x = -\frac{eB}{m_b}p_y - \frac{p_x}{\tau},$$

$$eE_y = \frac{eB}{m_b}p_x - \frac{p_y}{\tau},$$

where we have chosen the magnetic field in the z-direction. In the above expressions, one notices the appearence of a characteristic frequency,

$$\omega_C = \frac{eB}{m_b}, \tag{6.2}$$

which is called the *cyclotron frequency* because it characterizes the cyclotron motion of a charged particle in a magnetic field. With the help of the Drude conductivity,

$$\sigma_0 = \frac{n_{el}e^2\tau}{m_b}, \tag{6.3}$$

one may rewrite the above equations as

$$\sigma_0 E_x = -en_{el}\frac{p_x}{m_b} - en_{el}\frac{p_y}{m_b}(\omega_C\tau),$$

$$\sigma_0 E_y = en_{el}\frac{p_x}{m_b}(\omega_C\tau) - en_{el}\frac{p_y}{m_b},$$

or, in terms of the current density

$$\mathbf{j} = -en_{el}\frac{\mathbf{p}}{m_b}, \tag{6.4}$$

in matrix form as $\mathbf{E} = \rho\mathbf{j}$, with the resistivity tensor

$$\rho = \sigma^{-1} = \frac{1}{\sigma_0}\begin{pmatrix} 1 & \omega_C\tau \\ -\omega_C\tau & 1 \end{pmatrix} = \frac{1}{\sigma_0}\begin{pmatrix} 1 & \mu B \\ -\mu B & 1 \end{pmatrix}, \tag{6.5}$$

where we have introduced, in the last step, the mobility

$$\mu = \frac{e\tau}{m_b}. \tag{6.6}$$

From the above expression, one may immediately read off the Hall resistivity (the off-diagonal terms of the resistivity tensor ρ)

$$\rho_H = \frac{\omega_C \tau}{\sigma_0} = \frac{eB}{m_b}\tau \times \frac{m_b}{n_{el}e^2\tau} = \frac{B}{en_{el}}. \tag{6.7}$$

Furthermore, the conductivity tensor is obtained from the resistivity (6.5), by matrix inversion,

$$\sigma = \rho^{-1} = \begin{pmatrix} \sigma_L & -\sigma_H \\ \sigma_H & \sigma_L \end{pmatrix}, \tag{6.8}$$

with $\sigma_L = \sigma_0/(1+\omega_C^2\tau^2)$ and $\sigma_H = \sigma_0\omega_C\tau/(1+\omega_C^2\tau^2)$. It is instructive to discuss, based on these expressions, the theoretical limit of vanishing impurities, i.e. the limit $\omega_C\tau \to \infty$ of very long scattering times. In this case the resistivity and conductivity tensors read

$$\rho = \begin{pmatrix} 0 & \frac{B}{en_{el}} \\ -\frac{B}{en_{el}} & 0 \end{pmatrix} \quad \text{and} \quad \sigma = \begin{pmatrix} 0 & -\frac{en_{el}}{B} \\ \frac{en_{el}}{B} & 0 \end{pmatrix}, \tag{6.9}$$

respectively. Notice that if we had put under the carpet the matrix character of the conductivity and resistivity and if we had only considered the *longitudinal* components, we would have come to the counter-intuitive conclusion that the (longitudinal) resistivity would vanish at the same time as the (longitudinal) conductivity. The transport properties in the clean limit $\omega_C\tau \to \infty$ are therefore entirely governed, in the presence of a magnetic field, by the off-diagonal, i.e. transverse, components of the conductivity/resistivity. We will come back to this particular feature of quantum Hall systems when discussing the integer quantum Hall effect below.

Resistivity and resistance.   The above treatment of electronic transport in the framework of the Drude model allowed us to calculate the conductivity or resistivity of classical diffusive 2D electrons in a magnetic field. However, an experimentalist does not measure a conductivity or resistivity, i.e. quantities that are easier to calculate for a theoretician, but a *conductance* or a *resistance*. Usually, these quantities are related to one another but depend on the geometry of the conductor – the resistance R is thus related to the resistivity ρ by $R = (L/A)\rho$, where L is the length of the conductor and A its cross-section. From the scaling point of view of a d-dimensional conductor, the cross-section scales as L^{d-1}, such that the scaling relation between the resistance and the resistivity is

$$R \sim \rho L^{2-d}, \tag{6.10}$$

and one immediately notices that a 2D conductor is a special case. From the dimensional point of view, resistance and resistivity are the same in 2D, and the resistance is scale-invariant. Naturally, this scaling argument neglects the fact that the length L and the width W (the 2D cross-section) do not necessarily coincide: indeed, the resistance of a 2D conductor depends in general on the so-called *aspect ratio* L/W via some factor $f(L/W)$ (Akkermans and Montambaux, 2008). However, in the case of the transverse Hall resistance it is the length of the conductor itself that plays the role of the cross-section, such that the Hall resistivity and the Hall resistance truly coincide, i.e. $f = 1$. We will see in Section 6.3 that this conclusion also holds in the case of the quantum Hall effect and not only in the classical regime. Moreover, the quantum Hall effect is highly insensitive to the particular geometric properties of the sample used in the transport measurement, such that the quantization of the Hall resistance is surprisingly precise (on the order of 10^{-9}) and the quantum Hall effect is used nowadays in the definition of the resistance standard.

6.1.1.3 *Shubnikov–de Haas effect*

A first indication for the relevance of quantum phenomena in transport measurements of 2D electrons in a strong magnetic field was found in 1930 with the discovery of the Shubnikov–de Haas effect (Shubnikov and de Haas 1930). Whereas the classical result (6.5) for the resistivity tensor stipulates that the longitudinal resistivity $\rho_L = 1/\sigma_0$ (and thus the longitudinal resistance) is independent of the magnetic field, Shubnikov and de Haas found that above some characteristic magnetic field the longitudinal resistance oscillates as a function of the magnetic field. This is schematically depicted in Figure 6.2(a). In contrast to this oscillation in the longitudinal resistance, the Hall resistance remains linear in the B field, in agreement with the classical result from the Drude model (6.7).

The Shubnikov–de Haas effect is a consequence of the energy quantization of the 2D electron in a strong magnetic field, as has been shown by Landau at roughly the same moment. This so-called *Landau quantization* will be presented in great detail in Section 6.2. In a nutshell, Landau quantization consists of the quantization of the cyclotron radius, i.e. the radius of the circular trajectory of an electron in a magnetic field. This leads to the quantization of its kinetic energy into so-called Landau levels

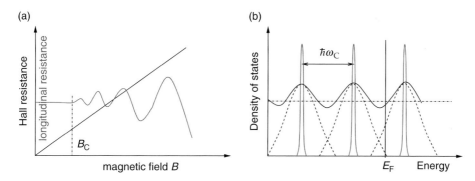

Fig. 6.2 (a) Sketch of the Shubnikov-de Haas effect. Above a critical field B_c, the longitudinal resistance (gray) starts to oscillate as a function of the magnetic field. The Hall resistance remains linear in B. (b) Density of states (DOS). In a clean system, the DOS consists of equidistant delta peaks (grey) at the energies $\epsilon_n = \hbar\omega_C(n + 1/2)$, whereas in a sample with a stronger impurity concentration, the peaks are broadened (dashed lines). The continuous black line represents the sum of overlapping peaks, and E_F denotes the Fermi energy.

(LLs), $\epsilon_n = \hbar\omega_C(n + 1/2)$, where n is an integer. In order for this quantization to be relevant, the magnetic field must be so strong that the electron performs at least one complete circular period without any collision, i.e. $\omega_C\tau > 1$. This condition defines the critical magnetic field $B_c \simeq m_b/e\tau = \mu^{-1}$ above which the longitudinal resistance starts to oscillate, in terms of the mobility (6.6). Notice that today's samples of highest mobility are characterized by $\mu \sim 10^7$ cm^2/V s $= 10^3$ m^2/V s such that one may obtain Shubnikov–de Haas oscillations at magnetic fields as low as $B_c \sim 1$ mT.

The effect may be understood within a slightly more accurate theoretical description of electronic transport (e.g. with the help of the Boltzmann transport equation) than the Drude model. The resulting Einstein relation then relates the conductivity to a diffusion equation, and the longitudinal conductivity

$$\sigma_L = e^2 D\rho(E_F) \tag{6.11}$$

turns out to be proportional to the density of states (DOS) $\rho(E_F)$ at the Fermi energy E_F rather than the electronic density.[2] Due to Landau quantization, the DOS of a clean system consists of a sequence of delta peaks at the energies $\epsilon_n = \hbar\omega_C(n + 1/2)$,

$$\rho(\epsilon) = \sum_n g_n\delta(\epsilon - \epsilon_n),$$

where g_n takes into account the degeneracy of the energy levels. These peaks are eventually impurity broadened in real samples and may even overlap [see Figure 6.2(b)], such that the DOS oscillates in energy with maxima at the positions of the energy levels ϵ_n. Consider a fixed number of electrons in the sample that fixes the zero-field Fermi energy the B-field dependence of which we omit in the argument.[3] When sweeping the magnetic field, one varies the energy distance between the LLs, and the DOS thus becomes maximal when E_F coincides with the energy of a LL and minimal if E_F lies between two adjacent LLs. The resulting oscillation in the DOS as a function of the magnetic field is translated, via the relation (6.11), into an oscillation of the longitudinal conductivity (or resistivity), which is the essence of the Shubnikov–de Haas effect.

6.1.1.4   *Integer quantum Hall effect*

An even more striking manifestation of quantum mechanics in the transport properties of 2D electrons in a strong magnetic field was revealed 50 years later with the discovery of the integer quantum Hall effect (IQHE) by v. Klitzing, Dorda, and Pepper in 1980 (v. Klitzing, *et al.* 1980). The Nobel Prize was attributed in 1985 to v. Klitzing for this extremely important discovery.

Indeed, the discovery of the IQHE was intimitely related to technological advances in material science, namely in the fabrication of high-quality field-effect transistors for the realization of 2D electron gases. These technological aspects will be briefly reviewed in a separate section (Section 6.1.2).

The IQHE occurs at low temperatures, when the energy scale set by the temperature k_BT is significantly smaller than the LL spacing $\hbar\omega_C$. It consists of a quantization of the Hall resistance, which is no longer linear in B, as one would expect from the classical treatment presented above, but reveals plateaus at particular values of the

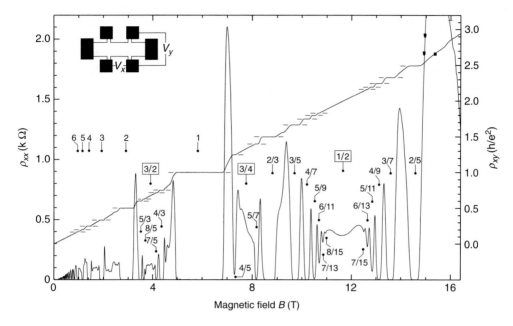

Fig. 6.3 Typical signature of the quantum Hall effect (measured by J. Smet, MPI-Stuttgart). Each plateau in the Hall resistance is accompanied by a vanishing longitudinal resistance. The classical Hall resistance is indicated by the dashed-dotted line. The numbers label the plateaus: integral n denote the IQHE and $n = p/q$, with integral p and q, indicate the FQHE.

magnetic field (see Figure 6.3). In the plateaus, the Hall resistance is given in terms of universal constants – it is indeed a fraction of the inverse quantum of conductance e^2/h, and one observes

$$R_H = \left(\frac{h}{e^2}\right)\frac{1}{n}, \tag{6.12}$$

in terms of an integer n. The plateau in the Hall resistance is accompanied by a vanishing longitudinal resistance. This is at first sight reminiscent of the Shubnikov–de Haas effect, where the longitudinal resistance also reveals minima although it never vanishes. The vanishing of the longitudinal resistance at the Shubnikov–de Haas minima may indeed be used to determine the crossover from the Shubnikov–de Haas regime to the IQHE.

It is worth mentioning that the quantization of the Hall resistance (6.12) is a *universal* phenomenon, i.e. independent of the particular properties of the sample, such as its geometry, the host materials used to fabricate the 2D electron gas and, even more importantly, its impurity concentration or distribution. This universality is the reason for the enormous precision of the Hall-resistance quantization (typically $\sim 10^{-9}$), which is nowadays – since 1990 – used as the resistance standard,[4]

$$R_{K-90} = h/e^2 = 25\,812.807\,\Omega, \tag{6.13}$$

also called the Klitzing constant (Poirier and Schopfer, 2009*a*,*b*). Furthermore, as already mentioned in Section 6.1.1.2, the vanishing longitudinal resistance indicates that the scattering time tends to infinity [see Eq. (6.9)] in the IQHE. This is another indication of the above-mentioned universality of the effect, i.e. that IQHE does not depend on a particular impurity (or scatterer) arrangement.

A detailed presentation of the IQHE, namely the role of impurities, may be found in Section 6.3.

6.1.1.5 *Fractional quantum Hall effect*

Two years after the discovery of the IQHE, an even more unexpected effect was observed in a 2D electron system of higher quality, i.e. of higher mobility: the *fractional quantum Hall effect* (FQHE). The effect ows its name to the fact that contrary to the IQHE, where the number n in Eq. (6.12) is an integer, a Hall-resistance quantization was discovered by Tsui, Störmer and Gossard with $n = 1/3$ (Tsui, *et al.* 1982). From a phenomenological point of view, the effect is extremely reminiscent of the IQHE: whereas the Hall resistance is quantized and reveals a plateau, the longitudinal resistance vanishes (see Figure 6.3, where different instances of both the IQHE and the FQHE are shown). However, the origins of the two effects are completely different: whereas the IQHE may be understood from Landau quantization, i.e. the kinetic-energy quantization of independent electrons in a magnetic field, the FQHE is due to strong electronic correlations, when a LL is only partially filled and the Coulomb interaction between the electrons becomes relevant. Indeed, in 1983 Laughlin showed that the origin of the observed FQHE with $n = 1/3$, as well as any $n = 1/q$ with q being an odd integer, is due to the formation of a *correlated incompressible electron liquid* with extremely exotic properties (Laughlin, 1983) that will be reviewed in Section 6.4. As for the IQHE, the discovery and the theory of the FQHE was awarded a Nobel Prize (1998 for Tsui, Störmer and Laughlin).

After the discovery of the FQHE with $n = 1/3$,[5] a plethora of other types of FQHE has been discovered and theoretically described. One should first mention the 2/5 and 3/7 states (i.e. with $n = 2/5$ and $n = 3/7$), which are part of the series $p/(2sp \pm 1)$, with the integers s and p. This series has found a compelling interpretation within the so-called *composite-fermion* (CF) theory according to which the *F*QHE may be viewed as an *I*QHE of a novel quasi-particle that consists of an electron that "captures" an even number of flux quanta (Jain, 1989; 1990). The basis of this theory is presented in Section 6.4.4. Another intriguing FQHE was discovered in 1987 by Willet, *et al.* with $n = 5/2$ and $7/2$ (Willett, *et al.* 1987) – it was intriguing insofar as up to that moment only states $n = p/q$ with *odd* denominators had been observed in monolayer systems. From a theoretical point of view, it was shown in 1991 by Moore and Read (Moore and Read, 1991) and by Greiter, Wilczek and Wen (Greiter, *et al.* 1991) that this FQHE may be described in terms of a very particular, so-called *Pfaffian*, wave function, which involves particle pairing and the excitations of which are anyons with non-Abelian statistics. These particles are intensively studied in today's research because they may play a relevant role in quantum computation. The physics of anyons will be introduced briefly in Section 6.4.3. Finally, we would mention in this brief (and naturally incomplete) historical overview a FQHE with $n = 4/11$ discovered in 2003 by Pan, *et al.* (Pan, *et al.* 2003): it does not fit into the above-mentioned CF series, but it would correspond to a FQHE of CFs rather than an IQHE of CFs.

6.1.1.6   *Relativistic quantum Hall effect in graphene*

Recently, quantum Hall physics experienced another unexpected boost with the discovery of a "relativistic" quantum Hall effect in graphene, a one-atom-thick layer of graphite (Novoselov, *et al.* 2005; Zhang, *et al.* 2005). Electrons in graphene behave as if they were relativistic massless particles. Formally, their quantum-mechanical behavior is no longer described in terms of a (non-relativistic) Schrödinger equation, but rather by a relativistic 2D Dirac equation (Castro Neto, *et al.* 2009). As a consequence, Landau quantization of the electrons' kinetic energy turns out to be different in graphene from that in conventional (non-relativistic) 2D electron systems, as we will discuss in Section 6.2. This yields a "relativistic" quantum Hall effect with an unusual series for the Hall plateaus. Indeed, rather than having plateaus with a quantized resistance according to $R_H = h/e^2 n$, with integer values of n, one finds plateaus with $n = \pm 2(2n' + 1)$, in terms of an integer n', i.e. with $n = \pm 2, \pm 6, \pm 10, ...$ The different signs in the series ($\pm$) indicate that there are two different carriers, electrons in the conduction band and holes in the valence band, involved in the quantum Hall effect in graphene. As we will briefly discuss in Section 6.1.2, one may easily change the character of the carriers in graphene with the help of the electric field effect.

Interaction effects may be relevant in the formation of other integer Hall plateaus, such as $n = 0$ and $n = \pm 1$ (Zhang, *et al.* 2006), which do not occur naturally in the series $n = \pm 2(2n' + 1)$ characteristic of the relativistic quantum Hall effect. Furthermore, a FQHE with $n = 1/3$ has very recently been observed, although in a simpler geometric (two-terminal) configuration than the standard one depicted in Figure 6.1(a) (Du, *et al.* 2009; Bolotin, *et al.* 2009).

6.1.2   **Two-dimensional electron systems**

As already mentioned above, the history of the quantum Hall effect is intimately related to technological advances in the fabrication of 2D electron systems with high electronic mobilities. The increasing mobility allows one to probe the fine structure of the Hall curve and thus to observe those quantum Hall states that are more fragile, such as some exotic FQHE states (e.g. the 5/2, 7/2 or the 4/11 states). This may be compared to the quest for high resolutions in optics: the higher the optical resolution, the better the chance of observing tinier objects. In this sense, electronic mobility means resolution and the tiny object is the quantum Hall state. As an order of magnitude, today's best 2D electron gases (in GaAs/AlGaAs heterostructures) are characterized by mobilities $\mu \sim 10^7$ cm^2/V s.

6.1.2.1   *Field-effect transistors*

The samples used in the discovery and in the first studies of the IQHE were so-called *metal-oxide-semiconductor field-effect transistors* (MOSFET). A metallic layer is seperated from a semiconductor (typically doped silicon) by an insulating oxide

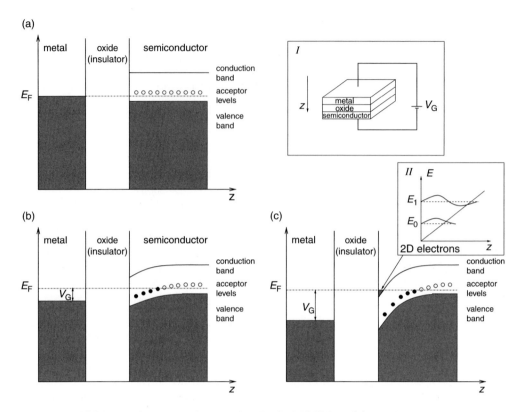

Fig. 6.4   MOSFET. The inset I shows a sketch of a MOSFET. (a) Level structure at $V_G = 0$. In the metallic part, the band is filled up to the Fermi energy E_F, whereas the oxide is insulating. In the semiconductor, the Fermi energy lies in the band gap (energy gap between the valence and the conduction bands). Close to the valence band, albeit above E_F, are the acceptor levels. (b) The chemical potential in the metallic part may be controled by the gate voltage V_G via the electric field effect. As a consequence of the introduction of holes in the metal, the semiconductor bands are bent downwards, and above a threshold voltage (c), the conduction band is filled in the vicinity of the interface with the insulator. One thus obtains a 2D electron gas. Its confinement potential is of triangular shape, the levels (electronic subbands) of which are represented in the inset II.

(e.g. SiO_2) layer (see inset I in Figure 6.4). The chemical potential in the metallic layer may be varied with the help of a gate voltage V_G. At $V_G = 0$, the Fermi energy in the semiconductor lies in the band gap below the acceptor levels of the dopants [Figure 6.4(a)]. When lowering the chemical potential in the metal with the help of a positive gate voltage $V_G > 0$, one introduces holes in the metal that attract, via the electric field effect, electrons from the semiconductor to the semiconductor insulator interface. These electrons populate the acceptor levels, and as a consequence, the semiconductor bands are bent downwards when they approach the interface, such that the filled acceptor levels now lie below the Fermi energy [Figure 6.4(b)].

Above a certain threshold of the gate voltage, the bending of the semiconductor bands becomes so strong that not only are the acceptor levels below the Fermi energy, but also the conduction band in the vicinity of the interface, which consequently gets filled with electrons [Figure 6.4(c)]. One thus obtains a confinement potential of triangular shape for the electrons in the conduction band, the dynamics of which is quantised into discrete electronic subbands in the perpendicular z-direction (see inset *II* in Figure 6.4). Naturally, the electronic wave functions are then extended in the z-direction, but in typical MOSFETs only the lowest electronic subband E_0 is filled, such that the electrons are purely 2D from a dynamical point of view, i.e. there is no electronic motion in the z-direction.

The typical 2D electronic densities in these systems are on the order of $n_{el} \sim 10^{11}$ cm^{-2}, i.e. much lower than in usual metals. This turns out to be important in the study of the IQHE and FQHE, because the effects occur, as we will show below, when the 2D electronic density is on the order of the density of magnetic flux $n_B = B/(h/e)$ threading the system, in units of the flux quantum h/e. This needs to be compared to metals where the surface density is on the order of 10^{14} cm^{-2}, which would require inaccessibly high magnetic fields (on the order of 1000 T) in order to probe the regime $n_{el} \sim n_B$.

6.1.2.2   *Semiconductor heterostructures*

The mobility in MOSFETs, which is typically on the order of $\mu \lesssim 10^6$ cm^2/V s, is limited by the quality of the oxide semiconductor interface (surface roughness). This technical difficulty is circumvented in semiconductor heterostructures – most popular are GaAs/AlGaAs heterostructures – which are grown by molecular-beam epitaxy (MBE), where high-quality interfaces with almost atomic precision may be achieved, with mobilities on the order of $\mu \sim 10^7$ cm^2/V s. These mobilities were necessary to observe the FQHE, which was indeed first observed in a GaAs/AlGaAs sample (Tsui, *et al.* 1982).

In the (generic) case of GaAs/AlGaAs, the two semiconductors do not possess the same band gap – indeed that of GaAs is smaller than that of AlGaAs, which is chemically doped by donor ions at a certain distance from the interface between GaAs and AlGaAs [Figure 6.5(a)]. The Fermi energy is pinned by these donor levels in AlGaAs, which may have a higher energy than the originally unoccupied conduction band in the GaAs part, such that it becomes energetically favorable for the electrons in the donor levels to occupy the GaAs conduction band in the vicinity of the interface. As a consequence, the energy bands of AlGaAs are bent upwards, whereas those of GaAs are bent downwards. Similarly to the above-mentioned MOSFET, one thus obtains a 2D electron gas at the interface on the GaAs side, with a triangular confinement potential.

6.1.2.3   *Graphene*

Graphene, a one-atom thick layer of graphite, presents a novel 2D electron system, which, from the electronic point of view, is either a zero-overlap semi-metal or a zero-gap semiconductor, where the conduction and the valence bands are no longer

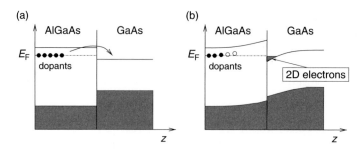

Fig. 6.5 Semiconductor heterostructure (GaAs/AlGaAs). (a) Dopants are introduced in the AlGaAs layer at a certain distance from the interface. The Fermi energy lies below the conduction band of AlGaAs, in the band gap, and is pinned by the dopant levels. The GaAs conduction band has an energy that is lower than that of the dopant levels, such that it is energetically favorable for the electrons in the dopant layer to populate the GaAs conduction band in the vicinity of the interface. (b) This polarization bends the bands in the vicinity of the interface between the two semiconductors, and thus a 2D electron gas is formed there on the GaAs side.

separated by an energy gap. Indeed, in the absence of doping, the Fermi energy lies exactly at the points where the valence band touches the conduction band and where the density of states vanishes linearly.

In order to vary the Fermi energy in graphene, one usually places a graphene flake on a 300-nm thick insulating SiO_2 layer that is itself placed on top of a positively doped metallic silicon substrate (see Figure 6.6). This sandwich structure, with the metallic silicon layer that serves as a backgate, may thus be viewed as a capacitor (Figure 6.6) the capacitance of which is

$$C = \frac{Q}{V_G} = \frac{\epsilon_0 \epsilon \mathcal{A}}{d},$$ (6.14)

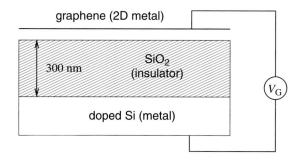

Fig. 6.6 Schematic view of graphene on a SiO_2 substrate with a doped Si (metallic) backgate. The system graphene–SiO_2 – backgate may be viewed as a capacitor the charge density of which is controled by a gate voltage V_G.

where $Q = en_{2D}\mathcal{A}$ is the capacitor charge, in terms of the total surface $\mathcal{A}$, V_G is the gate voltage, and $d = 300$ nm is the thickness of the SiO$_2$ layer with the dielectric constant $\epsilon = 3.7$. The field-effect induced 2D carrier density is thus given by

$$n_{2D} = \alpha V_G \quad \text{with} \quad \alpha \equiv \frac{\epsilon_0 \epsilon}{ed} \simeq 7.2 \times 10^{10} \; \frac{\text{cm}^{-2}}{\text{V}}. \tag{6.15}$$

The gate voltage may vary roughly between -100 and 100 V, such that one may induce maximal carrier densities on the order of 10^{12} cm^{-2}, on top of the intrinsic carrier density, which turns out to be zero in graphene, as will be discussed in the next section. At gate voltages above ± 100 V, the capacitor breaks down (electrical breakdown).

In contrast to 2D electron gases in semiconductor heterostructures, the mobilities achieved in graphene are rather low: they are typically on the order of $\mu \sim 10^4 - 10^5$ cm^2/V s. Notice, however, that these graphene samples are fabricated in the so-called exfoliation technique, where one simply "peals" thin graphite crystals, under ambient conditions, whereas the highest-mobility GaAs/AlGaAs laboratory samples are fabricated with a very high technological effort. The mobilities of graphene samples are comparable to those of commercial silicon-based electronic elements.

6.2   Landau quantization

The basic ingredient for the understanding of both the IQHE and the FQHE is Landau quantization, i.e. the kinetic-energy quantization of a (free) charged 2D particle in a perpendicular magnetic field. In this section, we give a detailed introduction to the different aspects of Landau quantization. We have chosen a very general presentation of this quantization in order to account for both a non-relativistic and a relativistic 2D particle some properties of which, such as the level degeneracy, are identical. In Section 6.2.1, we introduce the basic Hamiltonians for 2D particles in the absence of a magnetic field and discuss both Schrödinger- and Dirac-type particles. The case of a non-zero B-field is presented in Section 6.2.2, and Section 6.2.3 is devoted to the discussion of the LL structure of non-relativistic and relativistic particles.

6.2.1   Basic one-particle Hamiltonians for $B = 0$

In this section, we introduce the basic Hamiltonians that we treat in a quantum-mechanical manner in the following parts. Quite generally, we consider a Hamiltonian for a 2D particle[6] that is translation invariant, i.e. its momentum $\mathbf{p} = (p_x, p_y)$ is a constant of motion, in the absence of a magnetic field. In quantum mechanics, this means that the momentum operator commutes with the Hamiltonian, $[\mathbf{p}, H] = 0$, and that the eigenvalue of the momentum operator is a good quantum number.

6.2.1.1   Hamiltonian of a free particle

In the case of a free particle, this is a very natural assumption, and one has for the non-relativistic case,

$$H = \frac{\mathbf{p}^2}{2m}, \tag{6.16}$$

in terms of the particle mass m.[7] However, we are interested, here, in the motion of electrons in some material (in a metal or at the interface of semiconductors). It seems, at first sight, to be a very crude assumption to describe the motion of an electron in a crystalline environment in the same manner as a particle in free space. Indeed, a particle in a lattice is not described by the Hamiltonian (6.16) but rather by the Hamiltonian

$$H = \frac{\mathbf{p}^2}{2m} + \sum_i^N V(\mathbf{r} - \mathbf{r}_i), \tag{6.17}$$

where the last term represents the electrostatic potential caused by the ions situated at the lattice sites $\mathbf{r}_i$. Evidently, the Hamiltonian now depends on the position $\mathbf{r}$ of the particle with respect to that of the ions, and the momentum $\mathbf{p}$ is therefore no longer a constant of motion or a good quantum number.

This problem is solved with the help of Bloch's theorem: although an arbitrary spatial translation is not an allowed symmetry operation, in contrast to a free particle (6.16), the system is invariant under a translation by an arbitrary lattice vector if the lattice is of infinite extension – an assumption we make here.[8] In the same manner as for the free particle, where one defines the momentum as the generator of a spatial translation, one may then define a generator of a lattice translation. This generator is called the *lattice momentum* or also the *quasi-momentum*. As a consequence of the discreteness of the lattice translations, not all values of this lattice momentum are physical, but only those within the first Brillouin zone (BZ) – any vibrational mode, be it a lattice vibration or an electronic wave, with a wavevector outside the first BZ can be described by a mode with a wavevector within the first BZ. Since this chapter cannot include a full class on basic solid-state physics, we refer the reader to standard textbooks on solid-state physics (Ashcroft and Mermin, 1976); (Kittel, 2005).

The bottom line is that also in a (perfect) crystal, the electrons may be described in terms of a Hamiltonian $H(p_x, p_y)$ if one keeps in mind that the momentum $\mathbf{p}$ in this expression is a lattice momentum restricted to the first BZ. Notice, however, that although the resulting Hamiltonian may often be written in the form (6.16), the mass is generally not the free electron mass but a *band mass* m_b that takes into account the particular features of the energy bands[9] – indeed, the mass may even depend on the direction of propagation, such that one should write the Hamiltonian more generally as

$$H = \frac{p_x^2}{2m_x} + \frac{p_y^2}{2m_y}.$$

6.2.1.2 *Dirac Hamiltonian in graphene*

The above considerations for electrons in a 2D lattice are only valid in the case of a *Bravais* lattice, i.e. a lattice in which all lattice sites are equivalent from a crystallographic point of view. However, some lattices, such as the honeycomb lattice that describes the arrangement of carbon atoms in graphene due to the sp^2 hybridization of the valence electrons, are not Bravais lattices. In this case, one may describe the lattice as a Bravais lattice plus a particular pattern of N_s sites, called the *basis*. This is illustrated in Figure 6.7(a) for the case of the honeycomb lattice. When one compares a site A (full circle) with a site B (empty circle), one notices that the environment of these two sites is different: whereas a site A has nearest neighbors in the directions north-east, north-west and south, a site B has nearest neighbors in the directions north, south-west and south-east. This precisely means that the two sites are not equivalent from a crystallographic point of view – although they may be equivalent from a chemical point of view, i.e. occupied by the same atom or ion type (carbon in the case of graphene). However, all sites A form a triangular Bravais lattice as well as all sites B. Each subset of lattice sites forms one of the two *sublattices*, and the honeycomb lattice may thus be viewed as a triangular Bravais lattice with a two-atom basis, e.g. the pattern of two A and B sites connected by the vector $\boldsymbol{\delta}_3$.

In order to calculate the electronic bands in a lattice with N_s Bravais sublattices, i.e. a basis with N_s sites, one needs to describe the general electronic wave function as a superposition of N_s different wave functions, which satisfy each of Bloch's theorem for all sublattices (Ashcroft and Mermin, 1976; Kittel, 2005). Formally, this may be described in terms of a $N_s \times N_s$ matrix, the eigenvalues of which yield N_s different energy bands. In a lattice with N_s different sublattices, one therefore obtains one energy band per sublattice, and for graphene, one obtains two different bands for the conducting electrons, the valence band and the conduction band.

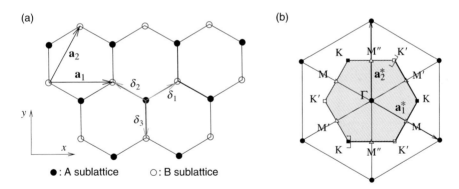

Fig. 6.7 (a) Honeycomb lattice. The vectors $\boldsymbol{\delta}_1$, $\boldsymbol{\delta}_2$, and $\boldsymbol{\delta}_3$ connect nearest-neighbor carbon atoms, separated by a distance $a = 0.142$ nm. The vectors $\mathbf{a}_1$ and $\mathbf{a}_2$ are basis vectors of the triangular Bravais lattice. (b) Reciprocal lattice of the triangular lattice. Its primitive lattice vectors are $\mathbf{a}_1^*$ and $\mathbf{a}_2^*$. The shaded region represents the first Brillouin zone (BZ), with its center Γ and the two inequivalent corners K (black squares) and K' (white squares). The thick part of the border of the first BZ represents those points that are counted in the definition such that no points are doubly counted. The first BZ, defined in a strict manner, is, thus, the shaded region plus the thick part of the border. For completeness, we have also shown the three inequivalent cristallographic points M, M', and M'' (white triangles).

The Hamiltonian for low-energy electrons in reciprocal space reads

$$H(\mathbf{k}) = t \begin{pmatrix} 0 & \gamma_{\mathbf{k}}^* \\ \gamma_{\mathbf{k}} & 0 \end{pmatrix}, \tag{6.18}$$

which is obtained within a tight-binding model, where one considers electronic hopping between nearest-neighboring sites with a hopping amplitude t. Because the nearest neighbor of a site A is a site B and *vice versa* [see Figure 6.7(a)], the Hamiltonian is off-diagonal, and the off-diagonal elements are related by complex conjugation due to time-reversal symmetry, $H(-\mathbf{k})^* = H(\mathbf{k})$. As already mentioned above, the lattice momentum $\mathbf{k}$ is restricted to the first BZ, which is of hexagonal shape and that we have depicted in Figure 6.7(b) for completeness. The precise form of the functions $\gamma_{\mathbf{k}}$ is derived in Appendix A [Eq. (A.9)]. The band structure is obtained by diagonalizing the Hamiltonian (6.18), and one finds the two bands, labelled by $\lambda = \pm$, $\epsilon_\lambda(\mathbf{k}) = \lambda t |\gamma_{\mathbf{k}}|$, which are plotted in Figure 6.8. The valence band ($\lambda = -$) touches the conduction band ($\lambda = +$) in the two inequivalent corners K and K' of the first BZ. Because there are as many electrons in the π-orbitals, that determine the low-energy conduction properties of graphene, as lattice sites, the overall energy band structure is half-filled. This is due to the two spin orientations of the electrons, which allow for a quantum-mechanical double occupancy of each π-orbital. As a consequence, the Fermi energy lies exactly in the contact points K and K' of the two bands unless the graphene sheet is doped, e.g. with the help of the electric field effect, as described in Section 6.1.2.

The inset in Figure 6.8 shows the band dispersion in the vicinity of the contact points K and K', the linearity of which is sufficient to describe the low-energy electronic properties in graphene, i.e. when all relevant energy scales are much smaller than the full band width.[10] The conical form of the two bands is reminiscent of that of

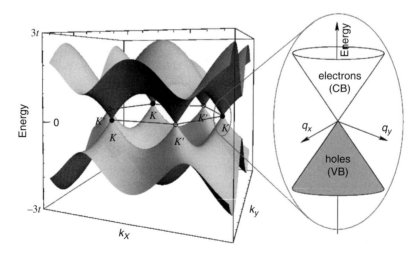

Fig. 6.8 Energy bands of graphene. The valence band touches the conduction band in the two inequivalent BZ corners K and K'. For undoped graphene, the Fermi energy lies precisely at the contact points, and the band dispersion in the vicinity of these points is of conical shape.

relativistic particles, the general dispersion of which is $E = \pm\sqrt{m^2c^4 + \mathbf{p}^2c^2}$, in terms of the light velocity c and the particle mass m. If the latter is zero, one obtains precisely $E = \pm c|\mathbf{p}|$, as in the case of low-energy electrons in graphene (inset of Figure 6.8), which may thus be treated as *massless Dirac fermions*. Notice that in the continuum description of electrons in graphene, we have two electron types – one for the K point and another one for the K' point. This doubling is called *valley degeneracy* that is two-fold here.

The analogy between electrons in graphene and massless relativistic particles is corroborated by a low-energy expansion of the Hamiltonian (6.18) around the contact points K and K', at the momenta $\mathbf{K}$ and $\mathbf{K'} = -\mathbf{K}$ [see Figures 6.7(b) and 6.8], $\mathbf{k} = \pm\mathbf{K} + \mathbf{p}/\hbar$, where $|\mathbf{p}/\hbar| \ll |\mathbf{K}|$. One may then expand the function $\gamma_{\pm\mathbf{K}+\mathbf{p}/\hbar}$ to first order, and one obtains formally[11]

$$H = t\begin{pmatrix} 0 & \nabla\gamma_K^* \cdot \mathbf{p} \\ \nabla\gamma_K \cdot \mathbf{p} & 0 \end{pmatrix} = v\begin{pmatrix} 0 & p_x - ip_y \\ p_x + ip_y & 0 \end{pmatrix} = v\mathbf{p} \cdot \boldsymbol{\sigma}, \qquad (6.19)$$

where $\boldsymbol{\sigma} = (\sigma^x, \sigma^y)$ in terms of the Pauli matrices

$$\sigma^x = \begin{pmatrix} 0 & 1 \\ 1 & 0 \end{pmatrix}, \qquad \sigma^y = \begin{pmatrix} 0 & -i \\ i & 0 \end{pmatrix} \qquad \text{and} \qquad \sigma^z = \begin{pmatrix} 1 & 0 \\ 0 & -1 \end{pmatrix}$$

and where we have chosen to expand the Hamiltonian (6.18) around the K point.[12] Here, the Fermi velocity v plays the role of the velocity of light c, which is though roughly 300 times larger, $c \simeq 300v$. The details of the above derivation may be found in Appendix A. The Hamiltonian (6.19) is indeed formally that of massless 2D particles, and it is sometimes called Weyl or Dirac Hamiltonian.

We will discuss, in the remainder of this section, how the two Hamiltonians

$$H_S = \frac{\mathbf{p}^2}{2m_b} \qquad \text{and} \qquad H_D = v\mathbf{p} \cdot \boldsymbol{\sigma}, \qquad (6.20)$$

for non-relativistic and relativistic particles, respectively, need to be modified in order to account for a non-zero magnetic field.

6.2.2 Hamiltonians for non-zero B fields

6.2.2.1 *Minimal coupling and Peierls substitution*

In order to describe free electrons in a magnetic field, one needs to replace the momentum by its gauge-invariant form (Jackson, 1999)

$$\mathbf{p} \to \boldsymbol{\Pi} = \mathbf{p} + e\mathbf{A}(\mathbf{r}), \qquad (6.21)$$

where $\mathbf{A}(\mathbf{r})$ is the vector potential that generates the magnetic field $\mathbf{B} = \nabla \times \mathbf{A}(\mathbf{r})$. This gauge-invariant momentum is proportional to the electron velocity $\mathbf{v}$, which must naturally be gauge-invariant because it is a physical quantity. Notice that because $\mathbf{A}(\mathbf{r})$ is not gauge-invariant, neither is the momentum $\mathbf{p}$, and remember that adding the gradient of an arbitrary derivable function $\lambda(\mathbf{r})$, $\mathbf{A}(\mathbf{r}) \to \mathbf{A}(\mathbf{r}) + \nabla\lambda(\mathbf{r})$, does not change the magnetic field because the rotational of a gradient is zero. Indeed, the momentum is transformed as $\mathbf{p} \rightarrow \mathbf{p} - e\nabla\lambda(\mathbf{r})$ under a gauge transformation in order to compensate the transformed vector potential, such that $\mathbf{\Pi}$ is gauge-invariant. The substitution (6.21) is also called *minimal substitution*.

In the case of electrons on a lattice, this substitution is more tricky because of the presence of several bands. Furthermore, the vector potential is unbound, even for a finite magnetic field; this becomes clear if one chooses a particular gauge, such as e.g. the Landau gauge $\mathbf{A}_L(\mathbf{r}) = B(-y, 0, 0)$, in which case the value of the vector potential may become as large as $B \times L_y$, where L_y is the macroscopic extension of the system in the y-direction. However, it may be shown that the substitution (6.21), which is called *Peierls substitution* in the context of electrons on a lattice, remains correct as long as the lattice spacing a is much smaller than the *magnetic length*

$$l_B = \sqrt{\frac{\hbar}{eB}}, \tag{6.22}$$

which is the fundamental length scale in the presence of a magnetic field. Because a is typically an atomic scale (~ 0.1 to 10 nm) and $l_B \simeq 26\,\mathrm{nm}/\sqrt{B[\mathrm{T}]}$, this condition is fulfilled in all atomic lattices for the magnetic fields, which may be achieved in today's high-field laboratories (~ 45 T in the continuous regime and ~ 80 T in the pulsed regime).[13]

With the help of the (Peierls) substitution (6.21), one may thus immediately write down the Hamiltonian for charged particles in a magnetic field if one knows the Hamiltonian in the absence of a magnetic field,

$$H(\mathbf{p}) \rightarrow H(\mathbf{\Pi}) = H(\mathbf{p} + e\mathbf{A}) = H^B(\mathbf{p}, \mathbf{r}).$$

Notice that because of the spatial dependence of the vector potential, the resulting Hamiltonian is no longer translation-invariant, and the (gauge-dependent) momentum $\mathbf{p}$ is no longer a conserved quantity. We will limit the discussion to the B-field Hamiltonians corresponding to the Hamiltonians (6.20)

$$H_S^B = \frac{[\mathbf{p} + e\mathbf{A}(\mathbf{r})]^2}{2m_b} \tag{6.23}$$

for non-relativistic and

$$H_D^B = v[\mathbf{p} + e\mathbf{A}(\mathbf{r})] \cdot \boldsymbol{\sigma} \tag{6.24}$$

for charged relativistic 2D particles, respectively.

6.2.2.2   *Quantum-mechanical treatment*

In order to analyze the one-particle Hamiltonians (6.23) and (6.24) in a quantum-mechanical treatment, we use the standard method, *canonical quantization* (Cohen-Tannoudji, *et al.* 1973), where one interprets the physical quantities as operators that act on state vectors in a Hilbert space. These operators do in general not commute with each other, i.e. the order matters in which they act on the state vector that describes the physical system. Formally, one introduces the *commutator* $[\mathcal{O}_1, \mathcal{O}_2] \equiv \mathcal{O}_1\mathcal{O}_2 - \mathcal{O}_2\mathcal{O}_1$ between the two operators $\mathcal{O}_1$ and $\mathcal{O}_2$, which are said to commute when $[\mathcal{O}_1, \mathcal{O}_2] = 0$ or else not to commute. The basic physical quantities in the argument of the Hamiltonian are the 2D position $\mathbf{r} = (x, y)$ and its canonical momentum $\mathbf{p} = (p_x, p_y)$, which satisfy the commutation relations

$$[x, p_x] = i\hbar, \quad [y, p_y] = i\hbar \quad \text{and} \quad [x, y] = [p_x, p_y] = [x, p_y] = [y, p_x] = 0, \qquad (6.25)$$

i.e. each component of the position operator does not commute with the momentum in the corresponding direction. This non-commutativity between the position and its associated momentum is the origin of the Heisenberg inequality according to which one cannot know precisely both the position of a quantum-mechanical particle and, at the same moment, its momentum, $\Delta x \Delta p_x \gtrsim h$ and $\Delta y \Delta p_y \gtrsim h$.

As a consequence of the commutation relations (6.25), the components of the gauge-invariant momentum no longer commute themselves,

$$[\Pi_x, \Pi_y] = [p_x + eA_x(\mathbf{r}), p_y + eA_y(\mathbf{r})] = e\left([p_x, A_y(\mathbf{r})] - [p_y, A_x(\mathbf{r})]\right)$$

$$= e\left(\frac{\partial A_y}{\partial x}[p_x, x] + \frac{\partial A_y}{\partial y}[p_x, y] - \frac{\partial A_x}{\partial x}[p_y, x] - \frac{\partial A_x}{\partial y}[p_y, y]\right),$$

where we have used the relation[14]

$$[\mathcal{O}_1, f(\mathcal{O}_2)] = \frac{df}{d\mathcal{O}_2}[\mathcal{O}_1, \mathcal{O}_2] \qquad (6.26)$$

between two arbitrary operators, the commutator of which is a c-number or an operator that commutes itself with both $\mathcal{O}_1$ and $\mathcal{O}_2$ (Cohen-Tannoudji, *et al.* 1973). With the help of the commutation relations (6.25), one finds that

$$[\Pi_x, \Pi_y] = -ie\hbar\left(\frac{\partial A_y}{\partial x} - \frac{\partial A_x}{\partial y}\right) = -ie\hbar\left(\nabla \times \mathbf{A}\right)_z = -ie\hbar B,$$

and, in terms of the magnetic length (6.22),

$$[\Pi_x, \Pi_y] = -i\frac{\hbar^2}{l_B^2}. \qquad (6.27)$$

This equation is the basic result of this section and merits some further discussion.

- As one would have expected for gauge-invariant quantities (the two components of $\mathbf{\Pi}$), their commutator is itself gauge-invariant. Indeed, it only depends on universal constants and the (gauge-invariant) magnetic field B, and not on the vector potential $\mathbf{A}$.
- The components of the gauge-invariant momentum $\mathbf{\Pi}$ are mutually *conjugate* in the same manner as x and p_x or y and p_y. Remember that p_x generates the translations in the x-direction (and p_y those in the y-direction). This is similar here: Π_x generates a "boost" of the gauge-invariant momentum in the y-direction, and similarly Π_y one in the x-direction.

- As a consequence, one may not diagonalize at the same time Π_x *and* Π_y, in contrast to the zero-field case, where the arguments of the Hamiltonian, p_x and p_y, commute.

For solving the Hamiltonians (6.23) and (6.24), it is convenient to use the pair of conjugate operators Π_x and Π_y and to introduce *ladder operators* in the same manner as in the quantum-mechanical treatment of the one-dimensional harmonic oscillator. Remember from your basic quantum-mechanics class that the ladder operators may be viewed as the complex position of the one-dimensional oscillator in the phase space, which is spanned by the position (x-axis) and the momentum (y-axis),

$$\tilde{a} = \frac{1}{\sqrt{2}}\left(\frac{x}{x_0} - i\frac{p}{p_0}\right) \qquad \text{and} \qquad \tilde{a}^\dagger = \frac{1}{\sqrt{2}}\left(\frac{x}{x_0} + i\frac{p}{p_0}\right),$$

where $x_0 = \sqrt{\hbar/m_b\omega}$ and $p_0 = \sqrt{\hbar m_b\omega}$ are normalization constants in terms of the oscillator frequency ω (Cohen-Tannoudji, *et al.* 1973). The fact that the position x and the momentum p are conjugate variables and the particular choice of the normalization constants yields the commutation relation $[\tilde{a}, \tilde{a}^\dagger] = 1$ for the ladder operators.

In the case of the 2D electron in a magnetic field, the ladder operators play the role of a *complex* gauge-invariant momentum (or velocity), and they read

$$a = \frac{l_B}{\sqrt{2}\hbar}\left(\Pi_x - i\Pi_y\right) \qquad \text{and} \qquad a^\dagger = \frac{l_B}{\sqrt{2}\hbar}\left(\Pi_x + i\Pi_y\right), \tag{6.28}$$

where we have chosen the appropriate normalization such as to obtain the usual commutation relation

$$[a, a^\dagger] = 1. \tag{6.29}$$

It turns out to be helpful for future calculations to invert the expression for the ladder operators (6.28),

$$\Pi_x = \frac{\hbar}{\sqrt{2}l_B}\left(a^\dagger + a\right) \qquad \text{and} \qquad \Pi_y = \frac{\hbar}{i\sqrt{2}l_B}\left(a^\dagger - a\right). \tag{6.30}$$

6.2.3 Landau levels

The considerations of the preceding section are extremely useful in the calculation of the level spectrum associated with the Hamiltonians (6.23) and (6.24) of both the non-relativistic and the relativistic particles, respectively. The understanding of this level spectrum is the issue of the present section. Because electrons not only possess a charge but also a spin, the Zeeman effect splits each level into two spin branches separated by the energy difference $\Delta_Z = g\mu_B B$, where g is the g-factor of the host material and $\mu_B = e\hbar/2m_0$ the Bohr magneton. In order to simplify the following presentation of the quantum-mechanical treatment and the level structure, we neglect the spin degree of freedom. Formally, this amounts to considering *spinless fermions*. Notice, however, that there exist interesting physical properties related to the spin degree of freedom, which will be treated separately in Section 6.5.

6.2.3.1　*Non-relativistic Landau levels*

In terms of the gauge-invariant momentum, the Hamiltonian (6.23) for non-relativistic electrons reads

$$H_S^B = \frac{1}{2m_b} \left(\Pi_x^2 + \Pi_y^2 \right).$$

The analogy with the one-dimensional harmonic oscillator is apparent if one notices that both conjugate operators Π_x and Π_y occur in this expression in a quadratic form. If one replaces these operators by the ladder operators (6.30), one obtains, with the help of the commutation relation (6.29),

$$H_S^B = \frac{\hbar^2}{4ml_B^2} \left[a^{\dagger 2} + a^\dagger a + aa^\dagger + a^2 - \left(a^{\dagger 2} - a^\dagger a - aa^\dagger + a^2 \right) \right]$$

$$= \frac{\hbar^2}{2ml_B^2} \left(a^\dagger a + aa^\dagger \right) = \hbar \omega_C \left(a^\dagger a + \frac{1}{2} \right),$$

where we have used the relation $\omega_c = \hbar/m_b l_B^2$ between the cyclotron frequency (6.2) and the magnetic length (6.22) in the last step.

As in the case of the one-dimensional harmonic oscillator, the eigenvalues and eigenstates of the Hamiltonian (6.31) are therefore those of the *number operator* $a^\dagger a$, with $a^\dagger a |n\rangle = n|n\rangle$. The ladder operators act on these states in the usual manner (Cohen-Tannoudji, *et al.* 1973)

$$a^\dagger |n\rangle = \sqrt{n+1}|n+1\rangle \qquad \text{and} \qquad a|n\rangle = \sqrt{n}|n-1\rangle, \tag{6.31}$$

where the last equation is valid only for $n > 0$ – the action of a on the ground state $|0\rangle$ gives zero,

$$a|0\rangle = 0. \tag{6.32}$$

This last equation turns out to be helpful in the calculation of the eigenstates associated with the level of lowest energy, as well as the construction of states in higher levels n (see Section 6.2.4.1)

$$|n\rangle = \frac{\left(a^\dagger \right)^n}{\sqrt{n!}} |0\rangle. \tag{6.33}$$

The energy levels of the 2D charged non-relativistic particle are therefore discrete and labelled by the integer n,

$$\epsilon_n = \hbar \omega_C \left(n + \frac{1}{2} \right). \tag{6.34}$$

These levels, which are also called *Landau levels* (LL), are depicted in Figure 6.9(a) as a function of the magnetic field. Because of the linear field dependence of the cyclotron frequency, the LLs disperse linearly themselves with the magnetic field.

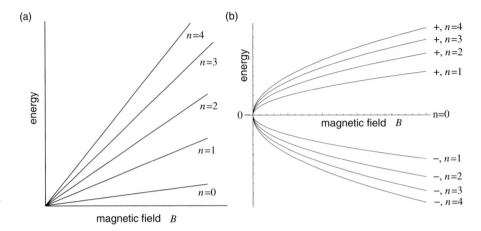

Fig. 6.9 Landau levels as a function of the magnetic field. (a) Non-relativistic case with $\epsilon_n = \hbar\omega_C(n + 1/2) \propto B(n + 1/2)$. (b) Relativistic case with $\epsilon_{\lambda,n} = \lambda(\hbar v/l_B)\sqrt{2n} \propto \lambda\sqrt{Bn}$.

6.2.3.2   *Relativistic Landau levels*

The relativistic case (6.24) for electrons in graphene may be treated exactly in the same manner as the non-relativistic one. In terms of the ladder operators (6.28), the Hamiltonian reads

$$H_D^B = v \begin{pmatrix} 0 & \Pi_x - i\Pi_y \\ \Pi_x + i\Pi_y & 0 \end{pmatrix} = \sqrt{2}\frac{\hbar v}{l_B} \begin{pmatrix} 0 & a \\ a^\dagger & 0 \end{pmatrix}. \tag{6.35}$$

One notices the occurence of a characteristic frequency $\omega' = \sqrt{2}v/l_B$, which plays the role of the cyclotron frequency in the relativistic case. Notice, however, that this frequency may not be written in the form eB/m_b because the band mass is strictly zero in graphene, such that the frequency would diverge.[15]

In order to obtain the eigenvalues and the eigenstates of the Hamiltonian (6.35), one needs to solve the eigenvalue equation $H_D^B \psi_n = \epsilon_n \psi_n$. Because the Hamiltonian is a 2×2 matrix, the eigenstates are 2-spinors,

$$\psi_n = \begin{pmatrix} u_n \\ v_n \end{pmatrix},$$

and we thus need to solve the system of equations

$$\hbar\omega' a\, v_n = \epsilon_n\, u_n \qquad \text{and} \qquad \hbar\omega' a^\dagger u_n = \epsilon_n\, v_n, \tag{6.36}$$

which yields the equation

$$a^\dagger a\, v_n = \left(\frac{\epsilon_n}{\hbar\omega'}\right)^2 v_n \tag{6.37}$$

for the second spinor component. One notices that this component is an eigenstate of the number operator $n = a^\dagger a$, which we have already encountered in the preceding subsection. We may therefore identify, up to a numerical factor, the second spinor component v_n with the eigenstate $|n\rangle$ of the *non-relativistic* Hamiltonian (6.31), $v_n \sim |n\rangle$. Furthermore, one observes that the square of the energy is proportional to this quantum number, $\epsilon_n^2 = (\hbar\omega')^2 n$. This equation has two solutions, a positive and a negative one, and one needs to introduce another quantum number $\lambda = \pm$, which labels the states of positive and negative energy, respectively. This quantum number plays the same role as the band index ($\lambda = +$ for the conduction and $\lambda = -$ for the valence band) in the zero-B-field case discussed in Section 6.2.1. One thus obtains the level spectrum (McClure, 1956)

$$\epsilon_{\lambda,n} = \lambda \frac{\hbar v}{l_B} \sqrt{2n} \tag{6.38}$$

the energy levels of which are depicted in Figure 6.9(b). These *relativistic Landau levels* disperse as $\lambda\sqrt{Bn}$ as a function of the magnetic field.

Once we know the second spinor component, the first one is obtained from Eq. (6.36), which reads $u_n \propto a\, v_n \sim a|n\rangle \sim |n-1\rangle$. One then needs to distinguish the zero-energy LL ($n = 0$) from all other levels. Indeed, for $n = 0$, the first component is zero as one may see from Eq. (6.32), and one obtains the spinor

$$\psi_{n=0} = \begin{pmatrix} 0 \\ |n = 0\rangle \end{pmatrix}. \tag{6.39}$$

In all other cases ($n \neq 0$), one has positive and negative energy solutions, which differ from each other by a relative sign in one of the components. A convenient representation of the associated spinors is given by

$$\psi_{\lambda,n\neq0} = \frac{1}{\sqrt{2}} \begin{pmatrix} |n-1\rangle \\ \lambda|n\rangle \end{pmatrix}. \tag{6.40}$$

Experimental observation of relativistic Landau levels. Relativistic LLs have been observed experimentally in transmission spectroscopy, where one irradiates the sample and measures the intensity of the transmitted light. Such experiments have been performed on so-called epitaxial graphene[16] (Sadowski, *et al.* 2006) and later on exfoliated graphene (Jiang, *et al.* 2007). When the monochromatic light is in resonance with a dipole-allowed transition from the (partially) filled LL (λ, n) to the (partially) unoccupied LL $(\lambda', n \pm 1)$, the light is absorbed due to an electronic excitation between the two levels [see Figure 6.10(a)]. Notice that, in a non-relativistic 2D electron gas, the only allowed dipolar transition is that from the last occupied LL n to the first unoccupied one $n + 1$. The transition energy is $\hbar\omega_C$, independently of n, and one therefore observes a single absorption line (cyclotron resonance). In graphene, however, there are many more allowed transitions due to the presence of two electronic bands, the conduction and the valence band, and the dipole transitions have the energies

$$\Delta_{n,\xi} = \frac{\hbar v}{l_B} \left[\sqrt{2(n+1)} - \xi\sqrt{2n} \right],$$

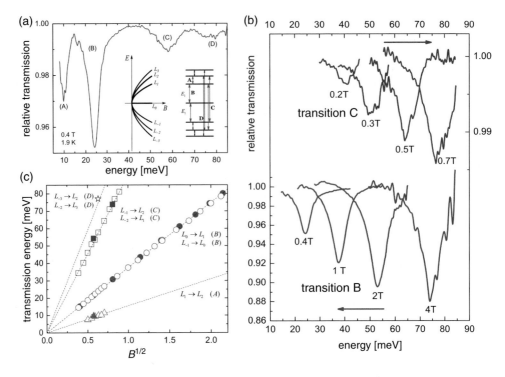

Fig. 6.10 LL spectroscopy in graphene (from Sadowski, *et al.* 2006). (a) For a fixed magnetic field (0.4 T), one observes resonances in the transmission spectrum as a function of the irradiation energy. The resonances are associated with allowed dipolar transitions between relativistic LLs. (b) These resonances are shifted as a function of the magnetic field. (c) If one plots the resonance energies as a function of the square root of the magnetic field, $\sqrt{B}$, a linear dependence is observed, as one would expect for relativistic LLs.

where $\xi = +$ denotes an intraband and $\xi = -$ an interband transition. One therefore obtains families of resonances the energy of which disperses as $\Delta_{n,\xi} \propto \sqrt{B}$, as has been observed in the experiments [see Figure 6.10(c), where we show the results from Sadowski, *et al.* (Sadowski, *et al.* 2006)]. Notice that the dashed lines in Figure 6.10(c) are fits with a single fitting parameter (the Fermi velocity v), which matches well all experimental points for different values of n.

6.2.3.3   *Level degeneracy*

In the preceding subsection, we have learnt that the energy of 2D (non-)relativistic charged particles is characterized by a quantum number n, which denotes the LLs (in addition to the band index λ for relativistic particles). However, the quantum system is yet underdetermined, as may be seen from the following dimensional argument. The original Hamiltonians (6.23) and (6.24) are functions that depend on *two* pairs of conjugate operators, x and p_x, and y and p_y, respectively, whereas when they are expressed in terms of the gauge-invariant momentum $\mathbf{\Pi}$ or else the ladder operators a

and $a^\dagger$ the Hamiltonians (6.31) and (6.35) depend only on a *single* pair of conjugate operators. From the original models, one would therefore expect the quantum states to be described by two quantum numbers (one for each spatial dimension). This is indeed the case in the zero-field models (6.20), where the quantum states are characterized by the two quantum numbers p_x and p_y, i.e. the components of the 2D momentum. For a complete description of the quantum states, we must therefore search for a second pair of conjugate operators, which necessarily commutes with the Hamiltonian and that therefore gives rise to the *level degeneracy* of the LLs – in addition to the degeneracy due to internal degrees of freedom such as the spin[17] or, in the case of graphene, the two-fold valley degeneracy.

In analogy with the gauge-invariant momentum, $\mathbf{\Pi} = \mathbf{p} + e\mathbf{A}(\mathbf{r})$, we consider the same combination with the *opposite* relative sign,

$$\tilde{\mathbf{\Pi}} = \mathbf{p} - e\mathbf{A}(\mathbf{r}), \tag{6.41}$$

which we call *pseudo-momentum* to give a name to this operator. One may then express the momentum operator $\mathbf{p}$ and the vector potential $\mathbf{A}(\mathbf{r})$ in terms of $\mathbf{\Pi}$ and $\tilde{\mathbf{\Pi}}$,

$$\mathbf{p} = \frac{1}{2}(\mathbf{\Pi} + \tilde{\mathbf{\Pi}}) \quad \text{and} \quad \mathbf{A}(\mathbf{r}) = \frac{1}{2e}(\mathbf{\Pi} - \tilde{\mathbf{\Pi}}). \tag{6.42}$$

Notice that, in contrast to the gauge-invariant momentum, the pseudo-momentum depends on the gauge and, therefore, does not represent a physical quantity.[18] However, the commutator between the two components of the pseudo-momentum turns out to be gauge-invariant,

$$\left[\tilde{\Pi}_x, \tilde{\Pi}_y\right] = i\frac{\hbar^2}{l_B^2}. \tag{6.43}$$

This expression is calculated in the same manner as the commutator (6.27) between Π_x and Π_y, as well as the mixed commutators between the gauge-invariant momentum and the pseudo-momentum,

$$\left[\Pi_x, \tilde{\Pi}_x\right] = 2ie\hbar\frac{\partial A_x}{\partial x},$$

$$\left[\Pi_y, \tilde{\Pi}_y\right] = 2ie\hbar\frac{\partial A_y}{\partial y}, \tag{6.44}$$

$$\left[\Pi_x, \tilde{\Pi}_y\right] = ie\hbar\left(\frac{\partial A_x}{\partial y} + \frac{\partial A_y}{\partial x}\right) = -\left[\tilde{\Pi}_x, \Pi_y\right].$$

These mixed commutators are unwanted quantities because they would induce unphysical dynamics due to the fact that the components of the pseudo-momentum would not commute with the Hamiltonian, $[\tilde{\Pi}_{x/y}, H] \neq 0$. However, this embarrassing situation may be avoided by choosing the appropriate gauge, which turns out to be the *symmetric* gauge

$$\mathbf{A}_S(\mathbf{r}) = \frac{B}{2}(-y, x, 0), \tag{6.45}$$

with the help of which all mixed commutators (6.44) vanish such that the components of the pseudo-momentum also commute with the Hamiltonian.

Notice that there exists a second popular choice for the vector potential, namely the *Landau* gauge, which we have already mentioned above,

$$\mathbf{A}_L(\mathbf{r}) = B(-y, 0, 0), \tag{6.46}$$

and for which the last of the mixed commutators (6.44) would not vanish. This gauge choice may even occur simpler: because the vector potential only depends on the y-component of the position, the system remains then translation invariant in the x-direction. Therefore, the associated momentum p_x is a good quantum number, which may be used to label the quantum states in addition to the LL quantum number n. For the Landau gauge, which is useful in the description of geometries with translation invariance in the y-direction, the wave functions are calculated in Section 6.2.4.2. However, the symmetric gauge, the wave functions of which are presented in Section 6.2.4.1, plays an important role in two different respects. First, it allows for a straightforward semi-classical interpretation, in contrast to the Landau gauge; and secondly, the wave functions obtained from the symmetric gauge happen to be the basic ingredient in the construction of trial wave functions *à la Laughlin* for the description of the FQHE, as we will see in Section 6.4.

The pseudo-momentum, with its mutually conjugate components $\tilde{\Pi}_x$ and $\tilde{\Pi}_y$, allows us to introduce, in the same manner as for the gauge-invariant momentum $\boldsymbol{\Pi}$, ladder operators,

$$b = \frac{l_B}{\sqrt{2}\hbar}\left(\tilde{\Pi}_x + i\tilde{\Pi}\right) \qquad \text{and} \qquad b^\dagger = \frac{l_B}{\sqrt{2}\hbar}\left(\tilde{\Pi}_x - i\tilde{\Pi}\right), \tag{6.47}$$

which again satisfy the usual commutation relations $[b, b^\dagger] = 1$ and that, in the symmetric gauge, commute with the ladder operators a and $a^\dagger$, $[b, a^{(\dagger)}] = 0$, as well as with the Hamiltonian, $[b^{(\dagger)}, H_B] = 0$. One may then introduce a number operator $b^\dagger b$ associated with these ladder operators, the eigenstates of which satisfy the eigenvalue equation

$$b^\dagger b|m\rangle = m|m\rangle.$$

One thus obtains a second quantum number, an integer $m \geq 0$, which is necessary to describe, as expected from the above dimensional argument, the full quantum states in addition to the LL quantum number n. The quantum states therefore become tensor products of the two Hilbert vectors

$$|n, m\rangle = |n\rangle \otimes |m\rangle \tag{6.48}$$

for non-relativistic particles. In the relativistic case, one has

$$\psi_{\lambda n, m} = \psi_{\lambda n} \otimes |m\rangle = \frac{1}{\sqrt{2}}\begin{pmatrix} |n-1, m\rangle \\ \lambda|n, m\rangle \end{pmatrix} \tag{6.49}$$

for $n \neq 0$ and

$$\psi_{n=0,m} = \psi_{n=0} \otimes |m\rangle = \begin{pmatrix} 0 \\ |n=0,m\rangle \end{pmatrix} \tag{6.50}$$

for the zero-energy LL.

6.2.3.4 Semi-classical interpretation of the level degeneracy

How can we illustrate this somewhat mysterious pseudo-momentum introduced formally above? Remember that, because the pseudo-momentum is a gauge-dependent quantity, any physical interpretation needs to be handled with care. However, within a semi-classical treatment, the symmetric gauge allows us to make a connection with a classical constant of motion that one obtains from solving the classical equations of motion for a massive electron in a magnetic field,

$$m_b \ddot{\mathbf{r}} = -e(\dot{\mathbf{r}} \times \mathbf{B}) \qquad \Leftrightarrow \qquad \begin{cases} \ddot{x} = -\omega_C \dot{y} \\ \ddot{y} = \omega_C \dot{x}, \end{cases} \tag{6.51}$$

which is simply the electron's acceleration due to the Lorentz force. These equations may be integrated, and one then finds

$$\left. \begin{aligned} \dot{x} &= \frac{\Pi_x}{m_b} = -\omega_C(y - Y) \\ \dot{y} &= \frac{\Pi_y}{m_b} = \omega_C(x - X) \end{aligned} \right\} \qquad \Leftrightarrow \qquad \begin{cases} y = Y - \frac{\Pi_x}{eB} \\ x = X + \frac{\Pi_y}{eB}, \end{cases} \tag{6.52}$$

where $\mathbf{R} = (X, Y)$ is an integration constant, which physically describes a constant of motion. This quantity may easily be interpreted: it represents the center of the electronic cyclotron motion (see Figure 6.11). Indeed, further integration of the equations (6.52) yields the classical cyclotron motion

$$x(t) = X - r\sin(\omega_C t + \phi) \qquad \text{and} \qquad y(t) = Y + r\cos(\omega_C t + \phi),$$

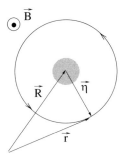

Fig. 6.11 Cyclotron motion of an electron in a magnetic field around the guiding center $\mathbf{R}$. The gray region indicates the quantum-mechanical uncertainty of the guiding-center position due to the non-commutativity (6.54) of its components.

where the phase ϕ is another constant of motion. The cyclotron motion itself is described by the velocities (or else the gauge-invariant momenta) Π_x/m and Π_y/m. Whereas the energy depends on these velocities that determine the radius r of the cyclotron motion, it is completely independent of the position of its center $\mathbf{R}$, which we call the *guiding center* from now on, as one would expect from the translational invariance of the equations of motion (6.51).

In order to relate the guiding center $\mathbf{R}$ to the pseudo-momentum $\tilde{\mathbf{\Pi}}$, we use Eq. (6.42) for the vector potential in the symmetric gauge,

$$e\mathbf{A}(\mathbf{r}) = \frac{eB}{2}\begin{pmatrix} -y \\ x \end{pmatrix} = \frac{1}{2}(\mathbf{\Pi} - \tilde{\mathbf{\Pi}}).$$

The postions x and y may then be expressed in terms of the momenta $\mathbf{\Pi}$ and $\tilde{\mathbf{\Pi}}$,

$$y = \frac{\tilde{\Pi}_x}{eB} - \frac{\Pi_x}{eB}$$

$$x = -\frac{\tilde{\Pi}_y}{eB} + \frac{\Pi_y}{eB}.$$

A comparison of these expresssions with Eq. (6.52) allows us to identify

$$X = -\frac{\tilde{\Pi}_y}{eB} \qquad \text{and} \qquad Y = \frac{\tilde{\Pi}_x}{eB}. \tag{6.53}$$

This means that, in the symmetric gauge, the components of the pseudo-momentum are simply, apart from a factor to translate a momentum into a position, the components of the guiding center, which are naturally constants of motion. In a quantum-mechanical treatment, these operators therefore necessarily commute with the Hamiltonian, as we have already seen above. Furthermore, the commutation relation (6.43) between the components of the pseudo-momentum, $[\tilde{\Pi}_x, \tilde{\Pi}_y] = i\hbar^2/l_B^2$ induces the commutation relation

$$[X, Y] = il_B^2 \tag{6.54}$$

between the components of the guiding-center operator. This means that there is a Heisenberg uncertainty associated with the guiding-center position of a quantum-mechanical state – one cannot know its x- and y-components simultaneously, and the guiding center is, therefore, smeared out over a surface

$$\Delta X \Delta Y = 2\pi l_B^2 \tag{6.55}$$

(see gray region in Figure 6.9).[19] This minimal surface plays the same role as the surface (action) h in phase space and therefore allows us to count the number of possible quantum states of a given (macroscopic) surface $\mathcal{A}$,

$$N_B = \frac{\mathcal{A}}{\Delta X \Delta Y} = \frac{\mathcal{A}}{2\pi l_B^2} = n_B \times \mathcal{A},$$

where we have introduced the flux density

$$n_B = \frac{1}{2\pi l_B^2} = \frac{B}{h/e},\qquad(6.56)$$

which is simply the magnetic field measured in units of the flux quantum h/e. Therefore, *the number of quantum states in a LL equals the number of flux quanta threading the sample surface A, and each LL is macroscopically degenerate.* We will show in a more quantitative manner than in the above argument based on the Heisenberg inequality that the number of states per LL is indeed given by N_B when discussing, in the next section, the electronic wave functions in the symmetric and the Landau gauges.

Similarly to the guiding-center operator, we may introduce the *cyclotron variable* $\boldsymbol{\eta} = (\eta_x, \eta_y)$, that determines the cyclotron motion and that fully describes the dynamical properties. The cyclotron variable is perpendicular to the electron's velocity and may be expressed in terms of the gauge-invariant momentum $\boldsymbol{\Pi}$,

$$\eta_x = \frac{\Pi_y}{eB} \quad \text{and} \quad \eta_y = -\frac{\Pi_x}{eB},\qquad(6.57)$$

as one sees from Eq. (6.52). The position of the electron is thus decomposed into its guiding center and its cyclotron variable, $\mathbf{r} = \mathbf{R} + \boldsymbol{\eta}$. Also, the components of the cyclotron variable do not commute, and one finds with the help of Eq. (6.27)

$$[\eta_x, \eta_y] = \frac{[\Pi_x, \Pi_y]}{(eB)^2} = -i l_B^2 = -[X, Y].\qquad(6.58)$$

Until now, we have only discussed a single particle and its possible quantum states. Consider now N independent quantum-mechanical electrons at zero temperature. In the absence of a magnetic field, electrons in a metal, due to their fermionic nature and the Pauli principle, which prohibits double occupancy of a single quantum state, fill all quantum states up to the Fermi energy, which thus depends on the number of electrons itself. The situation is similar in the presence of a magnetic field: the electrons preferentially occupy the lowest LLs, i.e. those of lowest energy. But once a LL is filled, the remaining electrons are forced to populate higher LLs. In order to describe the LL filling, it is therefore useful to introduce the dimensionless ratio between the number of electrons $N_{el} = n_{el} \times \mathcal{A}$ and that of flux quanta,

$$\nu = \frac{N_{el}}{N_B} = \frac{n_{el}}{n_B} = \frac{h n_{el}}{eB},\qquad(6.59)$$

which is called the *filling factor*. Indeed the integer part, $[\nu]$, of the filling factor counts the number of completely filled LLs. Notice that one may vary the filling factor either by changing the particle number or by changing the magnetic field. At fixed particle number, lowering the magnetic field corresponds to an increase of the filling factor.

6.2.4   Eigenstates

6.2.4.1   *Wave functions in the symmetric gauge*

The algebraic tools established above may be used calculate the electronic wave functions, which are the space representations of the quantum states $|n, m\rangle$, $\phi_{n,m}(x, y) = \langle x, y|n, m\rangle$.[20] Notice first that one may obtain all quantum states $|n, m\rangle$ from a single state $|n = 0, m = 0\rangle$, with the help of

$$|n, m\rangle = \frac{(a^\dagger)^n (b^\dagger)^m}{\sqrt{n!} \sqrt{m!}} |n = 0, m = 0\rangle, \tag{6.60}$$

which is a generalization of Eq. (6.33). Naturally, this equation translates into a differential equation for the wave functions $\phi_{n,m}(x, y)$.

A state in the lowest LL ($n = 0$) is characterized by the condition (6.32)

$$a|n = 0, m\rangle = 0, \tag{6.61}$$

which needs to be translated into a differential equation. Remember from Eq. (6.28) that $a = (l_B/\sqrt{2}\hbar)(\Pi_x - i\Pi_y)$ and, by definition, $\boldsymbol{\Pi} = -i\hbar\nabla + e\mathbf{A}(\mathbf{r})$, where we have already represented the momentum as a differential operator in position representation, $\mathbf{p} = -i\hbar\nabla$. One then finds

$$a = -i\sqrt{2} \left[\frac{l_B}{2} (\partial_x - i\partial_y) + \frac{x - iy}{4l_B} \right],$$

where ∂_x and ∂_y are the components of the gradient $\nabla = (\partial_x, \partial_y)$, and one sees from this expression that it is convenient to introduce *complex coordinates* to describe the 2D plane. We define $z = x - iy$, $z^* = x + iy$, $\partial = (\partial_x + i\partial_y)/2$ and $\bar{\partial} = (\partial_x - i\partial_y)/2$. The lowest LL condition (6.61) then becomes a differential equation,

$$\left(\frac{z}{4l_B} + l_B\bar{\partial} \right) \phi_{n=0}(z, z^*) = 0, \tag{6.62}$$

which may easily be solved by the complex function

$$\phi_{n=0}(z, z^*) = f(z)e^{-|z|^2/4l_B^2}, \tag{6.63}$$

where $f(z)$ is an *analytic* function, i.e. $\bar{\partial}f(z) = 0$, and $|z|^2 = zz^*$. This means that there is an additional degree of freedom because $f(z)$ may be *any* analytic function. It is not unexpected that this degree of freedom is associated with the second quantum number m, as we will now discuss.

The ladder operators b and $b^\dagger$ may be expressed in position representation in a similar manner as a, and one obtains the space representation of the different ladder operators,

$$a = -i\sqrt{2}\left(\frac{z}{4l_B} + l_B\bar{\partial}\right), \qquad a^\dagger = i\sqrt{2}\left(\frac{z^*}{4l_B} - l_B\partial\right)$$

$$b = -i\sqrt{2}\left(\frac{z^*}{4l_B} + l_B\partial\right), \qquad b^\dagger = i\sqrt{2}\left(\frac{z}{4l_B} - l_B\bar{\partial}\right). \tag{6.64}$$

In the same manner as for a state in the lowest LL, the condition for the reference state with $m = 0$ is $b|n, m = 0\rangle = 0$, which yields the differential equation

$$\left(z^* + 4l_B^2\partial\right)\phi'_{m=0}(z, z^*) = 0,$$

with the solution

$$\phi'_{m=0}(z, z^*) = g(z^*)e^{-|z|^2/4l_B^2},$$

in terms of an *anti-analytic* function $g(z^*)$ with $\partial g(z^*) = 0$. The wave function $\phi_{n=0,m=0}(z, z^*)$ must therefore be the Gaussian with a prefactor that is both analytic and anti-analytic, i.e. a constant that is fixed by the normalization. One finds

$$\phi_{n=0,m=0}(z, z^*) = \langle z, z^*|n = 0, m = 0\rangle = \frac{1}{\sqrt{2\pi l_B^2}}e^{-|z|^2/4l_B^2}, \tag{6.65}$$

and a lowest-LL state with arbitrary m may then be obtained with the help of Eq. (6.60),

$$\phi_{n=0,m}(z, z^*) = \frac{i^m\sqrt{2^m}}{\sqrt{2\pi l_B^2 m!}}\left(\frac{z}{4l_B} - l_B\bar{\partial}\right)^m e^{-|z|^2/4l_B^2}$$

$$= \frac{i^m}{\sqrt{2\pi l_B^2 m!}}\left(\frac{z}{\sqrt{2}l_B}\right)^m e^{-|z|^2/4l_B^2}. \tag{6.66}$$

The states within the lowest LL are therefore, apart from the Gaussian, given by the usual polynomial basis states z^m of analytic functions. In an arbitrary LL, the states may be obtained in a similar manner, but they happen to be more complicated because the differential operators (6.64) no longer act on the Gaussian only but also on the polynomial functions. They may be expressed in terms of Laguerre polynomials.

To conclude the discussion about the wave functions in the symmetric gauge, we calculate the average value of the guiding-center operator in the state $|n = 0, m\rangle$. With the help of Eqs. (6.47) and (6.53), one may express the components of the guiding-center operator in terms of the ladder operators b and $b^\dagger$,

$$X = \frac{l_B}{i\sqrt{2}}(b^\dagger - b) \qquad \text{and} \qquad Y = \frac{l_B}{\sqrt{2}}(b^\dagger + b), \tag{6.67}$$

and the ladder operators act, in analogy with Eq. (6.31), on the states $|n, m\rangle$ as

$$b^\dagger|n, m\rangle = \sqrt{m+1}|n, m+1\rangle \qquad \text{and} \qquad b|n, m\rangle = \sqrt{m}|n, m-1\rangle.$$

The average value of the guiding-center operator is therefore zero in the states $|n, m\rangle$,

$$\langle \mathbf{R} \rangle \equiv \langle n = 0, m | \mathbf{R} | n = 0, m \rangle = 0,$$

but we have

$$\langle |\mathbf{R}| \rangle = \left\langle \sqrt{X^2 + Y^2} \right\rangle = l_B \left\langle \sqrt{2b^\dagger b + 1} \right\rangle = l_B \sqrt{2m + 1}. \tag{6.68}$$

This means that the guiding center is situated, in a quantum state $|n, m\rangle$, somewhere on a circle of radius $l_B \sqrt{2m + 1}$ whereas its angle (or phase) is completely undetermined.

The symmetric gauge is the natural gauge to describe a sample in the form of a disc. Consider the disc to have a radius R_{max} (and a surface $\mathcal{A} = \pi R^2_{max}$). How many quantum states may be accomodated within the circle? The quantum state with maximal m quantum number, which we call M, has a radius $l_B \sqrt{2M + 1}$, which must naturally coincide with the radius R_{max} of the disc. One therefore obtains $\mathcal{A} = \pi l^2_B (2M + 1)$, and the number of states within the disc is then, in the thermodynamic limit $M \gg 1$,

$$M = \frac{\mathcal{A}}{2\pi l^2_B} = n_B \times \mathcal{A} = N_B, \tag{6.69}$$

in agreement with the result (6.56) obtained from the argument based on the Heisenberg uncertainty relation.

6.2.4.2   Wave functions in the Landau gauge

If the sample geometry is rectangular, the Landau gauge (6.46), $\mathbf{A}_L(\mathbf{r}) = B(-y, 0, 0)$, is more appropriate than the symmetric gauge to describe the physical system. As already mentioned above, the momentum $p_x = \hbar k$ is a good quantum number due to translational invariance in the x-direction. One may therefore use a plane-wave ansatz

$$\psi_{n,k}(x, y) = \frac{e^{ikx}}{\sqrt{L}} \chi_{n,k}(y),$$

for the wave functions. In this case, the Hamiltonian (6.23) becomes

$$H^B_S = \frac{(p_x - eBy)^2}{2m} + \frac{p^2_y}{2m} = \frac{p^2_y}{2m} + \frac{1}{2} m\omega_C (y - y_0)^2, \tag{6.70}$$

where we have defined

$$y_0 = k l^2_B. \tag{6.71}$$

The Hamiltonian (6.70) is simply the Hamiltonian of a one-dimensional oscillator centred around the position y_0, and the eigenstates are

$$\chi_{n,k}(y) = H_n \left(\frac{y - y_0}{l_B} \right) e^{-(y - y_0)^2 / 4 l^2_B},$$

in terms of Hermite polynomials $H_n(x)$ (Cohen-Tannoudji, *et al.* 1973). The coordinate y_0 plays the role of the guiding-centre component Y, the component X being smeared over the whole sample length L, as it is dictated by the Heisenberg uncertainty relation resulting from the commutation relation (6.54), $[X, Y] = il_B^2$.

Using periodic boundary conditions $k = m \times 2\pi/L$ for the wavevector in the x-direction, one may count the number of states in a rectangular surface of length L and width W (in the y-direction), similarly to the above arguments in the symmetric gauge. Consider the sample to range from $y_{min} = 0$ to $y_{max} = W$, the first corresponding via the above-mentioned condition (6.71) to $k = 0$ and the latter to a wavevector $k_{max} = M \times 2\pi/L$. Two neighboring quantum states are separated by the distance $\Delta y = \Delta k l_B^2 = \Delta m (2\pi/L) l_B^2 = 2\pi l_B^2/L$, and each state therefore occupies a surface $\sigma = \Delta y \times L = 2\pi l_B^2$, which agrees with the result (6.55) obtained above with the help of the Heisenberg uncertainty relation. The total number of states is, as in the symmetric gauge and the general argument leading to Eq. (6.56),

$$M = N_B = n_B \times LW = n_B \times \mathcal{A},$$

i.e. the number of flux quanta threading the (rectangular) surface $\mathcal{A} = LW$.

6.3 Integer quantum Hall effect

The quantum-mechanical treatment of the 2D electron in a perpendicular magnetic field is the key ingredient for the understanding of the quantum Hall effect. However, we need to relate the kinetic-energy quantization to the resistance quantization, which is the essential feature of the IQHE. In the present section, we discuss the transport properties of electrons in the IQHE, namely the somewhat mysterious role that disorder plays in this type of transport. Recall from the introduction that the Hall resistance is quantized with an astonishingly high precision (10^{-9}), such that it is now used as the standard of resistance [see Eq. (6.13)]. The resistance quantization in the IQHE therefore does reflect neither a particular disorder distribution nor a particular sample geometry. Nevertheless, disorder turns out to play an essential role in the occurence of the IQHE, as we will see in this section.

We will first consider, in Section 6.3.1, the motion of a 2D electron in a perpendicular magnetic field when an external electrostatic potential is also present, such as the one generated by disorder or the confinement potential that defines the sample boundaries. In Section 6.3.2, we then calculate the conductance of a single LL within a mesoscopic picture and discuss the difference between a two-terminal and a six-terminal transport measurement in Section 6.3.3. Furthermore, we discuss, in Section 6.3.4, the IQHE within a percolation picture and present some scaling properties that characterize the plateau transitions. We terminate this section with a short discussion of the peculiarities of the relativistic quantum Hall effect in graphene the understanding of which requires essentially the same ingredients as the IQHE in non-relativistic quantum Hall systems.

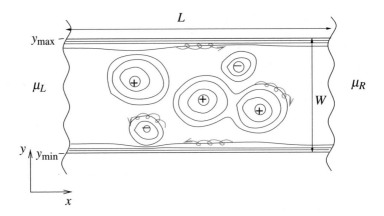

Fig. 6.12 Potential landscape of an electrostatic potential in a sample. The metallic contacts are described by the chemical potentials μ_L and μ_R for the left and right contacts, respectively. We consider $L \gg W \gg \xi \gg l_B$, where ξ is the typical length scale for the variation of the electrostatic potential. The sample is confined in the y-direction between y_{max} and y_{min}. The thin lines indicate the equipotential lines. When approaching one of the sample edges, they become parallel to the edge. The gray lines indicate the electronic motion with the guiding center moving along the equipotential lines. The electron turns counter-clockwise around a summit of the potential landscape, which is caused e.g. by a negatively charged impurity $(-)$, and clockwise around a valley $(+)$. At the sample edges, the equipotential lines due to the confinement potential connect the two contacts on the left- and on the right-hand side.

6.3.1 Electronic motion in an external electrostatic potential

We consider a system the length L of which is much larger than the width W (see Figure 6.12). This may be modelled by a confinement potential $V_{\mathrm{conf}}(y)$ that only depends on the y-direction, i.e. the system remains translation-invariant in the x-direction with respect to this potential.[21] In addition to the confinement, we consider a smoothly varying electrostatic potential $V_{\mathrm{imp}}(x,y)$ that is caused by the impurities in the sample. This impurity potential breaks the translation invariance in the x-direction as well as that in the y-direction, which is already broken by the confinement potential. The Hamiltonian of a 2D particle in a perpendicular magnetic field then needs to be completed by a potential term

$$V(\mathbf{r}) = V_{\mathrm{conf}}(y) + V_{\mathrm{imp}}(x,y), \qquad (6.72)$$

which creates a potential landscape that is schematically depicted in Figure 6.12.

6.3.1.1 Semi-classical treatment

In a first step, we consider a potential $V(\mathbf{r})$ that varies smoothly on the scale set by the magnetic length, i.e. $\xi \gg l_B$, where ξ describes the characteristic length scale for the variation of $V(\mathbf{r})$. Notice first that the external electrostatic potential lifts the LL degeneracy because the Hamiltonian $H = H_B + V(\mathbf{r} = \mathbf{R} + \boldsymbol{\eta})$ no longer commutes with the guiding-center operator $\mathbf{R}$, in contrast to the "free" Hamiltonian H_B, $[H, \mathbf{R}] = [V, \mathbf{R}] \neq 0$. Physically, this is not unexpected: the guiding center is a constant of motion due to translation invariance, i.e. it does not matter whether the electron performs its cyclotron motion around a point $\mathbf{R}_1$ or $\mathbf{R}_2$ in the 2D plane as long as the cyclotron radius is the same. However, the electrostatic potential $V(\mathbf{r})$ breaks this translation invariance and thus lifts the degeneracy associated with the guiding center.

In the case where the electrostatic potential varies smoothly on a length scale set by the magnetic length and does not generate LL mixing, i.e. when $|\nabla V| \ll \hbar \omega_C / l_B$, we may approximate the argument $\mathbf{r}$ in the potential (6.72) by the guiding-center variable $\mathbf{R}$,[22]

$$V(\mathbf{r}) \simeq V(\mathbf{R}). \tag{6.73}$$

Notice that this approximation may seem inappropriate when we consider the confinement potential in the y-direction that may vary abruptly when approaching the sample edges. The confinement potential with translation invariance in the x-direction will be discussed separately in the following subsection.

As a consequence of the non-commutativity of the potential term $V(\mathbf{R})$ with the guiding-center operator, the latter quantity acquires dynamics, as may be seen from the Heisenberg equations of motion

$$i\hbar \dot{X} = [X, H] = [X, V(\mathbf{R})] = \frac{\partial V}{\partial Y}[X, Y] = i l_B^2 \frac{\partial V}{\partial Y}$$

$$i\hbar \dot{Y} = [Y, V(\mathbf{R})] = -i l_B^2 \frac{\partial V}{\partial X}, \tag{6.74}$$

i.e. $\dot{\mathbf{R}} \perp \nabla V$. Here, we have used the commutation relation (6.54) for the guiding-center components and Eq. (6.26). The Heisenberg equations of motion are particularly useful in the discussion of the semi-classical limit because the averaged equations satisfy the classical equations of motion,

$$\langle \dot{\mathbf{R}} \rangle \perp \nabla V, \tag{6.75}$$

which means that, within the semi-classical picture, *the guiding centers move along the equipotential lines of the smoothly varying external electrostatic potential.* This feature, which is also called the *Hall drift*,

$$\mathbf{v}_D = \frac{\mathbf{E} \times \mathbf{B}}{B^2} = \langle \dot{\mathbf{R}} \rangle = \frac{\nabla V \times \mathbf{B}}{e B^2}, \tag{6.76}$$

in terms of the (local) electric field $\mathbf{E} = \nabla V / e$, is depicted in Figure 6.12 by the gray lines.

In the bulk, the potential landscape is created by the charged impurities in the sample, and the electrons turn counter-clockwise on an equipotential line around a summit that is caused by a negatively charged impurity and clockwise around a valley created by a positively charged impurity. If the equipotential lines are closed, as it is the case for most of the equipotential lines in a potential landscape,[23] an electron cannot move from one point to another one over a macroscopic distance, e.g. from one contact to the other one. An electron moving on a closed equipotential line can therefore not contribute to the electronic transport, and the electron is thus *localized*. Notice that this type of localization is different from other popular types. Anderson localization in 2D, e.g., is due to quantum interferences of the electronic wave functions (Abrahams, *et al.* 1979). Here, however, the localiation is a purely *classical* effect. The high-field localization is also different from the interaction-driven Mott insulator, where the electrons freeze out in order to minimize the mutual Coulomb repulsion between the electrons.

At the edge, the equipotential lines reflect the confinement potential, which is zero in the bulk but rapidly increases when approaching the sample edge at y_{min} and y_{max} (see Figure 6.12). In this case, the equipotential lines are open and therefore connect the two different electronic contacts. The electrons occupying quantum states at these equipotential lines then contribute to the electronic transport, in contrast to those on closed equipotential lines in the bulk. These quantum states are called *extended states*,[24] as opposed to the *localized states* discussed above. The difference between localized and extended states turns out to be essential in the understanding of the IQHE, as we will see below (Section 6.3.4).

6.3.1.2   *Electrostatic potential with translation invariance in the x-direction*

Although the above semi-classical considerations yield the correct physical picture of localized and extended states, it is based on the assumption that the electrostatic potential varies smoothly on the scale set by the magnetic length, such that we may replace the electron's position by that of its guiding center ([Eq. (6.73)]). This assumption is, however, problematic in view of the confinement potential, which varies strongly at the sample edges, i.e. in the vicinity of y_{min} and y_{max}. We will therefore treat the y-dependent confinement potential in a quantum treatment. Naturally, the appropriate gauge for the quantum-mechanical treatment is the Landau gauge (6.46), which respects the translation invariance in the x-direction, and the Hamiltonian (6.70) becomes

$$H = \frac{p_y^2}{2m} + \frac{1}{2}m\omega_C(y - y_0)^2 + V_{\text{conf}}(y). \tag{6.77}$$

Remember that for a fixed wavevector k in the x-direction, the position around which the one-dimensional harmonic oscillator is centered is fixed by Eq. (6.71), $y_0 = kl_B^2$. We may therefore expand the confinement potential, even in the case of a strong variation, around this position,

$$V_{\text{conf}}(y) \simeq V_{\text{conf}}(y_0 = kl_B^2) + eE(y_0)(y - y_0) + \mathcal{O}\left(\frac{\partial^2 V_{\text{conf}}}{\partial y^2}\right),$$

where the local electric field is given in terms of the first derivative of the potential at y_0,

$$eE(y_0) = \frac{\partial V_{\text{conf}}}{\partial y}\bigg|_{y_0}.$$

This expansion yields the Hamiltonian

$$H = \frac{p_y^2}{2m} + \frac{1}{2}m\omega_C(y - y_0')^2 + V_{\text{conf}}(y_0) - \frac{1}{2}mv_D^2(y_0),$$

where the local drift velocity reads $v_D = E(y_0)/B$ and the position of the harmonic oscillator is shifted, $y_0 \to y_0' = y_0 - eE(y_0)/m\omega_C^2$. Notice that the last term is quadratic in the electric field $E(y_0)$ and therefore a second-order term in the expansion of the confinement potential. We neglect this term in the following calculations. The final Hamiltonian then reads

$$H = \frac{p_y^2}{2m} + \frac{1}{2}m\omega_C(y - y_0')^2 + V_{\text{conf}}(y_0'), \tag{6.78}$$

where we have replaced the argument y_0 by the shifted harmonic-oscillator position y_0', which is valid at first order in the expansion of the confinement potential. One therefore obtains the energy spectrum

$$\epsilon_{n,y_0} = \hbar\omega_C\left(n + \frac{1}{2}\right) + V_{\text{conf}}(y_0), \tag{6.79}$$

where we have omitted the prime at the shifted harmonic-oscillator position to simplify the notation. One therefore obtains the same LL spectrum as in the absence of a confinement potential, apart from an energy shift that is determined by the value of the confinement potential at the harmonic-oscillator position, which may indeed vary strongly. This position y_0 plays the role of the guiding-center position, as we have already mentioned in the last section, where we have calculated the electronic wave functions in the Landau gauge (6.2.4.2). One thus obtains a result that is consistent with the semi-classical treatment presented above.

6.3.2 Conductance of a single Landau level

We now calculate the conductance of a completely filled LL for the geometry depicted in Figure 6.12, i.e. when all quantum states (described within the Landau gauge) of the nth LL are occupied. In a first step, we calculate the current of the nth LL, which flows from the left to the right contact, with the help of the formula (Büttiker, 1992)

$$I_n^x = -\frac{e}{L}\sum_k \langle n, k|v_x|n, k\rangle, \tag{6.80}$$

i.e. as the sum over all N_B quantum channels labelled by the wavevector $k = 2\pi m/L$, with the velocity

$$\langle n, k|v_x|n, k\rangle = \frac{1}{\hbar}\frac{\partial \epsilon_{n,k}}{\partial k} = \frac{L}{2\pi\hbar}\frac{\Delta\epsilon_{n,m}}{\Delta m},$$

in terms of the dispersion relation (6.79).[25] Notice that the velocity in the y-direction is zero because the energy does not disperse as a function of the y-component of the wave vector. The above expression is readily evaluated with $\Delta m = 1$, and one obtains

$$\langle n, k | v_x | n, k \rangle = \frac{L}{h} \left(\epsilon_{n,m+1} - \epsilon_{n,m} \right).$$

With the help of this expression, the current (6.80) of the nth LL becomes

$$I_n^x = -\frac{e}{L} \sum_m \frac{L}{h} \left(\epsilon_{n,m+1} - \epsilon_{n,m} \right),$$

and one notices that all terms in the sum cancel apart from the boundary terms $\epsilon_{n,m_{min}}$ and $\epsilon_{n,m_{max}}$, which correspond to the chemical potentials μ_{min} and μ_{max}, respectively. The difference between these two chemical potentials may be described in terms of the (Hall) voltage V between the upper and the lower edge, $\mu_{max} - \mu_{min} = -eV$. One thus obtains the final result

$$I_n^x = -\frac{e}{h} \left(\mu_{max} - \mu_{min} \right) = \frac{e^2}{h} V. \tag{6.81}$$

This means that each LL contributes one *quantum of conductance* $G_n = e^2/h$ to the electronic transport and n completely filled LLs contribute a conductance[26]

$$G = \sum_{n'=0}^{n-1} G_{n'} = n \frac{e^2}{h}. \tag{6.82}$$

This is a particular example of the Landauer–Büttiker formula of quantum transport

$$G_n = \frac{e^2}{h} T_n$$

through a conduction channel n, where T_n is the transmission coefficient of the channel (Büttiker, *et al.* 1985; Büttiker 1992; Datta 1995). Because $T_n + R_n = 1$, in terms of the reflexion coefficient, the above result (6.81) indicates that each filled LL may be viewed as a conduction channel with perfect transmission $T_n = 1$, i.e. where an injected electron is not reflected or backscattered.

6.3.2.1 *Edge states*

The astonishing feature of perfect transmission, which is independent of the length L (or more precisely of the aspect ratio L/W, see the discussion in Section 6.1.1.2 of the introduction) or the particular geometry of the sample, may be understood from the edge-state picture which we have introduced above (see Figure 6.13). Consider the upper edge, without loss of generality. The current-transporting edge state of the nth LL is the one situated at y_{max}^n, where the nth LL crosses the Fermi energy and where the filling factor jumps from $\nu = n + 1$ to n.[27] Due to the upward bend of the confinement potential a particular direction is imposed on the electronic motion, which is simply the Hall drift (see Figure 6.12). This uni-directional motion is also

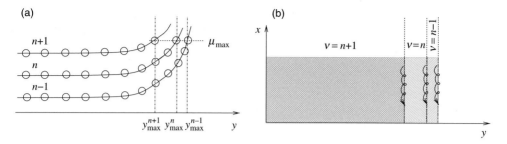

Fig. 6.13 Edge states. (a) The LLs are bent upwards when approaching the sample edge, which may be modelled by an increasing confinement potential. One may associate with each LL n a maximal value y^n_{max} of the y-component where the LL crosses the chemical potential μ_{max}. (b) At each position y^n_{max}, the filling factor decreases by a jump of 1. The nth edge state is associated with the jump at y^n_{max} and the gradient of the confinement potential imposes a direction to the Hall drift of this state (*chirality*). This chirality is the same for all edge states at the same edge.

called *chirality* of the edge state. Notice that this is the same chirality that one expects from the semi-classical expression (6.76) for the drift velocity. The chirality is the same for all edge states n at the same sample edge where the gradient of the confinement potential does not change its direction. Therefore, even if an electron is scattered from one channel n to another one n' at the same edge it does not change its direction of motion, and the electron cannot be backscattered unless it is scattered to the opposite edge with inverse chirality. However, in a usual quantum Hall system, the opposite edges are separated by a macroscopic distance $\sim W$, and backscattering processes are therefore strongly (exponentially) suppressed in the ratio W/l_B between the magnetic length, which determines the spatial extension of a quantum-mechanical state, and the macroscopic sample width W. Notice that the quantum Hall system at integer filling factors $\nu = n$ is therefore a very unusual electron liquid: it is indeed a bulk insulator with perfectly conducting (non-dissipative) edges.

6.3.3 Two-terminal versus six-terminal measurement

6.3.3.1 Two-terminal measurement

In the preceding Section (6.3.2), we have calculated the conductance of a single LL (and n filled LLs) within a so-called two-terminal measurement, where we inject a current in the left contact with chemical potential μ_L and collect the outcoming current at the right contact with μ_R. As a consequence of Eq. (6.81), this current builds up a voltage V between the upper and the lower sample edge. This voltage drop is therefore associated with a Hall resistance that is the inverse of the conductance $G = ne^2/h$,

$$R_H = G^{-1} = \frac{h}{e^2}\frac{1}{n},\tag{6.83}$$

and that coincides with the contact (or interface) resistance of a mesoscopic system (Datta, 1995). However, the voltage drop V_L between the left and the right contact is given by the chemical-potential difference in the contacts, $\mu_R - \mu_L = -eV_L$, and the associated longitudinal resistance V_L/I is non-zero, in contrast to what we have seen in section 6.1. This is due to the fact that the difference between the longitudinal and the Hall resistance is not clearly defined in such a two-terminal measurement.

This feature may be understood from Figure 6.14. Indeed, due to the above-mentioned absence of backscattering, the chemical potential is constant along a sample edge, but there is a potential difference between the two edges. This means that the chemical potential must change somewhere along the edge. Consider the upper edge that is fed with electrons by the left contact, i.e. the upper edge is in thermodynamic equilibrium with the left contact and the chemical potentials therefore coincide, $\mu_L = \mu_{max}$ (see Figure 6.14). When the upper edge touches the right contact that is at a different chemical potential μ_R, the chemical potential of the upper edge must rapidly relax to be in equilibrium with the right contact. In the same manner, the lower edge is in equilibrium with the right contact, $\mu_{min} = \mu_R$, and abruptly changes when touching the left contact. The rapid change in the chemical potential is associated with dissipation of energy (at so-called *hot spots*) that has been observed experimentally (Klaß, *et al.* 1991). In this experiment, the sample was put in liquid helium and the heating at the hot spots caused a local vaporization of the helium observable in the form of a fountain of gas bubbles.

Due to the equivalence of the chemical potentials $\mu_L = \mu_{max}$ and $\mu_{min} = \mu_R$, the voltage drops V, between the upper and the lower edge, and V_L between the current

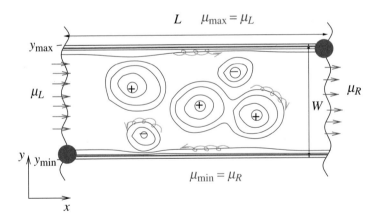

Fig. 6.14 Two-terminal measurement. The current is driven through the sample via the left and the right contacts, where one also measures the voltage drop and thus a resistance. The upper edge is in thermodynamic equilibrium with the left contact, whereas the lower one is in equilibrium with the right contact. The chemical potential drops abruptly when the upper edge reaches the right contact, and when the lower edge reaches the left contact. Dissipation occurs in these hot spots (dots). The measured resistance between the two contacts thus equals the Hall resistance.

contacts are equal, $V = V_L$. A perhaps unexpected consequence of this equation is that in a resistance measurement between the two contacts, in the two-terminal configuration, the two-terminal resistance equals the Hall resistance,

$$R_{R-L} = R_H = \frac{h}{e^2}\frac{1}{n}, \tag{6.84}$$

and *not* the (vanishing) longitudinal resistance, when the bulk is insulating (at $\nu = n$).

6.3.3.2   Six-terminal measurement

A more sophisticated geometry that allows for the simulaneous measurement of a well-defined longitudinal and Hall resistance is the six-terminal geometry, with two additional contacts at the upper and two at the lower edge (see Figure 6.15(a)). These additional contacts (2 and 3 at the upper and 5 and 6 at the lower edge, the left and the right contacts being labelled by 1 and 4, respectively) are used to measure a voltage, i.e. they have ideally an infinitely high internal resistance to prevent electrons

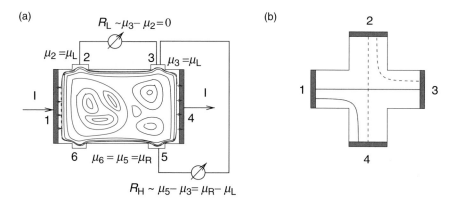

Fig. 6.15 (a) Six-terminal measurement. The current I is driven through the sample via the contacts 1 and 4. Between these two contacts the chemical potential on the upper edge μ_L does not vary because the electrons do not leak out or in at the contacts 2 and 3, where one measures the longitudinal resistance. In the same manner, the chemical potential μ_R remains constant between the contacts 5 and 6 on the lower edge. The longitudinal resistance measured between 2 and 3 as well as between 5 and 6 is therefore $R_L = (\mu_2 - \mu_3)/eI = (\mu_5 - \mu_6)/eI = 0$. The Hall resistance is determined by the potential difference between the two edges and thus measured, e.g. between the contacts 5 and 3, where $\mu_5 - \mu_3 = \mu_R - \mu_L$, and thus $R_H = (\mu_3 - \mu_5)/eI$. (b) Four-terminal measurement in the van-der-Pauw geometry. In a Hall-resistance measurement, one drives a current through the sample via the contacts 1 and 3 (connected by the continuous dark line) and measures the Hall resistance via the contacts 2 and 4 (dark dashed line). In a measurement of the longitudinal resistance, the current is driven through the sample via the contacts 1 and 4 (continuous gray line) and one measures a resistance between the contacts 2 and 3 (connected by the gray dashed line).

to leak out of or into the sample. The chemical potential therefore remains constant at the upper edge $\mu_L = \mu_2 = \mu_3$, as well as that at the lower edge $\mu_R = \mu_5 = \mu_6$, and one measures a zero-resistance, $R_L = (\mu_2 - \mu_3)/eI = (\mu_5 - \mu_6)/eI = 0$, as one expects from the calculation of the conductance through n LLs (see Section 6.3.2), which is entirely transverse. The conductance matrix is thus off-diagonal, as well as the resistance matrix,

$$G = \begin{pmatrix} 0 & n\frac{e^2}{h} \\ -n\frac{e^2}{h} & 0 \end{pmatrix} \quad \text{and} \quad R = \begin{pmatrix} 0 & -\frac{h}{e^2}\frac{1}{n} \\ \frac{h}{e^2}\frac{1}{n} & 0 \end{pmatrix}, \tag{6.85}$$

and one precisely measures the diagonal elements of the resistance matrix, the longitudinal resistance, between the contacts 3 and 2 (or 6 and 5). The off-diagonal elements, i.e. the Hall resistance, may e.g. be measured between the contacts 5 and 3 [as shown in Figure 6.15(a)], and one measures then the result $R_H = G_n^{-1} = h/e^2 n$ obtained from the calculation presented in Section 6.3.2 because of the voltage drop $V = (\mu_L - \mu_R)/e = (\mu_3 - \mu_5)/e$ between the upper and the lower edge.

Finally, we mention that there exists an intermediate geometry that consists of four terminals (van-der-Pauw geometry), where the resistance measurements are equally well defined [Figure 6.15(b)]. If one labels the contacts from 1 to 4 in a clockwise manner, one may measure a Hall resistance between the contacts 2 and 4 while driving a current through the sample by the contacts 1 and 3 (Figure 6.15(b)). In this case, one may use the clear topological definition mentioned at the beginning of Section 6.1. If one connects the contacts 2 and 3 by an imaginary line through the sample (dark dashed line) it necessarily crosses the imaginary line (dark continuous) that connects the current contacts 1 and 3 through the sample. This is precisely the topological definition of a Hall-resistance measurement.

Similarly, one may measure the longitudinal resistance between the contacts 2 and 3 if one drives a current through the sample via the contacts 1 and 4. In this case, the imaginary line (gray dashed) that connects the contacts 2 and 3 where one measures a resistance does not need to cross the line (dark lines in continuous) between the contacts 1 and 4 at which one injects and collects the current, respectively. As mentioned at the beginning of Section 6.1, this defines topologically a measurement of the longitudinal resistance.

These considerations show that a resistance measurement, although it does not depend on the microscopic details of the sample, depends nevertheless on the geometry in which the contacts are placed at the sample (Büttiker, 1988, 1992). This aspect is often not sufficiently appreciated in the literature, namely the fact that one measures, in a two-terminal geometry, a Hall resistance between the contacts that are used to inject and collect the current and not a longitudinal resistance, as one may have naively expected, when the system is in the IQHE condition.

6.3.4 The integer quantum Hall effect and percolation

Until now we have shown that the Hall resistance is quantized (Eq. (6.83) when n LLs are completely filled, i.e. when the filling factor is exactly $\nu = n$. However, we have not yet explained the occurence of plateaus in the Hall resistance, i.e. a Hall resistance that remains constant even if one varies the filling factor, e.g. by sweeping the magnetic field, around $\nu = n$.[28] In order to explain the constance of the Hall resistance over a rather large magnetic-field range around $\nu = n$, we need to take into account the semi-classical localization of additional electrons (or holes) described in Section 6.3.1. This is shown in Figure 6.16, where we represent the filling of the LLs (first line), the potential landscape of the last partially filled level (second line) and the resistances as a function of the magnetic field, measured in a six-terminal geometry (third line). We start with the situation of n completely filled LLs (column (a) of Figure 6.16), which we have extensively discussed above: the LL n (and its potential landscape) is unoccupied.[29] In a six-terminal measurement, one therefore measures the Hall resistance $R_H = h/e^2 n$ and a zero longitudinal resistance, as we have seen in Eq. (6.85).

In column (b) of Figure 6.16, we represent the situation where the LL n gets moderately filled by electrons when the magnetic field B is decreased. These electrons in n populate preferentially the valleys of the potential landscape, or more precisely the closed equipotential lines that enclose these valleys. The electrons in the LL n are thus (classically) localized somewhere in the bulk and do not affect the global

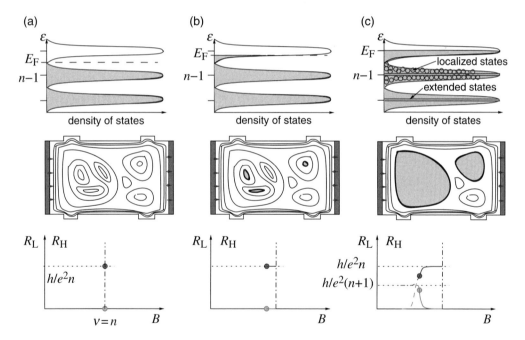

Fig. 6.16 Quantum Hall effect. The (impurity-broadened) density of states is shown in the first line for increasing fillings (a)–(c) described by the Fermi energy E_F. The second line represents the impurity-potential landscape the valleys of which become successively filled with electrons when increasing the filling factor, i.e. when lowering the magnetic field at fixed particle number. The third line shows the corresponding Hall (dark gray) and the longitudinal (light gray) resistance measured in a six-terminal geometry, as a function of the magnetic field. The first figure in column (c) indicates that the bulk extended states are close to the center of the DOS peaks, whereas the localized states are found in their tails.

transport characteristics, measured by the resistances, because they are not probed by the sample contacts. Therefore, the Hall resistance remains unaltered and the longitudinal resistance remains zero despite the change in the magnetic field. This is the origin of the plateau in the Hall resistance.

If one continues to lower the magnetic field, the regions of the potential landscape in the LL n occupied by electrons become larger, and they are eventually enclosed by equipotential lines that pass through the bulk and that connect the opposite edges. In this case, an electron injected at the left contact and travelling a certain distance at the upper edge may jump into the state associated with this equipotential line and thus reach the lower edge. Due to its chirality, the electron is then backscattered to the left contact, which causes an increase in the longitudinal resistance. Indeed, if one measures the resistance between the two contacts at the lower edge, a potential drop is caused by electrons that leak in from this equipotential line connecting the upper and the lower edge. It is this potential drop that causes a non-zero longitudinal resistance. At the same moment the Hall resistance is no longer quantized and jumps to the next (lower) plateau, a situation that is called *plateau transition*. This situation of electron-filled equipotential lines connecting opposite edges, which are thus *extended states* (see first line of Figure 6.16(c)) as opposed to the bulk *localized states*, arises when the LL n is approximately half-filled. Notice that these extended states, which are found in the center of the DOS peaks (see upper part of Figure 6.16(c)), are bulk states in contrast to the above-mentioned edge states, which are naturally also extended.

The clean jump in the Hall resistance at the plateau transition, accompanied by a peak in the longitudinal one, is only visible in the six- (or four-)terminal measurement. As we have argued in Section 6.3.3.2, there is no clear cut difference between the longitudinal and the Hall resistance in the two-terminal configuration, where the resistance measured between the current contacts is indeed quantized in the IQHE. At the plateau transition, however, the chemical potential at the edges is no longer constant because of backscattered electrons and the resistance is no longer quantized. One observes indeed the resistance peak associated with the *longitudinal* resistance in the six- or four-terminal configuration. As a consequence, one measures, at the plateau transition, the superposition of the Hall and the longitudinal resistances.

If one increases even more the filling of the LL n, the same arguments apply but now in terms of *hole states*. The Hall resistance is quantized as $R_H = h/e^2(n+1)$, and the holes (i.e. the lacking electrons with respect to $n+1$ completely filled LLs) get localized in states at closed equipotential lines around the potential summits. As a consequence, the longitudinal resistance drops to zero again.

6.3.4.1   *Extended and localized bulk states in an optical measurement*

The physical picture presented above, in terms of localized and extended bulk states, has recently been confirmed in scanning-tunnelling spectroscopy (STS) of a 2D electron system that was prepared on an n-InSb surface instead of the more common GaAs/AlGaAs heterostructure (Hashimoto, *et al.* 2008). Its advantage consists of its accessibility by an "optical" (surface) measurement that cannot be performed if the 2D

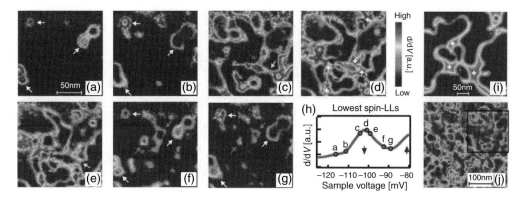

Fig. 6.17 STS measurements by Hashimoto, *et al.* 2008, on a 2D electron system on an *n*-InSb surface. The figures (a)–(g) show the local DOS at various sample voltages, around the peak obtained from a dI/dV measurement (h). Figure (i) shows a calculated characteristic LDOS, and figure (j) an STS result on a larger scale.

electron gas is buried deep in a semiconductor heterostructure. In an STS measurement one scans the sample and thus measures the *local* density of states at a certain energy that can be tuned via the voltage between the tip of the electron microscope and the sample. When measuring the differential conductance dI/dV, which is proportional to the DOS, one observes a peak that corresponds to the center of a LL (Figure 6.17(h)) where the extended states are capable of transporting a current between the different electric contacts, as mentioned above. Whereas the quantum states at energies corresponding to closed equipotential lines of the impurity landscape are clearly visible as closed orbits in Figure 6.17(a),(b) and (f),(g), the states in the vicinity of the peak are more and more extended, as shown by the spaghetti-like lines in Figures 6.17(c),(d) and (e), as one would expect from the arguments presented above.

6.3.4.2 Plateau transitions and scaling laws

The physical picture presented above suggests that the plateau transition in the Hall resistance is related to a *percolation* transition, where initially separated electron-filled valleys start to percolate between the opposite sample edges beyond a certain threshold of the filling. Because of the second-order character of a percolation transition, this scenario suggests that the plateau transition is a *second-order quantum phase transition* described by universal scaling laws, where the control parameter is just the magnetic field B. We provide, here, a brief overview of these scaling laws and refer the interested reader to the literature (Sondhi, *et al.* 1997; Sachdev 1999) and the class given by G. Batrouni at the same Singapore session of Les Houches Summer School 2009.[30]

The phase transition occurs at the critical magnetic field B_c and is characterized by an algebraically diverging correlation length

$$\xi \sim |B - B_c|^{-\nu}, \tag{6.86}$$

where ν is called the *critical exponent*.[31] In the same manner, the temporal fluctuations are described by a correlation "length" ξ_τ that is related to the spatial correlation length ξ by

$$\xi_\tau \sim \xi^z \sim |B - B_c|^{-z\nu},\tag{6.87}$$

where z is called *dynamical critical exponent*. It is roughly a measure of the anisotropy between the spatial and temporal fluctuations, and it is often encountered in non-relativistic condensed-matter systems.[32]

At the phase transition B_c, the longitudinal and transverse resistivities $\rho_{L/H}$ are described in terms of *universal* functions that are functions of the ratio τ/ξ_τ between the (imaginary) time τ, which is proportional to the inverse temperature, $\hbar/\tau = k_B T$ (Sondhi, *et al.* 1997; Sachdev 1999) and the temporal correlation length ξ_τ,

$$\rho_{L/H} = f_{L/H}\left(\frac{\tau}{\xi_\tau}\right) = f_{L/H}\left(\frac{\Delta B^{z\nu}}{T}\right),\tag{6.88}$$

where we have defined $\Delta B \equiv |B - B_c|$. In the case of an AC (alternating current) measurement at frequency ω, another dimensionless quantity, namely the ratio between the frequency and the temperature, $\hbar\omega/k_B T$, needs to be taken into account such that the universal function reads

$$\rho_{L/H}^{AC} = f_{L/H}\left(\frac{\tau}{\xi_\tau}, \frac{\hbar\omega}{k_B T}\right).$$

However, we do not consider an alternating current here. Equation (6.88) then yields the scaling of the width of the peak in the longitudinal resistance (or else the plateau transition)

$$\Delta B \sim T^{1/z\nu}.\tag{6.89}$$

A measurement of this width by Wei, *et al.* (Wei, *et al.* 1988) has confirmed such critical behavior with an exponent $1/z\nu = 0.42 \pm 0.04$ (see Figure 6.18).

Furthermore, one may distinguish between the two exponents ν and z within a measurement of the plateau-transition width as a function of the electric field E via current fluctuations. One may identify the energy fluctuation $eE\xi$ at the correlation length ξ with the energy scale $\hbar/\xi_\tau \sim \hbar/\xi^z$ set by the temporal fluctuation ξ_τ, which yields $E \sim \xi^{-(1+z)} \sim \Delta B^{\nu(1+z)}$, and thus

$$\Delta B \sim E^{1/\nu(1+z)}.\tag{6.90}$$

Other measurements by Wei, *et al.* (Wei, *et al.* 1994) have shown that these types of fluctuations yield $z \simeq 1$, i.e. $\nu \simeq 2.3$. The precision of the measured critical exponent has since been improved – more recent experiments (Li, *et al.* 2005; 2009) have revealed $\nu = 2.38 \pm 0.06$.

Theoretically, one knows that the critical exponent for classical 2D percolation is $\nu_{\text{class}} = 4/3$ and thus much smaller than the measured one. This discrepancy is due to the quantum nature of the percolation in quantum Hall systems. Indeed,

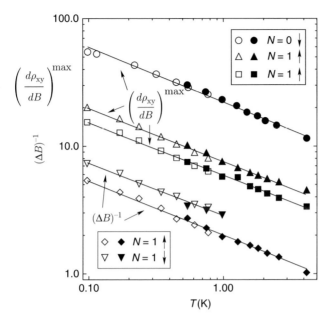

Fig. 6.18 Experiment by Wei, *et al.* 1988. The width of the transition ΔB and of the derivative of the Hall resisitivity $\partial \rho_{xy}/\partial B$, measured as a function of temperature, reveals a scaling law with an exponent $1/z\nu = 0.42 \pm 0.04$, for the transition between the filling factors $1 \to 2$ ($N = 0 \downarrow$), $2 \to 3$ ($N = 1 \uparrow$) and $3 \to 4$ ($N = 1 \downarrow$).

quantum-mechanical tunneling and the typical extension $\sim l_B$ of the wave functions associated with the equipotential lines enhance percolation, i.e. the electron puddles in the potential valleys may percolate *before* they touch each other in the classical sense. A model that takes into account this effect has been proposed by Chalker and Coddington (Chalker and Coddington, 1988), though with simplifying assumptions for the puddle geometry,[33] and one obtains a critical exponent $\nu = 2.5 \pm 0.5$ from numerical studies of this model (Chalker and Coddington, 1988; Huckestein, 1995), in quite a good agreement with the experimental data (Wei, *et al.* 1988; 1994).

In spite of the good agreement with experimental findings, these theoretical results need to be handled with care – indeed, analytical calculations have shown that the dynamical exponent should be exactly $z = 2$ for non-interacting electrons, whereas the measured value $z \simeq 1$ is obtained when interactions are taken into account on the level of the Hartree–Fock approximation (Huckestein and Backhaus, 1999). Furthermore, very recent numerical calculations within the Chalker–Coddington model have shown that the accurate value of the critical exponent is slightly larger ($\nu \simeq 2.59$) than the measured one when interactions are not taken into account (Slevin and Ohtsuki, 2009).

6.3.5 Relativistic quantum Hall effect in graphene

We finish this section on the IQHE with a short presentation of the relativistic quantum Hall effect (RQHE) in graphene, which is understandable in the same framework of

LL quantization and (semi-classical) one-particle localization as the IQHE in a non-relativistic 2D electron system. Indeed, the above arguments also apply to relativistic electrons in graphene, but we need to take into account the two different carrier types, electrons and holes, which carry a different charge. This is not so much a problem in the case of the impurity potential with its valleys and summits: in a particle – hole transformation, a valley becomes a summit and vice versa.[34] Furthermore, the direction of the Hall drift changes in this transformation. Because of the universality of the quantum Hall effect, both types of impurity distributions related by particle – hole symmetry yield the *same* quantization of the Hall resistance. The picture of semi-classical localization therefore applies also in the case of relativistic electrons in graphene.

The situation is different for the confinement potential. An ansatz of the form $V(y)\mathbb{1}$ – remember that the Hamiltonian of electrons in relativistic graphene is a 2×2 matrix that reflects the two different sublattices A and B – has the problem that an increase $V(y - y_{max/min}) \to \infty$ at the sample edge confines electrons but not the holes of the valence band for which we would need $V(y - y_{max/min}) \to -\infty$ for an efficient confinement. A possible confinement potential may be formed with the Pauli matrix σ^z,

$$V_{\text{conf}}(y) = V(y)\,\sigma^z = \begin{pmatrix} V(y) & 0 \\ 0 & -V(y) \end{pmatrix}, \qquad (6.91)$$

which, together with the Hamiltonian (6.24), yields the Hamiltonian that corresponds to the non-relativistic model (6.77). For a constant term $M = V(y)$ the contribution (6.91) plays the role of a mass of a relativistic particle (see also Appendix B). Therefore, the confinement (6.91) may also be called *mass confinement*. The corresponding energy spectrum, which one obtains within the same approximation as in Section 6.3.1 via the replacement $y \to y_0 = kl_B^2$ in the Landau gauge, reads (c.f. Eq. (B.8) in Appendix B)

$$\epsilon_{\lambda n, y_0} = \lambda \sqrt{M^2(y_0) + 2\frac{\hbar^2 v^2}{l_B^2}n}, \qquad (6.92)$$

and is schematically represented in Figure 6.19(a). Notice that Eq. (6.92) is only valid for $n \neq 0$ – indeed, the $n = 0$ acquires a non-zero energy $M(y_0)$, which is negative for our particular choice (see Appendix B). This feature is sometimes called *parity anomaly* in high-energy physics. Remember that in the case of graphene, one has two inequivalent low-energy points in the first BZ that give rise to a relativistic energy spectrum. The Dirac Hamiltonians (6.20) and (6.24) for the zero-B and magnetic-field case, respectively, applies principally only to one of the two valleys (say K), whereas that for the other valley is given by $-H_D$ (or $-H_D^B$) if one interchanges the A and B components (c.f. Eq. (A.16) in Appendix B). The confinement term (6.91) therefore reads $-V_{\text{conf}}(y)$ in the other valley, i.e. with a negative mass. The $n = 0$ LL thus shifts to positive energies in the second valley, and the two-fold valley degeneracy

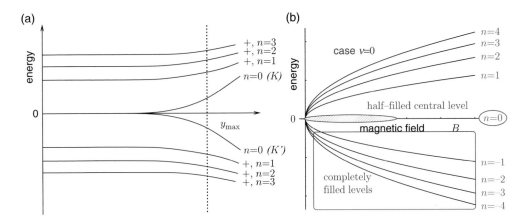

Fig. 6.19 (a) Mass confinement for relativistic Landau levels. Whereas the electron-like LLs ($\lambda = +$) are bent upwards when approaching the sample edge (y_{max}), the hole-like LLs ($\lambda = -$) are bent downwards. The fate of the $n = 0$ LL depends on the valley (parity anomaly) – in one valley (K), the level energy decreases, whereas it increases in the other valley (K'). (b) Filling of the bulk Landau levels at $\nu = 0$. All electron-like LLs ($\lambda = +$) are unoccupied, whereas all hole-like LLs ($\lambda = -$) are completely filled. The $n = 0$ LL is altogether half-filled.

is lifted in this level. A more detailed discussion of the mass confinement (6.91) in graphene may be found in the Appendix B.

This type of confinement may seem to be somewhat artificial, whereas the confinement in the non-relativistic case is easier to accept. Notice, however, that the whole model of massless Dirac fermions (second Hamiltonian in Eq. (6.20)) only describes the physical properties at length scales that are large compared to the lattice spacing (in graphene). In the true lattice model, the electrons are naturally confined because one does not allow for hopping from a lattice site at the edge into free space. The expression (6.91) is therefore only an *effective* model to describe confinement. We notice that, although the effective model yields a qualitatively correct picture, the fine structure of the dispersion at the edge depends on the edge geometry (Brey and Fertig, 2006). For further reading, we refer the interested reader to the literature (Castro Neto, *et al.* 2009).

With the help of these preliminary considerations, we are now prepared to understand the RQHE in graphene – the semi-classical localization is the same as in the non-relativistic case, and the confinement, which needed to be adopted to account for the simultaneous presence of electron- and hole-like LLs, yields the edge states that are responsible for the quantum transport and, thus, the resistance quantization. The RQHE was indeed discovered in 2005 by two different groups (Novoselov, *et al.* 2005; Zhang, *et al.* 2005), and the results are shown in Figure 6.20 (Zhang, *et al.* 2005). The phenomenology of the RQHE is the same as that of the IQHE in non-relativistic LLs: one observes plateaus in the Hall resistance while the longitudinal resistance vanishes. Notice that one may vary the filling factor either by changing the

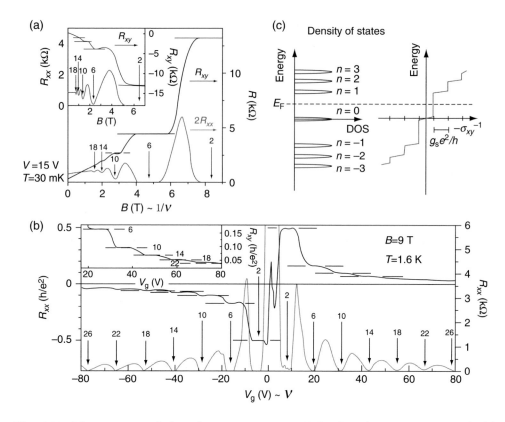

Fig. 6.20 Measurement of the relativistic quantum Hall effect (Zhang, *et al.* 2005). (a) RQHE at fixed carrier density ($V_G = 15$ V) at $T = 30$ mK. The filling factor is varied by sweeping the magnetic field. (b) Sketch of the DOS with the Fermi energy between the LLs $n = 0$ and $+, n = 1$. (c) RQHE at fixed magnetic field ($B = 9$ T) at higher temperatures, $T = 1.6$ K. The filling factor is now varied by changing the gate voltage.

B-field at fixed carrier density [Figure 6.20(a)] or one keeps the *B*-field fixed while changing the carrier density with the help of a gate voltage [Figure 6.20(c)]. The latter measurement is much easier to perform in graphene than in non-relativistic 2D electron gases in semiconductor heterostructures.

In spite of the similarity with the non-relativistic IQHE, one notices, in Figure 6.20, an essential difference: the quantum Hall effect is observed at the filling factors

$$\nu = \pm 2(2n + 1), \tag{6.93}$$

in terms of the LL quantum number n, whereas the IQHE is observed at $\nu = n$ (or $\nu = 2n$ if the LLs are spin-degenerate). The step in units of 4 is easy to understand: each relativistic LL in graphene is four-fold degenerate (in addition to the guiding-center degeneracy), due to the two-fold spin and the additional two-fold valley degeneracy. However, there is an "offset" of 2. This is due to the fact that the filling factor $\nu = 0$

corresponds to no carriers in the system, i.e. to a situation where the Fermi energy is exactly at the Dirac point (undoped graphene). In this case, one has a perfect electron – hole symmetry, and the $n = 0$ LL must therefore be *half-filled* (see Figure 6.19(b)), or else: there are as many electrons as holes in $n = 0$. According to the considerations presented in Section 6.3.4, this does not correspond to a situation where one observes a quantum Hall effect due to percolating extended states. Indeed, the system turns out to be *metallic* at $\nu = 0$ with a finite non-zero longitudinal resistance (Novoselov, *et al.* 2005; Zhang, *et al.* 2005). A situation, where one would expect a quantum Hall effect, arises when the central LL $n = 0$ is completely filled (or completely empty). As a consequence of the four-fold level degeneracy, one obtains the quantum Hall effect at $\nu = 2$ (or $\nu = -2$) observed in the experiments (see Figure 6.20). This is the origin of the particular filling-factor sequence (6.93) of the RQHE in graphene.

6.4 Strong correlations and the fractional quantum Hall effect

In the preceding section, we have seen that one may understand the essential features of the IQHE within a one-particle picture, i.e. in terms of Landau quantization; at integer filling factors $\nu = n$, which correspond to n completely filled LLs,[35] an additional electron is forced, as a result of the Pauli principle, to populate the next higher (unoccupied) LL (see Figure 6.21(a). It therefore needs to "pay" a finite amount of energy $\hbar\omega_C$ [or $\sqrt{2}(\hbar v/l_B)(\sqrt{n} - \sqrt{n-1})$ in the case of the RQHE in graphene] and is localized by the impurities in the sample, due to the classical Hall drift that forces the electron to move on closed equipotential lines. The system is said to be *incompressible* because one may not vary the filling factor and pay only an infinitesimal amount of energy – indeed in the case of a fixed particle number, consider an infinitesimal decrease of the magnetic field that amounts to an infinitesimal change of the surface $2\pi l_B^2$ occupied by each quantum state. Since the total surface of the system remains constant, the infinitesimal increase of $2\pi l_B^2$ may not be accomodated by an infinitesimal change in energy, due to the gap between the LL $n - 1$ and n where at least one electron must be promoted to. This gives rise to a zero compressibility.

(a)   (b)

Fig. 6.21   (a) Sketch of a completely occupied LL. An additional electron (gray circle) is forced to populate the same LL because of the Pauli principle. (b) Sketch of a partially filled LL. Because of the presence of unoccupied states in the LL (crosses), the Pauli principle does not prevent an additional electron (gray circle) from populating the same LL. The low-energy dynamical properties of the electrons are described by excitations within the same LL (no cost in kinetic energy), and inter-LL excitations are now part of the high-energy degrees of freedom.

In view of this picture of the quantum Hall effect, it was therefore a big surprise to observe a FQHE at a filling factor $\nu = 1/3$, with the corresponding Hall quantization $R_H = h/e^2\nu = 3h/e^2$ (Tsui, *et al.* 1982), and, later, at a large set of other fractional filling factors. Indeed, if only the kinetic energy is taken into account, the ground state at $\nu = 1/3$ is highly degenerate and there is no evident gap present in the system: the Pauli principle no longer prevents an additional electron from populating the next-higher LL, but it finds enough place in the lowest LL, which is only one-third filled.

Notice that we have neglected so far the mutual Coulomb repulsion between the electrons, which happens to be responsible for the occurence of the FQHE. The relevance of electronic interactions is discussed in the next section (Section 6.4.1). In Section 6.4.2, we present the basic results of Laughlin's theory of the FQHE, such as the ground-state wave functions, fractionally charged quasi-particles and the interpretation of Laughlin's wave function in terms of a 2D one-component plasma. The related issue of fractional statistics is introduced in a separate section (Section 6.4.3), and we close this section with a short discussion of different generalizations of Laughlin's wave function, such as CF theory or the Moore – Read wave function in half-filled LLs.

6.4.1   The role of Coulomb interactions

As already mentioned above, the situation of a partially filled LL is somewhat opposite to that of n completely occupied levels, where one observes the IQHE. This difference is summarized in Figure 6.21 and it is also the origin of the different role played by the Coulomb repulsion between the electrons. In the case of n completely filled LLs, one has a non-degenerate (Fermi-liquid-like) ground state, where the interactions may be treated within a perturbative approach. Indeed, any type of excitation involves a transition between two adjacent LLs that are separated by an energy gap of $\hbar\omega_C$ (see Figure 6.21(a)),[36] and we need to compare the Coulomb energy at the characteristic length scale $R_C = l_B\sqrt{2n+1}$ to this gap,

$$\frac{V_C}{\hbar\omega_C} \sim \frac{me^{3/2}}{\epsilon\hbar^{3/2}}(Bn)^{-1/2},$$

which turns out to be simply the usual dimensionless coupling constant

$$r_s = \frac{me^2}{\epsilon\hbar^2}n_{el}^{-1}$$

for the 2D Coulomb gas in Fermi liquid theory (Mahan, 1993; Giuliani and Vignale, 2005). The last expression is obtained by identifying the Fermi energy $E_F = \hbar^2 k_F^2/2m$, in terms of the Fermi wavevector k_F, with the energy of the last occupied LL $\hbar\omega_C n$. The perturbative approach allows one, e.g., to describe collective electronic excitations in the IQHE, such as magneto-plasmon modes (the 2D plasmon in the presence of a magnetic field) or magneto-excitons (inter-LL excitations that acquire a dispersion due to the Coulomb interaction) (Kallin and Halperin, 1984), or else the corresponding modes of the RQHE in graphene (Iyengar, *et al.* 2007; Roldán, *et al.* 2009).

In the case of a partially filled LL n the situation is inverted: for an electronic excitation, there are enough unoccupied states in the LL n for an electron of the same level to hop to. From the point of view of the kinetic energy, there is no energy cost associated with such an excitation (*low-energy* degrees of freedom) whereas an excitation to the next-higher (unoccupied) LL costs an energy $\hbar\omega_C$. Inter-LL excitations may then be neglected as belonging to *high-energy* degrees of freedom [Figure 6.21(b)]. Therefore, all possible distributions of N electrons within the same partially filled LL n have the same kinetic energy, which effectively drops out of the problem.

The macroscopic degeneracy may be lifted by phenomena due to other energy scales, such as those associated with the impurities in the sample or else the electron-electron interactions. The first hypothesis (impurities) may be immediately discarded as the driving mechanism of the FQHE because, in contrast to the IQHE, the FQHE only occurs in high-quality samples with low impurity concentrations. Indeed, the hierarchy of energy scales in the FQHE may be characterized by the succession

$$\hbar\omega_C \gtrsim V_C \gg V_{imp}, \tag{6.94}$$

and we therefore need to consider seriously the Coulomb repulsion, which governs the low-energy electronic properties in a partially filled LL.[37] Notice that we thus obtain a system of *strongly correlated* electrons for the description of which all perturbative approaches starting from the Fermi liquid are doomed to fail. The only hope one may have to describe the FQHE is then a well-educated guess for the ground state.

The most natural guess would be that the electrons in a partially filled LL behave as classical charged particles that form a crystalline state in order to minimize their mutual Coulomb repulsion. Such a state is also called a Wigner crystal (WC) because it was first proposed by Wigner (Wigner 1934). A WC has indeed been thought – before the discovery of the FQHE – to be the ground state of electrons in a partially filled LL (Fukuyama, *et al.* 1979). Even if the WC is the ground state at very low filling factors, as it has been shown experimentally (Andrei, *et al.* 1988), this state may not allow for an explanation of the FQHE. Indeed, the WC is a state that breaks a continuous spatial symmetry (translation invariance) and any such state has gapless long-wavelength excitations (*Goldstone modes*). The Goldstone mode of the WC (as of any other crystal) is the acoustic phonon the energy of which tends to zero at zero wavevector. One may thus compress the WC by changing the occupied surface in an infinitesimal manner or else by adding an electron without changing the macroscopic surface and pay only an infinitesimal amount of energy. The ground state is therefore compressible, i.e. it is not separated by an energy gap from its single-particle excitations, a situation that is at odds with the FQHE.

6.4.2   Laughlin's Theory

As a consequence of the above-mentioned considerations on the WC, one thus needs to search for a candidate for the ground state that does not break any continuous spatial symmetry and that has an energy gap. Such a state is the *incompressible quantum liquid* that was proposed by Laughlin (Laughlin 1983) the basic features of which we review in the present section. We consider, here, only the FQHE in the lowest LL (LLL), for simplicity. There are different prescriptions to generalize the associated wave functions to higher LLs, e.g. with the help of Eq. (6.33) (see MacDonald, 1984).

Experimentally, several FQHE states have been observed in the next-higher LL $n = 1$ although the majority of FQHE states is found in the LLL.[38]

6.4.2.1 Laughlin's guess from two-particle wave functions

In order to illustrate – one cannot speak of a derivation – Laughlin's wave function, we first need to remember the one-particle wave function of the LLL and then consider the corresponding two-particle wave function. We have already seen in Section 6.2.4.1 that a one-particle wave function in the LLL is described in terms of an analytic function times a Gaussian,[39]

$$\psi \sim z^{m'} e^{-|z|^2/4},$$

in terms of the integer $m' = 0, ..., N_B - 1$, where we have absorbed now (and in the remainder of this chapter) the magnetic length in the definition of the complex position, $z = (x - iy)/l_B$.

Consider, in a second step, an arbitrary two-particle wave function. This wave function must also be an analytic function of both postions z_1 and z_2 of the first and second particle, respectively, and may be a superposition of polynomials, such as, e.g., of the basis states

$$\psi^{(2)}(z, Z) \sim Z^M z^m e^{-(|z_1|^2 + |z_2|^2)/4}, \tag{6.95}$$

where we have defined the center of mass coordinate $Z = (z_1 + z_2)/2$ and the relative coordinate $z = (z_1 - z_2)$. The quantum number m plays the role of the *relative* angular momentum between the two particles, and M is associated with the *total* angular momentum of the pair. Because of the analyticity of the LLL wave functions, m must be an integer, and the exchange of the positions z_1 and z_2 constrains m to be *odd* because of the electrons' fermionic nature.

Laughlin's wave function (Laughlin, 1983) is a straightforward N-particle generalization of the two-particle wave function (6.95),

$$\psi_m^L\left(\{z_j, z_j^*\}\right) = \prod_{k<l} (z_k - z_l)^m e^{-\sum_j |z_j|^2/4}, \tag{6.96}$$

where we have omitted the normalization constants in order to simplify the notation and where all indices run from 1 to the total number of particles N. Notice that there is no dependence on the center of mass, but only on the relative coordinates between the particle pairs. Had there been such a dependence, described by a non-zero value of the total angular momentum quantum number $M \neq 0$, one would have broken a continuous spatial symmetry, in which case the state would describe a compressible rather than an incompressible state required for the FQHE, as we have mentioned above. We emphasize once again that Laughlin's wave function is not based on a mathematical derivation, although we will see below that there exist some mathematical models for which it describes the *exact ground state*, but it is more appropriately characterized as a *variational* wave function.

Variational parameter. The variational parameter in Laughlin's wave function (6.96) is simply the exponent m, with respect to which we would, in principle, need to optimize the wave function in order to approximate the true ground state of the system. Notice, however, that due to the LLL analyticity condition and fermionic statistics, the exponent is restricted to odd integers, $m = 2s + 1$, in terms of the integer s. Furthermore, this variational parameter turns out to be fully determined by the filling factor ν, as we will show with the following argument.[40]

Consider Laughlin's wave function as a function of the position z_k of some arbitrary but fixed electron k. There are $N - 1$ factors of the type $(z_k - z_l)^m$, one for each of the remaining $N - 1$ electrons, l, occuring in the ansatz (6.96), such that the highest power of z_k is $m(N - 1)$,

$$\prod_{k<l} (z_k - z_l)^m \sim z_k^{m(N-1)}.$$

Now, remember from Section 6.2.4.1 (see Eq. (6.69)) that the highest power of the complex particle position is fixed by the number of states N_B in each LL. This yields the relation

$$mN - \delta = N_B \tag{6.97}$$

between the number of particles N and the number of flux quanta N_B threading the system. Here, δ is some *shift* that is on the order of unity and that plays no role in the thermodynamic limit $N, N_B \to \infty$.[41] Because the ratio between the number of particles and that of flux quanta is simply the LL filling factor (6.59), $\nu = N/N_B$, one notices that, in the thermodynamic limit, the "variational parameter" is entirely fixed by the filling factor, i.e.

$$m = 2s + 1 = \frac{1}{\nu} \qquad \Leftrightarrow \qquad \nu = \frac{1}{m} = \frac{1}{2s + 1}, \tag{6.98}$$

and Laughlin's wave function is therefore a candidate wave function for the ground state at the filling factors

$$\nu = 1, 1/3, 1/5, ...$$

Remember that the odd value $m = 2s + 1$ is required by the fermionic nature of the electrons. Formally, one may though lift this restriction and generalize Laughlin's wave function to *bosonic* particles by choosing an even exponent $2s$. Such bosonic Laughlin wave functions have been studied theoretically in the context of rotating cold Bose gases in an optical trap (Cooper, 2008).

Laughlin's wave function at $\nu = 1$. It may seem, at first sight, astonishing that also the case of a completely filled LL for $\nu = 1$ is described in terms of a Laughlin wave function with $m = 1$ (or $s = 0$). Indeed, the state

$$\psi(\{z_j\}) = f_N(\{z_j\}) e^{-\sum_j |z_j|^2/4}$$

at $\nu = 1$ is non-degenerate and can thus be described in terms of a Slater determinant,

$$f_N(\{z_j\}) = \det \begin{pmatrix} z_1^0 & z_1^1 & \cdots & z_1^{N-1} \\ z_2^0 & z_2^1 & \cdots & z_2^{N-1} \\ \vdots & \vdots & & \vdots \\ z_N^0 & z_N^1 & \cdots & z_N^{N-1} \end{pmatrix}, \tag{6.99}$$

where we have omitted the ubiquitous Gaussian factor $\exp(-\sum_j |z_j|^2/4)$. Notice that the jth line in this determinant corresponds to all LLL states of the jth particle described in terms of the polynomials z_j^m. The determinant takes into account all permutations of the N particles over the N particle positions, $z_1, \ldots, z_N$, and may be rewritten in a compact manner with the help of the so-called Vandermonde determinant,

$$f_N(\{z_j\}) = \prod_{i<j} (z_i - z_j), \tag{6.100}$$

which is indeed simply the polynomial prefactor in Laughlin's wave function (6.96) with $m = 1$.

Until now we have obtained an N-particle wave function from some very general symmetry considerations (LLL analyticity condition, fermionic statistics, no broken continuous spatial symmetries), but we have not at all shown that it describes the ground state responsible of the FQHE. In the following, we will therefore discuss the basic physical properties of this, for the moment rather abstract, mathematical entity. In a first step, we will discuss some energy properties of the ground state and show that Laughlin's wave function is the exact ground state of a certain class of models that are qualitatively compared to the physical one (Coulomb interaction). We will then discuss the fractionally charged quasi-particle excitations of this wave function.

6.4.2.2 Haldane's pseudopotentials

In order to describe the energetic properties of Laughlin's wave function (6.96), we consider again the two-particle wave function (6.95). Notice that this wave function is an *exact* eigenstate for any central interaction potential that depends only on the relative coordinate z between particle pairs, such as is the case for the Coulomb interaction, $V = V(|z|)$. One may therefore decompose such an interaction potential in the relative angular momentum quantum numbers m,

$$v_m \equiv \frac{\langle m, M|V|m, M\rangle}{\langle m, M|m, M\rangle}, \tag{6.101}$$

where the denominator takes into account the fact that we have not properly normalized the two-particle wave functions (6.95), $\psi^{(2)}(z, Z) = \langle z, Z|m, M\rangle$.[42] The fact that there is no dependence on M is a direct consequence of the assumption that we deal with a central interaction potential, i.e. $\langle z, Z|V|z', Z'\rangle = V(|z|)\delta_{z,z'}\delta_{Z,Z'}$. Furthermore, there are no off-diagonal terms of the form $\langle m, M|V|m', M\rangle$, with $m' \neq m$, as one may show explicitly in the polar representation $z = \rho\exp(i\phi)$,

$$\langle m, M | V | m', M \rangle \propto \int_0^{2\pi} d\phi \int_0^{\infty} d\rho\, \rho^{m+m'+1} V(\rho) e^{-i(m-m')\phi} \propto \delta_{m,m'},$$

due to the integration over the polar angle. The potentials v_m obtained from the decomposition into relative angular momentum states are also called *Haldane's pseudopotentials* (Haldane, 1983). They fully characterize the two-particle energy spectrum because the kinetic energy is the same for all two-particle states $|m, M\rangle$, as described above. Notice that this is a very special case: normally any repulsive interaction potential yields unbound states with a continuous energy spectrum, such as the plane-wave states in scattering theory. Here, the energy spectrum is discrete even if the interaction is repulsive, due to the presence of a quantizing magnetic field. Notice further that Haldane's pseudopotentials are an image of the real-space form of the interaction potential. Indeed, if a pair of electrons is in a quantum state with relative angular momentum m, the average distance between the electrons is $|z| \sim l_B \sqrt{2m}$.[43] Haldane's pseudopotential v_m is therefore roughly the value of the original interaction potential at the relative distance $l_B \sqrt{2m}$,

$$v_m \simeq V\left(|z| = l_B\sqrt{2m}\right), \tag{6.102}$$

and the small-m components of Haldane's pseudopotentials correspond to the short-range components of the underlying interaction potential. Figure 6.22 shows the pseudopotential expansion for the Coulomb interaction in the lowest ($n = 0$) and the first excited ($n = 1$) LL.

Haldane's pseudopotentials are extremely useful in the description of the N-particle state as well. Indeed, the N-particle interaction Hamiltonian V may be rewritten in terms of pseudopotentials as

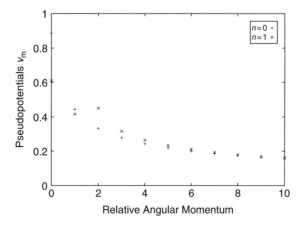

Fig. 6.22 Haldane's pseudopotentials for the Coulomb interaction in the LLs $n = 0$ and $n = 1$. Notice that we have plotted the pseudopotentials for both odd and even values of the relative angular momentum m even though only odd values matter in the case of fermions.

$$V = \sum_{i<j} V(|z_i - z_j|) = \sum_{i<j} \sum_{m'=0}^{\infty} v_{m'} \mathcal{P}_{m'}(ij), \tag{6.103}$$

where the operator $\mathcal{P}_{m'}(ij)$ projects the electron pair ij onto the relative angular momentum state m'. Notice that due to the factor $\prod_{k<l}(z_k - z_l)^m$ in Laughlin's wave function (6.96), no particle pair is in a relative angular momentum state $m' < m$. If one then chooses, though somewhat artificially, all pseudopotentials with a $m < m'$ to be positive (say 1) and all others zero,

$$v'_m = \begin{cases} 1 \text{ for } m' < m \\ 0 \text{ for } m' \geq m \end{cases} \tag{6.104}$$

one obtains $V\psi_m^L = 0$, i.e. Laughlin's wave function is the zero-energy eigenstate of the model (6.104). Since the model describes an entirely repulsive interaction, all possible states must have an energy $E \geq 0$. Therefore, Laughlin's wave function is even the *exact* ground state of the model (6.104). Furthermore, it is the only zero-energy state because if one keeps the total number of particles and flux fixed, any other state different from that described by Laughlin's wave function involves a particle pair in a state with an angular momentum quantum number different from m. If it is smaller than m, this particle pair is affected by the associated non-zero pseudo-potential m' and thus costs an energy on the order of $v_{m'} > 0$. If the particle pair is in a momentum state with $m' > m$, there is at least another pair with $m'' < m$ in order to keep the filling factor fixed, and this pair raises the energy. These general arguments show that any excited state involves a finite (positive) energy given by a pseudo-potential $v_{m'}$, with $m' < m$, which plays the role of an *energy gap*. In this sense, the liquid state described by Laughlin's wave function is indeed an *incompressible* state that already hints at the possibility of a quantum Hall effect if we can identify the correct quasi-particle of this N-particle state that becomes localized by the sample impurities.

Notice that the above considerations are based on an extremely artificial model interaction (6.104) that has, at first sight, very little to do with the physical Coulomb repulsion. However, the model is often used to generate numerically (in exact-diagonalization calculations) the Laughlin state, which may then be compared to the Coulomb potential decomposed in Haldane's pseudopotentials. This procedure has shown that the Laughlin state generated in this manner has an overlap of more than 99% with the state obtained from the Coulomb potential (Haldane and Rezayi 1985; Fano, *et al.* 1986), which is amazingly high for a wave function obtained from a well-educated guess. This high accuracy of Laughlin's wave function may be understood in the following manner: when one decomposes the Coulomb interaction potential in Haldane's pseudopotentials, one obtains a monotonically decreasing function when plotted as a function of m (see Figure 6.22). Furthermore, the component v_1 is much larger than v_3 and all other pseudopotentials v_m with higher values of m.[44] These higher terms may be treated in a perturbative manner and do not change the ground state, which is protected by the above-mentioned gap on the order of $v_1 > v_m$, with $m > 1$.

Furthermore, we mention that, apart from its successful verification by exact-diagonalization calculations (Haldane and Rezayi 1985; Fano, *et al.* 1986), Laughlin, in his original paper (Laughlin, 1983), showed within a variational calculation that the quantum liquid described by his wave function (6.96) has indeed a lower energy than the previously proposed WC. Again, the reason for this unexpected feature is the capacity of Laughlin's wave function, the modulus square of which varies as r^{2m} when two particles i and j approach each other with $r = |z_i - z_j|$, to screen the short-range components of the interaction potential. Notice that for a WC of fermions, the modulus square of the corresponding N-particle wave function decreases as r^2, as dictated by the Pauli principle.

6.4.2.3 *Quasi-particles and quasi-holes with fractional charge*

Until now, we have discussed some ground-state properties of Laughlin's wave function. We have seen that the Laughlin state at $\nu = 1/m$ is insensitive to the short-range components of the interaction potential described by Haldane's pseudopotentials $v_{m'}$ with $m' < m$, whereas excited states must be separated from the ground state by a gap characterized by these short-range pseudopotentials. However, we have not characterized so far the nature of the excitations.

There are two different sorts of excitations: (i) elementary excitations (quasi-particles or quasi-holes) that one obtains by adding or removing charge from the system, and (ii) collective excitations at fixed charge. The latter are simply a charge-density-wave excitation that consist of a superposition of particle – hole excitations at a fixed wave vector $\mathbf{q}$ (the momentum of the pair) and that may be shown to be gapped at all values of $\mathbf{q}$. Its dispersion reveals a minimum (called *magneto-roton minimum*) at a non-zero value of the wave vector that indicates a certain tendency to form a ground state with modulated charge density, such as a WC. The characteristic dispersion relation of these collective excitations is shown in Figure 6.23(a). However, we do not discuss collective excitations here and refer the interested reader to the literature for a more detailed discussion (Girvin, *et al.* 1986; Prange and Girvin 1990; Girvin 1999), but concentrate here on a presentation of the elementary excitations.

Quasi-holes. Elementary excitations are obtained when sweeping the filling factor slightly away from $\nu = 1/m$. Remember that there are two possibilities for varying the filling factor: adding charge to the system by changing the electronic density or adding (or removing) flux by varying the magnetic field. Remember further (see Eq. (6.97)) that the number of flux is intimitely related to the number of zeros in Laughlin's wave function. We therefore consider the ansatz

$$\psi_{qh}\left(z_0, \{z_j, z_j^*\}\right) = \prod_{j=1}^{N}(z_j - z_0)\,\psi_m^L\left(\{z_j, z_j^*\}\right) \tag{6.105}$$

for an excited state. Each electron at the positions z_j thus "sees" an additional zero at z_0. In order to verify that this wave function indeed adds another flux quantum to the system, we may expand Laughlin's wave function (6.96) formally in a polynomial,

$$\psi_m^L(\{z_j, z_j^*\}) = \sum_{\{m_i\}} \alpha_{m_1,\dots,m_N}\, z_1^{m_1} \dots z_N^{m_N} e^{-\sum_j |z_j|^2/4},$$

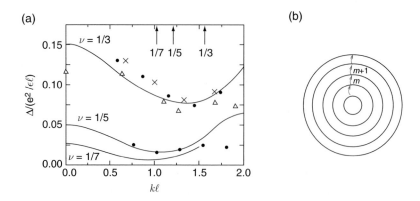

Fig. 6.23 (a) Dispersion relation for collective charge-density-wave excitations (Girvin, et al, 1986; Girvin, 1999). The continuous lines have been obtained in the so-called single-mode approximation (Girvin, et al, 1986) for the Laughlin states at $\nu = 1/3$, $1/5$ and $1/7$, whereas the points are exact-diagonalization results (Haldane and Rezayi, 1985; Fano, et al, 1986). The arrows indicate the characteristic wavevector of the WC state at the corresponding densities. (b) Quasi-hole excitation. Each electron jumps from the state m to the next-higher angular momentum state $m + 1$.

where the $\alpha_{m_1,...,m_N}$ describe the expansion coefficients. We now choose the position z_0 at the center of the disc, in which case the wave function of the excited state (6.105) simply reads

$$\psi_{qh}(\{z_j, z_j^*\}) = \sum_{\{m_i\}} \alpha_{m_1,...,m_N} z_1^{m_1+1} ... z_N^{m_N+1} e^{-\sum_j |z_j|^2/4},$$

i.e. each exponent is increased by one, $m_i \rightarrow m_i + 1$. This may be illustrated in the following manner: each electron jumps from the angular momentum state m to a state in which the angular momentum is increased by one (see Figure 6.23), leaving behind an empty state at $m = 0$. The excitation is therefore called a *quasi-hole* as we have already suggested by the subscript in Eq. (6.105). This also affects the quantum state with highest angular momentum M, i.e. we have increased the sample size by the surface occupied by one flux quantum, while keeping the number of electrons fixed.[45] Furthermore, this quasi-hole is associated with a *vorticity* if one considers the phase of the additional factor in Eq. (6.105),

$$\psi_{qh}(z_0 = 0, \{z_j, z_j^*\}) \propto \prod_j^N e^{-i\theta_j} \times \psi_m^L(\{z_j, z_j^*\}),$$

i.e. each particle that circles around the origin $z_0 = 0$ experiences an additional phase shift of 2π as compared to the original situation described in terms of Laughlin's wave function (6.96). This is reminiscent of the vortex excitation in a type-II superconductor (Tinkham, 2004).

We have seen above that one can create a quasi-hole excitation at the postion z_0 by introducing one additional flux quantum, $N_B \to N_B + 1$, which lowers the filling factor by a tiny amount. However, we have not yet determined the charge associated with this elementary excitation. This charge may be calculated by considering the filling factor fixed, i.e. we need to add some (negative) charge to compensate the extra flux quantum in the system. From Eq. (6.97), we notice that the relation between the extra flux ΔN_B and the compensating extra charge ΔN is simply given by

$$ m\Delta N = \Delta N_B \qquad \Leftrightarrow \qquad \Delta N = \frac{\Delta N_B}{m} . \tag{6.106} $$

This very important result is somewhat unexpected: in order to compensate one additional flux quantum ($\Delta N_B = 1$), one would need to add the *m*th *fraction of an electron*. The charge deficit caused by the quasi-hole excitation is therefore

$$ e^* = \frac{e}{m} , \tag{6.107} $$

i.e. the quasi-hole carries *fractional charge*.

Quasi-particles. In the preceding paragraph, we have considered a quasi-hole excitation that is obtained by introducing an additional flux quantum in the system [or, mathematically, an additional zero in the Laughlin wave function, see Eq. (6.105)]. Naturally, one may also *lower* the number of flux quanta by one, in which case one obtains a *quasi-particle* excitation with opposite vorticity as compared to that of the quasi-hole excitation. This opposite vorticity suggests that we use a prefactor $\prod_{j=1}^{N}(z_j^* - z_0^*)$, instead of $\prod_{j=1}^{N}(z_j - z_0)$ in the expression (6.105), in order to create a quasi-particle excitation at the position z_0. Remember, however, that the resulting wave function would have unwanted components in higher LLs because the analyt- icity condition of the LLL is no longer satisfied. In order to heal the quasi-particle expression, one formally projects it into the LLL,

$$ \psi_{qp}\left(z_0, \{z_j, z_j^*\}\right) = \mathcal{P}_{LLL} \prod_{j=1}^{N}(z_j^* - z_0^*)\, \psi_m^L\left(\{z_j, z_j^*\}\right) . \tag{6.108} $$

There are several ways of taking into account this projection $\mathcal{P}_{LLL}$. A common one consists of replacing each occurence of the non-analytic variables z_j^* (and powers of them) in the polynomial part of the wave function by a derivative with respect to z_j in the same polynomial (Girvin and Jach, 1984). By partial integration, this amount to deriving the Gaussian factor by $(\partial_{z_j})^m$ which, up to a numerical prefactor, yields exactly the non-analytic polynomial factor z_j^{*m}. We will encounter this projection scheme again in the discussion of the CF generalization of Laughlin's wave function (Section 6.4.4.1).

6.4.2.4 *Experimental observation of fractionally charged quasi-particles*

That the fractional charge of Laughlin quasi-particles[46] is not only a mathematical concept but a physical reality has been proven in a spectacular manner in so-called *shot-noise* experiments on the $\nu = 1/3$ FQHE state (de Picciotto, *et al.* 1997; Samina-dayar, *et al.* 1997).[47] In these experiments, one constrains the quantum Hall system with the help of side gates (see Figure 6.24) that are used to deplete the region in their vicinity via the application of a gate voltage V_{sg}. As a consequence of this depletion the quantum Hall system has a bottleneck where the corresponding edge states are brought into spatial vicinity (6.24(a)). In the first case, an injected charge may be backscattered in a tunnelling event at the bottleneck over a region filled by the $\nu = 1/3$ liquid (weak-backscattering limit). If one increases the side-gate voltage, the incompressible liquid becomes eventually cut into two parts separated by a completely depleted barrier, and one obtains the strong-backscattering limit [Figure 6.24(b)].

In a shot-noise measurement, one does not only measure the average current $\bar{I}$ (over a certain time interval) but simultaneously the (square of the) current fluctuation $\overline{(\Delta I)^2}$ which is proportional to the carrier charge. If the elementary charged excitations are $e^* = e/3$ quasi-particles and not electrons, one may expect to measure this particular charge. The experiments (de Picciotto, *et al.* 1997; Saminadayar, *et al.* 1997) have indeed shown that the charge measured in the shot noise is $e^* = e/3$ if the tunneling process takes place at a bottleneck filled with the incompressible quantum liquid (Figure 6.24(a)), whereas it is the usual elementary charge e in the case of a tunnelling process over a depleted region (Figure 6.24(b)).

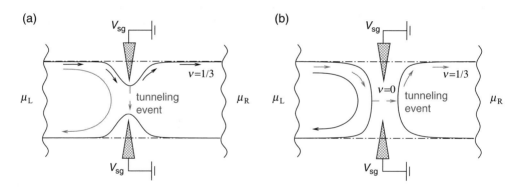

Fig. 6.24 Experimental setup for the observation of fractionally charged quasi-particles. In addition to the usual geometry, one adds, at the upper and the lower edges, side gates that are used to deplete the region around the gates by the application of a voltage V_{sg}. The filling factor is chosen to be $\nu = 1/3$. As a result, the edge states at the opposite edges are brought into close vicinity. (a) Weak-backscattering limit. The incompressible liquid has a *bottleneck* at the side gates, i.e. the edges are so close to each other that a tunnelling event between them has a finite probability. A particle injected at the left contact may thus be backscattered (gray arrow) in a region filled by the incompressible Laughlin liquid, although the majority of the particles reaches the right contact (black arrows). (b) Strong-backscattering limit. If one increases the side-gate voltage V_{sg}, the incompressible $\nu = 1/3$ liquid is eventually cut into two parts separated by a fully depleted region ($\nu = 0$). In this case, backscattering is the majority process (black arrow), and a tunnelling event may occur over the depleted region such that a particle injected at the left contact may still reach the right one (gray arrows).

6.4.2.5 Laughlin's plasma analogy

A compelling physical picture of Laughlin's wave function (6.96) and the properties of its elementary excitations (6.105) and (6.108) with fractional charge has been provided by Laughlin himself (Laughlin, 1983), in terms of an analogy with a *classical 2D one-component plasma*. In the present subsection, we present the basic ideas and results of this plasma analogy, for completeness and pedagogical reasons. However, no new results will come out of this analogy here, as compared to those derived above.

Remember from basic *quantum mechanics* that the modulus square of a quantum-mechanical wave function may be interpreted as a statistical probability distribution. For Laughlin's wave function (6.96), one obtains the probability distribution

$$\left|\psi_m^L\left(\{z_j\}\right)\right|^2 = \prod_{i<j}|z_i - z_j|^{2m}\, e^{-\sum_j |z_j|^2/2}.$$

Now, remember from *classical statistical mechanics* that a probability distribution in the canonical ensemble is the Boltzmann weight, $\exp(-\beta\mathcal{H})$, of some Hamiltonian $\mathcal{H}$ and that the classical partition function, which encodes all relevant statistical information, is obtained from a sum over the Boltzmann weights of all possible configurations $\mathcal{C}$, $\mathcal{Z} = \sum_{\mathcal{C}} \exp[-\beta\mathcal{H}(\mathcal{C})]$. Laughlin's plasma analogy consists precisely of the identification of the modulus square of his wave function with the Boltzmann weight of some *mock* Hamiltonian U_{cl}.[48] The mock Hamiltonian may be obtained exactly from this identification,

$$-\beta U_{cl} = \ln\left|\psi_m^L\left(\{z_j\}\right)\right|^2,\qquad(6.109)$$

and one obtains, by choosing somewhat artificially $\beta = 2/q$,[49]

$$U_{cl} = -q^2 \sum_{i<j} \ln|z_i - z_j| + q\sum_j \frac{|z_j|^2}{4}.\qquad(6.110)$$

This is simply the classical Hamiltonian of a 2D one-component plasma, in terms of the plasma-particle charge

$$q = m = 2s + 1.\qquad(6.111)$$

The first term of Eq. (6.110) reflects the interactions between the charged plasma particles, whereas the second term describes their interaction with a neutralizing background of positive charge, as in the case of the jellium model of the Coulomb gas (Mahan, 1993; Giuliani and Vignale, 2005). This may be seen best with the help of Poisson's equation, $-\Delta\phi = 2\pi qn_q(\mathbf{r})$, for an electrostatic 2D potential due to the charge density qn_q. The first term then indeed describes particles with charge q interacting via the 2D Coulomb interaction potential $\phi(\mathbf{r}) = -\ln(|\mathbf{r}|/l_B)$, and the second term is the interactions with the neutralizing background because $\Delta|\mathbf{r}|^2/4l_B^2 = 1/l_B^2 = 2\pi n_B$, where the flux density n_B may thus be viewed as the charge density of the positively charged background.

In order to minimize the energy of the mock Hamiltonian U_{cl}, which corresponds to a distribution of highest weight, the 2D plasma thus needs to be charge-neutral, i.e. the charge density of the plasma particles qn_{el} must be compensated by that of the background n_B,

$$n_B - qn_{el} = 0, \tag{6.112}$$

which, together with Eq. (6.111), yields simply the relation between the filling factor ν and the exponent in Laughlin's wave function (6.98), $\nu = n_{el}/n_B = 1/m$.

The plasma analogy does not only apply to the ground-state wave function (6.96) but also to the quasi-hole excitation (6.105). The additional factor $\prod_{j=1}^{N}(z_j - z_0)$ in the quasi-hole wave function (6.105) yields, within the plasma analogy (6.109), an additional term

$$V = -q \sum_{j=1}^{N} \ln|z_j - z_0| \tag{6.113}$$

to the mock Hamiltonian (6.110), $U_{cl} \to U_{cl} + V$. This additional term may be interpreted as the interaction of the plasma particles with an "impurity" of unit charge at the position z_0. In order to maintain charge neutrality, the impurity needs to be screened by the plasma particles. Since the charge of each plasma particle is $q = m = 2s + 1$ and thus greater than unity, one needs $1/q$ plasma particles to screen the impurity of charge one. Remember that each plasma particle represents one electron of unit charge in the original Laughlin liquid. One therefore obtains the same charge fractionalization of the Laughlin quasi-particle (6.107), $e^* = e/m$, as in the original quantum model.

6.4.3 Fractional statistics

6.4.3.1 Bosons, fermions and anyons – an introduction

One of the most exotic consequences of charge fractionalization in 2D quantum mechanics, exemplified by Laughlin quasi-particles, is *fractional statistics*. Remember that, in three space dimensions, the quantum-mechanical treatment of two and more particles yields a *superselection rule* according to which quantum particles are, from a statistical point of view, either *bosons* or *fermions*. This superselection rule is no longer valid in 2D (two space dimensions), and one may find intermediate statistics between bosons and fermions. The corresponding particles are called *anyons*, because the statistics may be *any*. The present section is meant to illustrate these amazing aspects of 2D quantum mechanics, and we try to avoid a too formal or mathematical treatment. We refer, again, the interested reader to the more detailed literature (Nayak, *et al.* 2008).

In order to illustrate the different statistical (i.e. exchange) properties of two quantum particles in three and two space dimensions, consider a particle A that moves adiabatically on a closed path C in the xy-plane around another one B of the same species (see Figure 6.25). We choose the path to be sufficiently far away from particle

(a)   (b)

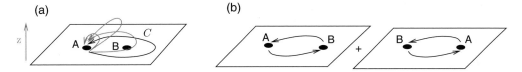

Fig. 6.25 (a) Process in which a particle A moves on a path $\mathcal{C}$ around another particle B. In three space dimensions, one may profit from the third direction (z-direction) to lift the path over particle B and thus to shrink the path into a single point. (b) Process equivalent to moving A on a closed path around B that consists, apart from a topologically irrelevant translation, of two successive exchanges of A and B.

B and the two particles to be sufficiently localized such that we can neglect corrections due to the overlap between the two corresponding wave functions. Notice first that such a process $\mathcal{T}$ is equivalent, apart from a topologically unimportant translation, to two successive exchange processes $\mathcal{E}$, in which one exchanges the positions of A and B. Algebraically, this may be expressed in terms of the corresponding operators as

$$\mathcal{E}^2 = \mathcal{T} \quad \text{or} \quad \mathcal{E} = \pm\sqrt{\mathcal{T}}, \tag{6.114}$$

modulo a translation.

Let us discuss first the three-dimensional case. Because of the presence of the third direction (z-direction), one may elevate the closed path in this direction while keeping the position of particle A fixed in the xy-plane. We call the elevated path $\mathcal{C}'$. Furthermore, one may now shrink the closed loop $\mathcal{C}'$ into a single point at the position A without crossing the position of particle B, which remains in the xy-plane. This final (point-like) path is called $\mathcal{C}''$. Although this procedure may seem somewhat formal, a quantum-mechanical exchange process does principally not specify the precise geometry of the exchange path in the distinction between a boson and a fermion, but only its *topological* properties. From a topological point of view, all paths that can be continuously deformed into each other define a *homotopy class* (Mermin, 1979). Equation (6.114) must therefore be viewed as an equation for homotopy classes in which a simple translation and an allowed deformation are irrelevant operations. As a consequence of these considerations, the simple point-like path $\mathcal{C}''$ at the position of particle A, which may be formally described by $\mathcal{C}'' = 1$, is in the same homotopy class as the original path $\mathcal{C}$. Therefore, the associated processes are the same, and one has

$$\mathcal{T} = \mathcal{T}(\mathcal{C}) = \mathcal{T}(\mathcal{C}'') = 1 \quad \text{and thus} \quad \mathcal{E} = \sqrt{1}, \tag{6.115}$$

where the last equation is symbolic in terms of the identity operator. It indicates that the quantum-mechanical operator $\mathcal{E}$, corresponding to particle exchange, has two eigenvalues that are the two square roots of unity, $e_B = \exp(2i\pi) = 1$ and $e_F = \exp(i\pi) = -1$. This is precisely the above-mentioned superselection rule, according to which all quantum particles in three space dimensions are either bosons ($e_B = 1$) or fermions ($e_F = -1$).

In two space dimensions, this topological argument yields a completely different result. It is not possible to shrink a path $\mathcal{C}$ enclosing the second particle B into a single point at the position of A, without passing by B itself. This means that the position of B must be an element of the path at a certain moment of the shrinking process, which cannot profit from a third dimension in order to elevate the loop on which it moves above the xy-plane. The single point still represents a homotopy class of paths, but these paths do not enclose another particle, and $\mathcal{C}$ is therefore an element of another homotopy class, i.e. that of all paths starting from A and enclosing only the particle B. If there are more than two particles present, the homotopy classes are described by the integer number of particles enclosed by the paths in this class. From an algebraic point of view, the exchange processes are no longer described by the two roots of unity, 1 and -1, but by the so-called *braiding group*, and the classification into bosons and fermions is no longer valid. In the simplest case of Abelian statistics,[50] one needs to generalize the commutation relation

$$\psi(\mathbf{r}_1)\psi(\mathbf{r}_2) = \pm\psi(\mathbf{r}_2)\psi(\mathbf{r}_1), \qquad (6.116)$$

for bosons and fermions, respectively, to

$$\psi(\mathbf{r}_1)\psi(\mathbf{r}_2) = e^{i\alpha\pi}\psi(\mathbf{r}_2)\psi(\mathbf{r}_1), \qquad (6.117)$$

where α is also called the *statistical angle*. One has $\alpha = 0$ for bosons and $\alpha = 1$ for fermions, and all other values of α in the interval between 0 and 2 for *anyons*. Sometimes anyonic statistics is also called *fractional statistics* – indeed all physical quasi-particles, such as those relevant for the FQHE, have an angle that is a fractional (or rational) number, but there is no fundamental objection that irrational values of the statistical angle should be excluded.

Before discussing the anyonic nature of Laughlin quasi-particles, we need to mention an important issue in these statistical considerations. We know that fermions are forced to satisfy Pauli's principle, which excludes double occupancy of a single quantum state, whereas the number of bosons per quantum state is unrestricted. What about anyons then? In the context of quantum fields the Pauli principle yields, via Eq. (6.116) for $\mathbf{r} = \mathbf{r}_1 = \mathbf{r}_2$,

$$\psi(\mathbf{r})\psi(\mathbf{r}) = 0.$$

For an arbitrary statistical angle, one obtains in the same manner, from Eq. (6.117),

$$\left(1 - e^{i\alpha\pi}\right)\psi(\mathbf{r})\psi(\mathbf{r}) = 0, \qquad (6.118)$$

which may be viewed as a *generalized Pauli principle for 2D anyons* (Haldane, 1991). Only if $\alpha = 0$ modulo 2, we can satisfy Eq. (6.118) with $\psi(\mathbf{r})\psi(\mathbf{r}) \neq 0$. Otherwise, when $\alpha \neq 0$ modulo 2, we necessarily have $\psi(\mathbf{r})\psi(\mathbf{r}) = 0$. Anyons are, thus, from an exclusion-principle point of view more similar to fermions than to bosons.

6.4.3.2 *Statistical properties of Laughlin quasi-particles*

We may now apply the above general statistical considerations to the case of Laughlin quasi-particles. The basic idea is to describe the statistical angle as an Aharonov-Bohm phase due to some gauge field that is generated by the flux bound to the charges included in a closed loop $\partial\Sigma$. This closed loop, around which a quasi-particle moves adiabatically, encloses a surface Σ. The gauge field is not to be confunded with the one the generates the true magnetic field B – it is rather a *mock* (or fake) field $\mathbf{A}_M$ (with $\mathbf{B}_M = \nabla \times \mathbf{A}_M$) that generates the flux bound, e.g., by the electrons in the Laughlin liquid via the relation (6.97). We consider the case where the area Σ is filled with $N_{el}(\Sigma)$ electrons condensed in an incompressible quantum liquid described by Laughlin's wave function (6.96) and $N_{qh}(\Sigma)$ quasi-hole excitations (6.105), such that there are two contributions to $B_M = |\mathbf{B}_M|$,

$$B_M \Sigma = N_{\text{flux}} \frac{h}{e} = [m N_{el}(\Sigma) + N_{qh}(\Sigma)] \frac{h}{e}. \tag{6.119}$$

The corresponding Aharonov–Bohm phase, which the quasi-particle picks up when turning around the area Σ on the boundary path $\partial\Sigma$, is given by

$$\Gamma_{A-B} = 2\pi \frac{e^*}{h} \oint_{\partial\Sigma} d\mathbf{r} \cdot \mathbf{A}_M(\mathbf{r}) = 2\pi \frac{e^*}{h} \int_{\Sigma} d^2r \, B_M(\mathbf{r}),$$

where $e^* = e/m$ is the charge of the quasi-particle and where we have used Stoke's theorem to convert the line integral of $\mathbf{A}_M$ on $\partial\Sigma$ into a surface integral of B_M over the area Σ. The Aharonov–Bohm phase has therefore two contributions, one Γ_{el} that stems from the electrons condensed in the Laughlin liquid and the other one Γ_{qh} that is due to the enclosed quasi-holes. One obtains from Eq. (6.119)

$$\Gamma_{el} = 2\pi \frac{e^*}{e} m N_{el} = 2\pi N_{el}, \tag{6.120}$$

for the enclosed electrons, i.e. an integer times 2π. Notice that this contribution to the Aharonov–Bohm phase may not be interpreted in terms of a statistical angle because it does not describe a true exchange process: the involved particles are not of the same type – we have chosen a quasi-particle to move on a path enclosing condensed electrons. However, had we chosen an *electron* rather than a *quasi-hole* to move along the path $\partial\Sigma$, the Aharonov–Bohm phase,

$$\Gamma_{el-el} = 2\pi \frac{e}{e} m N_{el}(\Sigma),$$

would give rise to a statistical angle $\alpha = m N_{el}(\Sigma)$.[51] If we have only one electron enclosed by the path, $N_{el}(\Sigma) = 1$, the statistical angle is simply the odd integer m, which is equal to 1 (modulo 2), as it should be for fermions.

A more interesting situation arises when the path encloses Laughlin quasi-holes, the contribution of which to the Aharonov–Bohm phase reads

$$\Gamma_{qh} = 2\pi \frac{e^*}{e} N_{qh} = 2\pi \frac{N_{qh}}{m}. \tag{6.121}$$

Consider a single quasi-hole in the area Σ, $N_{qh} = 1$: one encounters the rather unusual situation in which the Aharonov-Bohm phase is a *fraction* of 2π, and the associated statistical angle is $\alpha = 1/m$. This illustrates that Laughlin quasi-holes are indeed anyons with fractional statistics, as we have argued above.

6.4.4   Generalizations of Laughlin's wave function

Although Laughlin's wave function (6.96) has been extremely successful in the description of the FQHE at $\nu = 1/3$ and $1/5$, it is not capable of describing all observed FQHE states. Indeed, there are e.g. FQHE states at $\nu = 2/5, 3/7, 4/9, ...$ corresponding to the series $p/(2p+1)$, or more generally to $p/(2sp+1)$, in terms of the integers s and p, which may be accounted for within composite-fermion (CF) theory, which we present below. Furthermore, even-denominator FQHE states have been observed at $\nu = 5/2$ and $7/2$ (Willett, *et al.* 1987), in the first excited LL ($n = 1$), and, in wide quantum wells or bilayer quantum Hall systems, at $\nu = 1/2$ and $\nu = 1/4$ (Luhman, *et al.* 2008; Shabani, *et al.* 2009). Whereas the latter may be understood within a multi-component picture, which we will briefly introduce in Section 6.5, the states at $\nu = 5/2$ and $7/2$ may find their explanation in terms of a so-called *Pfaffian* wave function. Both the CF and the Pfaffian wave functions are sophisticated generalizations of Laughlin's original idea.

6.4.4.1   *Composite fermions*

Soon after the discovery of the most prominent FQHE state at $\nu = 1/3$, a lot of other states have been observed at the filling factors $\nu = p/(2sp+1)$. In a first theoretical approach, these states were interpreted in the framework of a hierarchy scheme (Haldane, 1983; Halperin, 1984) according to which the quasi-particles of the Laughlin (parent) state, such as $\nu = 1/3$, condense themselves into a Laughlin-type (daughter) state, due to their residual Coulomb repulsion – remember that the Laughlin quasi-particles carry an electric charge $e^* = e/m$. In this picture, the 2/5 state would be the daughter state formed of Laughlin quasi-particle excitations of the 1/3 state.

An alternative picture, though related to the above-mentioned hierarchy scheme, was proposed by Jain (Jain 1989, 1990). The basic idea consists of a reinterpretation of Laughlin's wave function (6.96): consider only the polynomial part, the Gaussian $\exp(-\sum_j^N |z_j|^2/4)$ being an ubiquitous factor that finally needs to be multiplied with the polynomial wave function,

$$\psi_m^L(\{z_j\}) = \prod_{k<l}(z_k - z_l)^{2s+1} = \prod_{k<l}(z_k - z_l)^{2s}\prod_{k<l}(z_k - z_l). \qquad (6.122)$$

In the last step of this equation, we have split the product into two parts, one with the exponent $2s$, which we call the *vortex* part, and another one with the exponent 1.

Before introducing Jain's generalization, let us interpret the above wave function in terms of the statistical properties introduced in the last section. Quite generally, one may express any LLL N-particle wave function ψ_{LLL} as a product of such a vortex factor and another (residual) wave function ψ_{res},

$$\psi_{LLL}(\{z_j\}) = \prod_{k<l}(z_k - z_l)^{m'}\psi_{res}(\{z_j\}).$$

If the original wave function is fermionic, i.e. antisymmetric with respect to an exchange process of an arbitrary particle pair, the symmetry properties of ψ_{res} depend on the parity of m'. If it is odd, ψ_{res} must be a symmetric (bosonic) wave function, and if it, is even, both the original and the residual wave functions are antisymmetric (fermionic). In terms of the above-mentioned gauge field $\mathbf{A}_M(\mathbf{r})$, the statistical angle associated with the vortex factor is just given by the parity of m', which may be viewed as the number of *flux quanta attached to each particle* at the positions z_j. Flux attachment may thus be used, in 2D quantum mechanics, to transform fermions into bosons and vice versa.

For the above decomposition (6.122) of Laughlin's wave function, the vortex part attaches s pairs of flux quanta to each particle position and therefore does not affect the statistical properties of the wave function. The second factor

$$\chi_{\nu^*=1}(\{z_j\}) = \prod_{k<l}(z_k - z_l)$$

is indeed fermionic and corresponds, as we have mentioned in Section 6.4.2.1, to a completely filled LL at a virtual (CF) filling factor of $\nu^* = 1$, the true filling factor being still $\nu = 1/(2s + 1)$. This is schematically represented in Figure 6.26.

Jain's generalization consists of replacing the term $\prod_{k<l}(z_k - z_l)$ by any other Slater determinant $\chi_{\nu^*=p}(\{z_j, z_j^*\})$ of p completely filled LLs, with a CF filling factor $\nu^* = p$,

$$\psi^J(\{z_j, z_j^*\}) = \mathcal{P}_{LLL} \prod_{k<l}(z_k - z_l)^{2s} \chi_{\nu^*=p}(\{z_j, z_j^*\}), \tag{6.123}$$

where we need to take into account the same projection $\mathcal{P}_{LLL}$ to the LLL as in the case of quasi-particle excitations (6.108) because, contrary to the $\nu^* = 1$ case,

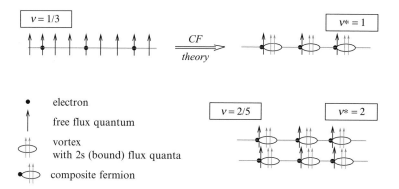

Fig. 6.26 Schematic view of composite fermions. The electronic state at $\nu = 1/3$ may be interpreted as a CF state at an integer CF filling factor $\nu^* = 1$, where each vortex bound to an electron carries $2s$ (here $s = 1$) flux quanta. In the same manner a CF filling factor $\nu^* = 2$ gives rise to an (electronic) FQHE state at $\nu = 2/5$.

the wave function $\chi_{\nu^*=p}(\{z_j, z_j^*\})$ has by construction non-analytic components, i.e. components in higher (CF) LLs.

Jain's wave function (6.123) may be illustrated in the following manner. Via the first factor $\prod_{k<l}(z_k - z_l)^{2s}$, we have effectively bound $2s$ flux quanta to each of the electrons, as we have already mentioned above. This novel type of particle is what we call the *composite fermion* (CF). The residual (free) flux quanta now determine the effective number of states per (CF) LL,

$$N_B \to N_B^* = N_B - 2sN_{el},$$

which correspond to a renormalized magnetic field

$$B \to B^* = B - 2s\left(\frac{h}{e}\right)n_{el}. \tag{6.124}$$

Similarly, the CF filling factor is defined with respect to the renormalized number of flux quanta,

$$\nu^* = \frac{N_{el}}{N_B^*} \qquad \Rightarrow \qquad \nu^{*-1} = \nu^{-1} - 2s, \tag{6.125}$$

which leads to the relation

$$\nu = \frac{\nu^*}{2s\nu^* + 1} \tag{6.126}$$

between the CF filling factor and the usual one ν (Eq. (6.59)). For completely filled LLs, $\nu^* = p$, this yields the above-mentioned series

$$\nu = \frac{p}{2sp + 1} \tag{6.127}$$

for the FQHE states, which may thus be interpreted as IQHE states of CFs. To be explicit, the physical picture of CF theory is the following: the ground state is described by the wave function (6.123), which describes an incompressible quantum liquid in the same manner as Laughlin's wave function does. The elementary excitation in the CF theory consists of a CF promoted to the next higher CF LL, which is separated from the ground state by an energy gap, in analogy with the electron added to n completely filled (electronic) LLs in the IQHE.[52] Again, these elementary CF excitations become localized by the sample impurities, and one therefore obtains a plateau in the Hall resistance that is thus quantized.

Numerically, Jain's CF wave function (6.123) has been successful in the description of the series (6.127) of FQHE states: even if the overlap with the exact ground states decreases when the quantum number p increases, the overlap is still reasonably high (above 95%) for the number of particles accessible in state-of-the-art exact-diagonalization calculations. Notice, however, that the physical interpretation is more involved as compared to Laughlin's wave function because the LLL projection $\mathcal{P}_{LLL}$ is rather complicated to implement in analytical as well as numerical calculations. For a further review of CF theory, we refer the interested reader to the literature. The above-mentioned wave-function approaches are thoroughly reviewed in Jain's recent book (Jain, 2007). Furthermore, there have been field-theoretical approaches beyond the numerical wave-function description presented above, such as in terms of Chern–Simons theories (Lopez and Fradkin 1991; Halperin, *et al.* 1993) or in terms of a Hamiltonian theory (Murthy and Shankar, 2003). For a review of these complementary theories we refer the reader to the book edited by Heinonen (Heinonen, 1998) or the excellent pedagogical review by Murthy and Shankar (Murthy and Shankar, 2003).

6.4.4.2 Half-filled LLs and Pfaffian states

Within the CF picture, we have seen that the effective magnetic field becomes renormalized due to flux attachment (Eq. (6.124)). An interesting situation arises when the filling factor is $\nu = 1/2$, which corresponds to the limit $p \to \infty$ in Eq. (6.127). In this limit the effective magnetic field (6.124) vanishes, $B^* = 0$, and one may then expect the corresponding phase to be described in terms of a metallic state, such as a Fermi liquid that one would obtain for electrons when the magnetic field vanishes. A natural ansatz for the N-particle wave function of such a Fermi-liquid state is given by the Slater determinant

$$\psi_{FL} = \det\left(e^{i\mathbf{k}_i \cdot \mathbf{r}_j}\right),$$

where the N electrons occupy the states described by the wavevectors $\mathbf{k}$, $i = 1, ..., N$, the modulus of which is delimited by the Fermi wavevector $|\mathbf{k}_i| \leq k_F$, and $\mathbf{r}_j$ is the position of the jth particle. Notice that this state is nevertheless unappropriate in the description of a state in the LLL. Indeed, if the scalar product in the exponent is rewritten in terms of complex variables, $\mathbf{k}_i \cdot \mathbf{r}_j = (k_i z_j^* + k_i^* z_j)/2$, one realizes that the Fermi-liquid state violates the LLL condition of analyticity. Formally, one may again avoid this problem by projecting the Fermi-liquid state into the LLL, and one thus obtains a state,

$$\psi_{FL}^{\nu=1/2} = \mathcal{P}_{LLL} \prod_{k<l} (z_k - z_l)^2 \det\left(e^{i\mathbf{k}_i \cdot \mathbf{r}_j}\right), \tag{6.128}$$

that was proposed by Rezayi and Read for the description of a compressible metallic state at $\nu = 1/2$ (Rezayi and Read, 1994). The first term is the same factor as in CF theory, which attaches 2 flux quanta to each particle and that thus cancels the external magnetic field, $B^* = B - 2(h/e)n_{el} = 0$.

Because the wave function (6.128) describes a compressible state, one should not observe a quantized Hall resistance, in agreement with most experimental data. A FQHE at $\nu = 1/2$ (and 1/4) has only been observed in very wide quantum wells (Luhman, *et al.* 2008; Shabani, *et al.* 2009), which are likely to be described by two-component wave functions (Papić, *et al.* 2009) that we will briefly introduce in Section 6.5.

In contrast to the LLL, the half-filled LL $n = 1$ reveals, in both spin branches, a FQHE (5/2 and 7/2 states). The difference between the half-filled LL $n = 0$ and

$n = 1$ is due to a different *effective* interaction potential that takes into account the wave function overlap between two (interacting) particles, which we do not discuss in detail here. Indeed, the Fermi-liquid-like state (6.128) turns out to be quite unstable with respect to particle pairing. This is reminiscent of the BCS (Bardeen–Cooper–Schrieffer) instability of a conventional Fermi liquid that gives rise to superconductivity (Mahan, 1993; Tinkham, 2004), although the glue between the particles is no longer a phonon-mediated attractive interaction, but only the *repulsive* Coulomb interaction in a strong magnetic field. As we have already mentioned in Section 6.4.2.2, such an interaction may yield a discrete two-particle spectrum, in contrast to a repulsive interaction in the absence of a magnetic field. As a consequence, pairing may occur at certain relative angular momenta for particular pseudopotential sequences and for sufficiently high filling factors.[53] In the present case, one may exclude s-wave pairing, i.e. in the relative angular momentum state with $m = 0$, due to the Pauli principle, and the most natural candidate would therefore be p-wave pairing in the relative angular momentum state $m = 1$ (Greiter, *et al.* 1991).

A wave function that accounts for p-wave pairing was proposed by Moore and Read in 1991 (Moore and Read, 1991),

$$\psi_{MR}\left(\{z_j\}\right) = \mathrm{Pf}\left(\frac{1}{z_i - z_j}\right) \prod_{k<l}(z_k - z_l)^2, \tag{6.129}$$

where we have again omitted the ubiquitous Gaussian factor. As for the CF wave functions (6.123) and the Rezayi–Read wave function (6.128), the factor $\prod_{k<l}(z_k - z_l)^2$ attaches two flux quanta to each electron and therefore does not change the statistical properties of the wave function. If the wave function consisted only of this factor (times the Gaussian), one would have a *bosonic* Laughlin wave function that describes an incompressible quantum liquid at the desired filling factor $\nu = 1/2$. However, it does not have the correct statistical properties. This problem is healed by the first factor $\mathrm{Pf}[1/(z_i - z_j)]$ that represents the *Pfaffian* of the $N \times N$ matrix $\mathcal{M}_{ij} = 1/(z_i - z_j)$. The Pfaffian may be viewed as the square root of the more familiar determinant, $\mathrm{Pf}(\mathcal{M}) = \sqrt{\det(\mathcal{M})}$, and has the same anti-symmetric properties as the determinant in an exchange of two particles i and j, such that it generates a fermionic wave function. Notice, furthermore, that this Pfaffian seems, at first sight, to take away some of the zeros such that one could expect the filling factor to increase. However, the function $\prod_{k<l}(z_k - z_l)^2$ is a product of $N(N-1) \sim N^2$ terms, whereas the Pfaffian is a sum of products of $N/2 \sim N$ terms. Therefore, the number of zeros, and thus the filling factor, is unchanged in the thermodynamic limit, $N \to \infty$.

A particularly interesting feature of the Pfaffian state are the quasi-particle excitations of charge $e/4$ that satisfy non-Abelian anyonic statistics (Moore and Read, 1991), in contrast to the corresponding excitations of Laughlin's (6.96) or Jain's (6.123) wave functions. These non-Abelian quasi-particles are currently investigated in detail within the proposal of topologically protected quantum computation (Kitaev, 2003). A more detailed discussion of this issue is beyond the scope of these lecture notes, and we refer the reader to the review article by Nayak, *et al.* (Nayak, *et al.* 2008).

6.5 Brief overview of multicomponent quantum Hall systems

6.5.1 The different multi-component systems

6.5.1.1 The role of the electronic spin

In the preceding section, we have completely neglected the physical consequences of possible internal degrees of freedom, apart from an occasional degeneracy factor that has been smuggled in to account for experimental data. This choice has been made simply for pedagogical reasons, but it is clear that one prominent internal degree of freedom – the electronic spin – may not so easily be put under the carpet. Naively, one may expect that each LL is split into two distinct spin-branches separated by the energy gap Δ_Z due to the Zeeman effect. If this gap is large, one may use the same one-particle arguments as in the case of the IQHE, but now for each spin branch separately: once the lowest spin branch of a paticular LL is completely filled, additional electrons must overcome an energy gap that is no longer given by the LL separation but by Δ_Z. This would indeed not change the presented explanation of the IQHE – instead of a localized electron in the next higher LL, one simply needs to invoke localization in the upper spin branch.

Also in the case of the FQHE, the explanation would need to be modified only in the fine structure if the Zeeman gap is sufficiently large. If the electrons fill partially the lower spin branch of the lowest (or any) LL, one may omit all transitions to the upper spin branch and argue that they constitute the high-energy degrees of freedom, in the same manner as inter-LL excitations in the case of the "spinless" fermions that we have discussed in Section 6.4.1.

However, the situation is not so easy as the above picture might suggest. Indeed, already in 1983 Halperin pointed out (Halperin, 1983) that the Zeeman gap in GaAs, with a g-factor of $g = -0.4$, is $\Delta_Z = g\mu_B B = g(\hbar e/2m_0)B \simeq 0.33B[\text{T}]$ K and therefore much smaller than both the LL separation $\hbar\omega_C = (\hbar e/m)B \simeq 24B[\text{T}]$ K, due to the rather small band mass ($m = 0.068m_0$, in terms of the bare electron mass m_0, in GaAs), and the Coulomb energy scale $V_C = e^2/\epsilon l_B \simeq 50\sqrt{B[\text{T}]}$ K with a dielectric constant of $\epsilon \simeq 13$. For a characteristic field of 6 T, for which one typically reaches the LLL condition $\nu = 1$, one therefore has the energy scales

$$\Delta_Z \simeq 2\,\text{K} \quad \ll \quad \frac{e^2}{\epsilon l_B} \simeq 120\,\text{K} \quad \lesssim \quad \hbar\omega_C \simeq 140\,\text{K}, \tag{6.130}$$

in GaAs. The situation is qualitatively the same in graphene, where one finds for a field[54] of 6 T

$$\Delta_Z \simeq 7\,\text{K} \quad \ll \quad \frac{e^2}{\epsilon l_B} \simeq 620\,\text{K} \quad \lesssim \quad \sqrt{2}\frac{\hbar v}{l_B} \simeq 1000\,\text{K}, \tag{6.131}$$

for $g \simeq 2$ and $\epsilon \simeq 2.5$, which are the appropriate values for graphene on a SiO_2 substrate.[55]

The inevitable consequence of these considerations is that, even if one may integrate out the kinetic energy scale in a low-energy description of a partially filled LL, one cannot do so with the Zeeman energy scale. One must therefore take into account the electron spin within a two-component picture in which each quantum state $|n, m\rangle$ is doubled, $|n, m; \sigma\rangle$ with $\sigma = \uparrow$ and $\downarrow$.

6.5.1.2 *Graphene as a four-component quantum Hall system*

Another multi-component system that we have already discussed is precisely graphene, not only because of the tiny Zeeman gap that requires to take into account the electronic spin, but also because of its double valley degeneracy due to the two inequivalent Dirac points situated at the corners K and K' in the first BZ. Each quantum state $|n, m\rangle$ therefore occurs in *four* copies, $|n, m; \sigma\rangle$ with $\sigma = (K, \uparrow)$, $(K, \downarrow)$, $(K', \uparrow)$ and $(K', \downarrow)$. Formally, this four-fold degeneracy may be described with the help of an SU(4) spin, whereas the two-fold spin degeneracy in GaAs, e.g., is represented by the usual SU(2) spin. Notice that it is very difficult in graphene to lift the valley degeneracy, and the associated energy scale is expected to be on the same order of magnitude as the Zeeman gap, i.e. it is tiny with respect to the one set by the Coulomb interactions.

6.5.1.3 *Bilayer quantum Hall systems*

A third multi-component system that we would like to mention consists of a double quantum well (see Figure 6.27(a)). These bilayer systems, which are fabricated by molecular-beam epitaxy, consist of two quantum wells spatially separated by an insulating barrier that is on the same order of magnitude as the width of each of the wells. Formally, each of the wells (layers) may be described in terms of an SU(2) *pseudo-spin*, $\sigma = \uparrow$ for an electron in the left well and $\sigma = \downarrow$ for one in the right well. In contrast to the true electron spin, the Coulomb interaction does not respect this

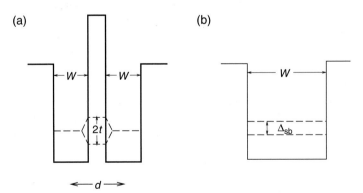

Fig. 6.27 (a) Profile of a double quantum well. The two wells are separated by a distance d that is typically on the same order of magnitude as the well width W, $d \sim W \sim 10$ nm. In the presence of a tunnelling term t between the two wells, the electronic subband is split into a symmetric and an antisymmetric combination, separated by the energy scale $\Delta_{SAS} = 2t$. (b) Wide quantum well. In a wide quantum well the energy gap between the occupied lowest electronic subband and the unoccupied first excited subband, Δ_{sb}, is decreaased as compared to a narrow quantum well.

SU(2) symmetry – indeed, the repulsion is stronger between particles within the same layer (i.e. with the same pseudo-spin orientation) than between particles in different layers (with opposite pseudo-spin orientation) because, in the second case, electrons cannot be brought together closer than the distance d between the layers. In order to minimize the interaction energy, it is therefore favorable to charge both layers equally. Alternatively, this may be viewed as a capacitive energy, if one interprets the two-layer system in terms of a capacitor, that favors an equal charge distribution between the two layers as compared to a charging of only one layer. Notice, furthermore, that tunnelling, with the tunnelling energy t, between the two quantum wells lifts the pseudo-spin degeneracy: whereas the symmetric superposition $|+\rangle = (|\uparrow\rangle + |\downarrow\rangle)/\sqrt{2}$ of the layer pseudo-spin lowers the energy, the antisymmetric superposition $|-\rangle = (|\uparrow\rangle - |\downarrow\rangle)/\sqrt{2}$ describes anti-binding. The energy separation between the associated subbands is given by $\Delta_{SAS} = 2t$ (see Figure 6.27(a)), but it may be strongly reduced experimentally with the help of a high potential barrier separating the two wells. The term Δ_{SAS}, which plays the role of a Zeeman gap (though in the x-quantization axis), may become the lowest energy scale in the system, such that the SU(2) pseudo-spin symmetry breaking only stems from the difference in the Coulomb interaction between particles in the same and in different layers.

6.5.1.4 Wide quantum wells

Another quantum Hall system that may be characterized as a multi-component system is a wide quantum well (Figure 6.27(b)). Indeed, the samples that reveal the highest mobilities are those fabricated in wide quantum wells, where the well width w is often much larger than the magnetic length l_B. As compared to a narrow quantum well, the energy difference between the lowest and the first excited electronic subbands, which are the energy levels of the confinement potential in the z-direction, is strongly decreased. Although the Fermi level still resides in the lowest electronic subband (pseudo-spin $\sigma = \uparrow$), the energy gap to the next unoccupied one (pseudo-spin $\sigma = \downarrow$) may then become smaller than the relevant Coulomb energy scale. In the same manner as for the electronic spin, one must therefore no longer discard higher electronic subbands. In a first approximation one may restrict the calculations to these two lowest subbands (Abolfath, *et al.* 1997; Papić, *et al.* 2009) although the next-higher subbands also shift to lower energies and need eventually to be taken into account. Similarly to the quantum Hall bilayer, which is sometimes also used in the description of the large quantum well, the Coulomb interaction decomposed in these electronic subband states is not pseudo-spin SU(2)-symmetric.

   In the remainder of this section, we discuss some general aspects of correlated states that one encounters in multi-component quantum Hall systems, starting (Section 6.5.2) with the completely spin-polarized state at $\nu = 1$ (quantum Hall ferromagnet) and its various manifestations in the different quantum Hall systems described above. We will not discuss, for reasons of space limitation, the amazing physical properties of the elementary excitations of the quantum Hall ferromagnet, which are topological spin-texture states *(skyrmions)*, and refer the interested reader to the literature

(Sondhi, *et al.* 1993; Moon, *et al.* 1995; Girvin 1999; Ezawa 2000). In the line of the preceding section, we have chosen to discuss a generalization of Laughlin's wave function, which we owe to Halperin (Halperin, 1983), in order to account for the electronic spin (Section 6.5.3). These wave functions are further generalized to even more components than two, and we close this section with a discussion of their possible use in the description of multi-component FQHE states.

6.5.2 The state at $\nu = 1$

If one takes into account internal degrees of freedom, the state at $\nu = 1$ is no longer simply a Slater determinant of all occupied quantum states in the lowest LL, but one must take into account the macroscopic degeneracy due to the fact that each state $|n, m\rangle$ may now be occupied by 0, 1 or 2 particles. In this sense the situation at $\nu = 1$ is much more similar to the FQHE in a partially filled LL than to the IQHE that one obtains for completely filled LL (Sondhi, *et al.* 1993), and the macroscopic degeneracy is again lifted by the Coulomb interactions between the electrons.

6.5.2.1 *Quantum Hall ferromagnetism*

We first consider the generic case of electrons at $\nu = 1$ in the conventional monolayer quantum Hall system while taking into account their physical spin. In view of the above-mentioned energy arguments, we completely neglect the Zeeman effect, which would otherwise trivially lift the macroscopic degeneracy at $\nu = 1$ by polarizing all electron spins. Because of the fact that two electrons, with opposite spin, may now occupy the same quantum state $|n, m\rangle$, the electron pair may in principle be in a relative angular momentum state with $m = 0$ – the Pauli principle, which only applies to fermions of the same species, no longer prevents this quantum number from being zero. Indeed, such an electron pair is described by a two-particle wave function with the rather unspectacular polynomial factor $(z_{i,\uparrow} - z_{j,\downarrow})^0 = 1$, where $z_{i,\uparrow}$ is the position of an arbitrarily chosen spin-$\uparrow$ electron and $z_{j,\downarrow}$ that of a spin-$\downarrow$ electron. Such an electron pair therefore interacts via the Haldane pseudopotential v_0, which is the largest pseudopotential in the case of a repulsive Coulomb interaction because it characterizes the interaction at the shortest possible length scale (see Figure 6.22).[56] Since $v_0 \simeq 2v_1$, the system thus tends to avoid double occupancy, and the ground state is described by the fully antisymmetric (orbital) wave function (6.100) regardless of whether the electron at the position z_j is spin-$\uparrow$ or spin-$\downarrow$.

Notice that, although both spinless and spin-1/2 electrons are described by the same wave function, the physical origin of these ground states is different: in the case of spinless fermions, it is simply the non-degenerate wave function described by a Slater determinant, whereas in the case of electrons with spin, the state is formed in order to minimize the mutual Coulomb repulsion.

Because the orbital wave function (6.100) for electrons with spin at $\nu = 1$ is fully antisymmetric, the spin wave function describing the internal degrees of freedom must be fully symmetric, e.g.

$$\chi_{FM} = |\uparrow_1, \uparrow_2, ..., \uparrow_N\rangle, \tag{6.132}$$

in order to form an overall wave function that is antisymmetric. The subscript indicates the index of the particle that the spin is associated with. The global wave function, therefore, reads

$$\psi_{\nu=1,FM} = \prod_{k<l}(z_k - z_l) \otimes |\uparrow_1, \uparrow_2, ..., \uparrow_N\rangle. \tag{6.133}$$

This is simply a (spin) wave function of a *quantum ferromagnet*, similar to ferromagnetism in a usual Fermi liquid. Indeed, the spontaneous spin polarization in a Fermi liquid is also due to a minimization of the Coulomb repulsion by the formation of an antisymmetric orbital wave function. Notice, however, that the spin polarization in a Fermi liquid comes with an energy cost as a consequence of the mismatch between the Fermi energies of spin-↑ and spin-↓ electrons. The competition between the gain in interaction energy and the cost in kinetic energy determines the final polarization of the system, which is never perfect. In the case of the quantum Hall ferromagnet, there is no cost in kinetic energy when the system is fully polarized because all quantum states have the same kinetic energy, and the system is therefore *fully* polarized.

Collective excitations. Because the spontaneous spin polarization in the quantum Hall ferromagnet chooses, in the absence of a Zeeman effect, an arbitrary direction in the three-dimensional spin space, one is confronted with a spontaneous SU(2) symmetry breaking. As a consequence of this broken continuous symmetry, there exists a gapless collective excitation (Goldstone mode) the energy of which tends to zero in the long-wavelength limit. Indeed, even if we have chosen the ferromagnet in Eq. (6.132) to be oriented in the z-direction, any other orientation, such as these described by the wave functions

$$|\downarrow_1, \downarrow_2, ..., \downarrow_N\rangle \quad \text{or} \quad \bigotimes_{j=1}^{N} |+_j\rangle = |+_1, +_2, ..., +_N\rangle,$$

where the $+_j$ sign indicates the symmetric superposition $|+_j\rangle = (|\uparrow_j\rangle + |\downarrow_j\rangle)/\sqrt{2}$ of both spin orientations of the jth electron, would also describe a ground state. The Goldstone mode in the large wavelength limit may then be viewed as a global rotation of all spins into another ground-state configuration, which naturally does not imply an energy cost.

In the case of a ferromagnet, the Goldstone mode is simply the spin-density wave,[57] which disperses as $\omega \propto q^2$ in the small-wavevector limit, $ql_B \ll 1$. At first sight, this mode seems in contradiction with the observation of a quantum Hall effect at $\nu = 1$, even in the absence of a Zeeman effect, which requires a gap as we have seen above. Notice, however, that this gap needs to be a transport gap in which a quasi-particle moves independently from a quasi-hole in order to transport a current. This is not the case in a spin wave with $ql_B \ll 1$, but one obtains freely moving quasi-particles and quasi-holes only in the limit $ql_B \gg 1$. In this limit, the spin-wave dispersion tends to a finite value that is given by the exchange energy between particles of different spin orientation and that is proportional to the interaction energy scale $e^2/\epsilon l_B$, as in the case of the FQHE (Moon, *et al.* 1995).

There are more exotic spin-texture excitations (skyrmions), which are described by a topological quantum number associated with the winding of the spin-texture. These are gapped excitations that carry an electric charge related to this topological quantum number. As mentioned above, a detailed discussion of these amazing excitations is beyond the scope of the present chapter.

6.5.2.2   *Exciton condensate in bilayer systems*

The $\nu = 1$ in a bilayer system is remarkably different from the quantum Hall ferromagnet described in the preceding subsection. Although the electronic interactions still favor a fully antisymmetric orbital wave function (6.100) and thus a symmetric, i.e. ferromagnetic, pseudo-spin wave function, the interaction potential is no longer $SU(2)$ symmetric in the pseudo-spin degree of freedom.[58] As we have already mentioned above, a charge imbalance Q between the two quantum wells (layers) is penalized by a charging energy, $E_C = Q^2/2C$, in terms of the capacitance $C = \epsilon \mathcal{A}/d$, where $\mathcal{A}$ is the area of the 2D system. Because $Q = -en_{el}\mathcal{A} = -e\nu n_B \mathcal{A} = -e\nu \mathcal{A}/2\pi l_B^2$ when all electrons reside in a single layer and $Q = 0$ if they are equally distributed between the two layers, one obtains an energy cost

$$\frac{E_C}{N_{el}} \sim \nu \frac{e^2}{\epsilon l_B} \frac{d}{l_B},$$

per particle in the charge-imbalanced state, in agreement with a more sophisticated microscopic calculation (Moon, *et al.* 1995). In terms of the pseudo-spin magnetization, this means that in the ground-state configuration, with a homogeneous charge distribution over both layers, all pseudo-spins are oriented in the xy-plane. Remember that a pseudo-spin $\uparrow$ corresponds to an electron in the upper layer and $\downarrow$ to one in the lower layer, and a configuration as the one described in Eq. (6.132) is therefore excluded, whereas the symmetric and antisymmetric combinations

$$\chi_+ = \bigotimes_{j=1}^{N} |+_j\rangle \qquad \text{and} \qquad \chi_- = \bigotimes_{j=1}^{N} |-_j\rangle,$$

with $|\pm_j\rangle = (|\uparrow_j\rangle \pm |\downarrow_j\rangle)/\sqrt{2}$ are not. These two states, which correspond to a ferromagnet in the x- and the $-x$-direction, respectively, may be generalized by choosing any other direction described by the angle ϕ in the xy-plane,

$$\chi_\phi = \bigotimes_{j=1}^{N} |\phi_j\rangle, \tag{6.134}$$

where $|\phi_j\rangle \equiv [|\uparrow\rangle + \exp(i\phi)|\downarrow\rangle]/\sqrt{2}$. The states χ_+ and χ_- are obtained for $\phi = 0$ and $\phi = \pi$ (modulo 2π), respectively.

Contrary to the case of the spin ferromagnet with full $SU(2)$ symmetry, where a general state would be described in terms of two angles θ and ϕ, the different possible *easy-plane* pseudo-spin ferromagnets are characterized by the angle ϕ that may vary between 0 and 2π. The low-energy degrees of freedom are therefore described by a different *universality class* that turns out to be the same as the one that describes superfluidity or superconductivity. The relation between superfluidity and the easy-plane pseudo-spin ferromagnet in bilayer systems at $\nu = 1$ may indeed be understood in the following manner: on average, the filling factor per layer is $\nu_\uparrow = \nu_\downarrow = 1/2$ in order to minimize the charging energy due to the capacitive term, i.e. there are as many electrons as holes in the LLL of each layer. Naturally, because of the Coulomb interaction between the particles in the two different layers, an electron in one layer wants to be bound to a hole in the other one. Since the number of electrons in each layer equals, on average, that of holes in the other one, all particles find their appropriate partner in the opposite layer. The electron-hole pair in the two layers may be viewed as a charge-neutral *interlayer exciton* that satisfies bosonic statistics (Figure 6.28(a)). Below a certain temperature, these bosons condense into a collective state that is simply the *exciton superfluid* (Fertig 1989; Wen and Zee 1992a; Ezawa and Iwazaki 1993; Moon, *et al.* 1995). The phase coherence between the different excitons is precisely described by the angle ϕ.

The first experimental indication of excitonic superfluidity in bilayer quantum Hall systems was a zero-bias anomaly in tunnelling experiments (Spielman, *et al.* 2000). Indeed, if one injects a charge in a tunnelling experiment into one of the layers and collects it in a contact at the other layer, the tunnelling conductance dI_z/dV is expected to be weak in the case of uncorrelated electrons because of the Coulomb repulsion between electrons in the opposite layers. However, below a critical value of d/l_B, where one expects the interlayer correlations to be sufficiently strong to form a phase-coherent excitonic condensate, the injected electron systematically finds a hole in the other layer, such that tunnelling between the layers is strongly enhanced. This strong enhancement, which, due to its reminiscence with the Josephson effect in superconductors (Tinkham, 2004), is also called the *quasi-Josephson effect*,[59] has indeed been observed experimentally (Spielman, *et al.* 2000).

Another strong indication for excitons in bilayer quantum Hall systems stems from transport measurements in the counterflow configuration, where the current in the upper layer $I_\uparrow = I$ flows in the opposite direction as compared to that in the lower layer $I_\downarrow = -I$ (see Figure 6.28(a)). From a technical point of view, it is indeed possible to contact the two layers separately such that one may measure the Hall resistance (and also the longitudinal resistance) in both layers independently. In the case of exciton condensation, the charges involved in transport are *zero* because the excitons are charge-neutral objects, which are not coupled to the magnetic field and are thus not affected by the Lorentz force. In addition to a vanishing longitudinal resistance, one would therefore expect a vanishing Hall resistance because no density gradient between opposite edges is built up to compensate the Lorentz force (Wen and Zee, 1992a; Ezawa and Iwazaki, 1993). This is schematically shown in Figure 6.28(b). The simultaneous vanishing of the Hall and longitudinal resistances was indeed observed in 2004 by two different experimental groups (Kellogg, *et al.* 2004; Tutuc, *et al.* 2004).

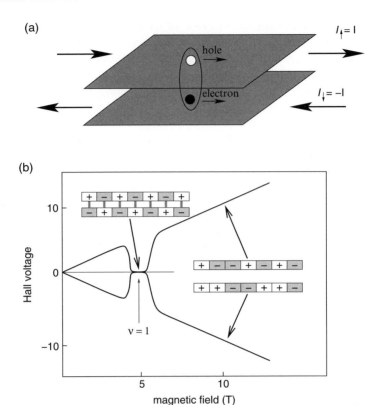

Fig. 6.28   Hall-resistance measurement used to detect excitonic condensation, adopted from (Eisenstein and MacDonald, 2004). (a) Counterflow configuration, in which one drives a current $I_\uparrow = I$ through the upper layer that is flowing in the opposite direction as that, $I_\downarrow = -I$ in the lower layer. The hole component of the excitonic quantum state in one layer thus moves in the same direction as the electron component in the other one. (b) The two curves schematically represent, when taking into account only excitonic superfluidity, the Hall resistance in both layers within the counterflow configuration. Because of the relative sign between the currents in the two layers, the measured Hall resistances are of opposite sign. Electrons with no interlayer correlations yield the usual linear B-field dependence of the Hall resistance in order to compensated the Lorentz force acting on them individually. In the case of exciton condensation (around $B = 5$ T), charge tranport is due to a uniform current of charge-neutral excitons, which are not affected by the Lorentz force, and the Hall resistance vanishes, as it has been observed in the experiments (Kellogg, *et al.* 2004; Tutuc, *et al.* 2004).

6.5.2.3   *SU(4) ferromagnetism in graphene*

The arguments in favor of a quantum Hall ferromagnetism may easily be generalized to graphene, where the Coulomb interaction respects to great accuracy the four-fold spin-valley degeneracy, as we have described above. In order to avoid confusion about the filling factor, one first needs to remember that the filling factor ν_G in graphene is defined with respect to the charge-neutral point, which happens to be in the middle of the central $n = 0$ LL (see Section 6.3.5). Two of the four (degenerate) spin-valley branches are therefore completely filled at $\nu_G = 0$, which in non-relativistic quantum Hall systems would correspond rather to a filling factor $\nu = 2$. Similarly, the filling factor $\nu = 1$ would correspond to a graphene filling factor $\nu_G = -1$, whereas $\nu_G = 1$ implies three completely filled spin-valley branches ($\nu = 3$).

Let us first consider the filling factor $\nu_G = -1$ and see how the above considerations apply to graphene with its SU(4) symmetry.[60] In the same manner as for the spin quantum Hall ferromagnet at $\nu = 1$, the short-range component v_0 of the Coulomb potential is screened in the completely antisymmetric orbital wave function (6.100), and the spin part of the wave function must therefore be completely symmetric. Notice, however, that one may now distribute the electron over the four internal states $|m; K, \uparrow\rangle$, $|m; K, \downarrow\rangle$, $|m; K', \uparrow\rangle$ and $|m; K', \downarrow\rangle$. The general spin wave function is therefore a superposition of all these states

$$
\chi_{\mathrm{SU}(4)} = \bigotimes_{m=1}^{N} \left(u_{m,1}|m; K, \uparrow\rangle + u_{m,2}|m; K, \downarrow\rangle + u_{m,3}|m; K', \uparrow\rangle + u_{m,4}|m; K', \downarrow\rangle \right),
$$

$$(6.135)$$

where the complex coefficients $u_{m,i}$ satisfy the normalization condition $\sum_{i=1}^{4} |u_{m,i}|^2 = 1$. In the case of global coherence, all coefficients are independent of the guiding-center quantum number m, $u_{m,i} = u_i$, and one thus obtains the spin wave function of an *SU(4) ferromagnetism* (Nomura and MacDonald 2006; Goerbig, *et al.* 2006; Alicea and Fisher 2006; Yang, *et al.* 2006). These arguments may also be generalized to the case of $\nu_G = 0$, where two branches are completely filled (Yang, *et al.* 2006), but the ground state does not reveal the same degeneracy as the SU(4) ferromagnet at $\nu_G = \pm 1$. Indeed, a general argument on K-component quantum Hall system shows that one has generalized ferromagnetic states at all integer values of the filling factor $\nu = 1, \ldots, K - 1$ (Arovas, *et al.* 1999).

As a consequence of the SU(4) quantum Hall ferromagnet, one may expect a quantum Hall effect in graphene at the unusual filling factors $\nu_G = 0, \pm 1$. Remember that these states do not belong to the series (6.93), $\nu_G = \pm 2, \pm 6, \ldots$ of the RQHE, which may be explained by LL quantization within the picture of non-interacting relativistic particles. In the same manner as for the spin quantum Hall ferromagnet, the gapless spin-density-wave modes, which reveal a higher degeneracy due to the larger SU(4) symmetry, do not imply that the charged modes are also gapless. Indeed, the elementary charged excitations of the SU(4) quantum Hall ferromagnet are generalized skyrmions (Yang, *et al.* 2006; Douçot, *et al.* 2008) that are separated by a gap from the ground state, which therefore describes an incompressible quantum liquid that displays the quantum Hall effect. A quantum Hall effect has indeed been observed at these unusual filling factors (Zhang, *et al.* 2006), in agreement with the formation of an SU(4) quantum Hall ferromagnet. However, there exist alternative scenarios to describe the appearance of a quantum Hall effect at these filling factors (Gusynin, *et al.* 2006; Fuchs and Lederer 2007; Herbut 2007) and a clear indication of SU(4) quantum Hall ferromagnetism is yet lacking.

We finally emphasize that an SU(4) description is not restricted to graphene. Indeed, if one takes into account the electron spin, the bilayer quantum Hall system and its excitations may also be treated within the SU(4) framework (Arovas, *et al.* 1999; Ezawa 1999, 2000; Douçot, *et al.* 2008) although the interaction does not respect the full SU(4) symmetry because of the asymmetry in the layer pseudo-spin described above.

6.5.3 Multi-component wave functions

Until now, we have considered a multi-component quantum Hall effect at the integer filling factor $\nu = 1$ (or other integer fillings in the case of graphene) that is described in terms of the Vandermonde determinant (6.100) $\prod_{k<l}(z_k - z_l)$ regardless of whether the particle at the position z_k is in a state ↑ or ↓. The spin orientation has only been taken into account within a spin wave function that is multiplied to the Vandermonde determinant. One may naturally ask the question whether one can also describe other filling factors than $\nu = 1$.

A simple generalization of the quantum Hall ferromagnetism to other filling factors consists of replacing the Vandermonde determinant by, e.g., the Laughlin (6.96) at $\nu = 1/(2s+1)$ or the Jain wave function (6.123) at $\nu = p/(2sp+1)$ and of multiplying it again with a spin wave function. This spin wave function is naturally ferromagnetic because the orbital wave function remains antisymmetric. There are, however, more general states for which the orbital wave function is not fully antisymmetric, but only in the intra-component parts as it is required by the Pauli principle. These states are described in terms of wave functions proposed by Halperin in 1983 (Halperin, 1983) that we present in this section, as well as a natural generalization to systems with more components than $K = 2$.

6.5.3.1 *Halperin's wave function*

Halperin's wave function for spin-1/2 electrons is a straightforward generalization of Laughlin's proposal (6.96). We consider the particle positions to be separated into two sets $\{z_1^\uparrow, z_2^\uparrow, ..., z_{N_\uparrow}^\uparrow\}$ for spin-↑ particles and $\{z_1^\downarrow, z_2^\downarrow, ..., z_{N_\downarrow}^\downarrow\}$ for spin-↓ particles. If the particles with different spin orientation could be treated as independent from one another, i.e. in the absence of an interaction between spin-↑ and spin-↓ particles, one would simply write down a product ansatz

$$\psi_{\uparrow,m_1}^L(\{z_j^\uparrow\}) \times \psi_{\downarrow,m_2}^L(\{z_j^\downarrow\}) = \prod_{k<l}^{N_\uparrow} \left(z_k^\uparrow - z_l^\uparrow\right)^{m_1} \prod_{k<l}^{N_\downarrow} \left(z_k^\downarrow - z_l^\downarrow\right)^{m_2} \tag{6.136}$$

of two independent Laughlin wave functions that need not necessarily be described by the same exponent m. The total filling factor would then be simply the sum $\nu = \nu_\uparrow + \nu_\downarrow$ of the filling factors $\nu_\uparrow = 1/m_1$ and $\nu_\downarrow = 1/m_2$ for spin-↑ and spin-↓ particles, respectively.

Apart from the fact that this situation is not particularly interesting, it is also unphysical because the Coulomb interaction does not depend on the spin orientation of the particle pairs. In the wave function (6.136), two particles of opposite spin orientation may be at the same position, i.e. the wave function does not vanish in general for $z_k^\uparrow = z_l^\downarrow$. Remember that such a double occupancy of the same position would be penalized by an energy cost on the order of the short-range component v_0 in a pseudo-potential expansion.

In order to account for these inter-component correlations, Halperin proposed to add a factor $\prod_{k=1}^{N_\uparrow} \prod_{l=1}^{N_\downarrow} (z_k^\uparrow - z_l^\downarrow)^n$ to the wave function (6.136) the exponent of which does not necessarily need to be odd because particles of opposite spin orientation are not constrained by the Pauli principle. Halperin's wave function

$$\psi_{m_1,m_2,n}^{H}(\{z_j^\uparrow, z_j^\downarrow\}) = \prod_{k<l}^{N_\uparrow} \left(z_k^\uparrow - z_l^\uparrow\right)^{m_1} \prod_{k<l}^{N_\downarrow} \left(z_k^\downarrow - z_l^\downarrow\right)^{m_2} \prod_{k=1}^{N_\uparrow}\prod_{l=1}^{N_\downarrow} \left(z_k^\uparrow - z_l^\downarrow\right)^{n} \qquad (6.137)$$

is therefore characterized by the set (m_1, m_2, n) of three exponents.

In analogy with Laughlin's wave function, for which we have $\nu = 1/m$, the exponents fix the (component) filling factors, as one may see from the power-counting argument (see Section 6.4.2). According to this argument, the maximal exponent for a particular particle position cannot exceed the number of flux quanta N_B threading the area $\mathcal{A}$ of the 2D electron system. Apart from the shift δ that vanishes in the thermodynamic limit, one obtains the two equations

$$N_B = m_1 N_\uparrow + n N_\downarrow \qquad \text{and} \qquad N_B = m_2 N_\downarrow + n N_\uparrow. \qquad (6.138)$$

This means that, contrary to the simpler case of Laughlin's wave function, the number of zeros in one component is not simply given by the corresponding exponent times the number of particles in this component (first term in the above expressions). Instead, it is also affected by the particles in the other component that each contribute a zero of order n (second term) due to the mixed term in Halperin's wave function (6.137). In terms of the component filling factors,

$$\nu_\sigma = \frac{N_\sigma}{N_B}, \qquad (6.139)$$

Eq. (6.138) may be rewritten in matrix form

$$\begin{pmatrix} 1 \\ 1 \end{pmatrix} = \begin{pmatrix} m_1 & n \\ n & m_2 \end{pmatrix} \begin{pmatrix} \nu_\uparrow \\ \nu_\downarrow \end{pmatrix}, \qquad (6.140)$$

from which one obtains the component filling factors by matrix inversion

$$\begin{pmatrix} \nu_\uparrow \\ \nu_\downarrow \end{pmatrix} = \frac{1}{m_1 m_2 - n^2} \begin{pmatrix} m_2 & -n \\ -n & m_1 \end{pmatrix} \begin{pmatrix} 1 \\ 1 \end{pmatrix}, \qquad (6.141)$$

and one finds

$$\nu = \nu_\uparrow + \nu_\downarrow = \frac{m_1 + m_2 - 2n}{m_1 m_2 - n^2} \qquad (6.142)$$

for the total filling factor.

One first notices that, in Eq. (6.141), not only the filling factors are fixed by the exponents but also, for a given magnetic field (i.e. a given number of flux quanta), the number of particles per component. Contrary to what one could have expected from the expression of Halperin's wave function (6.137), the numbers N_σ, namely the ratio between them, cannot be chosen arbitrarily.

Furthermore, the above expressions (6.141) and (6.142) for the filling factors are ill-defined if the exponent matrix in Eq. (6.140) is not invertible, i.e. when its determinant is zero, $m_1 m_2 - n^2 = 0$. The only physically relevant situation arises when all exponents are equal odd integers $m_1 = m_2 = n$. However, this result should not surprise us: we are then confronted again with a completely antisymmetric wave function, actually a Laughlin wave function, which requires a ferromagnetic spin wave function. As we have seen above, in the discussion of the quantum Hall ferromagnetism, the ground-state manifold comprises states with different polarization along the z-axis: the state with $N_\uparrow = N$ and $N_\downarrow = 0$ is an equally valid ground state as a state with $N_\uparrow = N_\downarrow = N/2$ or $N_\uparrow = 0$ and $N_\downarrow = N$, where $N = N_\uparrow + N_\downarrow$ is the total number of particles. The component filling factor is therefore not well defined and depends on the polarization

$$p_z = \frac{N_\uparrow - N_\downarrow}{N} = \frac{\nu_\uparrow - \nu_\downarrow}{\nu}, \tag{6.143}$$

whereas the total filling factor is simply given by $\nu = 1/m$, in terms of the common odd exponent m. Notice that contrary to the quantum Hall ferromagnet, a state with an invertible exponent matrix has a polarization that is completely fixed,

$$p_z = \frac{m_2 - m_1}{m_1 m_2 - n^2}. \tag{6.144}$$

We finally mention that not all states that can be written down in terms of Halperin's wave function are good candidates for the description of the ground state chosen by the system. One may show, e.g. within a generalization of Laughlin's plasma analogy (presented in Section 6.4.2.5) to two or more components, that several of Halperin's wave functions do not describe a homogeneous liquid but a liquid in which the different components phase separate (de Gail, *et al.* 2008). For two components, the condition for a homogeneous state is simply that both the exponents m_1 and m_2, which describe the intra-component correlations, must be larger than n for the inter-component correlations. As an example, we may study the states $(3, 3, 1)$ and $(1, 1, 3)$, which would both be candidates for a possible two-component FQHE at $\nu = 1/2$ and which have indeed been investigated in the literature (MacDonald, *et al.* 1989). However, only the first one describes a homogeneous liquid, such that the second one may be discarded right from the beginning.

Furthermore, some of Halperin's wave functions, even if they satisfy the above-mentioned condition, turn out to be problematic if the interaction is SU(2) symmetric, such as for the true electron spin. In this case, one may show that (m, m, n) states are only eigenstates of the total-spin operator, which commutes with the interaction Hamiltonian, if $n = m$ (i.e. in the ferromagnetic state) or if $n = m - 1$ (Prange and Girvin 1990). However, this restriction may be omitted in bilayer quantum Hall systems or in wide quantum wells where the interaction Hamiltonian is not pseudo-spin SU(2)-symmetric.

Physical relevance of Halperin states. A physically relevant Halperin state is, e.g., the unpolarized $(3,3,2)$ state that would occur at a filling factor $\nu = 2/5$. Remember from the discussion of CF theory in Section 6.4.4 that there is also a (naturally polarized) CF candidate, with $p = 2$ completely filled CF LLs, to describe the ground state at this filling factor. Which of them is now the better one? This question could be answered within exact-diagonalization calculations, which showed that, in the absence of a Zeeman effect, the true ground state is described in terms of the unpolarized Halperin wave function $(3,3,2)$ (Chakraborty and Zhang, 1984). Notice, however, that the energy difference between the two states is quite small, as may be seen from variational calculations (Jain, 2007), such that the polarized CF state becomes the ground state above a critical value of the energy Δ_Z associated with the Zeeman effect. This critical value would therefore describe a phase transition between an unpolarized and a fully polarized FQHE state. Such transitions have indeed been observed in polarization experiments, where the strength of the Zeeman effect was varied by a simultaneous change in the magnetic field and in the electronic density (Kang, *et al.* 1997; Kukushkin, *et al.* 1999).

6.5.3.2 Generalized Halperin wave functions

We will finally mention that Halperin's wave function may easily be generalized to describe possible FQHE states in systems with a larger number of components, such as the four spin-valley components in graphene. This generalized wave function for K-component quantum Hall systems may be written as a product

$$\psi^{SU(K)}_{m_1,\dots,m_K;n_{ij}}\left(\left\{z^{(1)}_{j_1}, z^{(2)}_{j_2}, \dots, z^{(K)}_{j_K}\right\}\right) = \psi^{L}_{m_1,\dots,m_K} \times \psi^{inter}_{n_{ij}} \qquad (6.145)$$

of a product of Laughlin wave functions

$$\psi^{L}_{m_1,\dots,m_K} = \prod_{j=1}^{K} \prod_{k_j < l_j}^{N_j} \left(z^{(j)}_{k_j} - z^{(j)}_{l_j}\right)^{m_j}$$

for each of the components and a term

$$\psi^{inter}_{n_{ij}} = \prod_{i<j}^{K} \prod_{k_i}^{N_i} \prod_{k_j}^{N_j} \left(z^{(i)}_{k_i} - z^{(j)}_{k_j}\right)^{n_{ij}}$$

that takes into account the correlations between particles in different components (Goerbig and Regnault, 2007). Here, the indices i and j denote the component, $i,j = 1, \dots, K$, and $z^{(i)}_{k_i}$ is the complex position of the k_ith particle in the component i.

Although the wave function (6.145) may seem scary at the first sight, it is as easily manipulated as Halperin's original wave function (6.137). The component filling factors $\nu_j = N_j/N_B$ may be determined, in the same manner as in the two-component case

(6.140), with the help of the "exponent matrix" $\mathcal{M}$ the off-diagonal terms of which are the exponents $(\mathcal{M})_{ij} = n_{ij}$ (for $i \neq j$), whereas the diagonal terms are simply the exponents corresponding to the intra-component correlations, $(\mathcal{M})_{ii} = m_i$. The zero-counting argument yields the matrix equation

$$\begin{pmatrix} 1 \\ \vdots \\ 1 \end{pmatrix} = \mathcal{M} \begin{pmatrix} \nu_1 \\ \vdots \\ \nu_K \end{pmatrix} \tag{6.146}$$

relating the component filling factors to the exponents, and if $\mathcal{M}$ is invertible, all component filling factors are fixed by the inverse equation

$$\begin{pmatrix} \nu_1 \\ \vdots \\ \nu_K \end{pmatrix} = \mathcal{M}^{-1} \begin{pmatrix} 1 \\ \vdots \\ 1 \end{pmatrix}. \tag{6.147}$$

If the determinant $\det(\mathcal{M})$ is zero and the matrix thus not invertible, not all component filling factors can be determined. In analogy with the two-component case this hints at underlying ferromagnetic states. A perfect $SU(K)$ ferromagnetic state is obtained when all components are equal odd integers, $m_i = n_{ij} = m$, in which case one obtains again a simple (fully antisymmetric) Laughlin wave function for all particles regardless of to which component they belong. For $K = 4$ and $m = 1$, this is just the $SU(4)$ ferromagnetic state at $\nu = 1$ which we have already discussed in the context of the quantum Hall effect at $\nu_G = \pm 1$ in graphene (Section 6.5.2.3).

Notice, however, that contrary to a two-component system, where one only needs to distinguish between an invertible and a non-invertible matrix, the situation is much richer for $K > 2$. One may indeed have different "degrees" of invertibility that are described by the *rank* of the matrix. Consider, e.g., the fully antisymmetric wave function with $m_i = n_{ij} = m$. In this case, Eq. (6.146) actually consists only of one single equation relating the component filling factors, i.e. $1 = m(\nu_1 + \ldots + \nu_K) = m\nu$, and all other lines of the matrix equation are simply copies of the first one. The rank of this matrix is 1, i.e. only the total filling factor is fixed, $\nu = 1/m$ [$SU(K)$ ferromagnet], whereas in the case of an invertible matrix the rank is K, and the K lines in the matrix equation (6.146) represent (linearly) independent equations. If the rank of an exponent matrix is smaller than K but larger than 1, the resulting state is neither a full $SU(K)$ ferromagnet nor a state with completely fixed component filling factors (or polarizations) – it is rather a state with some intermediate ferromagnetic properties.

As for two-component Halperin wave functions (6.137), a generalization of Laughlin's plasma analogy allows one to distinguish between physical (i.e. homogeneous) and unphysical states (which show a phase separation of at least some of the components). Indeed, the exponent matrix $\mathcal{M}$ must have only positive eigenvalues in order to describe a homogeneous state (de Gail, *et al.* 2008). We finally mention that $\mathcal{M}$ encodes not only information concerning the filling factors (6.147), but fully describes the quantum Hall state (6.145), such as its topological degeneracy, the charges of its quasi-particle excitations as well as the statistical properties of the latter (Wen and Zee, 1992b).

Appendix A: Electronic band structure of graphene

In this appendix, we calculate the band structure of graphene in the tight-binding model (Wallace 1947), the results of which are summarized in Section 6.1.2.3. Because graphene's honeycomb lattice consists of two distinct sublattices A and B, the electronic wave function

$$\psi_{\mathbf{k}}(\mathbf{r}) = a_{\mathbf{k}}\psi_{\mathbf{k}}^{(A)}(\mathbf{r}) + b_{\mathbf{k}}\psi_{\mathbf{k}}^{(B)}(\mathbf{r}), \tag{A.1}$$

is a superposition of two wave functions, for the A and B sublattice, respectively, where $a_{\mathbf{k}}$ and $b_{\mathbf{k}}$ are complex functions of the quasi-momentum $\mathbf{k}$. Both $\psi_{\mathbf{k}}^{(A)}(\mathbf{r})$ and $\psi_{\mathbf{k}}^{(B)}(\mathbf{r})$ are Bloch functions with

$$\psi_{\mathbf{k}}^{(j)}(\mathbf{r}) = \sum_{\mathbf{R}_l} e^{i\mathbf{k}\cdot\mathbf{R}_l}\phi^{(j)}(\mathbf{r}+\boldsymbol{\delta}_j - \mathbf{R}_l), \tag{A.2}$$

in terms of the atomic wave functions $\phi^{(j)}(\mathbf{r}+\boldsymbol{\delta}_j - \mathbf{R}_l)$ centered around the position $\mathbf{R}_l - \boldsymbol{\delta}_j$, where $\boldsymbol{\delta}_j$ is the vector that connects the sites $\mathbf{R}_l$ of the underlying Bravais lattice with the site of the j atom within the unit cell. Typically, one chooses the sites of one of the sublattices, e.g. the A sublattice, to coincide with the sites of the Bravais lattice such that $\boldsymbol{\delta}_A = 0$.

With the help of these wavefunctions, we may now search the solutions of the Schrödinger equation

$$H\psi_{\mathbf{k}} = \epsilon_{\mathbf{k}}\psi_{\mathbf{k}},$$

where H is the full Hamiltonian for electrons on a lattice, which is of the type (6.17) mentioned in Section 6.2.1. Here, we have chosen an arbitrary representation, which is not necessarily that in real space.[61] Multiplication of the Schrödinger equation by $\psi_{\mathbf{k}}^*$ from the left yields the equation $\psi_{\mathbf{k}}^* H\psi_{\mathbf{k}} = \epsilon_{\mathbf{k}}\psi_{\mathbf{k}}^*\psi_{\mathbf{k}}$, which may be rewritten in matrix form with the help of Eqs. (A.1) and (A.2)

$$(a_{\mathbf{k}}^*, b_{\mathbf{k}}^*)\,\mathcal{H}_{\mathbf{k}}\begin{pmatrix} a_{\mathbf{k}} \\ b_{\mathbf{k}} \end{pmatrix} = \epsilon_{\mathbf{k}}\,(a_{\mathbf{k}}^*, b_{\mathbf{k}}^*)\,\mathcal{S}_{\mathbf{k}}\begin{pmatrix} a_{\mathbf{k}} \\ b_{\mathbf{k}} \end{pmatrix}. \tag{A.3}$$

Here, the Hamiltonian matrix is defined as

$$\mathcal{H}_{\mathbf{k}} \equiv \begin{pmatrix} \psi_{\mathbf{k}}^{(A)*} H \psi_{\mathbf{k}}^{(A)} & \psi_{\mathbf{k}}^{(A)*} H \psi_{\mathbf{k}}^{(B)} \\ \psi_{\mathbf{k}}^{(B)*} H \psi_{\mathbf{k}}^{(A)} & \psi_{\mathbf{k}}^{(B)*} H \psi_{\mathbf{k}}^{(B)} \end{pmatrix} = \mathcal{H}_{\mathbf{k}}^{\dagger}, \tag{A.4}$$

and the overlap matrix

$$\mathcal{S}_{\mathbf{k}} \equiv \begin{pmatrix} \psi_{\mathbf{k}}^{(A)*} \psi_{\mathbf{k}}^{(A)} & \psi_{\mathbf{k}}^{(A)*} \psi_{\mathbf{k}}^{(B)} \\ \psi_{\mathbf{k}}^{(B)*} \psi_{\mathbf{k}}^{(A)} & \psi_{\mathbf{k}}^{(B)*} \psi_{\mathbf{k}}^{(B)} \end{pmatrix} = \mathcal{S}_{\mathbf{k}}^{\dagger}. \tag{A.5}$$

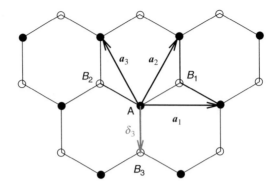

Fig. A.1 Tight-binding model for the honeycomb lattice.

accounts for the non-orthogonality of the trial wave functions. The eigenvalues $\epsilon_\mathbf{k}$ of the Schrödinger equation yield the energy bands, and they may be obtained from the secular equation

$$\det\left[\mathcal{H}_\mathbf{k} - \epsilon_\mathbf{k}^\lambda \mathcal{S}_\mathbf{k}\right] = 0, \tag{A.6}$$

which needs to be satisfied for a non-zero solution of the wavefunctions, i.e. for $a_\mathbf{k} \neq 0$ or $b_\mathbf{k} \neq 0$. The label λ denotes the energy bands, and it is clear that there are as many energy bands as solutions of the secular equation (A.6), i.e. two bands for the case of two atoms per unit cell.

From now on, we neglect the overlap of wave functions on neighboring sites, such that the overlap matrix (A.5) simply becomes the one matrix $\mathbb{1}$ times the number of particles N due to the normalization of the wave functions. The secular equation then tells us that the energy bands are just the eigenvalues of the Hamiltonian matrix (A.4). Furthermore, one notices that because the two sublattices are equivalent from a chemical point of view, we have $\psi_\mathbf{k}^{(A)*} H \psi_\mathbf{k}^{(A)} = \psi_\mathbf{k}^{(B)*} H \psi_\mathbf{k}^{(B)}$, and the diagonal terms therefore contribute just a constant shift to the band energies that we may set to zero. The only relevant terms are then the off-diagonal terms in Eq. (A.4), $\mathcal{H}_\mathbf{k}^{AB} \equiv \psi_\mathbf{k}^{(A)*} H \psi_\mathbf{k}^{(B)} = N t_\mathbf{k}^{AB}$, with the *hopping term*

$$t_\mathbf{k}^{AB} \equiv \sum_{\mathbf{R}_l} e^{i\mathbf{k}\cdot\mathbf{R}_l} \int d^2r\, \phi^{(A)*}(\mathbf{r}) H \phi_l^{(B)}(\mathbf{r} + \boldsymbol{\delta}_{AB} - \mathbf{R}_l), \tag{A.7}$$

where $\boldsymbol{\delta}_{AB}$ is a vector that connects an A site to a B site.

In order to obtain the basic band structure of graphene, it is sufficient to consider a hopping only between nearest-neighboring sites described by the *hopping amplitude*

$$t \equiv \int d^2r\, \phi^{A*}(\mathbf{r}) H \phi^B(\mathbf{r} + \boldsymbol{\delta}_3), \tag{A.8}$$

where we have chosen $\boldsymbol{\delta}_{AB} = \boldsymbol{\delta}_3$ (see Figure A.1). Notice that one may also take into account hopping to sites that are further away such as the next-nearest neighbors that turn out to be on the same sublattice and that would thus yield diagonal terms to the Hamiltonian matrix. However, whereas we have $t \sim 3$ eV, the hopping amplitude for next-nearest-neighbor hopping is roughly 10 times smaller (Saito, *et al.* 1998; Castro Neto, *et al.* 2009) and only marginally affects the low-energy properties of electrons in graphene.

If we now consider an arbitrary site A on the A sublattice (Figure A.1), we may see that the hopping term (A.7) consist of three terms corresponding to the nearest neighbors B_1, B_2, and B_3, all of which have the same hopping amplitude t. However, only the site B_3 is described by the same lattice vector (shifted by $\boldsymbol{\delta}_3$) as the site A and thus yields a zero phase to the hopping matrix. The sites B_1 and B_2 correspond to lattice vectors shifted by

$$\mathbf{a}_2 = \frac{\sqrt{3}a}{2}(\mathbf{e}_x + \sqrt{3}\mathbf{e}_y) \quad \text{and} \quad \mathbf{a}_3 \equiv \mathbf{a}_2 - \mathbf{a}_1 = \frac{\sqrt{3}a}{2}(-\mathbf{e}_x + \sqrt{3}\mathbf{e}_y),$$

respectively, where $a = |\boldsymbol{\delta}_3| = 0.142$ nm is the distance between nearest-neighbor carbon atoms. Therefore, they contribute a phase factor $\exp(i\mathbf{k} \cdot \mathbf{a}_2)$ and $\exp(i\mathbf{k} \cdot \mathbf{a}_3)$, respectively. The hopping term (A.7) may therefore be written as

$$t_{\mathbf{k}}^{AB} = t\gamma_{\mathbf{k}}^* = \left(t_{\mathbf{k}}^{BA}\right)^*,$$

where we have defined the sum of the nearest-neighbor phase factors

$$\gamma_{\mathbf{k}} \equiv 1 + e^{i\mathbf{k}\cdot\mathbf{a}_2} + e^{i\mathbf{k}\cdot\mathbf{a}_3}. \tag{A.9}$$

The band dispersion may now easily be obtained by solving the secular equation (A.6),

$$\epsilon_\lambda(\mathbf{k}) = \lambda \left|t_{\mathbf{k}}^{AB}\right| = \lambda t \left|\gamma_{\mathbf{k}}\right|, \tag{A.10}$$

and is plotted in Figure 6.8. The band dispersion is obviously particle – hole symmetric, and the valence band ($\lambda = -$) touches the conduction band ($\lambda = +$) in the inequivalent points

$$\pm\mathbf{K} = \pm\frac{4\pi}{3\sqrt{3}a}\mathbf{e}_x,$$

which one determines by setting $\gamma_{\pm\mathbf{K}} = 0$ and that coincide with the two inequivalent BZ corners K and K'. Because the whole band structure is half-filled in undoped graphene, as we have mentioned in Section 6.2.1, the Fermi energy lies exactly in these points K and K'.

Continuum Limit

The low-energy electronic properties may be obtained by expanding the band structure in the vicinity of these points, and the low-energy Hamiltonian is obtained simply by expanding the sum of the phase factors (A.9) around K and K',

$$\gamma_{\mathbf{p}}^{\pm} \equiv \gamma_{\mathbf{k}=\pm\mathbf{K}+\mathbf{p}} = 1 + e^{\pm i\mathbf{K}\cdot\mathbf{a}_2}e^{i\mathbf{p}\cdot\mathbf{a}_2} + e^{\pm i\mathbf{K}\cdot\mathbf{a}_3}e^{i\mathbf{p}\cdot\mathbf{a}_3}$$

$$\simeq 1 + e^{\pm i2\pi/3}\left[1 + i\mathbf{p}\cdot\mathbf{a}_2\right] + e^{\mp i2\pi/3}\left[1 + i\mathbf{p}\cdot\mathbf{a}_3\right]$$

$$= \gamma_{\mathbf{p}}^{\pm(0)} + \gamma_{\mathbf{p}}^{\pm(1)}.$$

By definition of the Dirac points and their position at the BZ corners K and K', we have $\gamma_{\mathbf{p}}^{\pm(0)} = \gamma_{\pm\mathbf{K}} = 0$. We limit the expansion to first order in $|\mathbf{p}|a$. Notice that, in order to simplify the notations, we have used a system of units with $\hbar = 1$, i.e. where the momentum has the same units as the wavevector.

The first-order term is given by

$$\gamma_{\mathbf{p}}^{\pm(1)} = i\frac{\sqrt{3}a}{2}\left[(p_x + \sqrt{3}p_y)e^{\pm i2\pi/3} + (-p_x + \sqrt{3}p_y)e^{\mp i2\pi/3}\right]$$

$$= \mp\frac{3a}{2}(p_x \pm ip_y), \tag{A.11}$$

which is obtained with the help of $\sin(\pm 2\pi/3) = \pm\sqrt{3}/2$ and $\cos(\pm 2\pi/3) = -1/2$. This yields the effective low-energy Hamiltonian

$$H_{\mathbf{p}}^{\xi} = \xi v(p_x\sigma^x + \xi p_y\sigma^y), \tag{A.12}$$

in terms of the Fermi velocity

$$v \equiv \frac{3ta}{2\hbar}. \tag{A.13}$$

The index $\xi = \pm$ denotes the valleys K and K', and one obtains at the K point the Dirac Hamiltonian mentioned in Eq. (6.20)

$$H_D = v\mathbf{p}\cdot\boldsymbol{\sigma}, \tag{A.14}$$

whereas the low-energy Hamiltonian at the K' point reads

$$H_D' = -v\mathbf{p}\cdot\boldsymbol{\sigma}^*, \tag{A.15}$$

with $\boldsymbol{\sigma}^* = (\sigma^x, -\sigma^y)$. Both Hamiltonians yield the same energy spectrum, which is therefore *two-fold valley degenerate*.

Notice that if one prefers to avoid the complex conjugation in the Hamiltonian (A.15), one simply changes the representation by interchanging the A and B sublattices, in which case one may write the Hamiltonians for the two valleys K ($\xi = +$) and K' ($\xi = -$) in a compact form,

$$H_D^{\xi} = \xi H_D = \xi v\mathbf{p}\cdot\boldsymbol{\sigma}. \tag{A.16}$$

Appendix B: Landau levels of massive Dirac particles

Mass confinement of Dirac fermions at $B = 0$

Even in the absence of a magnetic field, electronic confinement in graphene turns out to be quite tricky because a simple-minded approach in terms of a potential $V_{\text{conf}} = V(y)\mathbb{1}$ cannot confine Dirac electrons. This fact is due to an intrinsically relativistic effect that is called the *Klein paradox*, according to which a (massless) relativistic particle may transverse a potential barrier without being backscattered (Klein, 1929). This effect may be understood in the following manner: consider an incident electron in the region with $V = 0$ the energy of which is slightly above the Fermi energy. In the potential barrier, the Dirac point is shifted to a higher energy that corresponds to the barrier height, and the Fermi energy lies now in the valence band, where the electron may still find a quantum state (with the same velocity v) – instead of moving as an electron in the conduction band, it thus simply moves in the same direction as an electron in the valence band (Figure B.1(a)). This is in stark contrast with quantum-mechanical tunnelling of a non-relativistic particle, for which the transmission probability through a potential barrier is exponentially suppressed because of a lacking quantum state at the same energy as that of the incident electron.

The problem is circumvented by a so-called *mass confinement*

$$V_{\text{conf}} = V(y)\,\sigma^z = \begin{pmatrix} V(y) & 0 \\ 0 & -V(y) \end{pmatrix}, \tag{B.1}$$

and we discuss first the simpler case of a constant mass term $M\sigma^z$ that needs to be added to the Dirac Hamiltonian. That this term indeed yields a mass may be seen from the Dirac Hamiltonian at $B = 0$

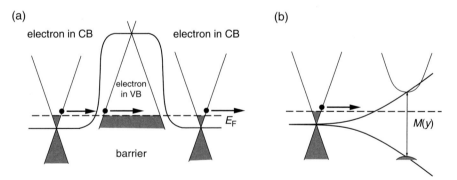

Fig. B.1 (a) Klein tunnelling through a barrier. An incident electron in the conduction band (CB) above the Fermi energy, which is at the Dirac point before the barrier, transverses the barrier as en electron above the Fermi energy in the valence band (VB). The valence band is partially emptied because the Dirac point has shifted to a higher energy corresponding to the barrier height. (b) Mass confinement. A gap opens when the particle approaches the edge, which becomes a forbidden region where no quantum state can be found at the energy corresponding to that of the incident electron.

$$H_D^m = v\mathbf{p} \cdot \boldsymbol{\sigma} + M\sigma^z = \begin{pmatrix} M & v(p_x - ip_y) \\ v(p_x + ip_y) & -M \end{pmatrix}, \tag{B.2}$$

the diagonalization of which yields the energy spectrum

$$\epsilon_\lambda(\mathbf{p}) = \lambda\sqrt{v^2|\mathbf{p}|^2 + M^2},$$

which is gapped at zero momentum. This is simply the dispersion relation of a relativistic particle[62] with mass m such that $M = mv^2$. Qualitatively, one may see from Figure B.1(b) why a mass confinement is more efficient than a potential barrier. Indeed, when the particle approaches the edge with $M(y) \neq 0$ a gap opens. An electron slightly above the Dirac point may then only propagate in the region with $M = 0$, whereas at the edge its energy lies in the gap that is a forbidden region, and the electron is thus confined.

Similarly to the $B = 0$ case, one may find the energy spectrum of the massive Dirac Hamiltonian (B.2) in a perpendicular magnetic field, which reads, in terms of the ladder operators a and $a^\dagger$,

$$H_D^B = \begin{pmatrix} M & v(\Pi_x - i\Pi_y) \\ v(\Pi_x + i\Pi_y) & -M \end{pmatrix} = \begin{pmatrix} M & \sqrt{2}\frac{\hbar v}{l_B}a \\ \sqrt{2}\frac{\hbar v}{l_B}a^\dagger & -M \end{pmatrix}. \tag{B.3}$$

Its eigenvalues may be obtained in the same manner as in the $M = 0$ case (c.f. Section 6.2.3.2), and one obtains

$$\epsilon_{\lambda n} = \lambda\sqrt{M^2 + 2\frac{\hbar^2 v^2}{l_B^2}n} \tag{B.4}$$

for the massive relativistic LLs, $n \neq 0$.

Special care needs to be taken in the discussion of the central LL $n = 0$, which necessarily shifts away from zero energy. The associated quantum state (6.39) is zero in the first component u_0, whereas the second component is given by $v_0 = |0\rangle$. In order to satisfy the second line in the eigenvalue equation

$$H_D^B\psi_0 = \epsilon_0\psi_0 \qquad \Leftrightarrow \qquad \begin{pmatrix} M & \sqrt{2}\frac{\hbar v}{l_B}a \\ \sqrt{2}\frac{\hbar v}{l_B}a^\dagger & -M \end{pmatrix}\begin{pmatrix} 0 \\ |0\rangle \end{pmatrix} = \epsilon_0\begin{pmatrix} 0 \\ |0\rangle \end{pmatrix},$$

one needs to fulfil

$$\sqrt{2}\frac{\hbar v}{l_B}a^\dagger u_0 = (\epsilon_0 + M)v_0 \qquad \Leftrightarrow \qquad 0 = (\epsilon_0 + M)|0\rangle, \tag{B.5}$$

such that the only solution is $\epsilon_0 = -M$. The relativistic $n = 0$ LL is therefore shifted to negative energies and no longer satisfies particle – hole symmetry. This effect is called *parity anomaly* and depends on the *sign of the mass*.

In the case of graphene, we need to remember that there are two copies of the energy spectrum, one at the K point and one at the K' point. As we have disussed in Appendix 2, the Hamiltonian (B.3) describes the low-energy properties at the K

point whereas we need to interchange the A and B sublattices at the K' point and add a global sign in front of the off-diagonal terms (see Eq. (A.16)),

$$H_D'^B = \begin{pmatrix} -M & -\sqrt{2}\frac{\hbar v}{l_B} a \\ -\sqrt{2}\frac{\hbar v}{l_B} a^\dagger & M \end{pmatrix} = -H_D^B. \tag{B.6}$$

Naturally, the eigenstates of this Hamiltonian are the same as those of the Hamiltonian (B.3) at the K point, but the eigenvalues change their sign. Due to the particle – hole symmetry of the levels (B.4), the global sign does not affect the energy spectrum for $n \neq 0$. However, the $n = 0$ LL, which does not respect particle – hole symmetry, must again be treated separately, and one finds in the same manner as for the K point the condition corresponding to Eq. (B.5),

$$-\sqrt{2}\frac{\hbar v}{l_B} a^\dagger u_0 = (\epsilon_0 - M) v_0 \qquad \Leftrightarrow \qquad 0 = (\epsilon_0 - M)|0\rangle. \tag{B.7}$$

One notices that the $n = 0$ LL level at the K' point shifts to positive energies as a function of the mass, such that the overall level spectrum for graphene, when one takes into account *both* valleys, is again particle – hole symmetric, but the valley degeneracy is lifted for $n = 0$.

The case of a mass term that varies in the y-direction, such as for the mass confinement potential, may finally be treated in the same manner as we have discussed in Section 6.3.1.2: the system remains translation-invariant in the x-direction, such that the Landau gauge is the appropriate gauge and the wavevector k in this direction is a good quantum number. Because this wavevector determines the position of the eigenstate in the y-direction, $y_0 = kl_B^2$, the energy spectrum is given by the expression (6.92),

$$\epsilon_{\lambda n, y_0; \xi} = \lambda\sqrt{M^2(y_0) + 2\frac{\hbar^2 v^2}{l_B^2} n}, \tag{B.8}$$

for $n \neq 0$ and both valleys $\xi = \pm$, whereas the $n = 0$ LL is found at

$$\epsilon_{n=0, y_0; \xi} = -\xi M(y_0). \tag{B.9}$$

Notes

1. These lectures are also available on the preprint server, http://arxiv.org/abs/cond-mat/9907002
2. Notice, however, that the Fermi energy and thus the DOS is a function of the electronic density. Furthermore, we mention that in a fully consistent treatment the diffusion constant D also depends on the density of states and eventually on the magnetic field. This affects the precise form of the oscillation but not its periodicity.

3. Naturally, this is a crude assumption because if the density of states $\rho(\epsilon, B)$ depends on the magnetic field, so does the Fermi energy via the relation

$$\int_0^{E_F} d\epsilon\, \rho(\epsilon, B) = n_{el}.$$

However, the basic features of the Shubnikov–de Haas oscillation may be understood when keeping the Fermi energy constant.

4. The subscript K honours v. Klitzing and 90 stands for the date since which the unit of resistance is defined by the IQHE.

5. The quantity n determines the filling of the LLs, usually described by the Greek letter ν, as we will discuss in Section 6.2.

6. All vector quantities (also in the quantum-mechanical case of operators) $\mathbf{v} = (v_x, v_y)$ are hence 2D, unless stated explicitly.

7. The statement that $\mathbf{p}$ is a constant of motion remains valid also in the case of a relativistic particle. However, the Hamiltonian description depends on the frame of reference because the energy is not Lorentz-invariant, i.e. invariant under a transformation into another frame of reference that moves at constant velocity with respect to the first one. For this reason a Lagrangian rather than a Hamiltonian formalism is often prefered in relativistic quantum mechanics.

8. Although this may seem to be a typical "theoretician's assumption", it is a very good approximation when the lattice size is much larger than all other relevant length scales, such as the lattice spacing or the Fermi wavelength.

9. In GaAs, e.g., the band mass is $m_b = 0.068 m_0$, in terms of the free electron mass m_0.

10. Indeed, in graphene, the relevant low-energy scales are in the $10 - 100$ meV regime, whereas non-linear corrections of the band dispersion become relevant in the eV regime.

11. Notice that $\gamma_{\pm\mathbf{K}} = 0$ by symmetry.

12. One obtains a similar result at the K' point, see Eq. (A.15) in Appendix B.

13. Higher magnetic fields may be obtained only in *semi-destructive experiments*, in which the sample survives the experiment but not the coil that is used to produce the magnetic field.

14. More precisely we have used a gradient generalization of this relation to operator functions that depend on several different operators,

$$[\mathcal{O}_0, f(\mathcal{O}_1, ..., \mathcal{O}_J)] = \sum_{j=1}^{J} \frac{\partial f}{\partial \mathcal{O}_j} [\mathcal{O}_0, \mathcal{O}_j],$$

which is valid if $[[\mathcal{O}_0, \mathcal{O}_j], \mathcal{O}_0] = [[\mathcal{O}_0, \mathcal{O}_j], \mathcal{O}_j] = 0$ for all $j = 1, ..., N$.

15. Sometimes, a *cyclotron mass* m_C is formally introduced via the equality $\omega' \equiv eB/m_C$. However, this mass is a somewhat artificial quantity, which turns out to depend on the carrier density. We will therefore not use this quantity in the present chapter.

16. Epitaxial graphene is obtained from a thermal graphitization process of an epitaxially grown SiC crystal (Berger, *et al.* 2004).

17. The quantum states are naturally only degenerate if one neglects the Zeeman effect.

18. We will nevertheless try to give a physical interpretation to this operator below, within a semi-classical picture.

19. Mathematicians speak of a *non-commutative geometry* in this context, and the charged 2D particle in a magnetic field may be viewed as a paradigm of this concept.

20. We limit the discussion to the non-relativistic case. The spinor wave functions for relativistic electrons are then easily obtained with the help of Eqs. (6.49) and (6.50).

21. Naturally, the system is also confined in the x-direction, but since we consider a sample with $L \gg W$, the system appears as translation-invariant in the x-direction when one considers intermediate length scales. The latter may be taken into account with the help of periodic boundary conditions that discretize the wavevector in the x-direction, as we have seen in the preceding section within the quantum-mechanical treatment of the 2D electron in the Landau gauge (see Section 6.2.4.2).

22. This approximation may be viewed as the first term of an expansion of the electrostatic potential in the coherent (or vortex) state basis, where the states are maximally localized around the guiding-center position $\mathbf{R}$ (Champel and Florens 2007).

23. In order to illustrate this point, consider a hiking tour in the mountains, e.g. around Les Houches in the French Alps. To go from one point to another one at the same height, one usual needs go downhill as well as uphill. It is very rare to be able to stay on the same height unless one wants to turn in circles that are just the closed countour lines that correspond to closed equipotential lines in our potential landscape. For those who participated at the Les Houches session that was outsourced to Singapore, where there are no mountains and where the weather is anyway too hot for hiking, just look at a hiking map of some mountainous region. Then search for countour lines that connect one border of the map to the opposite border. It turns out to be very hard to find such lines as compared to a large number of closed countour lines.

24. In the semi-classical picture the extended states are also called *skipping orbits*.

25. This relation may be obtained from the Heisenberg equations of motion, $i\hbar \dot{x} = [x, H] = (\partial H/\partial p_x)[x, p_x] = i\partial H/\partial k$, where we have used Eq. (6.26) and $p_x = \hbar k$. One therefore obtains the operator equation

$$\dot{x} = \frac{1}{\hbar}\frac{\partial H}{\partial k},$$

which we evaluate in the state $|n, k\rangle$. In the last step we have used the periodic boundary conditions.

26. Because the lowest LL is labelled by $n' = 0$, the last one has the index $n - 1$ in the case of n completely filled levels.

27. Strictly speaking the filling factor does not jump abruptly when one takes interactions between the electrons into account. In this case, two incompressible strips,

of $\nu = n + 1$ and $\nu = n$ are separated by a *compressible* strip of finite width. The picture of chiral electron transport remains, however, essentially the same when considering such compressible regions.

28. Strictly speaking, we have not gained anything because the quantum treatment allows us only to determine the Hall resistance at certain points of the Hall curve, those at the magnetic fields corresponding to $\nu = hn_{el}/eB = n$. If we substitute the filling factor in Eq. (6.83), we see immediately that $R_H = h/e^2\nu = B/en_{el}$, i.e. one retrieves the *classical* result for the Hall resistance.

29. Remember that due to the label 0 for the lowest LL, all LLs with $n' = 0, ..., n - 1$ are the completely filled and the LL n is then the lowest *unoccupied* level.

30. The lecture notes for this class are available on the School's program webpage: http://www.ntu.edu.sg/ias/upcomingevents/LHSOPS09/Pages/programme.aspx

31. Although we use the same Greek letter ν for the critical exponent, it must not be confused with the filling factor, which plays no role in this subsection.

32. Notice that in relativity, time is considered as the "fourth" dimension, and Lorentz invariance would require that spatial and temporal fluctuations be equivalent, i.e. $z = 1$.

33. Notice, however, that due to the universality of the scaling laws and the fluctuations at all length scales, the results are expected to be independent on these *microscopic* assumptions.

34. The particle – hole transformed landscape corresponds to an impurity distribution in which one interchanges negatively and positively charged impurities.

35. As before, we neglect the electron spin to render the discussion as simple as possible. The role of the spin will be discussed briefly in the last section on multi-component systems.

36. In order to simplify the discussion, we consider only the IQHE in non-relativistic quantum Hall systems, but the arguments apply also to the RQHE in graphene.

37. As for the IQHE, impurities play nevertheless an important role in the localization of quasi-particles, which we need to invoke later in this chapter in order to explain the transport properties of the FQHE.

38. There is even some slight indication for a 1/5 FQHE state in the next excited LL $n = 2$ (Gervais, *et al.* 2004).

39. We neglect, here, the numerical prefactors that account for the normalization of the wave functions.

40. We are therefore confronted with the somewhat bizarre situation where we dispose of a variational wave function with no possible variation.

41. Notice, however, that this shift plays an important role in numerical calculations, such as exact diagonalization, when performed on special geometries, such as on a sphere.

42. In order to simplify the notations, we have omitted the LL quantum number $n = 0$, which is the same for both particles in this wave function.

43. This is similar to the average value of the radius at which the electron's guiding center is placed in the symmetric gauge (see Section 6.2.4.1). Remember (e.g. from classical mechanics) that the decomposition of a two-particle wave function into

relative and center-of-mass coordinates maps the two-body problem to an effective one-body problem.

44. One has $v_1/v_3 \simeq 1.6$ in the LLL. Remember that pseudopotentials with even angular momentum quantum number m do not play any physical role because of the fermionic nature of the electrons.

45. Naturally, the total surface of the quantum Hall system remains constant, but physically we have slightly *increased* the B-field. Each quantum state occupies then an infinitesimally smaller surface $2\pi l_B^2$, such that the system may accomodate for one more quantum state, $M = N_B \to N_B + 1$.

46. From now on, we use the term "(Laughlin) quasi-particles" generically in order to denote quasi-particles *and* quasi-holes.

47. Later, this kind of experiment was repeated for other FQHE states.

48. mock: *singlish* for fake; mainly used in the description of Singaporean catering food.

49. Notice that Laughlin's wave function describes a system at $T = 0$, such that temperature does not intervene in the expressions. The choice is purely formal.

50. There are more complicated cases of non-Abelian statistics, in which the exchange processes of more than two different particles no longer commute, but we do not discuss this case here and refer the reader to the review by Nayak, *et al.* (Nayak, *et al.* 2008).

51. Remember that the statistical angle is defined with respect to an exchange process $\mathcal{E}$ that is the square root of the process $\mathcal{T}$ considered here [Eq. (6.114)]. The relation between the statistical angle and the Aharononv–Bohm phase is therefore $\Gamma = 2\pi\alpha$ and not $\pi\alpha$.

52. Remember, however, that the energy scale of this gap is not given in terms of a kinetic energy $\hbar eB/m$, but in terms of the Coulomb interaction $e^2/\epsilon l_B$.

53. There have been attempts in the literature to formalize this point (Haldane and Rezayi 1985; Wójs and Quinn 2000).

54. Remember that this field is somewhat arbitrary because the situation $\nu = 1$ may also be obtained easily for other fields by varying the gate voltage V_G.

55. Naturally, the dielectric constant depends on the dielectric environment around the graphene sheet and thus also on the substrate.

56. This pseudo-potential, as well as any other with an even value of m, does not play any physical role due to the Pauli principle if one considers only spinless electrons, as we have mentioned in Section 6.4.2.2.

57. Remember that for a crystaline ground state (WC), the Goldstone mode is the acoustic phonon, as we have briefly discussed in Section 6.4.1.

58. Naturally, such an antisymmetric orbital wave function is only physical if the layer separation d is not too large (as compared to the magnetic length) – otherwise one would simply have completely decoupled layers.

59. Contrary to the Josephson effect, only the tunnelling conductance dI_z/dV is strongly enhanced, whereas the tunnelling *current* remains zero in the quasi-Josephson effect in bilayer systems.

60. The filling factor $\nu_G = 1$ is related to $\nu_G = -1$ by particle – hole symmetry and therefore does not require a separate discussion.

61. The wavefunction $\psi_{\mathbf{k}}(\mathbf{r})$ is, thus, the real-space representation of the Hilbert vector $\psi_{\mathbf{k}}$.
62. The sign $\lambda = -$ corresponds to the anti-particle.

References

Abolfath, M., Belkhir, L., and Nafari, N. (1997). *Phys. Rev. B*, **55**, 10643.

Abrahams, E., Anderson, P. W., Licciardello, D. C., and Ramakrishnan, T. V. (1979). *Phys. Rev. Lett.*, **42**, 673.

Akkermans, E. and Montambaux, G. (2008). *Mesoscopic physics of electrons and photons.* Cambridge University Press, Cambridge.

Alicea, J. and Fisher, M. P. A. (2006). *Phys. Rev. B*, **74**, 075422.

Andrei, E. Y., Deville, G., Glattli, D. C., Williams, F. I. B., Paris, E., and Etienne, B. (1988). *Phys. Rev. Lett.*, **60**, 2765.

Arovas, D. P., Karlhede, A., and Lilliehöök, D. (1999). *Phys. Rev. B*, **59**, 13147.

Ashcroft, N. W. and Mermin, N. D. (1976). *Solid state physics.* Harcourt, Orlando.

Berger, C., Song, Z., Li, T., Ogbazghi, A. Y., Feng, R., Dai, Z., Marchenkov, A. N., Conrad, E. H., First, P. N., and de Heer, W. A. (2004). *J. Phys. Chem.*, **108**, 19912.

Bolotin, K. I., Ghahari, F., Shulman, M. D., Stormer, H. L., and Kim, P. (2009). Nature **462**, 196.

Brey, L. and Fertig, H.A. (2006). *Phys. Rev. B*, **73**, 195408.

Büttiker, M. (1988). *Phys. Rev. B*, **38**, 9375.

Büttiker, M. (1992). *The quantum Hall effect in open conductors*, in M. Reed (ed.) *Nanostructured systems (Semiconductors and Semimetals*, **35**, 191). Academic Press, Boston.

Büttiker, M., Imry, Y., Landauer, R., and Pinhas, S. (1985). *Phys. Rev. B*, **31**, 6207.

Castro Neto, A. H., Guinea, F., Peres, N. M. R., Novoselov, K. S., and Geim, A. K. (2009). *Rev. Mod. Phys.*, **81**, 109.

Chakraborty, T. and Zhang, F. C. (1984). *Phys. Rev. B*, **29**, 7032.

Chalker, J. T. and Coddington, P. D. (1988). *J. Phys. C*, **21**, 2665.

Champel, Th. and Florens, S. (2007). *Phys. Rev. B*, **75**, 245326.

Cohen-Tannoudji, C., Diu, B., and Laloë, F. (1973). *Quantum mechanics.* Hermann, Paris.

Cooper, N. R. (2008). *Adve. Phys.*, **57**, 539.

Datta, S. (1995). *Electronic transport in mesoscopic systems.* Cambridge University Press, Cambridge.

de Gail, R., Regnault, N., and Goerbig, M. O. (2008). *Phys. Rev. B*, **77**, 165310.

de Picciotto, R., Reznikov, M., Heidblum, M., Umansky, V., Bunin, G., and Mahalu, D. (1997). *Nature*, **389**, 162.

Douçot, B., Goerbig, M. O., Lederer, P., and Moessner, R. (2008). *Phys. Rev. B*, **78**, 195327.

Du, X., Skachko, I., Duerr, F., Luican, A., and Andrei, E. Y. (2009). *Nature*, **462**, 192.

Ezawa, Z. F. (1999). *Phys. Rev. Lett.*, **82**, 3512.

Ezawa, Z. F. (2000). *Quantum Hall effects – field theoretical approach and related topics.* World Scientific, Singapore.

Ezawa, Z. F. and Iwazaki, A. (1993). *Phys. Rev. B*, **47**, 7295.

Fano, G., Ortolani, F., and Colombo, E. (1986). *Phys. Rev. B*, **34**, 2670.

Fertig, H. A. (1989). *Phys. Rev. B*, **40**, 1087.

Fuchs, J.-N. and Lederer, P. (2007). *Phys. Rev. Lett.*, **98**, 016803.

Fukuyama, H., Platzman, P. M., and Anderson, P. W. (1979). *Phys. Rev. B*, **19**, 5211.

Gervais, G., Engel, L. W., Stormer, H. L., Tsui, D. C., Baldwin, K. W., West, K. W., and Pfeiffer, L. N. (2004). *Phys. Rev. Lett.*, **93**, 266804.

Girvin, S. M. (1999). *The quantum Hall effect: Novel excitations and broken symmetries,* in A. Comptet, T. Jolicoeur, S. Ouvry and F. David (eds.) *Topological aspects of low-dimensional systems – École d'Éte de Physique Théorique LXIX.* Springer, Berlin.

Girvin, S. M. and Jach, T. (1984). *Phys. Rev. B*, **29**, 5617.

Girvin, S. M., MacDonald, A. H., and Platzman, P. M. (1986). *Phys. Rev. B*, **33**, 2481.

Giuliani, G. F. and Vignale, G. (2005). *Quantum theory of electron liquids.* Cambridge UP, Cambridge.

Goerbig, M. O., Douçot, B., and Moessner, R. (2006). *Phys. Rev. B*, **74**, 161407.

Goerbig, M. O. and Regnault, N. (2007). *Phys. Rev. B*, **75**, 241405.

Greiter, M., Wen, X.-G., and Wilczek, F. (1991). *Phys. Rev. Lett.*, **66**, 3205.

Gusynin, V. P., Miransky, V. A., Sharapov, S. G., and Shovkovy, I. A. (2006). *Phys. Rev. B*, **74**, 195429.

Haldane, F. D. M. (1983). *Phys. Rev. Lett.*, **51**, 605.

Haldane, F. D. M. (1991). *Phys. Rev. Lett.*, **67**, 937.

Haldane, F. D. M. and Rezayi, E. H. (1985). *Phys. Rev. Lett.*, **54**, 237.

Halperin, B. I. (1983). *Helv. Phys. Acta*, **56**, 75.

Halperin, B. I. (1984). *Phys. Rev. Lett.*, **52**, 1583.

Halperin, B. I., Lee, P. A., and Read, N. (1993). *Phys. Rev. B*, **47**, 7312.

Hashimoto, K., Sohrmann, C., Wiebe, J., Inaoka, T., Meier, F., Hirayama, Y., Römer, R. A., Wiesendanger, R., and Morgenstern, M. (2008). *Phys. Rev. Lett.*, **101**, 256802.

Heinonen, O., Ed. (1998). *Composite fermions.* World Scientific, Singapore.

Herbut, I. F. (2007). *Phys. Rev. B*, **75**, 165411.

Huckestein, B. (1995). *Rev. Mod. Phys.*, **67**, 357.

Huckestein, B. and Backhaus, M. (1999). *Phys. Rev. Lett.*, **82**, 5100.

Iyengar, A., Wang, J., Fertig, H. A., and Brey, L. (2007). *Phys. Rev. B*, **75**, 125430.

Jackson, J. D. (1999). *Classical electrondynamics.* Wiley, 3rd edn., New York.

Jain, J. K. (1989). *Phys. Rev. Lett.*, **63**, 199.

Jain, J. K. (1990). *Phys. Rev. B*, **41**, 7653.

Jain, J. K. (2007). *Composite fermions.* Cambridge University Press, Cambridge.

Jiang, Z., Henriksen, E. A., L. C. Tung, Y.-J. Wang, Schwartz, M. E., Han, M. Y., Kim, P., and Stormer, H. L. (2007). *Phys. Rev. Lett.*, **98**, 197403.

Kallin, C. and Halperin, B. I. (1984). *Phys. Rev. B*, **30**, 5655.

Kang, W., Young, J. B., Hannahs, S. T., Palm, E., Campman, K. L., and Gossard, A. C. (1997). *Phys. Rev. B*, **56**, R12776.

Kellogg, M., Eisenstein, J. P., and an K. W. West, L. N. Pfeiffer (2004). *Phys. Rev. Lett.*, **036801**.

Kitaev, A. Y. (2003). *Ann. Phys. (N.Y.)*, **303**, 2.

Kittel, C. (2005). *Introduction to solid state physics.* 8th edn., Wiley, New York.

Klaß, U., Dietsche, W., v. Klitzing, K., and Ploog, K. (1991). *Z. Phys. B: Cond. Matt.*, **82**, 351.

Klein, O. (1929). *Z. Phys.*, **53**, 157.

Kukushkin, I. K., v. Klitzing, K., and Eberl, K. (1999). *Phys. Rev. Lett.*, **82**, 3665.

Laughlin, R. B. (1983). *Phys. Rev. Lett.*, **50**, 1395.

Li, W., Csathy, A., Tsui, D. C., Pfeiffer, L. N., and West, K. W. (2005). *Phys. Rev. Lett.*, **94**, 206807.

Li, W., Vicente, C. L., Xia, J. S., Pan, W., Tsui, D. C., Pfeiffer, L. N., and West, K. W. (2009). *Phys. Rev. Lett.*, **102**, 216801.

Lopez, A. and Fradkin, E. (1991). *Phys. Rev. B*, **44**, 5246.

Luhman, D. R., Pan, W., Tsui, D. C., Pfeiffer, L. N., Baldwin, K. W., and West, K. W. (2008). *Phys. Rev. Lett.*, **101**, 266804.

MacDonald, A. H., Yoshioka, D., and Girvin, S. M. (1989). *Phys. Rev. B*, **39**, 8044.

Mahan, G. D. (1993). *Many-particle physics.* 2nd edn., Plenum Press, New York.

McClure, J. W. (1956). *Phys. Rev.*, **104**, 666.

Mermin, N. D. (1979). *Rev. Mod. Phys.*, **51**, 591.

Moon, K., Mori, H., Yang, K., Girvin, S. M., MacDonald, A. H., Zheng, I., Yoshioka, D., and Zhang, S.-C. (1995). *Phys. Rev. B*, **51**, 5143.

Moore, G. and Read, N. (1991). *Nucl. Phys. B*, **360**, 362.

Murthy, G. and Shankar, R. (2003). *Rev. Mod. Phys.*, **75**, 1101.

Nayak, Ch., Simon, S. H., Stern, A., Friedman, M., and Das Sarma, S. (2008). *Rev. Mod. Phys.*, **80**, 1083.

Nomura, K. and MacDonald, A. H. (2006). *Phys. Rev. Lett.*, **96**, 256602.

Novoselov, K. S., Geim, A. K., Morosov, S. V., Jiang, D., Katsnelson, M. I., Grigorieva, I. V., Dubonos, S. V., and Firsov, A. A. (2005). *Nature*, **438**, 197.

Pan, W., Stormer, H. L., Tsui, D. C., Pfeiffer, L. N., Baldwin, K. W., and West, K. W. (2003). *Phys. Rev. Lett.*, **90**, 016801.

Papić, Z., Möller, G., Milovanović, M., Regnault, N., and Goerbig, M. O. (2009). *Phys. Rev. B*, **79**, 245327.

Poirier, W. and Schopfer, F. (2009a). *Eur. Phys. J. Special Topics*, **172**, 207.

Poirier, W. and Schopfer, F. (2009b). *Int. J. Mod. Phys. B*, **23**, 2779.

Prange, R. and Girvin, S. M., Eds. (1990). *The quantum Hall effect.* Springer, New York.

Rezayi, E. H. and Read, N. (1994). *Phys. Rev. Lett.*, **72**, 100.

Roldán, R., Fuchs, J.-N., and Goerbig, M. O. (2009). *Phys. Rev. B*, **80**, 085408.

Sachdev, S. (1999). *Quantum phase transitions.* Cambridge University Press, Cambridge.

Sadowski, M. L., Martinez, G., Potemski, M., Berger, C., and de Heer, W. A. (2006). *Phys. Rev. Lett.*, **97**, 266405.

Saito, R., Dresselhaus, G., and Dresselhaus, M. S. (1998). *Physical properties of carbon nanotubes.* Imperial College Press, London.

Saminadayar, L., Glattli, D. C., Jin, Y., and Etienne, B. (1997). *Phys. Rev. Lett.*, **79**, 2526.

Shabani, J., Gokmen, T., and Shayegan, M. (2009). *Phys. Rev. Lett.*, **103**, 046805.

Shubnikov, L. W. and de Haas, W. J. (1930). *Proceedings of the Royal Netherlands Society of Arts and Science*, **33**, 130 and 163.

Slevin, K. and Ohtsuki, T. (2009). *Phys. Rev. B*, **80**, 041304.

Sondhi, S. L., Girvin, S. M., Carini, J. P., and Shahar, D. (1997). *Rev. Mod. Phys.*, **69**, 315.

Sondhi, S. L., Karlhede, A., Kivelson, S. A., and Rezayi, E. H. (1993). *Phys. Rev. B*, **47**, 16419.

Spielman, I. B., Eisenstein, J. P., Pfeiffer, L. N., and West, K. W. (2000). *Phys. Rev. Lett.*, **84**, 5808.

Tinkham, M. (2004). *Introduction to superconductivity.* 2nd edn., Dover Publications, Dover.

Tsui, D. C., Störmer, H., and Gossard, A. C. (1982). *Phys. Rev. Lett.*, **48**, 1559.

Tutuc, E., Shayegan, M., and Huse, D. A. (2004). *Phys. Rev. Lett*, **93**, 036802.

v. Klitzing, K., Dorda, G., and Pepper, M. (1980). *Phys. Rev. Lett.*, **45**, 494.

Wallace, P. R. (1947). *Phys. Rev.*, **71**, 622.

Wei, H. P., Engel, L. W., and Tsui, D. C. (1994). *Phys. Rev. B*, **50**, 14609.

Wei, H. P., Tsui, D. C., Paalanen, M. A., and Pruisken, A. M. M. (1988). *Phys. Rev. Lett.*, **61**, 1294.

Wen, X.-G. and Zee, A. (1992*a*). *Phys. Rev. Lett*, **69**, 1811.

Wen, X.-G. and Zee, A. (1992*b*). *Phys. Rev. B*, **46**, 2290.

Wigner, E. (1934). *Phys. Rev.*, **102**, 46.

Willett, R. L., Eisenstein, J. P., Stormer, H. L., Tsui, D. C., Gossard, A. C., and English, J. H. (1987). *Phys. Rev. Lett.*, **59**, 1776.

Wójs, A. and Quinn, J. J. (2000). *Philos. Mag. B*, **80**, 1405.

Yang, K., Das Sarma, S., and MacDonald, A. H. (2006). *Phys. Rev. B*, **74**, 075423.

Yoshioka, D. (2002). *The quantum Hall effect.* Springer, Berlin.

Zhang, Y., Jiang, Z., Small, J. P., Purewal, M. S., Tan, Y.-W., Fazlollahi, M., Chudow, J. D., Jaszczak, J. A., Stormer, H. L., and Kim, P. (2006). *Phys. Rev. Lett.*, **98**, 197403.

Zhang, Y., Tan, Y.-W., Stormer, H. L., and Kim, P. (2005). *Nature*, **438**, 201.

7
Quantum phase transitions

G. G. BATROUNI[1] and R. T. SCALETTAR[2]

[1]Institut Non Linéaire de Nice, Université of Nice Sophia, CNRS;
1361 route des Lucioles, 06560 Valbonne, France.
[2]Physics Department, University of California, Davis, CA 95616, USA

7.1   Introduction

Thermal phase transitions, usually triggered by fluctuations caused by tuning the temperature, T, close to some special value T_c, are very familiar daily phenomena. As the temperature is lowered, thermal fluctuations decrease and eventually will cease as $T \to 0$. As is well known, however, quantum fluctuations do not stop at zero temperature. These quantum fluctuations can, under certain conditions, trigger phase transitions known as quantum phase transitions (QPT). For this to happen, the amplitude of these fluctuations needs to be controlled, which can be accomplished by tuning a parameter in the Hamiltonian governing the system. In what follows, we will present an introduction to these phase transitions with the help of two concrete examples that will serve to bring out some of the most important features and make clear the relation to classical statistical mechanics and thermal phase transitions.

We will start in this section with a brief review of thermal phase transitions, scaling relations and their generalization to QPT. We will then consider dynamics and its relation to excitation spectra. In Section 7.2 we will present in some detail the specific example of the one-dimensional Ising model in a transverse magnetic field. In Section 7.3 we will present the Hubbard model with a brief discussion of the main features of its phase diagram in the bosonic case. In addition, while this is not the focus of this chapter, we will, nonetheless, make contact between the formalism we will develop and the methods of quantum Monte Carlo (QMC).

Our goal here is to present only a first pedagogical introduction to this interesting and important subject. For further study and application to other systems, we refer the reader to Refs. (Sachdev, 1999) and (Sondhi *et al.*, 1997).

7.1.1   Critical points and scaling relations

Consider a lattice system at temperature $T = 0$ governed by a Hamiltonian of the form

$$\mathcal{H} = H_0 + gH_1, \tag{7.1}$$

where g is a tunable dimensionless parameter. If the lattice is finite and $[H_0, H_1] \neq 0$, then as g is tuned, there could be an avoided level crossing between the ground state and an excited state. As the system size increases, such an avoided level crossing can sharpen and eventually, in the thermodynamic limit, become an actual level crossing. When this happens, the energy, $E(g)$ (which is equal to the free energy, $F(g)$, since $T = 0$), becomes non-analytic at that value of g, signalling a phase transition. This is not a thermal but a quantum phase transition. A qualitative physical picture of this QPT, which will be made more precise below, is as follows. H_0 and H_1 are competing terms that favor different quantum states. When $g \ll 1$, the ground state is dominated by H_0 with some perturbative quantum fluctuations caused by H_1. When $g \gg 1$, the ground state is essentially that of H_1. As g is tuned between these two extreme limits, the competition between H_0 and H_1 will determine the ground state and, at some value, g_c, the above-mentioned level crossing may happen, indicating a QPT and a change in the relative dominance of H_0 and H_1.

As for thermal phase transitions, a QPT can be continuous (for example second order) or discontinuous (first order). In this chapter we will focus exclusively on the case of second-order transitions. It is well known that in the thermodynamic limit, second-order thermal phase transitions in classical statistical physics are characterized by diverging quantities (Stanley, 1971; Ma, 1985; LeBellac *et al.*, 2004) such as the specific heat and the magnetic susceptibility. In particular, a very important diverging quantity is the correlation length, ξ, which exhibits the power law

$$\xi \sim |T - T_c|^{-\nu}, \tag{7.2}$$

where T_c is the critical temperature and ν is the *correlation length critical exponent*. The correlation length plays a crucial role near T_c since it sets the distance scale. For example, experiments show that near the transition, the Fourier transform of the correlation function (i.e. the structure factor) has the form (Stanley, 1971; Ma, 1985; LeBellac *et al.*, 2004)

$$\tilde{G}(|\vec{q}|) = \int d^D r \, e^{i\vec{q}\cdot\vec{r}} G(\vec{r}), \tag{7.3}$$

$$= \frac{1}{q^{2-\eta}} f(q\xi), \tag{7.4}$$

which defines the critical exponent η. Dimensional arguments then give the long-distance behavior of the correlation function as

$$G(r) = \frac{1}{r^{D+\eta-2}} g\left(\frac{r}{\xi}\right). \tag{7.5}$$

The meaning of Eqs. (7.4) and (7.5) is that, near the critical point, the only characteristic length scale is the diverging correlation length, ξ. Equation (7.4) shows that in order for $\tilde{G}(|\vec{q}|)$ to be defined for $q = 0$, $f(q\xi) \sim (q\xi)^{2-\eta}$ as $q \to 0$ yielding

$$\tilde{G}(0) \sim \xi^{2-\eta} \sim |T - T_c|^{-\nu(2-\eta)}. \tag{7.6}$$

An example of $G(r)$ is the spin–spin correlation function. In such a case the magnetic susceptibility, which diverges as $\chi \sim |T - T_c|^{-\gamma}$, is proportional to $\tilde{G}(0)$ this gives the following relation between the critical exponents,

$$\gamma = \nu(2 - \eta). \tag{7.7}$$

Relation (7.7) is known as a *scaling relation*. Arguments similar to the above can be applied to other quantities such as the specific heat, $C \sim |T - T_c|^{-\alpha}$, the order parameter, M, at T_c as a function of an external field, h, $M \sim h^{1/\delta}$ and the order parameter near T_c at $h = 0$, $M \sim |T - T_c|^{\tilde{\beta}}$. Such an analysis leads to the following scaling relations (Stanley, 1971; Ma, 1985; LeBellac *et al.*, 2004),

$$\alpha = 2 - \nu D, \qquad 2\tilde{\beta} = \nu(D - 2 + \eta), \qquad \delta = \frac{D + 2 - \eta}{D - 2 + \eta}. \tag{7.8}$$

The six critical exponents $(\alpha, \tilde{\beta}, \gamma, \delta, \nu, \eta)$ obey four scaling relations. Remark: the critical exponent $\tilde{\beta}$ traditionally has no tilde, which we have added to avoid confusion with the inverse temperature β.

How is this modified by a QPT? To answer this question, we start with the partition function for a quantum system (Sachdev, 1999; LeBellac *et al.*, 2004),

$$Z = \text{Tr } e^{-\beta \mathcal{H}}, \tag{7.9}$$

where the Hamiltonian, $\mathcal{H}$, describes a D-dimensional quantum system. The expectation value of a quantum operator, $\mathcal{O}$, is given by

$$\langle \mathcal{O} \rangle = \frac{1}{Z} \text{Tr } \mathcal{O} e^{-\beta \mathcal{H}}. \tag{7.10}$$

Since the trace is an invariant, it can be evaluated in any representation and Eq. (7.9) can be written as

$$Z = \sum_{\{\phi\}} \langle \{\phi\} | e^{-\beta \mathcal{H}} | \{\phi\} \rangle, \tag{7.11}$$

where the $|\{\phi\}\rangle$ is a complete set of states. The operator $e^{-\beta \mathcal{H}}$ has the form of a quantum-mechanical time-evolution operator, $e^{-it\mathcal{H}}$, with $t \to -i\beta$. Equation (7.11) then lends itself to the following physical interpretation: The operator $e^{-\beta \mathcal{H}}$ evolves the state $|\{\phi\}\rangle$ by an imaginary time β; the partition function is the sum of the amplitudes, matrix elements, of all "paths" that bring the system back to the original state. The return to the original state, *i.e.* periodic boundary conditions in the imaginary time direction, is imposed by the trace in Eq. (7.9). Taking the thermodynamic limit of the quantum system means that the extent of the system in the D space dimensions tends to infinity. Since we are interested in quantum phase transitions, we want to take $T \to 0$, in other words $\beta \to \infty$. This means that the system now has infinite extent in the original D space dimensions plus the new imaginary time direction: It has become effectively a $(D+1)$-dimensional (classical) model. The details of this mapping will be worked out in the next section, for now we are only concerned with generalities. When the quantum critical point is approached, $g \to g_c$, the system will exhibit diverging quantities like in the thermal critical case. In particular, the correlation length in the space directions will diverge as (Sachdev, 1999)

$$\xi \sim |g - g_c|^{-\nu}. \tag{7.12}$$

The correlation "length", ξ_τ, along the new imaginary time direction will also diverge but there is no *a priori* reason why it should diverge with the same exponent ν. In general, for a second-order QPT, we have (Sachdev, 1999; Hohenberg and Halperin, 1989)

$$\xi_\tau \sim \xi^z \sim |g - g_c|^{-\nu z}, \tag{7.13}$$

which defines the dynamic critical exponent z. We did not encounter this exponent in the thermal transitions of classical systems because they have no intrinsic dynamics.

Equations (7.12) and (7.13) mean that the length scales in the space and imaginary time directions can, in general, be different, $z \neq 1$. Noting that ξ_τ has the same units as β, inverse energy, we see that the divergence of ξ_τ implies the vanishing of a characteristic energy. This is what happens when energy levels cross, as discussed above. We will see in Section 7.2 an explicit example of such a vanishing energy scale.

The presence of the additional imaginary time dimension and a diverging correlation length along it will modify the scaling relations Eq. (7.8), which involve the dimensionality, D. Perhaps the first impulse is simply to replace D by $(D+1)$ but this is not justified since the length scales are not necessarily the same in the new dimension. In fact, the same arguments that led to Eq. (7.5) will now lead to the following scaling form for the correlation function,

$$G(r, \tau) \sim \frac{1}{s^{(D+z-2+\eta)}} g\left(\frac{r}{\xi}, \frac{\tau}{\xi_\tau}\right),$$

$$\sim \frac{1}{s^{(D+z-2+\eta)}} g\left(\frac{r}{\xi}, \frac{\tau}{\xi^z}\right), \tag{7.14}$$

where $s = \sqrt{r^2 + \tau^2}$ is the distance in space-imaginary time. We see that the effect of the added dimension is to add z rather than 1 to D; this is reasonable since distances along the imaginary time direction are not measured in units of ξ but of ξ^z. Carrying these arguments forward generalizes the scaling relations Eq. (7.8) to yield (Fisher et al., 1989),

$$\alpha = 2 - \nu(D + z), \qquad 2\beta = \nu(D + z - 2 + \eta), \qquad \delta = \frac{D + z + 2 - \eta}{D + z - 2 + \eta}. \tag{7.15}$$

The scaling relation Eq. (7.7) does not involve D and remains valid without any change. We see then that the commonly used statement "a D-dimensional quantum system is equivalent to a $(D+1)$-dimensional classical system", while true, should be treated with care near a quantum critical point.

7.1.2 Dynamics and excitation energies

Consider a system in its ground state $|0\rangle$. The *dynamic correlation function* of an observable represented by an operator $\mathcal{O}(\vec{r}, t)$ is defined by

$$S(\vec{r}, \vec{r}', t) = \langle 0|\mathcal{O}^\dagger(\vec{r}, 0)\mathcal{O}(\vec{r}', t)|0\rangle,$$

$$= \sum_m \langle 0|\mathcal{O}^\dagger(\vec{r}, 0)|m\rangle\langle m|\mathcal{O}(\vec{r}', t)|0\rangle, \tag{7.16}$$

where we inserted a complete set of energy eigenstates $1 = \sum|m\rangle\langle m|$ between the two operators. Recalling that

$$\mathcal{O}(\vec{r}', t) = e^{-i\mathcal{H}t}\mathcal{O}(\vec{r}', 0)e^{i\mathcal{H}t}, \tag{7.17}$$

where we take $\hbar = 1$, Eq. (7.16) becomes

$$S(\vec{r}, \vec{r}', t) = \sum_m \langle 0|\mathcal{O}^\dagger(\vec{r}, 0)|m\rangle\langle m|e^{-i\mathcal{H}t}\mathcal{O}(\vec{r}', 0)e^{i\mathcal{H}t}|0\rangle,$$

$$= \sum_m e^{-i(E_m - E_0)t}\langle 0|\mathcal{O}^\dagger(\vec{r}, 0)|m\rangle\langle m|\mathcal{O}(\vec{r}', 0)|0\rangle. \qquad (7.18)$$

We immediately see that the dynamic correlation function gives information about the excitation energies due to the appearance of $(E_m - E_0)$ in the exponential. Assuming that $S(\vec{r}, \vec{r}', \tau) = S(|\vec{r} - \vec{r}'|, \tau)$ and calculating its Fourier transform in space and time gives the dynamic structure factor

$$S(\vec{k}, \omega) = \sum_m \delta(\omega - E_m + E_0)|\langle 0|\tilde{\mathcal{O}}(\vec{k}, 0)|m\rangle|^2. \qquad (7.19)$$

We see that, for a given $\vec{k}$, $S(\vec{k}, \omega)$ is non-zero only when $\omega = E_m - E_0$. The dynamic structure factor is, therefore, a series of spikes whose height is given by $|\langle 0|\tilde{\mathcal{O}}(\vec{k}, 0)|m\rangle|^2$ and whose position gives the excitation spectrum of the system.

However, as discussed above and shown in Eq. (7.11), the additional dimension we have here is imaginary time $t \to -i\beta$. Consequently, the dynamic structure factor will not exhibit oscillating complex exponentials as in Eq. (7.18) but rather exponential decay,

$$S(\vec{k}, \tau) = \sum_m e^{-(E_m - E_0)\tau}|\langle 0|\tilde{\mathcal{O}}(\vec{k}, 0)|m\rangle|^2. \qquad (7.20)$$

The first excited state is the slowest decaying one and, therefore, will dominate the exponential approach to the ground-state contribution,

$$S(\vec{k}, \tau) \approx |\langle 0|\tilde{\mathcal{O}}(\vec{k}, 0)|0\rangle|^2 + e^{-(E_1 - E_0)\tau}|\langle 0|\tilde{\mathcal{O}}(\vec{k}, 0)|1\rangle|^2. \qquad (7.21)$$

We see here a novel feature not present in classical thermal phase transitions namely the presence of a special direction, β, along which correlations can yield dynamical information. However, to access the excitation specturm in this way, one needs the "Fourier transform in imaginary time" of the dynamical correlation function, i.e. its Laplace transform. If one has the functional form of the dynamical correlation function, it may be possible to calculate the Laplace transform without too much difficulty. However, one normally does not have this information. This is a situation where quantum Monte Carlo (QMC) can help. If one evaluates $S(\vec{r}, \vec{r}', \tau)$ by means of QMC (see Sections 7.2 and 7.3), one can then implement a numerical Laplace transform, for example, using the maximum entropy method (Gubernatis *et al.*, 1991). The details of QMC and maximum entropy are beyond our scope here, they are mentioned as an example of how to exploit features of quantum critical phenomena absent in the classical case.

To end this section, we note that the exponential decay in Eq. (7.19) occurs only in the presence of an energy gap: $E_0 < E_1$. At a quantum critical point, where we have a level crossing, E_1 becomes equal to the ground-state energy, E_0, the gap and exponential decay disappear and power laws appear. We also note that the exponential decay in Eq. (7.21) can be written as $e^{-\tau/\xi_\tau}$, which makes explicit that the diverging correlation length in the imaginary time direction is due to a vanishing energy scale, $\xi_\tau \sim (E_1 - E_0)^{-1}$.

7.2   Quantum Ising model

7.2.1   Generalities

The Ising model is, probably, the most familiar classical statistical model of magnetism (Stanley, 1971; Ma, 1985). Consider the sites i of a regular lattice on which we have classical "spins" $S_i = \pm 1$. The energy and partition function of the Ising model are

$$E = -J \sum_{\langle i,j \rangle} S_i S_j, \tag{7.22}$$

$$Z = \sum_{\{S_i\}} e^{\beta J \sum_{\langle i,j \rangle} S_i S_j}, \tag{7.23}$$

where β is the inverse temperature (we take Boltzmann's constant $k_B = 1$), J is the coupling between the Ising spin variables and $\langle i,j \rangle$ denotes nearest neighbors. The sum over $\{S_i\}$ denotes a sum over all spin configurations. The properties of this model are well known and we will only discuss those we need as we proceed to the quantum case.

Now consider a system governed by the quantum Hamiltonian

$$\mathcal{H} = H_0 + H_1, \tag{7.24}$$

$$H_0 = -J \sum_{\langle i,j \rangle} \sigma_i^z \sigma_j^z, \tag{7.25}$$

$$H_1 = -h \sum_i \sigma_i^x, \tag{7.26}$$

where σ^x and σ^z are the usual Pauli matrices,

$$\sigma^x = \begin{pmatrix} 0 & 1 \\ 1 & 0 \end{pmatrix}, \quad \sigma^y = \begin{pmatrix} 0 & -i \\ i & 0 \end{pmatrix}, \quad \sigma^z = \begin{pmatrix} 1 & 0 \\ 0 & -1 \end{pmatrix}. \tag{7.27}$$

Equation (7.24) is the same as Eq. (7.1) but now with explicit forms for the non-commuting H_0 and H_1. When $h = 0$ all the operators in the problem commute, and can be replaced by their eigenvalues, resulting in the classical Ising model. However, when $h \neq 0$, the non-commutativity $[H_0, H_1] \neq 0$, requires that we treat $\mathcal{H}$ as a fully quantum system. While H_0 by itself is just the classical Ising model, H_1 describes an external magnetic field in the x-direction and therefore transverse to the spin projections in H_0. For this reason, Eq. (7.25) is referred to as the Ising model in a transverse field.

We will now consider the one-dimensional case of Eq. (7.24) in some detail to illustrate many of the points discussed in the previous section. We write

$$\mathcal{H} = -J \sum_{i=1}^{N} \sigma_{i+1}^z \sigma_i^z - h \sum_{i=1}^{N} \sigma_i^x, \tag{7.28}$$

where N is the total number of sites and we take periodic boundary conditions, $\sigma_{N+1}^z = \sigma_1^z$. The partition function is given by Eq. (7.9). Consider first the limiting case $h = 0$,

$$Z = \mathrm{Tr}\, e^{\beta J \sum_{i=1}^{N} \sigma_{i+1}^z \sigma_i^z}. \tag{7.29}$$

To evaluate the trace, we use σ^z eigenvectors:

$$|+\rangle = \begin{pmatrix} 1 \\ 0 \end{pmatrix}, \quad |-\rangle = \begin{pmatrix} 0 \\ 1 \end{pmatrix}, \tag{7.30}$$

with

$$\sigma^z|+\rangle = |+\rangle, \quad \sigma^z|-\rangle = -|-\rangle, \quad \sum_{S^z=\pm1} |S^z\rangle\langle S^z| = 1. \tag{7.31}$$

The partition function becomes,

$$Z = \left(\prod_{i=1}^{N} \sum_{S_i^z} \right) \left(\prod_{i=1}^{N} \langle S_i^z| \right) e^{\beta J \sum_{i=1}^{N} \sigma_{i+1}^z \sigma_i^z} \left(\prod_{i=1}^{N} |S_i^z\rangle \right)$$

$$= \left(\prod_{i=1}^{N} \sum_{S_i^z} \right) \left(\prod_{i=1}^{N} \langle S_i^z| \right) e^{\beta J \sum_{i=1}^{N} S_{i+1}^z S_i^z} \left(\prod_{i=1}^{N} |S_i^z\rangle \right)$$

$$= \sum_{\{S_i^z\}} \left(\prod_{i=1}^{N} \langle S_i^z|S_i^z\rangle \right) e^{\beta J \sum_{i=1}^{N} S_{i+1}^z S_i^z}$$

$$= \sum_{\{S_i^z\}} e^{\beta J \sum_{i=1}^{N} S_{i+1}^z S_i^z}. \tag{7.32}$$

This is simply the partition function of the classial one-dimensional Ising model which can be easily solved and that has no phase transition at any finite temperature (LeBellac *et al.*, 2004). The other limiting case is to take $J = 0$,

$$Z = \mathrm{Tr}\, e^{\beta h \sum_{i=1}^{N} \sigma_i^x}. \tag{7.33}$$

Following the same steps that led to Eq. (7.32) but this time using eigenstates of σ^x,

$$|\pm\rangle = \frac{1}{\sqrt{2}} \begin{pmatrix} 1 \\ \pm1 \end{pmatrix}, \quad \sigma^x|+\rangle = |+\rangle, \quad \sigma^x|-\rangle = -|-\rangle, \tag{7.34}$$

leads to

$$Z = \prod_{i=1}^{N} \left[\sum_{S_i^x = \pm 1} e^{\beta h S_i^x} \right], \tag{7.35}$$

$$= [2\cosh(\beta h)]^N, \tag{7.36}$$

which is simply the partition function of N independent spins. We, therefore, see that the two extreme limits, $h = 0$ and $J = 0$ lead back to classical cases.

7.2.2   Mapping to classical system

Having both h and J non-zero leads to the interesting quantum critical behavior we seek. Several approaches to this problem are possible and here we demonstrate a general approach that can be applied essentially to any quantum model. The partition function of the full Hamiltonian governing the one-dimensional quantum system will be mapped onto that of a $(1 + 1)$ classical system as follows,

$$Z = \text{Tr} e^{-\beta \mathcal{H}},$$

$$= \text{Tr} \left[e^{-\Delta \tau \mathcal{H}} e^{-\Delta \tau \mathcal{H}} e^{-\Delta \tau \mathcal{H}} \dots e^{-\Delta \tau \mathcal{H}} e^{-\Delta \tau \mathcal{H}} \right], \tag{7.37}$$

where there are L exponentials in the product and $\beta = L \Delta \tau$ defines L and $\Delta \tau$. The product inside the trace can be viewed as a succession of imaginary time evolution operators. Note that no approximations were made in Eq. (7.37). Now, between each pair of exponentials, insert a complete set of σ_i^z eigenstates,

$$1 = \prod_{i=1}^{N} \left[\sum_{S_i^z = \pm 1} |S_i^z\rangle\langle S_i^z| \right], \tag{7.38}$$

$$\equiv \sum_{\{S_i^z\}} |S^z\rangle\langle S^z|. \tag{7.39}$$

Note that in Eq. (7.39) there is a complete set of eigenstates for every site that we then abbreviate with the notation of Eq. (7.39). Since we will make this insertion at several places along the string of exponentials, we will use a new index, ℓ, to label that position,

$$1 = \sum_{\{S_{i,\ell}^z\}} |S_\ell^z\rangle\langle S_\ell^z|. \tag{7.40}$$

The partition function then becomes,

$$Z = \sum_{\{S_{i,\ell} = \pm 1\}} \langle S_1^z | e^{-\Delta \tau \mathcal{H}} | S_L^z \rangle \langle S_L^z | e^{-\Delta \tau \mathcal{H}} | S_{L-1}^z \rangle \langle S_{L-1}^z | e^{-\Delta \tau \mathcal{H}} | S_{L-2}^z \rangle \dots \tag{7.41}$$

$$\dots \langle S_3^z | e^{-\Delta \tau \mathcal{H}} | S_2^z \rangle \langle S_2^z | e^{-\Delta \tau \mathcal{H}} | S_1^z \rangle. \tag{7.42}$$

Since Z is a trace, we have the same state at the beginning and end of the chain. As we discussed for Eq. (7.11), $e^{-\Delta\tau\mathcal{H}}$ is an evolution operator in imaginary time for a step of $\Delta\tau$. The partition function has become a product of matrix elements each representing the evolution of the quantum-mechanical state from "imaginary time slice" ℓ to "imaginary time slice" $\ell+1$. It now remains to evaluate these matrix elements, which is where we will make our first approximation,

$$\langle S_{\ell+1}^z|e^{-\Delta\tau\mathcal{H}}|S_\ell^z\rangle = \langle S_{\ell+1}^z|e^{-\Delta\tau H_1-\Delta\tau H_0}|S_\ell^z\rangle$$

$$\approx \langle S_{\ell+1}^z|\left(e^{-\Delta\tau H_1}e^{-\Delta\tau H_0} + O(\varepsilon)\right)|S_\ell^z\rangle. \qquad (7.43)$$

$O(\varepsilon)$ is the Trotter error and is of order $(\Delta\tau)^2 Jh$, as can be seen from

$$\varepsilon = [\Delta\tau H_0, \Delta\tau H_1] = (\Delta\tau)^2[H_0, H_1] = O\left((\Delta\tau)^2 Jh\right). \qquad (7.44)$$

This approximation, called the Trotter–Suzuki approximation, is controlled in the sense that one can get as close to the exact result as one desires by requiring that

$$(\Delta\tau)^2 Jh \ll 1,$$

$$L^2 \gg \beta^2 Jh,$$

$$L \gg \beta\sqrt{Jh}. \qquad (7.45)$$

In other words, we have divided β into L time slices to make $\Delta\tau$ small, justifying the Trotter–Suzuki approximation and allowing us to evaluate the matrix elements. It is very important to keep in mind that, regardless of the value of L, this is *not* a perturbative calculation: Non-perturbative effects are all present in this formalism.

The matrix element, Eq. (7.43), can now be evaluated easily by noting that the exponential of H_0 (see Eq. (7.25)) acts on σ^z eigenstates on the right giving,

$$\langle S_{\ell+1}^z|e^{-\Delta\tau H_1}e^{-\Delta\tau H_0}|S_\ell^z\rangle = \langle S_{\ell+1}^z|e^{-\Delta\tau H_1}e^{\Delta\tau J\sum_{i=1}^N S_{i,\ell}^z S_{i+1,\ell}^z}|S_\ell^z\rangle$$

$$= e^{\Delta\tau J\sum_{i=1}^N S_{i,\ell}^z S_{i+1,\ell}^z}\langle S_{\ell+1}^z|e^{\Delta\tau h\sum_{i=1}^N \sigma_i^x}|S_\ell^z\rangle. \qquad (7.46)$$

To calculate the remaining matrix element we note that since $\sigma_x^2 = 1$, the identity matrix, then

$$e^{\Delta\tau h\sigma_x} = 1\cosh(\Delta\tau h) + \sigma_x\sinh(\Delta\tau h). \qquad (7.47)$$

We require that the matrix element be expressed in the form

$$\langle S_z'|e^{\Delta\tau h\sigma_x}|S_z\rangle \equiv \Lambda e^{\gamma S_z' S_z}, \qquad (7.48)$$

which defines Λ and γ. To determine Λ and γ, we calculate Eq. (7.48), using Eq. (7.47), for the two cases $S_z = S_z'$ and $S_z = -S_z'$,

$$\langle S_z|e^{\Delta\tau h\sigma_x}|S_z\rangle = \cosh(\Delta\tau h) = \Lambda e^\gamma, \qquad (7.49)$$

$$\langle -S_z|e^{\Delta\tau h\sigma_x}|S_z\rangle = \sinh(\Delta\tau h) = \Lambda e^{-\gamma}, \qquad (7.50)$$

which yield,

$$\gamma = -\frac{1}{2}\ln\tanh(\Delta\tau h).$$ (7.51)

$$\Lambda^2 = \sinh(\Delta\tau h)\cosh(\Delta\tau h).$$ (7.52)

Equation (7.46) then becomes,

$$\langle S_{\ell+1}^z | e^{-\Delta\tau H_1} e^{-\Delta\tau H_0} | S_\ell^z \rangle = \Lambda^N e^{\Delta\tau J \sum_{i=1}^N S_{i,\ell}^z S_{i+1,\ell}^z + \gamma \sum_{i=1}^N S_{i,\ell}^z S_{i,\ell+1}^z}.$$ (7.53)

Note that in Eq. (7.53), H_0 led to Ising spin couplings at the same imaginary time slice but near-neighbor spatial sites, while H_1 led to couplings between Ising spins on the same spatial site but "near-neighbor" (consecutive) imaginary time slices. Substituting Eq. (7.53) in Eq. (7.42) finally yields,

$$Z = \Lambda^{NL} \sum_{\{S_{i,\ell}=\pm1\}} e^{\Delta\tau J \sum_{i=1}^N \sum_{\ell=1}^L S_{i,\ell}S_{i+1,\ell} + \gamma \sum_{i=1}^N \sum_{\ell=1}^L S_{i,\ell}S_{i,\ell+1}},$$ (7.54)

where $\gamma > 0$ is given by Eq. (7.51) and where we dropped the z superscript on the spins. The prefactor, Λ^{NL}, is not important because it does not affect the spins and therefore does not affect the transition. In addition, this prefactor cancels out when average values of physical quantities are calculated. Equation(7.54) is simply the partition function of the classical $D = 2$ Ising model, Eq. (7.23), with different couplings in the i and ℓ directions,

$$Z_{cl} = \sum_{\{S_{i,\ell}=\pm1\}} e^{\beta_{cl}J_x \sum_{i=1}^{N_x} \sum_{j=1}^{N_y} S_{i,j}S_{i+1,j} + \beta_{cl}J_y \sum_{i=1}^{N_x} \sum_{j=1}^{N_y} S_{i,j}S_{i,j+1}}.$$ (7.55)

We then identify,

$$N_x = N$$
$$N_y = L$$
$$\beta_{cl}J_x = \Delta\tau J$$
$$\beta_{cl}J_y = \gamma.$$ (7.56)

The one-dimensional Ising model in a transverse field is thus mapped onto a classical $(1 + 1)$-dimensional Ising model with anisotropic couplings. Note that β_{cl} is not the inverse temperature of the original quantum system, $\beta = L\Delta\tau$.

7.2.3 Quantum phase transition

The classical two-dimensional Ising model, Eq. (7.55), can be solved exactly in the absence of an external magnetic field (Onsager, 1944; Baxter, 1982). It is known that in the thermodynamic limit, $N_x \to \infty$ and $N_y \to \infty$, this system has a critical point given by the relation (which can also be obtained using the duality transformation (Baxter, 1982; Kramers and Wannier, 1941; Savit, 1980)

$$\sinh(2J_x\beta_{cl}^c)\sinh(2J_y\beta_{cl}^c) = 1, \tag{7.57}$$

where β_{cl}^c is the critical inverse temperature for the classical system. For low temperature, $\beta_{cl} > \beta_{cl}^c$, the system is ordered (magnetized), while for high temperature, $\beta_{cl} < \beta_{cl}^c$, the system is disordered. While the value of β_{cl}^c depends on the relative values of J_x and J_y, i.e. on the anisotropy, the critical exponents themselves do not (Onsager, 1944): The critical points for all J_x/J_y are in the same universality class. For $J_x = J_y$ Eq. (7.57) yields $2J_x\beta_{cl}^c = \ln(1+\sqrt{2})$.

With the mapping of the $D = 1$ quantum system to the $(1+1)$ classical system, Eq. (7.54), and the parameter identification Eq. (7.56), we, therefore, see that as $N \to \infty$ and $L \to \infty$ the quantum system will exhibit a critical point given by the condition

$$\sinh(2J^c\Delta\tau)\sinh(2\gamma^c) = 1, \quad \gamma^c = -\frac{1}{2}\ln\tanh(h^c\Delta\tau). \tag{7.58}$$

This, however, is not a thermal phase transition since it takes place when the quantum system is at zero temperature. We see this by noting that for the transition to take place, both $N \to \infty$ and $L \to \infty$. But when $L \to \infty$ at fixed $\Delta\tau$, which is the case here, it means that $\beta \to \infty$ and thus the system is at $T = 0$. We have here, therefore, a quantum phase transition reached by tuning J and h, the coupling parameters while the system is at zero temperature. Equation (7.58) serves to get the critical point. Substituting the expression for γ^c in the condition for criticality, Eq. (7.58), yields

$$1 = \frac{1}{2}\sinh(2J^c\Delta\tau)\left(\frac{\cosh(h^c\Delta\tau)}{\sinh(h^c\Delta\tau)} - \frac{\sinh(h^c\Delta\tau)}{\cosh(h^c\Delta\tau)}\right)$$

$$= \sinh(2J^c\Delta\tau)\left(\frac{\cosh^2(h^c\Delta\tau) - \sinh^2(h^c\Delta\tau)}{2\cosh(h^c\Delta\tau)\sinh(h^c\Delta\tau)}\right)$$

$$= \frac{\sinh(2J^c\Delta\tau)}{\sinh(2h^c\Delta\tau)}. \tag{7.59}$$

Putting $h = gJ$, we see that Eq. (7.59) is satisfied for any J^c as long as $g^c = 1$. We therefore conclude that the $D = 1$ Ising model in a transverse field, undergoes a quantum phase transition, i.e. at $T = 0$, at the critical coupling $g^c = 1$.

As we saw in the introduction, Eq. (7.13), the correlation length along the imaginary time direction diverges as $\xi_\tau \sim \xi^z$. But it is known for the anisotropic Ising model that rotational isotropy is restored at the critical point: The system looks and behaves the same in all directions. Therefore, in our effective $(1+1)$ model, $\xi_\tau \sim \xi$ at the quantum critical point. Consequently, for the $D = 1$ Ising model in transverse field we have the dynamic critical exponent $z = 1$. All the other exponents, α, $\tilde{\beta}$, γ, δ, η and ν have the same values as the classical $D = 2$ model. This is, perhaps, not hard to understand qualitatively: The couplings in Eq. (7.54) have the same form in the x- and β-directions, only the details are somewhat different, so when the correlations diverge it is not unreasonable that the system behaves in the same way in both directions. This is not true for all quantum models, as we will see in the next section.

Applying the above to the isotropic $D = 2$ Ising model in a transverse field at inverse temparature β, we obtain the mapping to the $(2+1)$- dimensional classical model with equal couplings in the x- and y-directions but different couplings in the β-direction. The QPT is obtained in the same way: $N_x \to \infty$, $N_y \to \infty$ and $L \to \infty$ (with $\Delta\tau$ constant). This system will undergo a quantum phase transition, in the thermodynamic and $\beta \to \infty$ limits, at parameter values that are not known exactly but that can be estimated, for example, numerically.

Again, near the transition, isotropy is restored, which gives $z = 1$.

Practice problem: A good practice problem is to apply the above to the Hamiltonian

$$\mathcal{H} = -J\sigma^z - h\sigma^x, \tag{7.60}$$

which represents a $(D = 0)$ single Ising variable in a transverse field. This Hamiltonian will lead to the $(0+1)$-dimensional classical Ising model that can be solved trivially. Although this model does not have a phase transition, the exercise is, nonetheless, very instructive.

7.2.4 Discussion

- It is important not to confuse β, the inverse temperature of the quantum system, with β_{cl}; see for example Eqs. (7.54), (7.55), (7.57) and (7.58). β is the true inverse temperature of the quantum system and, after the mapping to the $(D+1)$ classical system, manifests itself only through $\Delta\tau$ and L.
- The following point, which was discussed above, is worth re-iterating. The fact that we needed to take $\Delta\tau$ "very small" (actually $(\Delta\tau)^2 Jh \ll 1$) does not mean that the results are perturbative. We needed this condition only to be able to evaluate the matrix elements at each imaginary time slice. All non-perturbative effects are included.
- The mapping to take the one-dimensional quantum system to a two-dimensional "classical" system is rather general. Details of the mapping, such as evaluation of the matrix elements, will depend on the model, but the general approach is the same: Any D-dimensional quantum system can, through a similar succession of steps, be mapped onto an equivalent $(D+1)$-dimensional classical model. The additional dimension, β, is interpreted as the "imaginary time" direction. We will see another example below.
- If one wants to simulate numerically a quantum system, one can first map it onto its equivalent $(D+1)$-dimensional classical equivalent and perform a classical simulation on the resulting system. This is known as the quantum Monte Carlo method. One should, however, be careful with the observables: It is not always obvious what form a quantum observable takes in the equivalent classical model in $(D+1)$ dimensions. However, the new form can be found by perfoming the same mapping on the expectation value, Eq. (7.10), as on the partition function (LeBellac *et al.*, 2004; Hirsch *et al.*, 1992; Batrouni and Scalettar, 1992; Tobochnik *et al.*, 1992).

- We have seen that the path integral for the $D = 1$ Ising model in a transverse field led to the $D = 2$ classical Ising model for which we have Onsager's exact solution. It should not be surprising, then, that the transverse Ising model should have a direct analytic solution.

  Indeed, an elegant and direct solution is found with the Jordan–Wigner transformation (see appendix) which can map a 1-dimensional spin system onto a 1-dimensional non-interacting fermionic system. The Hamiltonian for a non-interacting system can be easily diagonalized to yield the energy spectrum. For the Hamiltonian Eq. (7.24) one obtains

$$\epsilon(k) = 2J \left(1 + g^2 - 2g\cos(k)\right)^{1/2}, \tag{7.61}$$

  where k is the wavevector, $p = \hbar k$. The lowest excitation (i.e. gap) energy is for $k = 0$, $\epsilon(0) = 2J|1 - g|$, which gives $g_c = 1$, in agreement with Eq. (7.59). We can then write,

$$\epsilon(0) = 2J|g - g_c|. \tag{7.62}$$

  At the end of Section 7.1 we said that $\xi_\tau \sim (E_1 - E_0)^{-1}$, which, because of Eq. (7.62), becomes $\xi_\tau \sim |g - g_c|^{-1}$. Therefore, Eq. (7.13) gives $\nu z = 1$ and, since $\nu = 1$ for the classical two-dimensional Ising model, we obtain again $z = 1$.
  The Jordan–Wigner transformation applies only in one dimension, so this method cannot be used, for example, to solve the 2-dimensional Ising model in a transverse field.

- What happens to the QPT at finite temperature? At finite T there are thermal fluctuations that tend to drive the phase transition, *if there is one*.
  For example, the $D = 1$ Ising model in a transverse field has a QPT at $T = 0$ for $g_c = 1$. We also saw, Eqs.(7.29) to (7.32), that, when $h = 0$ (and therefore $g = 0$), this model is simply the classical one-dimensional Ising chain that exhibits no phase transition at any temperature. At finite T, L is finite and therefore when $N \to \infty$, the partition function Eq. (7.54) describes an infinite strip and consequently the system is effectively one-dimensional and will not exhibit any phase transitions. Its phase diagram will, therefore, be like Figure 7.1(a).
  Now consider the $D = 2$ Ising model in a transverse field. When $T = 0$, there is a QPT at g_c, as discussed above. In addition, however, when $h = 0$ (i.e. $g = 0$) the system becomes the two-dimensional classical Ising model that exhibits a thermal second-order phase transition at $T/J = 2.2691\ldots$. So, there is a quantum phase transition at $(g_c, T = 0)$ and a *thermal* one at $(g = 0, T_c)$. These two transitions are typically connected by a line of thermal critical points as illustrated in Figure 7.1(b).

7.3   Hubbard model

What has become known as "the Hubbard model" covers a rather large class of models, both bosonic and fermionic. These models play a central role in the study of strongly

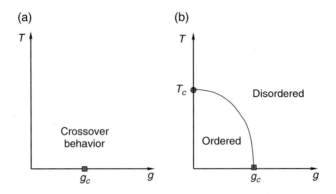

Fig. 7.1 (a) Schematic of the phase diagram of the 1-dimensional quantum Ising model in the temperature-coupling space. The square denotes the quantum critical point. (b) The phase diagram for the 2-dimensional case. The square denote a quantum phase transition and the circle a thermal one; the two are connected by a line of thermal critical points.

correlated systems exhibiting a very wide range of quantum phases and quantum phase transitions. A partial list of phases exhibited by these systems include high-temperature superconductivity, superfluidity, supersolids, antiferromagnetism, Mott insulator, Bose glasses, ultracold atoms loaded on optical lattices, etc. (Fazekas, 1999; Rasetti, 1991).

The original model (Gutzwiller, 1963; Hubbard, 1964; Kanamori, 1963), used to study magnetic transitions, described fermions on a square lattice governed by the Hamiltonian

$$\mathcal{H} = -t \sum_{\langle ij \rangle \sigma} (c_{i\sigma}^\dagger c_{j\sigma} + c_{j\sigma}^\dagger c_{i\sigma}) + U \sum_i \left(\hat{n}_{i\uparrow} - \frac{1}{2} \right) \left(\hat{n}_{i\downarrow} - \frac{1}{2} \right) - \mu \sum_i (\hat{n}_{i\uparrow} + \hat{n}_{i\downarrow}).$$
(7.63)

$c_{i\sigma}$ and $c_{i\sigma}^\dagger$ are destruction and creation operators of fermions with spin $\sigma = \uparrow, \downarrow$ on site i and they satisfy the anti-commutation relations

$$\{c_{i\sigma}^\dagger, c_{j\sigma'}\} = \delta_{i,j} \delta_{\sigma,\sigma'}, \quad \{c_{i\sigma}, c_{j\sigma'}\} = \{c_{i\sigma}^\dagger, c_{j\sigma'}^\dagger\} = 0.$$
(7.64)

Therefore, the first term in Eq. (7.63) is the kinetic energy term governing how fermions jump between near neighbor sites, $\langle ij \rangle$. t is called the hopping parameter and by making contact with the kinetic energy in the Schödinger equation, we can identify $t = \hbar^2/2m$. The contact interaction is described by the second term where U is the interaction strength and $\hat{n}_{i\sigma} = c_{i\sigma}^\dagger c_{i\sigma}$ is the number operator on site i. The chemical potential μ controls the particle population in the grand canonical ensemble.

A natural generalization of the above model is to consider the properties of a system of bosons on a lattice governed by a similar Hamiltonian (Fisher *et al.*, 1989),

$$\mathcal{H} = -t \sum_{\langle i,j \rangle} \left(a_i^\dagger a_j + a_j^\dagger a_i \right) + \frac{U}{2} \sum_i \hat{n}_i \left(\hat{n}_i - 1 \right) - \mu \sum_i \hat{n}_i,$$
(7.65)

where the bosonic operators satisfy

$$[a_i, a_j^\dagger] = \delta_{i,j}, \quad [a_i^\dagger, a_j^\dagger] = [a_i, a_j] = 0. \tag{7.66}$$

In Eq. (7.65) we considered the simplest case of spin-0 bosons but one can also consider bosons of higher spins. In both Eqs. (7.63) and (7.65) we only included contact interactions. However, both models have been extended to include nearest and next-nearest neighbors (Niyaz *et al.*, 1991, 1994; Batrouni and Scalettar, 2000) and even longer-range interactions and disorder (Fisher *et al.*, 1989). Such additional interactions vastly enrich the phase diagrams of these models and introduce new exotic phases of which we mentioned a few at the begining of this section.

Although, initially, interest in these models was focused on condensed-matter issues, this focus recently widened considerably with the experimental realization of atomic Bose–Einstein condensates (BEC). It was shown that if one produces a standing wave with pairs of counter-propagating laser beams (called an "optical lattice") and then loads ultracold fermionic or bosonic atoms on this optical lattice, the resulting system is, indeed, governed by Eqs. (7.63) or (7.65) with *experimentally tunable* interaction strength and range (Jaksch *et al.*, 1998). In addition, the optical lattices can be made in 1, 2, and 3 dimensions. This high degree of experimental flexibility and control allows detailed comparisons between the physics of strongly correlated systems governed by Hamiltonians of the Hubbard type, Eq. (7.63, 7.65), and those realized experimentallly. This is currently a very active field with new experimental and theoretical/numerical results appearing very frequently.

In this chapter we will concentrate on the phase diagram and the quantum phase transitions of the spin-0 bosonic Hubbard model governed by Eq. (7.65).

7.3.1   Mapping to classical system

The mapping of the Hubbard model onto a classical model proceeds along a path similar to that for the Ising model in a transverse field. We will, nonetheless, show it in some detail in order to fix ideas and to emphasize that, unlike the Ising case, the classical model can be unfamiliar and not correspond to a well-studied model.

To implement the mapping, we consider the one-dimensional version of Eq. (7.65),

$$\mathcal{H} = -t \sum_{i=1}^{N} \left(a_i^\dagger a_{i+1} + a_{i+1}^\dagger a_i \right) + \frac{U}{2} \sum_{i=1}^{N} \hat{n}_i \left(\hat{n}_i - 1 \right) - \mu \sum_{i=1}^{N} \hat{n}_i, \tag{7.67}$$

where N is the number of sites and we take periodic boundary conditions, $a_{N+1} = a_1$. As for the Ising case, we start with the partition function,

$$Z = \text{Tr}\, e^{-\beta \mathcal{H}},$$
$$= \text{Tr} \left[e^{-\Delta\tau\mathcal{H}} e^{-\Delta\tau\mathcal{H}} e^{-\Delta\tau\mathcal{H}} \ldots e^{-\Delta\tau\mathcal{H}} e^{-\Delta\tau\mathcal{H}} \right], \tag{7.68}$$

where there are L terms in the product and, as before, $\beta = L\Delta\tau$. There are several representations possible to calculate the trace. We will choose the occupation number representation, $|\mathbf{n}\rangle \equiv |n_1, n_2, \ldots, n_i, \ldots, n_N\rangle$ where

$$a_i^\dagger |n_1, n_2, \ldots, n_i, \ldots, n_N\rangle = \sqrt{n_i + 1} \, |n_1, n_2, \ldots, n_i + 1, \ldots, n_N\rangle, \qquad (7.69)$$

$$a_i |n_1, n_2, \ldots, n_i, \ldots, n_N\rangle = \sqrt{n_i} \, |n_1, n_2, \ldots, n_i - 1, \ldots, n_N\rangle, \qquad (7.70)$$

$$\hat{n}_i |n_1, n_2, \ldots, n_i, \ldots, n_N\rangle = n_i \, |n_1, n_2, \ldots, n_i, \ldots, n_N\rangle, \qquad (7.71)$$

and the completeness condition,

$$\sum_{\{\mathbf{n}\}} |\mathbf{n}\rangle\langle\mathbf{n}| \equiv \left(\prod_{i=1}^{N} \sum_{n_i=0}^{\infty}\right) |n_1, n_2, \ldots, n_N\rangle\langle n_1, n_2, \ldots, n_N| = 1. \qquad (7.72)$$

This representation has the advantage of being very familiar and will lead to a partition function amenable to intuitive physical representation. Before inserting complete sets of state in Eq. (7.68), we will choose to work in the canonical ensemble where the total number of particles is fixed and, therefore, the chemical potential term is not present. We will also split the Hamiltonian, as follows,

$$\mathcal{H} = H_1 + H_2$$

$$H_1 = -t \sum_{i=odd} \left(a_i^\dagger a_{i+1} + a_{i+1}^\dagger a_i\right) + \frac{U}{4} \sum_{i=1}^{N} \hat{n}_i \left(\hat{n}_i - 1\right),$$

$$H_2 = -t \sum_{i=even} \left(a_i^\dagger a_{i+1} + a_{i+1}^\dagger a_i\right) + \frac{U}{4} \sum_{i=1}^{N} \hat{n}_i \left(\hat{n}_i - 1\right). \qquad (7.73)$$

This splitting of $\mathcal{H}$ means that the hopping term in H_1 connects sites $(1, 2)$, $(3, 4)$, $(5, 6)$, etc. while H_2 connects sites $(2, 3)$, $(4, 5)$, $(6, 7)$, etc. It is called the "checkerboard decomposition" for reasons that will become clear shortly (Hirsch *et al.*, 1992). Notice that the interaction term was divided equally between H_1 and H_2. We then have,

$$e^{-\Delta\tau H} \approx e^{-\Delta\tau H_1} e^{-\Delta\tau H_2} + O(\varepsilon), \qquad (7.74)$$

where

$$\varepsilon = [\Delta\tau H_1, \Delta\tau H_2] = (\Delta\tau)^2 [H_1, H_2] = O\left((\Delta\tau)^2 tU\right). \qquad (7.75)$$

Once again, we find that for this procedure to be accurate, one needs to take a large enough number of time slices, L, to make $(\Delta\tau)^2 Ut \ll 1$. Now insert Eq. (7.74) in Eq. (7.68) and insert between each pair of exponentials a complete set of states, Eq. (7.72). This yields,

$$Z = \sum_{\{\mathbf{n}\}} \langle \mathbf{n}^1 | e^{-\Delta\tau H_2} | \mathbf{n}^{2L} \rangle \langle \mathbf{n}^{2L} | e^{-\Delta\tau H_1} | \mathbf{n}^{2L-1} \rangle \ldots \langle \mathbf{n}^3 | e^{-\Delta\tau H_2} | \mathbf{n}^2 \rangle \langle \mathbf{n}^2 | e^{-\Delta\tau H_1} | \mathbf{n}^1 \rangle.$$

$$(7.76)$$

The partition function is now expressed as a path integral, Equation (7.76), with an appealing intuitive geometrical interpretation in terms of *world lines*, *i.e.* the paths

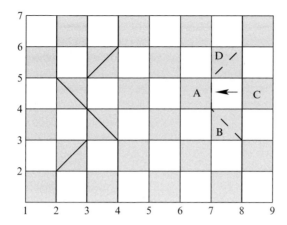

Fig. 7.2 The decomposition of the full Hamiltonian $\mathcal{H}$ into H_1 and H_2 results in a "checkerboard" pattern for the space-imaginary time lattice. Periodic boundary conditions are used in space, so that on this 8-site lattice, site 1 is identified with site 9. Furthermore, the trace in the partition function imposes periodic boundary conditions in the imaginary time direction so that world lines have to close on themselves. Only for even time slices are bosons allowed to hop between sites $(2,3)$, $(4,5)$, $(6,7)$, $\ldots$,, while only for odd time slices can they hop between sites $(1,2)$, $(3,4)$, $\ldots$ The bold paths are boson world lines, the arrow shows a typical allowed deformation.

traced by the bosons as they evolve in space-imaginary time (Hirsch *et al.*, 1992). To see this, consider, for example, the matrix element $\langle \mathbf{n}^2|e^{-\Delta\tau H_1}|\mathbf{n}^1\rangle$: As is clear from its definition, Eqs. (7.75), the hopping term (kinetic energy) in H_1 couples only the pairs $(1,2)$, $(3,4)$, $(5,6)$ *etc.* but not the pairs $(2,3)$, $(4,5)$, $(6,7)$, etc., which are connected by H_2. Consequently, the matrix elements of $e^{-\Delta\tau H_1}$ are products of *independent* two-site problems: $(1,2)$, $(3,4)$, $(5,6)\ldots$ Similarly, a matrix element of $\exp(-\Delta\tau H_2)$ is a product of independent two-site problems $(2,3)$, $(4,5)$, $(6,7)\ldots$ This is shown in Figure 7.2 where the shaded squares represent the imaginary-time evolution of connected pairs. The thick lines represent a possible configuration of boson world lines traced by the particles initially present at the first time slice as the imaginary time evolution operators evolve the state from one time slice to the next. Because of the checkerboard division of $\mathcal{H} = H_1 + H_2$, a boson can jump only across a shaded square.

The matrix element represented by a shaded square can now be calculated easily. Consider such a square at a time slice where, for example, H_1 acts (see Figure 7.3),

$$\langle n'_j, n'_{j+1}|e^{-\Delta\tau H_1}|n_j, n_{j+1}\rangle \approx \langle n'_j, n'_{j+1}|e^{-\Delta\tau\hat{\mathcal{U}}/2}\, e^{\Delta\tau t(a_j^\dagger a_{j+1}+a_{j+1}^\dagger a_j)}\, e^{-\Delta\tau\hat{\mathcal{U}}/2}|n_j, n_{j+1}\rangle,$$

$$= e^{-\Delta\tau\mathcal{U}'/2}e^{-\Delta\tau\mathcal{U}/2}$$

$$\times\langle n'_j, n'_{j+1}|e^{\Delta\tau t(a_j^\dagger a_{j+1}+a_{j+1}^\dagger a_j)}|n_j, n_{j+1}\rangle, \tag{7.77}$$

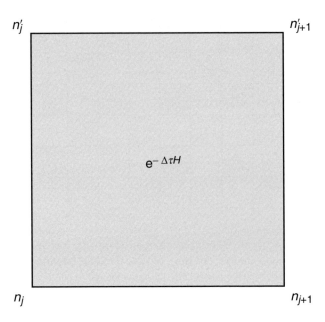

Fig. 7.3   An elementary matrix element represented by shaded squares in Figure 7.2.

where

$$\mathcal{U} = \frac{U}{4}\left(n_j(n_j + 1) + n_{j+1}(n_{j+1} + 1)\right),$$

$$\mathcal{U}' = \frac{U}{4}\left(n'_j(n'_j + 1) + n'_{j+1}(n'_{j+1} + 1)\right). \tag{7.78}$$

The remaining matrix element in Eq. (7.77) is easy to calculate by expanding the exponential in powers of $(\Delta\tau)$, but can also be calculated exactly (LeBellac *et al.*, 2004). It vanishes unless $n_j + n_{j+1} = n'_j + n'_{j+1}$ and is positive when this condition is satisfied.

The partition function of the one-dimensional bosonic Hubbard model is, therefore, expressed as a "path integral", which is the sum over all conformations and deformations of N_b lines where N_b is the number of bosons. The partition function of the two-dimensional system can be represented in a similar manner but in this case there are more intricate configurations possible. For example, boson world lines can wrap around each other forming a helical structure.

We remark that all the matrix elements in Eq. (7.76) are real and positive (for the bosonic model in the absence of frustration). Consequently, this world line representation forms the basis for many quantum Monte Carlo algorithms applied to a wide class of bosonic and magnetic systems (Hirsch *et al.*, 1992; Batrouni and Scalettar, 1992; Tobochnik *et al.*, 1992; Prokof'ev *et al.*, 1998).

This mapping of the quantum system onto a classical system of interacting extended objects (lines) can be exploited by using known properties of the quantum system to understand those of clasical extended objects. For example, this was used to understand the properties of interstitials and vacancies and argue for supersolid order in vortex crystals (Frey *et al.*, 1994). In this case, understanding the quantum system shed light on the behavior of a classical system.

Notice that it is clear with this representation that, unlike the example of the quantum Ising model, the space and imaginary time directions are not equivalent. One way to see this is as follows: World lines can hop in the positive or negative space directions, but they can only hop in the positive imaginary time direction!

Practice problem: There are other useful representations for expressing the partition function as a path integral. A very useful representation is in terms of coherent states, eigenstates of the destruction operator,

$$a_i|\{\Phi\}\rangle = \phi(i)|\{\Phi\}\rangle, \quad \langle\{\Phi\}|a_i^\dagger = \langle\{\Phi\}|\phi^\star(i), \tag{7.79}$$

where the eigenvalues $\phi(i)$ are complex numbers that are defined on the sites, i, of the lattice. In terms of the more familiar occupation number representation vacuum $(a|0\rangle = 0, a^\dagger|0\rangle = |1\rangle)$ the coherent state $|\Phi\rangle$ is defined as follows (Negele and Orland, 1998; Gardiner, 2009).

$$|\{\Phi\}\rangle = \exp\left(\sum_i (-\frac{1}{2}\phi^\star(i)\phi(i) + \phi(i)a_i^\dagger)\right)|0\rangle. \tag{7.80}$$

With this normalization we obtain the inner product of two coherent states

$$\langle\{\Psi\}|\{\Phi\}\rangle = \exp\left(\sum_i \left(\psi^\star(i)\phi(i) - \frac{1}{2}\phi^\star(i)\phi(i) - \frac{1}{2}\psi^\star(i)\psi(i)\right)\right), \tag{7.81}$$

and the resolution of unity

$$1 = \int \prod_i \frac{d^2\phi(i)}{2\pi}|\{\Phi\}\rangle\langle\{\Phi\}|. \tag{7.82}$$

Use this representation to evaluate the trace in Eq. (7.68) and express the partition function as the path integral,

$$Z = \int \prod_{i,\ell} \frac{d^2\phi(i,\ell)}{\pi} e^{-S(\phi^\star,\phi)}, \tag{7.83}$$

where the action is given by (Batrouni and Mabilat, 1999)

$$S = \sum_{i,\ell} \phi^\star(i,\ell)\Delta_{-\tau}\phi(i,\ell) + \Delta\tau \sum_\ell \mathcal{H}[\phi^\star(i,\ell+1), \phi(i,\ell)], \tag{7.84}$$

and where i is the spatial index of the site and ℓ the imaginary time slice. In Eq. (7.84), $\mathcal{H}[\phi^\star(i,\ell), \phi(i,\ell-1)]$ means that at the imaginary time slice ℓ, we replace in the Hamiltonian, Eq. (7.65), the destruction operator a_i by the complex field $\phi(i,\ell)$, and the creation operator, $a_i^\dagger$, by $\phi^*(i, \ell+1)$. Furthermore, the forward and backward finite difference operators are

$$\Delta_\tau \phi(i, \ell) \equiv \phi(i, \ell+1) - \phi(i, \ell), \quad \Delta_{-\tau} \phi(i, \ell) \equiv \phi(i, \ell) - \phi(i, \ell-1). \tag{7.85}$$

Note: It can be shown (Batrouni and Mabilat, 1999) that the world-line representation is dual to the coherent state representation in the Kramers–Wannier sense (Kramers and Wannier, 1941).

7.3.2   Phase diagram

Mapping the bosonic Hubbard model onto a $(D+1)$ classical model did not lead to a well-known model whose phase diagram is already studied. However, one can form a good idea of its phase diagram by first looking at extreme cases. We recall first that this model has two independent control parameters, the density, ρ, and the interaction, U, or, in the grand canonical ensemble the chemical potential, μ, and the interaction, U. In what follows we shall examine the phase diagram in the $(t/U, \mu/U)$ plane (Fisher *et al.*, 1989).

One easy limiting case to study is the no-hopping limit, $t/U \to 0$. The Hamiltonian, in arbitrary dimensionality, reduces to the sum of single-site Hamiltonians,

$$\mathcal{H} = \sum_{i=1}^{N} H_i \tag{7.86}$$

$$H_i = \frac{U}{2} \hat{n}_i \left(\hat{n}_i - 1 \right) - \mu \hat{n}_i. \tag{7.87}$$

The ground state is obtained simply by minimizing H_i with respect to the site occupation for a fixed μ. It is easy to see (Fisher *et al.*, 1989) that for

$$(n-1) < \frac{\mu}{U} < n, \tag{7.88}$$

the site Hamiltonian, H_i, is minimized for a site occupation of n bosons/site. In other words,

$$0 < \frac{\mu}{U} < 1 \Longrightarrow n = 1, \tag{7.89}$$

$$1 < \frac{\mu}{U} < 2 \Longrightarrow n = 2, \tag{7.90}$$

$$2 < \frac{\mu}{U} < 3 \Longrightarrow n = 3. \tag{7.91}$$

This means that as μ/U is increased from 0, the system will be stuck at one boson/site for the interval $0 < \frac{\mu}{U} < 1$ and as μ/U goes into the next interval, $1 < \frac{\mu}{U} < 2$, the occupation will change everywhere to two bosons/site and so on. This change in the density as μ/U is tuned at $T = 0$ is a quantum phase transition! Within each interval, the system is incompressible since the compressibility, $\kappa = \partial\rho/\partial\mu = 0$, it is also an insulator because there is no particle transport when $t/U = 0$. This incompressible insulating phase at multiples of fulfilling is called the Mott insulator.

We have, thus, established, that in the limit $t/U = 0$, the system passes through a series of Mott insulating phases. One can demonstrate via perturbation theory that as t/U increases (the interaction, U, decreases), the Mott regions shrink and eventually vanish. We demonstrate this here, to first order in t/U, by considering the nth Mott phase, $|\mathsf{M}_n\rangle$. To take the system out of this phase, we can add one particle to it, creating a one-particle excitation $|\mathsf{P}_n\rangle$, or remove a particle, creating a one-hole excitation $|\mathsf{H}_n\rangle$. The properly normalized states are given by

$$|\mathsf{M}_n\rangle = |n, n, n, \ldots, n\rangle, \tag{7.92}$$

$$= \prod_{i=1}^{N} \frac{(a_i^{\dagger})^n}{\sqrt{(n+1)!}} |0\rangle,$$

$$|\mathsf{P}_n\rangle = \frac{1}{\sqrt{N}} \sum_{i=1}^{N} \frac{a_i^{\dagger}}{\sqrt{n+1}} |\mathsf{M}_n\rangle, \tag{7.93}$$

$$|\mathsf{H}_n\rangle = \frac{1}{\sqrt{N}} \sum_{i=1}^{N} \frac{a_i}{\sqrt{n}} |\mathsf{M}_n\rangle, \tag{7.94}$$

where $N = N_1 N_2 \ldots N_D$. To first order in the perturbation, t/U, the energies of these states are simply given by the expectation values of $\mathcal{H}$. Since we are fixing the particle numbers here, the chemical potential, μ, should not be included in the Hamiltonian for this calculation. The energies are easy to obtain and are given by,

$$E(\mathsf{M}_n) = \langle \mathsf{M}_n | \mathcal{H} | \mathsf{M}_n \rangle,$$

$$= \frac{U}{2} N n(n-1), \tag{7.95}$$

$$E(\mathsf{P}_n) = \langle \mathsf{P}_n | \mathcal{H} | \mathsf{P}_n \rangle,$$

$$= \frac{U}{2} N n(n-1) + Un - 2tD(n+1), \tag{7.96}$$

$$E(\mathsf{H}_n) = \langle \mathsf{H}_n | \mathcal{H} | \mathsf{H}_n \rangle,$$

$$= \frac{U}{2} N n(n-1) - U(n-1) - 2tDn. \tag{7.97}$$

Since $T = 0$ here, the internal energy is equal to the free energy that allows us to calculate the chemical potential for n particles as $\mu(n) = E(n) - E(n-1)$. The chemical potential where the $\mathsf{H}_n - \mathsf{M}_n$ transition takes place gives the lower boundary of the nth Mott region,

$$\frac{\mu_{lower}}{U} = \frac{1}{U} \left(E(\mathsf{M}_n) - E(\mathsf{H}_n) \right) = (n-1) + \frac{2Dt}{U} n, \tag{7.98}$$

while the upper boundary is given by the $P_n - M_n$ transition,

$$\frac{\mu_{upper}}{U} = \frac{1}{U}\left(E(P_n) - E(M_n)\right) = n - \frac{2Dt}{U}(n+1). \tag{7.99}$$

This gives the nth Mott gap, Δ_n,

$$\Delta_n = \frac{\mu_{upper}}{U} - \frac{\mu_{lower}}{U} = 1 - \frac{2Dt}{U}(1 + 2n), \tag{7.100}$$

which vanishes at the critical coupling

$$\left(\frac{2Dt}{U}\right)_c = \frac{1}{2n+1}. \tag{7.101}$$

Equation (7.100) shows that in the $t/U \to 0$ limit the gap has unit width in agreement with Eq. (7.88) and that the effect of hopping is to make the Mott regions protrude into the finite t/U region while at the same time getting narrower. Equation (7.101) shows that these Mott regions eventually end at a critical value, $(2Dt/U)_c$, giving the Mott regions lobe-like shapes in the $(\mu/U, t/U)$ plane (see Figure 7.4 (Batrouni *et al.*, 1990)). See reference (Freericks and Monien, 1996) for a more elaborate treatment of perturbation to third order in t/U.

What is the phase of the system when the Mott insulator disappears? It is well known that a non-interacting bosonic system in two (three) dimensions undergoes Bose–Einstein condensation (BEC) at zero (finite) temperature. In the presence of weak interaction, this BEC is also known to become a superfluid (SF). In addition, at $T = 0$, the one-dimensional system is superfluid but without BEC.

We therefore form (Fisher *et al.*, 1989) the following qualitative picture of the phase diagram: At small enough t/U, the system is in a Mott insulating phase when the density is an integer multiple of the number of sites. As t/U increases, the bosons can hop around the lattice more easily and the Mott insulators disappear to be replaced by a superfluid phase. The system is always superfluid when the number of particles is incommensurate with the number of sites. The phase diagram of the one-dimensional system, determined from quantum Monte Carlo simulations, is shown in Figure 7.4 (Batrouni *et al.*, 1990) clearly showing the Mott lobes. Going from the superfluid to the Mott phase, the system undergoes a quantum phase transition.

This qualitative picture will serve as the basis for a mean field calculation. But first, how does one determine if a phase is superfluid? Superfluidity is a quantum phenomenon where the phase of the wavefunction is coherent over macroscopic length scales. In such a case, the phase is said to be rigid or stiff: A twist in the phase at one end of the system makes itself felt throughout. The average monentum in quantum mechanics is given by

$$\vec{p} = \langle \Psi | -i\hbar\vec{\nabla} | \Psi \rangle, \tag{7.102}$$

$$= \langle \Psi | \hbar\vec{\nabla}\phi(\vec{r}) | \Psi \rangle, \tag{7.103}$$

$$\equiv m\vec{v}_s, \tag{7.104}$$

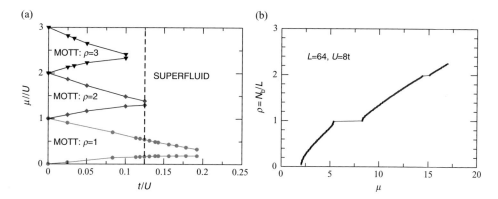

Fig. 7.4 Left: The ground-state phase diagram of the one-dimensional bosonic Hubbard model, Eq. (7.65), from quantum Monte Carlo simulations. The system is always superfluid for incommensurate filling. For commensurate filling, $\rho = 1, 2, 3, \ldots$, the systems is a Mott insulator for strong coupling, small t/U, and superfluid for small coupling, large t/U. Right: A plot of ρ versus μ at fixed t/U along the dashed vertical in the left panel. In the horizontal plateaux $\kappa = \partial \rho / \partial \mu = 0$ indicating the incompressible Mott lobes. The transitions from the superfluid to the Mott insulator lobes are quantum phase transitions.

where $\phi(\vec{r})$ is the phase of the wave function and $\vec{v}_s$ is, by definition, the superfluid velocity. We see that $\vec{p} = 0$ if the phase of the wave function is not coherent. In other words, if the phase fluctuates with very short correlation length (called coherence length in this context), then the average of the gradient will vanish even over short length scales. If, on the other hand, the phase is coherent over long length scales, one can have $\vec{p} \neq 0$. One way, then, to probe for superfluidity, is to calculate the free energy of the system under normal periodic boundary conditions, $F(0)$, and then to measure the free energy with an externally imposed phase gradient, $\delta \phi$ (see below), $F(\delta \phi)$. The difference is non-zero only if the phase is rigid and is interpreted as the kinetic energy of the superfluid component due to the imposed velocity field (phase gradient) (Fisher *et al.*, 1973),

$$F(\delta \phi) = F(v_s) \approx F(0) + \rho_s \frac{V m}{2} \vec{v}_s^2, \tag{7.105}$$

where V is the volume, m the mass of the bosons and ρ_s the number density of the superfluid component. Note that, since the wave function must be single valued, then in a system with periodic boundary conditions, *e.g.* superfluid helium in a toroidal container, the total phase change when the entire length of the system is traversed must be an integer multiple of 2π. Consequently,

$$\langle \vec{\nabla} \phi(\vec{r}) \rangle = \frac{m}{\hbar} \vec{v}_s = 2\pi n / L, \tag{7.106}$$

where L is the length of the system and n is an integer. The topological number n counts the number of times the phase winds as the system is traversed. This is the remarkable Onsager velocity quantization condition (Onsager, 1949) for superfluids and has been verified experimentally! Therefore, the superfluid density can be obtained from

$$\rho_s = \frac{1}{Vm} \frac{\partial^2 F(v_s)}{\partial v_s^2}. \tag{7.107}$$

In the world-line representation discussed earlier, we can show that the superfluid density is given by (Pollock and Ceperley, 1987),

$$\rho_s = \frac{\langle W^2 \rangle}{tD\beta L^{D-2}}, \tag{7.108}$$

where W, the "winding number", is an integer giving the number of times the world lines wind around the lattice in the space direction. The configuration in Figure 7.2 has only $W = 0$. Equation (7.108) leads to an appealing physical picture for the phases of the bosonic Hubbard model in terms of world lines. In Figure 7.5 we show three possible configurations of world lines. The left panel shows the world-line configuration with one

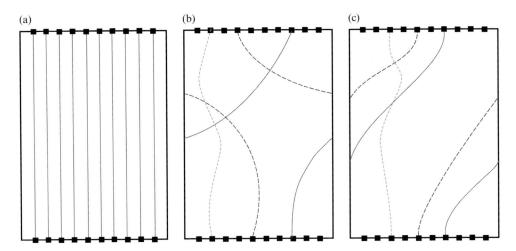

Fig. 7.5 Possible configurations of boson world lines. The horizontal axis is the space direction with the squares denoting lattice sites. The vertical axis is β, the imaginary time. Left panel shows a Mott insulator: The energy cost of double occupancy of a site is too high, the boson occupation does not fluctuate and the bosons go straight along the imaginary time direction. Large quantum fluctuations are allowed for in dilute systems and systems where the contact repulsion is not very large. The middle panel depicts such a case where the fluctuations are not coherent: The solid world line undergoes a large fluctuation giving it $W = 1$, the long dashed line gets $W = -1$ while the short dashed line has only small fluctuations. The total winding of this configuration is $W = 0$. The right panel shows a coherent fluctuation: The total winding is $W = 2$. A phase where, as a function of real time, the fluctuations tend to be like those of the center panel, the system is a normal (non-superfluid) quantum liquid. If the fluctuations are like those of the right panel, the system is in the superfluid phase.

boson/site and large U/t, in other words the Mott insulator. In this phase, the system is incompressible and the bosons cannot hop among the sites and therefore their world lines are straight lines in the imaginary time direction. The winding is strictly vanishing and $\rho_s = 0$. The center panel shows a normal (non-superfluid) bosonic quantum liquid. We choose a dilute system to keep the figure uncluttered. In this situation, there may be large enough quantum fluctuations to produce a world line with $W \neq 0$. However, the fluctuations are not coherent and, typically, there is an equal number of positive and negative windings at any given moment that keeps $W = 0$. However, note the differences between this situation and that of the Mott insulator. In the superfluid phase, the fluctuations are coherent and, typically, when a boson undergoes a large fluctuation it does this collectively with other particles leading to configurations such as the one depicted in the right panel of Figure 7.2. When averaged over configurations, the situation depicted in the right panel will also give $\langle W \rangle = 0$ but it will have $\langle W^2 \rangle \neq 0$, whereas the situations depicted in the left and center panels will give $\langle W \rangle = \langle W^2 \rangle = 0$.

Scaling arguments (Fisher *et al.*, 1989) lead to the critical exponents $\nu = 1/2$ and $z = 2$ for all dimensions. These values lead to the prediction that as the system goes from the SF phase into a Mott lobe in Figure 7.5, the superfluid density vanishes as $\rho_s \sim |\rho - \rho_c|$, where $\rho_c = 1$ for the first Mott lobe, $\rho_c = 2$ for the second, etc. Figure 7.6 (Batrouni *et al.*, 1990) shows QMC results that confifirm this prediction.

Note that, like the quantum Ising model discussed above, we have here $z\nu = 1$. However, in contradistinction with the quantum Ising model where the correlation

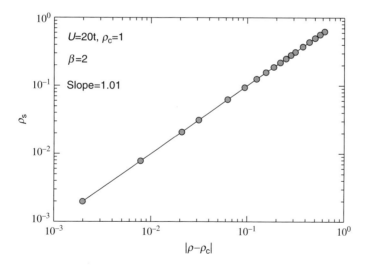

Fig. 7.6 The superfluid density, ρ_s, as a function of $|\rho - \rho_c|$, on a log-log scale, for the one-dimensional bosonic Hubbard model. This confirms that $\rho_s \sim |\rho - \rho_c|$. Note that although $\beta = 2$ does not seem to be very large, the system is essentially in its ground state because the thermal energy, β^{-1}, is much smaller than U. The systems sizes used ranged from $N = 16$ to $N = 256$.

length in the imaginary time direction diverged like the spatial correlation length, $\xi_\tau \sim \xi$ since $z = 1$ in that case, in the present case we have $\xi_\tau \sim \xi^2$: The space and imaginary time directions are not equivalent.

We now illustrate some of the critical properties of the bosonic Hubbard model using a simple but instructive method, namely mean field. We begin with a word of caution, however. The mean-field method is only as good as the guess for the mean-field state and it is not reliable at low dimensions. Furthermore, this is not a controlled approximation in the sense that the error cannot be made aribtrarily small by including higher corrections. There are several ways, of varying sophistication, to implement a mean field calculation for the model Eq. (7.65) (Oosten *et al.*, 2001). Here we shall illustrate the simplest method applied to the hard-core bosonic Hubbard model. In this model, the contact interaction between bosons is infinitely repulsive that, consequently, strictly forbids multiple occupation of a site. The Hamiltonian is given by

$$\mathcal{H} = -t \sum_{\langle i,j \rangle} \left(a_i^\dagger a_j + a_j^\dagger a_i \right) - \mu \sum_i \hat{n}_i, \tag{7.109}$$

with the condition that no multiple occupation is allowed. We now propose a mean-field state,

$$|\Psi_{MF}\rangle = \prod_{i=1}^N (u + v a_i^\dagger)|\mathbf{0}\rangle, \tag{7.110}$$

where $|\mathbf{0}\rangle$ is the empty state, $a_i|\mathbf{0}\rangle = 0$. In Eq. (7.110), u (v) is the amplitude that site i is vacant (occupied) and, consequently, we have $u^2 + v^2 = 1$ that ensures that $\langle \Psi_{MF}|\Psi_{MF}\rangle = 1$. This can be written as

$$u = \sin(\theta/2), \quad v = \cos(\theta/2), \tag{7.111}$$

The angle, θ, will be determined by minimizing the grand potential,

$$\Gamma = \langle \Psi_{MF}|\mathcal{H}|\Psi_{MF}\rangle. \tag{7.112}$$

Keeping in mind the hard-core condition of no multiple occupancy ($a_i^\dagger a_i^\dagger|\mathbf{0}\rangle = 0$), it is easy to show that,

$$\Gamma = -t \sum_{\langle i,j \rangle} \langle \mathbf{0}|(u + v a_i)(u + v a_j)(a_i^\dagger a_j + a_j^\dagger a_i)(u + v a_i^\dagger)(u + v a_j^\dagger)|\mathbf{0}\rangle$$
$$- \mu \sum_i \langle \mathbf{0}|(u + v a_i)\hat{n}_i(u + v a_i^\dagger)|\mathbf{0}\rangle, \tag{7.113}$$
$$= -2tDu^2v^2N - \mu v^2 N,$$
$$= -2tDN \sin^2(\theta/2) \cos^2(\theta/2) - \mu N \cos^2(\theta/2),$$
$$= -\frac{tDN}{2} \sin^2(\theta) - \frac{\mu N}{2} (1 + \cos(\theta)), \tag{7.114}$$

where D is the number of space dimensions and N is the number of sites. Minimizing Γ with respect to θ, $\partial\Gamma/\partial\theta = 0$, gives

$$\cos(\theta) = \frac{\mu}{2tD},\tag{7.115}$$

and consequently,

$$\Gamma = -\frac{tDN}{2}\left(1 + \frac{\mu}{2tD}\right)^2.\tag{7.116}$$

The density as a function of the chemical potential is given by

$$\rho = -\frac{1}{N}\frac{\partial\Gamma}{\partial\mu} = \frac{1}{2} + \frac{\mu}{4tD}.\tag{7.117}$$

Note that from Eq. (7.115) we have the condition that $-1 \le \mu/2tD \le 1$ and, therefore, Eq. (7.117) gives $0 \le \rho \le 1$ as required by the hard-core condition.

The condensate, which is the order parameter, is given by the number of particles in the $k = 0$ quantum level,

$$N_0 = \langle\Psi_{MF}|\tilde{a}_0^\dagger\tilde{a}_0|\Psi_{MF}\rangle,\tag{7.118}$$

where $\tilde{a}_k$ is the Fourier transform of a_i,

$$\tilde{a}_{\vec{k}} = \frac{1}{\sqrt{N}}\sum_{\vec{r}} a_{\vec{r}}\, e^{i\vec{k}.\vec{r}}.\tag{7.119}$$

The periodic boundary conditions we took impose,

$$\vec{k} = (\frac{2\pi n_1}{N_1}, \frac{2\pi n_2}{N_2} \ldots \frac{2\pi n_D}{N_D}),\tag{7.120}$$

where N_i is the number of sites along the ith direction, $-N_i/2 \le n_i \le N_i/2 - 1$, and $N = N_1 N_2 \ldots N_D$. The condensate is, therefore, given by

$$N_0 = \frac{1}{N}\sum_{i,j}\langle\Psi_{MF}|a_i^\dagger a_j|\Psi_{MF}\rangle,$$

$$= \frac{1}{N}\sum_{i\ne j}\langle 0|(u + va_i)(u + va_j)a_i^\dagger a_j(u + va_i^\dagger)(u + va_j^\dagger)|0\rangle$$

$$+\frac{1}{N}\sum_i\langle 0|(u + va_i)a_i^\dagger a_i(u + va_i^\dagger)|0\rangle,\tag{7.121}$$

$$= (N - 1)u^2 v^2 + v^2,\tag{7.122}$$

$$\approx \frac{N}{4}\left(1 - \left(\frac{\mu}{2tD}\right)^2\right),\tag{7.123}$$

$$= N\rho(1 - \rho),\tag{7.124}$$

where, going from Eq. (7.122) to Eq. (7.123) we kept only the term of order N and used Eq. (7.115). We see that as the Mott lobe, $\rho_c = 1$, is approached, the condensate fraction, $\rho_0 = N_0/N$, vanishes as a power with exponent equal to 1, $(\rho_c - \rho)$.

In addition to the order parameter, ρ_0, we can calculate the superfluid density, ρ_s, by imposing an external phase gradient on the system (Bernardet et $al.$, 2002).

$$\mathcal{H}_{\delta\phi} = -t \sum_{\langle i,j \rangle} \left(a_i^\dagger e^{i\delta\phi} a_j + a_j^\dagger e^{-i\delta\phi} a_i \right) - \mu \sum_i \hat{n}_i, \tag{7.125}$$

where the constant phase gradient, $\delta\phi$, is non-zero along only one direction and vanishes along the other directions. Taking the expectation value of this Hamiltonian in the mean-field state gives,

$$\Gamma_{\delta\phi} = -2t(D-1)Nu^2v^2 - 2tNu^2v^2\cos(\delta\phi) - \mu Nv^2, \tag{7.126}$$

$$\approx -2tDNu^2v^2 + tNu^2v^2(\delta\phi)^2 - \mu Nv^2, \tag{7.127}$$

$$= \Gamma + tNu^2v^2(\delta\phi)^2, \tag{7.128}$$

$$= \Gamma + \frac{m}{2} Nu^2v^2v_s^2. \tag{7.129}$$

To get the last equation we used Eq. (107) and the fact that the hopping parameter correspnds to $t = \hbar^2/2m$. The superfluid density is therefore given by,

$$\rho_s = \frac{1}{Nm} \frac{\partial^2 \Gamma_{\delta\phi}}{\partial v_s^2}, \tag{7.130}$$

$$= u^2v^2, \tag{7.131}$$

$$= \rho(1-\rho). \tag{7.132}$$

We see that at this level of mean-field approximation, the condensate fraction and the superfluid density are equal. Furthermore, ρ_s vanishes linearly with $(1-\rho)$ as predicted by the scaling arguments mentioned above and from QMC. This mean-field result is shown (dot-dash line) in Figure 7.7 up to $\rho = 1/2$; the curve is symmetric with respect to that point.

One can improve on this approximation by taking fluctuations into account. Doing that lifts the degeneracy between ρ_0 and ρ_s. The result of this calculation (Bernardet et $al.$, 2002) in two dimensions is shown in Figure 7.7 and compared with exact QMC results. The very close agreement between the exact QMC results and the mean field plus fluctuations is quite remarkable. Such close quantitative agreement between mean-field and exact results is rather unusual. Note, in addition, that $\rho_s > \rho_0$. This emphasizes the important fact that the condensate and the superfluid fraction should not be considered to be the same: They are not. Quantum fluctuations can knock bosons out of the $k = 0$ level, reducing ρ_0, and bosons in the higher-momentum states can participate in superfluid transport, thus increasing ρ_s.

Similar close agreement is obtained in three dimensions. Not surprisingly, this calculation fails in one dimension. One aspect of this failure is that one finds, in

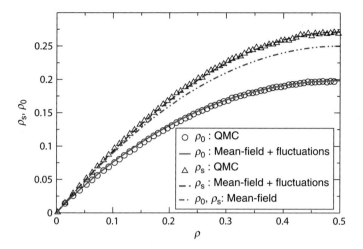

Fig. 7.7 The solid line shows the condensate fraction, ρ_0, and the dahsed line shows the superfluid density, ρ_s, for the $D=2$ hard-core bosonic Hubbard model using the corrected mean-field approximation (see text). The dot-dash line is the lowest-order mean-field result for both ρ_s and ρ_0 (Eq. (7.132)).

$D=1$, that the condensate fraction, ρ_0, is still given by Eq. (7.124), whereas it is well known that $\rho_0 = 0$ in this case.

Practice problem: Hard-core bosons are somewhat awkward to handle. They satisfy the commutation relations,

$$[a_i, a_j] = [a_i^\dagger, a_j^\dagger] = 0, \tag{7.133}$$

and for $i \neq j$,

$$[a_i, a_j^\dagger] = 0. \tag{7.134}$$

But since for hard-core bosons we have $a^\dagger|1\rangle = 0$, it is easy to show that, instead of the usual commutaion relation $[a_i, a_i^\dagger] = 1$, we have

$$a_i a_i^\dagger + a_i^\dagger a_i = 1. \tag{7.135}$$

In fact, hard-core bosons can be easily mapped to a spin-1/2 system via $a_i \to \sigma_i^-$, $a_i^\dagger \to \sigma_i^+$ and $a^\dagger a_i = n_i \to \sigma_i^z + 1/2$. This mapping is valid in any dimension (Bernardet *et al.*, 2002), but for $D=1$ we can use in addition the Jordan–Wigner transformation (see the appendix) to diagonalize the resulting Hamiltonian Eq. (7.109). Show that

$$1 - 2c_j^\dagger c_j = e^{i\pi c_j^\dagger c_j}, \tag{7.136}$$

where c_i and $c_i^\dagger$ are fermionic operators. Use the Jordan–Wigner transformation to express Eq. (7.125) as

$$\mathcal{H}_{\delta\phi} = -t \sum_{i=1}^{N} \left(c_i^\dagger e^{i\delta\phi} c_{i+1} + c_{i+1}^\dagger e^{-i\delta\phi} c_i \right) - \mu \sum_{i=1}^{N} c_i^\dagger c_i. \qquad (7.137)$$

By using Fourier transformation, show that

$$\mathcal{H}_{\delta\phi} = \sum_{k=1}^{N} \epsilon_k c_k^\dagger c_k, \qquad (7.138)$$

$$\epsilon_k = -2t \cos(\delta\phi + k) - \mu. \qquad (7.139)$$

In the ground state (i.e. at zero temperature) there is one fermion per value of k and terminates at N_b, the number of original bosons present,

$$\mathcal{H}_{\delta\phi} = \sum_{k=1}^{N_b} \epsilon_k. \qquad (7.140)$$

Show that the superfluid density is given by

$$\rho_s = \frac{1}{N} \sum_{n=1}^{N_b} \cos\left(\frac{2\pi}{N} n\right), \qquad (7.141)$$

and perform the sum. Show that $\rho_s = 0$ for $\rho = N_b/N = 1$ and that for very dilute systems $\rho_s = \rho$.

Practice problem: In this problem we will perform a rudimentary mean-field calculation for the soft-core bosonic Hubbard model, Eq. (7.65), which has the virtue of simplicity and ease of calculation and which will give a qualitative picture of the phase diagram. On the other hand, due to its simplicity, it will not be very accurate quantitatively. This calculation is based on a straightforward generalization of the method used for hard-core bosons where we took the mean-field state,

$$|\Psi_{MF}\rangle = \prod_{i=1}^{N} (u + v a_i^\dagger)|0\rangle, \qquad (7.142)$$

with the condition $a_i^\dagger a_i^\dagger |0\rangle = 0$. In this case, the calculation decribed the system for $0 \le \rho \le 1$. Now we propose the mean-field state

$$|\Psi_{MF}^n\rangle = \prod_{i=1}^{N} \left(u + \frac{v}{\sqrt{n+1}} a_i^\dagger\right)|n\rangle, \qquad (7.143)$$

where the state $|n\rangle$ has n bosons/site. In analogy with the hard-core case, u is the amplitude to have n particles on a site, and v is the amplitude to create a particle and have $(n + 1)$ particles on that site; $u^2 + v^2 = 1$. Furthermore, we impose the condition $a_i^\dagger a_i^\dagger |n\rangle = 0$: We cannot create two particles on top of the n particles already present, only one creation is allowed. Consequently, $|\Psi_{MF}^n\rangle$ can only describe the system for $n \le \rho \le n + 1$ It is clear that this mean-field state is a poor choice when t/U is large because at weak coupling large fluctuations are not uncommon.

Show that,

$$\frac{\Gamma_n}{N} = \frac{\langle \Psi_{MF}^n | \mathcal{H} | \Psi_{MF}^n \rangle}{N} = -2tDu^2v^2(n + 1) + u^2 \left[\frac{U}{2}n(n - 1) - \mu n \right]$$

$$+ v^2 \left[\frac{U}{2}n(n + 1) - \mu(n + 1) \right]. \quad (7.144)$$

Writing $u = \sin(\theta/2)$ and $v = \cos(\theta/2)$, minimize Γ_n with respect to θ and show that

$$\cos(\theta) = \frac{\mu - Un}{2tD(n + 1)}. \quad (7.145)$$

Substituting in Eq. (7.144), show that

$$\frac{\Gamma_n}{N} = -\frac{t}{2}D(n + 1) \left[1 + \frac{\mu - Un}{2tD(n + 1)} \right]^2 + \frac{1}{2} \left[Un^2 - Un - 2\mu n \right]. \quad (7.146)$$

Calculate

$$\rho^n = -\frac{1}{N} \frac{\partial \Gamma_n}{\partial \mu}, \quad (7.147)$$

and show that $n \le \rho^n \le n + 1$. Hint: Look at Eq. (7.145).

Now we want to calculate the boundaries of the Mott lobes. Since $n \le \rho^n \le n + 1$, the lower boundary, μ_n^-/U, of the nth Mott lobe comes from $\rho_{max}^{(n+1)} = n$ and the upper boundary, μ_n^+/U, comes from $\rho_{min}^n = n$. Show that

$$\frac{\mu_n^-}{U} = (n - 1) + 2Dn\frac{t}{U}, \quad (7.148)$$

$$\frac{\mu_n^+}{U} = n - 2D(n + 1)\frac{t}{U}. \quad (7.149)$$

Draw the phase diagram in the $(t/U, \mu/U)$ plane and compare with Figure 7.4. The Mott gap, the width of the Mott lobe at t/U, can be calculated from $(\mu_n^+ - \mu_n^-)/U$. Show that the nth lobe terminates at the criticial value

$$\frac{t}{U_c} = \frac{1}{2D(2n + 1)}. \quad (7.150)$$

This shows that the tip of the nth lobe recedes as n^{-1} for large n and agrees with more elaborate mean-field calculations. Following the calculation in the hard-core case, show that the condensate fraction is given by,

$$\rho_0 = (n+1)(\rho - n)(1 + n - \rho). \tag{7.151}$$

This expression holds between two Mott lobes, the nth and $(n+1)$th. So, the critical densities are $\rho_c = n$ and $\rho_c = (n+1)$ and we, therefore, see that the condensate vanishes as $\rho_0 \sim |\rho - \rho_c|$. The superfluid density can be calculated by imposing a gradient, as before. Show that ρ_s is given by Eq. (7.151).

Note that due to the constraint that only one particle can be created, this approximation gives nonsensical results beyond the tips of the Mott lobes. Also note that Eqs. (7.148–7.150) agree with the perturbation results Eqs. (7.98–7.101). This is because we used the same states in both cases and because in both cases the energies are just the expectation values of the Hamiltonian. In the perturbation case we fixed the particle number and then calculated the chemical potential; in the mean-field case, we fixed the chemical potential that we used as a variational parameter. Higher-order perturbations and more elaborate mean field do not give identical results.

7.3.3 Dynamics

The mapping of D-dimensional quantum systems onto $(D+1)$-dimensional classical systems has yielded insights into the behavior of both classical and quantum systems. One property of quantum systems not shared by classical ones is their dynamics: Once the Hamiltonian is written, the dynamics of the quantum system are determined. As discussed above in Section 7.2, the additional dimension introduced because of the mapping onto a classical model can be interpreted as the (imaginary) time axis and can, thus, be exploited to study dynamic effects. For example, excitation spectra can be obtained by performing a numerical Laplace transform on Eq. (7.20) using the maximum entropy method (Gubernatis *et al.*, 1991).

We now discuss briefly the properties of the dynamic density–density correlation function of the bosonic Hubbard model (in $D = 1$) and what it tells us about the excitation spectrum and transport properties. The dynamic density–density correlation function is obtained by taking $\tilde{\mathcal{O}}(\vec{k}, 0) = \tilde{n}(\vec{k}, 0)$ in Eq. (7.20).

We will consider a 1-dimensional lattice with $L = 20$ sites on which we place $N_b = 20$ bosons. We are interested in the ground-state properties of the system and to that end we take $\beta = 20t$, which corresponds to a low enough temperature. We study two cases, $U = 2t$ which, according to Figure 7.4, is in the superfluid phase and $U = 8t$ which is in the first Mott insulator lobe.

The left panel of Figure 7.8 (Batrouni *et al.*, 2005) shows $S(k, \omega)$ for $\pi/10 \le k \le 9\pi/10$ obtained from the Laplace transform of $S(k, \tau)$ when $U = 2t$. For each k value, we see a peak corresponding to the excitation energy at that momentum. As k increases, the peaks get wider but remain relatively narrow. In fact, exact diagonalization of smaller systems (Roth and Burnett, 2004) shows that the wider peaks consist of several narrow peaks bunched close together that maximum entropy cannot resolve. Plotting the location of each peak, i.e. energy, versus the momentum, k, yields the dispersion relation of the excitations and is shown in the right panel of Figure 7.8. We see that the dispersion is linear for the smaller values of k, indicating

(a)   (b)

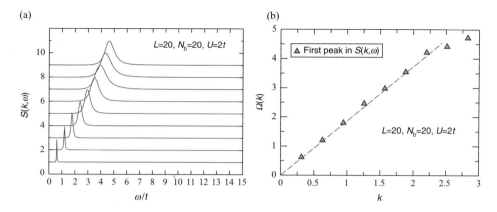

Fig. 7.8 The superfluid phase. Left: $S(k,\omega)$ for the bosonic Hubbard model in $D=1$ obtained by using the maximum-entropy method to calculate the Laplace transform with respect to τ of $S(k,\tau)$. The horizontal axis is the frequency (i.e. energy), each curve corresponds to a different k value. The lowest curve is $k = 2\pi/L$ and moves up to $k = 2\pi(L/2-1)/L \approx \pi$. Right: The dispersion relation of excitations is obtained from the positions of the peaks of $S(k,\omega)$ as a function of k. We see that the dispersion in the superfluid phase is linear for small k. The linear dispersion indicates phonon excitations and a stable superfluid phase.

that the excitations are phonons and the superfluid is stable. Note that as $k \to 0$, the excitation energy $\Omega(k) \to 0$.

The behavior is different when $U = 8t$ and the system is in the Mott insulating phase. The left panel in Figure 7.9 (Batrouni *et al.*, 2005) shows $S(k,\omega)$ that exhibits distinct features not present for $U = 2t$. For example, all the peaks are rather wide, indicating that each is a collection of peaks representing a pack of closely spaced excited states. This means that the precise location of the lowest excitation energy is not easy to determine. In the right panel is shown the location of the peak maximum versus the momentum. It is clear that as $k \to 0$, $\Omega(k \to 0)$ is non-vanishing: There is an energy gap to produce the lowest excited state. This is one of the characteristics of the Mott insulator. However, note that while the energy gap calculated here is $\Omega(k \to 0) \approx 4t$, the value obtained from ρ versus μ in Figure 7.4 (right panel) is $3t$. The discrepancy is easy to understand: The gap in Figure 7.4 is the energy needed to reach the first excited state, whereas the value obtained from Figure 7.9 corresponds to a higher excited state in the middle of a group of excited states. However, note that, for the smallest k value, the value of ω/t where $S(k,\omega)$ rises appreciably does correspond to a value $\omega/t \approx 3t$.

The above example illustrates the utility of the formal interpretation of the β direction as imaginary time and how it can can be exploited to obtain information not easily accessible otherwise. Here we only showed particle excitations obtained from the dynamic density–density correlation function. Other excitations can be calculated by using different operators for $\mathcal{O}$; for example $\langle 0|a(\vec{r}',\tau)a^\dagger(\vec{r},0)|0\rangle$ yields the density of states.

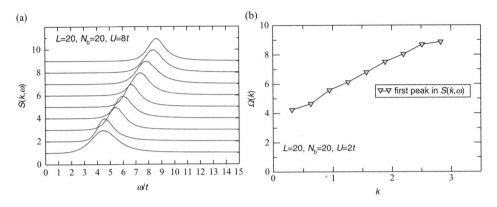

Fig. 7.9 The Mott insulating phase. Left: $Sk(k,\omega)$ from the maximum-entropy method. Note that the peaks are wider. In fact, exact diagonalization of small systems reveals that each wide bump is composed of several peaks where the maximum entropy merges into one. In addition, as $k \to 0$, the excitation energy does not go to zero: It takes a finite energy to produce excitations. This is the Mott gap and is seen clearly in the right panel where $\Omega(k = 0) \approx 4t$. Figure 7.4 gives a value of $3t$ for this value of U. The uncertainty is due to the width of the peaks in the right panel.

7.4   Conclusions

We have presented in this chapter a pedagogical introduction to quantum phase transitions and some of the ideas and methods used to study this fascinating and very active subject. The experimental realization of trapped ultracold boson and fermion atomic systems loaded on optical lattices, provides actual systems governed by the type of Hamiltonians we studied here. These systems are very versatile because the Hamiltonian parameters are tunable and thus allow the experimental exploration of the many exotic phases and phases transitions expected in these models.

7.5   Appendix: The Jordan–Wigner transformation

The Jordan–Wigner transformation is not within the scope of this chapter but we present it in this appendix because it is useful and interesting and to point out why it does not work in more than one dimension.

Consider the following relations,

$$\sigma_i^z = 1 - 2c_i^\dagger c_i, \tag{7.152}$$

$$\sigma_i^+ = \prod_{j<i}(1 - 2c_j^\dagger c_j)c_i, \tag{7.153}$$

$$\sigma_i^- = \prod_{j<i}(1 - 2c_j^\dagger c_j)c_i^\dagger, \tag{7.154}$$

where $\sigma^\pm = (\sigma^x \pm i\sigma^y)$ are raising $(+)$ and lowering $(-)$ operators for the z-component of the spin. c_i and $c_i^\dagger$ are operators whose commutation relations are determined from those of the Pauli matrices,

$$[\sigma_i^+, \sigma_j^-] = \delta_{i,j}\sigma_i^z, \quad [\sigma_i^z, \sigma_j^\pm] = \pm\delta_{i,j}\sigma_i^\pm. \tag{7.155}$$

Inverting Eqs. (7.153) and (7.154) gives

$$c_i = \left(\prod_{j<i}\sigma_j^z\right)\sigma_i^+,$$

$$c_i^\dagger = \left(\prod_{j<i}\sigma_j^z\right)\sigma_i^-, \tag{7.156}$$

with which it can be verified easily that

$$\{c_i, c_j^\dagger\} = \delta_{i,j}, \quad \{c_i, c_j\} = \{c_i^\dagger, c_j^\dagger\} = 0. \tag{7.157}$$

Therefore, c_i and $c_i^\dagger$ are *fermionic* annihilation and creation operators. Notice that central to this transformation is the product for $j < i$ of σ_j^z in Eq. (7.157), which, in effect, gives the operators $\sigma_i^\pm$ tails extending all the way to the edge of the system. In one dimension, there is only one way to lay this tail and the transformation works. But in higher dimensions, the path taken by the tail is not unique: There is an infinity of tails (in the thermodynamic limit) and the Jordan–Wigner transformation fails for $D > 1$ (see below).

The above transformation is the conventional Jordan–Wigner transformation, but for the Hamiltonian Eq. (7.28), it is more convenient to rotate the spin axis,

$$\sigma^z \to \sigma^x, \quad \sigma^x \to -\sigma^z. \tag{7.158}$$

Equations (7.153) and (7.154) then become,

$$\sigma_i^x = 1 - 2c_i^\dagger c_i, \tag{7.159}$$

$$\frac{1}{2}(-\sigma_i^z + i\sigma_i^y) = \prod_{j<i}(1 - 2c_j^\dagger c_j)c_i, \tag{7.160}$$

$$\frac{1}{2}(-\sigma_i^z - i\sigma_i^y) = \prod_{j<i}(1 - 2c_j^\dagger c_j)c_i^\dagger, \tag{7.161}$$

and therefore

$$\sigma_i^z = -\prod_{j<i}(1 - 2c_j^\dagger c_j)(c_i^\dagger + c_i). \tag{7.162}$$

We then have,

$$
\begin{aligned}
\sigma_{i+1}^z \sigma_i^z &= \prod_{k<i+1}(1 - 2c_k^\dagger c_k)(c_{i+1} + c_{i+1}^\dagger)\prod_{j<i}(1 - 2c_j^\dagger c_j)(c_i + c_i^\dagger),\\
&= \prod_{k<i}(1 - 2c_k^\dagger c_k)(1 - 2c_i^\dagger c_i)(c_{i+1} + c_{i+1}^\dagger)\prod_{j<i}(1 - 2c_j^\dagger c_j)(c_i + c_i^\dagger),\\
&= \left(\prod_{k<i}(1 - 2c_k^\dagger c_k)\right)^2 (1 - 2c_i^\dagger c_i)(c_{i+1} + c_{i+1}^\dagger)(c_i + c_i^\dagger),\\
&= (1 - 2c_i^\dagger c_i)(c_{i+1} + c_{i+1}^\dagger)(c_i + c_i^\dagger),\\
&= c_{i+1}c_i + c_i^\dagger c_{i+1}^\dagger + c_{i+1}^\dagger c_i + c_i^\dagger c_{i+1}.
\end{aligned}
\tag{7.163}
$$

Notice how the tails in the Jordan–Wigner transformation disappeared: The overlap of the tails gave the square of a term whose value is ± 1. For this transformation to work, it is crucial that the tails overlap in this way. That is why this fails in higher dimensions, there are infinitely many ways to thread the tail through the system without overlap. With Eq. (7.163), the Hamiltonian, Eq. (7.28) becomes

$$
\mathcal{H} = -J\sum_{i=1}^{N}\left(c_i^\dagger c_{i+1} + c_{i+1}^\dagger c_i + c_i^\dagger c_{i+1}^\dagger + c_{i+1}c_{i+1} - 2gc_i^\dagger c_i - g\right),
\tag{7.164}
$$

where we put $h = gJ$. To diagonalize Eq. (7.164), we first apply the Fourier transformation,

$$
c_k = \frac{1}{\sqrt{N}}\sum_{\ell=1}^{N}c_\ell e^{-ik\ell},
\tag{7.165}
$$

which gives,

$$
\mathcal{H} = J\sum_{k}\left(2[g - \cos(ka)]c_k^\dagger c_k - i\sin(ka)[c_{-k}^\dagger c_k^\dagger + c_{-k}c_k]\right),
\tag{7.166}
$$

where $k = 2\pi n/N$, $n = -N/2, -(N-1)/2, \ldots 0, 1, 2, \ldots (N-1)/2$ and a is the lattice constant that we will take to be unity. Now we apply the Bogoliubov transformation,

$$
\gamma_k = u_k c_k - i v_k c_{-k}^\dagger,
\tag{7.167}
$$

$$
c_k = u_k \gamma_k + i v_k \gamma_{-k}^\dagger,
\tag{7.168}
$$

where γ_k is a new fermionic operator. In order for γ_k and $\gamma_k^\dagger$ to satisfy the usual fermionic anticommutation relations, Eq. (7.157), one can show that u_k and v_k are real functions subject to the conditions $u_{-k} = u_k$, $v_{-k} = v_k$ and $u_k^2 + v_k^2 = 1$. This last condition can be expressed as $u_k = \sin(\theta_k)$, $v_k = \cos(\theta_k)$. Substituting Eq. (7.168) in Eq. (7.166) and requiring that the resulting Hamiltonian only contain terms of the form $\gamma_k^\dagger \gamma_k$ fixes the angle θ_k,

$$\tan\theta_k = \frac{\sin(k)}{\cos(k) - g}, \qquad (7.169)$$

and gives for the diagonalized Hamiltonian,

$$\mathcal{H} = \sum_k \epsilon_k (\gamma_k^\dagger \gamma_k - \frac{1}{2}), \qquad (7.170)$$

$$\epsilon_k = 2J \left(1 + g^2 - 2g \cos(k)\right)^{1/2} . \qquad (7.171)$$

As mentioned above, the Jordan–Wigner transformation is very useful in $D = 1$ but fails in higher dimensions due to the certainty of non-overlapping tails.

References

Batrouni, G.G., Assaad, F.F., Scalettar, R.T., and Denteneer, P.J.H. (2005). *Phys. Rev.*, **A72**, 031601(R).

Batrouni, G. G. and Mabilat, H. (1999). *Comp. Phys. Comm.*, **121-122**, 468.

Batrouni, G. G. and Scalettar, R. T. (1992). *Phys. Rev.*, **B46**, 9051.

Batrouni, G. G. and Scalettar, R. T. (2000). *Phys. Rev. Lett.*, **84**, 1599.

Batrouni, G. G., Scalettar, R. T., and Zimanyi, G. T. (1990). *Phys. Rev. Lett.*, **65**, 1765.

Baxter, R. J. (1982). *Exactly solved models in statistical mechanics*. Academic Press, New York.

Bernardet, K., Batrouni, G. G., Meunier, J.-L., Schmid, G., Troyer, M., and Dorneich, A. (2002). *Phys. Rev.*, **B65**, 104519.

Fazekas, P. (1999). *Lecture notes on electron correlation and magnetism*. World Scientific, Singapore.

Fisher, M. E., Barber, M. N., and Jasnow, D. (1973). *Phys. Rev.*, **A8**, 1111.

Fisher, M. P. A., Weichman, P. B., Grinstein, G., and Fisher, D. S. (1989). *Phys. Rev.*, **B40**, 546.

Freericks, J.K. and Monien, H. (1996). *Phys. Rev.*, **B53**, 2691.

Frey, E, Nelson, D. R., and Fisher, D. S. (1994). *Phys. Rev.*, **B49**, 9723.

Gardiner, C. (2009). *Stochastic methods: A handbook for the natural and social sciences*. Springer, Berlin.

Gubernatis, J. E., Jarrell, Mark, Silver, R. N., and Sivia, D. S. (1991). *Phys. Rev.*, **B44**, 6011.

Gutzwiller, M.C. (1963). *Phys. Rev. Lett.*, **10**, 159.

Hirsch, J. E., Sugar, R. L., Scalapino, D. J., and Blankenbecler, R. (1992). *Phys. Rev.*, **B46**, 9051.

Hohenberg, P. C. and Halperin, B. I. (1989). *Rev. Mod. Phys.*, **B40**, 546.

Hubbard, J. (1964). *Proc. Roy. Soc.*, **A276**, 238.

Jaksch, D., Bruder, C., Cirac, J.I., Gardiner, C.W., and Zoller, P. (1998). *Phys. Rev. Lett.*, **81**, 3108.

Kanamori, J. (1963). *Prog. Theor. Phys.*, **30**, 275.

Kramers, H. A. and Wannier, G. H. (1941). *Phys. Rev.*, **60**, 252.

LeBellac, M., Mortessagne, F., and Batrouni, G. G. (2004). *Equilibrium and non-equilibrium statistical thermodynamics*. Cambridge University Press, Cambridge.

Ma, S.-K. (1985). *Statistical mechanics*. World Scientific, Singapore.

Negele, J. W. and Orland, H. (1998). *Quantum many-particle systems*. Westview Press, Boulder.

Niyaz, P., Scalettar, R. T., Fong, C.Y., and Batrouni, G. G. (1991). *Phys. Rev.*, **B44**, 7143.

Niyaz, P., Scalettar, R. T., Fong, C.Y., and Batrouni, G. G. (1994). *Phys. Rev.*, **B50**, 362.

Onsager, L. (1944). *Phys. Rev.*, **65**, 117.

Onsager, L. (1949). *Nuovo Cimento*, **6**, 249.

Oosten, D. Van, van der Straten, P., and Stoof, H. T. C. (2001). *Phys. Rev.*, **A63**, 053601.

Pollock, E. L. and Ceperley, D. M. (1987). *Phys. Rev.*, **B36**, 8343.

Prokof'ev, N. V., Svistunov, B. V., and Tupitsyn, I. S. (1998). *Phys. Lett.*, **A238**, 253.

Rasetti, M. (1991). *The Hubbard model: Recent results*. World Scientific, Singapore.

Roth, R. and Burnett, K. (2004). *J. Phys. B*, **37**, 3893.

Sachdev, S. (1999). *Quantum phase transitions*. Cambridge University Press, Cambridge.

Savit, R. (1980). *Rev. Mod. Phys.*, **52**, 453.

Sondhi, S. L., Girvin, S. M., Carini, J. P., and Shahar, D. (1997). *Rev. Mod. Phys.*, **69**, 315.

Stanley, H. E. (1971). *Introduction to phase transitions and critical phenomena*. Oxford University Press, Oxford.

Tobochnik, J., Batrouni, G. G., and Gould, H. (1992). *Comput.Phys.*, **6**, 673.

8
Interactions in quantum fluids

T. GIAMARCHI

DPMC-MaNEP, University of Geneva, 24 Quai Ernest Ansermet, 1211 Geneva 4, Switzerland

8.1 Introduction

The problem of interacting quantum particles is one of the most fascinating problems in physics, with a history nearly as long as quantum mechanics itself. From a purely fundamental point of view this is a staggering problem. Even in one gram of solid there are more particles interacting together than there are stars in the universe. In addition, these particles behave as waves, since quantum effects are important and must obey symmetrization or antisymmetrization principles. It it thus no wonder that despite nearly a century of efforts we still lack the tools to give a complete solution of this problem, and that sometimes even the concepts needed to describe such interacting systems need to be sharpened.

The problem is not simply an academic one, however. Indeed, with the discovery of quantum mechanics, came the understanding of the behavior of electrons in solids and the band-structure theory. One of the fantastic successes of such a theory was to understand why some materials are metals, insulators or semiconductors, an understanding that is at the root of the discovery of the transistor and all the modern electronic industry. An additional piece of knowledge was added by Landau, with the so-called Fermi liquid theory, where he showed that in most fermionic systems the effects of interactions could be essentially forgotten, and hidden in the redefinition of simple parameters such as the mass of the particles. Being able to forget about interactions paved the way to study the effects on the real electronic systems of much smaller perturbations than the electron–electron interactions, for example the electron and lattice vibrations. This is at the root of the understanding of many possible orders of the solids, such as superconductivity and magnetism, or the interplay of magnetism and conducting properties of the systems such as the giant magnetoresistance. These properties have also had a profound impact on our everyday life.

Now such systems have been intensively studied and the forefront of research has moved to materials for which we cannot hide the effects of interactions any longer. These include in particular the wide class of materials such as the oxides that have properties as varied as being the best superconductors known to date, or exhibit properties such as ferroelectricity. These compounds or others that remain to be discovered are clearly those that will be the materials of the future for applications, and mastering of their properties requires the deep understanding of the effects of interactions. In addition to problems in condensed matter, cold atomic gases have recently provided marvellous realizations of such strongly cor-related systems and have thus added both to the challenge we have to face, but also provided model systems that could help us to make progress in that difficult field.

In this chapter I review the basic concepts of the effects of interactions on quantum particles. I focus here mostly on the case of fermions, but several aspects of interacting bosons will be mentioned as well. The chapter has been voluntarily kept at an elementary level and should be suitable for students wanting to enter this field. I review the concept of Fermi liquid, and then move to a description of the interaction effects, as well as the main models that are used to tackle these questions. Finally,

I study the case of one-dimensional interacting particles that constitutes a fascinating special case.

8.2 Fermi liquids

This section is based on a master course given at the university of Geneva, over several years together with C. Berthod, A. Iucci, P. Chudzinski. Many more details can be found on the course notes on `http://dpmc.unige.ch/gr_giamarchi/`.

8.2.1 Weakly interacting fermions

Let me start by recalling briefly some well-known but important facts about non-interacting fermions. I will not recall the calculation since they can be found in every textbook on solid-state physics (Ashcroft and Mermin, 1976; Ziman, 1972), but just give the main results. These will be important for the case of interacting electrons.

We consider independent electrons described by the Hamiltonian

$$H_{\text{kin}} = \sum_{\vec{k}\sigma} \varepsilon(\vec{k}) c_{\vec{k}\sigma}^{\dagger} c_{\vec{k}\sigma}, \tag{8.1}$$

where the sum over $\vec{k}$ runs in general over the first Brillouin zone. One usually incorporates the chemical potential in the energy $\xi(\vec{k}) = \varepsilon(\vec{k}) - E_{\text{F}}$ to make sure that $\xi(\vec{k}) = 0$ at the Fermi level. The ground state of such a system is the unpolarized Fermi sea

$$|\text{F}\rangle = \prod_{\vec{k}, \xi(\vec{k}) \leq 0} c_{\vec{k}\uparrow}^{\dagger} c_{\vec{k}\downarrow}^{\dagger} |\varnothing\rangle. \tag{8.2}$$

At finite temperature, states are occupied with a probability

$$n(k) = \langle c_{\vec{k}}^{\dagger} c_{\vec{k}} \rangle = f_{\text{F}}(\xi(\vec{k})) = \frac{1}{e^{\beta\xi(\vec{k})} + 1} \tag{8.3}$$

given by the Fermi factor. A very important point, true for most solids, is that the order of magnitude of the Fermi energy is $E_{\text{F}} \sim 1\,\text{eV} \sim 12\,000\,\text{K}$. This means that the temperature, or most of the energies that are relevant for a solid (for example $30\,\text{GH}_z \sim 1\text{K}$) are extremely small compared to the Fermi energy. As a result, the broadening of the Fermi distribution is extremely small. The important states are thus the ones in a tiny shell close to the Fermi level, as shown in Figure 8.1. The other excitations are completely blocked by the Pauli principle. This hierarchy of energies is of course what confers to fermions in solids their unique properties and make them so different from a classical system. As a consequence, some of the response of such a fermion gas are rather unique. The specific heat is linear with temperature (contrarily to the case of a classical gas for which it would be a constant)

$$C_V(T) \propto k_{\text{B}}^2 \mathcal{N}(E_{\text{F}}) T, \tag{8.4}$$

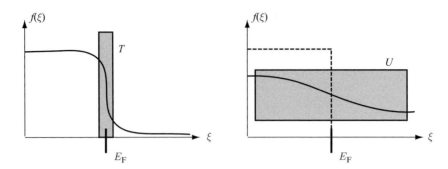

Fig. 8.1 (left) Broadening of the Fermi distribution due to the temperature T. Because in most solids the temperature T is much smaller than the Fermi energy E_F only a tiny fraction close to the Fermi level can be excited. These fermions control all the properties of the system. (right) If the interaction U was acting as a temperature and was broadening the Fermi distribution one would get an extremely large broadening. All the consequences of the sharp Fermi distribution (specific heat linear in temperature, compressibility and spin susceptibility going to a constant at low temperatures) would be lost.

where $\mathcal{N}(E_F)$ is the density of states at the Fermi level, and k_B the Boltzmann constant. The compressibility of the fermion gas goes to a constant in the limit $T \to 0$, and the same goes for the spin susceptibility, namely the magnetization M of the electron gas in response to an applied magnetic field H

$$\chi = \left(\frac{dM}{dH} \right)_T. \tag{8.5}$$

Note that a system made of independent spins would have had a divergent spin susceptibility when $T \to 0$ instead of a constant one. The slope of the specific heat, the compressibility and the spin susceptibility of the free fermion gas are all controlled, by the same quantity, namely the density of states at the Fermi level.

One could thus wonder what would be the effects of interactions on such behavior. Because of the interactions, the energy of a particle can now fluctuate since the particle can give or take energy from the others. Thus, one could naively imagine that the interactions produce an effect on the distribution function similar to that of a thermal bath, with an effective "temperature" of the order of the strength of the interaction. In order to determine the consequences of such a broadening, one needs to estimate the strength of the interactions. In a solid the interaction is mostly the Coulomb interaction. However, in a metal this interaction is screened beyond a length λ that one can easily compute in the Thomas–Fermi approximation (Ashcroft and Mermin, 1976; Ziman, 1972)

$$\lambda^{-2} = \frac{e^2 \mathcal{N}(E_F)}{\epsilon_0}, \tag{8.6}$$

where e is the charge of the particles and ϵ_0 the dielectric constant of the vacuum. To estimate λ we can use the fine structure constant

$$\alpha = \frac{e^2}{4\pi\epsilon_0\hbar c} = \frac{1}{137} \tag{8.7}$$

to obtain

$$\lambda^{-2} = 4\pi\alpha\hbar c \mathcal{N}(E_F) = 4\pi\alpha\hbar c \frac{3n}{2E_F}, \tag{8.8}$$

where n is the density of particles. Using the density of states per unit volume of free fermions in three dimensions

$$n(\varepsilon) = \frac{m}{2\pi^2\hbar^2} \left(\frac{2mE_F}{\hbar^2}\right)^{1/2} = \frac{3}{2}\frac{n}{E_F} \tag{8.9}$$

and $E_F = \hbar v_F k_F$, and $6\pi^2 n = k_F^3$ one gets

$$\lambda^{-2} = \frac{1}{\pi}\alpha\frac{c}{v_F}k_F^2. \tag{8.10}$$

Since $c/v_F \sim 10^2$ in most systems, one finds that $k_F\lambda \sim 1$. The screening length is of the order of the inverse Fermi length, i.e. essentially the lattice spacing in normal metals. This is a striking result: not only is the Coulomb interaction screened, but the screening is so efficient that the interaction is practically *local*. We will use this fact extensively in the definition of models below. Let us now estimate the order of magnitude of this screened interaction. The interaction between two particles can be written as

$$H_{\text{int}} = \frac{1}{2}\int d\vec{r}d\vec{r}'V(\vec{r}-\vec{r}')\rho(\vec{r})\rho(\vec{r}'). \tag{8.11}$$

Since the interaction is screened it is convenient to replace it by a local interaction. Given our previous result let us simply replace the screening length by a the fermion–fermion distance. The effective potential seen at point $\vec{r}$ by one particle is

$$\int d\vec{r}'V(\vec{r}-\vec{r}')\rho(\vec{r}'). \tag{8.12}$$

Due to screening we should only integrate within a radius a around the point r. Assuming that the density is roughly constant one obtains

$$\int_{|\vec{r}-\vec{r}'|<a} d\vec{r}\frac{e^2}{4\pi\epsilon_0|\vec{r}-\vec{r}'|}\rho_0 \sim \frac{e^2\rho_0 S_d a^{d-1}}{4(d-1)\pi\epsilon_0}, \tag{8.13}$$

where S_d is the surface of the sphere in d dimensions. Using $\rho_0 \sim 1/a^d$ and Eq. (8.7) one gets

$$\frac{S_d \alpha \hbar c}{(d-1)a}. \tag{8.14}$$

This potential acting on a particle has to be compared with the kinetic energy of this particle at the Fermi level, which is $E_F = \hbar v_F k_F$. Since $k_F \sim a^{-1}$ one has again to compare α and c/v_F. The two are about the same order of magnitude. The Coulomb energy, even if screened (i.e. even in a very good metal), is thus of the *same* order of magnitude as the kinetic energy. This means, for solids, typical energies of the order of the electron volt. If such an interaction was acting as a temperature in smearing the Fermi function this would lead to an enormous smearing, as shown in Figure 8.1.

This would be in complete contradiction with data on most of solids. The specific heat in real materials is found to be linear, albeit with a slope different from the naive free-electron picture at temperatures much smaller than the scale of the interactions (see e.g Figure 1.8 in Ashcroft and Mermin (1976)). This would be totally impossible if the interactions had smeared the Fermi distribution to a practically flat distribution. Similarly, spin susceptibility and compressibility are still found, e.g. in ^{3}He to be essentially constant at low temperature (Greywall, 1983; 1984), again implying that the Fermi distribution must remain quite sharp.

In addition, a remarkable experimental technique to look at the single-particle excitations, and momenta distributions is provided by the photoemission technique (Damascelli, *et al.* 2003). Pending some hypothesis this technique is a direct measure of the spectral function $A(\vec{k}, \omega)$, which is the probability of finding an excitation with the energy ω and a momentum $\vec{k}$. For free particles $A(\vec{k}, \omega) = \delta(\omega - \xi(\vec{k}))$. Naively, one would expect that, because an energy of the order of the interaction can be exchanged, these perfect peaks are broadened over an energy of the order of the interaction. As can be seen from Figure 8.2 this is clearly not the case. Very sharp peaks exist, and become sharper and sharper as one gets closer to the Fermi energy. The momentum distribution seems to be broadened uniquely by the temperature when one is at the Fermi surface.

One is thus faced with a remarkable puzzle: the "free-electron" picture seems, at least qualitatively, to work *much better* than it should, based on estimates of the interaction strength. This must hide a profound effect, and is thus a great theoretical challenge.

8.2.2 Landau–Fermi liquid theory

The solution to this puzzle was given by Landau and is known under the name of Fermi-liquid theory (Landau, 1957*a*; 1957*b*).

Let me first give a very qualitative description of the underlying ideas before embarking on a more rigorous definition and derivation. The main idea behind Fermi liquids (Nozieres, 1961) is to look at the excitations that would exist above the ground state of the system. In the absence of interactions the ground state is the Fermi sea (Eq. 8.2). Let us assume now that one turns on the interactions. This ground state will evolve into a very complicated object, that we will be unable to describe, but that does not interest us directly. What we need are the excitations that correspond to the addition or removal (creation of a hole) of a fermion in the ground state. In the absence of interactions one just adds an electron in an empty $\vec{k}$ state and such an excitation does not care about the presence of all the other electrons in

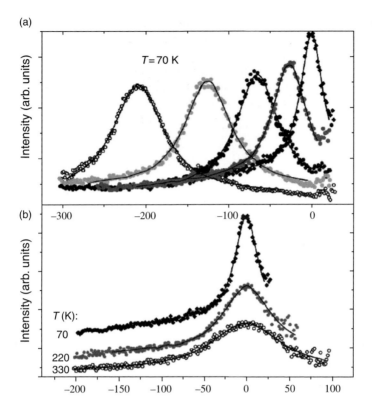

Fig. 8.2 Photoemission data from a Mo (110) surface. (Top) The spectral function $A(\vec{k}, \omega)$ is plotted as a function of ω. The zero denotes the Fermi level. Different peaks corresponds to different values of $\vec{k}$. One sees that, contrarily to naive expectations, the peaks in the spectral function become narrower and narrower as one gets closer to the Fermi energy. (Bottom) Width of a peak close to the Fermi level as a function of the temperature T. One sees that the width of the peak is controlled to a large part by the temperature, which corresponds to energies several orders of magnitude smaller than the typical energy of the interactions. (After Valla, Fedorov, Johnson and Hulbert 1999).

the ground state (otherwise than via the Pauli principle that prevents creating it in an already occupied state). In the presence of interactions this will not be the case and the added particle interacts with the existing particles in the ground state. For example, for repulsive interactions one can expect that this excitation repels other electrons in its vicinity. This is schematically represented in Figure 8.3. On the other hand, if one is at low temperature (compared to the Fermi energy) there are very few such excitations and one can thus neglect the interactions between them. This picture strongly suggests that the main interaction is between the excitation and the ground state. This defines a new composite object (fermion or hole surrounded by its own polarization cloud). This complex object essentially behaves as a particle, with the same quantum numbers (charge, spin) as the original fermion, albeit with

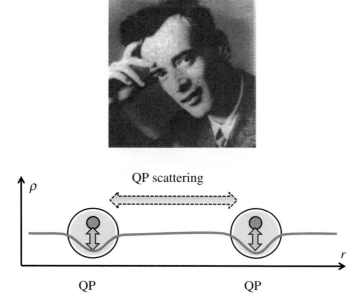

Fig. 8.3 (Top) Lev Landau, the man behind the Fermi-liquid theory (and many other things). (Bottom) In a Fermi liquid, the few excitations above the ground state can interact strongly with all the other electrons present in the ground state. The effect of such interactions is strong and leads to a strong change of the parameters compared to free electrons. The combined object behaves as a single particle, named a quasi-particle. The quasi-particles thus have characteristics depending strongly on the interactions. However, the scattering of the quasi-particles is blocked by the Pauli principle leaving a very small phase space of scattering. The lifetime of the quasi-particles is thus extremely large. This is the essence of the Fermi-liquid theory.

renormalized parameters, for example its mass. This image thus strongly suggests that even in the presence of interactions good excitations looking like free particles, still exist. These particles resemble free fermions but with a renormalized energy $E(\vec{k})$ and thus a renormalized mass. Since the interaction has been incorporated in the definition of such objects, it will not act as a source of broadening for their momentum distribution and the momentum distribution for the quasi-particles will remain very sharp, with only a small temperature broadening.

Of course the above is just a qualitative idea. Let me now give a more formal treatment. For that we can consider the retarded single correlation function (Mahan, 1981; Abrikosov, *et al.* 1963)

$$G(\vec{k}, t_2 - t_1) = -i\,\theta(t_2 - t_1)\langle[c_{\vec{k},t_2}, c^\dagger_{\vec{k},t_1}]_+\rangle. \tag{8.15}$$

This correlation represents the creation of a particle in a well-defined momentum state $\vec{k}$ at time t_1, let it propagate and then tries to destroy it in a well-defined momentum

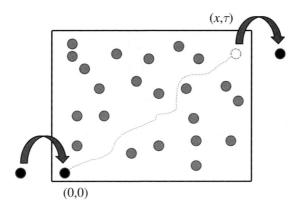

(x,τ)

$(0,0)$

Fig. 8.4 One way to understand the nature of the excitations in an interacting system is to inject a particle at point r_1 and time $t = 0$. The resulting system evolves, and then one destroys a particle at time t and point r_2. The amplitude of such a process indicates how much the propagation of the excitation between the two points and times is similar to the case of non-interacting electrons or not.

state $\vec{k}$ at time t_2. It thus measures how well in the interacting system the single-particle excitations still resemble plane or Bloch waves, i.e. independent particles. This corresponds to the gedanken experiment described in Figure 8.4. The imaginary part of the Fourier transform of this correlation function is just the spectral function

$$A(\vec{k}, \omega) = \frac{-1}{\pi} \operatorname{Im} G(\vec{k}, \omega).\tag{8.16}$$

The spectral function measures the probability to find a single-particle excitation with an energy ω and a momentum $\vec{k}$. It does obey the sum rule of probabilities $\int d\omega A(\vec{k}, \omega) = 1$. The general form of the retarded correlation is

$$G(\vec{k}, \omega) = \frac{1}{\omega - \xi(\vec{k}) - \Sigma(\vec{k}, \omega) + i\delta},\tag{8.17}$$

where $\Sigma(\vec{k}, \omega)$, called the self-energy, is a certain function of momenta and frequency. The relation (8.17) defines the function Σ. For non-interacting systems $\Sigma = 0$, and perturbative methods (Feynman diagrams) exist to compute Σ in powers of the interaction (Mahan, 1981; Abrikosov, *et al.* 1963). However, we will not attempt here to compute the self-energy Σ but simply to examine how it controls the spectral function. I incorporate the small imaginary part $i\delta$ in Σ for simplicity, since one can expect in general Σ to have a finite imaginary part as well. The spectral function is

$$A(\vec{k}, \omega) = -\frac{1}{\pi} \frac{\operatorname{Im} \Sigma(\vec{k}, \omega)}{(\omega - \xi(\vec{k}) - \operatorname{Re} \Sigma(\vec{k}, \omega))^2 + (\operatorname{Im} \Sigma(\vec{k}, \omega))^2}.\tag{8.18}$$

We see that $\operatorname{Im}\Sigma$ and $\operatorname{Re}\Sigma$ play very different roles in the spectral function. Note that quite generally Eq. (8.18) imposes that $\operatorname{Im}\Sigma(\vec{k},\omega) < 0$ to get a positive spectral function. In the absence of interactions $\Sigma = 0$ and one recovers

$$A(\vec{k},\omega) = \delta(\omega - \xi(\vec{k})). \tag{8.19}$$

The imaginary part of the self-energy gives the broadening of the peaks, which now acquire a lorentzian-like form, with a width of order $\operatorname{Im}\Sigma$ and a height of order $1/\operatorname{Im}\Sigma$. This can be readily seen from Eq. (8.18), for example by setting the real part of Σ to zero and taking a constant $\operatorname{Im}\Sigma$. Note that now the peaks are *not* centered around $\xi(\vec{k})$ any longer but have a maximum at a an energy $E(\vec{k})$, which is a solution $\omega = E(\vec{k})$ of

$$\omega - \xi(\vec{k}) - \operatorname{Re}\Sigma(\vec{k},\omega) = 0. \tag{8.20}$$

The imaginary part thus defines the broadening of the peaks, i.e. how sharply the excitations are defined, while the real part gives the *new energy* of the excitations. Let us look at these two elements in slightly more detail.

Lifetime:.   The imaginary part of the self-energy controls the spread in energy of the particles. This leads to the physical image of a particle with an average energy $\omega = E(\vec{k})$, related to its momentum, but with a certain spread $\operatorname{Im}\Sigma$ in energy. To understand the physics of this spread let us consider the Green function of a particle in real time:

$$G(\vec{k},t) = -i\,\theta(t)e^{-iE(\vec{k})t}e^{-t/\tau}. \tag{8.21}$$

The oscillatory part is the normal time evolution of a wave function of a particle with an energy $E(\vec{k})$. We have added the exponential decay that would be produced by a finite lifetime τ for the particle. The Fourier transform becomes

$$G(\vec{k},\omega) = \frac{1}{\omega - E(\vec{k}) + i/\tau}, \tag{8.22}$$

and the spectral function is

$$A(\vec{k},\omega) = \frac{1}{\pi}\frac{1/\tau}{(\omega - E(\vec{k}))^2 + (1/\tau)^2}, \tag{8.23}$$

which is essentially the one we are considering with the identification

$$\frac{1}{\tau} = \operatorname{Im}\Sigma. \tag{8.24}$$

We thus see from Eq. (8.21) that a Lorentzian-like spectral function corresponds to a particle with a well defined energy $E(\vec{k})$ that defines the center of the peak, but also with a finite *lifetime* τ. Of course the existence of such lifetime does not mean that the particle physically disappears, but simply that it does not exist as an excitation with the given quantum number $\vec{k}$. This is indeed an expected effect of the interaction since the particle exchanges momentum with the other particles and thus is able to change its quantum state.

With the more general form of the self-energy, which depends on $\vec{k}$ and ω this interpretation in terms of a lifetime, still holds if the peak is narrow enough. Indeed in that case the self-energy at the position of the peak $\operatorname{Im}\Sigma(\vec{k},\omega = E(\vec{k}))$ matters if one assumes that the self-energy varies slowly enough with ω compared to $\omega - E(\vec{k})$.

Effective mass and quasi-particle weight:.   Let us now turn to the real part. For simplicity, let me set the imaginary part to zero in Eq. (8.18) since in this section we will be mostly interested in the position and weight of the peak. This simplification replaces the Lorentzian peaks by sharp δ functions, but keeps the other characteristics unchanged. With this simplification the spectral function becomes

$$A(\vec{k},\omega) = \delta(\omega - \xi(\vec{k}) - \operatorname{Re}\Sigma(\vec{k},\omega)). \tag{8.25}$$

As we already pointed out, the role of the real part of the self-energy is to modify the position of the peak. One has now a new dispersion relation $E(\vec{k})$, which is defined by

$$E(\vec{k}) - \xi(\vec{k}) - \operatorname{Re}\Sigma(\vec{k},\omega = E(\vec{k})) = 0. \tag{8.26}$$

The interactions, via the real part of the self-energy are thus leading to a modification of the energy of single-particle excitations. Although we can in principle compute the whole dispersion relation $E(\vec{k})$, in practice we do not need it since the low-energy excitations close to the Fermi level control all the physical properties. At the Fermi level the energy, with a suitable subtraction of the chemical potential, is zero. One can thus expand it in powers of $\vec{k}$. For free electrons with $\xi(\vec{k}) = \frac{\vec{k}^2}{2m} - \frac{k_F^2}{2m}$ the corresponding expansion would give

$$\xi(\vec{k}) = \frac{k_F}{m}(\vec{k} - k_F). \tag{8.27}$$

A similar expansion for the new dispersion $E(\vec{k})$ gives

$$E(\vec{k}) = 0 + \frac{k_F}{m^*}(\vec{k} - k_F). \tag{8.28}$$

which defines the coefficient m^*. Comparing with Eq. (8.27) we see that m^* has the meaning of a mass. Close to the Fermi level we only need to compute the effective mass m^* to fully determine (at least for a spherical Fermi surface) the effects of the interactions on the energy of single-article excitations. To relate the effective mass to the self-energy one computes from Eq. (8.26)

$$\frac{dE(k)}{dk} = \frac{d\xi(k)}{dk} + \frac{\partial\operatorname{Re}\Sigma(k,\omega)}{\partial k}\bigg|_{\omega=E(k)} + \frac{\partial\operatorname{Re}\Sigma(k,\omega)}{\partial\omega}\bigg|_{\omega=E(k)}\frac{dE(k)}{dk}, \tag{8.29}$$

which can be solved to give

$$
\frac{k_F}{m^*} = \frac{\frac{k_F}{m} + \left. \frac{\partial \operatorname{Re} \Sigma(k,\omega)}{\partial k} \right|_{\omega=E(k)}}{1 - \left. \frac{\partial \operatorname{Re} \Sigma(k,\omega)}{\partial \omega} \right|_{\omega=E(k)}},
\tag{8.30}
$$

or in a more compact form

$$
\frac{m}{m^*} = \frac{1 + \frac{m}{k_F} \left. \frac{\partial \operatorname{Re} \Sigma(k,\omega)}{\partial k} \right|_{\omega=E(k)}}{1 - \left. \frac{\partial \operatorname{Re} \Sigma(k,\omega)}{\partial \omega} \right|_{\omega=E(k)}}.
\tag{8.31}
$$

To determine the effective mass these relations should be computed on the Fermi surface $E(k_F) = 0$. Equation (8.31) indicates how the self-energy changes the effective mass of the particles. This renormalization of the mass by interaction is well consistent with the experimental findings showing that in the specific heat one had something that was resembling the behavior of free electrons but with a different mass m^*.

However, he interactions have another more subtle effect. Indeed if we try to write the relation (8.25) in the canonical form $\delta(\omega - E(\vec{k}))$ that we would naively expect for a free particle with the dispersion $E(\vec{k})$ one obtains from Eq. (8.25)

$$
A(k, \omega) = Z_k \delta(\omega - E(k)),
\tag{8.32}
$$

with

$$
Z_k = \left[\left. \frac{\partial}{\partial \omega}(\omega - \xi(k) - \operatorname{Re} \Sigma(k, \omega)) \right|_{\omega=E(k)} \right]^{-1}
$$
$$
= \frac{1}{1 - \left. \frac{\partial \operatorname{Re} \Sigma(k,\omega)}{\partial \omega} \right|_{\omega=E(k)}}.
\tag{8.33}
$$

Because of the *frequency* dependence of the real part of the self-energy, the total spectral weight in the peak is no longer one, but the total weight is now Z_k, which is in general a number smaller than one. It is as if not the whole electron (or rather the total spectral weight of an electron) is converted into something that looks like a free particle with a new dispersion relation, but only a faction Z_k of it. With our crude approximation the rest of the spectral function has totally vanished. The fact that this violates the conservation of the probability to find an excitation is clearly an artefact of setting only the imaginary part to zero, while keeping the real part, since the real and imaginary parts of the self-energy are related by a Kramers–Kronig relation. However, the reduction of the quasi-particle weight that we found is quite real. What becomes of the remaining spectral weight will be described in the next section.

To conclude, we see that the real part of the self-energy controls the dispersion relation and the total weight of excitations that in the spectral function produce peaks exactly like free particles. The frequency and momentum dependence of the real part of the self-energy lead to the two independent quantities m^* the effective mass of the excitations and Z_k the weight. In the particular case when the momentum dependence of the self energy is small on can see from Eq. (8.33) and Eq. (8.31)

$$\frac{m}{m^*} = Z_{k_\mathrm{F}}.\tag{8.34}$$

Landau quasi-particles:. From the previous analysis of the spectral function and its connection with the self-energy we have a schematic idea of the excitations as summarized in Figure 8.5. Quite generally we can thus distinguish two parts in the spectral function. There is a continuous background, without any specific feature for which the probability to find a particle with an energy ω is practically independent of its momentum k. This part of the spectrum cannot be easily identified with excitations resembling free or quasi-free particles. On the other hand, in addition to this part, which carries a total spectral weight $1 - Z_k$, another part of the excitations gives a spectral weight with a lorentzian peak, well centered around a certain energy $E(k)$. This part of the spectrum can thus be identified with a "particle", called a Landau quasi-particle, with a well-defined relation between its momentum k and energy $\omega = E(k)$. This quasi-particle has a only a finite lifetime, determined by the inverse width and height of the peak. The dispersion relation and the total weight of the quasi-particle peak are controlled by the real part of the self-energy, while the lifetime

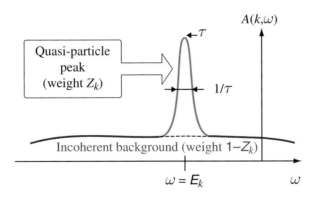

Fig. 8.5 A cartoon of the spectral function for interacting particle. One can recognize several features. There is a continuous background of excitations of total weight $1 - Z_k$. This part of the spectrum cannot be identified with excitations that resemble quasi-free particles. In addition to this continuous background there can be a quasi-particle peak. The total weight of the peak is Z_k determined by the real part of the self-energy. The center of the peak is at an energy $E(k)$, which is renormalized by the interactions compared to the independent electron dispersion $\xi(k)$. This change of dispersion defines an effective mass m^* determined also by the real part of the self-energy. The quasi-particle peak has a lorentzian lineshape that traduces the finite lifetime of the quasi-particles. The lifetime is inversely proportional to the imaginary part of the self-energy.

is inversely proportional to the imaginary part. Depending on the self-energy, and thus the interactions, we can still have objects that we could identify with "free" particles, solving our problem of why the free-electron picture works qualitatively so well with just a renormalization of the parameters such as the mass into an effective mass.

However, it is not clear that in the presence of interactions one can have sharp quasi-particles. In fact one would naively expect exactly the opposite. Indeed, we would like to identify the peak in the spectral function with the existence of a quasi-particle. The energy of this excitation is $E(k)$, which of course tends towards zero at the Fermi level, while the imaginary part of the self-energy is the inverse lifetime $1/\tau$. Since $E(k)$ gives the oscillations in time of the wave function of the particle $e^{-iE(k)t}$, in order to be able to identify properly a particle it is mandatory, as shown in Figure 8.6 that there are many oscillations by the time the lifetime has damped the wave function. This imposes

$$E(k)^{-1} \gg \tau. \tag{8.35}$$

Since $1/\tau$ is the imaginary part of the self-energy and controlled by energy scales of the order of the interactions, one would naively expect the life-time to be roughly constant close to the Fermi level. On the other hand, one has always $E(k) \to 0$ when $k \to k_F$, and thus the relation (8.35) is violated when one gets close to the Fermi level. This would mean that for weak interactions one has perhaps excitations that resemble particles far from the Fermi level, but that this becomes worse and worse as one looks at low-energy properties, with finally all the excitations close to the Fermi level being quite different from particles. Quite remarkably, as was first shown by Landau, this "intuitive" picture is totally incorrect and the lifetime has a quite different behavior when one approaches the Fermi level.

Quasi-particle scattering:.   In order to estimate the lifetime let us look at what excitations can lead to the scattering of a particle from a state k to another state.

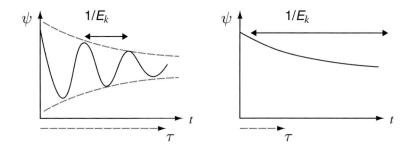

Fig. 8.6   For particles with an energy $E(k)$ and a finite lifetime τ, the energy controls the oscillations in time of the wave function. (left) In order to properly identify an excitation as a particle it is mandatory that the wave function can oscillate several times before being damped by the lifetime, otherwise it is impossible to precisely define the frequency of the oscillations. This is illustrated on the left part of the figure. (right) On the contrary, if the damping is too fast, one cannot define an average energy and thus identify the excitation with a particle.

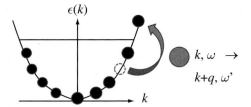

Fig. 8.7 Cartoon of the process giving the lifetime of a particle with energy w. It can interact with the ground state of the system, which has all single-particle states filled below the Fermi energy E_F. The excitations are thus particle–hole excitations where a particle is promoted from below the Fermi level to above the Fermi level. Due to the presence of the sharp Fermi level, the phase space available for making such particle–hole excitations is severely restricted.

Let us start from the non-interacting ground state in the spirit of a perturbative calculation in the interactions. As shown in Figure 8.7 a particle coming in the system with an energy w and a momentum k can excite a particle–hole excitation, taking a particle below the Fermi surface with an energy w_1 and putting it above the Fermi level with an energy w_2. The process is possible if the initial state is occupied and the final state is empty. One can estimate the probability of transition using the Fermi golden rule. The probability of the transition gives directly the inverse lifetime of the particle, and thus the imaginary part of the self-energy. We will not care here about the matrix elements of the transition, assuming that all possible transitions will effectively happen with some matrix element. The probability of transition is thus the sum over all possible initial states and final states that respect the constraints (energy conservation and initial state occupied, final state empty). Since the external particle has an energy w it can give at most w in the transition. Thus, $w_2 - w_1 \leq w$. This also implies directly that the initial state cannot go deeper below the Fermi level than w, otherwise the final state would also be below the Fermi level and the transition would be forbidden. The probability of transition is thus

$$P \propto \int_{-w}^{0} dw_1 \int_{0}^{w+w_1} dw_2 = \frac{1}{2}w^2. \tag{8.36}$$

One has thus the remarkable result that because of the *discontinuity* due to the Fermi surface and the Pauli principle that only allows the transitions from below to above the Fermi surface, the inverse lifetime behaves as w^2. This has drastic consequences since it means that contrarily to the naive expectations, when one considers a quasi-particle at the energy w, the lifetime grows much *faster* than the period $D \sim 1/w$ characterizing the oscillations of the wave function. In fact

$$\frac{\tau}{D} = \frac{1}{w} \to \infty \tag{8.37}$$

when one approaches the Fermi level. In other words, the Landau quasi-particles become *better and better defined* as one gets closer to the Fermi level. This is a remarkable result since it confirms that we can view the system as composed of single-particle excitations that resemble the original electrons, but with renormalized parameters (effective mass m^* and quasi-particle weight Z_k). Other quantum numbers are the same as those of an electron (charge, spin). Note that this does *not* mean that close to the Fermi level the interactions are disappearing from the system. They are present and can be extremely strong, and affect both the effective mass and quasi-particle weight very strongly. It is only the scattering of the quasi-particles that is going to zero when one is going close to the Fermi level. This is thus a very unusual situation, quite different from what would happen in a classical gas. In such a case diluting the gas would thus reduce both the interaction between the particles and also their scattering in essentially the same proportion. On the contrary, in a Fermi liquid there are many $\mathcal{N} \to \infty$ electrons in the ground state, which are in principle strongly affected by the interactions. Note again that computing the ground state would be a very complicated task. However, there are very few excitations above this ground state at low energy. These excitations can interact strongly with the other electrons in the soup of the ground state, leading to a very strong change of the characteristics compared to free electron excitations. This can lead to very large effective masses or small quasi-particle weight. On the other hand, the lifetime of the quasi-particles is controlled by a totally different mechanism since it is blocked by the Pauli principle, as shown in Figure 8.3. Thus, even if the interaction is strong the *phase space* available for such a scattering is going to zero close to the Fermi level, making the quasi-particle in practice infinitely long-lived particles, and allowing to use them to describe the system. The image of Figure 8.7 also gives us a description of what a quasi-particle is: this is an electron that is surrounded by a cloud of particle–hole excitations, or in other words density fluctuations since $c^\dagger_{k+q}c_k$ is typically the type of terms entering the density operator. Such density fluctuations are of course neutral and do not change the spin. This composite object electron+density fluctuation cloud, thus represents a tightly bound object (just like an electron dresses with a cloud of photons in quantum electrodynamics), that is the Landau quasi-particle. Since the electron when moving must carry with it its polarization cloud, one can guess that its effective mass will indeed be affected.

The Fermi-liquid theory is a direct explanation of the fact that "free" electrons theory works very well *qualitatively* (such as the specific heat linear in temperature) even when the change of parameters can be huge. We show in Figure 8.8 the case of systems where the renormalization of the mass is about $m^* \sim 10^3 m$ indicating very strong interactions effects. Nevertheless, we see that the specific heat varies linearly with temperature just like for free electrons. The prediction for the quasi-particle peaks fits very well with the photoemission data of Figure 8.2, in which one clearly sees the peaks becoming sharper as one approaches the Fermi level. There is another direct consequence of the prediction for the lifetime. At finite temperature one can expect the lifetime to vary as $\tau \sim 1/T^2$, since T is the relevant energy scale when $T \gg \omega$. If we put such a lifetime in the Drude formula for the conductivity we get

$$\sigma(T) = \frac{ne^2\tau}{m} \propto \frac{1}{T^2}. \tag{8.38}$$

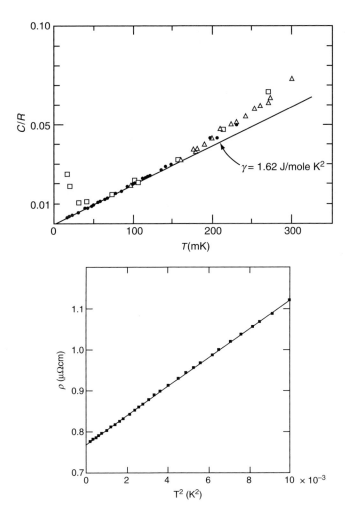

Fig. 8.8 Physical properties of the compound CeAl$_3$. (Top) The specific heat is linear in temperature T, but the slope gives an effective mass of about $10^3 m$ showing extremely strong interaction effects, showing that the Fermi liquid theory applies even when the interaction effects are strong. (Bottom) The resistivity varies as T^2 in very good agreement with the Fermi liquid theory. [After Andres, Graebner and Ott (1975).]

This result can be confirmed by a full calculation. This shows that the electron–electron interactions give an intrinsic contribution to the resistivity that varies as $\rho(T) \sim T^2$, and that also can be taken as one of the characteristics of Fermi-liquid behavior. This is, however, difficult to test since this temperature dependence can easily be masked by other scattering phenomena (impurities, scattering by the phonons, etc.) that must be added to the electron–electron scattering and that have a quite different temperature dependence. Nevertheless, there are some materials where the T^2 law can be well observed, as shown in Figure 8.8. Another interesting consequence can be deduced by looking at the occupation factor $n(k)$, which can be expressed from the spectral function by

$$n(k) = \int d\omega' A(k, \omega') f_{\mathrm{F}}(\omega').$$

(8.39)

For free electrons one recovers the step function at $k = k_{\mathrm{F}}$. For a Fermi liquid, if we represent the spectral function as

$$A(k, \omega) = Z_k \delta(\omega - E(k)) + A_{\mathrm{inc}}(k, \omega),$$

(8.40)

where the incoherent part is a smooth flattish function without any salient feature, then $n(k)$ becomes

$$n(k) = Z_k f_{\mathrm{F}}(E(k)) + Cste.$$

(8.41)

Thus, even in the presence of interaction there is still a *discontinuity* at the Fermi level, that is only rounded by the temperature. Contrarily to the case of free electrons the amplitude of the singularity at $T = 0$ is no longer one but is now $Z_{k_{\mathrm{F}}} < 1$. The existence of this discontinuity *if* a quasi-particle exists tells us directly that the Fermi-liquid theory is internally consistent since the very existence of the quasi-particles (namely the large lifetime) was heavily resting on the existence of such a discontinuity at the Fermi level. One can thus in a way consider that the existence of a sharp discontinuity at the Fermi level is a good order parameter to characterize the existence of a Fermi liquid.

One important question is when does the Fermi-liquid theory apply. This is of course a very delicate issue. One can see both from the arguments given above, and from direct perturbative calculations that when the interactions are weak the Fermi-liquid theory will in general be valid. There are some notable exceptions that we will examine in the next section, and for which the phase-space argument given above fails. However, the main interest of the Fermi-liquid theory is that it does not rest on the fact that the interactions are small and, as we have seen through examples, works also remarkably well for the case of strong interactions, even when all perturbation theory fails to be controlled. This is specially important for realistic systems since, as we showed, the interaction is routinely of the same order as the kinetic energy even in very good metals. The Fermi-liquid theory has thus been the cornerstone of our description of most condensed-matter systems in the last 50 years or so. Indeed it tells us that we can "forget" (or easily treat) the main perturbation, namely the interaction among fermions, by simply writing what is essentially a free-fermion Hamiltonian with some parameters changed. It is not even important to compute microscopically these parameters since one can simply extract them from one experiment and then use them consistently in the others. This allows as to go much further and treat effects caused by much smaller perturbations that one would otherwise have been totally unable to take into account. One of the most spectacular examples is the possibility to now look at the very tiny (compared to the electron–electron interactions) electron–phonon coupling, and to obtain from that the solution to the phenomenon of superconductivity, or other instabilities such as magnetic ordering.

8.3   Beyond Fermi liquid

Of course not all materials follow the Fermi-liquid theory. There are cases when this theory fails to apply. In that case the system is commonly referred to as strongly correlated or a "non-Fermi liquid" a term that hides our poor knowledge of their properties. For such systems, the question of the effects of interactions becomes again a formidable problem. As discussed in the introduction, most of the actual research in now devoted to such non-Fermi-liquid systems.

There are fortunately some situations where one can understand the physics and we will examine such cases in this section as well as define the main models that are at the heart of the study of these systems.

8.3.1   Instabilities of the FL

The Fermi liquid can become unstable for a variety of reasons. Some of them are well known.

The simplest instability consists, at low temperature, for the system to go into an ordered state. Many types of order are possible, the most common are spin order such as ferromagnetism, antiferromagnetism, charge order such as a charge density wave, or superconductivity. In general, analyzing such instabilities can be done by computing the corresponding susceptibility and looking for divergences as the temperature is lowered. When the normal system is well described by a Fermi liquid it is in general relatively easy to compute these susceptibilities by a mean-field decoupling of the interaction. This is of course much more complex when one starts from a normal phase that is a non-Fermi liquid.

Another important ingredient in the stability of the Fermi-liquid phase is the dimensionality of the system. Intuitively, we can expect that the lower the dimension the more important the effects of the interactions will be since the particles have a harder time to avoid each other. The ultimate case in that respect is the one-dimensional situation where one particle moving will push the particle in front and so on, as anybody queuing in a line has already had chance to notice. This effect of dimensionality is confirmed by a direct perturbative calculation of the self energy (Mahan, 1981; Abrikosov, *et al.* 1963)

$$\Sigma_{3D}(\omega) \propto U^2 \omega^2$$

$$\Sigma_{2D}(\omega) \propto U^2 \omega^2 \log(\omega) \tag{8.42}$$

$$\Sigma_{1D}(\omega) \propto U^2 \omega \log(\omega),$$

where U is the strength of the interaction. In particular, one immediately sees that for the one-dimensional case, the self-energy becomes dominant compared to the mean energy ω. Using Eq. (8.33) one sees that for one dimension the quasi-particle weight is zero at the Fermi level. This means the whole argument we used in the previous section to justify the existence of sharper and sharper peaks, and thus the existence of Landau quasi-particles fails *regardless* of the strength of the interaction. In one dimension a Fermi liquid always fails. This leads to a remarkable physics that we examine in more detail in Section 8.4.

With the exception of the one-dimensional case, one can thus expect the Fermi liquid to be perturbatively valid in two dimensions and above. Of course, if the interaction becomes large whether a non-Fermi-liquid state can appear is one of the major challenges of today's research.

8.3.2   Tight binding and Hubbard Hamiltonian

One ingredient of special importance is the presence of a lattice on which the fermions can move. This is the normal situation in solids, with the presence of the ionic lattice, and can be realized very well in cold atomic systems by imposing an optical lattice. Even for free electrons the lattice is an essential ingredient, since it leads to the existence of bands of energy.

One very simple description of the effects of a lattice is provided by the tight-binding model (Ziman, 1972). This model is of course an approximation of the real band structure in solids, but contains the important ingredients. Recently, cold atomic systems in optical lattices have provided excellent realization of such a model, and we recall here its main features.

Let us consider a system of cold atoms of fermions or bosons (Anglin and Ketterle, 2002). Such a system can be described by

$$H = \int dr \frac{\hbar^2 (\nabla \psi)^\dagger (\nabla \psi)}{2m} + \frac{1}{2} \int dr \, dr' \, V(r - r')\rho(r)\rho(r') - \int dr \, \mu(r)\rho(r). \quad (8.43)$$

The first term is the kinetic energy, the second term is the interaction V between the particles and the last term is the chemical potential. The three-dimensional interaction is characterized by a scattering length (Pitaevskii and Stringari, 2003) a_s

$$V(x, y, z) = V_0 \delta(x)\delta(y)\delta(z) = \frac{4\pi\hbar^2 a_s}{m}\delta(x)\delta(y)\delta(z). \quad (8.44)$$

The chemical potential takes into account the confining potential $V_c(r)$ keeping the particles in the trap

$$\mu(\vec{r}) = \mu_0 - V_c(r) = \mu_0 - \frac{1}{2}\omega_0 r^2, \quad (8.45)$$

and is thus in general position dependent. In the presence of interactions the effect of the confining potential can usually be taken by making a local density approximation. If one neglects the kinetic energy (so-called Thomas–Fermi approximation; for other situations see Pitaevskii and Stringari (2003)) the density profile is obtained by minimizing Eqs. (8.43) and (8.45), leading to

$$V_0 \rho(r) + [V_c(r) - \mu_0] = 0, \quad (8.46)$$

the density profile is thus an inverted parabola, reflecting the change of the chemical potential. In the following we will ignore for simplicity the confining potential and consider the system as homogeneous.

If one adds an optical lattice, produced by counter-propagating lasers, to the system it produces a periodic potential of the form $V_L(x)$ coupled to the density (Bloch, *et al.* 2008)

$$H_L = \int dx \, V_L(x)\rho(x). \tag{8.47}$$

This term, which favors certain points in space for the position of the bosons, mimics the presence of a lattice of period a, the periodicity of the potential $V_L(x)$. We take the potential as

$$V_L(x) = V_L \sum_{\alpha=1}^{d} \sin^2(k_\alpha r_\alpha) = \frac{V_L}{2} \sum_{\alpha=1}^{d} [1 - \cos(2k_\alpha r_\alpha)], \tag{8.48}$$

one has thus $a = \pi/k_\alpha$ as the lattice spacing.

If the lattice amplitude V_L is large then it is possible to considerably simplify the full Hamiltonian. In that case one can focus on the solution in one of the wells of the optical lattice, as shown in Figure 8.9.

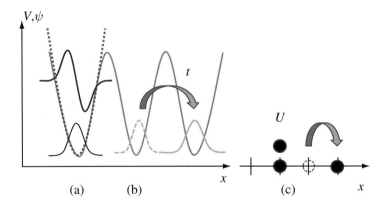

Fig. 8.9 Optical lattices and tight-binding model. (a) If the optical lattice is deep enough one can consider the wells as practically independent and obtain the states in each well. Since the well can be approximated by a parabolic potential one has harmonic-oscillator states. (b) Some level of overlap exists between the wave functions of different wells. This provides a hopping amplitude t for a particle to move from one well to the next. Similarly, the interaction between particles is only efficient if the two particles are in the same well. (c) The system can be mapped on a tight-binding model, describing particles leaving the sites of a lattice and hopping from one site to the next with an amplitude t. If interactions are included they can be modelled by local interactions of strength U. The resulting Hamiltonian is the famous Hubbard model.

The well can be approximated by a parabola $\frac{1}{2}(4V_L)k^2r^2$ and the solutions are those of an harmonic oscillator. The ground-state one is

$$\psi_0(x) = \left(\frac{m\omega_0}{\hbar\pi}\right)^{1/4} e^{-\frac{m\omega_0}{2\hbar}x^2}, \tag{8.49}$$

and higher excited states are schematically indicated on Figure 8.9. Typical values for the above parameters are $a_s \sim 5\,\text{nm}$, while, $a \sim 400\,\text{nm}$ (Stöferle, *et al.* 2004).

If V_L is large the energy levels in each well are well separated and one can retain only the ground-state wave-function in each well, higher excited states being well separated in energy. One can thus fully represent the state of the system by a creation (or destruction) of a particle in the ith well (i.e. with the wave function (8.49)). One thus has the energy

$$H = E_0 \sum_u c_i^\dagger c_i. \tag{8.50}$$

This is just a chemical-potential term and can be absorbed in the chemical potential. Of course if the lattice is not infinite, as indicated on Figure 8.9 the wave functions in different wells overlap, and there is a hybridation term t between the wave function in two adjacent wells. This term allows the particle to go from one well to the next. For an optical lattice one gets (Zwerger, 2003)

$$t/E_r = (4/\sqrt{\pi})(V_L/E_r)^{(3/4)} \exp\left(-2\sqrt{V_L/E_r}\right). \tag{8.51}$$

Here, $E_r = \hbar^2 k^2/(2m)$ is the so-called recoil energy, i.e. the kinetic energy for a momentum of order π/a. In the presence of such a term the energy becomes (we have written it for two spin species)

$$H_t = -t \sum_{\langle ij\rangle, \sigma} c_{i\sigma}^\dagger c_{j\sigma} - \sum_i \mu_{i\sigma} c_{i\sigma}^\dagger c_{i\sigma}, \tag{8.52}$$

where $\langle\rangle$ denotes nearest neighbors. In Fourier space one recovers Eq. (8.1) with

$$\xi(\vec{k}) = -2t \sum_{j=1}^d \cos(k_j a), \tag{8.53}$$

where d is the dimension of space, a the lattice spacing (I assumed here for simplicity a square lattice). The momentum $\vec{k}$ is restricted to the first Brillouin zone

$$k_j \in [-\pi/a, \pi/a]. \tag{8.54}$$

Thus, one particle per site corresponds to a half-filled zone (half of the available states are occupied). For example, in one dimension this gives $k_F = \pi/(2a)$. For independent fermions this is the best situation to have a metallic state since the Fermi level is far from a band edge. An empty band and a completely filled band (two particles per site) correspond to an insulating state. For a filled band all the excitations are blocked by the Pauli principle. The tight-binding description of the kinetic energy thus contains the essential ingredient of quantum periodic systems namely the existence of bands, and the existence of a Brillouin zone.

One can add to this description, the effects of interaction. In a solid, as we discussed, the interaction is screened. A good approximation is thus to take a local interaction. For optical lattice such an approximation is truly excellent. Indeed, since atoms are neutral the interaction (8.44) has a range much smaller than the size of the well. Thus, atoms only interact if they are in the same well of the optical lattice. One can thus represent the interaction by a purely local term

$$H_U = U \sum_i n_{i\uparrow} n_{i\downarrow}. \tag{8.55}$$

The effective potential U can be easily computed by using the shape of the on-site wave function (8.49) with $\omega_0^2 = 4V_L k^2$

$$U = \int dx dy dz |\psi(x, y, z)|^4, \tag{8.56}$$

where $\psi(x, y, z) = \psi_0(x)\psi_0(y)\psi_0(z)$.

The full Hamiltonian of the system is thus

$$H = H_t + H_U, \tag{8.57}$$

and describes particle hopping on a lattice with a purely local repulsion. This is the famous Hubbard model. This model, although extremely simple contains the essential ingredients of interacting fermions. Optical lattices are thus excellent realizations of this model. They offer in addition a powerful control on the ratio of U/t the interaction measured compared to the kinetic energy, since we see that by increasing the height of the optical lattice one essentially modifies the overlap integral t between two different sites (Jaksch, *et al.* 1998; Greiner, *et al.* 2002). Recently, cold atomic systems have also provided a direct control over the interaction U by using a Feshbach resonance (Bloch, *et al.* 2008).

8.3.3   Mott insulators

Let us now study the properties of interacting fermions on a lattice as described in the previous section. For non-interacting fermions, the system remains metallic unless the band is completely filled. The situation changes drastically in the presence of interactions. The combination of lattice and interactions leads to a striking effect of interactions known as the Mott transition, a phenomenon predicted by Sir Nevil Mott (Mott, 1949) (see Figure 8.10)

To understand the phenomenon, let us consider a simple variational calculation of the energy of the Hubbard model (8.57), assuming that the system is described by a FL-like ground state. We will be even more primitive and take free electrons. In that case the ground state is simply the Fermi sea (Eq. 8.2). The energy of such a state $E_0 = \langle \psi_0 | H | \psi_0 \rangle$ is simply given by

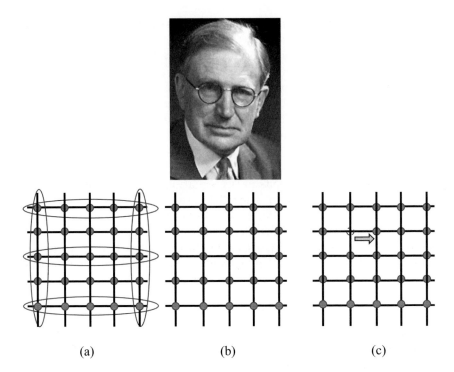

(a)   (b)   (c)

Fig. 8.10 (Top) Sir Nevil Mott. (a) If the repulsion is weak the particles prefer to gain kinetic energy and are essentially in a delocalized (plane-wave) state. This leads to a metallic state, and usually to a Fermi liquid. (b) For one particle per site $n = 1$ and strong interactions the system prefers to localize the particles, not to pay the large repulsion energy. This leads to a Mott insulating site. Residual kinetic energy tends to favor antiferromagnetic exchange between nearest neighbors, to avoid Pauli-principle blocking. Depending on the lattice this can lead to various types of magnetic order. (c) If there is less than one particle per site $n < 1$ but the interaction is large, a state with only one particle per site is still favored. The conduction is only provided from the holes present in the system that can hop at no interaction cost. Although this is still a metallic state this state has quite different properties from the weakly correlated one described in (a):

$$E_0 = 2 \sum_{|k| < k_F} \epsilon(k) + U \sum_i \langle \psi_0 | n_{i\uparrow} n_{i\downarrow} | \psi_0 \rangle, \tag{8.58}$$

where the dispersion relation is Eq. (8.53). The first term in Eq. (8.58) is simply the average of the kinetic energy

$$\int_{-W}^{\mu_0} d\epsilon \mathcal{N}(\epsilon)\epsilon = \Omega \int_{-W}^{\mu_0} d\epsilon n(\epsilon)\epsilon = \Omega \epsilon_0, \tag{8.59}$$

where Ω is the number of sites in the systems, $n(\epsilon)$ the density of states per site, W the minimum of energy in the band and μ_0 the chemical potential. For the dispersion relation (8.53), ϵ_0 is a filling-dependent negative number (for $n \leq 1$). The second term in (8.58) depends on the average number of doubly occupied sites. Because the plane waves of the ground state have a uniform amplitude on each site

$$\langle n_{i\uparrow} n_{i\downarrow} \rangle = \langle n_{i\uparrow} \rangle \langle n_{i\downarrow} \rangle = \frac{n^2}{4}, \tag{8.60}$$

and thus

$$E_0 = \Omega[\epsilon_0 + U\frac{n^2}{4}]. \tag{8.61}$$

One can notice two interesting points in this equation: i) in the absence of interaction ($U = 0$) the system gains energy by delocalizing the particles. The maximum gain in energy, so in a way the best metallic state, is obtained when the band is half-filled $n = 1$; ii) if the system has a FL (free-electron-like) ground state, the energy cost due to the interactions is growing. Thus, one could naively expect that for large U a better state could occur. Let us consider for example a variational function in which each particle is localized on a given site (for simplicity let us only consider $n \leq 1$)

$$|\psi_M\rangle = c^\dagger_{i_1,\uparrow} \cdots c^\dagger_{i_N,\downarrow} |\varnothing\rangle, \tag{8.62}$$

Note that we can put the spins in an arbitrary order for the moment. There are thus many such states $|\psi_M\rangle$ differing by their spin orientation. What is the best spin order will be discussed in the next section. One has obviously

$$\langle \psi_M | H | \psi_M \rangle = 0 \tag{8.63}$$

since there is no kinetic energy and no repulsion given the fact that at most one particle is on a given site. We can now take these two states as variational estimates of what could be the best representation of the ground state of our system, the idea being that the wave function with the lower energy is the one that has probably the largest overlap with the true ground state of the system. At a pure variational level, this calculation would strongly suggest that the wave function $|\psi_0\rangle$ is a good approximation of the ground state at small interaction, while $|\psi_M\rangle$ would be the best approximation of the ground state at large U. One could thus naively expect a crossover or even a quantum phase transition at a critical value of U. The nature of the two "phases" can be inferred from the two variational wavefunction. For small U one expects a good FL, while for large interactions one can expect a state where the particles are localized. For $n = 1$, as shown in Figure 8.10, such a state would be an insulator. One could thus expect a metal–insulator transition, driven by the interactions at a critical value of the interactions U of the order of the kinetic energy per site ϵ_0 of the non-interacting system. Of course our little variational argument is far too primitive to allow as to seriously conclude on the above points, and one must perform more sophisticated calculations (Imada, *et al.* 1998). This includes physical arguments, more refined variational calculations (Gutzwiller, 1965, Brinkman and Rice 1970; Yokoyama and Shiba 1987), slave boson techniques (Kotliar and Ruckenstein 1986) and refined mean-field theories (Kotliar and Vollhardt 2004).

As it turns out our little calculation already gives the right physics, originally understood by Mott. An important ingredient is the filling of the band. Indeed, the situation is very different depending on whether the filling is one particle per site or not. If the filling is one particle per site (half-filled band), then our little variational state, with one particle per site is indeed an insulator, as shown on Figure 8.10. Applying the kinetic-energy operator on this state would force it to go through a state with at least a doubly occupied site, with an energy cost of U, so such excitations could not propagate. Such an insulating state created by interactions has been nicknamed a Mott insulator. From the point of view of band theory and free electrons this is a remarkable state, since a half-filled system would normally give the best type of metallic state possible. We see that interactions are able to transform this state into an insulating state.

If the filling is not exactly one particle per site, something remarkable occurs again when $U > U_c$. In that case, as shown in Figure 8.10, the function whese one has only singly occupied sites is obviously still a very good starting point. However, if there are now empty sites ($n < 1$) these holes can now propagate freely at no cost in U. Such a state is thus not an insulator but still a metal. There are important differences with the metal one would get for small U. Indeed in this strongly correlated metal, only the holes can propagate freely, and one can expect that the number of carriers will be proportional to the doping $n - 1$ and not to the total number of particles n in the system. Other properties are also affected. I will not address in detail these points but refer the reader to the literature for more details.

To finish, let me mention three important points: the first one is that some additional phenomena can shift the Mott transition to $U = 0$ for special lattices. Indeed if the lattice is bipartite, i.e. can be separated into two sublattices that are connected only by the hopping operator, special properties occur. This is particularly the case of the square or the hexagonal lattice. On such lattices, for one particle per site the dispersion relation has a special property known as the nesting property. Namely, there is a wavevector Q such that for all k

$$\xi(k + Q) = -\xi(k). \tag{8.64}$$

For the tight-binding relation on a square lattice, nesting occurs for $n = 1$ and $Q = (\pi, \pi, \cdots)$. When a Fermi surface is nested the charge and spin susceptibilities have a logarithmic divergence due to the nesting. Typically

$$\chi(Q, T) \simeq \mathcal{N}(E_F) \log[\beta W]. \tag{8.65}$$

This divergent susceptibility provokes an instability of the metallic state for arbitrary interactions. As a result, nested systems become Mott insulators as soon as the interactions become repulsive, and order antiferromagnetically (at least in $d = 3$ where quantum fluctuations cannot destroy the magnetic order).

The second remark is that Mott insulators are not limited to the case of one particle per site. Any commensurate filling can potentially lead to an insulating state, depending on the *range* of the interactions. For example, one can expect an insulating state at $1/4$ filling if both on-site and nearest-neighbor interactions are present and of sufficient strength.

The third important point is that nothing in such a mechanism limits Mott insulators to fermionic statistics. As we saw in the simple derivation, interactions are simply avoiding double occupancy, which turns the system into an insulator. So we can also expect Mott insulators to exist for bosons and even for Bose–Fermi mixtures. The Mott phase will be essentially the same in each case (only one particle per site), but of course the low-interaction phase will strongly depend on the precise properties of the system (e.g. for bosons one can expect to have a superfluid phase at small interactions).

Mott insulators are one of the most spectacular effects of strong correlations. They are ubiquitous in nature and occur in oxides, cuprates, organic conductors and of course are realized in cold atomic systems both for bosons and fermions.

8.3.4   Magnetic properties of Mott insulators

Let me concentrate now on the Mott phase ($n = 1$) at large interactions $U > U_c$. In that case, at low temperature all charge excitations are gapped, with a gap $\Delta \sim U$ and the system is an insulator. One could thus naively think that one has properties very similar to these of a band insulator. Although this naive expectation is, to some extent, correct for the charge properties, it is completely incorrect for the spin ones, and very interesting spin physics occurs. Indeed, a band insulators corresponds to either zero or two particles per site. This means that in both cases the on-site spin is zero. The Pauli principle forces two particles on the same site to be in a singlet spin state to get a fully antisymmetric wave function. Thus, a band, insulator has no interesting charge nor spin properties. This is not the case for a Mott insulator since each site is singly occupied. There is thus a spin $1/2$ degree of freedom on each site, and it is thus important to understand the corresponding magnetic properties. This was worked out by Anderson (Anderson, 1959) (see Figure 8.11).

To do so let us examine the case of two sites. The total Hilbert space is

$$|\uparrow, \downarrow\rangle, |\downarrow, \uparrow\rangle, |\uparrow\downarrow, 0\rangle, |0, \uparrow\downarrow\rangle, |\uparrow, \uparrow\rangle, |\downarrow, \downarrow\rangle. \tag{8.66}$$

Since the states are composed of fermions one should be careful with the order of operators to avoid minus signs. Let us take the convention that

$$|\uparrow, \downarrow\rangle = c_{1\uparrow}^\dagger c_{2\downarrow}^\dagger |\varnothing\rangle \quad , \quad |\downarrow\uparrow\rangle = c_{1\downarrow}^\dagger c_{2\uparrow}^\dagger |\varnothing\rangle$$
$$|\uparrow\downarrow, 0\rangle = c_{1\uparrow}^\dagger c_{1\downarrow}^\dagger |\varnothing\rangle \quad , \quad |0, \uparrow\downarrow\rangle = c_{2\uparrow}^\dagger c_{2\downarrow}^\dagger |\varnothing\rangle. \tag{8.67}$$

The states with two particles per site are states of energy $\sim U$ and therefore strongly suppressed. We thus need to find what is the form of the Hamiltonian when restricted to the states with only one particle per site. It is easy to check that the two states $|\uparrow\uparrow\rangle$ and $|\downarrow\downarrow\rangle$ are eigenstates of H

$$H |\uparrow\uparrow\rangle = 0, \tag{8.68}$$

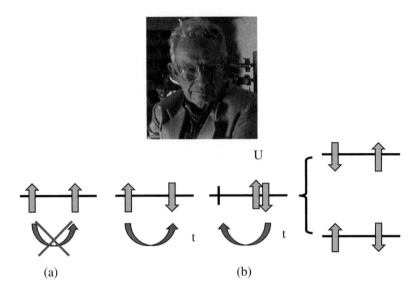

Fig. 8.11 (Top) P. W. Anderson. (a) The hopping between two neighboring identical spins is totally blocked by the Pauli principle. The system cannot gain some residual kinetic energy in this case. (b) If the spins are antiparallel then the kinetic energy can allow virtual hopping through an intermediate state of energy U. This state comes back either to the original state or to a configuration in which the spins have been exchanged. This is the mechanism of superexchange leading to dominant antiferromagnetic correlations for fermionic Mott insulators.

and a similar equation for $|\downarrow\downarrow\rangle$. The reason why the kinetic energy does not act on such a state is shown in Figure 8.11. The Pauli principle blocks the hopping if the two spins are equal. On the contrary, if the two spins are opposite the particles can make a virtual jump on the neighboring site. Since this state is of high energy U the particles must come back to the original position, or the two particles can exchange leading to a term similar to a spin exchange (see Figure 8.11). The *kinetic energy* thus leads to a *magnetic* exchange, named superexchange, that will clearly favor configurations with opposite neighboring spins, namely it will be antiferromagnetic.

Let us now quantify the mechanism. The Hamiltonian can be written in the basis (8.66) (only the action on the first four states is shown since $|\uparrow\uparrow\rangle$ and $|\downarrow\downarrow\rangle$ are eigenstates and thus are uncoupled to the other ones)

$$H = \begin{pmatrix} 0 & 0 & -t & -t \\ 0 & 0 & t & t \\ -t & t & U & 0 \\ -t & t & 0 & U \end{pmatrix}. \tag{8.69}$$

This Hamiltonian couples the low-energy states with one particle per site to high-energy states of energy U. In order to find the restriction of H to the low-energy sector, let us first make a canonical transformation of H

$$H' = e^{iS} H e^{-iS} \simeq H + i[S, H] + \frac{i^2}{2}[S, [S, H]] + \cdots , \tag{8.70}$$

where the matrix S is expected to be perturbative in t/U. In this transformation one wants to chose the matrix S such that

$$H_t + i[S, H_U] = 0. \tag{8.71}$$

This will ensure that H' has no elements connecting the low-energy sector with the sector of energy U. The restriction of H' to the low-energy sector will thus be the Hamiltonian we need to diagonalize to find the spin properties.

Using the condition (8.71) and keeping only terms up to order t^2/U it is easy to check that

$$H' = H_U + \frac{i}{2}[S, H_t]. \tag{8.72}$$

Since H is block diagonal one can easily determine S to be

$$S = \frac{i}{U} \begin{pmatrix} 0 & 0 & -t & -t \\ 0 & 0 & t & t \\ t & -t & 0 & 0 \\ t & -t & 0 & 0 \end{pmatrix}. \tag{8.73}$$

This leads, using Eq. (8.72) to

$$H' = \frac{4t^2}{U}\frac{1}{2}[|\uparrow\downarrow\rangle \langle\downarrow\uparrow| + |\downarrow\uparrow\rangle \langle\uparrow\downarrow|] - \frac{1}{2}[|\uparrow\downarrow\rangle \langle\uparrow\downarrow| + |\downarrow\uparrow\rangle \langle\downarrow\uparrow|]. \tag{8.74}$$

This Hamiltonian must be complemented by the fact that $H'|\uparrow\uparrow\rangle = 0$. Since there is one particle per site, it can now be represented by a spin $1/2$ operator $S^\alpha = \frac{1}{2}\sigma^\alpha$, where $\alpha = x, y, z$ and the σ are the Pauli matrices. Using $S^+ = S^x + iS^y$ for the raising spin operator one obtains

$$H' = \frac{4t^2}{U}\frac{1}{2}[S_1^+ S_2^- + S_1^- S_2^+] + S_1^z S_2^z - \frac{1}{4}]. \tag{8.75}$$

Up to a constant that is a simple shift of the origin of energies, the final Hamiltonian is thus the Heisenberg Hamiltonian

$$H = J \sum_{\langle ij \rangle} \vec{S}_i \cdot \vec{S}_j, \tag{8.76}$$

where the magnetic exchange is $J = 4t^2/U$.

As we saw for fermions, the exchange J is positive and thus the system has dominant antiferromagnetic correlations. This explains naturally the remarkable fact that antiferromagnetism is quite common in solid-state physics. Of course, the magnetic properties depend on the lattice, and the Hamiltonian (8.76) can potentially lead to very rich physics Auerbach (1998). It is important to note that we are dealing here with quantum spins for which $[S_x, S_y] = iS_z$ and thus the three component of the spins cannot be determined simultaneously. Quantum fluctuation will thus drastically affect the possible spin orders. Depending on the lattice, various ground states are possible ranging from spin liquids to ordered states.

8.3.5 Summary of the basic models

Let me summarize in this section the basic models that are commonly used to tackle the properties of strongly correlated systems. Of course these are just the simplest possible case, and many extensions and interesting generalizations have been proposed. There is currently a great interest in realizing these models with cold atomic gases.

Fermions. The simplest interacting model on a lattice is the Hubbard model

$$H = -t \sum_{\langle ij \rangle, \sigma} c_{i\sigma}^{\dagger} c_{i\sigma} + U \sum_{i} n_{i\uparrow} n_{i\downarrow}, \tag{8.77}$$

where $\langle \rangle$ denote the nearest neighbors, which of course depends on the topology of the lattice (square, triangular, etc.). Having only on-site interactions this model can only stabilize a Mott insulating phase at half-filling ($n = 1$). If the lattice is bipartite (e.g. square), the Mott phase is stable for any $U > 0$ and the ground state is an antiferromagnetic state. If the lattice is not bipartite, then a critical value of U is normally requested to reach the Mott insulating state. The spin order then depends on the lattice (see the spin section below).

For $U < 0$ this model has a BCS-type instability leading to a superconducting ground state. There are interesting symmetries for the Hubbard model. In particular, for bipartite lattices there is a particle-hole symmetry for one spin species that can map the repulsive Hubbard model onto the attractive one, and exchange the magnetic field and the chemical potential. This symmetry can be exploited in several contexts, but in particular can be very useful in the cold-atom context for probing the phases of the repulsive Hubbard model in a much more convenient way (Ho, *et al.* 2009).

The canonical Hubbard model can be easily extended in several ways. Longer-range hopping can be added, and of course longer-range interactions than on-site can also be included. In that case, insulating phases can in general be stabilized at other commensurate fillings. Finally, one can also increase the number of states per site (so called multi-orbital Hubbard model), a situation that has been useful in condensed matter and also is potentially relevant for cold atoms in optical lattices, for example if one needs to take into account the higher states in each well of Figure 8.9.

Bosons. Here also the canonical model is the Hubbard model (nicknamed the Bose–Hubbard model in that case). The main difference is that the simplest case does not need spins for the bosons. The simplest Bose–Hubbard model is thus

$$H_B = -t \sum_{\langle ij \rangle} [b_i^{\dagger} b_j + b_j^{\dagger} b_i] + U \sum_{i} n_i(n_i - 1). \tag{8.78}$$

This model has also in general a Mott transition for sufficiently large interaction when the filling is commensurate $n = 1$ (1 boson per site) (Haldane 1981*a*; Fisher, *et al.*

1989). In a similar way as for the fermions the model can be extended to the case of longer-range hopping or longer-range interactions, allowing for more insulating phases at other commensurabilities than for $n = 1$.

The case of the Hubbard model with a nearest-neighbor interaction

$$U \sum_i n_i(n_i - 1) + V \sum_{\langle ij \rangle} n_i n_j \tag{8.79}$$

has an interesting property that is worth noting. When U and V are of the same order of magnitude, the system can have fluctuations of charge on a site going between zero, one and two bosons per site, since putting two bosons per site is a way to escape paying the repulsion V. The system is thus very close to a system with three states per site, i.e. of a system that can be mapped onto a spin 1. In one dimension spin 1 are known to have very special properties, and thus similar properties are expected for such an extended Hubbard model (Berg, *et al.* 2009). In particular, they can have a topologically ordered phase (Haldane phase (Haldane, 1983) for spin one). In higher dimensions such models can have complex orders, and there is in particular a debate on whether such models can sustain simultaneously a crystalline order (Mott phase for the bosons) and superfluidity, the so-called supersolid phase (Niyaz, *et al.* 1994; Wessel and Troyer 2005). Various extensions of the canonical Bose–Hubbard are worth noting. The most natural one, in connection with cold atomic systems is to consider a model with more than one species, in other word re-introducing a kind of "spin for the bosons". This will correspond to bosonic mixtures. In that case one can generally expect interactions of the form

$$H = U_{\uparrow\uparrow} n_\uparrow (n_\uparrow - 1) + U_{\downarrow\downarrow} n_\downarrow (n_\downarrow - 1) + U_{\uparrow\downarrow} n_\uparrow n_\downarrow. \tag{8.80}$$

Note that the $U_{\sigma\sigma}$ interactions did not exist for the fermionic Hubbard model because of the Pauli principle. For bosons, their presence allows for a very rich physics. In particular, the nature of the superexchange interactions will depend on the difference (Duan, *et al.* 2003) between the inter, and the intra-species interactions. If $U_{\uparrow\uparrow}, U_{\downarrow\downarrow} \gg U_{\uparrow\downarrow}$ one is in a situation very similar to the case of fermions, with a dominant antiferromagnetic exchange. The ground state of such a model will be antiferromagnetic. On the contrary, if one is in the opposite situation then it is more favorable for the kinetic energy to have parallel spins nearby and the superexchange will be ferromagnetic. This leads to the possibility of new ground states, and even new phases in one dimension (Zvonarev, *et al.* 2007).

Spin systems.   If the charge degrees of freedom are localized, usually but not necessarily for one particle per site, the resulting Hamiltonian only involves the spin degrees of freedom. The simplest one is for fermions with spin 1/2, and is the Heisenberg Hamiltonian

$$H = J \sum_{\langle ij \rangle} \vec{S}_i \cdot \vec{S}_j. \tag{8.81}$$

Depending on the spin exchange a host of magnetic phases can exist. As discussed above fermions lead mostly to antiferromagnetic exchanges, while bosons with two degrees of freedom would mostly be ferromagnetic.

Let me mention a final mapping that is quite useful in connecting the spins and itinerant materials. Spin 1/2 can in fact be mapped back onto bosons. The general transformation has been worked out by Holstein and Primakov for a spin S, but let me specialize here to the case of spin 1/2, which has been worked out by Matsubara and Matsuda (Matsubara and Matsuda 1956) and is particularly transparent. Since the Hilbert space of spin 1/2 has only two states, they can be mapped onto the presence and absence of a boson by the mapping

$$S^+ = b^\dagger$$
$$S^z = n_i - 1/2 = b_i^\dagger b_i - 1/2,$$
(8.82)

because only two states are possible for the spins, it is important to put a hard-core constraint on the bosons, imposing that at most one boson can exist on a given site.

The Hamiltonian (8.81), where we have introduced two coupling constants J_{XY} and J_Z for the two corresponding terms, can be rewritten as

$$H = \frac{J_{XY}}{2} \sum_{\langle ij \rangle} [S_i^+ S_j^- + S_i^- S_j^+] + J_Z \sum_{\langle ij \rangle} S_i^z S_j^z,$$
(8.83)

which becomes using the mapping (8.82)

$$H = \frac{J_{XY}}{2} \sum_{\langle ij \rangle} [b_i^\dagger b_j + b_i b_j^\dagger] + J_Z \sum_{\langle ij \rangle} (n_i - 1/2)(n_j - 1/2).$$
(8.84)

The hard-core constraint can be imposed by adding an on-site interaction U and letting it go to infinity. It is thus possible to map a spin model onto

$$H = \frac{J_{XY}}{2} \sum_{\langle ij \rangle} [b_i^\dagger b_j + b_i b_j^\dagger] + U \sum_i n_i(n_i - 1) + J_Z \sum_{\langle ij \rangle} (n_i - 1/2)(n_j - 1/2)$$
(8.85)

in the limit $U \to \infty$. One can recognize the form of an extended Hubbard model.

This mapping allows not only to solve certain bosonic problems by borrowing the intuition on magnetic order or vice versa, but it also allows as to use spin systems to experimentally realize Bose–Einstein condensate and study their critical properties. This has been a line of investigation that has been very fruitfully pursued recently (Giamarchi, *et al.* 2008).

8.4   One-dimensional systems

Let us now turn to one-dimensional systems. As we discussed in the previous sections the effect of interactions is maximal there. For fermions this leads to the destruction of the Fermi liquid state. For bosons, it is easy to see that simple BEC states are likewise impossible. Due to quantum fluctuations it is impossible to break a continuous symmetry (here the phase symmetry of the wave function) in one dimension. One has thus to face a radically different physics than for their higher-dimensional counterparts. Fortunately, the one-dimensional character brings new physics but also new methods to tackle the problem. This allows for quite complete solutions to be obtained, revealing remarkable physics phenomena and challenges (Giamarchi, 2004).

I will not cover here all these developments and physical realizations. I have written a whole book on the subject (Giamarchi, 2004) where the interested reader can find this information in a much more detailed and pedagogical way than the size of this chapter allows me to present. I will, however, present the very basic ideas here.

Before we embark on the description of the physics, a short historical note. To solve one-dimensional systems, crucial theoretical progress was made, mostly in the 1970s allowing a detailed understanding of the properties of such systems. This culminated in the 1980s with a new concept of interacting one-dimensional particles, analogous to the Fermi liquid for interacting electrons in three dimensions: the Luttinger liquid (Haldane, 1981*b,c*). Since then many developments have enriched further our understanding of such systems (Giamarchi, 2004), ranging from conformal field theory to important progress in the exact solutions such as the Bethe ansatz. In addition to these important theoretical progress, experimental realizations have prodused comparably spectacular developments. One-dimensional systems were initially a theorist's toy. Experimental realizations started to appear in the 1970s with polymers and organic compounds. But in the last 20 years or so we have seen a real explosion of realization of one-dimensional systems. The progress in material research made it possible to realize bulk materials with one-dimensional structures inside. The most famous ones are the organic superconductors (Lebed, 2007) and the spin and ladder compounds (Dagotto and Rice, 1996). At the same time, the tremendous progress in nanotechnology allowed to obtain realizations of isolated one-dimensional systems such as quantum wires (Fisher and Glazman, 1997), Josephson junction arrays Fazio and van der Zant (2001), edge states in quantum all systems (Wen, 1995), and nanotubes (Dresselhaus, *et al.* 1995). Last but not least, the recent progress in Bose condensation in optical traps has allowed an unprecedented way to probe for strong interaction effects in such systems (Pitaevskii and Stringari, 2003; Greiner, *et al.* 2002; Bloch, *et al.* 2008).

8.4.1 Realization of one-dimensional systems

Let us first discuss how one can obtain "one-dimensional" objects in a real three-dimensional world. All the one-dimensional systems are characterized by a confining potential forcing the particles to be in a localized state. The wave function of the system is thus of the form

$$\psi(x, r_\perp) = e^{ikx}\phi(r_\perp), \tag{8.86}$$

where ϕ depends on the precise form of the confining potential For an infinite well, as show in Figure 8.12, ϕ is $\phi(y) = \sin((2n_y + 1)\pi y/l)$, whereas it would be a gaussian function (8.49) for an harmonic confinement. The energy is of the form

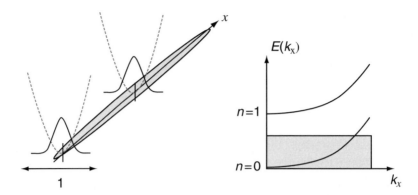

Fig. 8.12 (left) Confinement of the electron gas in a one-dimensional tube of transverse size l. x is the direction of the tube. Only one transverse direction of confinement has been shown for clarity. Due to the transverse confining potential the transverse degrees of freedom are strongly quantized. (right) Dispersion relation $E(k)$. Only half of the dispersion relation is shown for clarity. k is the momentum parallel to the tube direction. The degrees of freedom transverse to the tube direction lead to the formation of minibands, labelled by a quantum number n. If only one miniband is populated, as represented by the gray box, the system is equivalent to a one-dimensional system where only longitudinal degrees of freedom can vary.

$$E = \frac{k^2}{2m} + \frac{k_y^2}{2m},\tag{8.87}$$

where for simplicity I have taken hard-wall confinement. Due to the narrowness of the transverse channel l, the quantization of k_y is sizeable. Indeed, the change in energy by changing the transverse quantum number n_y is at least (e.g. $n_y = 0$ to $n_y = 1$)

$$\Delta E = \frac{3\pi^2}{2ml^2}.\tag{8.88}$$

This leads to minibands as shown in Figure 8.12. If the distance between the minibands is larger than the temperature or interactions energy one is in a situation where only one miniband can be excited. The transverse degrees of freedom are thus frozen and only k matters. The system is a one-dimensional quantum system.

8.4.2 Bosonization dictionary

Treating interacting particles in one dimension is quite a difficult task. One very interesting technique is provided by the so-called bosonization. It has the advantage of giving a very simple description of the low-energy properties of the system, and of being completely general and very useful for many one-dimensional systems. This section describe its vary basic features. For more details and physical insights on this technique both for fermions and bosons I refer the reader to (Giamarchi, 2004).

The idea behind the bosonization technique is to re-express the excitations of the system in a basis of collective excitations. Indeed, in one dimension it is easy to realize that single-particle excitations cannot really exit. One particle when moving will push its neighbors and so on, which means that any individual motion is converted into a collective one. Collective excitations should thus be a good basis to represent a one-dimensional system.

To exploit this idea, let us start with the density operator

$$\rho(x) = \sum_i \delta(x - x_i), \tag{8.89}$$

where x_i is the position operator of the ith particle. We label the position of the ith particle by an "equilibrium" position R_i^0 that the particle would occupy if the particles were forming a perfect crystalline lattice, and the displacement u_i relative to this equilibrium position. Thus,

$$x_i = R_i^0 + u_i. \tag{8.90}$$

If ρ_0 is the average density of particles, $d = \rho_0^{-1}$ is the distance between the particles. Then, the equilibrium position of the ith particle is

$$R_i^0 = di. \tag{8.91}$$

Note that at that stage it is not important whether we are dealing with fermions or bosons. The density operator written as Eq. (8.89) is not very convenient. To rewrite it in a more pleasant form we introduce a labelling field $\phi_l(x)$ (Haldane, 1981a). This field, which is a continuous function of the position, takes the value $\phi_l(x_i) = 2\pi i$ at the position of the ith particle. It can thus be viewed as a way to number the particles. Since in one dimension, contrary to higher dimensions, one can always number the particles in a unique way (e.g. starting at $x = -\infty$ and processing from left to right), this field is always well defined. Some examples are shown in Figure 8.13. Using this labelling field and the rules for transforming δ functions

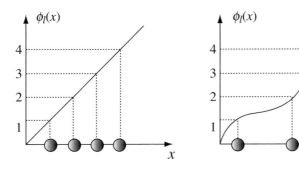

Fig. 8.13   Some examples of the labelling field $\phi_l(x)$. If the particles form a perfect lattice of lattice spacing d, then $\phi_l^0(x) = 2\pi x/d$, and is just a straight line. Different functions $\phi_l(x)$ allow to put the particles at any position in space. Note that $\phi(x)$ is always an increasing function regardless of the position of the particles. [From Giamarchi (2004).]

$$\delta(f(x)) = \sum_{\text{zeros of } f} \frac{1}{|f'(x_i)|} \delta(x - x_i) \tag{8.92}$$

one can rewrite the density as

$$\rho(x) = \sum_i \delta(x - x_i)$$

$$= \sum_n |\nabla \phi_l(x)| \delta(\phi_l(x) - 2\pi n). \tag{8.93}$$

It is easy to see from Figure 8.13 that $\phi_l(x)$ can always be taken as an increasing function of x, which allows as to drop the absolute value in Eq. (8.93). Using the Poisson summation formula this can be rewritten

$$\rho(x) = \frac{\nabla \phi_l(x)}{2\pi} \sum_p e^{ip\phi_l(x)}, \tag{8.94}$$

where p is an integer. It is convenient to define a field ϕ relative to the perfect crystalline solution and to introduce

$$\phi_l(x) = 2\pi \rho_0 x - 2\phi(x). \tag{8.95}$$

The density becomes

$$\rho(x) = \left[\rho_0 - \frac{1}{\pi} \nabla \phi(x)\right] \sum_p e^{i2p(\pi \rho_0 x - \phi(x))}. \tag{8.96}$$

Since the density operators at two different sites commute it is normal to expect that the field $\phi(x)$ commutes with itself. Note that if one averages the density over distances large compared to the interparticle distance d all oscillating terms in (8.96) vanish. Thus, only $p = 0$ remains and this smeared density is

$$\rho_{q \sim 0}(x) \simeq \rho_0 - \frac{1}{\pi} \nabla \phi(x). \tag{8.97}$$

We can now write the single-particle creation operator $\psi^\dagger(x)$. Such an operator can always be written as

$$\psi^\dagger(x) = [\rho(x)]^{1/2} e^{-i\theta(x)}, \tag{8.98}$$

where $\theta(x)$ is some operator. In the case where one would have Bose condensation, θ would just be the superfluid phase of the system. The commutation relations between the ψ impose some commutation relations between the density operators and the $\theta(x)$. For bosons, the condition is

$$[\psi_B(x), \psi_B^\dagger(x')] = \delta(x - x'). \tag{8.99}$$

Using Eq. (8.98) the commutator gives

$$e^{+i\theta(x)}[\rho(x)]^{1/2}[\rho(x')]^{1/2}e^{-i\theta(x')} - [\rho(x')]^{1/2}e^{-i\theta(x')}e^{+i\theta(x)}[\rho(x)]^{1/2}. \qquad (8.100)$$

If we assume quite reasonably that the field θ commutes with itself ($[\theta(x), \theta(x')] = 0$), the commutator (8.100) is obviously zero for $x \neq x'$ if (for $x \neq x'$)

$$[[\rho(x)]^{1/2}, e^{-i\theta(x')}] = 0. \qquad (8.101)$$

A sufficient condition to satisfy Eq. (8.99) would thus be

$$[\rho(x), e^{-i\theta(x')}] = \delta(x - x')e^{-i\theta(x')}. \qquad (8.102)$$

It is easy to check that if the density were only the smeared density (8.97) then Eq. (8.102) is obviously satisfied if

$$[\frac{1}{\pi}\nabla\phi(x), \theta(x')] = -i\delta(x - x'). \qquad (8.103)$$

One can show that this is indeed the correct condition to use (Giamarchi, 2004). Equation (8.103) proves that θ and $\frac{1}{\pi}\nabla\phi$ are canonically conjugate. Note that for the moment this results from totally general considerations and does not rest on a given microscopic model. Such commutation relations are also physically very reasonable since they encode the well-known duality relation between the superfluid phase and the total number of particles. Integrating by parts Eq. (8.103) shows that

$$\pi\Pi(x) = \hbar\nabla\theta(x), \qquad (8.104)$$

where $\Pi(x)$ is the canonically conjugate momentum to $\phi(x)$.

To obtain the single-particle operator one can substitute Eq. (8.96) into Eq. (8.98). Since the square root of a delta function is also a delta function, up to a normalization factor, the square root of ρ is identical to ρ up to a normalization factor that depends on the ultraviolet structure of the theory. Thus,

$$\psi_B^\dagger(x) = [\rho_0 - \frac{1}{\pi}\nabla\phi(x)]^{1/2}\sum_p e^{i2p(\pi\rho_0 x - \phi(x))}e^{-i\theta(x)}, \qquad (8.105)$$

where the index B emphasizes that this is the representation of a *bosonic* creation operator.

How to modify the above formulas if we have fermions instead of bosons? The density can obviously be expressed in the same way in terms of the field ϕ. For the single-particle operator one has to satisfy an anticommutation relation instead of Eq. (8.99). We thus have to introduce in representation (8.98) something that introduces the proper minus sign when the two fermions operators are commuted. This is known as a Jordan–Wigner transformation. Here, the operator to add is easy to guess. Since the field ϕ_l has been constructed to be a multiple of 2π at each particle, $e^{i\frac{1}{2}\phi_l(x)}$ oscillates between ± 1 at the location of consecutive particles. The Fermi field can thus be easily constructed from the boson field (8.98) by

$$\psi_F^\dagger(x) = \psi_B^\dagger(x)e^{i\frac{1}{2}\phi_l(x)}. \tag{8.106}$$

This can be rewritten in a form similar to Eq. (8.98) as

$$\psi_F^\dagger(x) = [\rho_0 - \frac{1}{\pi}\nabla\phi(x)]^{1/2} \sum_p e^{i(2p+1)(\pi\rho_0 x - \phi(x))}e^{-i\theta(x)}. \tag{8.107}$$

The above formulas are a way to represent the excitations of the system directly in terms of variables defined in the continuum limit.

The fact that all operators are now expressed in terms of variables describing *collective* excitations is at the heart of the use of such representation, since as already pointed out, in one dimension excitations are necessarily collective as soon as interactions are present. In addition, the fields ϕ and θ have a very simple physical interpretation. If one forgets their canonical commutation relations, order in θ indicates that the system has a coherent phase as indicated by Eq. (8.105), which is the signature of superfluidity. On the other hand, order in ϕ means that the density is a perfectly periodic pattern as can be seen from Eq. (8.96). This means that the system has "crystallized". For fermions note that the least oscillating term in Eq. (8.107) corresponds to $p = \pm 1$. This leads to two terms oscillating with a period $\pm\pi\rho_0$, which is simply $\pm k_F$. These two terms thus represent the Fermions leaving around their respective Fermi points $\pm k_F$, also known as right movers and left movers.

8.4.3   Physical results and Luttinger liquid

To determine the Hamiltonian in the bosonization representation we use Eq. (8.105) in the kinetic energy of bosons. It becomes

$$H_K \simeq \int dx \frac{\hbar^2\rho_0}{2m}(\nabla e^{i\theta})(\nabla e^{-i\theta}) = \int dx \frac{\hbar^2\rho_0}{2m}(\nabla\theta)^2, \tag{8.108}$$

which is the part coming from the single-particle operator containing fewer powers of $\nabla\phi$ and thus the most relevant. Using Eqs. (8.43) and (8.96), the interaction term becomes

$$H_{int} = \int dx V_0 \frac{1}{2\pi^2}(\nabla\phi)^2 \tag{8.109}$$

plus higher-order operators. Keeping only the above lowest order shows that the Hamiltonian of the interacting bosonic system can be rewritten as

$$H = \frac{\hbar}{2\pi} \int dx[\frac{uK}{\hbar^2}(\pi\Pi(x))^2 + \frac{u}{K}(\nabla\phi(x))^2], \tag{8.110}$$

where I have put back the $\hbar$ for completeness. This leads to the action

$$S/\hbar = \frac{1}{2\pi K} \int dx\, d\tau[\frac{1}{u}(\partial_\tau\phi)^2 + u(\partial_x\phi(x))^2]. \tag{8.111}$$

This Hamiltonian is a standard sound-wave one. The fluctuation of the phase ϕ represent the "phonon" modes of the density wave as given by Eq. (8.96). One immediately sees that this action leads to a dispersion relation, $\omega^2 = u^2 k^2$, i.e. to a linear spectrum. u is the velocity of the excitations. K is a dimensionless parameter whose role will become apparent below. The parameters u and K are used to parameterize the two coefficients in front of the two operators. In the above expressions they are given by

$$uK = \frac{\pi \hbar \rho_0}{m}$$
$$\frac{u}{K} = \frac{V_0}{\hbar \pi}. \tag{8.112}$$

This shows that for weak interactions $u \propto (\rho_0 V_0)^{1/2}$, while $K \propto (\rho_0/V_0)^{1/2}$. In establishing the above expressions we have thrown away the higher-order operators, that are less relevant. The important point is that these higher-order terms will not change the form of the Hamiltonian (like making cross-terms between ϕ and θ appears, etc.) but *only* renormalize the coefficients u and K (for more details see Giamarchi, 2004). For fermions it is easy to check that one obtains a similar form. The important difference is that since the single-particle operator contains already ϕ and θ at the lowest order (see Eq. (8.107)) the kinetic energy alone leads to $K = 1$ and interactions perturb around this value, while for bosons non-interacting bosons correspond to $K = \infty$.

The low-energy properties of interacting quantum fluids are thus described by an Hamiltonian of the form (8.110) *provided* the proper u and K are used. These two coefficients *totally* characterize the low-energy properties of massless one-dimensional systems. The bosonic representation and Hamiltonian (8.110) play the same role for one-dimensional systems that the Fermi liquid theory plays for higher-dimensional systems. It is an effective low-energy theory that is the fixed point of any massless phase, regardless of the precise form of the microscopic Hamiltonian. This theory, which is known as the Luttinger liquid theory Haldane (1981b,a), depends only on the two parameters u and K. Provided that the correct value of these parameters are used, *all* asymptotic properties of the correlation functions of the system then can be obtained *exactly* using Eqs. (8.96) and (8.105) or Eq. (8.107).

Computing the Luttinger liquid coefficient can be done very efficiently. For small interaction, perturbation theory such as Eq. (8.112) can be used. More generally, one just needs two relations involving these coefficients to obtain them. These could be, for example, two thermodynamic quantities, which makes it easy to extract from either Bethe-ansatz solutions if the model is integrable or numerical solutions. The Luttinger-liquid theory thus provides, coupled with the numerics, an incredibly accurate way to compute correlations and physical properties of a system (see, e.g., (Klanjsek, *et al.* 2008) for a remarkable example). For more details on the various procedures and models see (Giamarchi, 2004). But, of course, the most important use of Luttinger liquid theory is to justify the use of the boson Hamiltonian and fermion–boson relations as starting points for any microscopic model. The Luttinger parameters then become effective parameters. They can be taken as input, based on general rules (e.g. for bosons $K = \infty$ for non-interacting bosons and K decreases as the repulsion increases, for other general rules see (Giamarchi, 2004)), without any reference to a particular microscopic model. This removes part of the caricatural aspects of any modellization of a true experimental system. This use of the Luttinger liquid is reminiscent of the one where perturbations (impurity, electron–phonon interactions, etc.) are added on the Fermi-liquid theory. The calculations in $d = 1$ proceed in the same spirit with the Luttinger liquid replacing the Fermi liquid. The Luttinger-liquid theory is thus an invaluable tool to tackle the effect of perturbations on an interacting one-dimensional electron gas (such as the effect of lattice, impurities, coupling between chains, etc.). I refer the reader to (Giamarchi, 2004) for more on those points.

8.4.4 Correlations

Let us now examine in detail the physical properties of such a Luttinger liquid. For this we need the correlation functions. I briefly show here how to compute them using the standard operator technique. More detailed calculations and functional integral methods are given in (Giamarchi, 2004).

To compute the correlations we absorb the factor K in the Hamiltonian by rescaling the fields (this preserves the commutation relation)

$$\phi = \sqrt{K} \quad , \quad \theta = \frac{1}{\sqrt{K}} \tilde{\theta}. \tag{8.113}$$

The fields $\tilde{\phi}$ and $\tilde{\theta}$ can be expressed in terms of bosons operator $[b_q, b_{q'}^\dagger] = \delta_{q,q'}$. This ensures that their canonical commutation relations are satisfied. One has

$$\phi(x) = -\frac{i\pi}{L} \sum_{p\neq 0} \left(\frac{L|p|}{2\pi} \right)^{1/2} \frac{1}{p} e^{-\alpha|p|/2 - ipx} (b_p^\dagger + b_{-p})$$

$$\theta(x) = \frac{i\pi}{L} \sum_{p\neq 0} \left(\frac{L|p|}{2\pi} \right)^{1/2} \frac{1}{|p|} e^{-\alpha|p|/2 - ipx} (b_p^\dagger - b_{-p}), \tag{8.114}$$

where L is the size of the system and α a short distance cutoff (of the order of the interparticle distance) needed to regularize the theory at short scales. The above expressions are in fact slightly simplified and zero modes should also be incorporated (Giamarchi, 2004). This will not affect the remaining of this section and the calculation of the correlation functions.

It is easy to check by a direct substitution of Eq. (8.114) in Eq. (8.110) that Hamiltonian (8.110) with $K = 1$ is simply

$$\tilde{H} = \sum_{p\neq 0} u|p| b_p^\dagger b_p. \tag{8.115}$$

The time (or imaginary time Mahan (1981)) dependence of the field can now be easily computed from Eqs. (8.115) and (8.114). This gives

$$\phi(x,\tau) = -\frac{i\pi}{L} \sum_{p\neq 0} \left(\frac{L|p|}{2\pi} \right)^{1/2} \frac{1}{p} e^{-\alpha|p|/2 - ipx} (b_p^\dagger e^{u|p|\tau} + b_{-p} e^{-u|p|\tau}), \tag{8.116}$$

and a similar expression for θ. In order to compute physical observable we need to get correlations of exponentials of the fields ϕ and θ. To do so one simply uses that for an operator A that is *linear* in terms of boson fields and a quadratic Hamiltonian one has

$$\langle T_\tau e^A \rangle = e^{\frac{1}{2}\langle T_\tau A^2 \rangle}, \tag{8.117}$$

where T_τ is the time ordering operator. Thus, for example

$$\langle T_\tau e^{i2\phi(x,\tau)} e^{-i2\phi(0,0)} \rangle = e^{-2\langle T_\tau [\phi(x,\tau)-\phi(0,0)]^2 \rangle}. \tag{8.118}$$

Using these rules it is easy to compute the correlations (Giamarchi, 2004). If we want to compute the fluctuations of the density

$$\langle T_\tau \rho(x,\tau)\rho(0)\rangle \tag{8.119}$$

we obtain, for bosons or fermions, using Eq. (8.96)

$$\langle T_\tau \rho(x,\tau)\rho(0)\rangle = \rho_0^2 + \frac{K}{2\pi^2}\frac{y_\alpha^2 - x^2}{(x^2+y_\alpha^2)^2} + \rho_0^2 A_2 \cos(2\pi\rho_0 x)\left(\frac{\alpha}{r}\right)^{2K}$$

$$+ \rho_0^2 A_4 \, \cos(4\pi\rho_0 x)\left(\frac{\alpha}{r}\right)^{8K} + \cdots, \tag{8.120}$$

where $r = \sqrt{x^2+y^2}$ and $y = u\tau$. Here, the lowest distance in the theory is $\alpha \sim \rho_0^{-1}$. The amplitudes A_i are non-universal objects. They depend on the precise microscopic model, and even on the parameters of the model. Contrary to the amplitudes A_n, which depend on the precise microscopic model, the power-law decay of the various terms are *universal.* They *all* depend on the unique Luttinger coefficient K. Physically, the interpretation of the above formula is that the density of particles has fluctuations that can be sorted compared to the average distance between particles $\alpha \sim d = \rho_0^{-1}$. This is shown in Figure 8.14. The fluctuations of long-wavelength decay with a universal power law. These fluctuations correspond to the hydrodynamic modes of the interacting quantum fluid. The fact that their fluctuations decay very slowly is the signature that there are massless modes present. This corresponds to the sound waves of density described by Eq. (8.110). However, the density of particles has also higher fourier harmonics. The corresponding fluctuations also decay very slowly but this time with a non-universal exponent that is controlled by the LL parameter K. This is also the signature of the presence of a continuum of gaples modes, that exists for Fourier components around $Q = 2n\pi\rho_0$ as shown in Figure 8.14. For bosons K goes to infinity when the interaction goes to zero, which means that the correlations in the density decays increasingly faster with smaller interactions. This is consistent with the idea that the system becoming more and more superfluid smears more and more its density fluctuations. For fermions, the non-interacting point corresponds to $K = 1$ and one recovers the universal $1/r^2$ decay of the Friedel oscillations in a free-electron gas. For repulsive interactions $K < 1$ and density correlations decay more

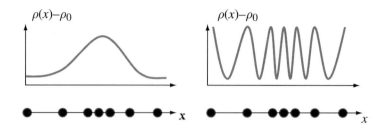

Fig. 8.14 The density $\rho(x)$ can be decomposed in components varying with different Fourier wavevectors. The characteristic scale to separate these modes is the interparticle distance. Only the two lowest harmonics are represented here. Although they have very different spatial variations both these modes depend on the *same* smooth field $\phi(x)$. (left) the smooth variations of the density at a lengthscale larger than the lattice spacing. These are simply $-\nabla\phi(x)/\pi$. (right) The density wave corresponding to oscillations of the density at a wavevector $Q = 2\pi\rho_0$. These modes correspond to the operator $e^{i\pm 2\phi(x)}$.

slowly, while for attractive interactions $K > 1$ they will decay faster, being smeared by the superconducting pairing.

Let us now turn to the single-particle correlation function

$$G(x,\tau) = -\langle T_\tau \psi(x,\tau)\psi^\dagger(0,0)\rangle. \tag{8.121}$$

For bosons, at equal time this correlation function is a direct measure on whether a true condensate exists in the system. Its Fourier transform is the occupation factor $n(k)$. In the presence of a true condensate, this correlation function tends to $G(x \to \infty, \tau = 0) \to |\psi_0|^2$ the square of the order parameter $\psi_0 = \langle \psi(x,\tau)\rangle$ when there is superfluidity. Its Fourier transform is a delta function at $q = 0$, as shown in Figure 8.15. In one dimension, no condensate can exist since it is impossible to break a continuous symmetry even at zero temperature, so this correlation must always go to zero for large space or time separation. Using Eq. (8.105) the correlation function can easily be computed. Keeping only the most relevant term ($p = 0$) leads to (I have also put back the density result for comparison)

$$\langle T_\tau \psi(r)\psi^\dagger(0)\rangle = A_1 \left(\frac{\alpha}{r}\right)^{\frac{1}{2K}} + \cdots$$

$$\langle T_\tau \rho(r)\rho(0)\rangle = \rho_0^2 + \frac{K}{2\pi^2}\frac{y_\alpha^2 - x^2}{(y_\alpha^2 + x^2)^2} + A_3 \cos(2\pi\rho_0 x)\left(\frac{1}{r}\right)^{2K} + \cdots, \tag{8.122}$$

where the A_i are the non-universal amplitudes. For the non-interacting system $K = \infty$ and we recover that the system possesses off-diagonal long-range order since the single-particle Green's function does not decay with distance. The system has condensed in the $q = 0$ state. As the repulsion increases (K decreases), the correlation function decays faster and the system has less and less tendency towards superconductivity. The occupation factor $n(k)$ has thus no delta function divergence but a power-law one, as shown in Figure 8.15. Note that the presence of the condensate or not is not directly

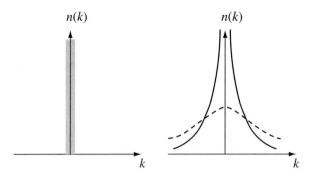

Fig. 8.15 Momentum distribution $n(k)$ for the bosons as a function of the momentum k. (left) For non-interacting bosons all bosons are in $k = 0$ state, thus $n(k) \propto \delta(k)$ (thick line). (right) As soon as interactions are introduced a true condensate cannot exist. The δ function is replaced by a power-law divergence with an exponent $\nu = 1 - 1/(2K)$ (solid line). At finite temperature, the superfluid correlation functions decay exponentially leading to a rounding of the divergence and a lorentzian-like shape for $n(k)$. This is indicated by the dashed line.

linked to the question of superfluidity. The fact that the system is a Luttinger liquid with a finite velocity u, implies that in one dimension an interacting boson system always has a linear spectrum $\omega = uk$, contrary to a free boson system where $\omega \propto k^2$. Such a system is thus a *true* superfluid at $T = 0$ since superfluidity is the consequence of the linear spectrum Mikeska and Schmidt (1970). Of course when the interaction tends to zero $u \to 0$, as it should to give back the quadratic dispersion of free bosons.

For fermions the single-particle correlation function contains the terms $p = \pm 1$, corresponding to fermions close to $\pm k_F$, respectively. If we compute the correlation for the right movers we get

$$G_R(x, \tau) = -e^{ik_F x} \langle T_\tau e^{i(\theta(x,\tau) - \phi(x,\tau))} e^{-i(\theta(0,0) - \phi(0,0))} \rangle$$
$$= e^{ik_F x} e^{-[\frac{K+K^{-1}}{2} \log(r/\alpha) - i\mathrm{Arg}(y+ix)]}.$$

(8.123)

The single-particle correlation thus decays as a non-universal power law whose exponent depends on the Luttinger-liquid parameter. For free particles $(K = 1)$ one recovers

$$G_R(r) = -e^{ik_F x} e^{-\log[(y_\alpha - ix)/\alpha]} = -ie^{ik_F x} \frac{1}{x + i(v_F \tau + \alpha \, \mathrm{Sign}(\tau))},$$

(8.124)

which is the normal function for ballistic particles with velocity u. For interacting systems $K \neq 1$ the decay of the correlation is always faster, which shows that single-particle excitations do not exist in the one-dimensional world. One important consequence is the occupation factor $n(k)$, which is given by the Fourier transform of the equal-time Green's function

$$n(k) = \int dx \, e^{-ikx} G_R(x, 0^-) = -\int dx \, e^{i(k_F - k)x} \left(\frac{\alpha}{\sqrt{x^2 + \alpha^2}} \right)^{\frac{K+K^{-1}}{2}} e^{i \, \text{Arg}(-\alpha + ix)}.$$

$$(8.125)$$

The integral can be easily determined by simple dimensional analysis. It is the Fourier transform of a power law and thus

$$n(k) \propto |k - k_F|^{\frac{K+K^{-1}}{2} - 1}.$$

$$(8.126)$$

The occupation factor is shown in Figure 8.16. Instead of the discontinuity at k_F that signals in a Fermi liquid that fermionic quasi-particles are sharp excitations, one thus finds in one dimension an essential power-law singularity. Formally, this corresponds to $Z = 0$, another signature that all excitations are converted to collective excitations and that new physics emerges compared to the Fermi-liquid case.

8.5   Conclusions

This concludes this brief tour of interacting quantum fluids. Of course we have just scratched the surface in this chapter. The problem is an active research field, with extensive efforts both from the theoretical and the experimental points of view.

The Fermi-liquid concept is still one of the cornerstones of our understanding of such systems, and is certainly the reference to which all novel properties must be compared. Models allowing to deal with interactions are still resisting our best attempts to fully solve them and in particular the Hubbard model, after about 50 years still remains a challenge, in particular in two dimensions. Cold atomic systems in optical lattices have provided a remarkable realization of such models and it is certain that the stimulation of those novel experimental realization should help drive the field forward.

Last but not least, the one-dimensional world, with its own properties and challenges is now at a stage where it has the corresponding concept of the Fermi liquid, the

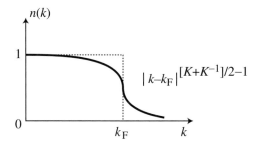

Fig. 8.16   The occupation factor $n(k)$. Instead of the usual discontinuity at k_F for a Fermi liquid, it has a power-law essential singularity. This is the signature that fermionic quasi-particles do not exist in one dimension. Note that the position of the singularity is still at k_F. This is a consequence of Luttinger's theorem stating that the volume of the Fermi surface cannot be changed by interactions.

so-called Luttinger liquid. This allows as to move forward and understand the effects of various perturbations, such as lattice, disorder, effect of coupling of one-dimensional chains, which are at the heart of the several realizations of the one-dimensional systems that we have today. We thus see that the one-dimensional world can be connected by that route to its higher-dimensional counterpart, and might also provide a way to tackle it. Considerable development can thus be expected in the coming years.

Acknowledgments

The research described in this work was supported in part by the Swiss NSF under MaNEP and Division II.

References

Abrikosov, A. A., Gorkov, L. P., and Dzyaloshinski, I. E. (1963). *Methods of quantum field theory in statistical physics*. Dover, New York.

Anderson, P. W. (1959). *Phys. Rev.*, **115**, 2.

Andres, K., Graebner, J. E., and Ott, H. R. (1975). *Phys. Rev. Lett.*, **35**, 1779.

Anglin, J. R. and Ketterle, W. (2002). *Nature (London)*, **416**, 211.

Ashcroft, N. W. and Mermin, N. D. (1976). *Solid state physics*. Saunders College, Philadelphia.

Auerbach, A. (1998). *Interacting electrons and quantum magnetism*. Springer, Berlin.

Berg, E., Dalla Torre, E. G., Giamarchi, T., and Altman, E. (2009). *Phys. Rev. B*, **77**, 245119.

Bloch, I., Dalibard, J., and Zwerger, W. (2008). *Rev. Mod. Phys.*, **80**, 885.

Brinkman, W. F. and Rice, T. M. (1970). *Phys. Rev. B*, **2**, 4302.

Dagotto, E. and Rice, T. M. (1996). *Science*, **271**, 5249.

Damascelli, A., Hussain, Z., and Shen, Z.-X. (2003). *Rev. Mod. Phys.*, **75**, 473.

Dresselhaus, M. S., Dresselhaus, G., and Eklund, P. C. (1995). *Science of fullerenes and carbon nanotubes*. Academic Press, San Diego, CA.

Duan, L.-M., Demler, E., and Lukin, M. D. (2003). *Phys. Rev. Lett.*, **91**, 090402.

Fazio, R. and van der Zant, H. (2001). *Phys. Rep.*, **355**, 235.

Fisher, M. P. A. and Glazman, L. I. (1997). In *Mesoscopic electron transport* (ed. L. Kowenhoven, *et al.*) Kluwer Academic Publishers Dordrecht cond-mat/9610037.

Fisher, M. P. A., Weichman, P. B., Grinstein, G., and Fisher, D. S. (1989). *Phys. Rev. B*, **40**, 546.

Giamarchi, Thierry (2004). *Quantum physics in one dimension*. Volume 121, International series of monographs on physics. Oxford University Press, Oxford, UK.

Giamarchi, T., Ruegg, C., and Tchernyshyov, O. (2008). *Nature Phys.*, **4**, 198.

Greiner, M., Mandel, O., Esslinger, T., Hänsch, T. W., and Bloch, I. (2002). *Nature*, **415**, 39.

Greywall, D. S. (1983). *Phys. Rev. B*, **27**, 2747.

Greywall, D. S. (1984). *Phys. Rev. B*, **29**, 4933.

Gutzwiller, M. C. (1965). *Phys. Rev.*, **137**, A1726.

Haldane, F. D. M. (1981*a*). *Phys. Rev. Lett.*, **47**, 1840.

Haldane, F. D. M. (1981*b*). *J. of Phys. C*, **14**, 2585.

Haldane, F. D. M. (1981*c*). *Phys. Rev. Lett.*, **47**, 1840.

Haldane, F. D. M. (1983). *Phys. Rev. Lett.*, **50**, 1153.

Ho, A. F., Cazalilla, M. A., and Giamarchi, T. (2009). *Phys. Rev. A*, **79**, 033620.

Imada, M., Fujimori, A., and Tokura, Y. (1998). *Rev. Mod. Phys.*, **70**, 1039.

Jaksch, D., Bruder, C., Cirac, J. I., Gardiner, C. W., and Zoller, P. (1998). *Phys. Rev. Lett.*, **81**, 3108.

Klanjsek, *et al.* M. (2008). *Phys. Rev. Lett.*, **101**, 137207.

Kotliar, Gabriel and Ruckenstein, Andrei E. (1986). *Phys. Rev. Lett.*, **57**(11), 1362.

Kotliar, G. and Vollhardt, D. (2004). *Phys. Today*, **57**, 53.

Landau, L. D. (1957*a*). *Sov. Phys. JETP*, **3**, 920.

Landau, L. D. (1957*b*). *Sov. Phys. JETP*, **5**, 101.

Lebed, A. (ed.) (2007). *The physics of organic superconductors*. Springer.

Mahan, G. D. (1981). *Many particle physics*. Plenum, New York.

Matsubara, T. and Matsuda, H. (1956). *Prog. of Theoret. Phys.*, **16**, 569.

Mikeska, H. J. and Schmidt, H. (1970). *J. Low Temp. Phys*, **2**, 371.

Mott, N. F. (1949). *Proc. Phys. Soc. Sect. A*, **62**, 416.

Niyaz, P., Scalettar, R. T., Fong, C. Y., and Batrouni, G. G. (1994). *Phys. Rev. B*.

Nozieres, P. (1961). *Theory of interacting Fermi systems*. W. A. Benjamin, New York.

Pitaevskii, L. and Stringari, S. (2003). *Bose-Einstein condensation*. Clarendon Press, Oxford.

Stöferle, T., Moritz, H., Schori, C., Köhl, M., and Esslinger, T. (2004). *Phys. Rev. Lett.*, **92**, 130403.

Valla, T., Fedorov, A. V., Johnson, P. D., and Hulbert, S. L. (1999). *Phys. Rev. Lett.*, **83**, 2085.

Wen, X. G. (1995). *Adv. In Phys.*, **44**, 405.

Wessel, S. and Troyer, M. (2005). *Phys. Rev. Lett.*, **95**, 127205.

Yokoyama, H. and Shiba, H. (1987). *J. Phys. Soc. Jpn.*, **56**, 3582.

Ziman, J. M. (1972). *Principles of the theory of solids*. Cambridge University Press, Cambridge.

Zvonarev, M., Cheianov, V. V., and Giamarchi, T. (2007). *Phys. Rev. Lett.*, **99**, 240404.

Zwerger, W. (2003). *J. Opt. B: Quantum Semiclass. Opt.*, **5**, 9.

9

Disorder and interference: localization phenomena

Cord A. MÜLLER* and Dominique DELANDE†

*Physikalisches Institut, Universität Bayreuth, 95440 Bayreuth, Germany
and Centre for Quantum Technologies, National University of Singapore,
3 Science Drive 2, Singapore 117543, Singapore
†Laboratoire Kastler-Brossel, Université Pierre et Marie Curie, Ecole Normale
Supérieure, CNRS; 4 Place Jussieu, F-75005 Paris, France

9.1   Introduction

Although complex systems are ubiquitous in nature, physicists tend to prefer "simple" systems. The reason is of course that simple systems obey simple laws, which can be represented by simple mathematical equations, as expressed by Goldenfeld and Kadanoff (Goldenfeld and Kadanoff, 1999):

One of the most striking aspects of physics is the simplicity of its laws. Maxwell's equations, Schrödinger's equation, and Hamiltonian mechanics can each be expressed in a few lines. The ideas that form the foundation of our worldview are also very simple indeed: The world is lawful, and the same basic laws hold everywhere. Everything is simple, neat, and expressible in terms of everyday mathematics, either partial differential or ordinary differential equations.

   Everything is simple and neat—except, of course, the world.

Even though the world obviously is not simple, many systems can be split, be it only in *Gedanken*, into simple components, each obeying simple laws. This is the viewpoint of standard reductionism, upon which modern science has been built (Anderson, 1972):

The reductionist hypothesis may still be a topic for controversy among philosophers, but among the great majority of active scientists I think it is accepted without question. The workings of our minds and bodies, and of all the animate or inanimate matter of which we have any detailed knowledge, are assumed to be controlled by the same set of fundamental laws, which except under certain extreme conditions we feel we know pretty well.

Reductionism was a key ingredient for the development of physics in the nineteenth and the first half of the twentieth century, when (classical and quantum) mechanics, electromagnetism, relativity, thermodynamics, etc. came to huge success. But is there anything fundamental beyond the simple laws of physics, or can one always reconstruct the properties of composed systems from the workings of their parts? This credo of "constructivism" has been challenged by P. W. Anderson, the author of the preceding quotation and one of the physicists who played a major role in the analysis of complex physical systems:

The ability to reduce everything to simple fundamental laws does not imply the ability to start from those laws and reconstruct the universe. In fact, the more the elementary particle physicists tell us about the nature of the fundamental laws, the less relevance they seem to have to the very real problems of the rest of science, much less to those of society.

   The constructionist hypothesis breaks down when confronted with the twin difficulties of scale and complexity. The behavior of large and complex aggregates of elementary particles, it turns out, is not to be understood in terms of a simple extrapolation of the properties of a few particles. Instead, at each level of complexity, entirely new properties appear, and the understanding of the new behaviors requires research which I think is as fundamental in its nature as any other.

which he summarized briefly with:

More Is Different.

It is one aim of this chapter to show—in a restricted context—how intricate the interplay between the small and the large can be in complex systems. We did not yet give a precise definition of the word "complex". It turns out that the very same word may be used in different contexts with different meanings. In this chapter, we will address a specific example of complexity, namely, *disorder*. In physics as in everyday life, disorder is associated with some lack of regularity. In a disordered material, atoms are not arranged in crystalline periodic patterns, but appear in more or less *random* positions. Randomness occurs because some agents, called "degrees of freedom" by physicists, are not under control, either because we cannot or chose not to control them. It means that we have to learn to deal not with a single specific complex system—that is called a single *realization* of the disorder—but with a whole family of systems whose properties are described in terms of distribution laws, correlation functions, etc. The goal of the game is not to describe as accurately as possible a single system, but rather to predict global properties shared by (almost) all systems, i.e. to acquire knowledge of universal features independent of the precise realization of the disorder. These are instances of the "new behaviors" mentioned by Anderson. And this is also the viewpoint taken long ago by classical thermodynamics where one forfeits the microscopic description of a gas in terms of all positions and momenta, concentrating instead on new concepts like entropy and temperature, which prove to be the relevant, and therefore fundamental, concepts at this level of complexity.

The specific problem we address in this chapter is the problem of transport and localization in disordered systems, when *interference* is present, as characteristic for waves. A wave propagates in some medium (be it vacuum), and interference occurs when different waves overlap, for example scattered from different positions with various wavevectors. The "simple laws" are the wave equation in the homogeneous medium together with a microscopic description of scattering by the impurities. The complex behavior we want to describe is, for example, the propagation of the wave over long distances and for long times. Physical situations of this type cover the propagation of sound in a concert hall with complicated shape, seismic waves multiply scattered inside the earth, electronic matter waves in dirty semiconductor crystals, atomic matter waves in the presence of a disordered potential, etc. In this context, it is good to keep in mind a warning issued by (Thirring, 1978):

It is notoriously difficult to obtain reliable results for quantum mechanical scattering problems. Since they involve complicated interference phenomena of waves, any simple uncontrolled approximation is not worth more than the weather forecast.

9.1.1 Anderson localization with atomic matter waves

To start with a specific, state-of-the-experimental-art example, imagine a one-dimensional non-relativistic particle evolving in a potential $V(z)$ as depicted in Figure. 9.1. The evolution of the wavefunction $\psi(z,t)$ is given by Schrödinger's equation:

$$i\hbar\partial_t\psi(z,t) = H\psi(z,t), \tag{9.1}$$

with the single-particle Hamiltonian

$$H = \frac{p^2}{2m} + V(z). \tag{9.2}$$

Let us assume that the particle is initially prepared in a Gaussian wavepacket. In the absence of any potential, the Gaussian wavepacket will show ballistic motion, where the center of mass moves at constant velocity while the width increases linearly with time at long times. In the presence of a certain realization $V(z)$ of the disorder, the wave function will take a certain form $\psi(z, t)$. For different realizations, different wave functions will be obtained. But we are not interested in the fine details of each wave function. Rather, we wish to understand the generic, if not universal, properties of the final stationary density distribution $|\psi(z)|^2$ obtained at long times. We will see that not only averages, but also their fluctuations contain important information.

Let us forget for a moment interference effects and try to guess what happens to a classical particle. If its kinetic energy is much larger than the typical strength of the disorder V_0, the particle will fly above the potential landscape, and the motion is likely to be ballistic on the average. If, on the other hand, V_0 is larger than the kinetic energy, the particle will be trapped inside a potential well and transport over long distance is suppressed, i.e. localization takes place.

Quantum mechanics modifies this simple picture fundamentally: waves can both tunnel through potential hills higher than the kinetic energy and be reflected even by small potential fluctuations. So the initial wave packet will split on each potential fluctuation into a transmitted part and a reflected part, no matter how large the kinetic energy with respect to the potential strength may be in detail. After many scattering instances, this looks like a random walk and one naively expects that, on average, the motion at long times will be diffusive, with a diffusion constant depending on some microscopic properties of particle and potential.

This simple model system has been recently realized experimentally (Billy *et al.*, 2008) using a quasi-one-dimensional atomic matter wave, interacting with an effective optical potential created by a speckle pattern, see Figure 9.1. The experimental result is the following: at short times, the wavepacket spreads as expected, but at long times, its average dynamics freeze, and the wave packet takes a characteristic exponential shape:

$$|\psi(z)|^2 \propto \exp\left(-\frac{|z|}{\xi_{\text{loc}}}\right), \tag{9.3}$$

where ξ_{loc} is called the localization length.[1] Moreover, if a different realization of the disorder is used (i.e. a microscopically different, but statistically equivalent speckle pattern), an almost identical shape is obtained, meaning that the phenomenon is robust versus a change of the microscopic details.

This surprising phenomenon is known as Anderson localization, sometimes also called strong localization. Although it was predicted on theoretical grounds in the late 1950s—most famously by Anderson himself (Anderson, 1958)—it has only been

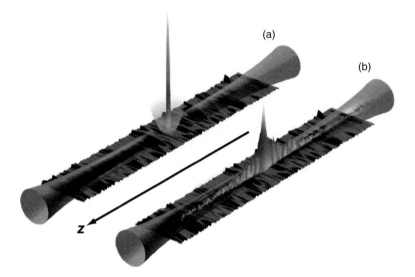

Fig. 9.1 Direct experimental observation of one-dimensional Anderson localization of an atomic matter wave in a disorder potential. The disorder potential (represented in dark gray in the lower part of the figure) is created by a speckle pattern. (a) An initially localized wave packet (prepared in a harmonic trap at the center) evolves freely, diffuses and eventually freezes at long times in a characteristic exponential shape (b). The tube represents the transverse-confinement laser beam that ensures an effectively one-dimensional dynamics. Reprinted from (Billy *et al.*, 2008) (courtesy of Ph. Bouyer).

observed directly rather recently. Cold atoms, where an *in situ* direct observation of the wave function is possible, are from that point of view highly valuable.

In this chapter, we present an introduction to transport properties in disordered systems, with a strong emphasis on Anderson localization. As an appetizer, we show in Section 9.2 how the one-dimensional case can be exactly solved, providing us with useful physical pictures. We then introduce in Section 9.3 the scaling theory of localization, a typical illustration of the appearance of new concepts and parameters relevant at the long distance and long timescale. After reviewing some of the most important experimental and numerical results in Section 9.4, we develop a microscopic description of quantum transport in Section 9.5, several applications of which are discussed in Sections 9.6–9.8.

Many other, interesting questions will not be touched upon, foremost the impact of *interaction* between several identical particles. However, technically speaking averages over disorder introduce an effective interaction. The relevant diagrammatic approach, originally introduced in the context of quantum electrodynamics, is quite versatile and used equally well to describe, e.g., interacting electrons in solid-state samples or interacting atoms in Bose–Einstein condensates. In this chapter, we restrict the discusion to non-interacting particles in a disordered medium, but the general framework and the technical tools introduced should provide our readers with solid foundations to follow also more advanced developments. Understanding the combined effects of interaction and disorder has been, still is, and doubtlessly will remain the subject of fascinating research for a long time to come.

9.2   Transfer-matrix description of transport and Anderson localization in 1D systems

In order to develop some intuition on transport in disordered systems, it is useful to study solvable models, where one can identify relevant phenomena and mechanisms. It turns out that a one-dimensional system, with the specific choice of δ point scatterers put at random positions, provides such a solvable model, that is moreover sufficiently rich to teach us useful lessons for more realistic disorder and higher dimensions.

Consider therefore a spinless particle confined to a 1D waveguide geometry by tight transverse trapping to its ground state. Free propagation along the z-direction with wavevector k is described by amplitudes $\psi_{\pm k}(z) = \exp\{\pm ikz\}$. In the following, we discuss the transmission of a single, fixed k-component through a series of obstacles $j = 1, 2, \ldots, N$, well separated and placed at randomly chosen distances $\Delta z_j = z_j - z_{j-1}$ as pictured in Figure 9.2.

9.2.1   Scattering matrix

Each obstacle shall be described by a potential $V_j(z)$. To fix ideas, we may assume that it is sufficiently short ranged to be well approximated by a δ-type impurity, $V_j(z) = V(z - z_j) = \sigma_0 V_0 \delta(z - z_j)$, with an internal length scale σ_0 that is not resolved by the propagating wave, $k\sigma_0 \ll 1$. Furthermore, we suppose that the obstacles are well separated, i.e. their density $n = N/L$ is small compared to the wavelength, $n \ll k$.

Consider first scattering by a single impurity at $z = 0$. We can decompose the wave functions to the left (L, $z < 0$) and right (R, $z > 0$) into left- and right-moving components:

$$\psi_{\mathrm{L}}(z) = \psi_{\mathrm{L}}^{\mathrm{in}} e^{+ikz} + \psi_{\mathrm{L}}^{\mathrm{out}} e^{-ikz} \tag{9.4}$$

$$\psi_{\mathrm{R}}(z) = \psi_{\mathrm{R}}^{\mathrm{out}} e^{+ikz} + \psi_{\mathrm{R}}^{\mathrm{in}} e^{-ikz}. \tag{9.5}$$

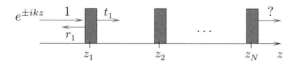

Fig. 9.2   One-dimensional waveguide with randomly placed scatterers. We know the reflection and transmission coefficients r_j and t_j of each scatterer. What is the total transmission T_N across the whole ensemble?

The outgoing amplitudes are linked to the incident amplitudes by the reflection and transmission coefficients r and t from the left, and r', t' from the right:

$$
\begin{aligned}
\psi_{\mathrm{L}}^{\mathrm{in}} \quad & \quad \psi_{\mathrm{R}}^{\mathrm{out}} = t\,\psi_{\mathrm{L}}^{\mathrm{in}} + r'\,\psi_{\mathrm{R}}^{\mathrm{in}} \\
r\,\psi_{\mathrm{L}}^{\mathrm{in}} + t'\,\psi_{\mathrm{R}}^{\mathrm{in}} = \psi_{\mathrm{L}}^{\mathrm{out}} \quad & \quad \psi_{\mathrm{R}}^{\mathrm{in}}
\end{aligned}
\tag{9.6}
$$

Writing these relations in matrix form introduces the scattering or S-Matrix:

$$
\begin{pmatrix} \psi_{\mathrm{L}}^{\mathrm{out}} \\ \psi_{\mathrm{R}}^{\mathrm{out}} \end{pmatrix} = \mathsf{S} \begin{pmatrix} \psi_{\mathrm{L}}^{\mathrm{in}} \\ \psi_{\mathrm{R}}^{\mathrm{in}} \end{pmatrix} \qquad \text{with} \qquad \mathsf{S} = \begin{pmatrix} r & t' \\ t & r' \end{pmatrix}.
\tag{9.7}
$$

For the present setting of a single-mode waveguide, the reflection and transmission coefficients are complex numbers, and the probabilities for reflection and transmission from the left are $R = |r|^2$ and $T = |t|^2$, respectively, and similarly from the right. In a more general setting of multimode scattering with m modes or "channels" on the left and m' modes on the right, r and t are matrices with $m \times m$ and $m' \times m$ entries, respectively. And for example, the total transmission probability "all channels in to all channels out" then reads $T = \sum_{m,m'} t_{m'm} t_{m'm}^* = \sum_{m'} (tt^\dagger)_{m'm'} =: \mathrm{tr}'\{tt^\dagger\}$. Within this section, we have only use for the single-channel notation and refer to the literature for the general case (Mello and Kumar, 2004; Imry, 2002; Datta, 2002; Beenakker, 1997)

Probability flux conservation requires that S be unitary, $\mathsf{S}^\dagger = \mathsf{S}^{-1}$. From $\mathsf{S}^\dagger\mathsf{S} = \mathbb{1}$, it follows directly that reflection and transmission probabilities add up to unity: $R + T = 1$ and $R' + T' = 1$. One also finds $r^* t' + t^* r' = 0$ and its complex conjugate $rt'^* + tr'^* = 0$. From this, it follows for the single-channel case that $R = R'$, $T = T'$: the reflection and transmission probabilities are the same from both sides.

Time-reversal exchanges the roles of "in" and "out" states. For a time-reversal invariant potential $V(z)$, this implies $\mathsf{S}^* = \mathsf{S}^{-1}$. With unitarity, this is equivalent to $\mathsf{S}^t = \mathsf{S}$ or $t = t'$, a symmetry called *reciprocity*. This setting defines the so-called "orthogonal" symmetry class of random matrix theory. Reciprocity is typically violated in presence of an external magnetic field or magnetic impurities (see Sections 9.6.3 and 9.7 below).

Exercise 9.1 Consider an elementary impurity with $V(z) = \sigma_0 V_0 \delta(z)$. Solve the Schrödinger eigenvalue equation $-\psi'' + (2m/\hbar^2)V\psi = k^2\psi$ at fixed k (use the continuity of the free wave function and compute its derivative discontinuity at $z = 0$) and show that the S-matrix in terms of $f = m\sigma_0 V_0/\hbar^2 k$ is given by (Mello, 1995);

$$
\mathsf{S} = \frac{1}{1 + if} \begin{pmatrix} -if & 1 \\ 1 & -if \end{pmatrix}.
\tag{9.8}
$$

9.2.2 Transfer matrix

If we now have several impurities in series, in principle the total transmission can be calculated from the S-matrix of the whole system. But the total S-matrix is not simply linked to the individual S-matrices. Since the transmission depends on the incident amplitudes from both left and right, adding a scatterer requires to recompute the entire sequence. So instead of distinguishing in/out amplitudes, one prefers to decompose the wave function into right-/left-moving amplitudes, respectively: $\psi(z) = \psi^+ e^{+ikz} + \psi^- e^{-ikz}$, and this on both sides R/L of the obstacles. The transfer matrix M then maps the amplitudes from the left side of the obstacle to the right:

$$\text{or} \qquad \begin{pmatrix} \psi_R^+ \\ \psi_R^- \end{pmatrix} = M \begin{pmatrix} \psi_L^+ \\ \psi_L^- \end{pmatrix}. \qquad (9.9)$$

One can easily determine its matrix elements in terms of t, t', r, r'. For instance, $\psi_L^{\text{out}} = r\psi_L^{\text{in}} + t'\psi_R^{\text{in}}$ rewrites as $\psi_L^- = r\psi_L^+ + t'\psi_R^-$, which we can immediately solve for $\psi_R^- = \frac{1}{t'}\psi_L^- - \frac{r}{t'}\psi_L^+$, and similarly for ψ_R^+. Eliminating r', t' with unitarity relations in favor of r, t and their complex conjugates yields a simple form:

$$M = \begin{pmatrix} 1/t^* & -r^*/t^* \\ -r/t & 1/t \end{pmatrix}. \qquad (9.10)$$

Exercise 9.2 Check the following interesting properties of the transfer matrix:

(*i*) $\det M = 1$.

(*ii*) Current conservation (unitarity of S) implies now that $M\sigma_z M^\dagger = \sigma_z$, with $\sigma_z = \begin{pmatrix} 1 & 0 \\ 0 & -1 \end{pmatrix}$ the third Pauli matrix.

(*iii*) Equivalently, $M^{-1} = \sigma_z M^\dagger \sigma_z$.

(*iv*) $(M^\dagger M)^{-1} = \sigma_z M^\dagger M \sigma_z$. Thus, the hermitian matrices $(M^\dagger M)^{-1}$ and $M^\dagger M$ have the same (real) eigenvalues. Verify this property by computing the eigenvalues directly using Eq. (9.10). Since these eigenvalues must also be each other's inverses, they can only be of the form $\lambda_+ = 1/\lambda_- = e^{2x}$.

(*v*) As a matrix, $2 + (M^\dagger M)^{-1} + M^\dagger M = 4/T$. Thus, the total transmission probability is $T = 1/(\cosh x)^2$.

9.2.2.1 Chaining transfer matrices

By construction, the transfer matrix maps the amplitudes from left to right across each scatterer. Therefore, the total transfer matrix across N scatterers is obtained by multiplying them:

$$M_{12\dots N} = M_N \dots M_2 M_1. \qquad (9.11)$$

Consider the simplest case of two obstacles $j = 1, 2$ in series, for which $M_{12} = M_2 M_1$. After matrix multiplication, one finds the transmission coefficient

$$t_{12} = \frac{t_1 t_2}{1 - r_1' r_2}. \tag{9.12}$$

This transmission amplitude contains the entire series of repeated internal reflection between the two scatterers: $t_{12} = t_2 t_1 + t_2 r_1' r_2 t_1 + t_2 (r_1' r_2)^2 t_1 + \ldots$. The transmission probability reads

$$T_{12} = \frac{T_1 T_2}{|1 - \sqrt{R_1 R_2} e^{i\theta}|^2}, \tag{9.13}$$

where θ is the total phase accumulated during one complete internal reflection. Since the scatterers are placed with a random distance $k\Delta z \gg 2\pi$, the phase θ will also be randomly distributed in $[0, 2\pi]$, independently of the details of the random distribution of distance between consecutive scatterers or their reflection phases. One can calculate expectation values of any function of θ by

$$\langle f(\theta) \rangle = \int_0^{2\pi} \frac{d\theta}{2\pi} f(\theta). \tag{9.14}$$

But as we will see in the following, a very important question is: "what quantity $f(\theta)$ should be averaged?"

9.2.2.2 Incoherent transmission: Ohm's law

The most natural thing seems to average directly the transmission probability (9.13). One finds

$$\langle T_{12} \rangle = \frac{T_1 T_2}{1 - R_1 R_2}. \tag{9.15}$$

The same transmission probability is obtained for a purely classical model where only reflection and transmission *probabilities* are combined:

Exercise 9.3 Show that the rule (9.15) is also obtained if one uses an S-matrix propagating probabilities instead of amplitudes, $\mathring{S} = \begin{pmatrix} R & T \\ T & R \end{pmatrix}$, by determining the corresponding transfer matrix $\mathring{M}$ and chaining it.

In this case, the distance Δz of ballistic propagation between the two scatterers is completely irrelevant, and T_{12} is given by Eq. (9.15). This classical description applies to systems that are subject to strong decoherence, where the phase of the particle is completely scrambled by coupling to an external degree of freedom, while travelling between scatterers 1 and 2.

The so-called element resistance of the obstacles, $(1-T)/T$, calculated with the classical transmission, is additive:

$$\frac{1-T_{12}}{T_{12}} = \frac{1-T_1}{T_1} + \frac{1-T_2}{T_2}. \tag{9.16}$$

Therefore, the classical resistance across N identical impurities distributed with linear density $n = N/L$ along a wire of length L grows like

$$\frac{R}{T}(L) = N\frac{R_1}{T_1} =: \frac{L}{l_1}, \tag{9.17}$$

where $l_1 = T_1/(nR_1)$ is a length characterizing the backscattering strength of a single impurity.

Within the context of electronic conduction, the result Eq. (9.17) is known as Ohm's law, stating that the total classical resistance of a wire grows linearly with its length L. Obviously, averaging the transmission itself at each step has wiped out completely the phase coherence and left us with a purely classical transport process. This process can be formulated equivalently as a persistent random walk on a lattice where the particle has uniform probability T_1 to continue in the same direction at each time step and probability R_1 to make a U-turn. For long times, this random walk leads to diffusive motion with diffusion constant (Godoy, 1997)

$$D = vl_0\frac{T_1}{2R_1}. \tag{9.18}$$

Here, v is the velocity of the particle and $l_0 = 1/n$ the distance between consecutive scatterers. Since the diffusion constant is related to the transport mean free path l by the general relation $D = vl/d$, with $d = 1$ the dimension of the system, one can identify l_1 as twice the transport mean free path, $l_1 = 2l$.

9.2.2.3   *Phase-coherent transmission: strong localization*

The relation (9.15) cannot be easily generalized to more than two scatterers. Indeed, already for three scatterers, the average of the complicated product of transmission matrices even over independent, random phases θ_{12} and θ_{23} becomes very complicated. In order to predict the behavior of transmission across long samples, it is advantageous to find a quantity that is additive as new scatterers are added to the wire. There is such a quantity that becomes additive under ensemble-averaging, namely the so-called extinction coefficient $\kappa = -\ln T = |\ln T|$. When averaging the logarithm of Eq. (9.13), the denominator drops out since

$$\int_0^{2\pi} \frac{d\theta}{2\pi} \ln\left|1 - \sqrt{R_1 R_2}e^{i\theta}\right| = 0 \tag{9.19}$$

due to the analyticity of the complex logarithm for all $0 \le R_1 R_2 < 1$. Thus, one immediately finds that the average extinction across two consecutive scatterers is strictly additive:

$$\langle \ln T_{12} \rangle = \ln(T_1) + \ln(T_2). \tag{9.20}$$

The generalization to many scatterers is now easy because $\langle \ln T \rangle$ is additive: the total extinction of a channel of length L grows on average like $|\langle \ln T \rangle| = nL|\ln T_1|$. With this scaling behavior, one obtains that the log-averaged transmission

$$\exp\{\langle \ln T \rangle\} = e^{-L/\xi_{\text{loc}}} \tag{9.21}$$

drops exponentially fast with increasing sample length L. In the absence of absorption, this is a hallmark of strong localization by disorder, and we have found the localization length $\xi_{\text{loc}} = 1/(n|\ln T_1|)$. In a weak-scattering situation where $nl_1 = T_1/R_1 \gg 1$, we approximate $|\ln T_1| \approx 1/nl_1$ and thus find the localization length as $\xi_{\text{loc}} = l_1 = 2l$.

What is the meaning of the log-averaged transmission (9.21)? It is important to realize that the transmission T as a function of the microscopic realization of disorder is a *random variable*. We will show in the following sections that T itself is not a self-averaging quantity, meaning that its average $\langle T \rangle$ has no resemblance to the most likely found value of T, called the *typical* transmission T_{typ}. We will shortly see that for long samples, the probability distribution of $\ln T$ is very close to a normal distribution (see Eq. (9.32) below), which is centered at $\langle \ln T \rangle = \ln T_{\text{typ}}$. And since the logarithm is monotonic, the *most probable value* of the transmission is indeed $T_{\text{typ}} = \exp\{\langle \ln T \rangle\} = e^{-L/\xi_{\text{loc}}}$.

How does one go about identifying a properly self-averaging quantity in the first place?—By observing that the transmission matrices of several obstacles multiply, see Eq. (9.11). Thus, their logarithm is additive: $\ln \mathsf{M}_{12...N} = \sum_{j=1}^{N} \ln \mathsf{M}_j$. Summing a large number of these random log-transmission matrices therefore realizes a variant of the central limit theorem, from which we know that the limiting probability distribution is a normal (Gaussian) distribution with a width that decreases with the number of addends. And really, many of the rigorous results available for (quasi-)1D Anderson localization make use of Furstenberg's theorem on products of random matrices (Furstenberg, 1963). Thus, the extinction or log-averaged transmission is indeed a good candidate for a self-averaging quantity.

9.2.3 Scaling equations

In order to substantiate the previous arguments, we should find the full distribution function $P(T, L)$ that permits us to derive expectation values for arbitrary functions of T at length L. This distribution function can be found exactly by solving recursion relations that describe how the transmission is changed when a small bit ΔL is added to a sample of length L:

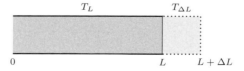

We know already that the transfer matrices multiply, $\mathsf{M}_{L+\Delta L} = \mathsf{M}_{\Delta L}\mathsf{M}_L$. The composition law (9.13) then gives

$$T_{L+\Delta L} = \frac{T_L T_{\Delta L}}{|1 - e^{i\theta}\sqrt{R_L R_{\Delta L}}|^2}. \tag{9.22}$$

The idea is now to study the change in the expectation values of T as the "time" $t = L/l_1$ grows. For this, we assume that the added part ΔL is long enough such that an independent disorder average $\langle\ldots\rangle_{\Delta L}$ in this section is meaningful. We also assume that the scatterers are weak such that the backscattering probability remains small:

$$R_{\Delta L} = \Delta L/l_1 =: \Delta t \ll 1. \tag{9.23}$$

These assumptions are verified if ΔL is of the order of $n^{-1} = l_1 R_1$ with a single weak scatterer on average. We may now expand Eq. (9.22) using $T_{\Delta L} = 1 - \Delta t$ to leading order in Δt, finding for the change

$$\Delta T = T_{L+\Delta L} - T_L = T_L \left[2\sqrt{R_L\Delta t}\cos\theta + (4R_L\cos^2\theta - 1 - R_L)\Delta t\right]. \tag{9.24}$$

Thus, to order Δt, we find by averaging $\langle\ldots\rangle_{\Delta L} = \langle\ldots\rangle_\theta$ that

$$\frac{\langle\Delta T\rangle_\theta}{\Delta t} = -T_L^2, \qquad \frac{\langle(\Delta T)^2\rangle_\theta}{\Delta t} = 2T_L^2(1 - T_L), \qquad \frac{\langle(\Delta T)^n\rangle_\theta}{\Delta t} = 0 \quad (n \geq 3). \tag{9.25}$$

So only the first two moments of fluctuations contribute. From the first relation we can read off that the transmission decreases with increasing length, an expected result. But the second relation shows that the relative fluctuations, while small in the beginning where $1 - T = R \ll 1$, will grow for long samples where $T \ll 1$.

9.2.4 Fokker–Planck equation and log-normal distribution

The above equations (9.25) describe a random quantity $T(t)$ whose first two moments obey the equations $\partial\langle T\rangle/\partial t = \langle A(T)\rangle$ and $\partial\langle T^2\rangle/\partial t = \langle 2TA(T) + B(T)\rangle$ with $A(T) = -T^2$ and $B(T) = 2T^2(1 - T)$. Here, we use the continuous notations $\Delta t \to dt$ and $\Delta T \to dT$, while bearing in mind the coarse-grained character of the averaged quantities. The theory of Brownian motion (van Kampen, 2007) then teaches us that the probability distribution $P(T, t)$ of T at time t obeys the Fokker–Planck equation

$$\partial_t P = -\partial_T[AP] + \frac{1}{2}\partial_T^2[BP]. \tag{9.26}$$

This equation of motion can be seen as a continuity equation, $\partial_t P + \partial_T J = 0$, for the locally conserved probability density with current $J = AP - \frac{1}{2}\partial_T(BP)$. Here, A describes the drift, whereas $B/2$ plays the role of a diffusion constant.

By standard terminology, the Fokker–Planck equation (9.26) is called "non-linear", because $B(T)$ and $A(T)$ depend non-linearly on T. By changing the variable, one can try to simplify these coefficients. A first option consists in using $T = 1/(\cosh x)^2$ (remember exercise 9.2(v)). In the remainder of this section, let us explore the consequences of this choice. Knowing that the change of variables for a probability density requires

$$P(T,t)\mathrm{d}T = P((\cosh x)^{-2}, t)\left|\frac{\mathrm{d}T}{\mathrm{d}x}\right|\mathrm{d}x = \tilde{P}(x,t)\mathrm{d}x, \tag{9.27}$$

we find the corresponding Fokker–Planck equation:

$$\partial_t \tilde{P}(x,t) = -\frac{1}{2}\partial_x\left[\coth(2x)\tilde{P}\right] + \frac{1}{4}\partial_x^2\tilde{P}, \tag{9.28}$$

with initial condition $\tilde{P}(x,0) = \delta(x)$. With this choice of variable, the second derivative describing the fluctuations has become most simple. Although this equation is still (too) difficult to solve exactly, we can extract the limiting distribution for long samples in the limit $t \gg 1$ as follows. First, we rewrite the equation as

$$\left[\partial_t + \frac{1}{2}\coth(2x)\partial_x\right]\tilde{P}(x,t) = \frac{1}{(\sinh 2x)^2}\tilde{P} + \frac{1}{4}\partial_x^2\tilde{P}. \tag{9.29}$$

We may interpret the derivatives on the left-hand side as a Lagrangian derivative $\partial_t + \dot{x}_0\partial_x = D_t$ in a co-moving frame defined by $\dot{x}_0(t) = \frac{1}{2}\coth(2x_0)$, which is solved by $x_0(t) = \frac{1}{2}\mathrm{arcosh}(e^t) \approx \frac{1}{2}t$, for large t. Developing all terms for small deviations $\Delta = x - x_0(t)$ from this point of reference, we find a very simple equation for $\tilde{F}(\Delta,t) = \tilde{P}(x_0(t) + \Delta, t)$:

$$\partial_t \tilde{F}(\Delta,t) = \frac{1}{4}\partial_\Delta^2\tilde{F}(\Delta,t). \tag{9.30}$$

This is the elementary diffusion equation, and the solution, perhaps best known as the heat kernel, is readily obtained by Fourier transformation:

$$\tilde{F}(\Delta,t) = \frac{1}{2\sqrt{\pi t}}\exp\left\{-\frac{\Delta^2}{4t}\right\}. \tag{9.31}$$

Going back to the transmission using $\ln T = -2x$ valid for large x, we thus find the limiting distribution

$$F_{\text{log-norm}}(\ln T, t) = \frac{1}{2\sqrt{\pi t}}\exp\left\{-\frac{(\ln T + t)^2}{4t}\right\} \tag{9.32}$$

for small deviations around the most probable value $\ln T_0(t) = -2x_0(t) = -t$. We have successfully demonstrated that indeed the logarithm of the transmission is a normally distributed random quantity. It is characteristic for disordered channels that the two defining moments

$$|\langle \ln T\rangle| = t, \qquad \mathrm{var}(\ln T) = \langle(\ln T)^2\rangle - \langle \ln T\rangle^2 = 2t \tag{9.33}$$

are determined by a single parameter, namely the length $t = L/2l = L/\xi_{\text{loc}}$ of the one-dimensional wire in units of the localization length. Clearly, the relative fluctuations $\mathrm{var}(\ln T)/\langle \ln T\rangle^2 = 2/t$ decay with system size. Thus, we are assured that the transmission logarithm is a self-averaging quantity, with moreover a normal probability distribution whose most probable value is equal to the mean.

But careful! Even in the limit $t \gg 1$, this does unfortunately *not* imply that one may use the log-normal distribution (9.32) indiscriminately to calculate moments of the transmission. A striking counterexample is

$$\langle T \rangle_{\text{log-norm}} = \langle e^{\ln T} \rangle_{\text{log-norm}} = \int dy F_{\text{log-norm}}(y, t) e^y = 1 \qquad \text{(wrong)}, \qquad (9.34)$$

and quite obviously so. What goes wrong here? We will have a second look at the end of the next section once we know the exact solution.

9.2.5 Full distribution function

Another choice of variable is $\rho = T^{-1}$, the dimensionless total resistance of the channel. The Fokker–Planck equation (9.26) for its probability distribution $W(\rho, t) = P(\rho^{-1}, t)\rho^{-2}$ reads

$$\partial_t W = \partial_\rho \left[\rho(\rho - 1) \partial_\rho W \right], \qquad (9.35)$$

with initial condition $W(\rho, 0) = \delta(\rho - 1)$, i.e. a wire of zero length has perfect transmission. The solution can be calculated in closed form (Abrikosov, 1981):

$$W(\rho, t) = \frac{\exp\{-t/4\}}{\sqrt{\pi} t^{3/2}} \int_{\text{arcosh}\sqrt{\rho}}^{\infty} \frac{\exp\{-y^2/t\} d(y^2)}{\sqrt{(\cosh y)^2 - \rho}}. \qquad (9.36)$$

Figure 9.3 shows how the distribution function $F(\kappa, t) = W(e^\kappa, t)e^\kappa$ for the extinction $\kappa = \ln \rho = -\ln T$ moves from a δ-distribution with growing system size t to the log-normal distribution (9.32), drawn as a dashed line at $t = 10$.

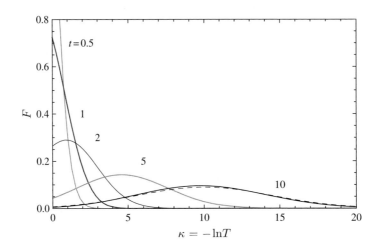

Fig. 9.3 Probability distribution function $F(\kappa, t) = W(e^\kappa, t)e^\kappa$ for the extinction $\kappa = \ln \rho = -\ln T$ across a one-dimensional channel of length $t = L/\xi_{\text{loc}} = 0.5, 1, 2, 5, 10$. Dashed line at $t = 10$: log-normal distribution (9.32).

The full distribution function permits to calculate all moments $\langle \rho^{-n} \rangle = \langle T^n \rangle$ of the transmission (Abrikosov, 1981). For $t \gg 1$, one finds the asymptotic expression

$$\langle T^n \rangle = \frac{\pi^{3/2} \Gamma(n - \frac{1}{2})^2}{2 \Gamma(n)^2} t^{-3/2} e^{-t/4}, \tag{9.37}$$

showing that *all* moments decay with the same dependence $t^{-3/2} e^{-t/4}$. The first two moments are then

$$\langle T \rangle = \frac{\pi^{5/2}}{2} t^{-3/2} e^{-t/4}, \qquad \langle T^2 \rangle = \frac{1}{4} \langle T \rangle \ll 1. \tag{9.38}$$

Although the average transmission and its fluctuations decay exponentially, as expected for a strongly localizing system, the *relative* fluctuations of the transmission itself grow very quickly: $\mathrm{var}(T)/\langle T \rangle^2 \propto t^{3/2} e^{t/4}$.

And why does the blind application (9.34) of the limiting log-normal distribution (9.32) predict $\langle T \rangle_{\text{log-norm}} = 1$ instead of the correct decrease (9.38)? Well, in Eq. (9.34), we have used the normal distribution on the entire real axis for $y = \ln T$, without paying attention to the constraint that the physically admissible transmission is $T \leq 1$. To take this into account, a popular recipe consists in using a *truncated* log-normal distribution on the half-line $\kappa \geq 0$ (Beenakker, 1997; Evers and Mirlin, 2008)

$$F_{\text{log-norm}}^+ (\kappa, t) = \frac{\Theta(\kappa)}{C(t) \sqrt{\pi t}} \exp \left\{ -\frac{(\kappa - t)^2}{4t} \right\}, \tag{9.39}$$

with a normalization $C(t) = 1 + \mathrm{erf}(\sqrt{t}/2) \approx 2$. The moments of this distribution depend entirely on the value at truncation, $\lim_{\kappa \to 0^+} F_{\text{log-norm}}^+ (\kappa, t) = (2\sqrt{\pi})^{-1} t^{-1/2} e^{-t/4}$, that is the probability for perfect transmission $T = 1$. The transmission moments for $t \gg 1$ are predicted to be

$$\langle T^n \rangle_{\text{log-norm}}^+ \approx \frac{1}{(2n - 1)\sqrt{\pi}} t^{-1/2} e^{-t/4}. \tag{9.40}$$

Comparing them with the exact moments (9.37), we see that the truncated log-normal distribution can describe the *leading exponential decay*, but fails to capture the algebraic dependence correctly. Mathematically, this is due to the fact that the log-normal distribution overestimates the probability of perfectly transmitting channels $T = 1$, which is really only $W(0, t) = (\pi^{3/2}/2) t^{-3/2} e^{-t/4}$.

On physical grounds, this limited applicability of the log-normal distribution emphasizes the difficulties one faces when dealing with broad distributions. Figure 9.4 shows the transmission probability distributions at different lengths $t = L/l$. It is instructive to look at the last curve for the altogether moderate system size $t = 10$. The *most probable* or "typical" value for the transmission is quite small, $T_{\text{typ}} = \exp\{\langle \ln T \rangle\} = e^{-10} \approx 4.5 \times 10^{-5}$. The exact *average value* is much bigger, $\langle T \rangle \approx 1.06 \times 10^{-2}$, indicating that this average is to a large extent determined by very rare events with anomalously large transmission. The truncated log-normal

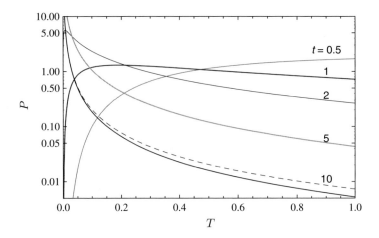

Fig. 9.4 Probability distribution function $P(T,t)$ for the transmission T across a one-dimensional channel of length $t = L/\xi_{\text{loc}}$. Dashed line at $t = 10$: log-normal distribution (9.32).

distribution, shown in dashed, overestimates the frequency of large transmissions and predicts $\langle T \rangle \approx 1.28 \times 10^{-2}$.

Let us close this section by emphasizing once more that the typical transmission—averaged over all possible relative phases accumulated between consecutive scatterers—displays an exponential decay at large distance as stated in Eq. (9.21). This is in sharp contrast to the classical transmission (9.17), where classical probabilities, not amplitudes, are combined to an algebraic decay. Clearly, averaging over disorder implies averaging over quantum-mechanical phases globally, but is not equivalent to removing phase-coherence and interference effects locally from the very start. This is a striking example of a *mesoscopic* effect, a rather counter-intuitive phenomenon, where microscopic phase coherence has macroscopic physical consequences that survive averaging over quenched disorder.

9.3   Scaling theory of localization

9.3.1   What is a scaling theory?

A scaling theory describes the relevant properties of physical systems by considering their behavior under changes of size $L \mapsto bL$. Quantitative scaling arguments were invented in quantum field theory in the context of renormalization. Scaling arguments became widely popular in statistical physics by the mid-1960s for describing phase transitions and critical phenomena (Kadanoff *et al.*, 1967). The immense success of renormalization-group techniques developed in the 1970s (Wilson, 1983) rapidly radiated to the field of disorder-induced phase transitions that Anderson's celebrated paper had founded (Anderson, 1958). After pioneering work by Wegner (Wegner, 1976), a scaling theory of localization was formulated by Abrahams, Anderson, Licciardello, and Ramakrishnan (Abrahams *et al.*, 1979), a quartet that became known as the "gang of four".

A scaling theory can hope to capture those features that are important on macroscopic scales, but will be insensitive to microscopic details. This means that its predictions are only semi-quantitative, in the sense that it cannot furnish the precise location of a critical point in parameter space nor provide any system-specific data. In return, if one feeds it with the microscopic data (such as the transport mean free path), it can give general, and surprisingly accurate, predictions of universal character.

Let us have a look at the different lengths characterizing quantum transport in a disordered material:

As short-scale lengths (in the left dotted box) one has the wavelength $\lambda = 2\pi/k$ of the propagating object and the correlation length ζ of the disorder. If $\zeta \ll \lambda$, the details of the disorder are unimportant, and models of δ-correlated scatterers are appropriate. If $\zeta \gg \lambda$, the disorder correlation can be resolved by the wave; this is typically the case for our example of optical speckle potentials probed with ultracold atoms (Billy *et al.*, 2008).

As larger scales (in the right dotted box) one has the transport mean free path l and the localization length ξ_{loc}. We have already seen in Section 9.2.2.3 that in $d = 1$ these two lengths are practically identical, $\xi_{\mathrm{loc}} = 2l$. In $d = 2$ the localization length is much larger than the transport length, as will be discussed in Sections 9.3.6 and 9.7.4 below, and in $d = 3$ it may well be infinite. Depending on the system size L, one can distinguish three basic transport regimes: *ballistic* transport through small samples with $L < l$, *diffusive* transport for $l < L < \xi_{\mathrm{loc}}$ with, possibly, weak-localization corrections, and finally *strong localization* for large samples with $\xi_{\mathrm{loc}} < L$. In $d = 1$, there is no room for diffusion between l and ξ_{loc}, and strong localization is basically a single-scattering effect.[2] The scaling theory of localization has the purpose of describing the transition between these regimes as function of system size L (Anderson *et al.*, 1980).

9.3.2 Dimensionless conductance

Traditionally, the scaling theory of localization is formulated in terms of a channel's *proper conductance*, a dimensionless parameter defined as $g = T/R$ by transmission and reflection probabilities. Equivalently, one may consider the channel's *proper resistance* g^{-1}. A perfectly transmitting channel $T = 1$ has a proper conductance of $g = \infty$, and a perfectly resisting channel with $T = 0$ has $g = 0$, which seems a rather sensible definition. Moreover, we have seen in Section 9.2.2.2 that this resistance is additive when classical subsystems are chained in series. Alternatively, one could define the total resistance as $\rho = 1/T = 1 + g^{-1}$, where the additional 1 represents the "contact resistance" due to the leads connecting the sample to the external world.

In the previous section, we also learned that the transmission of a disordered channel is a random variable with a broad distribution around a most probable, typical value $T_{\text{typ}} = \exp \langle \ln T \rangle$. Therefore, also $g = T/(1 - T)$ is a broadly distributed random variable, fluctuating around the *typical conductance* $g_{\text{typ}} = T_{\text{typ}}/(1 - T_{\text{typ}})$. In all of the following, we discuss the behavior of g_{typ}, but in order not to overburden the notation, we will simply write $g_{\text{typ}} = g$.

In order to get used to this vocabulary, let us reformulate the results of Section 9.2 for the typical conductance. The exact exponential behavior (9.21) of the typical transmission translates into

$$g(L) = \frac{1}{\exp\{L/2l\} - 1} = \begin{cases} 2l/L, & L \ll l, & (9.41) \\ \exp\{-L/2l\} & L \gg l. & (9.42) \end{cases}$$

This conductance together with the resistance g^{-1} is plotted in Figure 9.5. Only the conductance of short, ballistic channels is given by the classical expression (9.41), that we have already encountered as Ohm's law in Section 9.2.2.2.

The results of scaling for the conductance can be reformulated for other quantities if those seem more convenient. One of the most popular, and useful, quantities is the *diffusion constant* $D = vl/d$, the product of velocity and transport mean free path divided by the number of dimensions d, a convention whose rationale will become clearer below. For matterwaves with wavevector k, the velocity is $v = \hbar k/m$, and the diffusion constant can also be written $D = \frac{\hbar}{dm}kl$, i.e. the product of an elementary diffusion constant $(\hbar/dm)$ by the dimensionless quantity $kl = 2\pi l/\lambda$. This ratio describes the effective disorderedness of the medium: $kl \gg 1$ means that the wave can travel over many periods before suffering scattering. We will see in Section 9.7 below that kl is a crucial parameter for transport and localization properties.

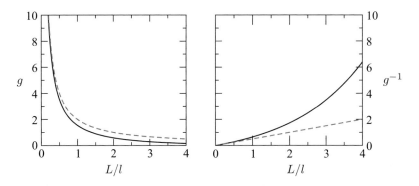

Fig. 9.5 Typical conductance (left panel) and resistance (right panel) of a 1D channel as a function of sample length L/l. Short channels have a resistance that grows linearly as expected by Ohm's law (green dashed, Eq. (9.41)), whereas long channels show exponentially large resistance (full line, Eq. (9.42)).

In a metallic sample with usual electrical conductivity, the Drude formula $\bar{\sigma} = ne^2\tau/m$ establishes the direct proportionality between the conductivity $\bar{\sigma}$ and the classical diffusion constant $D = v^2\tau/d$ of charge carriers with mean free path $l = v\tau$. We are thus led to define a dimensionless classical conductivity[3]

$$\mathring{\sigma} = \frac{2mD}{\hbar}. \tag{9.43}$$

The conductance $\mathring{g}$ of a sample of linear size L in d dimensions is the ratio of $\mathring{\sigma}$ to L^{2-d}. To see this, picture a metallic block where the voltage U is applied along one dimension to give $E = U/L$, whereas the current density $j = \bar{\sigma}E$ over the transverse area L^{d-1} yields the total current $I = L^{d-1}j$, which results in the dimension-full conductance $G = I/U = L^{d-2}\bar{\sigma}$. In order to define a dimensionless conductance $\mathring{g}$, one has to compensate the factors of L with another length scale. The simplest choice is the inverse of the wave number k. One can thus define a classical dimensionless conductance as

$$\mathring{g}(L) = (kL)^{d-2}\mathring{\sigma}. \tag{9.44}$$

In 1D, this definition gives $\mathring{g}(L) = 2l/L$, i.e. is fully compatible with the known exact result at short distance, Eq. (9.41) (this is the reason for the factor 2 introduced in Eq. (9.43)). It is of course no accident that the two definitions of dimensionless conductance—through the diffusion constant or through the transmission across a sample—coincide. The Landauer formula (Landauer, 1970) makes the connection explicit.

In any dimension, the classical dimensionless conductance can be rewritten as:

$$\mathring{g}(L) = \frac{2kl}{d}(kL)^{d-2}. \tag{9.45}$$

In particular, in dimension 2 we have $\mathring{g} = kl$ itself, independently of the system size.

9.3.3 Scaling in 1D systems

Since we wish to follow how the dimensionless conductance g evolves with system size, we make use of the β-*function*,

$$\beta = L\frac{\mathrm{d}\ln g}{\mathrm{d}L} = \frac{\mathrm{d}\ln g}{\mathrm{d}\ln(L/L_0)}. \tag{9.46}$$

$\beta = 0$ means that g does noot change with L. Actually, $\beta = cst$ implies a purely algebraic dependence $g(L) \propto L^\beta$. The celebrated function $\beta(g)$ has been introduced originally by Callan and Symanzik to describe the change of a coupling constant under a change of scale within quantum field theory (Peskin and Schroeder, 1995). Let us familiarize ourselves with the β-function, arguably the most important single object of scaling theory, in the case $d = 1$, for which we already know everything exactly. It is a matter of elementary calculus to find

$$\beta(L) = -\frac{L}{2l} \frac{1}{1 - \exp\{-L/2l\}}. \tag{9.47}$$

Since $g(L)$ is a monotonous function of $L/2l$, one can easily invert this dependence and express β as function of the conductance alone:

$$\beta(g) = -(1 + g) \ln \left[1 + g^{-1}\right]. \tag{9.48}$$

This result can also be derived directly as follows: the linear scaling of the typical-transmission logarithm implies $T_{\text{typ}}(bL) = [T_{\text{typ}}(L)]^b = \exp\{b \ln T_{\text{typ}}(L)\}$. Writing $T_{\text{typ}}^{-1} = 1 + g^{-1}$, differentiating with respect to b, and setting $b = 1$ at the end leads to Eq. (9.48). The fact that $\beta(g)$ can be expressed as function of g instead of the original length scale L does not seem very profound in $d = 1$ (Abrikosov, 1981). However, in field theory this property is vital for renormalizability (Peskin and Schroeder, 1995), and in statistical physics it guarantees that $\beta(g)$ can describe universal behavior close to a phase transition.

Figure 9.6 shows the β-function for $d = 1$, plotted as function of $\ln(g)$ together with its asymptotics. For short samples $L \ll l$, the conductance $g \propto L^{-1}$ is large, and $\lim_{g \to \infty} \beta(g) = -1$. More precisely, one has the following asymptotic behavior:

$$\beta(g) = -1 - \frac{1}{2g} + O(g^{-2}) \tag{9.49}$$

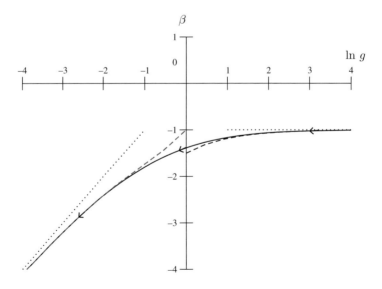

Fig. 9.6 Conductance scaling β-function in $d = 1$, Eq. (9.48). Arrows show the flow from the Ohmian behavior (9.49) for $g \gg 1$ in short samples to the exponential localization (9.50) for $g \ll 1$ in long samples.

which shows a *weak-localization* correction (see Section 9.7). In the opposite limit of a large sample, $g \ll 1$ is exponentially small, and

$$\beta(g) = \ln g - g(|\ln g| + 1) + O(g^2). \tag{9.50}$$

The transition between the two asymptotic regimes occurs around $g = 1$. The function $\beta(g)$ entirely describes how the dimensionless conductance evolves with system size L. Indeed, if g is known for some small size L, it can be deduced for any other size by solving the differential equation (9.46), moving along the arrows shown on the curve of figure 9.7. This is the so-called "renormalization flow" followed by the system when its size is increased towards macroscopic scales. For $d = 1$, β is always negative, implying that g always decreases with increasing system size, and thus the renormalization flow is unidirectional from right to left. Characteristically, when g decreases, β becomes more negative, which makes g decrease even faster, until it finishes by dropping exponentially fast.

A crucial asset of scaling theory is that its predictions are valid for an arbitrary 1D system, although the specific form of $\beta(g)$ was deduced using a specific model. Suppose you have a large-size complex disordered system and that you want to study its conductance. You may start with a small system for which you can calculate the conductance microscopically using a method of your choice. By following the renormalization flow, you are then provided, almost magically, with the conductance at any scale. Moreover, you find the localization length ξ_{loc} as the system size for which $g(\xi_{\text{loc}}) = O(1)$. Of course, there is no real magic here: your initial calculation yields the mean free path l and, as this is the only macroscopic length scale relevant for transport in a 1D system, you finally have everything.

9.3.4 Quasi-1D systems

The previous results may also be applied to quasi-one-dimensional systems that consist of several parallel channels $i = 1, \ldots, N_\perp$. One may either have in mind channels that are literally built parallel to each other (Lahini *et al.*, 2009) or a multimode waveguide with spatially overlapping transverse modes. If there is no coupling between channels, then the purely 1D description of Section 9.2 applies. For weakly coupled channels, which arises naturally by the disorder present, an equation of motion for the full distribution function of transmission eigenvalues very similar to Eq. (9.35) has been derived by Dorokhov and independently by Mello, Pereyra, and Kumar, known as the DMPK equation (Beenakker, 1997). Also, the scaling picture remains essentially the same. Keeping the number of transverse modes $N_\perp$ fixed, the short-scale conductance is $g = N_\perp 2l/L$, as expected for parallel resistors. Thus, the initial condition for the scaling flow on the curve $\beta(g)$ is changed, but the transition to the localized regime is the same. Since the crossover again occurs at $L = \xi_{\text{loc}}$ with $g(\xi_{\text{loc}}) = O(1)$, we simply find that the localization length is increased toward $\xi_{\text{loc}} = 2N_\perp l$.

9.3.5 Scaling in any dimension

In arbitrary dimension d, one changes the system size $L \mapsto bL$ in *all* directions, but still looks at the transmission along one chosen direction. In the ballistic regime $L \ll l$, we start again from the classical behavior, Eq. (9.45), where $g(L) \propto L^{d-2}$. One therefore expects to find $\lim_{g \to \infty} \beta(g) = d - 2$, and

$$\beta(g) = d - 2 - \frac{c_d}{g} + O(g^{-2}),\tag{9.51}$$

where a microscopic calculation is required to find the coefficient c_d that describes weak localization corrections.

In the strongly localized regime $L \gg \xi_{\text{loc}}$, exponential localization prevails. And since adding parts to the system beyond the localization length in the perpendicular direction cannot change its longitudinal transport, we still expect the power law $T_{\text{typ}}(bL) = [T_{\text{typ}}(L)]^b$ to hold in each channel. Thence follows the asymptotic behavior

$$\beta(g) = \ln(g/g_d)\tag{9.52}$$

in the strongly localized regime in any dimension, with a constant g_d of order unity.

Taking into account that the number of transverse channels scales as b^{d-1}, we would obtain the simple scaling relation $T_{\text{typ}}(bL) \approx b^{d-1}[T_{\text{typ}}(L)]^b$, if there were strictly no coupling between the channels. Then, the same calculation as for 1D would give

$$\beta(g) = (d-1) - (1+g)\ln\left[1 + g^{-1}\right],\tag{9.53}$$

that is a simple vertical shift of Eq. (9.48) by $d - 1$. In particular, this would imply that the weak-localization correction $-c_d/g$ is the same in all dimensions, a result known to be wrong, see Section 9.7. It nevertheless remains true that the shape of the true $\beta(g)$ curves, interpolating smoothly between the known asymptotics, is qualitatively given by Eq. (9.53), see also Figure 9.8. Although the scaling description encompasses arbitrary dimensions, its consequences are radically different in $d = 2$ and $d = 3$, meriting a separate discussion.

9.3.6   d = 2

In the ballistic limit of short samples with typical conductance $g \gg 1$, $\beta(g) \approx 0$ describes scale-independent conductance of $N_\perp \propto L$ transverse channels, each with element conductance $g \propto L^{-1}$. But then, $\beta(g)$ is not exactly zero. Starting the flow at the finite conductance g_0 of a sample of length L_0, a slightly negative $\beta(g) = -c_2/g$ makes g decrease with size (it will be shown in Section 9.7 below that $c_2 = 2/\pi$). We can integrate the flow equation

$$\beta(g) = -\frac{c_2}{g} = \frac{1}{g}\frac{dg}{d\ln(L/L_0)}\tag{9.54}$$

by elementary means to find

$$g(L) = g_0 - c_2 \ln(L/L_0).\tag{9.55}$$

To fix ideas, we can chose $L_0 = l$, a scale on which transport is classical, such that, from Eq. (9.45), $g(L_0) = \mathring{g} = kl \gg 1$. The transition to the strong localization regime

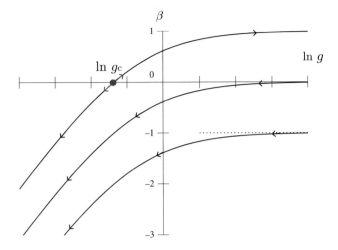

Fig. 9.7   Schematic plot of the β-function in $d = 1, 2, 3$, showing a smooth interpolation from the metallic regime (9.51) for $g \gg 1$ in short samples to the localized regime (9.52) for $g \ll 1$ in long samples. Note the existence of an unstable fix-point at critical g_c in $d = 3$.

occurs at $g(\xi_{\mathrm{loc}}) = O(1)$. Together with Eq. (9.55), this predicts an exponentially large localization length

$$\xi_{\mathrm{loc}} \sim L_0 \exp\{g_0/c_2)\} = l \exp\{kl/c_2\}. \tag{9.56}$$

The prediction of scaling theory for noninteracting particles in $d = 2$ is therefore that all states are localized. This transcends also from the scaling flow depicted in figures 9.7. However, the localization length can be extremely large if the system is only weakly disordered with $kl \gg 1$. Let us take some figures from the Orsay experiment (Billy *et al.*, 2008). With $l = 100\,\mu\mathrm{m}$ and $k = 2.5\,\mu\mathrm{m}^{-1}$, one finds the rather large localization length $\xi_{\mathrm{loc}} \approx le^{400}$, which would surely overstretch the possibilities of even the most capable experimentalist. The take-home message here is: In order to observe 2D localization, kl should be chosen as close to unity as possible. In turn, this implies that the classical diffusion constant $D = \hbar kl/2m$ must be of the order of $\hbar/m$, which for a typical cold atomic gas is a rather small quantity of the order of $10^{-9}\,\mathrm{m}^2/\mathrm{s}$. In order to observe 2D Anderson localization with atomic matterwaves, the experimentalist must be capable to observe the dynamics for a long time while keeping phase coherence, a challenging task indeed.

9.3.7 $d = 3$

In $d = 3$, we encounter a qualitatively new situation: the β-function is positive for large g. So if we start with some $g \gg 1$, the conductance flow will take us to even larger values of g. In renormalization-group terms, the behavior of the system in the thermodynamic limit $L \to \infty$ is described by the "infrared stable fix-point" $g = \infty$ (as in statistical physics, but in contrast to quantum field theory, we are interested in the large-distance behavior, i.e. the infrared asymptotics with respect to momentum). Since large conductance characterizes a good conductor, this is also known as the "metallic" fix-point.

By contrast, if we start with some $g \ll 1$, the negative β-function will drive the system towards the stable insulating fix-point $g = 0$ with exponentially small conductance at finite length. Between these two extrema, the β-function, assumed to be continuous, must have a zero at some g_c. A zero of $\beta(g) \propto \mathrm{d}g/\mathrm{d}L$ is also a fix-point, but in this case an unstable one. This unstable fix-point $\beta(g_c) = 0$ marks the critical point and shows the possibility of a *metal–insulator phase transition* at some critical strength of disorder. Although a scaling theory, with its roughly interpolating β-function, cannot predict the precise position of the critical point, it can give a semi-quantitative estimate. Indeed, a microscopic calculation of the transport mean free path l provides us with the dimensionless conductance $g(l)$. At such a scale, interference effects are unimportant, and thus, Eq. (9.45) can be used, giving $g(l) \approx 2(kl)^2/3$. As the critical point is such that $g_c \sim 1$, we obtain that the threshold for Anderson localization is given by:

$$kl \sim 1, \tag{9.57}$$

an equation known as the Ioffe–Regel criterion for localization. The precise value of the critical kl depends on microscopic details and is thus not universal.

Let us assume that the microscopic physics involves disorder whose strength is measured by some parameter W, typically the width of the disorder probability distribution (cf. Section 9.4.1.4). Even though scaling theory does not predict the precise position of the critical point, the behavior of the β-function around the critical point yields precious information about the large-scale physics: it permits us to calculate *critical exponents* that are the hallmark of universality. In their 1979 paper (Abrahams *et al.*, 1979), Abrahams et al. showed that the localization length diverges close to the transition for $W > W_c$ as

$$\xi_{\mathrm{loc}} \sim (W - W_c)^{-\nu}, \tag{9.58}$$

where the critical exponent $\nu = 1/s$ is determined by the slope of the β-function at the transition, $s = [\mathrm{d}\beta/\mathrm{d}\ln g]_{g_c}$.

The calculation leading to this prediction is elementary, but quite instructive in order to appreciate the power of a scaling description. Let us start at some length L_0 with some value $g_0 < g_c$ on the localized side of the fix-point. The β-function always allows us to calculate any other $g(L)$ implicitly by integration:

$$\ln\left(\frac{L}{L_0}\right) = \int_{\ln g_0}^{\ln g} \frac{\mathrm{d}\ln g'}{\beta(g')}. \tag{9.59}$$

Using the linearized form $\beta(g) = s\ln(g/g_c)$ around the fix-point leads to

$$\left(\frac{L}{L_0}\right)^s = \frac{\ln(g_c/g)}{\ln(g_c/g_0)}. \tag{9.60}$$

Now we are free to choose $L_0 = \xi_{\text{loc}}$ for which $g_0 = O(1)$ such that $\xi_{\text{loc}} \sim L \left[\ln(g_c/g) \right]^{-1/s}$. Because the microscopic physics on small scales ignores the critical behavior on large scales and can involve only smooth dependencies, one can always write $\ln(g/g_c) \approx (g - g_c)/g_c \propto (W_c - W)$ close enough to the critical conductance g_c, and we finally end up with Eq. (9.58).

For the simplest possible interpolation (9.53), one finds $\nu \approx 1.68$. This value is not disastrously far from the true value $\nu = 1.58 \pm 0.01$ that is known today from extensive numerical simulations (Slevin and Ohtsuki, 1999; Lemarié *et al.*, 2009), cf. Section. 9.4.3.

The "metallic" side of the transition can also be studied using a similar approach, but following the metallic branch $\beta > 0$ of the renormalization flow. For smaller-than-critical disorder strength $W < W_c$, the microscopically computed g at some size L_0 will be slightly larger than g_c. It is left as an exercise for the reader to show that this results at large scale in a diffusive (i.e. metallic) behavior with a diffusion constant

$$D \propto (W_c - W)^{\nu}. \tag{9.61}$$

The continuous (algebraic) vanishing of diffusion constant and conductance on the metallic side of the Anderson transition is characteristic of a continuous second-order phase transition.

9.3.8 $d > 3$

The Anderson transition is expected to take place in any dimension $d \geq 3$. According to the simple scaling theory sketched above, the transition point will shift to lower and lower g_c, requiring a more strongly scattering medium to observe localization, and thus a Ioffe–Regel criterion, Eq. (9.57), with a smaller constant.

Contrary to conventional phase transitions, the Anderson metal–insulator transition does not have a finite upper critical dimension above which fluctuations would be unimportant and critical exponents simply given by their mean-field values (Evers and Mirlin, 2008). This is compatible with the observation that as the dimension d increases, the zero of the β-function must shift more and more to the asymptotic $\ln(g)$-wing where the slope tends towards $s = 1$. Thus, from the scaling description it is tempting to surmise that the critical exponent tends towards $\nu = 1$ only continuously as $d \to \infty$. We will see in section 9.8 below that this observation is not only a theoretician's spleen but may be put to experimental testing.

9.4 Key numerical and experimental results

Over the past 50 years, a wealth of numerical and experimental results has been accumulated on localization phenomena, especially on Anderson localization in dimension 1, 2, 3 and beyond. In the following, we present a selection, necessarily subjective and limited, of the most remarkable results.

9.4.1 $d = 1$

Anderson localization is a generic feature in phase-coherent 1d and quasi-1d systems, as explained in Section 9.2.2.3 above. Any amount of disorder, even very small, will eventually localize a wavepacket, independently of how large its energy is. Of course, the localization length can be huge if the energy is large compared to the disorder; see Section 9.5.2.4 for a quantitative estimate.

9.4.1.1   Localization of cold atoms

Concerning the experiment described in Section 9.1.1, there is thus no surprise that a quasi-1d atomic wavepacket displays localization in an optical speckle potential. Figure 9.8(a) shows the experimentally measured spatial shape of the wave packet at various times. One clearly distinguishes an exponential decrease in the wings, from which a localization length is extracted by a fit to $\exp\{-2|z|/L_{\text{loc}}\}$. As shown in Figure 9.8(b), this localization length first increases with time, then settles for a stationary value after about 500 ms. In the stationary regime, the wavepacket displays spatial fluctuations that are different for each single realization of the disorder. In addition, there remains a large fraction of the atoms still trapped near the original location of the wave packet.

How can we understand these experimental results? The initial wave packet is not monochromatic at all: it contains plane waves with a large dispersion in the wavevector k and consequently in the kinetic energy $\hbar^2 k^2/2m$ (the added optical potential also contributes to the total energy, but is a small correction here). The initial, free expansion of an interacting Bose–Einstein condensate released from a harmonic trap leads to a population of the various k classes that is given by an inverted parabola (Sanchez-Palencia *et al.*, 2007):

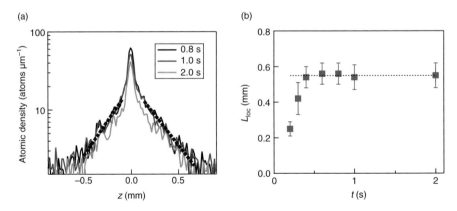

Fig. 9.8   (a) The atomic density of a BEC expanding in a quasi-1d optical speckle potential, shown in logarithmic scale at various times, displays clear exponential localization in the wings, from which the localization length is extracted by a fit (dashed line). (b) The localization length first increases linearly with time, then saturates in the stationary regime. Reprinted from (Billy *et al.*, 2008) (courtesy of Ph. Bouyer).

$$\Pi_0(k) = \frac{3(k_{\text{max}}^2 - k^2)}{4k_{\text{max}}^3} \Theta(k_{\text{max}} - |k|), \tag{9.62}$$

where k_{max} is the maximum k value, related to the initial chemical potential by $\mu = \hbar^2 k_{\text{max}}^2 / 2m$.

Because the disordered is "quenched" or stationary, energy is conserved, and each k-component of the wave packet evolves independently. When averaged over time, the interference terms between different energy components will be smoothed out, leaving the averaged wave packet as the *incoherent* superposition of all energy components. From Section 9.2.2.3, we know that each k-component localizes with a localization length $\xi_{\text{loc}}(k)$ equal to twice the transport mean free path. For the fastest atoms, this localization length is much larger than the initial spatial extension of the wave packet. Thus, we predict the stationary spatial distribution once localization sets in to be roughly given by

$$\langle |\psi(z)|^2 \rangle = \int_{-k_{\text{max}}}^{k_{\text{max}}} \frac{\Pi_0(k)\xi_{\text{loc}}(k)}{2} \exp\left(-\frac{|z|}{\xi_{\text{loc}}(k)}\right) dk. \tag{9.63}$$

Since $\xi_{\text{loc}}(k)$ is an increasing function of $|k|$, see Section 9.5.2.4, the asymptotic decrease at large distance is dominated by the largest $|k|$ values, such that $\langle |\psi(z)|^2 \rangle \propto \exp[-|z|/\xi_{\text{loc}}(k_{\text{max}})]$. The low-$k$ components have short localization lengths and thus produce the large bump near the origin in the final density. Using only the wings of the experimentally measured density, it is possible to estimate the localization length, for which we derive in Section 9.5.2.4 a theoretical prediction.

9.4.1.2 *Localization of light: a ten-Euro experiment*

The reasoning in Section 9.2.2.3 is entirely based on the construction of a 2×2 transfer matrix that can be chained; any randomness in the transfer matrix then leads to localization. The fact that our starting point was a quantum matterwave and the underlying wave–particle duality of quantum mechanics are not central to this argument. Indeed, the transfer-matrix description applies to all physical situations governed by a 1D (or quasi-1D) linear wave equation. Consequently, Anderson localization has been observed for many different types of non-quantum waves: microwaves, elastic waves in solids, optical waves, to cite a few.

Even a poor man's experiment using viewgraph transparencies, i.e. plastic films made of polyester carbonate, allows one to observe Anderson localization. A stack of several transparencies parallel to each other, separated by air layers of randomly varying thickness, realizes the simple model shown in figure 9.2. The transmission and reflection coefficients for an individual film can be computed from its index of refraction and its thickness. If the randomness in the film spacing is larger than an optical wavelength, we have a truly disordered system. Transport and localization can be observed by illuminating the stack of transparencies with a plane light wave (or rather a good approximation of a plane wave, namely, the light from a simple commercial He-Ne laser) and recording the transmission. Figure 9.9 shows the transmission vs. the number N of films or thickness of the sample. It displays a clear exponential decay,

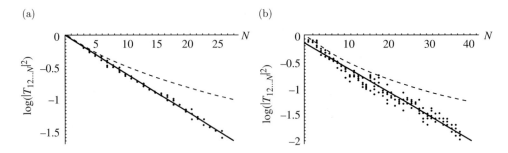

Fig. 9.9 Logarithmic transmitted intensity across stacks of N plastic films with mean thickness (a) 0.25 mm; (b) 0.1 mm. Dots: experimental data. Full curve: best fit to the data. Dashed curve: prediction of incoherent transmission (Ohm's law). Reprinted from (Berry and Klein, 1997) (courtesy of M.V. Berry).

one of the signatures of Anderson localization, and markedly differs from the linear decay (Ohm's law) predicted for the incoherent transport of intensities.

It is important to realize that absorption in the transparencies would also result in an exponential decay of the transmitted intensity. In all Anderson localization experiments, it is crucial to ensure that absorption is negligible, especially when working with electromagnetic waves. In this specific case, the bulk absorption coefficient of polyester carbonate is known to be negligible here. In principle, one could also double-check that no photon is lost by measuring the reflection coefficient of the sample and verifying that $R + T = 1$. In condensed-matter experiments with electrons or cold atoms, number conservation of massive particles makes absorption less relevant and simpler to monitor.

Another crucial requirement in the experiment is that it remains a 1D system, i.e. that there is a single transverse electromagnetic mode involved. Light polarization is not an issue (for perpendicular incidence, scattering is independent of polarization), but surface roughness or lack of parallelism can cause scattering into other transverse modes. This coupling into loss channels eventually destroys localization. Some indication of this loss is visible in Figure 9.9(b): the decay is not really exponential, but bent upwards, towards the prediction for incoherent transmission.

The experimental data show fluctuations of the transmission for various realizations of the experiment. This is not surprising, on the contrary, according to Section 9.2.5, the fluctuations are even expected to be large. However, it turns out that the observed fluctuations are smaller than predicted: in the plot, the transmission logarithm should appear as a cloud of points whose variance, Eq.(9.33), increases like N, which is clearly not the case. Most probably, this is due to the experimental imperfections mentioned above that couple several transverse modes and consequently attenuate the fluctuations.

9.4.1.3 *Fluctuations*

As already emphasized several times, the existence of large fluctuations of the transmission in a characteristic feature of Anderson localization. A key advantage of measuring relative fluctuations is that they are not much affected by absorption, which merely induces a global decay of the whole transmission distribution. Consequently, in the last few years, much progress has been made in calculating and measuring fluctuations in diffusive and localized systems. Fluctuations provide us with an unambiguous way of characterizing Anderson localization, even in the presence of absorption.

In order to illustrate this claim, we show in Figure 9.10 the transmission of microwaves across a quasi-1D sample composed of aluminum spheres randomly disposed in a long copper tube (cooled with liquid nitrogen so that absorption is negligible), as a function of the microwave frequency (Genack and Chabanov, 2005). The transport mean free path depends on the resonant scattering cross-section of the aluminum spheres, and thus varies strongly with frequency, implying large changes in the relative sample length $t = L/l$. Plot (a) is obtained in the diffusive regime, where the localization length is longer than the sample size: there, the transmitted intensity fluctuates in an apparently random way, but the fluctuations are relatively small, the rms deviation being comparable to the mean. This is expected in the diffusive regime for relatively short samples, where the transmission amplitude itself is expected to behave like a complex random number, whose real and imaginary parts are independent, normally distributed variables. In contrast, in the localized regime shown in plot (b), the fluctuations are much larger, the transmission being most of the time very small with some rare events of exceptionally high transmission, as predicted in Section 9.2.5.

Visual inspection reveals immediately that plots (a) and (b) are obtained in different regimes. While plot (a) has relatively small fluctuations, characteristic of a diffusive regime, where the fluctuations are comparable to the mean, plot (b) suggests some kind of huge (log-normal) fluctuations, typically associated with the localized or critical regime. The take-home message here is: don't rely solely on exponential decay

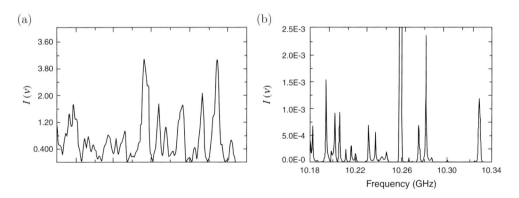

Fig. 9.10 Microwave intensity transmitted across a copper tube filled with scattering aluminum spheres vs. the microwave frequency. In the diffusive regime (a) (frequency around 17 GHz), fluctuations are comparable to the average value. In the localized regime (b), huge fluctuations are visible, a hallmark of Anderson localization. Reprinted from (Genack and Chabanov, 2005) (courtesy of A. Z. Genack).

to prove the existence of localization, look also at the fluctuations, they are better indicators. Since the relative fluctuations are insensitive to moderate absorption, they may even provide a quantitative criterion whether the strong localization threshold has been reached or not, and this under circumstances when the exponentially decreasing transmission alone could not be a reliable signature (Chabanov et al. , 2000; Chabanov and Genack, 2001).

9.4.1.4   The Anderson model: a free (numerical) experiment

Although the Anderson model was originally introduced as a tight-binding model for electrons in a disordered crystal, it is of broader interest and has become a paradigm for one-body localization effects.

   Up to now, we have considered continuous models where a wave propagates along a continuous 1D axis, encountering a set of discrete objects that scatter the wave backward and forward. The transfer-matrix game is just to combine the discrete scatterers with the proper phases. One can even go one step further, disregard the ballistic propagation, and build a completely discrete model, where the wave lives on a 1D lattice. One of the simplest discrete models is certainly the Anderson model whose Hamiltonian is

$$H = \sum_{n=-\infty}^{+\infty} (w_n |n\rangle\langle n| + t|n\rangle\langle n+1| + t|n+1\rangle\langle n|), \tag{9.64}$$

where the state $|n\rangle$ is the occupation amplitude of site n with w_n its on-site energy. t is the so-called "tunnelling" matrix element coupling neighboring sites, traditionally taken with numerical value $t = 1$. In a concrete physical realization, typically also the ts are random variables and thus define "off-diagonal disorder". However, it is enough to take diagonal disorder to observe localization. The precise value of t is irrelevant, as long as it is not zero, and may be used to define the energy scale of the problem. If $w_n = 0$, it is easy to check that the eigenstates are discrete plane Bloch waves $\psi_n = \exp ikn$, for $k \in [-\pi, \pi[$, and the energy is given by the dispersion relation $E(k) = 2\cos k$ of this single-band model.

   Disorder is introduced by allowing the on-site energies w_n to be random variables. The standard choice is to take w_n to be uncorrelated random variables uniformly distributed in the interval $[-W/2, W/2]$, with $W^2 = 12\langle w_n^2\rangle$ measuring the disorder strength. A noteworthy property—simplifying analytic calculations—is that the spatial correlation length of the disorder is zero (see Section 9.5.1.4 for a discussion of spatial correlations arising in optical speckle potentials). Actually, the 1D tight-binding Anderson model can be solved for almost any distribution (Luck, 1992), with simple closed expressions for the Cauchy-Lorentz on-site distribution. The techniques developed in Section 9.5 permit us to show that the localization length is at lowest order in W given by

$$\xi_{\text{loc}} = \frac{4\sin^2 k}{\langle w_n^2\rangle} = \frac{12(4 - E^2)}{W^2}. \tag{9.65}$$

Exercise 9.4 Few properties of the Anderson model.

(*i*) Show that the equations obeyed by an eigenstate of the Anderson model at energy E can be put in the following matrix form:

$$\begin{pmatrix} \psi_{n+1} \\ \psi_n \end{pmatrix} = T_n \begin{pmatrix} \psi_n \\ \psi_{n-1} \end{pmatrix}, \tag{9.66}$$

with a transfer matrix:

$$T_n = \begin{pmatrix} E - w_n & -1 \\ 1 & 0 \end{pmatrix}. \tag{9.67}$$

Show that the amplitudes of the left and right propagating plane waves in a disorder-free region can be expressed as simple linear combinations of ψ_n and ψ_{n+1}. Show that, in consequence, it is possible to construct a transfer matrix M as in Eq. (9.9). This shows that the general results of Section 9.2.2 can be used and that exponential localization is expected.

(*ii*) Consider a continuous model of a 1D particle in a disordered potential $V(z)$. By discretizing the Schrödinger equation on a lattice with sufficiently small spacing (much shorter than the de Broglie wavelength and the correlation length of the potential), show than one recovers the Anderson model, but with spatially correlated w_n.

Numerical simulations of the Anderson model are extremely easy, at least in dimension 1. Indeed, the previous exercise shows that the time-independent Schrödinger equation reduces, for an eigenstate $|\psi\rangle = \sum_n \psi_n |n\rangle$ with energy E, to the three-term recurrence relation

$$\psi_{n+1} + (w_n - E)\psi_n + \psi_{n-1} = 0 \tag{9.68}$$

which can be solved recursively. In Figure 9.11, we give an example of a simple script, written in the Perl language, that solves this equation at some arbitrary energy across a random sample of arbitrary length.

Note that the boundary condition used, $\psi_N = 1$ and $\psi_{N+1} = e^{ik}$, describes a purely outgoing wave with wavevector k and amplitude 1 on the right end of the sample. The boundary condition on the left is actually more complicated, because there the incident wave interferes with the reflected wave, whose amplitude depends on the microscopic realization of disorder of the entire sample. The Schrödinger recursion equation is thus better solved backwards from the right end of the sample, yielding on average an exponentially *increasing* solution toward the left. This is in agreement with the fact, shown in Exercise 9.2(*iv*) above, that the two eigenvalues of the transfer matrix are of the form $\lambda_\pm = e^{\pm 2x}$. So starting with this boundary condition and an arbitrary value of k (and thus E), one has, with probability one, a finite overlap with the eigenvectors of the larger eigenvalue and therefore numerically picks up an exponentially growing solution. This solution is then at the same time physically acceptable for the trans-mission experiment, namely, decreasing on average exponentially from left to right. Since Eq. (9.68) is linear, one can always normalize the solution to unit incoming

```perl
#!/usr/bin/perl
use Math::Trig;
use Math::Complex;
# compute_log_psi2_and_log_T.pl
# Author: Dominique Delande
# Release date: Feb, 24, 2010
# License: GPL3
# —————————————————————————————————————————————————————————————————————————————
# This script models Anderson localization in the Anderson model of disordered 1d systems.
# It computes the transmission across a sample of size $system_size at energy $energy.
# Without disorder, the spectrum is [−2,2], the dispersion relation $energy=2*cos($k).
# Disorder is given by the on−site energies $w_n,
# uncorrelated and uniformly distributed in [−$W/2,$W/2].
# The equations are:  \psi_{n+1} + \psi_{n−1} + (w_n−E) \psi_n = 0
# They are solved in the backward direction, starting from a normalized outgoing wave.
# |\psi|**2 is printed in file logpsi2.dat and −log(T) in logT.dat
# for $number_of_realizations independent realizations.
# The localization length is 12*(4−$energy**2)/$W**2 at lowest order in $W.
# —————————————————————————————————————————————————————————————————————————————
$system_size=1250;
$W=0.6;
$energy=0.5;
$k=acos(0.5*$energy);
$exp_i_k=cplx(cos($k),sin($k));
$number_of_realizations=3; # put 1000 or 10000 for a decent histogram
open(LOGT,">_logT.dat");   open(LOGPSI2,">_logpsi2.dat");
for ($j=1;$j<=$number_of_realizations;$j++) {
  $psi_n_plus_1=$exp_i_k;
  $psi_n=1.0;
  select(LOGPSI2);
  for ($n=$system_size;$n>0;$n−−) {
    $w_n=$W*(rand()−0.5);
    $psi_n_minus_1=($energy−$w_n)*$psi_n−$psi_n_plus_1;
    $logpsi2= 2.0*log(abs($psi_n));
    print "$n_$logpsi2\n";
    $psi_n_plus_1=$psi_n;
    $psi_n=$psi_n_minus_1;
  }
  print "_\n";
  $psi_n_minus_1=$energy*$psi_n−$psi_n_plus_1;
  $reflected=(0.5*abs($psi_n_minus_1*$exp_i_k−$psi_n)/sin($k))**2;
  $incident=(0.5*abs($psi_n_minus_1−$exp_i_k*$psi_n)/sin($k))**2;
  $minus_log_T=log($incident);
  select(LOGT);   print "$j_$minus_log_T\n";
}
```

Fig. 9.11 This perl script, available also at http://www.spectro.jussieu.fr/-Systemes-desordonnes-, can be run in a few seconds on any reasonable computer with a perl interpreter, preferably running a Unix-based operating system, or using a free perl interpreter for the non-free M$-Windoze operating system, such as http://strawberryperl.com/. Just type `perl compute_log_psi2_and_log_T.pl` in a shell window. The logarithm of the squared wave functions will be output (vs. position) in the file `logpsi2.dat` and can be visualized with your favorite plotting tool. (Minus) the logarithms of the transmissions (one for each realization of the disorder) are in the file `logT.dat`; use again your favorite plotting tool to build a histogram.

flux and thus finally find the transmission probability as the outgoing flux on the right side, calculated as a linear combination of ψ_N and ψ_{N+1} (cf. the perl script and Exercise 9.4(i)).

Figure 9.12 (left plot) shows the intensity $\ln|\psi_n|^2$ for three different realizations of the disorder at energy $E = 0.5$ and disorder strength $W = 0.6$ for a moderately

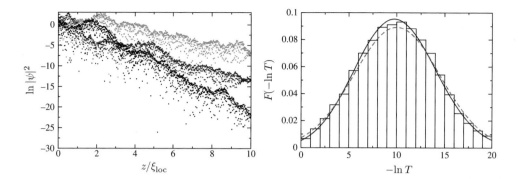

Fig. 9.12 Results of numerical experiments on the 1D Anderson model, with uncorrelated uniform distribution of disorder. The left plot shows the intensity $|\psi(z)|^2$ inside the disordered medium—on a logarithmic scale—for three different realizations of the disorder. Note the overall exponential decrease, decorated by huge fluctuations: the transmission across a sample of size $L = 10\xi_{\rm loc}$ fluctuates by more than 3 orders of magnitude. This illustrates why the typical transmission differs from the average one. The right plot shows the full probability distribution $F(-\ln(T), t)$ (histogram over 10 000 realizations) for $t = z/\xi_{\rm loc} = 10$, together with the prediction Eq. (9.36) (full red line), which is close to a truncated Gaussian (dashed).

large sample of 1250 sites, corresponding to a length $L = 10\xi_{\rm loc}$. Although the decay is on average exponential, huge fluctuations from one realization to another are visible with the naked eye. Moreover, even for this pure transmission experiment from left to right, the intensity is not monotonously decreasing at all, sometimes increasing by factors larger than 10. The histogram of the extinction (or transmission logarithm) over 10 000 realizations is shown in the right plot. Its width clearly visualizes the huge fluctuations characteristic for the localized regime. The agreement with the predicted distribution function, Eq. (9.36), shown with a red full line, is excellent. One also observes the convergence toward the truncated normal distribution (dashed), implying the truncated log-normal distribution (9.39) for T itself.

The reader is strongly encouraged to play with this script to experiment personally with Anderson localization. We recommend the following numerical experiments:

Exercise 9.5 Numerical study of 1D Anderson localization

(i) Use a single realization of the disorder and look at the wave function (or rather $|\psi|^2$) inside the medium. Try different sample lengths and different energies, but avoid the band center $E = 0$. Indeed, the Anderson model is singular at this value. There is still Anderson localization, but the localization length slightly differs from Eq. (9.65), a so-called Kappus–Wegner singularity (Luck, 1992).

(ii) Using a few hundred or thousand realizations, compute the statistical distribution of the transmission. Compare with the exact prediction Eq. (9.36) as well as with a truncated normal distribution.

(*iii*) Modify the script, for example for a Gaussian or Cauchy distribution of disorder and run additional numerical experiments. You may also introduce correlated disorder to simulate e.g. cold atoms in a speckle potential (see (Lugan *et al.*, 2009) for generation of a realization of the disorder with proper correlation functions).

A slightly different approach consist in diagonalizing the Hamiltonian for a large system numerically. Choosing strict boundary conditions on both ends of the sample yields normalizable eigenstates, centered at random positions within the sample and decreasing from there in both directions (similar to Figure 9.8(a)), together with the discrete set of corresponding eigenvalues. In the thermodynamic limit, these eigenenergies form a dense, but still pure-point spectrum.

An important message to keep in mind is that there is very little difference between a continuous and a discrete system, as far as localization on large spatial scales is concerned. For example, the exercise 9.4.(ii) shows how a particle in a continuous 1d random potential (e.g., an optical speckle) can be mapped to a variant of the Anderson model. This universality of the Anderson model cannot really surprise because localization is an asymptotic property taking place at large distance; whether the underlying configuration space is discrete or continuous plays only a minor role.

9.4.2 $d = 2$

Scaling theory predicts $d = 2$ to be the lower critical dimension for Anderson localization. In dimension $d = 2 + \epsilon$ (which can be numerically studied by constructing an Anderson model on a fractal set), a critical point should exist where $\beta(g_c) = 0$, separating a diffusive phase from an insulating one. Strictly at $d = 2$, scaling theory predicts localization provided there is a weak localization correction with $c_2 > 0$, see Section 9.3.6. We will see in Section 9.7 that this is indeed what a microscopic approach predicts in spinless time-reversal invariant systems. Scaling theory does not pretend to be an exact theory, there is thus a real interest in knowing whether there is localization in 2 dimensions for specific systems.

Experiments with cold atoms are expected to be much more difficult than in 1D. Indeed, the localization length, Eq. (9.56), is predicted to increase *exponentially* with the parameter kl, instead of linearly in 1D. Detailed theoretical studies (Kuhn *et al.*, 2007) have shown that experimental observation requires at the same time a speckle potential with a very short correlation length (comparable to what has been done in 1D, but in 2 directions) and a long atomic de Broglie wavelength, that is very cold atoms. Altogether, satisfying all conditions is far from easy, making 2d Anderson localization of ultracold atoms an interesting challenge.

A metal–insulator transition has been observed for electrons in clean semiconductor samples (Kravchenko *et al.*, 1994). It is generally acknowledged that the Coulomb electron–electron interaction—much stronger than the atom-atom interaction in a dilute cold atomic gas—plays a major role in this transition, which is thus qualitatively different from the pure Anderson transition and sometimes referred to as the Mott–Anderson transition (Belitz and Kirkpatrick, 1994).

Other types of waves have been successfully used in 2D systems. For example, using conveniently engineered optical fibers, one can create a 2D "photonic lattice" composed of parallel optical guides along which the light can freely propagate. Thanks to the photorefractive material used, its index of refraction can be adjusted by an external light source. Also, the transverse coupling between the optical guides can be adjusted at will, as well as the disorder due to small variations of the index of refraction in each guide. As the light propagates at roughly constant velocity along the guides, the spatial propagation mimics the temporal evolution of the Anderson model, each guiding mode playing the role of a site. Using such a device, the evolution from ballistic motion (on a scale shorter than the mean free path) to diffusive motion and eventually to strong localization has been experimentally observed (Schwartz *et al.*, 2007), see Figure 9.13.

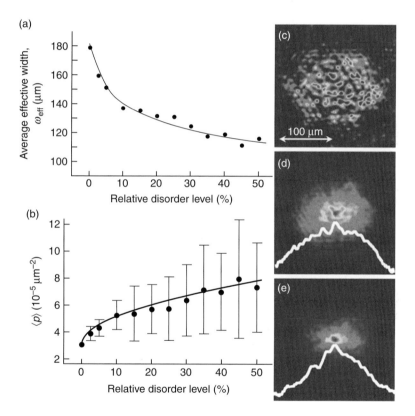

Fig. 9.13 Experimental results on the propagation of light across a transversally disordered 2D lattice of photonic waveguides, mimicking the 2D Anderson model. As the strength of the disorder is increased, the dynamics evolves from ballistic (a and c) to diffusive (b and d), characterized by a Gaussian shape of the wave packet, and eventually to Anderson localization, with a wave packet of characteristic exponential shape (e), when the disorder is sufficiently strong to make the localization length comparable to the extension of the wave packet. Reproduced from (Schwartz *et al.*, 2007) (courtesy of S. Fishman).

The Anderson model itself, described in Section 9.4.1.4, can be trivially extended to any dimension by adding hopping terms to nearest neighbors in a (hyper)cubic lattice. The numerical study is slightly more difficult than in 1D. The basic idea is to study first the quasi-1D propagation on a strip with a fixed number M of transverse sites, imposing for example periodic boundary conditions along this direction. One can write a $2M \times 2M$ transfer matrix for this quasi-1D system and calculate its asymptotic properties as the length N goes to infinity, extracting the quasi-1d localization length $\xi_{\mathrm{loc}}(M)$. Next, one studies the behavior of $\xi_{\mathrm{loc}}(M)$ as M is sent to infinity. If $\xi_{\mathrm{loc}}(M)$ diverges without bounds, one concludes that the system is not localized. If, on the other hand, $\xi_{\mathrm{loc}}(M)$ tends to a finite limiting value, one concludes that the system localizes with $\xi_{\mathrm{loc}} = \lim_{M \to \infty} \xi_{\mathrm{loc}}(M)$.

Powerful numerical techniques, such as finite-size scaling (Fisher and Barber, 1972), make it possible to extrapolate properties of the infinite system from numerical experiments on limited systems. In particular, the scaling function $\beta(g)$ can be reconstructed, see Figure 9.14. The fact that various data, computed for various values of the system parameters (energy, disorder strength, system size), lead to the very same $\beta(g)$ strongly indicates that the scaling approach is valid, and thus corroborates the existence of universal properties independent of the microscopic details. In 2D, the numerically computed $\beta(g)$ is always negative, as expected and its shape is in good agreement with the naive prediction, Eq. (9.53).

9.4.3 $d = 3$

Dimension 3 is arguably the most interesting, because scaling theory there predicts a transition between diffusive behavior for small disorder and Anderson localized behavior at large disorder. Consequently, much experimental and numerical effort has been spent to observe this Anderson transition. Numerical simulations of the 3D Anderson model are a very valuable tool, especially to locate the critical point where $\beta(g_c) = 0$ and to characterize its vicinity. The results in Figure 9.14 very clearly show the existence of the two regimes and the fact that $\beta(g)$ behaves smoothly across the transition. This constitutes a clear-cut proof that the Anderson transition is a continuous phase transition of second order. Note the absence of data for g just below g_c; this corresponds to localized systems with a localization length too large to be reliably measured in the numerical simulations. As shown in Section 9.3.7, the slope $\mathrm{d}\beta/\mathrm{d}\ln g|_{g_c}$ at the critical point is the inverse of the critical exponent ν of the Anderson transition. Although the slope at the critical point cannot be accurately measured on these data, it is without any doubt smaller than unity—the value of the asymptotic slope in the deep localized regime $\ln g \to -\infty$. This implies that the critical exponent ν is larger than unity. Recent numerical studies on much larger systems fully confirm this point, the current best estimate being $\nu = 1.58 \pm 0.01$ (Slevin and Ohtsuki, 1999; Lemarié *et al.*, 2009).

Direct experimental observation of Anderson localization in 3D is even more difficult than in 2D, because it requires an even more strongly scattering system (kl smaller than 1 from the Ioffe-Regel criterion, Eq. (9.57), instead of kl of the order of few units). Moreover, creation of a sufficiently disordered potential can be technically

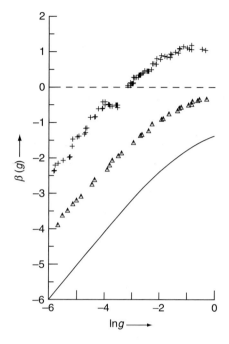

Fig. 9.14 Scaling function $\beta(g)$ reconstructed from numerical simulations of the Anderson model. The solid line is for dimension 1, the triangles for dimension 2 and the crosses for dimension 3. Different sets of microscopic parameters produce data lying on the same curve, which can be considered an "experimental proof" that the scaling approach is valid. In 2D, $\beta(g)$ is always negative, proving that the system is in the localized regime. In 3D, depending on the disorder strength, the system may be localized ($\beta(g) < 0$, strong disorder) or diffusive ($\beta(g) > 0$, weak disorder). Reprinted from (MacKinnon and Kramer, 1981) (courtesy of A. McKinnon and B. Kramer).

much more difficult in 3D: for a speckle potential, this would require to send plane waves with random phases from a large solid angle. Thus, Anderson localization of atomic matterwaves in a disordered potential has not yet been observed. However, using the equivalence of a quasi-periodically kicked rotor with a 3D Anderson model, the Anderson transition with atomic matterwaves has been observed, and its critical exponent experimentally measured, as discussed in Section 9.8.

Electronic transport in solids, the field where localization theory was originally developed, provides also interesting experimental results. Metal–insulator transitions can be observed in solid-state samples, but it is never easy to identify the microscopic mechanism. This is because electron–electron interactions play an essential role. Whether the observed transition is a one-body effect like the Anderson transition or a many-body one like the Mott transition (Greiner, 2002) is not easily proved. We are not aware of any unambiguous observation of the pure Anderson transition. Figure 9.15 shows the experimentally measured conductivity of a Si-doped AlGaAs

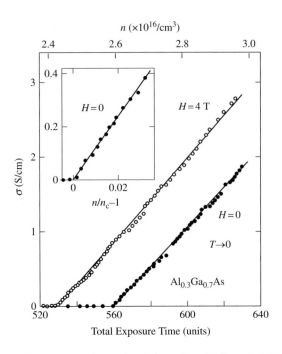

Fig. 9.15 Experimentally measured conductivity of a Si-doped AlGaAs 3D crystal vs. electron concentration (upper horizontal scale), showing a clear metal–insulator transition. The critical exponent is very close to unity, a value not compatible with the universal value $\nu = 1.58$ of the pure 3D Anderson transition. Electron–electron interaction is probably responsible for the difference. Adapted from (Katsumoto *et al.*, 1987) (courtesy of S. Katsumoto).

crystal vs. a parameter essentially representing the electronic Fermi energy. A clear insulator-to-metal transition is observed. It seems that the curve is almost linear in the metallic regime, which—because conductivity is essentially a measure of the diffusion constant—means that the measured critical exponent is $\nu \approx 1$. This markedly differs from the exponent of the pure Anderson transition, indicating that interaction effects are probably important.

One may also turn to other types of waves, for example ultrasonic waves (Hu *et al.*, 2008) or electromagnetic waves. Direct measurement of the electromagnetic field inside the disordered medium is not straightforward, and transmission experiments are easier. As mentioned earlier, absorption induces an exponential decay of the intensity, which must be carefully discriminated from the same effect being produced by Anderson localization. Thus, experimentalists have turned to measuring tell-tale properties right at the critical point. There, according to scaling theory, the dimensionless conductance has the constant value g_c, independently of the system size, whereas the classical dimensionless conductance, Eq. (9.45), increases linearly with the system size L. This additional power of L makes the total transmission across the sample evolve from a $1/L$ behavior (Ohm's law) in the diffusive regime to a $1/L^2$ scaling law at the critical point, and eventually to the exponential decay in the localized regime. Any spurious absorption is likely to transform the critical $1/L^2$ behavior into an exponential decrease. Thus, the existence of an $1/L^2$ may be considered a sensitive test of observing the Anderson transition. Figure 9.16 shows the experimental result obtained on the propagation of microwaves in a disordered medium, in the diffusive and critical regimes. The existence of a range with $1/L^2$ power law—before absorption wins at even larger size—is a convincing proof.

Similar results have been obtained in the optical regime (Wiersma *et al.*, 1997), where strong scattering is provided by oxide powders, but the role of absorption has been discussed controversially (Scheffold *et al.*, 1999). Recently, time-resolved transmission experiments, where absorption has less impact, have shown a slowing down of classical transport (Störzer *et al.*, 2006), which gives strong evidence for Anderson localization.

In the last few years, several numerical and laboratory experiments have characterized the fluctuations appearing in the vicinity of the Anderson transition. In particular, numerical experiments on the 3D Anderson model have shown that the critical eigenstates have a multi-fractal structure, implying the coexistence of regions where the wave function is exceptionally large together with regions where it is exceptionally small. This is presently a very active field of research (Faez *et al.*, 2009), whose description is beyond the scope of this chapter. The reader may refer to the recent review paper of Evers and Mirlin (Evers and Mirlin, 2008). In the near future, it is very likely that experiments on localization of atomic matterwaves will concentrate on the existence and properties of fluctuations.

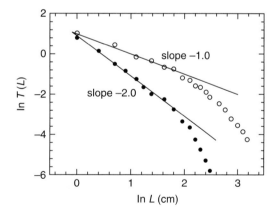

Fig. 9.16 Experimentally measured transmission of microwaves through a 3D strongly scattering medium vs. the size of the medium (doubly logarithmic scale). In the usual diffusive regime (open symbols), a $1/L$ decrease is observed, in agreement with classical transport theory (Ohm's law). At the critical point of the Anderson transition (filled symbols) a characteristic $1/L^2$ behavior is observed, in agreement with the scaling theory of localization. Note that, because of residual absorption, the signal drops at large size. Reprinted from (Genack, 2001) (courtesy of A.Z. Genack).

9.5   Microscopic description of quantum transport

9.5.1   Diagrammatic perturbation theory

In Section 9.2, we have seen that localization in (quasi-)one-dimensional systems can be very efficiently described by a transfer-matrix approach. In higher dimensions, we have resorted to the scaling arguments presented in Section 9.3. We now wish to give an introduction to a microscopic description of quantum transport in disordered systems. The main advantage of diagrammatic perturbation theory lies in its versatility. It applies in arbitrary dimensions d and to any model with a Hamiltonian of the form

$$H = H_0 + V, \tag{9.69}$$

in which H_0 describes regular propagation in an ordered substrate, and V is the disorder potential that breaks translational invariance. On a microscopic level, "disorder" refers to degrees of freedom whose detailed dynamics are not of interest and whose properties are only known statistically. In the following, we will consider *static* or *quenched* disorder that remains frozen on the timescale of wave propagation under study (as sole exception of this rule, we mention in Section 9.6.3 the dephasing effect of moving impurities).

A first model of type (9.69) describes a single quantum particle in an external potential,

$$H = \frac{p^2}{2m} + V(r), \tag{9.70}$$

with direct bearing on experiments with non-interacting matterwaves like (Billy *et al.*, 2008), but equally applicable to other massive particles like electrons, neutrons, etc. Note that the plain Hamiltonian (9.70) operating in Hilbert space describes the same single-particle physics as the more fanciful many-body version

$$H = \int d^d r \Psi^\dagger(r) \left[-\frac{\hbar^2}{2m} \nabla^2 + V(r) \right] \Psi(r), \tag{9.71}$$

defined in terms of particle creation and annihilation operators $\Psi^{(\dagger)}$ in Fock space.

Since we assume that H_0 is translation-invariant (if only by discrete translation on a lattice), the following equivalent formulation in Fourier space is also useful:

$$H = \sum_k \varepsilon_k^0 a_k^\dagger a_k + \sum_{k,q} V_q a_{k+q}^\dagger a_k. \tag{9.72}$$

Here, wavevectors k are used as good quantum numbers labelling the eigenstates of H_0. If there is an underlying lattice, one has to include also a Bloch-band index. ε_k^0 is the free dispersion relation; for matterwaves, $\varepsilon_k^0 = \hbar^2 k^2 / 2m$. Particle annihilators $a_k = L^{-d/2} \int d^d r e^{ik \cdot r} \Psi(r)$ and creators $a_k^\dagger$ fulfill the canonical commutation relations $[a_k, a_{k'}^\dagger]_\pm = \delta_{kk'}$.

The disorder potential breaks translation invariance by scattering particles $k \to k'$ with an amplitude given by its Fourier component $V_q = \langle k+q|V|k \rangle = L^{-d} \int d^d r e^{-iq \cdot r} V(r)$, conveniently represented by

$$V_q = \quad k \quad \overbrace{\qquad}^{q} \quad k+q \, . \qquad (9.73)$$

This model Hamiltonian (9.72) is not limited to matter waves. By appropriate changes in ε_k^0, different physical systems can be described. For instance, photons and other massless excitations have a linear dispersion $\varepsilon = \hbar c k$ with characteristic speed c. Yet another realization is provided by the elementary excitations of Bose–Einstein condensates, featuring the Bogoliubov dispersion $\varepsilon_k = \sqrt{\varepsilon_k^0 (\varepsilon_k^0 + 2\mu)}$, that interpolates between a linear sound-wave dispersion at low energy and a quadratic particle-like dispersion $\varepsilon_k^0 = \hbar^2 k^2 / 2m$ at high energy. The general formalism to be introduced below applies to all these cases, provided the scattering potential V_q is known.

The basic model can be made richer, depending on the circumstances and effects one wishes to describe. For example, spin often plays an important role, for instance via spin-orbit effects, due to coupling of spin and direction of propagation. Also, spin-flip processes can be of interest, as in electronic spin-flips induced by magnetic impurities or photon polarization flips induced by Zeeman-degenerate atomic dipole transitions. A typical spin-flip process $(m, \sigma) \mapsto (m', \sigma')$ changes the spin of the propagating object from σ to σ', while the impurity spin undergoes $m \mapsto m'$. Processes of this type can store information about the path travelled, and generally act as a source of strong decoherence (as discussed in Section 9.6.3).

The main drawback of the diagrammatic Green function approach is its perturbative character. Most results are obtained from an expansion in powers of V and are valid only for small enough potential strength. If one is interested in truly strong disorder effects, it may be worthwhile to start from the opposite situation where the propagation described by H_0 is small; Anderson's original method of a "locator expansion" (Anderson, 1958) is an example of such a weak-coupling perturbative approach. In any case, it takes considerable effort to derive non-perturbative results using controlled approximations. Yet, the basic diagrammatic technique is a prerequisite for more powerful, field-theoretic methods involving, e.g., replica methods, renormalization-group analysis and supersymmetry (Efetov, 1983).

9.5.1.1 *Quantum propagator*

Let us then start by calculating the Green function for the single-particle Hamiltonian (9.70) that determines the time evolution of a state $|\psi\rangle$ in Hilbert space according to the Schrödinger equation $i\hbar\partial_t |\psi\rangle = H|\psi\rangle$. For $t > 0$, the forward-time evolution operator $G^{\mathrm{R}}(t) = -\frac{i}{\hbar}\theta(t)\exp\{-iHt/\hbar\}$ solves the differential equation

$$[i\hbar\partial_t - H]\, G^{\mathrm{R}}(t) = \delta(t). \qquad (9.74)$$

Obviously, $G^{\mathrm{R}}(t)$ is the retarded Green operator for the Schrödinger equation. It encodes the same information as its many-body version $G_{kk'}^{\mathrm{R}}(t) = -\frac{i}{\hbar}\theta(t)\left\langle \left[a_k(t), a_{k'}^\dagger\right]\right\rangle$ that one would use starting from Eq. (9.72); in the following, we stick to the simpler form, referring the reader to the literature for the more advanced presentation (Bruus and Flensberg, 2004; Negele and Orland, 1998; Mahan, 2000;

Altland and Simons, 2006). Going from time to energy by Fourier transformation, one defines the *resolvent*

$$G^{\mathrm{R}}(E) = \lim_{\eta \to 0^+} \int dt e^{i(E+i\eta)t/\hbar} G^{\mathrm{R}}(T) = \lim_{\eta \to 0^+} [E - H + i\eta]^{-1} =: [E - H + i0]^{-1}. \tag{9.75}$$

The limiting procedure $\eta \to 0^+$ guarantees that indeed the retarded Green operator is obtained, different from zero only for $t > 0$. The advanced Green operator $G^{\mathrm{A}}(t)$ is obtained by taking $\eta \to 0^-$. In the basis where H is diagonal, $H|n\rangle = \varepsilon_n|n\rangle$, the resolvent is also diagonal and thus admits the spectral decomposition $G(z) = \sum_n |n\rangle[z - \varepsilon_n]^{-1}\langle n|$ for any argument $z \in \mathbb{C}$ outside the spectrum of H. The resolvent's matrix elements are called "propagators". For example, in the position representation $\langle r|n\rangle = \psi_n(r)$,

$$G^{\mathrm{R}}(r, r'; E) = \langle r'|G^{\mathrm{R}}(E)|r\rangle = \sum_n \frac{\psi_n(r')\psi_n^*(r)}{E - \varepsilon_n + i0} = \quad r \longrightarrow r' . \tag{9.76}$$

This propagator contains precious information: As function of E, it has singularities on the real axis that are precisely the spectrum of H and thus encode all possible evolution frequencies. Furthermore, the residues at these poles provide information about the eigenfunctions.

The total Hamiltonian (9.70) contains the disorder potential so that we cannot write down its eigenfunctions and eigenvalues analytically (numerically, one may of course calculate eigenfunctions and eigenvalues for each realization of disorder). We start therefore with the free Hamiltonian H_0. Its resolvent $G_0(z) = [z - H_0]^{-1}$ is diagonal in momentum representation, $\langle k'|G_0(z)|k\rangle = \delta_{kk'}G_0(k, z)$ with

$$G_0^{\mathrm{R}}(k; E) = \frac{1}{E - \varepsilon_k^0 + i0} = \xrightarrow{\quad k \quad}. \tag{9.77}$$

Now we are ready to describe the perturbation due to the potential V: Using $G(E) = [E - H_0 - V]^{-1} = [(E - H_0)\{1 - (E - H_0)^{-1}V\}]^{-1}$, we express

$$G(E) = [1 - G_0 V]^{-1} G_0 \tag{9.78}$$

$$= G_0 + G_0 V G_0 + G_0 V G_0 V G_0 + \ldots \tag{9.79}$$

as the Born series in powers of V. For notational brevity, we have already suppressed the energy argument on the right-hand side. Still, if one tries to write out a matrix element $\langle k'|G(E)|k\rangle$, the operator products convert into cumbersome expressions that tend to obscure the series' simple structure:

$$\langle k'|G(E)|k\rangle = \delta_{kk'}G_0(k) + G_0(k')V_{k'-k}G_0(k)$$

$$+ \sum_{k''} G_0(k')V_{k'-k''}G_0(k'')V_{k''-k}G_0(k) + \ldots . \tag{9.80}$$

At this point, we are well advised to use the graphical representation known as "Feynman diagrams", for which we have already all ingredients at hand:

$$\langle k'|G(E)|k\rangle = \delta_{kk'}\underset{k}{\longrightarrow} + \underset{k}{\longrightarrow}\underset{k'}{\longrightarrow} + \underset{k}{\longrightarrow}\underset{k''}{\longrightarrow}\underset{k'}{\longrightarrow} + \dots \qquad (9.81)$$

Already, we achieve a much more compact notation, aided by the fact that we do not need to label the dangling impurity lines, defined in Eq. (9.73), since their momentum is automatically determined by the incident and scattered momenta. Also, we henceforth use the prescription that all internal momenta have to be summed over, here k'' in the last contribution.

9.5.1.2   Ensemble average

In principle, the Born series (9.81) permits us to calculate the full propagator perturbatively. However, the result will be different for each realization of disorder. We are really only interested in suitable expectation values and thus have to understand how to perform the ensemble average over the disorder distribution.

The potential $V(r)$ as a function fluctuating in space is a *random process*. As such, it can be completely characterized by its moments or correlation functions $\langle V_1\rangle$, $\langle V_1V_2\rangle$, $\langle V_1V_2V_3\rangle$, etc., with the short-hand notation $V_i = V(r_i)$. We will assume that the process is *stationary* or, preferring the spatial dictionary, *statistically homogeneous*, which means that correlation functions can only depend on coordinate differences $r_{ij} = r_i - r_j$. We can therefore define the following correlation functions and corresponding diagrams:

$$\langle V_1\rangle = \langle V\rangle \qquad (9.82)$$

$$\langle V_1V_2\rangle = P(r_{12}) \quad = \quad \underset{1}{\bullet}\quad\underset{2}{\bullet} \qquad (9.83)$$

$$\langle V_1V_2V_3\rangle = T(r_{12}, r_{23}) = \underset{1}{\bullet}\quad\underset{2}{\bullet}\quad\underset{3}{\bullet} \qquad (9.84)$$

and so on for arbitrary n-point correlation functions. Without loss of generality, one may always take $\langle V\rangle = 0$ by defining a centered potential $V \mapsto V - \langle V\rangle$ while redefining the zero of energy $E - \langle V\rangle \mapsto E$. In Fourier representation, these correlation functions are

$$P(q) = \overset{q}{\underset{}{\bullet}}\quad\bullet, \qquad T(q, q') = \overset{q\quad q'}{\underset{}{\bullet}}\quad\bullet\quad\bullet, \qquad \text{etc.} \qquad (9.85)$$

Depending on the specific type of disorder, these general correlation functions can take different forms, and it may be instructive to discuss two of them in detail.

9.5.1.3   Gaussian disorder

As a first example, let us consider Gaussian-distributed disorder that is completely defined by its first two moments $\langle V \rangle = 0$ and $P(r) = V_0^2 C(r)$. Here, one conveniently factorizes the one-point variance $V_0^2 = \langle V_1^2 \rangle$ from the spatial correlation function $C(r)$ that obeys $C(0) = 1$ by construction. The characteristic property of a Gaussian process is that all higher-order correlation functions completely factorize into pair correlations. Indeed, a simple property of the Gaussian integral implies that the moments of a normally distributed, centered scalar random variable X are $\langle X^{2n} \rangle = C_n \langle X^2 \rangle^n$, where $C_n = (2n)!/(2^n n!)$ is the number of pairs that can be formed out of $2n$ individuals. Similarly, the Gaussian moment theorem applies to a Gaussian-distributed random potential:

$$\langle V_1 \cdots V_{2n} \rangle = \frac{1}{2^n n!} \sum_\pi \langle V_{\pi(1)} V_{\pi(2)} \rangle \cdots \langle V_{\pi(2n-1)} V_{\pi(2n)} \rangle \tag{9.86}$$

where π denotes the $(2n)!$ permutations. Pictorially, this implies also a complete factorization of $2n$-point potential correlation into products of pair correlations. The first interesting example is $n = 2$ with

$$\tag{9.87}$$

and so on for higher orders.

Such a Gaussian potential can be constructed with arbitrary spatial correlation $C(r)$. A popular choice here is often to model it as a Gaussian as well, $C(r) = \exp\{-r^2/2\sigma^2\}$, such as in (Hartung *et al.*, 2008), because this is easy to implement numerically (it suffices to draw uncorrelated random variables V_i on a discrete grid and convolute by a Gaussian correlation function afterwards). Moreover, this choice leads to simple analytical calculations because the k-space pair correlator is also Gaussian, $P(q) = V_0^2 \sigma^d (2\pi)^{d/2} \exp\{-q^2\sigma^2/2\}$. In the limit of low momenta $q\sigma \ll 1$, the potential details cannot be resolved and it appears δ-correlated. Then, everything can be expressed in terms of $P(0) = (2\pi)^{d/2} \sigma^d V_0^2$.

9.5.1.4   Speckle

A slightly more interesting example is provided by the optical speckle potential used recently for matterwave Anderson localization (Billy *et al.*, 2008; Clément *et al.*, 2006). The atoms are subject to an optical dipole potential $V(r) = K|E(r)|^2$ created by the local field intensity of far-detuned laser light. K contains the frequency-dependent atomic polarizability besides some constants (Allen and Eberly, 1987). With a laser beam that is blue-detuned from the optical resonance, one has $K > 0$ and thus expels atoms from high-intensity regions. This potential landscape features repulsive peaks with $\langle V \rangle > 0$. Conversely, a red-detuned laser leads to $K < 0$, and one finds a potential landscape with attractive wells and $\langle V \rangle < 0$. To create a disorder potential, the laser beam is focused through a diffuse glass plate, whose randomly positioned individual grains act as elementary sources for the emitted field. The electric field $E(r)$ at some far point then is the sum of a large number of complex amplitudes. By virtue of the central limit theorem, it is a *complex Gaussian random variable* with normalized pair correlator

$$\gamma_{ij} = \gamma(r_i - r_j) = \frac{\langle E^*(r_i)E(r_j)\rangle}{\langle |E|^2\rangle} = \overset{\cdots\cdots}{\underset{i \qquad j}{\circ}}, \tag{9.88}$$

with the obvious properties $\gamma_{ij}^* = \gamma_{ji}$ and $\gamma_{ii} = 1$. In a 1D-geometry, the pair correlator takes its simplest form in Fourier components:

$$\gamma(q) = \pi\zeta\Theta(1 - |q|/k_\zeta), \tag{9.89}$$

where $k_\zeta = k\alpha$ is the maximum wavevector that can be built from a monochromatic laser source with wavevector k seen under an optical aperture α. This Eq. (9.89) simply says that the random field contains all wavevectors inside the allowed interval with equal weight. In real space, this Fourier transforms to $\gamma(r) = \sin(r/\zeta)/(r/\zeta)$ with the correlation length $\zeta = 1/k_\zeta = 1/(\alpha k)$.

Other pair correlations such as $E_i E_j$ and $E_i^* E_j^*$ have uncompensated random phases and average to zero. The Gaussian moment decomposition now applies to arbitrary moments of the speckle disorder potential $V_i = K E_i^* E_i$. An n-point potential correlation is really a $(2n)$-field correlation, which decomposes into all possible pair correlations (9.88). As in a conventional ballroom dancing situation involving n couples, all possible heterosexual pairings between the E_i^*s and E_js are allowed. This gives for the 2-point potential correlator

$$\langle V_1 V_2\rangle = K^2\langle E_1^* E_1 E_2^* E_2\rangle = \langle V\rangle^2 \left[\gamma_{11}\gamma_{22} + \gamma_{12}\gamma_{21}\right]. \tag{9.90}$$

Setting $r_1 = r_2$ shows that $\langle V^2\rangle = 2\langle V\rangle^2$, which means that the potential variance is equal to its mean square, $\text{var}(V) = \langle V^2\rangle - \langle V\rangle^2 = \langle V\rangle^2$.

The shift $V \mapsto V - \langle V\rangle$ to the centered potential removes the first term in the bracket in Eq. (9.90). The same applies to all diagrams with field self-contractions:

$$\overset{\cdots}{\circledast} = 0. \tag{9.91}$$

So henceforth, we can neglect those diagrams by considering a centered potential $\langle V\rangle = 0$. Altogether, we have as a first building block the speckle potential pair correlator

$$\langle V_1 V_2\rangle = P(r_{12}) = V_0^2 C(r_{12}) = V_0^2 \underset{1 \qquad 2}{\circledast\overset{\cdots\cdots}{}\circledast}. \tag{9.92}$$

Here, the potential strength $V_0^2 = \text{var}(V)$ is factorized from the dimensionless correlation function $C(r) = |\gamma(r)|^2$ that is normalized to $C(0)=1$. In $d = 1$, from Eq. (9.89), we have the real-space intensity correlator $C(r) = [\sin(r/\zeta)/(r/\zeta)]^2$. In higher dimensions and in an isotropic setting, the Fourier transformation of the simple k-space field correlator yields $C(r) = [2J_1(r/\zeta)/(r/\zeta)]^2$ in $d = 2$ and $C(r) = [\sin(r/\zeta)/(r/\zeta)]^2$ again in $d = 3$ (Kuhn *et al.*, 2007).

An interesting effect occurs for potential correlations of odd order $(2n + 1)$. They are really field correlations of twice the order, which is even and thus different from zero. The first example of this kind is (since the fields $*$ and $\circ$ will always appear together, we note $\circledast = \bullet$ from now on)

$$\langle V_1 V_2 V_3 \rangle = V_0^3 2\text{Re}\{\gamma_{12}\gamma_{23}\gamma_{31}\} = V_0^3 \overset{\displaystyle \frown}{\underset{1 \quad 2 \quad 3}{\bullet \quad \bullet \quad \bullet}} . \tag{9.93}$$

Diagrams of this type can only contain closed loops of field correlations (because field self-contractions no longer appear). Since the loops can be closed both clockwise and counterclockwise, there are two contributions that are complex conjugates of each other.

9.5.1.5   *Average propagator: self-energy*

Now we are in position to take the ensemble average of the single-particle propagator (9.79):

$$\langle G \rangle = G_0 + G_0 \langle V G_0 V \rangle G_0 + G_0 \langle V G_0 V G_0 V \rangle G_0 + \dots, \tag{9.94}$$

or

$$\langle G \rangle = \text{——} + \text{——}\overset{\frown}{\bullet \quad \bullet}\text{——} + \text{——}\overset{\frown}{\bullet \quad \bullet \quad \bullet}\text{——} + \dots. \tag{9.95}$$

The precise form of potential correlations depends on the model of disorder. As shown by the example of the Gaussian model (9.87), starting from the fourth-order term there appear completely factorized contributions. Before writing all possible combinations down, we had better introduce one of the cornerstones of diagrammatic expansions: the *self-energy* $\Sigma(E)$ defined by the *Dyson equation*

$$\langle G \rangle = G_0 + G_0 \Sigma \langle G \rangle. \tag{9.96}$$

Introducing the self-energy invariably prompts the following frequently asked questions:

1. Why is the self-energy convenient for perturbation theory?
2. How do I calculate Σ?
3. What is the physical meaning of Σ?
4. Is there a simple example?

Let us answer them in turn.

1. By iterating the Dyson equation (9.96), one finds that the average propagator expands as

$$\langle G \rangle = \text{——} + \text{——}\textcircled{Σ}\text{——} + \text{——}\textcircled{Σ}\text{——}\textcircled{Σ}\text{——} + \text{——}\textcircled{Σ}\text{——}\textcircled{Σ}\text{——}\textcircled{Σ}\text{——} + \dots. \tag{9.97}$$

By construction, there are no disorder correlations between the different self-energies appearing here. In return, this implies that the self-energy contains *exactly all* correlations that cannot be completely factorized by removing a free propagator G_0

in between. These non-factorizable terms are called "one-particle irreducible" (1PI). Moreover, the self-energy contains only the correlations and internal propagators, but is stripped off the external propagator lines ("amputated"). This makes the self-energy the simplest object describing all relevant disorder correlations.

2. Due to statistical homogeneity, the self-energy is diagonal in momentum and thus only depends on k and E. The self-energy matrix element $\Sigma(k, E)$ is calculated by applying so-called Feynman rules to evaluate the diagrams. As a specific example, let us give the Feynman rules for the self-energy of the retarded single-particle propagator $\langle G^{\mathrm{R}}(k, E) \rangle$ in momentum representation for the case of the speckle potential:

(i) Draw all amputated 1PI diagrams with incident momentum k:

$$\Sigma(k, E) = \text{⚬⚬⚬⚬⚬} + \text{⚬⚬⚬⚬⚬} + \ldots \qquad (9.98)$$

(ii) Convert straight black lines to free propagators

$$\underset{k}{\longrightarrow} = G_0^{\mathrm{R}}(k, E) = [E - \varepsilon_k^0 + i0]^{-1}.$$

(iii) Convert disorder correlation lines to $\text{⚬⋯}\overset{q}{\dashrightarrow}\text{⋯⚬} = \gamma(q)$. The precise functional dependence $\gamma(q)$ depends on dimension and geometry.

(iv) For each scattering vertex, multiply by one power of the potential strength and the conservation of momentum:

$$\underset{k \quad k'}{\overset{q \diagdown \diagup q'}{\longrightarrow \quad \longrightarrow}} = V_0 \delta_{k+q, k'+q'}. \qquad (9.99)$$

(v) Sum over all free momenta after respecting momentum conservation.

For other types of disorder, these rules have to be adapted slightly to the precise shape of diagrams, correlation functions and vertex factors. But in all cases, the general idea of writing all possible combinations, respecting momentum conservation and integrating out the free momenta is the same.

3. One can rewrite the Dyson equation (9.90) as $[1 - G_0\Sigma]\langle G \rangle = G_0$ and solve formally for the average propagator: $\langle G \rangle = [1 - G_0\Sigma]^{-1}G_0 = [G_0^{-1} - \Sigma]^{-1}$. Thus, its matrix elements are

$$\langle G^{\mathrm{R}}(k, E) \rangle = \frac{1}{E - \varepsilon_k^0 - \Sigma(k, E)}. \qquad (9.100)$$

We recognize that the self-energy modifies the free dispersion relation. Generally, the self-energy is a complex quantity with a real as well as an imaginary part. The *modified dispersion relation*

$$E_k = \varepsilon_k^0 + \mathrm{Re}\Sigma(k, E_k) \qquad (9.101)$$

is an implicit equation for the new eigen-energy E_k of the mode k. So one effect of the disorder is to shift the energy levels.[4] But plane waves with fixed k are no longer proper eigenstates of the disordered system. This is encoded in the imaginary part. Writing $\Gamma_k = -2\text{Im}\Sigma(k, E_k)$ and using the fact that the self-energy varies smoothly with k and E, one finds a *spectral density*

$$A(k, E) = -2\text{Im}\langle G^{\text{R}}(k, E)\rangle = \frac{\Gamma_k}{(E - E_k)^2 + \Gamma_k^2/4}. \tag{9.102}$$

This spectral function is the probability density that an excitation k has energy E. Its wave-number integral is the average density of states per unit volume,

$$N(E) = \frac{1}{2\pi} \int \frac{d^d k}{(2\pi)^d} A(k, E). \tag{9.103}$$

For the free Hamiltonian, $A_0(k, E) = 2\pi\delta(E - \varepsilon_k^0)$. The disorder introduces a finite spectral width Γ_k, which translates into a finite lifetime $\hbar\Gamma_k^{-1}$. Equivalently, this finite lifetime translates into a finite scattering mean-free path l_s for the spatial matrix elements of the average propagator,

$$\langle G(r - r', E)\rangle = \int \frac{d^d k}{(2\pi)^d} e^{ik\cdot(r'-r)} \langle G(k, E)\rangle = G_0(r - r', E) e^{-|r'-r|/2l_\text{s}}, \tag{9.104}$$

showing an exponential decay with $l_\text{s} = k\Gamma_k/(2E)$ evaluated at $k = \sqrt{2mE}/\hbar$.

4. The simplest possible example is the calculation of the lifetime from the lowest-order, so-called *Born approximation*

$$\Sigma(k, E) = \underset{k \qquad k' \qquad k}{\overset{k - k'}{\longrightarrow\!\!\cdots\!\!\cdots\!\!\longrightarrow}} \tag{9.105}$$

for some potential with correlation function $P(q)$. To lowest order in V_0, we can use $E_k = \varepsilon_k^0$ and thus find

$$\frac{\Gamma_k}{2\varepsilon_k^0} = \frac{1}{kl_\text{s}} = \pi \int \frac{d^d k'}{(2\pi)^d \varepsilon_k^0} P(k - k')\delta(\varepsilon_k^0 - \varepsilon_{k'}^0) = \frac{\pi P(0) N_0(\varepsilon_k^0)}{\varepsilon_k^0} \tag{9.106}$$

in terms of the free density of states $N_0(\varepsilon)$ and the low-k limit $P(0)$ of potential correlation, (9.85), that is appropriate for the δ-correlated limit. This is precisely the result that one gets from a straightforward application of Fermi's golden rule for the average probability of scattering out of the mode k by the external potential V_q. The interest of the full-fledged diagrammatic expansion is of course that one is in principle able to calculate corrections to the lowest-order estimate, and to tackle more complicated potentials. There exist literally hundreds of other applications in the most diverse physical systems. Let us mention two examples from our own experience.

For two-dimensional Gaussian correlated potentials such as the one introduced in Section 9.5.1.3, the scattering rate evaluates to

$$\frac{1}{kl_s} = \frac{\Gamma_k}{2\varepsilon_k^0} = \frac{2\pi V_0^2}{k^2\sigma^2 E_\sigma^2}e^{-k^2\sigma^2}I_0(k^2\sigma^2) \quad (\text{Gauss}, d = 2), \tag{9.107}$$

where $E_\sigma = \hbar^2/m\sigma^2$ is a characteristic correlation energy and I_0 a modified Bessel function.

For matterwaves in a one-dimensional speckle potential, we can use Eqs. (9.105) and (9.106) with the speckle potential correlation function (9.92). In $d = 1$, the only contributions can come from forward scattering $k' = k$ and backward scattering $k' = -k$, such that

$$\frac{1}{kl_s} = \frac{\Gamma_k}{2\varepsilon_k^0} = \frac{V_0^2 k}{\varepsilon_k^0{}^2}[P(0) + P(2k)] \quad (\text{speckle}, d = 1) \tag{9.108}$$

in terms of the k-space pair correlator $P(2k) = \pi\zeta(1 - |k\zeta|)\,\Theta(1 - |k\zeta|)$.

The estimates (9.107) and (9.108) can only be trusted if $\Gamma_k/\varepsilon_k^0 \ll 1$ or equivalently $kl_s \gg 1$, otherwise the assumption of a small correction to the free dispersion is no longer valid. Since the scattering rates diverge at low k, we find that the perturbative approach breaks down at low energy. A closer analysis shows that a sufficient criterion for weak disorder is $E_k \gg V_0^2/E_\sigma$ (Kuhn *et al.*, 2007).

Sometimes, also the real part of the self-energy is of importance. For example, one can calculate the speed of sound in interacting Bose–Einstein condensates, and especially the shift due to correlated disorder by the same Green-function formalism (Gaul *et al.*, 2009). Incidentally, for sound waves the scattering mean free path grows as $k \to 0$, and the perturbative approach stays valid even at very low energy.

9.5.2   Intensity transport

We would like to calculate the ensemble-averaged density $n(r, t) = \langle\langle r|\rho(t)|r\rangle\rangle$ (or its many-body form $\langle\langle\Psi^\dagger(r)\Psi(r)\rangle\rangle$) in the limit of long time. In the Schrödinger picture, the state evolves as $\rho(t) = U^\dagger(t)\rho_0 U(t)$. After transforming the time evolution operators to Green functions as in Eq. (9.75), we need a theory for the ensemble-averaged product $\langle G^A(E)G^R(E')\rangle$. This is known as the average intensity propagator. In most experimental situations—be it with electromagnetic or matter waves—one measures intensities (see for example the average transmission through a 1D disordered system of length L studied in Section 9.2); the average intensity propagator is thus the fundamental quantity of interest. Before going into details, we propose to have a look at what we should expect to be the result.

9.5.2.1   *Density response*

The generic behavior that one may expect for transport in a disordered environment is *diffusion*. Indeed, diffusion follows from two very basic and rather innocuous hypotheses. First, one generally has a local conservation law, for instance for particle number, taking the form of a *continuity equation*:

$$\partial_t n + \nabla \cdot j = s, \tag{9.109}$$

where $j(r,t)$ is the current density associated with $n(r,t)$, and $s(r,t)$ is some source function. Secondly, one assumes a *linear response* in the form of Fourier's law

$$j = -D\nabla n, \tag{9.110}$$

saying that a density gradient induces a current that tries to re-establish global equilibrium. The diffusion constant D appears here as a linear response coefficient. Inserting Eq. (9.110) into Eq. (9.109), we immediately find as a consequence the *diffusion equation*

$$[\partial_t - D\nabla^2]n(r,t) = s(r,t). \tag{9.111}$$

This equation can be solved by Fourier transformation.[5] The solution for a unit source $s(r,t) = \delta(r)\delta(t)$ is the Green function for this problem, namely, the density relaxation kernel

$$\Phi_0(q,\omega) = \frac{1}{-i\omega + Dq^2}. \tag{9.112}$$

Its temporal version

$$\Phi_0(q,t) = \int \frac{d\omega}{2\pi} e^{-i\omega t} \Phi_0(q,\omega) = \theta(t)\exp\{-Dq^2 t\} \tag{9.113}$$

shows that the relaxation $\exp\{-t/\tau_q\}$ with characteristic time $\tau_q = 1/Dq^2$ becomes very slow in the large-distance limit $q \to 0$ because of the local conservation law. In real space and time, the relaxation kernel reads

$$\Phi_0(r,t) = \int \frac{d^d q}{(2\pi)^d} e^{iq\cdot r} \Phi_0(q,\omega) = \theta(t)[4\pi Dt]^{-d/2}\exp\{-r^2/4Dt\}. \tag{9.114}$$

This relaxation kernel describes diffusive spreading with $\langle r^2 \rangle = 2dDt$.

This is the "hydrodynamic" description of dynamics on large distances and for long times, accessed by small momentum q and frequency ω. A microscopic theory is then only required to calculate the linear response coefficient D.

9.5.2.2   *Quantum intensity transport*

In complete analogy to the Dyson equation (9.96) for the average single-particle propagator, one may write a structurally similar equation for the intensity propagator $\Phi = \langle G^R G^A \rangle$, known as the Bethe–Salpeter equation:

$$\Phi = \langle G^R \rangle \langle G^A \rangle + \langle G^R \rangle \langle G^A \rangle U\Phi. \tag{9.115}$$

Here, one splits off the known evolution with uncorrelated, average amplitudes

$$\langle G^{\mathrm{R}}(k, E)\rangle\langle G^{\mathrm{A}}(k', E')\rangle = \qquad\qquad . \tag{9.116}$$

The upper part of intensity diagrams describes the retarded propagator, called the "particle channel" in condensed-matter jargon, whereas the lower part contains the advanced propagator or "hole channel". All scattering events that couple these amplitudes are contained in the intensity scattering operator U. By construction, this "particle–hole irreducible" vertex contains exactly all diagrams that cannot be factorized by removing a propagator pair (9.116). Its detailed form again depends on the model of disorder. In all cases, $U_{kk'}(E)$ is essentially the differential cross-section for scattering from k to k' and generally has the following structure:

$$U(k, k'; E) = \qquad = \quad \vdots \quad + \quad \times \quad + \quad \vdots \quad + \quad \vdots \quad + \ldots \tag{9.117}$$

Linear-response theory shows that this scattering vertex permits us to calculate the transport mean free path l, in close analogy to the calculation of the scattering mean-free path l_{s} from the self-energy. Their ratio is expressed as

$$\frac{l_{\mathrm{s}}}{l} = 1 - \langle\cos\theta\rangle_U, \tag{9.118}$$

where θ is the scattering angle between k and k', and the brackets $\langle.\rangle_U$ indicate an average over the scattering cross-section U. The physical interpretation of the transport mean free path l is the following: while the scattering mean free path l_{s} measures the distance after which the memory of the initial phase of the wave is lost, l is the distance over which the direction of propagation is randomized.

9.5.2.3 Diffusion

The scattering processes encoded in U are perhaps more easily visualized in real space. We will draw a full line for every amplitude ψ propagated by G^{R} (upper lines in Eq. (9.117)) and a dashed line for every ψ^* propagated by G^{A} (lower lines in Eq. (9.117)). Impurities are represented by black dots as before. Then, the first contribution to U describes the single-scattering process

$$U_{\mathrm{B}}: \qquad \tag{9.119}$$

in which both ψ and ψ^* are being scattered by the same impurity at position r_1. This process is insensitive to phase variations and could just as well take place for classical particles. So, this Boltzmann contribution U_B describes classical diffusion with diffusion constant

$$D_B = \frac{v l_B}{d}. \tag{9.120}$$

The Boltzmann transport mean free path is calculated by inserting $U_{Bkk'} = V_0^2 P(k - k')$ into Eq. (9.118):

$$\frac{l_s(k)}{l_B(k)} = 1 - \frac{\int d\Omega_d \cos\theta P(2k|\sin(\theta/2)|)}{\int d\Omega_d P(2k|\sin(\theta/2)|)}. \tag{9.121}$$

Depending on the microscopic scattering process, l_B can be longer than l_s, if forward scattering is dominant, $\langle \cos\theta \rangle_{U_B} > 0$. This is the case for matterwaves in spatially correlated potentials. For isotropic scattering with $\langle \cos\theta \rangle_{U_B} = 0$, these two length scales coincide, $l_s = l_B$.

By combining Eq. (9.121) with Eq. (9.106) giving the scattering mean free path, one can easily compute, in the Born approximation, the transport mean free path and consequently the classical Boltzmann diffusion constant, using only microscopic ingredients: the dispersion relation of the free wave and the correlation function of the scattering potential.

9.5.2.4   Localization length in 1D systems

As we have already seen repeatedly in previous sections, in 1D the transport mean free path is (up a factor 2) equal to the localization length, $\ell = \xi_{loc}/2$, an identity that can also be verified microscopically (Thouless, 1973), at least to lowest order V_0^2 in perturbation theory.[6] So now we are in a position to give a microscopic prediction for the 1D localization length for arbitrarily correlated potentials, namely taking twice the backscattering contribution from Eq. (9.108), selected by the $(1 - \cos\theta)$-factor in Eq. (9.121):

$$\frac{1}{k\xi_{loc}} = \frac{V_0^2 k}{4\varepsilon_k^{0^2}} P(2k). \tag{9.122}$$

Figure 9.18, taken from (Billy *et al.*, 2008), shows this prediction for $L_{loc} = 2\xi_{loc}$ as a dashed line together with the results of a fit to the intensity measured in the real experiment (see Figure 9.8). Here, $P(2k) = \pi\zeta(1 - k_{max}\zeta)\Theta(1 - k_{max}\zeta)$ with k_{max} the largest k-value present in the expanding wave packet, resulting in:

$$\xi_{loc} = \frac{\hbar^4 k_{max}^2}{\pi m^2 V_0^2 \zeta(1 - k_{max}\zeta)}. \tag{9.123}$$

There is no adjustable parameter, and the agreement is rather satisfactory. Significant deviations are visible both for small disorder (there the localization length becomes too large and experimental limitations start to show) and for large disorder, where the lowest-order theoretical estimate, or Born approximation (9.123) becomes insufficient.

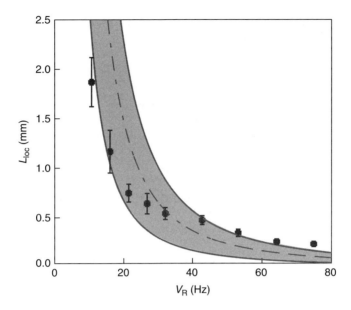

Fig. 9.17 Comparison between the experimentally measured localization length for a quasi-1D atomic wave-packet launched in the disordered optical potential created by a speckle pattern, and the theoretical prediction, Eq. (9.123), when the strength of the disordered potential is varied. There is no adjustable parameter. Reprinted from (Billy *et al.*, 2008) (courtesy of Ph. Bouyer).

Moreover, for strong disorder, atom–atom interaction for the strongly localized cloud may no longer be negligible and induce some delocalization.

An interesting scenario occurs in speckle potentials when the fastest atoms have a wavevector k_{max} larger than the most rapid spatial fluctuations, with wavevector $1/\zeta$. Then, $P(2k_{\mathrm{max}}) = 0$, and Eq. (9.123) predicts *prima facie* $\xi_{\mathrm{loc}} = \infty$ or absence of localization, which would signal the existence of a mobility edge in this 1D random potential, contradicting rigorous mathematical theorems stating that all states are exponentially localized (Kotani and Simon, 1987). In fact, exponential localization still prevails, but requires more than a single scattering event by the smooth random potential. Going to higher orders in perturbation theory, beyond the Born approximation, one can show in excellent quantitative agreement with numerics, that the localization length is always finite (Lugan *et al.*, 2009; Gurevich and Kenneth, 2009). However, for weak disorder, the localization length can become much larger than the system size, such that numerical or experimental results show an *apparent mobility edge*. For other types of long-range correlations, one does find mobility edges (de Moura and Lyra, 1998). There seems to be no obvious way of deciding, for a certain class of potentials, whether true exponential localization exists or not, and correlated potentials are still actively investigated in different contexts, see (Izrailev and Makarov, 2005) and references therein.

Despite the nice agreement shown in Figure 9.18, some caution is indicated. The experimental observation involves averaging over k, and also over several different realizations of the disorder (single-shot results look similar, just more noisy). This means that the experimental data resembles the average transmission $\langle T(z) \rangle$ as a function of sample thickness z.[7] In section 9.2.2.3, we showed that it is the *typical* transmission $T_{\rm typ}(z) = \exp(\langle \ln T(z) \rangle)$ that decays exponentially, not the average transmission. At very large z, this can make a huge difference, see Section 9.2.5. Fortunately, for z of the order of the localization length (t or order unity in the language of 9.2.5), the fluctuations have not yet built up, and the difference between the typical and the average value is still small, $\ln \langle T(z) \rangle \approx -z/\xi_{\rm loc}$, making the pure exponential decay an acceptable approximation. Further in the wings, one expects deviations of the average density from a pure exponential decay, see Eq. (9.38). This takes place, however, in the region where fluctuations are huge, so that a typical experiment may not measure the *average* value of the density, but rather its *typical* value.

9.5.2.5   *Weak-localization correction*

The first corrections to the classical, incoherent scattering process (9.119) shown in Eq. (9.117) involve one more scatterer and several possibilities of intermediate propagation. The most well-known type of correction stems from the diagram with two crossed lines. In real space, the scattering process is

$$(9.124)$$

This is an interference correction with a phase shift $\Delta\varphi$ between ψ and ψ^* that depends on the impurity positions r_1 and r_2. Contributions of this type are ensemble-averaged to zero—or rather, almost averaged to zero. Indeed, if the starting and final point of propagation come close, $r \approx r'$, the phase shift picked up by the two counter-propagating amplitudes becomes smaller:

$$(9.125)$$

At exact backscattering $r = r'$ and in the absence of any dephasing mechanisms, the phase difference is exactly zero. Vanishing phase difference means constructive interference and therefore enhanced backscattering probability to stay at the original position. This holds true no matter how many scatterers are visited on the path. One is led to consider all maximally crossed diagrams:

$$U_{\mathrm{C}} = \quad + \quad + \dots \tag{9.126}$$

These diagrams were first considered in the electronic context (Langer and Neal, 1966) and became known as the *Cooperon* contribution. This contribution is peaked around backscattering $k = -k'$. Therefore, one may resort to a diffusion approximation and sum up all contributions with the help of the diffusion kernel (9.112):

$$\frac{1}{l} = \frac{1}{l_{\mathrm{B}}} \left[1 + \frac{1}{\pi N_0} \int \frac{\mathrm{d}^d q}{(2\pi)^d} \frac{1}{-i\omega + D_{\mathrm{B}}q^2} \right]_{\omega \to 0}. \tag{9.127}$$

Writing this in terms of the diffusion constant, one arrives at the weak-localization correction

$$\frac{1}{D} = \frac{1}{D_{\mathrm{B}}} \left[1 + \frac{1}{\pi N_0 D_{\mathrm{B}}} \int \frac{\mathrm{d}^d q}{(2\pi)^d} \frac{1}{q^2 - i0} \right]. \tag{9.128}$$

The quantum correction of the Cooperon makes $D < D_{\mathrm{B}}$, and we have thus found the microscopic reason for the weak-localization correction that was first mentioned in the scaling Section 9.3.5. Before looking in more detail at this correction in Section 9.7, we should like to understand it better by selectively probing the Cooperon contribution. In optics, this is indeed possible and is developed in the next section.

9.6 Coherent backscattering (CBS)

One can probe the specific geometry of scattering paths like Eq. (9.124) by using a source of plane waves together with a collection of randomly positioned scatterers in a half-space geometry (figure 9.19). Hereafter, we suppose normal incidence and detection close to the backscattering direction; generalizing to arbitrary incident and detection angles changes nothing in the central argument.

9.6.1 Theory

The picture in Figure 9.19 shows scattering by four impurities, contributing to the incoherently transported intensity. The corresponding intensity diagram is

$$\tag{9.129}$$

The sum of all such diagrams with a distinct ladder topology yields the intensity propagator or *diffuson*, whose long-distance and long-time form is precisely the diffusion kernel (9.112), evaluated with the Boltzmann diffusion constant,

$$\Phi_{\mathrm{B}} = \frac{1}{-i\omega + D_{\mathrm{B}}q^2}. \tag{9.130}$$

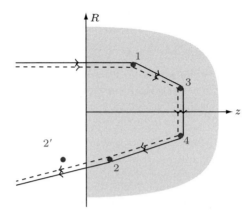

Fig. 9.18 Half-space geometry of a backscattering experiment with cylindrical coordinates $r = (R, z)$. $2'$ is the image point of the exit scatterer 2 used to construct the half-space propagator for the incoherent intensity. As an example, the contribution of scattering from four scatterers is depicted.

The total backscattered diffuse intensity per unit surface is given by summing contributions from all possible starting and end points:

$$I_L \propto \int dz_1 \, e^{-z_1/l_s} \int dz_2 \, e^{-z_1/l_s} \int d^2R \, \Phi_B(r_1, r_2). \tag{9.131}$$

The exponential attenuation factors describe the propagation of intensities from the surface to the first scatterer and back out again with average propagator $\langle G^R(z_i) \rangle$, (9.104), featuring the scattering mean free path l_s. Moreover, translation invariance along the surface direction has been used, leaving only the surface integral over the lateral distance $R = R_1 - R_2$.

  The propagation inside an infinite disordered medium would occur with the bulk kernel (9.130) and thus have a time-integrated diffusion probability of $\Phi_B(r) = \int dt\Phi_B(r, t) = [4\pi D_B r]^{-1}$. This expression leads to a diverging integral over R in Eq. (9.131). But the starting and end points r_1 and r_2 lie rather close to the surface, namely typically one scattering mean free path l_s away from it. So, for calculating the backscattered intensity (9.131), we have to worry about appropriate boundary conditions. The complete integral equation for intensity propagation in a half-space geometry of a scalar wave and isotropic scatterers, known by the name Milne equation, can be solved exactly (Morse and Feshbach, 1953; Nieuwenhuizen and Luck, 1993), albeit with considerable mathematical effort. For a simple solution involving the diffusive bulk propagator valid far from the boundary, one can employ the method of images that is often used in electrostatics. Since photons reaching the surface would escape prematurely from the medium, one can exclude these events by subtracting the contribution of propagation to an image point $r_{2'} = (R_2, -z_2)$ mirrored to the outside of the sample:

$$\Phi_B(r_1, r_2) = \frac{1}{4\pi D_B}\left[\frac{1}{r_{12}} - \frac{1}{r_{12'}}\right].$$ (9.132)

This half-space propagator behaves like R^{-3} at large R and thus permits us to carry out the integration. The final result is some number I_L and gives the incoherent background on top of which we now study the interference contribution.

Each multiple-scattering diagram like Eq. (9.129) has an interference-correction counterpart such as

(9.133)

in which the conjugate amplitude travels along the same scatterers, but in the opposite direction. The contribution of such maximally crossed diagrams to the backscattered intensity can be accounted for along the same lines. First, incident and scattered amplitudes now pick up a phase between the surface and the scattering end points. Namely, the amplitude ψ picks up $\exp\{ik \cdot r_1\}$ at the entrance and $\exp\{ik' \cdot r_2\}$ at the exit of the medium. The path-reversed complex conjugate amplitude picks up $\exp\{-i[k \cdot r_2 + k' \cdot r_1]\}$. So after all, there is a total phase difference of $\Delta\varphi = (k + k') \cdot (r_1 - r_2)$. Exactly toward the backscattering direction, $k' = -k$, this phase difference vanishes. Close to backscattering, for a small angle $\theta \ll 1$, one has $|k + k'| \approx k_\perp = k\sin\theta \approx k\theta$, and the phases differ by $\Delta\varphi = kR\theta$.

Thus, each path acts like a Young double-slit interferometer with the two end-point scatterers playing the role of the two slits. The larger the transverse distance R between the scatterers, the finer the interference fringes. The only point where all fringes are bright is the symmetry point $\theta = 0$ toward backscattering. Sufficiently far away from this direction, the sum of random fringe patterns averages out to zero. The sum of all interference terms is again the integral over all end points with the appropriate weight furnished by the intensity propagator (9.132) (which must be modified if some additional dephasing processes are at work, see Section 9.6.3 below). This simple calculation predicts a relative interference enhancement over the background

$$\frac{I_C(\theta)}{I_L} \approx \frac{1}{(1 + kl|\theta|)^2}.$$ (9.134)

The interference-induced enhancement, shown in Figure 9.20 as a dashed line, survives in an angular range $\Delta\theta = 1/kl = \lambda/(2\pi l)$ around backscattering. Very characteristically, this peak features a triangular cusp at backscattering (plotted using a dotted line),

$$\frac{I_C(\theta)}{I_L} = 1 - 2|q| + O(q^2),$$ (9.135)

where $q = kl\theta$ is the reduced momentum transfer.

The exact solution for scalar waves and isotropic point scatterers can be calculated solving the Milne equation of intensity transport. The CBS profile can then be expressed as the integral (Nieuwenhuizen and Luck, 1993)

(a)   (b)

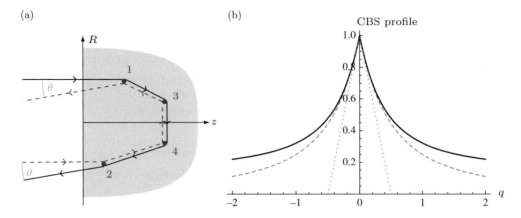

Fig. 9.19 Coherent backscattering (CBS) (a) Schematic picture of a four-scatterer path; the constructive interference of path-reversed amplitudes in the backscattering direction $\theta = 0$ leads to an observable intensity enhancement. Away from backscattering, the phase differences average out and leave only the background intensity. (b) CBS profile as a function of reduced scattering angle $q = kl\theta$, normalized to the value at $\theta = 0$. Solid black: exact solution (9.136). Dashed: diffusive solution (9.134). Dotted: linear solution (9.135) with characteristic slope discontinuity at backscattering.

$$\frac{I_C(\theta)}{I_L} = \frac{1}{C} \exp\left\{ -\frac{2}{\pi} \int_0^{\pi/2} d\beta \ln\left[1 - \frac{\arctan\sqrt{q^2 + \tan^2\beta}}{\sqrt{q^2 + \tan^2\beta}} \right] \right\}, \qquad (9.136)$$

where $q = kl\theta$ and a constant $C = \exp\left\{ -\frac{2}{\pi} \int_0^{\pi/2} d\beta \ln\left[1 - \beta \cot\beta \right] \right\} \approx 8.455$ such that at the origin $I_C(0) = I_L$. This profile is plotted as a solid curve in Figure 9.20. It becomes apparent that the diffusive solution (9.134) gives a very good description for small scattering angles. Notably, its slope at $\theta = 0$ is precisely equal to the exact value that can be extracted from Eq. (9.136). This was to be expected since diffusion should be valid for long-distance bulk propagation, and long scattering paths have widely separated end points that contribute to the small transverse momenta making up the top of the CBS peak. Indeed, the diffusion prediction (9.134) is precisely recovered by replacing the exact propagation kernel under the logarithm by its diffusion approximation:

$$A(Q) = \arctan(Q)/Q \approx 1 - \frac{1}{3}Q^2 \qquad (9.137)$$

valid at small $Q = \sqrt{q^2 + \tan^2\beta}$.

The agreement between the diffusive profile (9.134) and the exact result (9.136) deteriorates at larger angle $q = kl\theta$. This discrepancy is due to low scattering orders that are not accurately captured by the diffusion approximation and the imaging method used to mimic the exact boundary conditions. Indeed, only the very tip of the CBS peak stems from long paths reaching far into the bulk. The larger part of the total signal is due to contributions from rather short paths, for which scattering inside the surface skin layer is crucial.

One can calculate the contribution of scattering orders $n = 1, 2, \ldots$ to the total incoherently backscattered intensity, measured in units of the incident flux by the so-called bistatic coefficient $\gamma(\cos \theta', \cos \theta)$ (Ishimaru, 1978) that depends on the angles θ' and θ of incidence and observation. For exact backscattering and normal incidence ($\theta = \theta' = 0$), one has $\gamma = \sum_{n \geq 1} \gamma_n$. The largest contribution comes from single scattering with $\gamma_1 = 1/2$ followed by double scattering with $\gamma_2 = \ln(2)/2 \approx 0.35$ and so on, with an asymptotic decrease as $\gamma_n \sim n^{-3/2}$ (Nieuwenhuizen and Luck, 1993).

When the CBS peak was first observed in the beginning of the 1980s (Kuga and Ishimaru, 1984; Albada and Lagendijk, 1985; Wolf and Maret, 1985), the diffusive theory used an image point placed at $z_{2'} = -(2z_0 + z_2)$ such that the diffuse propagator vanishes at a distance $z_0 = 2/3$ outside the sample. This translates to the boundary condition that the total incident diffusive flux on the surface vanishes (Morse and Feshbach, 1953; Akkermans and Montambaux, 2007) and leads to a diffusive CBS peak shape of

$$\frac{I_C(q)}{I_L} = \frac{1}{(1 + |q|)^2} \frac{1}{1 + 2z_0} \left[1 + \frac{1 - \exp\{-2z_0|q|\}}{|q|} \right]. \tag{9.138}$$

This diffusive solution predicts a different slope at the origin, namely, $-2[1 + z_0^2/(1 + 2z_0)]$, which is off by more than 20% from the exact value, although one would expect the diffusion solution to get this value right (Nieuwenhuizen and Luck, 1993). This is all the more disturbing as fits to the diffuse CBS peak shape are generally used to measure the transport mean free path. Also at larger angles this solution cannot convince because the diffusion profile decays as q^{-2}, whereas the exact solution decreases like $|q|^{-1}$. This asymptotic behavior is known to come from the double-scattering contribution.

Van Tiggelen (van Tiggelen, 1992) has noticed that the diffusion approximation becomes virtually exact if single- and double scattering are included separately since

$$\frac{1}{1 - A(q)} = 1 + A(a) + \frac{A(q)^2}{1 - A(q)} \approx 1 + A(q) + \frac{3\alpha}{q^2}, \tag{9.139}$$

both *for small and large q*, with a numerical coefficient $\alpha = 1$ for $q \to 0$ and $\alpha = \pi^2/12 \approx 0.822$ for $q \to \infty$. Therefore, the best approximation to the exact solution is obtained by first taking the exact double-scattering profile (Nieuwenhuizen and Luck, 1993; Müller *et al.*, 2001)

$$\gamma_2(q) = \frac{1}{\pi} \int_0^{\pi/2} d\beta A \left(\sqrt{q^2 + \tan^2 \beta} \right) = \frac{2 \cosh^{-1}(1/|q|) - \cosh^{-1}(1/q^2)}{2\sqrt{1 - q^2}} \tag{9.140}$$

where $\cosh^{-1}(x)$ is the inverse hyperbolic cosine function, then adding the diffusive solution

$$\gamma_{\text{diff}}(q) = \frac{3\alpha}{2(1 + |q|)^2} \left[1 + \frac{1 - \exp\{-2z_0|q|\}}{|q|} \right], \tag{9.141}$$

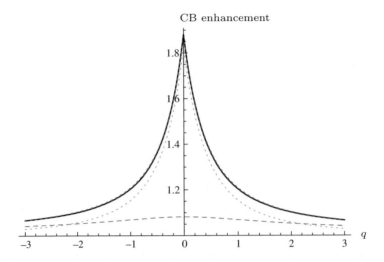

Fig. 9.20 CBS intensity enhancement in units of the background intensity as a function of reduced scattering angle $q = kl\theta$. Solid black: exact solution (9.136) minus the single-scattering value $\gamma_1 = 1/2$. Thin dashed line: double-scattering contribution (9.140). Thick dashed line: sum of double scattering and best diffusive solution, (9.141) with $\alpha^* \approx 0.86$ and $z_0^* \approx 0.81$. Dotted: Traditional diffusive CBS peak, Eq. (9.141), with $\alpha = 1$ and $z_0 = 2/3$ shown for comparison. Even for this time-reversal invariant case, the backscattering enhancement is slightly smaller than 2 because of the single-scattering contribution to the background that is absent in the CBS signal.

and finally fitting the extrapolation length z_0 and diffusion-constant multiplicator α such that height and slope are equal to the exact values at the origin. Doing this, we find $\alpha^* \approx 0.86$ within the expected interval $[0.822, 1]$ and $z_0^* \approx 0.81$. Figure 9.21 shows the exact CBS profile (with the single-scattering contribution subtracted as required) together with the double-scattering contribution plus the full approximated diffusive CBS profile that turns out to be in excellent agreement, both for small and large angles.

The full width at half-height of the CBS profile is $\Delta q \approx 0.73kl \approx 4.59l/\lambda$, and observing the CBS peak can be used to measure the transport mean free path quite accurately. The explicit occurrence of the wavelength λ emphasizes that CBS is a genuine interference effect. In many circumstances, the mean free path is much longer than the wavelength, such that $kl \sim 10^2 - 10^3$, and $\Delta\theta$ is at most a couple of mrad. This makes CBS difficult to observe with the naked eye, together with the constraint that one has to look exactly toward the backscattering direction, but it can be easily imaged using standard optics, as schematically shown in Figure 9.21.

9.6.2 Live experiment

Because the CBS cone is typically very narrow and its maximum height at best equal to the average background, a source with large angular dispersion will broaden the

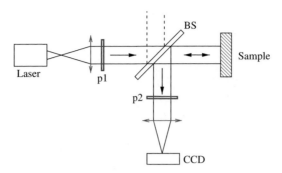

Fig. 9.21   Schematic view of a table-top CBS experiment. Light from a wide laser beam with small angular dispersion is directed onto a disordered sample. The diffuse retroreflected intensity is sent by a beamsplitter (BS) in the focal plane of a lens and recorded by a CCD camera, thus imaging the angular distribution. Polarization elements (p1 and p2) select a suitable polarization channel; generally, the helicity-preserving channel of opposite circular polarization is recommended.

signal too much and reduce enhanced backscattering. Thus, it is highly desirable to use a quasi-parallel beam obtained from a laser source, with angular divergence smaller than a fraction of mrad. This in turn requires a large spot, with a diameter larger than the mean free path, which is easily obtained by expanding the output beam of a commercial diode laser with a telescope. The scattering medium should scatter efficiently and must not absorb the light: a bright white object is thus chosen. A piece of ordinary paper turns out to give the best results. A sheet of paper is about 100 μm thick and obviously scatters most of the incoming beam, meaning that the mean free path does not exceed a few tens of μm. A piece of teflon could also be used, but the mean free path is significantly larger, meaning a narrower CBS cone, much harder to detect. White paint or milk also make good samples, with the advantage that the concentration and thus the mean free path can be varied and that the thermal motion of scatterers inside the solvent provides us with configuration averaging for free; however, these samples must be put inside some transparent container whose surface can produce specular reflection that is easily confounded with the CBS signal.

A semitransparent plate (beamsplitter) can be used to send the backreflected light into a 1280×1024 pixel CCD camera with pixel size around $5 \,\mu m$, located about 20 cm from the scattering medium, thus ensuring a 0.025-mrad angular resolution. Figure 9.22 (left) shows the image recorded from a fixed piece of paper. This situation corresponds to a single realization of the disorder. The electric field on each pixel is the coherent sum of the field amplitudes radiated by each point of the sheet of paper. Because of its disordered nature, each contribution picks a random amplitude and phase, resulting in a characteristic speckle pattern on the CCD camera, the "optical fingerprint" of the paper. The angular size of the speckle grain is of the order of $1/kL$ where L is the size of the illuminated spot on the sheet of paper. For our case, it is about 0.1 mrad, i.e. slightly larger than the pixel size, in agreement with the experimental observation. The attentive reader may notice that the bright spots

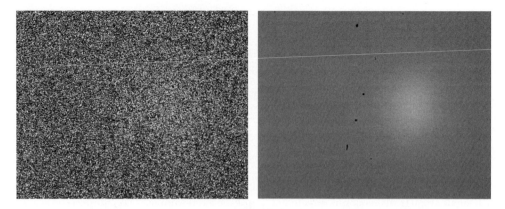

Fig. 9.22 Intensity around the backscattered direction for a piece of paper exposed to a parallel laser beam. For a fixed paper (left figure), one observes a characteristic speckle pattern, due to random interference of phase-coherent light scattered by a single configuration of disorder. By averaging over various parts of the paper, small-scale random variations are averaged out, but an enhanced intensity is clearly visible around exact backscattering. Original data from an experiment performed during the Les Houches Summer School in Singapore on July, 15th, 2009. Special thanks to David Wilkowski, Kyle Arnold, and Lu Yin from the Centre for Quantum Technologies, National University of Singapore, for generous support and invaluable help in setting up the experiment.

look slightly brighter in a roughly circular area on the right side of the figure. In order to *see* the CBS cone, one should perform configuration averaging. This is easily done by mounting the piece of paper on a rotating device (in our case a battery-powered computer fan). On the time-averaged intensity, shown in Figure 9.22 (right), the fluctuating speckle pattern has been washed out, leaving a uniform background, on top of which appears a smooth bright spot of approximate width 10 mrad due to coherent backscattering. The effect is perhaps not dramatic, as the enhancement factor cannot be larger than 2 (it is 1.6 in this live experiment), but clearly present and visible with the naked eye.

A cut across the spot center is presented in Figure 9.23 together with a fit to the simplest theoretical formula, Eq. (9.134). The fit is quite good in the wings and allows us to extract the mean free path inside the piece of paper, in our case 25 μm. The fact that the top of the CBS peak is rounded can be attributed to various experimental imperfections such as the finite angular resolution, geometrical aberrations, finite thickness and residual absorption of the piece of paper, but could in principle also highlight the presence of a decoherence mechanism.

9.6.3 Dephasing/decoherence

The CBS phenomenon presented so far relies on perfect phase coherence of the multiply scattered wave. What happens if some external agent—such as some degree of freedom inside the paper coupled to the wave—affects the scattered amplitude in

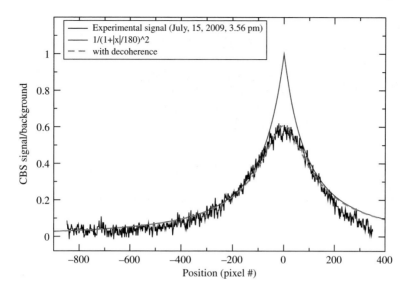

Fig. 9.23 Experimentally measured coherent backscattering signal, together with a fit to the simplest theoretical formula, (9.134) (solid gray line), and to Eq. (9.146) with phenomenological decoherence included (dashed line). From the angular width of the CBS signal, one can extract the transport mean free path inside the sheet of paper, here 25 μm. Experimental imperfections limit the coherence of the phenomenon and are responsible for the deviation from the peaked shape at the center of the cone. When properly taken into account (dashed curve), the agreement is very good.

an uncontrolled way? Qualitatively, it is clear that amplitudes of long scattering paths are more fragile than those of shorter paths. As very long paths are responsible for the characteristic triangular shape of the CBS cone around the exact backscattering direction, it is important to understand the effect of decoherence on the CBS signal. In return, the CBS enhancement factor can serve as a sensitive measure for phase coherence.

There is no universal way of breaking phase coherence, and the effect on CBS can be different depending on the specific mechanism at work. Nevertheless, a simple phenomenological approximation may often be used, and we will see below that several physical processes are well described by this approximation. This assumption is that phase coherence is lost at a constant rate, characterized by a phase-coherence time τ_ϕ, also called the dephasing time. Then, interference terms associated with paths who are visited in a time t have to be multiplied by a factor $\exp(-t/\tau_\phi)$. An example of such a situation is provided by a Michelson interferometer operated with a classical light source, where the interference disappears once the optical path length difference exceeds $c\tau_\phi$, with τ_ϕ the longitudinal coherence time of the source. Note that these phenomena are typically called "dephasing" in the context of classical waves, and "decoherence" for quantum-mechanical matterwaves. The bottom line is simply that interference is lost by coupling to some external degree of freedom.

The exponential attenuation of interference as $\exp(-t/\tau_\phi)$ applies especially often to the Cooperon contribution. In Fourier space, the effect is simply tantamount to the replacement

$$\omega \mapsto \omega + \frac{i}{\tau_\phi}, \tag{9.142}$$

or, in the diffusive propagator (9.112):

$$\frac{1}{-i\omega + D_B q^2} \mapsto \frac{1}{-i\omega + \frac{1}{\tau_\phi} + D_B q^2} = \frac{1}{-i\omega + D_B \left(q^2 + \frac{1}{D_B \tau_\phi}\right)}, \tag{9.143}$$

which can be also be obtained via the replacement

$$q^2 \mapsto q^2 + \frac{1}{L_\phi^2}. \tag{9.144}$$

The phase-coherence length,

$$L_\phi = \sqrt{D_B \tau_\phi}, \tag{9.145}$$

is the average distance over which the wave propagates *diffusively* before losing its phase coherence.

This simple replacement can be used to calculate the shape of the CBS cone in the presence of decoherence effects. Indeed, Section 9.6.1 discusses several approximate expressions for the shape, all expressed as a function of the transverse momentum $k_\perp \approx k|\theta|$, which is simply the sum of the incoming and outgoing momenta (in the limit of small angles $\theta \ll 1$). The substitution $k_\perp^2 \mapsto k_\perp^2 + 1/D_B\tau_\phi$ in Eq. (9.134) yields

$$\frac{I_C(\theta)}{I_L} \approx \frac{1}{\left[1 + \sqrt{(kl\theta)^2 + l^2/L_\phi^2}\right]^2}. \tag{9.146}$$

This expression is now a smooth function of θ (no cusp at $\theta = 0$ any longer). In the limiting case $l \ll L_\phi$, one recovers the previous expression, only slightly perturbed near the tip. In the opposite limit $L_\phi \ll l$, the CBS cone disappears completely, which is quite natural as interference effects are washed out before the wave travels a single mean free path. The relative height of the CBS peak, compared to the background at $kl|\theta| \gg 1$, is

$$\frac{I_C(0)}{I_L} = \frac{1}{[1 + l/L_\phi]^2} \approx 1 - 2\frac{l}{L_\phi} = 1 - 2\sqrt{\frac{\tau_l}{\tau_\phi}}, \tag{9.147}$$

where the last two expressions are valid in the limit of weak decoherence $l \ll L_\phi$. Here, $\tau_l = \sqrt{D_B/l^2}$ is the mean free time separating two consecutive scattering events. This expression emphasizes the sensitivity of the CBS cone to dephasing effects. Indeed, if the dephasing time is say 10 times larger than the mean free time, its effect on the CBS cone is still very noticeable, reducing its height by almost 50%. For example, the

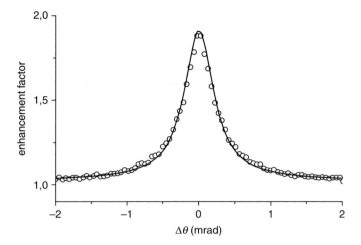

Fig. 9.24 CBS of light by cold strontium atoms (Bidel *et al.*, 2002). The backscattering enhancement is close to 2, which indicates full phase coherence. Signal observed in the helicity-preserving channel (incoming light circularly polarized, detection in the opposite circular polarization). Solid line: the result of a calculation taking into account the finite geometry and inhomogeneous density of the atomic cloud.

experimentally observed CBS cone in the live experiments is well fitted by Eq. (9.146), with a decoherence time $\tau_\phi = 12.5\tau_l$.

Several other types of decoherence have been studied in great detail in connection with light and cold atoms. In the following, we present a few of them qualitatively, referring to the literature for more details.

9.6.3.1 *Polarization*

For simplicity, we have up to now considered a scalar complex wave, describing e.g. a spinless atomic matterwave. Many real atoms, all electrons and also electromagnetic waves are more complicated because of their spin/polarization. In particular, light scattering does not preserve polarization, as is obvious from the requirement of transversality. Thus, the light backscattered along the direct and reverse paths generically emerges from the medium with different polarization. But orthogonal polarizations do not interfere, and thus one may expect a reduced enhancement factor.

Technically, one has to dress the multiple-scattering Cooperon contribution with the polarization structure. Two independent polarizations of the propagating intensity must be taken into account (one in the retarded, one in the advanced Green function), leading to a tensorial structure for diffuson and Cooperon alike. The complete calculation of this effect is possible by decomposing the intensity kernel into irreducible tensor moments (Müller and Miniatura, 2002). To make a long story short, it is enough to say that each contribution has a kernel of the type (9.143), with its own τ_ϕ of the order of τ_l. The physical interpretation is clear: because scattering will on average lead to depolarization, all channels associated with specific polarization correlations must decay during propagation. Only the one channel measuring the total intensity is protected by conservation of energy, with $1/\tau_\phi = 0$, and propagates diffusively.

If the system is additionally time-reversal invariant, the same conserved intensity channel also exists for the Cooperon. The population of the various contributions depends on the specific choices for the incoming and outgoing polarizations used for recording the CBS signal. Using the same linear polarization for excitation and analysis populates the conserved mode and ensures an optimal interference contrast for long scattering paths, at least for classical point-like objects (such as dye molecules) acting as Rayleigh scatterers (we will discuss in Section 9.6.3.4 the more general case). The same is true if the incident field has circular polarization and the opposite circular polarization is used for detection (helicity-preserving channel), with the additional advantage that the single scattering background of the diffuson is filtered out, allowing in principle the observation of a a perfect CBS enhancement by a factor of 2 (Bidel *et al.*, 2002; Wiersma *et al.*, 1995).

9.6.3.2 Residual velocity of the scatterers

The previous derivation assumed quenched disorder, i.e. scatterers at fixed positions. Moving scatterers are a cause of decoherence: as light travels along two reciprocal paths, it visits the same scatterers, but in opposite order, i.e. at different times. If, during the time delay separating the scattering events on the direct and reversed path, the atom has moved by at least one wavelength, the phase coherence between the two paths will be lost. This phenomenon can alternatively be interpreted in the frequency domain, where moving scatterers induce a Doppler shift of the scattered photon that is different along the direct and reversed path. Although this phenomenon does not lead to a strict exponential decay of the phase coherence (Golubentsev, 1984), it reduces the enhancement factor, which has been notably observed with cold atoms (Labeyrie *et al.*, 2006).

In general, interference of waves is suppressed once the environment has acquired knowledge of the path taken by the scattered object. This can be most simply seen in experiments of the Young's double-slit type (Itano *et al.*, 1998), but applies equally to the CBS by light from moving atoms, where moreover the storage of which-path information in the atomic recoil has been studied (Wickles and Müller, 2006).

9.6.3.3 Non-linear atom-light interaction

Because atoms have extremely narrow resonance lines, they have large polarizabilities and already quite low laser intensities can saturate an atomic transition, in which case the atom scatters photons inelastically. It is easy to understand that such a non-linear inelastic process will reduce the phase coherence of the scattered light and the enhancement factor. This indeed has been observed (Chanelière *et al.*, 2004). A full quantitative understanding of multiple inelastic scattering is still not available. For a model system of two atoms driven by a powerful laser field, a rather complete understanding of the CBS signal has been achieved (see Shatokhin *et al.*, 2007a,b and references therein).

In the context of matterwaves, one may study coherent backscattering of interacting matterwaves, obeying a non-linear equation such as the Gross–Pitaevskii equation, evolving in a random optical potential. It has been shown that already a moderate non-linearity induces a phase shift between the direct and reversed paths and thus a decrease of the height of the CBS peak, and may in some cases even create a negative contribution in the backward direction (Hartung *et al.*, 2008). These theoretical predictions still await experimental realization.

9.6.3.4 *Internal atomic structure/spin-flip*

The preceding description treats atoms as Rayleigh point scatterers that radiate a purely dipolar electromagnetic field, with an induced dipole directly proportional to the incoming electric field. This is an excellent approximation for atoms with a non-degenerate electronic ground state, such as strontium. The situation is radically different if the atomic ground state is degenerate: indeed, when scattering a photon, the atom may stay in the same atomic state (Rayleigh transition) or change to another state with the same energy (degenerate Raman transition), see Figure. 9.25(a).

The basic rules of quantum mechanics imply that orthogonal final states cannot interfere. In other words, two multiple scattered paths will interfere only if they are associated with *the same initial and final states of all atoms*.[8] Note that there is no need for the initial and final states to be identical, so that Raman transitions can very well contribute to interference terms, if and only if the same Raman transitions occur along the interfering paths.[9] A commonly encountered situation is that the degeneracy of the atomic ground state is due to its non-zero total angular momentum, see Figure 9.25(a): Raman transitions then involve different Zeeman substates. The detailed calculation of the scattering vertex requires us to incorporate also the angular momentum, i.e. the polarization, of the light. Then, the whole structure of the diffuson and the Cooperon boils down to various kernels of type (9.143), where the various depolarization/decoherence rates are rotational invariants that depend only on the angular momenta F_g, F_e (Müller *et al.*, 2001; 2005)

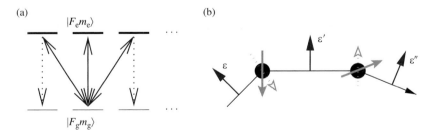

Fig. 9.25 (a) Light scattering by a degenerate atomic dipole transition $F_g \leftrightarrow F_e$ can either preserve spin (Rayleigh transition, full arrows) or change the spin (degenerate Raman transition, dotted arrows). (b) Multiple light scattering by randomly placed atoms with internal spin states involves depolarization/decoherence from both spin-orbit (transversality) and spin-flip effects.

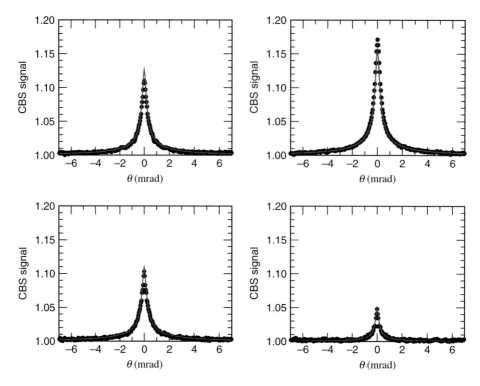

Fig. 9.26 CBS by a cloud of cold rubidium atoms, in four polarization channels (upper/lower left: parallel/perpendicular linear polarization; upper/lower right: circular polarization with non-preserving/preserving helicity). Solid line: calculation taking into account the geometry of the atomic cloud. (Labeyrie *et al.*, 2003). The enhancement factor is strongly reduced compared to ideal Rayleigh scatterers such as strontium, Figure 9.25. This is due to the internal Zeeman structure of the rubidium atom, Figure 9.26(a). Note in particular that the helicity-preserving channel—where the largest enhancement is found for point scatterers such as strontium—gives here the smallest enhancement.

For a typical alkali atom like ^{85}Rb with a $F_{\mathrm{g}} = 3 \to F_{\mathrm{e}} = 4$ resonance line, the longest decoherence time for the Cooperon is $\tau_\phi = \frac{19}{21}\tau_l$ (Müller and Miniatura, 2002), meaning that the CBS interference is quite efficiently killed already by very few scattering events. An immediate consequence is that the CBS cone observed on a cold Rb gas has a much reduced enhancement factor (Jonckheere *et al.*, 2000). A less trivial feature is that the best Rb CBS signal is not observed in the same channels as with Sr. Detailed calculations can be performed and an excellent agreement between the measured and the calculated CBS signals is observed (Labeyrie *et al.*, 2003), see Figure 9.26.

The internal atomic degrees of freedom are here responsible for the loss of coherence. Information flows from the light to the atoms; as long as we do not precisely measure the internal state of each atom, this information is lost and the interference contrast is reduced. This information-theoretic argument can be made quantitative by investigating how much which-path information is stored within the atomic internal degrees of freedom. A quantitative measure of this wave–particle duality, developed originally in the context of Mach–Zehnder-type interferometers (Englert, 1996), can be investigated analytically in the simplest cases and highlights the role of which-path information in the loss of CBS interference visibility (Miniatura *et al.*, 2007).

A simple way to restore phase coherence is to lift the atomic degeneracy by applying an external magnetic field. Fields as small as a few Gauss are enough to detune some of the atomic transitions far from resonance, thus reducing the effect of Raman transitions. It has been experimentally observed and theoretically explained how this can increase the enhancement factor (Sigwarth *et al.*, 2004). We here face a seemingly paradoxical situation (in view of the negative magnetoresistance discussed in Section 9.7), where adding an external magnetic field, which should break the time-reversal symmetry, has the effect of increasing the interference between time-reversed paths! Similarly, strong magnetic fields in electronic samples have been used to align free magnetic impurities, thus reduced spin-flip effects and restore Aharonov–Bohm interference (Washburn and Webb, 1986; Pierre and Birge, 2002).

9.7 Weak localization (WL)

As discussed in the preceding section, the Cooperon is responsible for enhanced backscattering, which implies an increased probability to return to the starting point. In the bulk of a disordered system, diffusive transport is thus hindered. This phenomenon, known as weak localization, is quantitatively expressed by a reduction of the diffusion constant (or dimensionless conductance/conductivity) with respect to the classical diffusion constant expected for phase-incoherent transport.

The weak-localization effect of the Cooperon is expressed by Eq. (9.128). For the sake of concreteness, we will take in the following quantitative estimates the example of atomic matter waves with a quadratic dispersion relation $\varepsilon = \hbar^2 k^2 / 2m$. The free density of states (9.103) is

$$N_0(\varepsilon) = \frac{S_d}{(2\pi)^d} \frac{mk^{d-2}}{\hbar^2}, \qquad (9.148)$$

where $S_d = 2\pi^{d/2}/\Gamma(d/2)$ is the area of the unit sphere in dimension d: $S_1 = 2$, $S_2 = 2\pi$, $S_3 = 4\pi$. The Boltzmann diffusion constant $D_{\mathrm{B}} = \hbar k l / dm$ is directly proportional to the transport mean free path l. Because the Cooperon is isotropic, the d-dimensional integral in Eq.(9.128) can be reduced to a trivial $(d-1)$-dimensional angular integral and a radial integral over momentum q, such that

$$\frac{1}{D} = \frac{1}{D_{\mathrm{B}}} \left(1 + \frac{\hbar}{\pi m k^{d-2} D_{\mathrm{B}}} \int_0^\infty \frac{q^{d-1} \mathrm{d}q}{q^2 - i0} \right). \qquad (9.149)$$

The result of the q-integral depends crucially on the dimensionality of the system. This is a consequence of the fact, well known from classical random walks, that the return probability to the origin is the higher, the lower the spatial dimension d. Therefore, weak—and consequently also strong—localizations are immediately seen to have the largest impact in low-dimensional systems.

9.7.1 $d = 1$

In dimension $d = 1$, the integral (9.149) diverges for small q. However, for a system of size L, the momentum q cannot take arbitrary small values, and a lower cutoff of the order of $1/L$ must be used. A simple way of implementing it—following the recipe of Section 9.6.3 for including decoherence effects—consists in replacing q^2 by $q^2 + 1/L^2$. One then gets:

$$\frac{1}{D} = \frac{1}{D_B}\left(1 + \frac{L}{2l}\right). \tag{9.150}$$

This expression is valid only if the weak-localization contribution is a small correction, i.e. for $L \ll l$. To lowest order, we recover $D \approx D_B(1 - L/2l)$, which is the exact result (9.19) already derived for 1D systems in Section 9.3.3. The interest of the present approach is that a full microscopic theory provides us with the weak-localization correction and thus puts the scaling theory of localization on firm grounds.

9.7.2 $d = 2$

In dimension 2, the integral diverges both for small and large q. A suitable cutoff at small q is again $1/L$, the inverse of the system size. Diffusive transport is a long-time, large-distance behavior. It is not expected to give an accurate description on a scale shorter than the mean free path. Performing the integral with a natural cutoff $1/l$ at large q thus leads to

$$D \approx D_B\left[1 - \frac{2}{\pi kl}\ln\left(\frac{L}{l}\right)\right]. \tag{9.151}$$

In terms of the dimensionless conductance $g = 2mD/\hbar$, this implies the following scaling relation:

$$\beta(g) = \frac{\mathrm{d}\ln g}{\mathrm{d}\ln L} = -\frac{2}{\pi g}. \tag{9.152}$$

Our microscopic calculation thus gives an explicit prediction that can be readily incorporated into scaling theory, as anticipated in Section 9.3.6.

How can weak localization be observed experimentally? *A priori*, any measured diffusion constant incorporates already all interference corrections to the classically expected value. Fortunately, the Cooperon contribution to weak localization is due to the constructive interference between a multiply scattered path and its time-reversal. If one breaks time-reversal symmetry on purpose, then the delicate interference is likely to disappear, and an enhancement of diffusive transport should be observed.

For charged particles—such as electrons in solid-state samples—the simplest way is to add a magnetic field perpendicular to the sample. In the presence of a vector potential $\vec{A}$, a charged particle picks an additional phase $\int e\vec{A}\cdot\mathrm{d}\vec{l}/\hbar$ along a closed

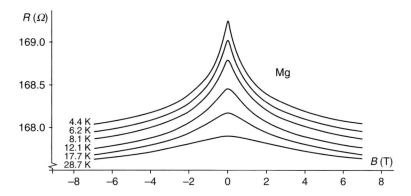

Fig. 9.27 Experimentally measured resistance of a thin Mg film exposed to a perpendicular magnetic field. The magnetic field breaks the constructive interference of waves counter-propagation along closed loops, and reduces the weak-localization effect, and thus results in *negative magnetoresistance*. Similarly, larger temperatures reduce the phase coherence of the electronic wave function, and also reduce weak-localization corrections. Adapted from (Bergmann, 1984) (courtesy of G. Bergmann).

loop. This is simply $e/\hbar$ times the enclosed magnetic flux. Along each closed loop, this additional phase appears in the Cooperon contribution. If this phase fluctuates largely from one loop to the other, the resulting interferential contribution will vanish. As the smallest area enclosed by a diffusive loop is l^2, the weak localization correction is expected to vanish above $B \approx \hbar/el^2$. For a typical mean free path of a fraction of µm, this is in the Tesla range. Figure 9.27 shows the measured resistance of a 2D Mg film vs. magnetic field at various temperatures (Bergmann, 1984). At the lowest temperature, propagation is almost fully phase coherent and one observes a decreasing resistance, i.e. a increasing conductance, when a magnetic field is applied. This *negative magnetoresistance* was a mystery when first observed and only later explained as a manifestation of weak localization. When temperature increases, the phase coherence of the electrons diminishes, and the weak localization correction gets smaller. This is, in a different context, analogous to decoherence phenomena discussed for coherent backscattering in Section 9.6.3. Note that, from such experimental data, it is possible to measure the transport mean free path (via the width of the weak localization peak) as well as the temperature dependence of the decoherence time. In recent times, weak-localization measurements have been used as very sensitive detectors for minute concentrations of magnetic impurities, which induce spin-flip decoherence and are responsible for finite decoherence times even at zero temperature (Müller, 2009; Pierre and Birge, 2002; Pierre *et al.*, 2003).

9.7.3 $d = 3$

In dimension 3, the integral in Eq. (9.149) requires only a cutoff at large q, which we take again as $1/l$ and obtain

$$D \approx D_{\rm B} \left(1 - \frac{3}{\pi(kl)^2} \right).$$ (9.153)

This diffusive Cooperon contribution to weak localization is found to scale as $1/(kl)^2$. However, this is not the whole story, because other diagrams, not included in the simple diffusive Cooperon, give contributions that are actually more important for small disorder $kl \gg 1$. Just as for the CBS cone discussed in Section 9.6.1, also here the double-scattering diagrams appearing in Eq. (9.117) contribute to leading order $1/kl$. In $d = 3$, the *static* electronic conductivity was found to be given by (Belitz and Kirkpatrick, 1994)

$$\frac{\sigma(\omega = 0)}{\mathring{\sigma}} = 1 - \frac{2\pi}{3kl} - \frac{\pi^2 - 4}{(kl)^2} \ln(kl) + O((kl)^{-2}).$$ (9.154)

As long as $kl \gg 1$, the weak localization is only a small correction, again providing us with a macroscopic ground for the scaling theory of localization. It also gives an approximate criterion for the onset of Anderson localization, which should set in approximately when the right-hand sides of Eq. (9.153) or Eq. (9.154) vanish, i.e. $(kl)_{\rm c} = O(1)$. This is Ioffe–Regel criterion, Eq. (9.57). However, the precise calculation of the critical point is a delicate endeavor. What precisely happens at the $1/l$ scale is not universal. The same is true for the Ioffe–Regel criterion, but the latter nonetheless yields a first estimate on where to expect the Anderson transition.

9.7.4   Self-consistent theory of localization

Weak localization describes how diffusive transport is affected by interference. In essence, however, weak localization is a perturbative result: first, because the Cooperon contribution is evaluated using a diffusive kernel valid in the absence of interference; secondly, because this simple approach takes into account only a specific type of diagrams. The first assumption is especially questionable in 1D, where diffusive transport actually never occurs, because localization appears at the very same scale (the localization length) than diffusion (the mean free path). Concerning the second point, the dominant role of the Cooperon in large systems was recognized already by (Gor'kov *et al.*, 1979) and (Abrahams *et al.*, 1979). However, a weak-disorder perturbation theory in powers of $1/kl$ alone would never be able to describe the Anderson transition (in 3D) for strong disorder, nor the crossover from weak to strong localization in 1D and 2D systems.

The self-consistent theory of localization, developed by Vollhardt and Wölfle in the 1980s (Vollhardt and Wölfle, 1980; 1982; 1992), is an attempt to escape this seemingly hopeless situation by applying a suitable self-consistency scheme, as often employed with success to describe phase transitions in statistical physics. Rather than a theory with rigorously controlled approximations, it must be thought of as a guess, albeit highly educated, about the most important contributions of diagrams to all orders. The basic observation is that the diffusive contribution of large closed loops in Eq. (9.149) must itself be modified by weak localization: inside a large loop, the wave explores smaller loops, leading to a decreased diffusion constant for propagation along the large loop. This argument can of course be repeated: one should take into account loops within loops within loops..., all the way down to the smallest loops, stopping at the scale of the transport mean free path.

The whole description must now be self-consistent, describing what happens at every scale from the mean free path up to the size of the system—or toward infinity in the bulk. The simplest idea would be to replace the static Boltzmann diffusion constant D_B in the integral of Eq. (9.149) by the renormalized diffusion constant D itself, thus providing us with an implicit equation for D. It turns out that this is not enough: indeed, a single number—the static diffusion constant D—cannot describe the full dynamics both for short times, where it is diffusive, and for long times where localization may eventually set in. So, we require a scale-dependent diffusion constant, and it turns out that it is simpler to consider various timescales rather than various spatial scales. We thus consider a diffusion constant $D(\omega)$ that depends on frequency ω. The self-consistent expression for $D(\omega)$ just derives from Eq. (9.149) by re-introducing the ω dependence and replacing D_B by $D(\omega)$ in the integral:

$$D(\omega) + \frac{\hbar}{\pi m k^{d-2}} \int \frac{q^{d-1} dq}{q^2 - (i\omega/D(\omega))} = D_B. \qquad (9.155)$$

In the short-time limit $\omega \to \infty$, the contribution of the integral vanishes and one gets back to classical Boltzmann diffusive propagation, as expected. The most interesting part takes place at long times, i.e. in the limit $\omega \to 0$, whose consequences again depend crucially on the dimension.

9.7.4.1 $d = 1$

At finite ω, the integral in Eq. (9.155) does not need any regularization, it is simply $\frac{\pi}{2}\sqrt{D(\omega)/(-i\omega)}$, and the implicit equation for $D(\omega)$ is easily solved (Lobkis and Weaver, 2005):

$$\frac{D(\omega)}{D_B} = \frac{\sqrt{1 - 16i\omega\tau_l} - 1}{\sqrt{1 - 16i\omega\tau_l} + 1}, \qquad (9.156)$$

where $\tau_l = l^2/D_B$ is the mean free time between two scattering events. This function is plotted in the left panel of Figure 9.28 as a function of $-i\omega$.[10] In the limit of small ω, it behaves linearly $D/D_B \approx -4i\omega\tau_l$. This in turns implies that the propagation kernel $1/(-i\omega + D(\omega)q^2)$ is just $1/(-i\omega) \times 1/(1 + 4l^2 q^2)$. When going back from momentum to configuration space by inverse Fourier transform, it implies that the intensity kernel is proportional to $\exp(-|z|/2l)$ at long times. It successfully describes exponential localization with the localization length $\xi_{loc} = 2l$, i.e. the exact result for the localization length! The elementary ingredients used for obtaining this important result are: quantum kinetic theory, microscopic calculation of the weak localization correction in the perturbative regime and its self-consistent extension. That the exact result is eventually obtained is a strong hint that the self-consistent approach catches an important part of the physics of localization.

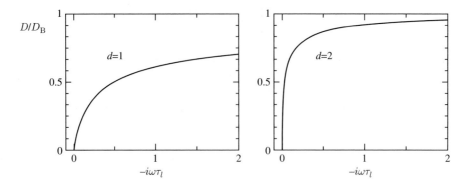

Fig. 9.28 Diffusion constant vs. (imaginary) frequency, in units of Boltzmann diffusion constant and mean free time, respectively, as predicted by the self-consistent theory of localization, in 1D (left) and 2D (right, for $kl=1.5$). At large ω (short time), one recovers the classical Boltzmann diffusive behavior. At small ω, the dependence is linear, which implies exponential localization in configuration space, in agreement with scaling theory, numerical and experimental observations (in 1D). In 2D, the localized regime is reached at much smaller frequency because the localization length and time are exponentially large.

But beware! This triumph is somewhat tarnished by the fact that it is the *typical* intensity that decays with ξ_{loc}, whereas the *average* intensity calculated here should asymptotically decay with $4\xi_{\mathrm{loc}}$, as shown in Section 9.2.5. The precise source for this discrepancy escapes our present understanding. Clearly, the self-consistent theory is built for the average intensity kernel $\langle G^{\mathrm{R}} G^{\mathrm{A}} \rangle$ and thus cannot describe the huge fluctuations in the localized regime. One lacks a diagrammatic expansion for the typical transmission, which would require us to calculate contributions of advanced and retarded Green functions to all orders.

Decoherence effects can be easily included in the self-consistent approach by the replacement $-i\omega \mapsto -i\omega + 1/\tau_\phi$ explained in Section 9.6.3. In Figure 9.28, this replacement simply translates the curve horizontally to the left. One immediately finds that the diffusion constant no longer vanishes at $\omega = 0$, but takes a finite value, implying diffusive motion at long times. In the limit of weak decoherence $\tau_l \ll \tau_\phi$, the residual diffusion constant is $D \approx 4\tau_l D_{\mathrm{B}}/\tau_\phi = \xi_{\mathrm{loc}}^2/\tau_\phi$. It is much smaller than the Boltzmann diffusion constant and allows for a simple physical interpretation: a phase-breaking event, occurring on average every τ_ϕ, destroys the delicate interference responsible for localization. This implies a restart of diffusion during time τ_l after which localization sets in again, until the next phase-breaking event, etc.

9.7.4.2 $d = 2$

In 2D, the integral in Eq. (9.155) diverges in the large-q limit, requiring a regularization. The natural short-distance cutoff is the mean free path l. Elementary manipulations show that $D(\omega)$ is implicitly determined by

$$\frac{D(\omega)}{D_{\mathrm{B}}} = 1 - \frac{1}{\pi kl} \ln\left(1 - \frac{D(\omega)}{D_{\mathrm{B}}} \frac{1}{2i\omega\tau_l}\right). \tag{9.157}$$

In contrast with the 1D case, D/D_{B} is not a universal function, it depends on the parameter kl. The right panel of Figure 9.28 plots it for $kl = 1.5$. It displays the classical diffusive behavior $D \approx D_{\mathrm{B}}$ at large ω (short times), and localization at long times. Indeed, for $\omega \to 0$, one finds $D(\omega) \approx -i\omega\xi_{\mathrm{loc}}^2$, i.e. exponential localization with the localization length

$$\xi_{\mathrm{loc}} = l\sqrt{\exp\left(\pi kl\right) - 1} \approx l\exp\left(\frac{\pi kl}{2}\right). \tag{9.158}$$

This provides us with a microscopic derivation of the result of scaling theory, Eq. (9.56). The self-consistent approach describes correctly the exponentially large localization length in 2D.[11] Note that, even for strong disorder with a rather small value $kl = 1.5$, the linear regime in Figure 9.28 is observed only at very small ω, i.e. for very long times.

Decoherence can be taken into account exactly like in 1D. Instead of a true metal–insulator transition, one observes a crossover from classical diffusion at large kl towards a residual diffusion (triggered by decoherence) at small kl. Explicit calculations have been carried out in (Miniatura *et al.*, 2009) for the case of atomic matterwaves in a speckle potential, where residual spontaneous emission is one source of decoherence that can be experimentally tuned. Figure 9.30 shows typical results. Because of the

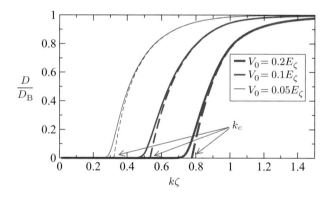

Fig. 9.29 Diffusion constant (normalized to the Boltzmann diffusion constant) computed from the self-consistent theory of localization, for atomic matterwaves with wavevector k exposed to a 2D speckle potential with correlation length ζ and different amplitudes V_0. A stronger potential means a smaller value of kl. Dashed lines are the prediction of the simple perturbative weak localization correction, Eq. (9.151), solid lines the result of the self consistent approach, Eq. (9.157), including residual decoherence due to spontaneous emission implemented via Eq. (9.142). A rather sharp crossover between the Boltzmann diffusive behavior at high energy and the quasi-localized behavior at low energy is observed around a critical value k_c (Miniatura *et al.*, 2009).

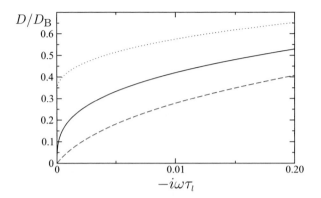

Fig. 9.30 Diffusion constant vs. (imaginary) frequency, in units of Boltzmann diffusion constant and mean free time, respectively, computed from the self-consistent theory of localization in 3D, Eq. (9.159). Dotted line: metallic regime $kl = 1.2$ with finite diffusion constant at long times. Dashed line: insulating regime $kl = 0.8$ with a finite localization length. Solid black line: critical point $kl = (kl)_c = \sqrt{3/\pi}$ of the metal–insulator Anderson transition, where the diffusion constant scales like $(-i\omega)^{1/3}$, implying an anomalous diffusion $\langle r^2(t) \rangle \propto t^{2/3}$ at long times.

exponential dependence in 2D, the crossover from quasi-localized behavior at small k to diffusive behavior at large k is rather rapid. In any case, a crucial requirement is to have very cold atoms, with de Broglie wavelength shorter than the speckle correlation length.

Considering an expanding BEC wave packet released from a harmonic trap, one can calculate the expected stationary (for negligible decoherence) density distribution along the lines of Eq. (9.63). Just as in 1D, the asymptotic decay is governed by the wavevector k_{max} of the fastest atoms, and the density is predicted to be $\langle |\psi(r)|^2 \rangle \approx Cr^{-5/2} \exp\{-r/\xi_{loc}(k_{max})\}$ (Miniatura *et al.*, 2009).

9.7.4.3 $d = 3$

In 3D, the same short-distance regularization as in 2D is necessary, leading to the following implicit equation, valid in the limit $\omega\tau_l \ll 1$:[12]

$$\frac{D(\omega)}{D_{\mathrm{B}}} + \frac{3}{\pi(kl)^2}\left(1 - \frac{\pi}{2}\sqrt{\frac{-3i\omega\tau_l}{D(\omega)/D_{\mathrm{B}}}}\right) = 1. \qquad (9.159)$$

The behavior of the solution, shown in Figure 9.31, depends on the Ioffe–Regel parameter kl and defines three distinct regimes:

Diffusive regime:.   For $kl > (kl)_c = \sqrt{3/\pi}$, D/D_{B} tends to a constant value in the limit $\omega \to 0$, which means that the system always behaves diffusively, albeit with a diffusion constant smaller than the Boltzmann diffusion constant. This is the regime of weak localization.

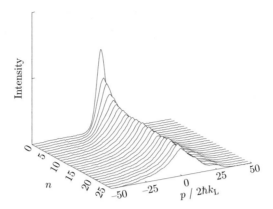

Fig. 9.31 Experimental time evolution of the momentum distribution of the atomic kicked rotor (Moore *et al.*, 1994), from the initial Gaussian distribution until the exponentially localized distribution at long time; N is the number of kicks (courtesy of M. Raizen).

Localized regime:. For $kl < (kl)_c$, it is easy to see that $D/D_B \to 0$ in the limit $\omega \to 0$. More precisely, one has exponential localization:

$$D(\omega) \approx -i\omega\xi_{\text{loc}}^2 \qquad \text{with} \qquad \xi_{\text{loc}} \sim \frac{1}{(kl)_c - kl} \qquad (9.160)$$

immediately below the Anderson transition, on the insulating side. This means, see Eq. (9.58), that the critical exponent deduced from the self-consistent approach is $\nu = 1$, quite far from the true value $\nu = 1.58$ known from numerical simulations (Slevin and Ohtsuki, 1999). The reason lies in the approximate character of the self-consistent approach, which disregards the huge fluctuations in the vicinity of the critical point. Field-theoretic approaches can in principle capture the effect of fluctuation and have been quantitatively tested in $d = 2 + \epsilon$ dimensions (Evers and Mirlin, 2008). For $d = 3$, however, to our knowledge no analytical theory is available that makes a better quantitative prediction than the self-consistent theory.

Critical regime:. At the critical point $kl = (kl)_c$, it is easy to see that the solution of Eq. (9.159) scales like $D(\omega) \sim (-i\omega)^{1/3}$. Consequently, the critical behavior is anomalous diffusion where the squared extension increases subdiffusively at long times: $\langle r^2(t) \rangle \propto t^{2/3}$. This anomalous diffusion has been experimentally observed with the quasi-periodically kicked rotor, see Section 9.8.

9.8   Kicked rotor

The physics of Anderson localization is, as amply discussed in the preceding sections, highly dependent on the dimension of the system. While the 1D situation is fairly well understood—localization is the generic behavior, the localization length is comparable to the mean free path, and the fluctuation properties in the localized regime are essentially well understood—the physics of higher dimensions is much richer still. Dimension 3 is especially interesting, as one expects a so-called mobility edge, separating, in the continuum case, localized states at low energy/strong disorder from extended states at high energy/weak disorder.

As explained in Section 9.4.3, it is very difficult to find a clean experimental system to observe this metal–insulator Anderson transition unambiguously. Cold atomic matterwaves are very attractive because they can be directly observed, and because most experimental imperfections as well as atom–atom interactions can be precisely controlled, if not reduced to a minimum. The main difficulty consists in reaching the Ioffe–Regel threshold $kl = O(1)$, Eq. (9.57), i.e. preparing a sufficiently small k (low energy, large de Broglie wavelength) and short mean free path l. Indeed, the latter cannot be shorter than the correlation length of the disordered potential, i.e. of the order of $1\,\mu$m for optical speckle.

This limitation on the mean free path can be overcome using a different approach, where disorder is not provided by an external potential in configuration space, but by classically chaotic dynamics in momentum space. This idea has been realized experimentally with the atomic kicked rotor, and Anderson localization in 1D has been observed as early as 1994 (Moore *et al.*, 1994), 14 years prior to the widely noticed Anderson localization in configuration space (Billy *et al.*, 2008)! A key advantage of the kicked rotor is that is does not require ultracold atoms from a Bose–Einstein condensate: a standard magneto-optical trap suffices to prepare the initial state. The kicked rotor also has permitted the clean observation of the metal-insulator Anderson transition in 3D, and the first experimental measurement of the critical exponent, with non-interacting matterwaves (Chabé *et al.*, 2008).

9.8.1 The classical kicked rotor

We consider a one-dimensional rotor whose position can be described by the angle x (defined modulo 2π) and the associated momentum p, and kick it periodically with a position-dependent amplitude. In properly scaled units, the Hamiltonian function can be written as

$$H = \frac{p^2}{2} - k \cos x \sum_{n=-\infty}^{+\infty} \delta(t - nT), \qquad (9.161)$$

where T and k are period and strength of the kicks, respectively.

Because of the time dependence, energy is not conserved, but thanks to the time periodicity, we can analyze the motion stroboscopically and build a Poincaré map picturing the evolution once every period. This map relates the phase-space coordinates just before kick $n + 1$ to the coordinates just before kick n:

$$\begin{cases} I_{n+1} = I_n + K \sin x_n \\ x_{n+1} = x_n + I_{n+1}, \end{cases} \qquad (9.162)$$

where $K = kT$ and $I_n = Tp_n$. This is simply the celebrated standard map (also known as the Chirikov map) that has been widely studied (Lichtenberg and Lieberman, 1983; Casati *et al.*, 1990): it is almost fully chaotic and ergodic around $K = 10$ and above.

When the stochasticity parameter K is very large, each kick is so strong that the positions of the consecutive kicks can be taken statistically uncorrelated. By averaging, one thus gets:

$$\langle p_{n+1}^2 \rangle \simeq \langle p_n^2 \rangle + k^2 \langle \sin^2 x_n \rangle \simeq \langle p_n^2 \rangle + \frac{k^2}{2}. \tag{9.163}$$

It follows that the motion in momentum space is diffusive ($\langle p^2 \rangle$ increases linearly with time) with diffusion constant

$$D = \frac{k^2}{2T}. \tag{9.164}$$

Numerical experiments (Lichtenberg and Lieberman, 1983) show that this expression works well for $K \geq 10$. Note that the kicked rotor is a perfectly deterministic system, without any randomness. It is the chaotic nature of the classical motion, and thus its extreme sensitivity to perturbations, which renders the deterministic classical motion diffusive *on average*.

9.8.2   The quantum kicked rotor

The quantum Hamiltonian is obtained from the classical one, Eq. (9.161), through the canonical replacement of p by $-i\hbar\partial_x$. The evolution operator over one period is the product of the free evolution operator and the instantaneous kick operator:

$$U = U(T,0) = \exp\left(-\frac{i}{\hbar} \frac{p^2 T}{2} \right) \exp\left(\frac{i}{\hbar} k \cos x \right). \tag{9.165}$$

The long-time dynamics is generated by successive iterations of U. Thus, one can use the eigenstates of U as a basis set. U being unitary, its eigenvalues are complex numbers with unit modulus:

$$U|\phi_i\rangle = \exp\left(-\frac{iE_i T}{\hbar} \right) |\phi_i\rangle, \tag{9.166}$$

with $0 < E_i \leq 2\pi\hbar/T$ are defined modulo $2\pi\hbar/T$. They are not exactly the energy levels of the system—the $|\phi_i\rangle$ are not stationary states of the time evolution, but are only periodic—and are called quasi-energy levels, the $|\phi_i\rangle$ being the Floquet eigenstates. This Floquet description is the time analog of the Bloch theorem that applies to spatially periodic potentials.

9.8.3   Dynamical Localization

The quantum dynamics of the kicked rotor can be quite simply studied numerically by repeated application of the one-period evolution operator U to the initial state, alternating free-propagation phases with instantaneous scattering events in momentum space induced by the kicks. The free evolution between kicks, $\exp\left(-ip^2 T/2\hbar\right)$, is diagonal in momentum representation, such that each momentum eigenstate, characterized by its momentum $m\hbar$ with integer m, picks up a different phase shift. The kick operator $\exp\left(ik\cos\theta/\hbar\right)$, in contrast, is diagonal in position representation and couples different momenta. Being unitary, it plays the role of a scattering matrix in momentum space and contains the quantum amplitude for changing an incoming initial momentum in an outgoing one, under the influence of one kick; k is the parameter controlling the scattering strength. The dynamics of the kicked rotor can be seen as a sequence of scattering events interleaved with free propagation phases.

For sufficiently large $K = kT$, the classical dynamics is diffusive in momentum space, but it should come as no surprise to the reader now familiar with 1D Anderson localization, that the quantum dynamics may be localized at long times. This localization was baptized "dynamical localization" when it was observed in numerical simulations (Casati *et al.*, 1979). Only later, people realized that it is simply the Anderson scenario of 1D localization, as explained below.

Dynamical localization has been experimentally observed in the dynamics of a Rydberg electron exposed to an external microwave field (Buchleitner *et al.*, 1995). Arguably the simplest observation uses a cold atomic gas, prepared in a standard magneto-optical trap with a typical velocity spread of few recoil velocities (Moore *et al.*, 1994; Chabé *et al.*, 2008; Ammann *et al.*, 1998). After the trap is switched off, a periodic train of laser pulses is applied to the atoms. Each pulse is composed of two far-detuned counter-propagating laser beams producing a spatially modulated optical potential. Each laser pulse thus produces a kick on the atom velocity, whose amplitude is proportional to the gradient of the optical potential.

If the kicks are infinitely short, we recover exactly the kicked rotor, Eq. (9.161), where the position of the atom in the standing wave plays the role of the x variable and its velocity is the p variable. The kick strength k is proportional to the laser intensity divided by the detuning. The spatial dimensions perpendicular to the laser beams do not play any role in the problem, so that we have an effectively one-dimensional time-dependent problem. The mapping of the dimensionfull Hamiltonian for cold atoms to the kicked rotor Hamiltonian, Eq. (9.161), shows that the effective Planck's constant of the problem (Lemarié *et al.*, 2009) is $\hbar_{\mathrm{eff}} = 4\hbar k_L^2 T/M = 8\omega_r T$, where k_L is the laser wave number and M the atomic mass. Up to a numerical factor, it is the ratio of the atomic recoil frequency ω_r to the pulse frequency, and can be easily varied in the experiment, from the semi-classical regime $\hbar_{\mathrm{eff}} \ll 1$ to the quantum regime $\hbar_{\mathrm{eff}} \sim 1$.

After the series of pulses is applied, the momentum distribution is measured either by a time-of-flight technique (Moore *et al.*, 1994) or velocity selective Raman transitions (Chabé *et al.*, 2008). Figure 9.32 shows the momentum distribution as a function of time. While, at short time, the distribution is Gaussian—as expected for a classical diffusion—its shape changes around the localization time and evolves toward an exponential shape $\exp(-|p|/\xi_{\mathrm{loc}})$ at long time, a clear-cut manifestation of Anderson/dynamical localization.

Adding decoherence on the system—either by adding spontaneous emission (Ammann *et al.*, 1998) or by weakly breaking the temporal periodicity (Klappauf *et al.*, 1998)—induces some residual diffusion at long time, in accordance with the discussion in Section 9.6.3 and 9.7.4. This is another proof that dynamical localization is based on delicate destructive interference.

9.8.4   Link between dynamical and Anderson localizations

So far, we have only made plausible that dynamical localization with the quantum kicked rotor is similar to Anderson localization in a spatially disordered medium. We now demonstrate the connection between the two phenomena, following (Grempel et al., 1984). Consider the evolution operator, Eq. (9.165), and the associated eigenstate $|\phi\rangle$ with quasi-energy E. The part of the evolution operator associated with the kick can be written as:

$$\exp\left(\frac{i}{\hbar}\, k \cos x\right) = \frac{1 + iW(x)}{1 - iW(x)} \tag{9.167}$$

where $W(x)$ is a periodic Hermitean operator that can be Fourier expanded:

$$W(x) = \sum_{r=-\infty}^{\infty} W_r \, \exp\left(irx\right). \tag{9.168}$$

Similarly, the kinetic part can be written as:

$$\exp\left[-\frac{i}{\hbar}\left(\frac{p^2}{2} - E\right) T\right] = \frac{1 + iV}{1 - iV}. \tag{9.169}$$

The operator V is diagonal is the eigenbasis of p, labelled by the integer m (see above). If one performs the following expansion in this basis set,

$$\frac{1}{1 - iW(x)}|\phi\rangle = \sum_{m} \chi_m \,|m\rangle, \tag{9.170}$$

it is straightforward to show that the eigenvalue equation (9.166) can be rewritten as

$$\epsilon_m \chi_m + \sum_{r \neq 0} W_r \chi_{m-r} = -W_0 \chi_m, \tag{9.171}$$

where

$$\epsilon_m = \tan\left[\left(E - \tfrac{1}{2}m^2\hbar^2\right) T/2\hbar\right]. \tag{9.172}$$

Equation (9.171) is the time-independent Schrödinger equation for a one-dimensional Anderson model, cf. Eq. (9.68), with site index m, on-site energy ϵ_m, coupling W_r to the nearest sites and total energy W_0. Compared to Eq. (9.68), there are two new ingredients: first, there are additional hopping amplitudes to other neighbors. But since they decrease sufficiently fast at large distance, they do not play a major role. Secondly, the ϵ_m values, determined deterministically by Eq. (9.172), are not really random variables, but only pseudo-random[13] with a Lorentzian distribution.[14] Still, localization is expected and indeed observed. The computation of the localization length follows the general lines explained in Section 9.5, and is in good agreement with experimental observations.

It should be emphasized that space and time play different roles in the Anderson model and in dynamical localization. What plays the role of the sites of the Anderson model are the momentum states. This is why dynamical localization is not observed in configuration space, but in momentum space.

9.8.5 The quasi-periodically kicked rotor

How can the kicked rotor be used to study Anderson localization in more than one dimension? The first idea is to use a higher-dimensional rotor with a classically chaotic dynamics and to kick it periodically. It turns out that this is not easily realized experimentally, as it requires building a specially crafted spatial dependence (Wang and Garcia-Garcia, 2009). Yet, remember that time and space have switched roles, and so a simpler idea is to use additional temporal dimensions rather than spatial dimensions. Instead of kicking the system periodically with kicks of constant strength, one may use a temporally quasi-periodic excitation. Various schemes have been used (Lignier *et al.*, 2005), but the one allowing to map on a multidimensional Anderson model uses a quasi-periodic modulation of the kick strength, the kicks being applied at fixed time interval (Casati *et al.*, 1989).

We will be interested in a 3D Anderson model, obtained by adding two quasi-periods to the system:[15].

$$\mathcal{H}_{\text{qp}} = \frac{p^2}{2} + \mathcal{K}(t) \cos x \sum_n \delta(t - n), \tag{9.173}$$

with

$$\mathcal{K}(t) = K \left[1 + \varepsilon \cos \left(\omega_2 t + \varphi_2 \right) \cos \left(\omega_3 t + \varphi_3 \right) \right]. \tag{9.174}$$

Now where is the three-dimensional aspect in the latter Hamiltonian? The answer lies in a formal analogy between this quasi-periodic kicked rotor and a 3D kicked rotor with the special initial condition of a "plane source", as follows.

Take the Hamiltonian of a 3D, periodically kicked rotor:

$$\mathcal{H} = \frac{p_1^2}{2} + \omega_2 p_2 + \omega_3 p_3 + K \cos x_1 \left[1 + \varepsilon \cos x_2 \cos x_3 \right] \sum_n \delta(t - n), \tag{9.175}$$

and consider the evolution of a wave function Ψ with the initial condition

$$\Psi(x_1, x_2, x_3, t = 0) \equiv \psi(x_1, t = 0) \delta(x_2 - \varphi_2) \delta(x_3 - \varphi_3). \tag{9.176}$$

This initial state, perfectly localized in x_2 and x_3 and therefore entirely delocalized in the conjugate momenta p_2 and p_3, is a "plane source" in momentum space (Lobkis and Weaver, 2005). A simple calculation shows that the stroboscopic evolution of Ψ under Eq. (9.175) coincides exactly with the evolution of the initial state $\psi(x = x_1, t = 0)$ under the Hamiltonian (9.173) of the quasi-periodically kicked rotor (for details, see (Lemarié *et al.*, 2009)). An experiment with the quasi-periodic kicked rotor can thus be seen as a localization experiment in a 3D disordered system, where localization is actually observed in the direction perpendicular to the plane source. In other words, the situation is comparable to a transmission experiment where the sample is illuminated by a plane wave and the exponential localization is only measured along the wavevector direction. Therefore, the behavior of the quasi-periodic kicked rotor (9.173) matches *all* dynamic properties of the quantum 3D kicked rotor.

The classical dynamics has been shown to be a chaotic diffusion, provided the parameter ε is sufficiently large to ensure efficient coupling between the 3 degrees of freedom (Lemarié *et al.*, 2010). As for the standard 3D kicked rotor (9.175), its quantum dynamics can be studied using the Floquet states via mapping to a 3D Anderson-like model:

$$\epsilon_{\mathbf{m}}\Phi_{\mathbf{m}} + \sum_{\mathbf{r}\neq 0} W_{\mathbf{r}}\Phi_{\mathbf{m-r}} = -W_0\Phi_{\mathbf{m}}, \tag{9.177}$$

where $\mathbf{m} \equiv (m_1, m_2, m_3)$ labels sites in a 3D cubic lattice, the on-site energy $\epsilon_{\mathbf{m}}$ is

$$\epsilon_{\mathbf{m}} = \tan\left\{\frac{1}{2}\left[\omega - \left(\hbar\frac{m_1^2}{2} + \omega_2 m_2 + \omega_3 m_3\right)\right]\right\}, \tag{9.178}$$

and the hopping amplitudes $W_{\mathbf{r}}$ are the Fourier expansion coefficients of

$$W(x_1, x_2, x_3) = \tan\left[K\cos x_1(1 + \varepsilon\cos x_2\cos x_3)/2\hbar\right]. \tag{9.179}$$

A necessary condition for localization is obviously that $\epsilon_{\mathbf{m}}$ not be periodic. This is achieved if $(\hbar, \omega_2, \omega_3, \pi)$ are incommensurate. When these conditions are verified, localization effects as predicted for the 3D Anderson model are expected, namely either a diffusive or a localized regime. Localized states would be observed if the disorder strength is large compared to the hopping. In the case of the model (9.177), the amplitude of the disorder is fixed, but the hopping amplitudes can be controlled by changing the stochasticity parameter K (and/or the modulation amplitude ε): $W_{\mathbf{r}}$ is easily seen to increase with K. In other words, the larger K, the smaller the disorder. One thus expects to observe diffusion for large stochasticity K and/or modulation amplitude ε (small disorder) and localization for small K and/or ε (large disorder). It should be emphasized that *stricto sensu* there is no mobility edge in our system that would separate localized from delocalized eigenstates. Depending on the parameters $K, \hbar, \varepsilon, \omega_2, \omega_3$, either *all* Floquet states are localized or all are delocalized. The boundary of the metal–insulator transition is in the $(K, \hbar, \varepsilon, \omega_2, \omega_3)$-parameter space. As seen below, K and ε are the primarily important parameters.

In the experiment performed at the University of Lille (Chabé *et al.*, 2008), kicks are applied to atoms with an initially narrow momentum distribution, and the final momentum distribution is measured using velocity-selective Raman transitions.[16] Figure 9.33 shows the experimental data. For large disorder, one clearly sees the initial diffusive phase and the freezing of the quantum dynamics in the localized regime (lower curve). In the diffusive regime (upper curve), $\langle p^2(t)\rangle$ is seen to increase linearly with time. The intermediate curve displays an anomalous diffusion $\langle p^2(t)\rangle \sim t^{2/3}$. The anomalous exponent $2/3$ is exactly the prediction of the self-consistent theory of localization, Section 9.7.4.3, which also fully agrees with the scaling theory of localization. Time here plays the role of the system size L in the scaling theory: going to longer times means following the renormalization flow in figure 9.7. Only exactly

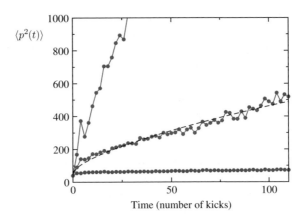

Fig. 9.32 Experimentally measured temporal dynamics of the quasi-periodically kicked rotor, for increasing values of the kick strength. The average kinetic energy $\langle p^2(t) \rangle$ tends to a constant in the localized regime (lower curve, $K = 4$, $\varepsilon = 0.1$), increases linearly with time in the diffusive regime (upper curve, $K = 9$, $\varepsilon = 0.8$). At the critical point $K = K_c \approx 6.4$ (middle curve), anomalous diffusion $\langle p^2(t) \rangle \sim t^{2/3}$ (dashed curve) is clearly observed.

at the unstable critical point will the anomalous diffusion subsist for arbitrarily long times. At slightly larger (respectively, smaller) K, the motion will eventually turn diffusive (respectively, localized) at long time. Experimental constraints prevent the observation beyond 150–200 kicks. Numerical simulations may extend much beyond: it has been checked that the anomalous diffusion with exponent 2/3 is followed for at least 10^8 kicks (Lemarié *et al.*, 2009).

Since in numerical or experimental practice one always works in finite-size systems, we should emphasize that there is an important difference between a true metal–insulator transition and a crossover between two limiting behaviors. For example, consider the simplest 1D situation where the dynamics eventually localizes for sure, with a localization time depending on the kick strength K. Over a finite experimental time, one may observe an apparently diffusive behavior if the localization time is longer than the duration of the experiment.[17] An intermediate situation with the localization time comparable to the duration of the experiment could produce data looking like anomalous diffusion. However, this could be only a transient behavior and a longer measurement will eventually show localization. In contrast, the $t^{2/3}$ behavior at the critical point of the Anderson transition is not a transient behavior, it extends to infinity, highlighting the scale-free behavior with fluctuations of all sizes present right at the critical point.

The unavoidable experimental limitation by finite size can also be turned into a powerful tool of analysis. It is known as finite-size scaling (Fisher and Barber, 1972) and has its roots in the scaling properties observed in the vicinity of the transition. The idea is that all results, obtained for various values of parameters and time, are described by a universal scaling law depending on a single parameter, namely the distance to the critical manifold. Close to the transition, there is only one characteristic length (which diverges at the critical point) and all details below this scale are irrelevant. Such an approach has been extremely successful to extract critical parameters from numerical simulations of the Anderson model for various system sizes. The approach has been transposed to the kicked rotor—see (Lemarié *et al.*, 2009; Lemarié, 2009) for details—and makes it possible to extract the localization length (in momentum space) from numerical or experimental data acquired over a restricted time interval.

The results are shown in Figure 9.33 for both numerical simulations and experimental observations. One clearly sees the divergence of the characteristic length (the localization length on the insulator side) in the vicinity of the transition. The divergence is smoothed by experimental imperfections and the finite duration of numerical and real experiments. The smoothing is much more important in the latter case than in the former one, because the duration of the real experiment (110 kicks maximum) is about 4 orders of magnitude shorter than in numerical experiments. It is nevertheless possible to extract the critical exponent of the transition. For the numerical experiments, one finds $\nu = 1.58 \pm 0.01$ in perfect agreement with the best determination on the Anderson model. Moreover, it has been checked that this exponent is universal, i.e. independent of the microscopic details such as the choice of the parameters $\hbar, \omega_2, \omega_3$ (Lemarié *et al.*, 2009). This is an additional confirmation that the transition observed is actually the metal–insulator Anderson transition.

The critical exponent can also be determined—albeit with reduced accuracy—from the experimental data (Chabé *et al.*, 2008). For the data of Figure 9.33, one obtains:

$$\nu_{\text{exp}} = 1.4 \pm 0.3. \tag{9.180}$$

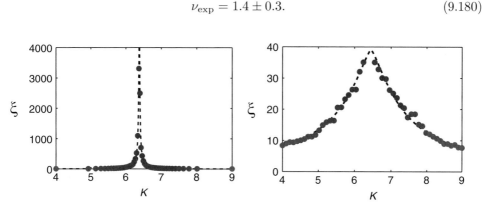

Fig. 9.33 Characteristic length (for localization in momentum space) extracted from numerical (left) and real (right) experiments on the quasi-periodically kicked rotor, in the vicinity of the metal-insulator Anderson transition (Chabé *et al.*, 2008). Finite-size scaling is used. The characteristic length is proportional to the localization length on the insulating side, and to the inverse of the diffusion constant on the metallic side. It has an algebraic divergence $1/|K - K_c|^{\nu}$ at the transition, smoothed by finite size and decoherence, which are more important in the real experiment (limited to 110 kicks) than in the numerical calculations (up to one million kicks). In both cases, it is possible to extract a rather precise estimate of the critical exponent $\nu \approx 1.5$.

These values are in excellent agreement with the numerical results. The key point is that the exponent significantly differs from unity, which is the value deduced from solid-state measurements, see Figure 9.15, and the prediction of the self-consistent approach.

Since the atom–atom contact interaction in a cold dilute gas is much smaller than the electron–electron Coulomb interaction in a solid sample, and since atoms are less easily lost than photons, cold atoms appear particularly suitable for precise measurements of the Anderson transition. Moreover, the possibility to picture wave functions directly opens the way to studies of fluctuations in the vicinity of the critical point (Lemarié *et al.*, 2010), and may even permit us to observe multi-fractal behavior (Faez *et al.*, 2009) with matterwaves. The flexibility of the kicked rotor could also be used to study the Anderson transition in lower dimensions (by reducing the number of quasi-periods) or, why not, even higher dimensions (by increasing it beyond 3). In any case, it is an attractive alternative to experiments on spatially disordered systems.

Notes

1. In the literature, the localization length is defined as the characteristic length for the decay of either $|\psi|^2$ or of $|\psi|$. The two quantities of course differ by a factor 2. Usage of one or the other definition depends on the community, but may also fluctuate from paper to paper. One has to live with this source of disorder.

2. See (Lugan *et al.*, 2009; Gurevich and Kenneth, 2009) for a situation where backscattering of an atom by a smooth speckle potential is zero at lowest order, but localization still prevails due to higher orders in perturbation theory.

3. This definition, as well as Eq. (9.44), uses $\hbar/m$ available for quantum matter waves. It should be adapted to any other specific transport problem under study, along the same lines. The somewhat arbitrary factor of 2 is included for future convenience.

4. Alternatively, one can also solve for k_E as the modified k-vector of an excitation with given energy E.

5. which was invented first for this purpose by Joseph Fourier, namesake of the French university in Grenoble hosting the Les Houches school.

6. Whether this holds to all orders in perturbation theory is to our knowledge an open, and also interesting question, especially in optical speckle potentials (Lugan *et al.*, 2009).

7. The situation is actually more complicated, because this expansion experiment strictly speaking does not measure the transmission across a sample. Instead, one starts with an atomic density inside the medium and see how it propagates. The boundary conditions are thus different from those of a transmission experiment with its connection to outside leads. Still, huge fluctuations must exist in the localized regime, implying that the average transmission deviates from a pure exponential, as discussed in Section 9.2.5.

8. Not taking this into account may lead to incorrect results, see for example (Assaf and Akkermans, 2007), corrected in (Grémaud *et al.*, 2008).

9. Thus, one must take statements like "Raman scattered light is incoherent", often made by quantum opticians, with great care. It is true that Raman scattered light does not interfere with the incoming reference beam—because the final states of the atom are different—but a single Raman-scattered photon along two different paths does interfere with itself very well.

10. Imaginary frequency is only used for convenience, as it makes the diffusion constant purely real and thus easier to plot. The most important transport property is the small-$|\omega|$ behavior, which is linear in the localized regime, both for real or imaginary ω.

11. The same caveats as in 1D exist, concerning average versus typical quantities.

12. This formula features only the static Cooperon contribution, which suffices for qualitative predictions, but should be extended to include the full interference terms appearing in Eq. (9.154) when quantitative precision is necessary.

13. As is well known, "random-number generators" implemented in computers also generate deterministic, merely pseudo-random sequences; most of them use formulae analogous to Eq. (9.172).

14. The non-random character appears for example, when the product $\hbar T/2$ is chosen as an integer multiple of 2π. Then all ϵ_m are equal, the motion is ballistic and localization is absent. Similarly, when $\hbar T$ is commensurate with π (the so-called quantum resonances), the sequence ϵ_m becomes periodic, and Anderson localization does not take place, giving way to Bloch-band transport.

15. In this section, we take the kicking period T as the unit of time.

16. Measuring the average $\langle p^2(t) \rangle$ is tedious and very sensitive to noise in the wings of the momentum distribution. It is much easier to measure the atomic population $\Pi_0(t)$ at zero momentum. Because of atom-number conservation, $\langle p^2(t) \rangle$ is roughly proportional to $1/\Pi_0^2(t)$. The proportionality factor depends on the shape of the distribution, but does not show large changes. As we are interested in scaling properties, $1/\Pi_0^2(t)$ or $\langle p^2(t) \rangle$ are essentially equivalent.

17. This also occurs for 1D Anderson localization in a speckle potential (Billy *et al.*, 2008), as already pointed out in Section 9.5.2.4: Because the localization length and time vary rapidly with energy, one observes localization at low energy and apparently diffusive behavior at high energy. In between, an apparent mobility edge appears (Lugan *et al.*, 2009), which should not be confounded with the true Anderson metal–insulator transition taking place in the thermodynamic limit, although the experimental signatures may be similar.

References

Abrikosov, A. A. "The paradox with the static conductivity of a one-dimensional metal", Solid State Commun. **37**, 997 (1981).

Abrahams, E., Anderson, P. W., Licciardello, D. C., and Ramakrishnan, T. V. "Scaling Theory of localization: Absence of quantum diffusion in two dimensions", Phys. Rev. Lett. **42**, 673 (1979).

Akkermans, E. and Montambaux, G. *Mesoscopic physics of electrons and photons*, (Cambridge University Press, 2007).

van Albada, M. P. and Lagendijk, A., "Observation of weak localization of light in a random medium", Phys. Rev. Lett. **55**, 2692 (1985).

Allen, L. and Eberly, J. H. *Optical resonance and two-level atoms* (Dover Publications, New York, 1987).

Altland, A. and Simons, B. *Condensed matter field theory* (Cambridge University Press, 2006).

Ammann, H., Gray, R., Shvarchuck, I., and Christensen, N. "Quantum Delta-kicked Rotor: Experimental Observation of Decoherence", Phys. Rev. Lett. **80**, 4111 (1998).

Anderson, P. W. "More Is Different", Science, **177**, 393 (1972).

Anderson, P. W. "Absence of Diffusion in Certain Random Lattices", Phys. Rev. **109**, 1492 (1958).

Anderson, P. W., Thouless, D. J., Abrahams, E., and Fisher, D. S. "New method for a scaling theory of localization", Phys. Rev. B **22**, 3519 (1980).

Assaf, O. and Akkermans, E. "Intensity correlations and mesoscopic fluctuations of diffusing photons in cold atoms", Phys. Rev. Lett. **98**, 083601 (2007); Phys. Rev. Lett. **100**, 199302 (2008).

Beenakker, C. W. J. "Random matrix theory of quantum transport", Rev. Mod. Phys. **69**, 731 (1997).

Belitz, D. and Kirkpatrick, T. R. "The Anderson-Mott transition", Rev. Mod. Phys. **66**, 261 (1994).

Bergmann, G. "Weak localization in thin films", Phys. Rep. **107**, 1–58 (1984).

Berry, M. V. and Klein, S. "Transparent mirrors: rays, waves and localization", Eur. J. Phys. **18**, 222 (1997).

Bidel, Y., Klappauf, B., Bernard, J.-C., Delande, D., Labeyrie, G., Miniatura, C., Wilkowski, D., and Kaiser, R. "Coherent light transport in a cold strontium cloud", Phys. Rev. Lett. **88**, 203902 (2002).

Billy, J., Josse, V., Zuo, Z., Bernard, A., Hambrecht, B., Lugan, P., Clément, D., Sanchez-Palencia, L., Bouyer, Ph., and Aspect, A. "Direct observation of Anderson localization of matter waves in a controlled disorder", Nature **453**, 891–894 (12 June 2008); Ph. Bouyer *et al.*, "Anderson localization of matter waves", Proceedings of the XXI ICAP Conference, R. Coté, Ph. L. Gould, M. Rozman and W. W. Smith, eds., World Scientific (2008).

Bruus, H. and Flensberg, K. *Many-body quantum theory in condensed matter physics* (Oxford University Press, 2004).

Buchleitner, A., Delande, D., and Gay, J.C. "Microwave ionization of 3-d hydrogen atoms in a realistic numerical experiment" J. Opt. Soc. Am. B **12**, 505 (1995).

Casati, G., Guarneri, I., and Shepelyansky, D. "Classical chaos, quantum localization and fluctuations: A unified view", Physica A **163**, 205 (1990).

Casati, G., Chirikov, B.V., Ford, J., and Izrailev, F.M. "Stochastic behavior of classical and quantum hamiltonian systems", Lecture Notes in Physics, **334**, G. Casati and J. Ford ed., (Springer, New York, 1979).

Casati, G., Guarneri, I., and Shepelyansky, D.L. "Anderson transition in a one-dimensional system with three incommensurable frequencies", Phys. Rev. Lett. **62**, 345–348 (1989).

Chabanov, A. A. and Genack, A. Z. "Photon localization in resonant media", Phys. Rev. Lett. **87**, 153901 (2001).

Chabanov, A. A., Stoytchev, M., and Genack, A. Z. "Statistical signatures of photon localization", Nature **404**, 850–853 (2000).

Chabé, J., Lemarié, G., Grémaud, B., Delande, D., Szriftgiser, P., and Garreau, J.C. "Experimental observation of the Anderson metal-insulator transition with atomic matter waves" Phys. Rev. Lett. **101**, 255702 (2008).

Chanelière, T., Wilkowski, D., Bidel, Y., Kaiser, R. and Miniatura, C. "Saturation-induced coherence loss in coherent backscattering of light", Phys. Rev. E **70**, 036602 (2004).

Clément, D., Varón, A.F., Retter, J.A., Sanchez-Palencia, L., Aspect, A., Bouyer, P., "Experimental study of the transport of coherent interacting matter-waves in a 1D random potential induced by laser speckle", New J. Phys. **8**, 165 (2006).

Datta, S. *Electronic transport in mesoscopic systems* (Cambridge University Press, 2002).

Efetov, K.B. "Supersymmetry and theory of disordered metals", Adv Phys. **32**, 53 (1983).

Englert, B.-G. "Fringe visibility and which-way information: an inequality", Phys. Rev. Lett. **77**, 2154 (1996).

Evers, F. and Mirlin, A. D. "Anderson transitions", Rev. Mod. Phys. **80**, 1355 (2008).

Faez, S., Strybulevych, A., Page, J.H., Lagendijk, A., and van Tiggelen, B.A. "Observation of multifractality at the Anderson localization transition of ultrasound in open three-dimensional media", Phys. Rev. Lett. **103**, 155703 (2009).

Fisher, M.E. and Barber, M.N. "Scaling theory for finite-size effects in the critical region", Phys. Rev. Lett. **28**, 1516 (1972).

Furstenberg, H. "Noncommuting Random Products", Trans. Am. Math. Soc. **108** 377 (1963).

Gaul, C., Renner, N., and Müller, C. A. "Speed of sound in disordered Bose-Einstein condensates", Phys. Rev. A **80**, 053620 (2009).

Genack, A.Z. "Statistical approach to photon localization" in Waves and Imaging through complex media, P. Sebbah (ed.) Kluwer (2001).

Genack, A.Z. and Chabanov, A.A. "Signatures of photon localization", J. Phys. A: Math. Gen. **38**, 10465 (2005).

Godoy, S. "Landauer diffusion coefficient: A classical result", Phys. Rev. E **56**, 4884 (1997).

Goldenfeld N. and Kadanoff L. P., "Simple Lessons from Complexity", Science, **284**, 87 (1999).

Golubentsev, A. A. "Suppression of interference effects in multiple scattering of light", JETP **59**, 26–32 (1984).

Gor'kov, L. P., Larkin, A. I., and Khmel'nitskiĭ, D. E. Pis'ma Zh. Eksp. Teor. Fiz. **30**, 248 (1979) [JETP Lett. **30**, 228 (1979)].

Greiner, M. *et al.,* "Quantum phase transition from a superfluid to a Mott insulator in a gas of ultracold atoms", Nature **415**, 39 (2002).

Grempel, D.R., Prange, R.E., and Fishman, S. "Quantum dynamics of a nonintegrable system", Phys. Rev. A **29**, 1639 (1984).

Grémaud, B., Delande, D., Müller, C. A., and Miniatura, C. "Comment on 'Intensity correlations and mesoscopic fluctuations of diffusing photons in cold atoms'", Phys. Rev. Lett. **100**, 199301 (2008).

Gurevich, E. and Kenneth, O. "Lyapunov exponent for the laser speckle potential: A weak disorder expansion", Phys. Rev. A **79**, 063617 (2009).

Hartung, M., Wellens, T., Müller, C. A., Richter, K., and Schlagheck, P. "Coherent backscattering of Bose-Einstein condensates in two-dimensional disorder potentials", Phys. Rev. Lett. **101**, 020603 (2008).

Hu, H., Strybulevych, A., Page, J.H., Skipetrov, S.E., and van Tiggelen, B.A. "Localization of ultrasound in a three-dimensional elastic network", Nature Phys. **4**, 945 (2008).

Imry, Y. *Introduction to mesoscopic physics* (Oxford University Press, 2002).

Ishimaru, A. *Wave propagation and scattering in random media* (Academic, New York, 1978), Vols. I and II.

Itano, W. M., Bergquist, J. C., Bollinger, J. J., Wineland, D. J., Eichmann, U. and Raizen, M. G. "Complementarity and Young's interference fringes from two atoms", Phys. Rev. A **57**, 4176 (1998).

Izrailev, F. M. and Makarov, N. M. "Anomalous transport in low-dimensional systems with correlated disorder", J. Phys. A: Math. Gen. **38**, 10613 (2005).

Jonckheere, T., Müller, C. A., Kaiser, R., Miniatura, C., Delande, D. "Multiple scattering of light by atoms in the weak localization regime", Phys. Rev. Lett. **85**, 4269 (2000).

Kadanoff et al., L. P. "Static phenomena near critical points: Theory and experiment", Rev. Mod. Phys. **39**, 395 (1967).

Katsumoto, S., Komori, F., Sano, N., and Kobayashi, S. "Fine tuning of metal-insulator transition in $Al_{0.3}Ga_{0.7}As$ using persistent photoconductivity", J. Phys. Soc. Jpn.. **56**, 2259 (1987).

van Kampen, N. G. *Stochastic proceses in physics and chemistry*, (Elsevier, Amsterdam, 2007).

Klappauf, B.G., Oskay, W.H., Steck, D.A., and Raizen, M.G. "Observation of Noise and Dissipation Effects on Dynamical Localization", Phys. Rev. Lett. **81**, 1203 (1998).

Kotani, S. and Simon, B. "Localization in general one-dimensional random systems", Commun. Math. Phys. **112**, 103 (1987).

Kravchenko, S.V., Kravchenko, G.V., Furneaux, J.E., Pudalov, V.M., and D'Iorio, M. "Possible metal-insulator transition at $B = 0$ in two dimensions", Phys. Rev. B **50**, 8039–8042 (1994).

Kuga, Y. and Ishimaru, A. "Retroreflectance from a dense distribution of spherical particles", J. Opt. Soc. Am. A **1**, 831-835 (1984).

Kuhn, R.C., Sigwarth, O., Miniatura, C., Delande, D., and Müller, C.A., "Coherent matter wave transport in speckle potentials" New J. Phys. **9**, 161 (2007).

Lahini, Y., Pugatch, R., Pozzi, F., Sorel, M., Morandotti, R., Davidson, N., and Silberberg, Y. "Direct observation of a localization transition in quasi-periodic photonic lattices", Phys. Rev. Lett. **103**, 013901 (2009).

Landauer, R. "Electrical resistance of disordered one-dimensional lattices", Phil. Mag. **21**, 863 (1970).

Langer, J. S. and Neal, T. "Breakdown of the concentration expansion for the impurity resistivity of metals", Phys. Rev. Lett. **16**, 984 (1966).

Labeyrie, G., Delande, D., Kaiser, R., and Miniatura, C. "Light transport in cold atoms and thermal decoherence", Phys. Rev. Lett. **97**, 013004 (2006).

Labeyrie, G., Delande, D., Müller, C.A., Miniatura, C., and Kaiser, R. "Coherent backscattering of light by cold atoms: Theory meets experiment", Europhys. Lett. **61**, 327 (2003).

Lemarié, G., Chabé, J., Szriftgiser, P., Garreau, J.C., Grémaud, B., and Delande, D. "Observation of the Anderson metal-insulator transition with atomic matter waves: Theory and experiment", Phys. Rev. A **80**, 043626 (2009).

Lemarié, G., Grémaud, B., and Delande, D. "Universality of the Anderson transition with the quasiperiodic kicked rotor", Europhys. Lett. **87**, 37007 (2009).

Lemarié, G., Lignier, H., Delande, D., Szriftgiser, P., and Garreau, J.C. "Critical state of the Anderson Transition: Between a Metal and an Insulator's Phys. Rev. Lett. **105**, 090601 (2010).

Lemarié, G. "Transition d'Anderson avec des ondes de matière atomiques", PhD thesis, Université Pierre et Marie Curie, Paris (2009), http://tel.archives-ouvertes.fr/tel-00424399/fr/

Lichtenberg, A.J. and Lieberman, M.A. *Regular and stochastic motion*, Springer-Verlag, New York (1983).

Lignier, H., Chabé, J., Delande, D., Garreau, J.C., and Szriftgiser, P. "Reversible destruction of dynamical localization", Phys. Rev. Lett. **95**, 234101 (2005).

Lobkis, O.I. and Weaver, R.L. "Self-consistent transport dynamics for localized waves", Phys. Rev. E **71**, 011112 (2005).

Luck, J.M. "Systèmes désordonnés unidimensionnels", Commissariat à l'énergie atomique (1992), in French.

Lugan, P., Aspect, A., Sanchez-Palencia, L., Delande, D., Grémaud, B., Müller, C.A., and Miniatura, C. "One-dimensional Anderson localization in certain correlated random potentials", Phys. Rev. A **80**, 023605 (2009).

MacKinnon, A. and Kramer, B. "One-parameter scaling of localization length and conductance in disordered systems", Phys. Rev. Lett. **47**, 1546 (1981).

Mahan, G. D. *Many-particle physics* (Kluwer, 2000).

Mello, P. A. and Kumar, N. *Quantum Transport in Mesoscopic Systems: Complexity and Statistical Fluctuations* (Oxford, 2004).

Mello, P. A. "Theory of Random Matrices: Spectral Statistics and Scattering Problems", in: E. Akkermans *et al.* (ed.), *Mesoscopic quantum physics*, Les Houches 1994 Session LXI (North-Holland, Elsevier, 1995).

Miniatura, C., Kuhn, R.C., Delande, D., and Müller, C. A. "Quantum diffusion of matter waves in 2D speckle potentials", Eur. Phys. J. B **68**, 353 (2009).

Miniatura, C., Müller, C. A., Lu, Y., Wang, G., Englert, B.-G. "Path distinguishability in double scattering of light by atoms", Phys. Rev. A **76**, 022101 (2007).

de Moura, F. A. B. F. and Lyra, M. L. "Delocalization in the 1D Anderson model with long-range correlated disorder", Phys. Rev. Lett. **81**, 3735 (1998).

Moore, F.L., Robinson, J.C., Bharucha, C., Williams, P.E., and Raizen, M.G. "Observation of dynamical localization in atomic momentum transfer: A New testing ground for quantum chaos", Phys. Rev. Lett. **73**, 2974 (1994); F. L. Moore, J. C. Robinson, C. F. Bharucha, B. Sundaram, and M. G. Raizen, "Atom optics realization of the quantum δ-kicked rotor", Phys. Rev. Lett. **75**, 4598 (1995).

Morse, P. M. and Feshbach, H. *Methods of theoretical physics* (McGraw-Hill, 1953).

Müller, C. A., Jonckheere, T., Miniatura, C. and Delande, D. "Weak localization of light by cold atoms: the impact of quantum internal structure", Phys. Rev. A **64**, 053804 (2001).

Müller, C. A. "Diffusive spin transport", in: Buchleitner, A., Viviescas, C., and Tiersch, M. (Eds.), *Entanglement and decoherence. Foundations and modern trends*, Lect. Notes Phys. **768**, 277–314 (Springer, 2009).

Müller, C. A., Miniatura, C., Akkermans, E., and Montambaux, G. "Mesoscopic scattering of spin *s* particles", J. Phys. A: Math. Gen. **38**, 7807 (2005).

Müller, C. A. and Miniatura, C. "Multiple scattering of light by atoms with internal degeneracy", J. Phys. A: Math. Gen. **35**, 10163 (2002).

Negele, J. W. and Orland, H., *Quantum many-particle systems* (Westview Press, 1998).

Nieuwenhuizen, T. M. and Luck, J. M. "Skin layer of diffusive media", Phys. Rev. E **48**, 569–588 (1993).

Peskin, M. E. and Schroeder, D. S. *An introduction to quantum field theory* (Addison-Wesley, 1995).

Pierre, F. and Birge, N.O. "Dephasing by extremely dilute magnetic impurities revealed by Aharonov-Bohm oscillations", Phys. Rev. Lett. **89**, 206804 (2002).

Pierre, F., Gougam, A.B., Anthore, A., Pothier, H., Estève, D., and Birge, N. "Dephasing of electrons in mesoscopic metal wires", Phys. Rev. B **68**, 085413 (2003).

Sanchez-Palencia, L., Clément, D., Lugan, P., Bouyer, P., Shlyapnikov, G. V., Aspect, A. "Anderson localization of expanding Bose-Einstein condensates in random potentials", Phys. Rev. Lett. **98**, 210401 (2007).

Scheffold, F., Lenke, R., Tweer, R., Maret, G., "Localization or classical diffusion of light?" Nature **398**, 206, (1999).

Schwartz, T., Bartal, G., Fishman, S., and Segev, M., "Transport and Anderson localization in disordered two-dimensional photonic lattices", Nature **446**, 52–55 (2007).

Shatokhin, V., Wellens, T., Grémaud, B., and Buchleitner, A. "Spectrum of coherently backscattered light from two atoms", Phys. Rev. A **76**, 043832 (2007b).

Shatokhin, V., Wellens, T., Müller, C. A., and Buchleitner, A. "Coherent backscattering of light from saturated atoms", Eur. Phys. J. Special Topics **151**, 51 (2007a).

Sigwarth, O., Labeyrie, G., Jonckheere, T., Delande, D., Kaiser, R., and Miniatura, C. "Magnetic field enhanced coherence length in cold atomic gases", Phys. Rev. Lett. **93**, 143906 (2004).

Slevin, K. and Ohtsuki, T. "Corrections to scaling at the Anderson transition", Phys. Rev. Lett. **82**, 382 (1999).

Störzer, M., Gross, P., Aegerter, G.M., and Maret, G. "Observation of the critical regime near Anderson localization of light", Phys. Rev. Lett. **96**, 063904 (2006).

Thirring, W. "Exact results for the scattering of three charged particles", in *Few Body Systems and Nuclear Forces II*, Lect. Notes Phys. **78**, 353–361 (Springer, 1978).

Thouless, D. J. "Localization distance and mean free path in one-dimensional disordered systems", J. Phys C.: Solid State Phys. **6**, L49 (1973).

van Tiggelen, B. A. *Multiple scattering and localization of light*, PhD thesis, University of Amsterdam (1992).

Vollhardt, D. and Wölfle, P. "Diagrammatic, self-consistent treatment of the Anderson localization problem in $d \leq 2$ dimensions", Phys. Rev. B **22**, 4666–4679 (1980).

Vollhardt, D. and Wölfle, P. "Scaling equations from a self-consistent theory of Anderson localization", Phys. Rev. Lett. **48**, 699 (1982)

Vollhardt, D. and Wölfle, P. "Self-consistent theory of Anderson localization", in: W. Hanke and Y. V. Kopaev, ed., *Electronic phase transitions* (Elsevier, Amsterdam, 1992).

Wang, J. and Garcia-Garcia, A.M. "The Anderson transition in a 3d kicked rotor", Phys. Rev. E **79**, 036206 (2009).

Washburn, S. and Webb, R. "Aharonov-Bohm effect in normal metal: Quantum coherence and transport", Adv. Phys. **35**, 375 (1986).

Wegner, F. " Electrons in disordered systems. Scaling near the mobility edge", Z. Phys. B **25**, 327 (1976).

Wiersma, D.S., Bartolini, P., Lagendijk, A., and Righini, R. "Localization of light in a disordered medium", Nature **390**, 671, (1997).

Wiersma, D. S., van Albada, M. P., van Tiggelen, B. A., and Lagendijk, A. "Experimental evidence for recurrent multiple scattering events of light in disordered media", Phys. Rev. Lett. **74**, 4193–4196 (1995).

Wickles, C. and Müller, C.A. "Thermal breakdown of coherent backscattering: a case study of quantum duality", Europhys. Lett. **74**, 240 (2006).

Wilson, K. G. "The renormalization group and critical phenomena", Rev. Mod. Phys. **55**, 583 (1983).

Wolf, P. E. and Maret, G. "Weak localization and coherent backscattering of photons in disordered media", Phys. Rev. Lett. **55**, 2696 (1985).

10

Quantum information processing and quantum optics devices

Christian KURTSIEFER
and
Antía LAMAS LINARES

Centre for Quantum Technologies and Physics Department,
National University of Singapore, 3 Science Drive 2, Singapore 117543, Singapore

10.1   Basic field quantization – the foundations

Quantum optics devices require, as a very first step, a reasonable understanding of what a quantum description of light actually covers. In this chapter, we will probably repeat elements of electrodynamics, with some specializations relevant to the optical domain, you may have seen several times already. The intent of this very basic introduction is to establish the notions, so one can perhaps more easily distinguish the quantum physics aspects from what is optics or classical electrodynamics.

Specifically, we consider here electromagnetic phenomena that take place on a time scale of $\tau \approx 10^{-15}$ s, or on wave phenomena with a characteristic length scale of $\lambda \approx 10^{-7}$ to 10^{-6} m, corresponding to energy scales of the order $E \approx \hbar/\tau \approx 10^{-19}$ J $\approx$ 1 eV.

An excellent reference book on the field quantization, which is the basis for the introductory part of this lecture, is Photons and atoms: Introduction to quantum electrodynamics (Cohen-Tannoudji, *et al.* 1997).

10.1.1   Recap of classical electrodynamics

10.1.1.1   *Maxwell equations*

We begin with understanding of the treatment of time-varying electromagnetic fields in classical physics. In a non-relativistic context (which will be the case for the scope of the matter interacting with the light field), the electromagnetic field is typically separated to two vectorial quantities with space and time as parameters, the electric and the magnetic field:

$$\mathbf{E}(\mathbf{x}, t) \qquad \mathbf{B}(\mathbf{x}, t). \tag{10.1}$$

Maxwell equations in vacuum. The dynamics of these two fields (in free space) are covered by a set of differential equations, commonly referred to as the Maxwell equations:

$$\nabla \cdot \mathbf{E}(\mathbf{x}, t) = \frac{1}{\epsilon_0} \rho(\mathbf{x}, t), \tag{10.2}$$

$$\nabla \times \mathbf{E}(\mathbf{x}, t) = -\frac{\partial}{\partial t} \mathbf{B}(\mathbf{x}, t), \tag{10.3}$$

$$\nabla \cdot \mathbf{B}(\mathbf{x}, t) = 0, \tag{10.4}$$

$$\nabla \times \mathbf{B}(\mathbf{x}, t) = \frac{1}{c^2} \frac{\partial}{\partial t} \mathbf{E}(\mathbf{x}, t) + \frac{1}{\epsilon_0 c^2} \mathbf{j}(\mathbf{x}, t) \tag{10.5}$$

Therein, the quantities $\rho(\mathbf{x}, t)$ represent a local charge density (this is a scalar quantity), and $\mathbf{j}(\mathbf{x}, t)$ a current density (this is a vector property), which describes the motion of charges, and is connected with the charge density through the charge current. Typically, we set $\rho = 0$ and $\mathbf{j} = 0$ in most of the space we consider, which allows us to understand the dynamics of a "free field".

Energy content in the electromagnetic field. An important notion for quantizing the electromagnetic field will be the total energy contained in the field of a given volume. For a system without dielectric media, the total energy of the electromagnetic field is given by

$$H = \frac{\epsilon_0}{2} \int d\mathbf{x} \, [E^2 + c^2 B^2]. \tag{10.6}$$

Here and subsequently, $\int d\mathbf{x}$ refers to volume integration over the relevant space. This is only the energy of the free field, without any charges and currents present; they can be added later.

10.1.2 Sorting the mess: Count your degrees of freedom

The Maxwell equations are a set of coupled differential equations where $\mathbf{E}$ and $\mathbf{B}$ (in the case of free space) are the variables describing the state of the system completely. More specifically, each point in the volume of interest has six scalar variables, and the state of the system is determined by describing each of these six field components at each point in space. However, this description is redundant. In the next subsection, we should find out how to eliminate that redundancy, and will arrive at the minimal set of variables we need to describe electromagnetic fields. Then, we will try to decouple the remaining equations of motion by transformation on normal coordinates.

10.1.2.1 Reducing degrees of freedom: Potentials and gauges

It can be shown that the electric and magnetic fields can always be written as derivatives of two fields $\mathbf{A}(\mathbf{x}, t)$ and $U(\mathbf{x}, t)$ called vector potential and scalar potential, respectively:

$$\mathbf{E}(\mathbf{x}, t) = -\frac{\partial}{\partial t}\mathbf{A} - \nabla U(\mathbf{x}, t), \tag{10.7}$$

$$\mathbf{B}(\mathbf{x}, t) = \nabla \times \mathbf{A}(\mathbf{x}, t). \tag{10.8}$$

We can then rewrite Maxwell equations in term of the potentials by substituting eqs. (10.7) and (10.8) into eqs. (10.2) and (10.5), leading to

$$\nabla^2 U(\mathbf{x}, t) = -\frac{1}{\epsilon_0}\rho(\mathbf{x}, t) - \nabla \cdot \frac{\partial}{\partial t}\mathbf{A}(\mathbf{x}, t), \tag{10.9}$$

$$\left(\frac{1}{c^2}\frac{\partial^2}{\partial t^2} - \nabla^2\right)\mathbf{A}(\mathbf{x}, t) = \frac{1}{\epsilon_0 c^2}\mathbf{j}(\mathbf{x}, t) - \nabla\left[\nabla \cdot \mathbf{A}(\mathbf{x}, t) + \frac{1}{c^2}\frac{\partial}{\partial t}U\right]. \tag{10.10}$$

All of these expressions are derivatives but completely describe the evolution of the field. By this trick, the number of variables describing the free field is already reduced to four scalar values for each point. There can be additive constants to the potentials U and $\mathbf{A}$ that leave the actual fields $\mathbf{E}$ and $\mathbf{B}$ and their dynamics unchanged. With the following transformation on the potentials

$$\mathbf{A}(\mathbf{x}, t) \rightarrow \mathbf{A}'(\mathbf{x}, t) = \mathbf{A}(\mathbf{x}, t) + \nabla F(\mathbf{x}, t), \tag{10.11}$$

$$U(\mathbf{x}, t) \rightarrow U'(\mathbf{x}, t) = U(\mathbf{x}, t) - \frac{\partial}{\partial t} F(\mathbf{x}, t), \tag{10.12}$$

where $F(\mathbf{x}, t)$ is any scalar field. This transformation of potentials has no physically observable consequences: electric and magnetic fields remain unchanged, and they are the quantities necessary to determine the forces on charged particles. Such a transformation is called a *gauge transformation*, and the property of the fields **E** and **B** being invariant under such a transformation is referred to as *gauge invariance*.

Therefore, we are free to choose the gauge function $F(\mathbf{x}, t)$ to come up with a particularly simple form of the equations of motion. There are two typical gauges used in electrodynamics. For many occasions, e.g. the free field, the so-called *Lorentz gauge* is very convenient. It is defined by

$$\nabla \cdot \mathbf{A}(\mathbf{x}, t) + \frac{1}{c^2} \frac{\partial}{\partial t} U(\mathbf{x}, t) = 0. \tag{10.13}$$

Equations (10.9) and (10.10) under Lorentz gauge take a particularly simple and symmetric form:

$$\Box U(\mathbf{x}, t) = \frac{1}{\epsilon_0} \rho(\mathbf{x}, t), \tag{10.14}$$

$$\Box \mathbf{A}(\mathbf{x}, t) = \frac{1}{\epsilon_0 c^2} \mathbf{j}(\mathbf{x}, t), \tag{10.15}$$

with the definition of the differential operator $\Box$ referred to as the d'Alembertoperator:

$$\Box := \frac{1}{c^2} \frac{\partial^2}{\partial t^2} - \nabla^2. \tag{10.16}$$

This form of the Maxwell equations is particularly suited for problems where a Lorentz invariance of the problem is important. However, for the purpose of optics or interaction with atoms of non-relativistic speed, another gauge is more favorable, the so-called Coulomb gauge:

$$\nabla \cdot \mathbf{A}(\mathbf{x}, t) = 0. \tag{10.17}$$

Equations (10.9) and (10.10) under Coulomb gauge take the form:

$$\nabla^2 U(\mathbf{x}, t) = -\frac{1}{\epsilon_0} \rho(\mathbf{x}, t), \tag{10.18}$$

$$\Box \mathbf{A}(\mathbf{x}, t) = \frac{1}{\epsilon_0 c^2} \mathbf{j}(\mathbf{x}, t) - \frac{1}{c^2} \nabla \frac{\partial}{\partial t} U(\mathbf{x}, t). \tag{10.19}$$

If we operate in a region without free charges, i.e. $\rho(\mathbf{x}, t) = 0$, we have $U(\mathbf{x}, t) = 0$ everywhere, and the free field is completely described by the three components of the vector potential $\mathbf{A}(\mathbf{x}, t)$. Its evolution in time is governed by a single equation of motion, a simplified version of eq. (10.19). Thus, the gauge invariance helps us to identify a redundancy in the combination of scalar and vector potential and their corresponding equations of motion.

10.1.2.2   *Decoupling of degrees of freedom: Fourier decomposition*

Now, we have to address another problem – the Maxwell equations form a set of coupled differential equations. The electromagnetic scenery in a region of space is thus described by the potentials $U(\mathbf{x}, t)$ and $\mathbf{A}(\mathbf{x}, t)$, but the values at different points (x, y, z) in space are coupled via the spatial differential operators. We therefore need to sort out this problem before we continue searching for further redundancies.

In order to arrive at the simplest (as in separable) description of the evolution of the field, we try to apply a mode decomposition of the field, very similar to the mode decomposition of mechanical oscillators in coupled systems like the lattice vibration or the vibration of a membrane. There, we try to express the local variables (like local displacement) as a function of normal coordinates, which have a completely decoupled evolution in time.

Such an attempt will be helpful for the field quantization. For most occasions, we choose the most common normal coordinate transformation for a system (as defined by the structure of equations of motion, and the boundary conditions) with translational invariance, namely the *Fourier transformation*. It has to be kept in mind that this is only a convenient choice, and by no means the only normal coordinate choice to make. There are many occasions in quantum optics where a different mode decomposition is appropriate, we will come back to that later in Section 10.1.4.

When using the Fourier transformation as the transformation to normal coordinates, plane waves (which are solutions of the homogenous Maxwell equations) form the basis for our solution. The amplitudes of various plane waves, characterized by a wavevector k, will be the new coordinates. We still want to arrive at a description of the electromagnetic field where we keep time as the parameter describing the evolution of the variables; this choice is suited to describe observations in typical non-relativistic lab environments. Therefore, we restrict the Fourier transformation only on the spatial coordinates. The transformation and their inverse for the electrical field reads explicitly:

$$\boldsymbol{\mathscr{E}}(\mathbf{k}, t) = \frac{1}{(2\pi)^{3/2}} \int d\mathbf{x}\, \mathbf{E}(\mathbf{x}, t) e^{-i\mathbf{k}\cdot\mathbf{r}}, \tag{10.20}$$

$$\mathbf{E}(\mathbf{x}, t) = \frac{1}{(2\pi)^{3/2}} \int d\mathbf{k}\, \boldsymbol{\mathscr{E}}(\mathbf{k}, t) e^{i\mathbf{k}\cdot\mathbf{r}}. \tag{10.21}$$

The quantity $\boldsymbol{\mathscr{E}}(\mathbf{k}, t)$ is a set of electric field amplitudes for every $\mathbf{k}$, which we will lead to a decoupled set of equation of motion, and thus form a suitable set of normal coordinates.

Similarly, we have transformations for all the other field quantities we have encountered so far:

$$\mathbf{E}(\mathbf{x}, t) \leftrightarrow \mathscr{E}(\mathbf{k}, t), \tag{10.22}$$

$$\mathbf{B}(\mathbf{x}, t) \leftrightarrow \mathscr{B}(\mathbf{k}, t), \tag{10.23}$$

$$\mathbf{A}(\mathbf{x}, t) \leftrightarrow \mathscr{A}(\mathbf{k}, t), \tag{10.24}$$

$$U(\mathbf{x}, t) \leftrightarrow \mathscr{U}(\mathbf{k}, t), \tag{10.25}$$

$$\rho(\mathbf{x}, t) \leftrightarrow \rho(\mathbf{k}, t), \tag{10.26}$$

$$\mathbf{j}(\mathbf{x}, t) \leftrightarrow \mathbf{j}(\mathbf{k}, t). \tag{10.27}$$

One thing to take note of is that, while the actual fields are real, the quantities in Fourier space can be complex. The reality of the electric field in real space, mathematically expressed by $\mathbf{E}^* = \mathbf{E}$, implies that

$$\mathscr{E}^*(\mathbf{k}, t) = \mathscr{E}(-\mathbf{k}, t) \tag{10.28}$$

for the electric field. This is a redundancy to keep in mind when counting the degrees of freedom of our system.

Two of the few identities that are useful to keep in mind are the Parseval–Plancherel identity and the Fourier transform of convolution product. The first one,

$$\int d\mathbf{x}\, F^*(\mathbf{x})G(\mathbf{x}) = \int d\mathbf{k}\, \mathscr{F}^*(\mathbf{k})\mathscr{G}(\mathbf{k}), \tag{10.29}$$

tells us that we can evaluate the integral over the whole space of a product of two functions (e.g. fields) also in a similar way in Fourier space. This will come in handy for evaluating the total field energy.

The convolution product of two functions is defined as

$$F(\mathbf{x}) \otimes G(\mathbf{x}) := \frac{1}{(2\pi)^{3/2}} \int d\mathbf{x}'\, F(\mathbf{x}')G(\mathbf{x} - \mathbf{x}'), \tag{10.30}$$

where the integration is carried out over a three-dimensional space. A similar definition of a convolution product can be defined in one dimension, with a properly adjusted normalization constant. Such a product typically appears in the evolution of correlation functions.

It can easily be shown that the Fourier transformation of the convolution product is just the product of the Fourier-transformed versions of the two functions:

$$F(\mathbf{x}) \otimes G(\mathbf{x}) \leftrightarrow \mathscr{F}(\mathbf{k})\mathscr{G}(\mathbf{k}). \tag{10.31}$$

This is a useful relation also in practical situations when correlations between functions need to be evaluated, as the numerical evaluation of a Fourier transformation is extremely efficient.

So far, we have not seen that Fourier transformation helps to decouple the Maxwell equations as the equations of motion for the fields, and that this is therefore actually a transformation to normal coordinates. For this, we use the transformation rules for differential operators,

$$\nabla \cdot \leftrightarrow i\mathbf{k}\cdot, \qquad \nabla \times \leftrightarrow i\mathbf{k} \times \qquad \text{etc.} \tag{10.32}$$

to rewrite the Maxwell equations (10.2)–(10.5) in terms of their Fourier-transformed fields $\mathscr{E}(\mathbf{k}, t)$ and $\mathscr{B}(\mathbf{k}, t)$:

$$i\mathbf{k} \cdot \mathscr{E}(\mathbf{k}, t) = \frac{1}{\epsilon_0}\rho(\mathbf{k}, t) \tag{10.33}$$

$$i\mathbf{k} \times \mathscr{E}(\mathbf{k}, t) = -\frac{\partial}{\partial t}\mathscr{B}(\mathbf{k}, t) \tag{10.34}$$

$$i\mathbf{k} \cdot \mathscr{B}(\mathbf{k}, t) = 0 \tag{10.35}$$

$$i\mathbf{k} \times \mathscr{B}(\mathbf{k}, t) = -\frac{1}{c^2}\frac{\partial}{\partial t}\mathscr{E}(\mathbf{k}, t) + \frac{1}{\epsilon_0 c^2}j(\mathbf{k}, t) \tag{10.36}$$

This is now a set of coupled differential equations *for each* $\mathbf{k}$, but the coupling extends only over the electric and magnetic field components for a given $\mathbf{k}$, not between those of different $\mathbf{k}$ any layer . Thus, the Fourier transformation helped to arrive at a decoupling between the field amplitudes at different locations $\mathbf{x}$ in space, and thus is a transformation on normal coordinates.

Similarly, the connections between fields and potentials in Fourier space are given by

$$\mathscr{B}(\mathbf{k}, t) = i\mathbf{k} \times \mathscr{A}(\mathbf{k}, t), \tag{10.37}$$

$$\mathscr{E}(\mathbf{k}, t) = -\frac{\partial}{\partial t}\mathscr{A}(\mathbf{k}, t) - i\mathbf{k}\mathscr{U}(\mathbf{k}, t), \tag{10.38}$$

and the gauge transformations turn into

$$\mathscr{A}(\mathbf{k}, t) \to \mathscr{A}'(\mathbf{k}, t) = \mathscr{A}(\mathbf{k}, t) + i\mathbf{k}\mathscr{F}(\mathbf{k}, t), \tag{10.39}$$

$$\mathscr{U}(\mathbf{k}, t) \to \mathscr{U}'(\mathbf{k}, t) = \mathscr{U}(\mathbf{k}, t) - \frac{\partial}{\partial t}\mathscr{F}(\mathbf{k}, t). \tag{10.40}$$

The equations of motion for the potentials transform into

$$k^2\mathscr{U}(\mathbf{k}, t) = \frac{1}{\epsilon_0}\rho(\mathbf{k}, t) + i\mathbf{k} \cdot \frac{\partial}{\partial t}\mathscr{A}(\mathbf{k}, t), \tag{10.41}$$

$$\frac{1}{c^2}\frac{\partial^2}{\partial t^2}\mathscr{A}(\mathbf{k}, t) + k^2\mathscr{A}(\mathbf{k}, t) = \frac{1}{\epsilon_0 c^2}j(\mathbf{k}, t) - i\mathbf{k}\frac{1}{c^2}\frac{\partial}{\partial t}\mathscr{U}(\mathbf{k}, t). \tag{10.42}$$

Again, these equations are simpler in Fourier space because partial differential equations were transformed into a set of ordinary differential equations for the different $\mathbf{k}$. Thus, the Maxwell equations are strictly local in the Fourier space.

10.1.2.3   *Longitudinal and transverse fields*

In an attempt to reduce the degrees of freedom we have to consider for a field quantization further, there is another – more subtle – redundancy we need to address.

It is connected with the fact that propagating plane waves of electromagnetic fields are *transverse fields*, and can be decomposed in only two polarization components.

As a mathematical definition, a vector field $\mathbf{V}_\parallel(\mathbf{x})$ is called a *longitudinal* vector field if and only if

$$\nabla \times \mathbf{V}_\parallel(\mathbf{x}) = 0. \tag{10.43}$$

This equation can be easier interpreted when written in Fourier space:

$$i\mathbf{k} \times \boldsymbol{\mathcal{V}}_\parallel(\mathbf{k}) = 0. \tag{10.44}$$

The vanishing cross-product between $\boldsymbol{\mathcal{V}}_\parallel$ and $\mathbf{k}$ simply means that both vectors are parallel. Thus, a longitudinal vector field has its components aligned with the wavevector $\mathbf{k}$.

Similarly, one can define a *transverse* vector field $\mathbf{V}_\perp(\mathbf{x})$ the following properties:

$$\nabla \cdot \mathbf{V}_\perp(\mathbf{x}) = 0, \tag{10.45}$$

$$i\mathbf{k} \cdot \boldsymbol{\mathcal{V}}_\perp(\mathbf{k}) = 0. \tag{10.46}$$

This time, the vector $\boldsymbol{\mathcal{V}}_\perp$ is perpendicular to the wavevector of a plane wave.

With these definitions, one may decompose any vector field $\mathbf{V}(\mathbf{x})$ into longitudinal and transverse part. This decomposition can be carried out conveniently from a representation in Fourier space:

$$\boldsymbol{\mathcal{V}}_\parallel(\mathbf{k}) = \boldsymbol{\kappa}[\boldsymbol{\kappa} \cdot \boldsymbol{\mathcal{V}}(\mathbf{k})], \tag{10.47}$$

$$\boldsymbol{\mathcal{V}}_\perp(\mathbf{k}) = \boldsymbol{\mathcal{V}}(\mathbf{k}) - \boldsymbol{\mathcal{V}}_\parallel(\mathbf{k}), \tag{10.48}$$

where $\boldsymbol{\kappa}$ is the unit vector in the direction of $\mathbf{k}$. The longitudinal and transverse fields ($\mathbf{V}_\parallel(\mathbf{x})$ and $\mathbf{V}_\perp(\mathbf{x})$) in real space can then be obtained via inverse Fourier transformation. This gives the decomposition of the following:

$$\mathbf{V}(\mathbf{x}) = \mathbf{V}_\parallel(\mathbf{x}) + \mathbf{V}_\perp(\mathbf{x}). \tag{10.49}$$

With the definitions of longitudinal and transverse fields, one can see that that the magnetic field is purely transverse. This is clear from Maxwell equation (10.35) that gives

$$\boldsymbol{\mathcal{B}}_\parallel(\mathbf{k}, t) = 0 = \mathbf{B}_\parallel(\mathbf{x}, t).^1 \tag{10.50}$$

Similarly, the expression for the source term of electrical field using eqs. (10.33) and (10.47), allows to isolate the parallel component of the electrical field $\boldsymbol{\mathcal{E}}_\parallel(\mathbf{k}, t)$ in Fourier space:

$$\boldsymbol{\mathcal{E}}_\parallel(\mathbf{k}, t) = -\frac{i}{\epsilon_0}\rho(\mathbf{k}, t)\frac{\mathbf{k}}{k^2}. \tag{10.51}$$

The expression for the electric field in real space can be obtained by an inverse Fourier transformation. The Fourier transformation relating convolution product in real space to product in Fourier space (eqn (10.31)) can be used to arrive at:

$$E_\parallel(\mathbf{x}, t) = \frac{1}{4\pi\epsilon_0} \int d\mathbf{x}' \, \rho(\mathbf{x}', t) \frac{\mathbf{x} - \mathbf{x}'}{|\mathbf{x} - \mathbf{x}'|^3} \tag{10.52}$$

This seemingly innocent expression looks just like an expression known from electrostatics, where the field from a charge distribution is obtained using the Green function of a point charge. However, keep in mind that $E_\parallel(\mathbf{x}, t)$ is the field created by the instantaneous position of charges $\rho(\mathbf{x})$ at time t at all locations $\mathbf{x}$, and no retardation effects are taken into account. However, this *does not* violate causality since $E_\parallel(\mathbf{x}, t)$ itself is not a physically observable quantity on its own. What is meaningful is the total electric field, which gets complemented by the transverse field, which in turn takes care of any information propagating around about charges that may have moved.

The dynamics of the transverse and longitudinal field components are still governed by the Maxwell equations, but we are left only with two equations not vanishing to zero:

$$\frac{\partial}{\partial t} \mathcal{B}_\perp(\mathbf{k}, t) = \frac{\partial}{\partial t} \mathcal{B}(\mathbf{k}, t) = -i\mathbf{k} \times \mathcal{E}(\mathbf{k}, t) = -i\mathbf{k} \times \mathcal{E}_\perp(\mathbf{k}, t), \tag{10.53}$$

$$\frac{\partial}{\partial t} \mathcal{E}_\perp(\mathbf{k}, t) = ic^2 \mathbf{k} \times \mathcal{B}(\mathbf{k}, t) - \frac{1}{\epsilon_0} \mathbf{j}_\perp(\mathbf{k}, t). \tag{10.54}$$

In terms of the transverse vector potential the equation of motion is given by

$$\frac{1}{c^2} \frac{\partial^2}{\partial t^2} \mathcal{A}_\perp(\mathbf{k}, t) + k^2 \mathcal{A}_\perp(\mathbf{k}, t) = \frac{1}{\epsilon_0 c^2} \mathbf{j}_\perp(\mathbf{k}, t). \tag{10.55}$$

For the longitudinal component, we are left with

$$k^2 \mathcal{U}(\mathbf{k}, t) = \frac{1}{\epsilon_0} \rho(\mathbf{k}, t) + i\mathbf{k} \cdot \frac{\partial}{\partial t} \mathcal{A}_\parallel(\mathbf{k}, t). \tag{10.56}$$

If the Coulomb gauge is chosen, we have $\mathcal{A}_\parallel(\mathbf{k}, t) = 0$. Then, the Maxwell equations leave us with two distinct problems: One governs the derivation of a scalar potential $U(\mathbf{x})$ from a given charge density $\rho(\mathbf{x})$, the other one governs the evolution of propagating free fields.

Referring to eq. (10.55), for a given mode index $\mathbf{k}$, we are left with two degrees of freedom corresponding to two perpendicular directions orthogonal to $\mathbf{k}$:

$$\mathcal{A}_\perp(\mathbf{k}, t) = \sum_\varepsilon a_\varepsilon(\mathbf{k}, t)\varepsilon, \tag{10.57}$$

where ε represents the two independent *polarization* directions. By using the concept of transverse fields, we finally arrived at two variables $a_\varepsilon(\mathbf{k}, t)$ for each point $\mathbf{k}$ describing completely the evolution of propagating electromagnetic fields.

10.1.2.4 *Normal coordinates – alternative approach*

With the two transverse components of the vector potential for each mode index $\mathbf{k}$ and their corresponding equation of motion (10.55) we have arrived at a minimal description of the electromagnetic field. Alternatively, we could arrive to a similar point by recalling eqns (10.53) and (10.54). Together with the assumption that $j_\perp(\mathbf{k}, t) = 0$, the two equations can be combined to:

$$\frac{\partial}{\partial t}(\boldsymbol{\mathscr{E}}_\perp[\mathbf{k}, t] \mp c\boldsymbol{\kappa} \times \boldsymbol{\mathscr{B}}(\mathbf{k}, t)] = \mp i\omega[\boldsymbol{\mathscr{E}}_\perp(\mathbf{k}, t) \mp c\boldsymbol{\kappa} \times \boldsymbol{\mathscr{B}}(\mathbf{k}, t)], \tag{10.58}$$

with $\omega = c|\mathbf{k}|$, making use of the time dependency $e^{-i\omega t}$ for the plane waves.

This suggests that we define the following two normal coordinates,

$$\boldsymbol{\alpha}(\mathbf{k}, t)) := -\frac{i}{2\mathscr{N}(k)}[\boldsymbol{\mathscr{E}}_\perp(\mathbf{k}, t) - c\boldsymbol{\kappa} \times \boldsymbol{\mathscr{B}}(\mathbf{k}, t)], \tag{10.59}$$

$$\boldsymbol{\beta}(\mathbf{k}, t)) := -\frac{i}{2\mathscr{N}(k)}[\boldsymbol{\mathscr{E}}_\perp(\mathbf{k}, t) + c\boldsymbol{\kappa} \times \boldsymbol{\mathscr{B}}(\mathbf{k}, t)], \tag{10.60}$$

where the normalization $\mathscr{N}(k)$ is somewhat arbitrary and will be chosen so that the Hamilton function has a nice form.

However, $\boldsymbol{\alpha}(\mathbf{k}, t)$ and $\boldsymbol{\beta}(\mathbf{k}, t)$ are not independent. Since $\mathbf{E}_\perp(\mathbf{x}, t)$ and $\mathbf{B}(\mathbf{x}, t)$ are real quantities, we have equations similar to eq. (10.28) for $\boldsymbol{\mathscr{E}}_\perp(\mathbf{k}, t)$ and $\boldsymbol{\mathscr{B}}(\mathbf{k}, t)$. These equations give the following relation between $\boldsymbol{\alpha}(\mathbf{k}, t)$ and $\boldsymbol{\beta}(\mathbf{k}, t)$:

$$\boldsymbol{\beta}(\mathbf{k}, t) = -\boldsymbol{\alpha}^*(-\mathbf{k}, t). \tag{10.61}$$

It is then sufficient to describe the electric and magnetic fields by one complex variable $\boldsymbol{\alpha}(\mathbf{k}, t)$ only.

Using eqs. (10.61), (10.59) and (10.60) can be solved for $\boldsymbol{\mathscr{E}}_\perp(\mathbf{k}, t)$ and $\boldsymbol{\mathscr{B}}(\mathbf{k}, t)$

$$\boldsymbol{\mathscr{E}}_\perp(\mathbf{k}, t) = i\mathscr{N}(k)[\boldsymbol{\alpha}(\mathbf{k}, t) - \boldsymbol{\alpha}^*(-\mathbf{k}, t)], \tag{10.62}$$

$$\boldsymbol{\mathscr{B}}_\perp(\mathbf{k}, t) = i\frac{\mathscr{N}(k)}{c}[\boldsymbol{\kappa} \times \boldsymbol{\alpha}(\mathbf{k}, t) + \boldsymbol{\kappa} \times \boldsymbol{\alpha}^*(-\mathbf{k}, t)]. \tag{10.63}$$

Therefore, the knowledge of $\boldsymbol{\alpha}(\mathbf{k}, t)$ for all $\mathbf{k}$ enables one to derive all physical quantities like $\boldsymbol{\mathscr{E}}_\perp(\mathbf{k}, t)$ and $\boldsymbol{\mathscr{B}}(\mathbf{k}, t)$. Since there is no restriction on the reality of $\boldsymbol{\alpha}(\mathbf{k}, t)$, they are really independent variables. The *complete* field is now described by the variables $\boldsymbol{\alpha}(\mathbf{k}, t)$.

Subtracting eq. (10.53) from eq. (10.54), with the definition of $\boldsymbol{\alpha}(\mathbf{k}, t)$ in mind, we have

$$\frac{\partial}{\partial t}\boldsymbol{\alpha}(\mathbf{k}, t) + i\omega\boldsymbol{\alpha}(\mathbf{k}, t) = \frac{i}{2\epsilon_0\mathscr{N}(k)}j_\perp(\mathbf{k}, t). \tag{10.64}$$

This is the equation of motion of the electromagnetic field, which is completely equivalent to eq. (10.55), this time formulated as a first- order differential equation in time, and the complex coefficients $\alpha_\varepsilon(\mathbf{k}, t)$ are the same as the ones used in the earlier derivation up to a normalization constant.

We note that $\boldsymbol{\alpha}(\mathbf{k}, t)$ is a *transverse* field because it is defined as a sum of two transverse fields in eq. (10.59). Therefore, there are only *two degrees of freedom* corresponding to the transverse direction rather than three degrees of freedom. This means that

$$\boldsymbol{\alpha}(\mathbf{k}, t) = \sum_\varepsilon \alpha_\varepsilon(\mathbf{k}, t)\boldsymbol{\varepsilon} = \alpha_{\varepsilon_1}(\mathbf{k}, t)\boldsymbol{\varepsilon}_1 + \alpha_{\varepsilon_2}(\mathbf{k}, t)\boldsymbol{\varepsilon}_2, \tag{10.65}$$

where the *mutually orthogonal* polarization vectors ε_1 and ε_2 are perpendicular to $\mathbf{k}$ for any given $\mathbf{k}$.

We now have the equation of motion as a decoupled set of equations

$$\frac{\partial}{\partial t}\alpha_\varepsilon(\mathbf{k}, t) + i\omega\alpha_\varepsilon(\mathbf{k}, t) = \frac{i}{2\epsilon_0 \mathcal{N}(k)} \mathbf{j}_\perp(\mathbf{k}, t) \cdot \boldsymbol{\varepsilon}, \tag{10.66}$$

and the pair $(\varepsilon, \mathbf{k})$ is an *index* of the different *modes* of the field.

10.1.2.5 Hamiltonian of the electromagnetic field

The total electromagnetic field energy in a propagating field[2] is given by

$$H = \frac{\epsilon_0}{2} \int d\mathbf{x}\, [E_\perp^2(\mathbf{x}, t) + c^2 B^2(\mathbf{x}, t)] = \frac{\epsilon_0}{2} \int d\mathbf{k}\, [\mathscr{E}_\perp^2(\mathbf{k}, t) + c^2 \mathscr{B}^2(\mathbf{k}, t)]. \tag{10.67}$$

From eqs. (10.62) and (10.63), we find

$$\mathscr{E}_\perp^* \cdot \mathscr{E}_\perp = \mathcal{N}^2(\boldsymbol{\alpha}^* \cdot \boldsymbol{\alpha} + \boldsymbol{\alpha}_- \cdot \boldsymbol{\alpha}_-^* - \boldsymbol{\alpha}^* \cdot \boldsymbol{\alpha}_-^* - \boldsymbol{\alpha}_- \cdot \boldsymbol{\alpha}), \tag{10.68}$$

$$c^2 \mathscr{B}_\perp^* \cdot \mathscr{B}_\perp = \mathcal{N}^2(\boldsymbol{\alpha}^* \cdot \boldsymbol{\alpha} + \boldsymbol{\alpha}_- \cdot \boldsymbol{\alpha}_-^* + \boldsymbol{\alpha}^* \cdot \boldsymbol{\alpha}_-^* + \boldsymbol{\alpha}_- \cdot \boldsymbol{\alpha}), \tag{10.69}$$

with $\boldsymbol{\alpha}_-^* = \boldsymbol{\alpha}^*(-\mathbf{k}, t)$. The first equation for $\mathscr{E}_\perp^* \cdot \mathscr{E}_\perp$ can be obtained easily. To go from eq. (10.63) to the second equation, the following identity is used, while keeping in mind that $\boldsymbol{\alpha}(\mathbf{k}, t)$ is transverse. The expression of the energy becomes[3]

$$H = \epsilon_0 \int d\mathbf{k}\, \mathcal{N}^2[\boldsymbol{\alpha}^* \cdot \boldsymbol{\alpha} + \boldsymbol{\alpha}_- \cdot \boldsymbol{\alpha}_-^*]. \tag{10.70}$$

The normalization coefficient $\mathcal{N}(k)$ is chosen to be $\sqrt{\frac{\hbar\omega}{2\epsilon_0}}$. A change of variable is performed for the second term in the equation above, where $\mathbf{k}$ is changed to $-\mathbf{k}$. Finally, we have

$$H = \int d\mathbf{k} \sum_\varepsilon \frac{\hbar\omega}{2}[\alpha_\varepsilon^*(\mathbf{k}, t)\alpha_\varepsilon(\mathbf{k}, t) + \alpha_\varepsilon(\mathbf{k}, t)\alpha_\varepsilon^*(\mathbf{k}, t)]. \tag{10.71}$$

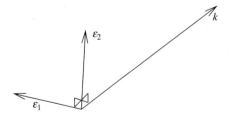

Fig. 10.1 Three mutually orthogonal vectors ε_1, ε_2 and $\mathbf{k}$.

Let's summarize the expressions for electric field, magnetic field and vector potential before we proceed to consider the quantization of the radiation field:

$$\mathbf{E}_\perp(\mathbf{x}, t) = i \int d\mathbf{k} \sum_\varepsilon \mathcal{E}_\omega [\alpha_\varepsilon(\mathbf{k}, t) e^{i\mathbf{k}\cdot\mathbf{x}} - \alpha_\varepsilon^*(\mathbf{k}, t) e^{-i\mathbf{k}\cdot\mathbf{x}}] \varepsilon, \tag{10.72}$$

$$\mathbf{B}(\mathbf{x}, t) = i \int \frac{d\mathbf{k}}{c} \sum_\varepsilon \mathcal{E}_\omega [\alpha_\varepsilon(\mathbf{k}, t) e^{i\mathbf{k}\cdot\mathbf{x}} - \alpha_\varepsilon^*(\mathbf{k}, t) e^{-i\mathbf{k}\cdot\mathbf{x}}] \kappa \times \varepsilon, \tag{10.73}$$

$$\mathbf{A}_\perp(\mathbf{x}, t) = \int d\mathbf{k} \sum_\varepsilon \frac{\mathcal{E}_\omega}{\omega} [\alpha_\varepsilon(\mathbf{k}, t) e^{i\mathbf{k}\cdot\mathbf{x}} + \alpha_\varepsilon^*(\mathbf{k}, t) e^{-i\mathbf{k}\cdot\mathbf{x}}] \varepsilon, \tag{10.74}$$

with $\mathcal{E}_\omega = \sqrt{\frac{\hbar\omega}{2\epsilon_0(2\pi)^3}} \cdot^4$

10.1.3 The works: Canonical quantization for dummies

Let's recall one form of the equation of motion (10.64) for the electromagnetic field in complex variables:

$$\frac{\partial}{\partial t} \boldsymbol{\alpha}(\mathbf{k}, t) + i\omega \boldsymbol{\alpha}(\mathbf{k}, t) = \frac{i}{2\epsilon_0 \mathcal{N}(k)} \mathbf{j}_\perp(\mathbf{k}, t), \tag{10.75}$$

This equation resembles a set of equations of motion for simple harmonic oscillators, each of the form

$$\frac{\partial}{\partial t} \alpha(t) + i\omega\alpha(t) = f(t). \tag{10.76}$$

Its quantization is well known, and this analogy suggests that a field quantization could be performed by interpretation of each single mode of the field as a harmonic oscillator following the standard harmonic oscillator quantization. This is, however, not completely according to the book. The proper way (as you probably have seen in your quantum-mechanics textbooks) would be to

1. Start with a Lagrange-density or a complete Langrage function,
2. Identify the coordinates of the system,
3. Find the canonically conjugated momenta,
4. Use the Hamilton–Jacobi formalism to express the energies of the field in terms of coordinates and conjugated momenta,
5. Express the physical quantities like **E** and **B** as a function of coordinates and momenta,
6. Use the Schroedinger or Heisenberg equation to describe the dynamics of the system in various pictures.

This procedure will lead to exactly the same result as the simple quantization approach obtained by simply using the analogy in the normal coordinates that will be done in this section.

10.1.3.1 Quantization

Now let's perform the quantization as according to the simple method outlined in the previous section. We use the transition

$$\alpha \to \hat{a}, \qquad \alpha^* \to \hat{a}^\dagger \tag{10.77}$$

similar to the case of harmonic oscillator. The field then turns into field operators as shown below:

$$\hat{\mathbf{A}}_\perp = \sum_j \mathscr{A}_{\omega_j} \left[\hat{a}_j \boldsymbol{\varepsilon}_j e^{i\mathbf{k}_j \cdot \mathbf{x}} + \hat{a}_j^\dagger \boldsymbol{\varepsilon}_j e^{-i\mathbf{k}_j \cdot \mathbf{x}} \right], \tag{10.78}$$

$$\hat{\mathbf{E}}_\perp = i \sum_j \mathscr{E}_{\omega_j} \left[\hat{a}_j \boldsymbol{\varepsilon}_j e^{i\mathbf{k}_j \cdot \mathbf{x}} - \hat{a}_j^\dagger \boldsymbol{\varepsilon}_j e^{-i\mathbf{k}_j \cdot \mathbf{x}} \right], \tag{10.79}$$

$$\hat{\mathbf{B}}_\perp = i \sum_j \mathscr{B}_{\omega_j} \left[\hat{a}_j (\boldsymbol{\kappa}_j \times \boldsymbol{\varepsilon}_j) e^{i\mathbf{k}_j \cdot \mathbf{x}} - \hat{a}_j^\dagger (\boldsymbol{\kappa}_j \times \boldsymbol{\varepsilon}_j) e^{-i\mathbf{k}_j \cdot \mathbf{x}} \right]. \tag{10.80}$$

The Hamilton operator of the field is given by

$$\hat{H} = \sum_j \frac{\hbar \omega_j}{2} (\hat{a}_j^\dagger \hat{a}_j + \hat{a}_j \hat{a}_j^\dagger), \tag{10.81}$$

which, as expected, resembles that of harmonic oscillators.

10.1.3.2 Harmonic-oscillator physics

The individual terms $\hat{H}_j$ of the Hamilton operator in eq. (10.81),

$$\hat{H} = \sum_j \hat{H}_j, \quad \hat{H}_j = \frac{\hbar \omega_j}{2} (\hat{a}_j^\dagger \hat{a}_j + \hat{a}_j \hat{a}_j^\dagger), \tag{10.82}$$

define the dynamics for the field state in each mode. We will come back to a few field states later, but mention that each mode j is associated with a Hilbert space $\mathcal{H}_j$ to capture every possible single-mode state $|\Psi_j\rangle$. A convenient way to characterize a in that space is its decomposition into the spectrum of energy eigenstates. We rewrite the $\hat{H}_j$ in the form

$$\hat{H}_j = \hbar \omega_j (\hat{n} + \frac{1}{2}) \quad \text{with} \quad \hat{n} = \hat{a}^\dagger \hat{a}, \tag{10.83}$$

with the so-called number operator $\hat{n}$ and using the commutator relation $[\hat{a}, \hat{a}^\dagger] = 1$. The energy eigenstates of the harmonic oscillator are given by the discrete set of eigenstates $|n\rangle$ of the number operator,

$$\hat{n}|n\rangle = n|n\rangle, n = 0, 1, 2, \ldots \tag{10.84}$$

Any state in this mode can now be expressed as a superposition of number states:

$$|\Psi\rangle = \sum_{n=0}^{\infty} c_n |n\rangle \quad \text{with coefficients} \quad c_n \in \mathbb{C}, \quad \sum_{n=0}^{\infty} |c_n|^2 = 1. \tag{10.85}$$

It is useful to see the analogy of a harmonic oscillator with a mass and a restoring force to the harmonic oscillator associated with an electromagnetic field mode. For the first case, we have a Hamilton operator of the form

$$\hat{H} = \frac{\hat{p}^2}{2m} + \frac{1}{2} m\omega^2 \hat{x}^2, \tag{10.86}$$

with operator $\hat{x}$ and $\hat{p}$ for position and momentum of the mass m. These operators can be expressed as the sum and difference of the ladder operators $\hat{a}$ and $\hat{a}^\dagger$:

$$\hat{x} = \sqrt{\frac{\hbar}{2m\omega}} \left(\hat{a}^\dagger + \hat{a} \right), \tag{10.87}$$

$$\hat{p} = \sqrt{\frac{\hbar m\omega}{2}} \, i \left(\hat{a}^\dagger - \hat{a} \right), \tag{10.88}$$

Since the position (or momentum) probability amplitude $\phi(x)$ is well known for some common harmonic oscillator states (e.g. the number states), this analogy can help to derive a distribution corresponding field quantities we will see in the next section.

10.1.3.3 Field operators

Often the field operators are again split up into

$$\hat{E} = \hat{E}^{(+)} + \hat{E}^{(-)}, \tag{10.89}$$

where

$$\hat{E}^{(+)}(\mathbf{x}, t) = i \sum_{j} \mathscr{E}_{\omega_j} \varepsilon_j \hat{a}_j e^{i\mathbf{k}\cdot\mathbf{x}}, \tag{10.90}$$

$$\hat{E}^{(-)}(\mathbf{x}, t) = -i \sum_{j} \mathscr{E}_{\omega_j} \varepsilon_j \hat{a}_j^\dagger e^{-i\mathbf{k}\cdot\mathbf{x}}. \tag{10.91}$$

They are referred to as positive and negative frequency contributions, corresponding to the evolutions in a Heisenberg picture; there $\hat{E}$ becomes time dependent, as well as the raising and lowering operators $\hat{a}$ and $\hat{a}^\dagger$.

For the lowering operator, it can be shown by remembering $[\hat{N}, \hat{a}] = -\hat{a}$ and $\hat{N} = \frac{1}{\hbar\omega}\hat{H} - \frac{1}{2}$ that

$$i\hbar \frac{\partial}{\partial t} \hat{a}(t) = \left[\hat{a}(t), \hat{H} \right] = \hbar\omega \hat{a}(t). \tag{10.92}$$

Solving this equation will result in

$$\hat{a}(t) = \hat{a}(0) e^{-i\omega t}. \tag{10.93}$$

The same can be done for the raising operator that leads us to the following equation of motion for the raising operator and its solution:

$$i\hbar \frac{\partial}{\partial t} \hat{a}^\dagger (t) = \left[\hat{a}^\dagger (t), \hat{H} \right] = -\hbar \omega \hat{a}^\dagger (t), \tag{10.94}$$

$$\hat{a}^\dagger (t) = \hat{a}^\dagger (0) e^{i\omega t}. \tag{10.95}$$

In the Heisenberg picture, the electric-field operator becomes time dependent and takes the form

$$\hat{E}_\perp (\mathbf{x}, t) = i \sum_j \mathcal{E}_{\omega_j} \varepsilon_j \left\{ \hat{a}_j e^{i(\mathbf{k} \cdot \mathbf{x} - \omega t)} - \hat{a}_j^\dagger e^{-i(\mathbf{k} \cdot \mathbf{x} - \omega t)} \right\}. \tag{10.96}$$

Next, we consider the following definitions of two operators,

$$\hat{a}_Q := \frac{1}{2} \left(\hat{a} + \hat{a}^\dagger \right), \tag{10.97}$$

$$\hat{a}_P := \frac{1}{2i} \left(\hat{a} - \hat{a}^\dagger \right). \tag{10.98}$$

These two operators corresponds to the $\hat{x}$ and $\hat{p}$ operators of the harmonic oscillator, which was discussed earlier. They are often referred to as *quadrature amplitudes*.

The two quadrature amplitudes are Hermitian operators, as can be shown:

$$\hat{a}_Q^\dagger = \frac{1}{2} \left(\hat{a}^\dagger + \hat{a} \right) = \hat{a}_Q, \tag{10.99}$$

$$\hat{a}_P^\dagger = -\frac{1}{2i} \left(\hat{a}^\dagger - \hat{a} \right) = \frac{1}{2i} \left(\hat{a} - \hat{a}^\dagger \right) = \hat{a}_P. \tag{10.100}$$

This means that the two operators can be associated with *observables*. In fact, they are associated with the sine and the cosine component of a particular field amplitude, as can be seen from the expression of the field operator in terms of these operators.

$$\hat{E}(\mathbf{x}, t) = -\sum_j 2\mathcal{E}_{\omega_j} \varepsilon_j \left\{ \hat{a}_{Q,j} \sin(\mathbf{k} \cdot \mathbf{x} - \omega_j t) + \hat{a}_{P,j} \cos(\mathbf{k} \cdot \mathbf{x} - \omega_j t) \right\}. \tag{10.101}$$

The commutation relation between $\hat{a}_P$ and $\hat{a}_Q$ can be shown easily as follows:

$$[\hat{a}_P, \hat{a}_Q] = \frac{1}{4i} \left[\hat{a} - \hat{a}^\dagger, \hat{a} + \hat{a}^\dagger \right] = \frac{1}{4i} \left([\hat{a}, \hat{a}^\dagger] - [\hat{a}^\dagger, \hat{a}] \right) = \frac{1}{2i}. \tag{10.102}$$

The reader can verify that for a single-modefield, the Hamilton operator can be written in terms of the quadrature amplitude as

$$\hat{H} = \hbar \omega \left(\hat{a}_Q^2 + \hat{a}_P^2 \right). \tag{10.103}$$

10.1.4 Different mode decompositions

So far, we have been using plane waves to decouple the partial differential equations, and to arrive at independent complex amplitudes $\alpha_{k,\epsilon}$ as independent variables. This is a convenient choice if we don't have any conductors or dielectrics in our volume of interest, but it is not the only possible mode decomposition. Before performing the quantization, let us review the case where we have a different set of boundary conditions.

For geometries where one wants to look at the interaction with an atom resting at the origin of a coordinate system, a set of spherical modes is more appropriate. Such a mode decomposition is, e.g., useful for understanding the spontaneous emission of light from an excited atom, or when dealing with spherical resonator geometries as sometimes found for microwaves.

Other common mode decompositions involve a cylindrical geometry, which will be important for describing the electromagnetic field, e.g., in optical fibers. There, the boundary conditions for guided modes lead to a field that is centered in the vicinity of a core in the fiber with high refractive index, while electric field decays with the radial distance according to a Bessel function or something similar, depending on the profile for the refractive index in the optical fiber.

A very common mode decomposition in experimental setups involves Gaussian beams, which have a radial field distribution following a Gaussian, and a well-defined dependency of the characteristic width along a light beam. These modes are eigensolutions for the paraxial wave equation, and are typically found in laser physics as eigenfunctions for arrangements of concave mirror cavities.

Many field geometries limit the volume of interest, thus leading to a modified mode structure altogether. A confinement of all three dimensions of space is typically referred to as a *cavity*. For simplicity, let's consider the simplest case and postulate a periodic boundary condition.

10.1.4.1 Periodic boundary conditions

The simplest step to make a transition from continuous variables to cavities is to enforce periodic boundary conditions. There, the wavevector $\mathbf{k}$ can only have integer multiples of the form:

$$k_x = \frac{2\pi n_x}{L}, \qquad k_y = \frac{2\pi n_y}{L}, \qquad k_z = \frac{2\pi n_z}{L}, \tag{10.104}$$

where L is the period.[5] Here, k_i refers to the component of the wavevector in the i direction.

We can write down the following correspondence between the sum in continuous case for the infinite space and the sum in discrete case for the finite cavities:

$$\int d\mathbf{k}\, f(\mathbf{k}) \leftrightarrow \sum_{k_{x,y,z}} \left(\frac{2\pi}{L}\right)^3 f(k_{x,y,z}), \tag{10.105}$$

We then arrive at expressions

$$H_{free,\perp} = \sum_{j=(k_{x,y,z},\varepsilon)} \frac{\hbar\omega_j}{2}(\alpha_j^*\alpha_j i + \alpha_j\alpha_j^*), \tag{10.106}$$

$$\mathbf{A}_\perp = \sum_j \mathscr{A}_{\omega_j}\left[\alpha_j\boldsymbol{\varepsilon}_j e^{i\mathbf{k}_j\cdot\mathbf{x}} + \alpha_j^*\boldsymbol{\varepsilon}_j e^{-i\mathbf{k}_j\cdot\mathbf{x}}\right], \tag{10.107}$$

$$\mathbf{E}_\perp = i\sum_j \mathscr{E}_{\omega_j}\left[\alpha_j\boldsymbol{\varepsilon}_j e^{i\mathbf{k}_j\cdot\mathbf{x}} - \alpha_j^*\boldsymbol{\varepsilon}_j e^{-i\mathbf{k}_j\cdot\mathbf{x}}\right], \tag{10.108}$$

$$\mathbf{B}_\perp = i\sum_j \mathscr{B}_{\omega_j}\left[\alpha_j(\boldsymbol{\kappa}_j\times\boldsymbol{\varepsilon}_j)e^{i\mathbf{k}_j\cdot\mathbf{x}} - \alpha_j^*(\boldsymbol{\kappa}_j\times\boldsymbol{\varepsilon}_j)e^{-i\mathbf{k}_j\cdot\mathbf{x}}\right], \tag{10.109}$$

with

$$\mathscr{E}_{\omega_j} = \left[\frac{\hbar\omega_j}{2\epsilon_0 L^3}\right]^{1/2}, \qquad \mathscr{B}_{\omega_j} = \frac{\mathscr{E}_{\omega_j}}{c}, \qquad \mathscr{A}_{\omega_j} = \frac{\mathscr{E}_{\omega_j}}{\omega_j}, \tag{10.110}$$

The mode index j now points to a discrete (but still infinite) set of modes, and the integration over all modes to obtain the electrical field strength $\mathbf{E}$ at any location $\mathbf{x}$ turns into a discrete sum.

For the particular case of periodic boundary conditions we still can get away with the simple exponentials describing plane waves; however, for realistic boundary conditions, this is no longer the case.

10.1.5   Realistic boundary conditions: Modes beyond plane waves

In a slightly more generalized mode decomposition, the operator for the electric field may be written as

$$\hat{\mathbf{E}}(\mathbf{x},t) = i\sum_j \mathscr{E}_{\omega_j}\left(\mathbf{g}_j(\mathbf{x})\hat{a}_j(t) - \mathbf{g}_j^*(\mathbf{x})\hat{a}_j^\dagger(t)\right), \tag{10.111}$$

where the spatial dependency and polarization property of a mode is covered by a mode function $\mathbf{g}_j(\mathbf{x})$, the time dependency is transferred into the ladder operators $\hat{a}, \hat{a}^\dagger$, and the dimensional components, together with some normalization of the mode function, is contained in the constant $\mathscr{E}_{\omega_j}$. The corresponding operator for the magnetic field can just be derived out of this quantity via one of the Maxwell equations, (10.4), taking into account the time dependency of the ladder operators:

$$\partial\hat{a}_j/\partial t = -i\omega_j\hat{a}_j, \qquad \partial\hat{a}_j^\dagger/\partial t = i\omega_j\hat{a}_j^\dagger. \tag{10.112}$$

With this, we end up with a magnetic field operator

$$\hat{\mathbf{B}}(\mathbf{x},t) = \sum_j \mathscr{B}_{\omega_j}\left(\nabla\times\mathbf{g}(\mathbf{x})\hat{a}(t) + \nabla\times\mathbf{g}^*(\mathbf{x})\hat{a}^\dagger(t)\right), \quad\text{with}\quad \mathscr{B}_{\omega_j} = \mathscr{E}_{\omega_j}/\omega_j. \tag{10.113}$$

In practice, the mode functions $\mathbf{g}_j(\mathbf{x})$ and the dispersion relation ω_j are known or chosen as an ansatz, and it remains to find the normalization constant $\mathcal{E}_{\omega_j}$ to make sure that the Hamilton operator, given as a volume integral in the form of eq. (10.67), corresponds to the standard harmonic oscillator Hamiltonian in eq. (10.81).

For a given mode function $\mathbf{g}(\mathbf{x})$ and dispersion relation compatible with the Maxwell equations, the normalization constant is given by:

$$\mathcal{E}_j = \sqrt{\frac{\hbar\omega_j}{\epsilon_0 V}} \quad \text{with} \quad V := \int d\mathbf{x} \left[|\mathbf{g}(\mathbf{x})|^2 + \frac{c^2}{\omega^2} |\nabla \times \mathbf{g}(\mathbf{x})|^2 \right]. \tag{10.114}$$

The choice of the mode function can now easily adapted to the symmetry of the problem or boundary condition. Depending on the problem, the mode indices j may be discrete, continuous, or a combination of both. In the following, we give a set of examples for various geometries.

Occasionally, it may be helpful to include a finite length or volume, artificially discretizing some continuous-mode indices. This may be helpful when evaluating the normalization constant $\mathcal{E}$ and keeping track of state densities for transition rates, but should not affect the underlying physics.

10.1.5.1 Square waveguide

This refers to very simple boundary conditions: The electromagnetic field is confined into a square pipe with ideally conducting walls (see Figure 10.2a). It is a mode decomposition suitable for TE modes as found in microwave waveguides. While not exactly of concern in the optics regime, it may become an important set of boundary conditions in the context of quantum circuit dynamics.

The generalized mode index $j \equiv (n, m, k)$ is formed by two discrete mode indices $n, m = 0, 1, 2, \ldots; n \cdot m \neq 0$ characterizing the nodes in the transverse direction across

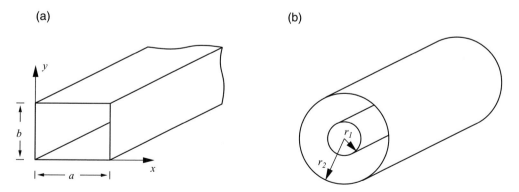

Fig. 10.2 Two simple boundary conditions for electromagnetic waves. (a) A waveguide with ideally conducting walls, as used in the microwave domain; (b) a coaxial waveguide, as found at low frequencies. Both geometries illustrate how to carry out field quantization in uncommon mode geometries.

the waveguide, and a continuous mode index k characterizing the wavevector along the waveguide.

The mode function $\mathbf{g}(\mathbf{x})$ of a TE_{nm} mode is given by

$$\mathbf{g}(\mathbf{x}) = \hat{e}_y e^{ikz} \sin\left(\frac{n\pi}{a}x\right) \cos\left(\frac{m\pi}{b}y\right), \tag{10.115}$$

and the dispersion relation is

$$\omega_j^2 = c_0^2 \left(k^2 + n^2\pi^2/a^2 + m^2\pi^2/b^2\right). \tag{10.116}$$

This dispersion relation is characteristic for waveguides, which in general have some discretized transverse mode structure and a continuous parameter k, which resembles the wavevector of the plane-wave solution. The second part poses a confinement term, leading to a dispersion just due to the geometry of the mode.[6]

To carry out the normalization, we introduce a "quantization length" L in the z direction. This discretizes the mode index k to $k = 2\pi l/L, l = 0, 1, 2, \dots$. The normalization constant for this mode is given by

$$\mathscr{E}_j = \sqrt{\frac{\hbar\omega_j}{\epsilon_0 V}}, \quad V = \frac{Lba}{2} \cdot \begin{cases} 1 & , m = 0 \\ 1/2 & , m > 0. \end{cases} \tag{10.117}$$

10.1.5.2   *Gaussian beams*

Many optical experiments work with light beams with a transverse Gaussian beam profile under a paraxial approximation. Such modes typically represent eigenmodes of optical resonators formed by spherical mirrors.

In a typical experiment, the transverse mode parameter (*waist, w_0*) is fixed, and the longitudinal mode index is a continuous wave number k. In a regime where there is no significant wave-front curvature, the mode function is given by

$$\mathbf{g}(\mathbf{x}) = \varepsilon e^{-\rho^2/w_0^2} e^{ikz}, \tag{10.118}$$

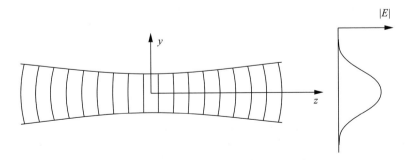

Fig. 10.3   Wave fronts and electrical field distribution of a Gaussian beam.

with a radial distance $\rho = \sqrt{x^2 + y^2}$, a transverse polarization vector $\boldsymbol{\varepsilon}$, and a position z along the propagation direction. The dispersion relation for this mode is given by

$$\omega^2 = \frac{c_0^2}{\epsilon_r}\left(k^2 + \frac{2}{w^2}\right). \tag{10.119}$$

This expression contains already the permittivity ϵ_r if the field is present in a dielectric medium. To cover the vacuum case, just set $\epsilon_r = 1$. The normalization constant $\mathscr{E}$ for the electric field operators is given by

$$\mathscr{E} = \sqrt{\frac{\hbar\omega}{\pi w^2 L \epsilon_0 \epsilon_r}}. \tag{10.120}$$

Herein, we introduced a "quantization length" L, assuming a periodic boundary condition in the propagation direction z. While not necessary, it avoids some confusion when counting over target modes.

This particular mode decomposition is also a good approximation when single-mode optical fibers are used to support the electromagnetic field. While the specific dielectric boundary conditions in optical fibers are more complicated and depend on the doping structure, the most common optical fibers have a transverse mode structure that resembles closely a Gaussian mode.

A discretized longitudinal mode index is actually a very common condition for optical resonators, assuming the leakage to the environment is relatively small. If the transition to a continuum is desired, the relevant observable quantity can be first evaluated with a discrete mode spectrum, followed by a transition $L \to \infty$. We will see an example of such a procedure in section 10.3.

A generalization of this mode function is obtained once the divergence of the Gaussian beam is taken into account. This is typically present in optical resonators with moderate focusing. There, the mode function becomes

$$\mathbf{g}(\mathbf{x}) = \boldsymbol{\varepsilon}\frac{w_0}{w(z)}e^{-\rho^2/2w^2(z)}e^{ikz+ik\frac{\rho^2}{2R(z)}+i\zeta(z)}, \tag{10.121}$$

with a beam waist w_0 other commonly used quantities

$$
\begin{array}{ll}
\text{beam parameter:} & w(z) = w_0\sqrt{1 + (z/z_R)^2} \\
\text{radius of curvature:} & R(z) = z + z_R^2/z \\
\text{Guoy phase:} & \zeta(z) = \tan^{-1} z/z_R \\
\text{Raleigh range:} & z_R^2 = \pi w_0^2/\lambda.
\end{array} \tag{10.122}
$$

Further extension of this concept includes higher-order transverse modes; similarly to the waveguides with conductive walls, these modes are characterized by the nodes in radial and angular components, or by nodes in a rectangular geometry (See, e.g., Saleh and Teich (Saleh and Teich, 2007)).

10.1.5.3   *Spherical harmonics*

This mode decomposition is particularly suited to adapt to electrical multipole transitions in atoms and molecules, since the problem has a rotational symmetry around the center, which contains the atom.

These modes fall into two classes, the transverse electrical (TE) or magnetic multipole fields, and the transverse magnetic (TM) or electrical multipole fields, each of them forming a complete set of modes for the electromagnetic field similarly to the plane waves we have seen earlier. Mode indices are given by a combination of two discrete indices L, M addressing the angular momentum, with $L = 0, 1, 2, \ldots$ and $M = -L, -L+1, \ldots, L-1, L$ and a radial wave index $k \in [-\infty, +\infty]$.

Their angular dependency involves the normalized vector spherical harmonics,

$$\mathbf{X}_{LM}(\theta, \phi) := \frac{1}{\sqrt{L(L+1)}} \frac{1}{i}(\mathbf{r} \times \nabla)Y_{LM}(\theta, \phi), \qquad (10.123)$$

with the usual spherical harmonics Y_{LM} known from atomic physics. For TM modes, associated with electrical multipole radiation, the mode function for the electrical field is given by

$$\mathbf{g}(\mathbf{x}) = \frac{i}{k}\nabla \times (h_L(kr)\mathbf{X}_{LM}(\theta, \phi)), \qquad (10.124)$$

with the spherical Hankel functions

$$h_L(kr) = \sqrt{\pi/2kr} \left[J_{L+1/2}(kr) \pm iN_{L+1/2}(kr) \right] \qquad (10.125)$$

governing the radial dependency. For electrical dipole modes, which are the most important modes for atomic transitions in the optical domain, $L = 1$, and $M = 0, \pm1$ corresponding to π and $\sigma^\pm$ polarized light. There, the radial part becomes

$$h_1(kr) = -\frac{e^{ikr}}{kr}\left(\pm1 + \frac{i}{kr}\right), \qquad (10.126)$$

where the $\pm$ reflects the sign of k, distinguishing between asymptotically incoming or outgoing solutions. The dispersion relation is simply $\omega = ck$, and asymptotically (i.e. for $r \gg \lambda$) the field resembles a locally transverse spherical wave. A comprehensive description of these modes can, e.g., be found in Jackson (Jackson, 1999).

Such modes are used to connect the spontaneous emission rate of atoms with their induced electrical dipole moment or susceptibility (Weisskopf and Wigner, 1930).

10.1.5.4   *Coaxial cable*

This is a somewhat textbook-like mode decomposition, which does not really reach into the optical domain, but is simple to solve analytically. It refers to the propagating modes in the usual cables used for signal transmission. These cables are formed by two concentric cylinders with radii r_1, r_2 that confine the electrical field as depicted in Figure 10.2(b). The most common (low frequency) mode function is indexed by a wavevector k and has a radial electrical field dependency:

$$\mathbf{g}(\mathbf{x}) = \begin{cases} \mathbf{e}_\rho \frac{e^{ikz}}{\rho} & \text{for} \quad r_1 < r < r_2, \\ 0 & \text{elsewhere} \end{cases} \tag{10.127}$$

with the radial unit vector $\mathbf{e}_\rho$. The dispersion relation has no confinement correction, i.e. $\omega = ck$. For a finite length L of the cable (periodic boundary conditions, i.e. $k = n \cdot 2\pi L, n \in \mathbb{N}$), one can simply calculate the "mode volume" V of the cable according to eq. (10.114) as

$$V = \int d\mathbf{x} \left[|\mathbf{g}|^2 + \frac{c^2}{\omega^2} |\nabla \times \mathbf{g}|^2 \right] = 4\pi L \log \frac{r_2}{r_1}, \tag{10.128}$$

leading to a normalization constant

$$\mathcal{E}_k = \sqrt{\frac{\hbar \omega_k}{\epsilon_0 V}} = \sqrt{\frac{\hbar \omega_k}{4\pi \epsilon_0 L \log(r_2/r_1)}}. \tag{10.129}$$

Keep in mind that the dimension of $\mathcal{E}_k$ is not that of the field strength, only the combined product of the mode function $\mathbf{g}$ and $\mathcal{E}_k$. Just to get a feel for orders of magnitude, we evaluate the normalization constant for a cable of length $L=1$ m and $k = 10\pi/L$, corresponding to $\omega/2\pi = 1.5$ GHz. With $\log r_2/r_1 = 1$, this results in a normalization constant $\mathcal{E} \approx 100$ nV – a quantity just about too small for contemporary microwave measurement, but not too far either. The currently developing area of quantum circuit dynamics is starting to explore this parameter regime.

10.2 A silly question: what is a photon?

So far, we only paid attention to the general field quantization – and did not consider specific states of the field. The intent of this section is to look at a few states of light, and attempt to get an idea of what we mean when we talk about photons. This is not as clean a question to answer as one would desire: much of the confusion and sometimes arguments between different camps in the community are based on different definitions, or perhaps conceptions that are carried over from a traditional mechanistic view.

To approach this problem, we quickly visit the history of how the concept of a photon came about. Then, we continue with the standard way of introducing field states in finite-size spaces, which perhaps allow the formally cleanest definition of what a photon is in the sense that these field states are energy eigenstates, and thus are stable excitations of a field. Such a set of boundary conditions is provided in practice by optical cavities or other resonators. This also allows a clean definition of typical field states, very similar to simple harmonic oscillator modes.

Another important approach of what a photon could be is inspired by the detection process, based on the photo-electric effect. Such a detection process is able to witness the smallest amount of energy, and usually results in a "detection event" localized in time. This is perhaps the most particle-like definition, and matches a wide range of experiments.

Another common definition of a photon is tied to a generation process, the most prominent one being the spontaneous emission of an excited state of an atom. Again, this leads to a field localized in time and space, not exactly reflecting an energy eigenstate.

10.2.1   History: The photon of Lewis and Lamb's rage

A somewhat polemic article of W.E. Lamb with a brief historical overview is not only entertaining, but helpful to get an idea where the problem lies (Lamb, 1995).

An investigation on the external photoelectric effect by Lenard in 1902 (Lenard, 1902) and others lead to the observation that the maximal kinetic energy of the electrons emitted from metals under illumination with light did not depend on the intensity of the light, but only on its frequency. The simple interpretation of this observation by Einstein, namely that the absorption of light can only take place in discrete quanta, and that the dependency of the electron's kinetic energy with frequency should be given by the Planck constant, is sometimes seen as the "birth" of the photon concept, assigning a corpuscular character to light in the framework of quantum physics (Einstein, 1905). The predicted linear dependency was then verified experimentally with a high accuracy by Millikan (Millikan, 1916).

The explanation of the photoelectric effect, however, does not really require a quantized treatment of the electromagnetic field, but can be understood by a quantum description of the electron with time-dependent electrical fields and first-order perturbation theory (Wentzel, 1926).

The original notion of a photon as a particle was introduced by Lewis in 1926 (Lewis, 1926) in a slightly different context, as a carrier for transmitting radiation between atoms that should obey a conservation law. That concept has long vanished.

10.2.2   Tying it to energy levels: Cavity QED

The perhaps cleanest situation can be found where one has a discrete set of modes: The electromagnetic field in each mode can be described as a harmonic oscillator, and many of the system states (e.g. states of the electromagnetic field) can be understood in terms of harmonic-oscillator states.

To be in a situation with a discrete set of modes, we need to have the electromagnetic field confined to a finite volume, preferably with distinct frequencies for the different modes. This is realized with optical cavities, or more recently with electronic resonators of a sufficiently high quality factor such that a coupling to an environment becomes only a small perturbation.

To understand the consequences of quantization of the electric and magnetic fields, we need to look into the expectation values and the variance of the fields using the tools of quantum mechanics. To simplify the treatment, we restrict the treatment to states that only involve a single mode, and come back to multimode fields later.

Here, we briefly review a few harmonic-oscillator states most relevant for quantum optics in cavities, namely the number states (or energy eigenstates, also known as Fock states), the thermal states, and the coherent states.

10.2.2.1 *Number states*

The number states (as energy eigenstates of the harmonic oscillator) have been discussed before. We now consider the expectation values and variance of the electric field for a single mode in a number state.

Assuming we focus our attention to the mode indexed by l, we find

$$\langle n|\hat{\mathbf{E}}_l|n\rangle = i\mathscr{E}_l\langle n|\hat{a}_l e^{i\mathbf{k}\cdot\mathbf{x}} - \hat{a}_l^\dagger e^{-i\mathbf{k}\cdot\mathbf{x}}|n\rangle = 0. \tag{10.130}$$

The equality is due to property of raising/lowering operator and the orthogonality of the number states:

$$\langle n|\hat{a}_l|n\rangle = \langle n|n-1\rangle\sqrt{n} = 0, \tag{10.131}$$

$$\langle n|\hat{a}_l^\dagger|n\rangle = \langle n|n+1\rangle\sqrt{n+1} = 0. \tag{10.132}$$

With this, the expectation value of $\hat{\mathbf{E}}_l^2$ can easily be obtained:

$$\langle n|\hat{E}_l^2|n\rangle = \mathscr{E}_l^2(2n+1). \tag{10.133}$$

Here, for the ground state $|0\rangle$, which we can identify as the vacuum state, the average of the field are zero, but there are fluctuations of the field, since

$$\langle \hat{E}_l^2\rangle = \mathscr{E}_l^2, \tag{10.134}$$

which is non-zero.

The *spread* or uncertainty of the electric field in a number state is given by

$$\Delta\hat{E}_l = \sqrt{\langle \hat{E}_l^2\rangle - \langle \hat{E}_l\rangle^2} = \mathscr{E}_l\sqrt{2n+1}. \tag{10.135}$$

Similarly, the expectation values of the quadrature amplitudes can be obtained:

$$\langle n|\hat{a}_Q|n\rangle = \langle n|\hat{a}_P|n\rangle = 0, \tag{10.136}$$

$$\langle n|\hat{a}_Q^2|n\rangle = \langle n|\hat{a}_P^2|n\rangle = \frac{2n+1}{4}. \tag{10.137}$$

10.2.2.2 *Thermal states*

In many cases, the exact state of a quantum mechanical system is not known, but can be described by an ensemble of possible states following a classical probability distribution; the classical thermal ensembles are a very typical example where one has a certain lack of classical knowledge about the exact state.

Such a partial knowledge about the state of a system can be expressed using a so-called density matrix ρ, which may be composed of projectors for a set of pure quantum states $|\psi\rangle_n$, where the probability of such a realization (of the system in state $|\psi_n\rangle$) is p_n:

$$\hat{\rho} = \sum_n p_n|\psi_n\rangle\langle\psi_n|. \tag{10.138}$$

The expectation values of operators $\hat{A}$ is a weighted average over the expectation values of the constituting states in the density matrix definition above, and may be obtained by tracing over these operators:

$$\langle \hat{A} \rangle = \sum_n p_n \langle \psi_n | \hat{A} | \psi_n \rangle \sum_n \langle \Psi_n | \hat{A} | \Psi_n \rangle =: tr \left(\hat{\rho} \hat{A} \right), \qquad (10.139)$$

where the sum is taken over a complete set of base vectors $|\Psi_n\rangle$. By definition, the trace of a meaningful density matrix has to be 1:

$$tr\,(\hat{\rho}) = 1. \qquad (10.140)$$

Sometimes, the definition

$$\langle \hat{A} \rangle = tr \left(\hat{\rho}^{1/2} \hat{A} \hat{\rho}^{-1/2} \right) \qquad (10.141)$$

is used, where the operator $\hat{\rho}^{1/2}$ is defined by the MacLaurin series expansion for operators.

For thermal states, the probability distribution is determined by the energy of each state, i.e.

$$p_n = \frac{1}{Z} e^{-\beta E_n}, \qquad (10.142)$$

where E_n is the energy of the state $|\psi_n\rangle$, $\beta = (k_B T)^{-1}$, and Z a normalization constant (partition function) so that the sum of probabilities equals 1:

$$Z = \sum_n e^{-\beta E_n}. \qquad (10.143)$$

For the harmonic oscillator in particular, we find

$$Z = \sum_{n=0}^{\infty} e^{-\beta E_n} = \sum_{n=0}^{\infty} e^{-\beta \hbar \omega (n + \frac{1}{2})} = e^{-\beta \hbar \omega / 2} \sum_{n=0}^{\infty} e^{-\beta \hbar \omega n}$$

$$= \frac{e^{-\beta \hbar \omega / 2}}{1 - e^{-\beta \hbar \omega}}. \qquad (10.144)$$

The probability p_n of being in the energy eigenstate $|psu_n\rangle$ can then be written as

$$p_n = \frac{e^{\beta \hbar \omega n}}{(1 - e^{-\beta \hbar \omega})^{-1}} = \frac{A^n}{(1 - A)^{-1}}, \qquad \text{with } A = e^{-\beta \hbar \omega}. \qquad (10.145)$$

In the density matrix formulation, one can write the thermal state as

$$\hat{\rho} = \sum_n \frac{1}{Z} e^{-\beta E_n} |n\rangle \langle n| = \sum_{n=0}^{\infty} \frac{A^n}{(1 - A)^{-1}} |n\rangle \langle n|. \qquad (10.146)$$

Coming back to the expectation values for the electrical field of a single mode and its variance, we now find

$$\langle \hat{\mathbf{E}}_l \rangle_{\text{thermal}} = 0, \tag{10.147}$$

$$\langle \hat{\mathbf{E}}_l^2 \rangle_{\text{thermal}} = \mathscr{E}_l^2 \frac{1+A}{1-A}. \tag{10.148}$$

In the high-temperature limit $(k_B T \gg \hbar\omega)$ we have

$$A = e^{-\beta\hbar\omega} \approx 1 - \beta\hbar\omega. \tag{10.149}$$

This means that the variance of the electrical field is given by

$$\langle \hat{\mathbf{E}}_l^2 \rangle_{\text{thermal}} \approx \mathscr{E}_l^2 \frac{2}{\beta\hbar\omega} = \frac{k_B T}{(2\pi)^3 \epsilon_0}. \tag{10.150}$$

On the other hand, for low temperature limit where $k_B T \ll \hbar\omega$, $A \ll 1$ we have

$$\langle \hat{\mathbf{E}}_l^2 \rangle_{\text{thermal}} \approx \mathscr{E}_l^2, \tag{10.151}$$

indicating that for low temperatures, the vacuum fluctuations dominate the distribution of possible measurement results for the electric field.

This is an important point for finding out at which temperatures experiments can be carried out. At room temperature $(T = 300\,\text{K})$, the characteristic thermal frequency to decide upon the high- or low- temperature limit is given by

$$\omega_{RT} = \frac{k_B T}{\hbar} \approx 2\pi \cdot 6.25\,\text{THz}, \tag{10.152}$$

corresponding to a vacuum wavelength of about $48\,\mu\text{m}$. Thus, in the optical domain with frequencies on the order of $10^{14}\,\text{Hz}$ or wavelengths on the order of $1\,\mu\text{m}$, electromagnetic fields are typically in the ground state or in close approximation thereof, while for experiments in the microwave domain (frequencies from a few $1\,\text{GHz}$ to a few $100\,\text{GHz}$), thermal occupation of modes is a problem, and the structures coupling to a thermal environment due to losses need to be cooled down to very low temperatures.

10.2.2.3 Coherent states

One of the perhaps most important class of states in quantum optics were introduced by Glauber in 1963 (Glauber, 1963). They are the closest quantum-mechanical analogon to classical motion of a harmonic oscillator, and – as any other pure oscillator state – are coherent superpositions of energy eigenstates of the oscillator.

For a formal approach to these states, consider the lowering operator $\hat{a}$ and one of its eigenstates $|\alpha\rangle$ corresponding to the eigenvalue α:

$$\hat{a}|\alpha\rangle = \alpha|\alpha\rangle. \tag{10.153}$$

There actually exists such a state for every $\alpha \in \mathbb{C}$. We leave it as an exercise to the reader to derive a representation of $|\alpha\rangle$ in the energy eigenbasis $\{|n\rangle\}$ in the form

$$|\alpha\rangle = \sum_{n=0}^{\infty} c_n |n\rangle. \tag{10.154}$$

The (normalized) result for that representation for a given α is given by

$$|\alpha\rangle = \sum_{n=0}^{\infty} e^{-|\alpha|^2/2} \frac{\alpha^n}{\sqrt{n!}} |n\rangle. \tag{10.155}$$

One particular state is obtained for $\alpha = 0$: all contributions except for the ground state of the oscillator vanish, so the corresponding coherent state is the ground state itself:

$$|\alpha = 0\rangle = |n = 0\rangle. \tag{10.156}$$

We now can find the expectation values for the electric field and its variance, as we have done before for number states and thermal states:

$$\langle\alpha|\hat{E}_l|\alpha\rangle = i\mathscr{E}_l\varepsilon_l \left(\alpha e^{i\mathbf{k}\cdot\mathbf{x}} - \alpha^* e^{-i\mathbf{k}\cdot\mathbf{x}} \right). \tag{10.157}$$

Here, the expression of $\hat{\mathbf{E}}$ in terms of $\hat{a}_Q$ and $\hat{a}_P$ comes in handy. With

$$\langle\alpha|\hat{a}_Q|\alpha\rangle = \frac{1}{2}\langle\alpha|(\hat{a} + \hat{a}^\dagger)|\alpha\rangle = \frac{1}{2}(\alpha + \alpha^*) = \mathrm{Re}(\alpha), \tag{10.158}$$

$$\langle\alpha|\hat{a}_P|\alpha\rangle = \frac{1}{2i}\langle\alpha|(\hat{a} - \hat{a}^\dagger)|\alpha\rangle = \frac{1}{2i}(\alpha - \alpha^*) = \mathrm{Im}(\alpha), \tag{10.159}$$

we can write the expectation value of the electrical field operator as

$$\langle\hat{\mathbf{E}}_l\rangle = -\mathscr{E}_l\varepsilon_l \{2\langle\hat{a}_Q\rangle \sin(\mathbf{k}\cdot\mathbf{x} - \omega t) + \langle\hat{a}_P\rangle \cos(\mathbf{k}\cdot\mathbf{x} - \omega t)\}$$
$$= -2\mathscr{E}_l\varepsilon_l \{\mathrm{Re}(\alpha) \sin(\mathbf{k}\cdot\mathbf{x} - \omega t) - \mathrm{Im}(\alpha) \cos(\mathbf{k}\cdot\mathbf{x} - \omega t)\}. \tag{10.160}$$

The real and imaginary parts of α are the expectation values of the sine and cosine component of something that looks like a classical field, e.g. an expectation value of the field strength that oscillates sinusoidally in time. Therefore, these states are also called quasi-classical states. Note that this property applies not only to the harmonic oscillators associated with an electromagnetic field mode, but any harmonic oscillator following a Hamiltonian with the same structure.

Let's now have a closer look at the variance of the electrical field, expressed both in the variance of $\langle\hat{\mathbf{E}}\rangle$ itself and its quadrature components. We start by finding the expectation value of the square of the electric field,

$$\langle\hat{E}_l^2\rangle = \langle\alpha| \left\{ i\mathscr{E}_l\varepsilon_l \left(\hat{a}_l e^{i\mathbf{k}\cdot\mathbf{x}} - \hat{a}_l^\dagger e^{-i\mathbf{k}\cdot\mathbf{x}} \right) \right\}^2 |\alpha\rangle$$
$$= -\mathscr{E}_l^2 \left\{ \alpha^2 e^{2i\mathbf{k}\cdot\mathbf{x}} - (2\alpha\alpha^* + 1) + \alpha^{*2} e^{-2i\mathbf{k}\cdot\mathbf{x}} \right\}. \tag{10.161}$$

For the variance, we also need

$$\langle \hat{\mathbf{E}}_l \rangle^2 = \left\{ i \mathscr{E}_l \left(\alpha e^{i\mathbf{k}\cdot\mathbf{x}} - \alpha^* e^{-i\mathbf{k}\cdot\mathbf{x}} \right) \right\}^2$$
$$= -\mathscr{E}_l^2 \left\{ \alpha^2 e^{2i\mathbf{k}\cdot\mathbf{x}} - 2\alpha\alpha^* + \alpha^{*2} e^{-2i\mathbf{k}\cdot\mathbf{x}} \right\}. \tag{10.162}$$

With both of these terms, we can evaluate the variance of the electric field,

$$(\Delta \hat{E}_l)^2 = \langle \hat{\mathbf{E}}_l^2 \rangle - \langle \hat{\mathbf{E}}_l \rangle^2 = \mathscr{E}_l^2. \tag{10.163}$$

Thus, the variance of the field is *independent* of the value of α, and equal to the variance of the field for a vacuum state since this is also a coherent state with $\alpha = 0$.

Now we turn to the variances in the quadrature components:

$$\langle \hat{a}_Q^2 \rangle = \frac{1}{4} \left(\alpha^2 + \alpha^{*2} + 2\alpha\alpha^* + 1 \right) \tag{10.164}$$

$$\langle \hat{a}_Q \rangle^2 = \frac{1}{4} \left(\alpha^2 + \alpha^{*2} + 2\alpha\alpha^* \right) \tag{10.165}$$

$$(\Delta \hat{a}_Q)^2 = \langle \hat{a}_Q^2 \rangle - \langle \hat{a}_Q \rangle^2 = \frac{1}{4}. \tag{10.166}$$

Similarly, the variance of the other quadrature component evaluates to

$$(\Delta \hat{a}_P)^2 = \langle \hat{a}_P^2 \rangle - \langle \hat{a}_P \rangle^2 = \frac{1}{4}. \tag{10.167}$$

Both quadrature amplitudes show the same variance, which is also the same as for the vacuum. Since there is an uncertainty relation between the quadratures, and the ground state is a minimal uncertainty state, this implies that the eigenstate of $\hat{a}$ is a minimum uncertainty state for the associated quadrature amplitudes of the electromagnetic field.

Without any proof, it should be noted that the coherent states represent the state typically emitted by a laser operating far above threshold, and assuming that the phase of the laser radiation is fixed by convention or a conditional measurement in an experiment.

It should also be noted that coherent states don't have a fixed number of photons in them, if the notion 'having n photons in a system' refers to the system being in the nth excited energy eigenstate of the discrete field mode.

10.2.3 Tying it to the detection processes

So far, we have discussed observables and measurements only from a very formal point of view. In this section, we will have a somewhat closer look into various measurement techniques for light, and try to get an idea what we really measure in a particular configuration – and how this connects to the various "observables".

10.2.3.1   The photo-electric process

Until very recently, all optical measurement techniques relevant for the domain of quantum optics were based on various versions of the photo-electric effect. The effect of electron emission upon irradiation of a metallic surface was essential in the development of a quantum-mechanical description of light.

The photo-electric effect refers to the phenomenon that upon exposure to light, electrons may be emitted from a metal surface and was experimentally observed in 1887 by Hertz (Hertz, 1887) as a change in a spark intensity upon exposure of electrodes to ultraviolet light. More quantitative studies were carried out 1899 by Thomspon, who observed together with the discovery of electrons that the emitted charge increases with intensity and frequency of the light. In 1902, von Lenard carried out more quantitative measurements on the electron energy emitted by light exposure in an experimental configuration symbolized in Figure 10.4 and found that the stopping potential V_r needed to suppress the observation of a photocurrent I_{ph} in a vacuum photodiode depended only on the wavelength of the light, and concluded that the kinetic energy of electrons after being liberated from the metal compound is determined by the frequency of the light, not its intensity. He also found a strong dependency of the liberation energy of the electrons, today referred to as the *work function*, which depended strongly on the preparation of the metal.

This led to the spectacular interpretation in 1905 by Einstein that the electron emission was due to an absorption process of electrons in the metal, and that absorption of light could only take place in well-defined packets or *quanta* of light, supplying another pillar in the foundation of a quantum-mechanical treatment of the electromagnetic radiation besides Planck's description of blackbody radiation.

The numeric expression for the kinetic energy of the emitted electrons,

$$E_{kin} = hf - \Phi, \tag{10.168}$$

with f being the frequency of a monochromatic light field and Φ a material constant suggested a linear dependency between the excess energy of the electrons and the light frequency. This linearity was then quantitatively observed in experiments of Millikan in 1915.

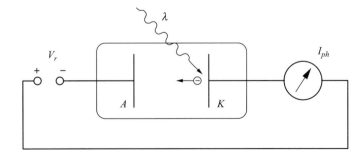

Fig. 10.4 Experimental configuration to observe the photo-electric effect: Light at wavelength λ causes electrons leaving the metal surface with an energy independent of the intensity of the light.

Photomultiplier. The charge of the single photo-electron liberated in an absorption process is very small, and it is technologically challenging to detect this single charge directly. Therefore, the metallic surface generating the primary photo-electron is often followed by an electron multiplier. This is an arrangement of subsequent metal surfaces, where electrons are accelerated towards these metal surfaces (*dynodes*) such that upon impact, a larger number of electrons are emitted, which are subsequently accelerated towards the next plate:

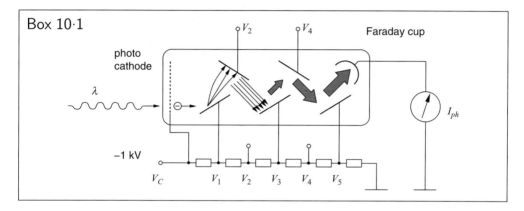

The photocathode, the dynode arrangement and a final Faraday cup to collect the secondary charge emission from the last dynode are kept in a small vacuum tube, and the cascaded accelerating potentials of the dynodes (a few 100 V) are derived via a voltage divider chain from a single high-voltage source.

The overall gain of such an electron-multiplication stage can be on the order of 10^6 to 10^8, leading to a charge pulse on the order of 10^{-11} A s. Such a charge can be conveniently detected, leading to a measurable signal from a single primary photo-electron.

The number of photo-electrons per unit time is proportional to the light power, as we will see later, so the photocurrent in such a device can be used to determine low light power levels.

A fundamental prerequisite for using the photo-electric effect to detect visible light is that the work function Φ of the photocathode is sufficiently small. As the binding energy of electrons in the metallic bulk can be on the order of a few eV, a careful choice of the photocathode material is necessary to observe the photo-electric effect with visible or infrared light. Typically, efficient photocathodes are made out of a combination of silver and several alkali metals and metal oxides.

Solid-state photodiodes. Another important effect used for light detection utilizes the internal photo-electric effect in semiconductors, light is absorbed by an electron in the valence band, and transported into the conduction band. There, only the energy to bridge the band gap needs to be provided.

Such electron–hole pairs can then be separated with an electrical field in the semiconductor, leading to a detectable electrical current. Such photodetectors typically have the geometry of a semiconductor diode, with a depleted region of low conduction

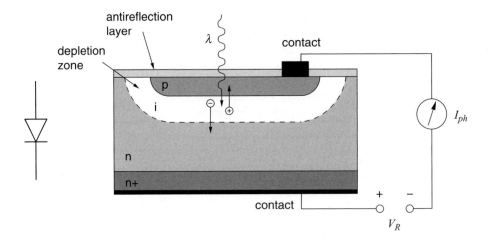

Fig. 10.5 Schematic of a $p - i - n$ photodiode. Electron–hole pairs are generated in a depletion region with a low charge carrier density i, and separated by an electrical field so they can be detected as a macroscopic current.

where the electron–hole pairs are generated by the absorbed light (see Figure 10.5). These pin devices have a diode characteristic, and are operated in a reverse-biased scheme. Typically, a large depletion volume is desired both to allow for an efficient absorption of the incoming light and to ensure a small parasitic capacity of the pn junction for a fast response of the photodetection process. The wavelength-dependent absorption coefficient is shown in Figure 10.6. For a wavelength of $\lambda = 600\,\text{nm}$, the absorption length is on the order of $5\,\mu\text{m}$, which gives some constraints to the construction of silicon photodiodes.

Various semiconductor materials are used for this type of photodetector, allowing us to construct photodetectors for a large range of wavelengths. In a wavelength regime

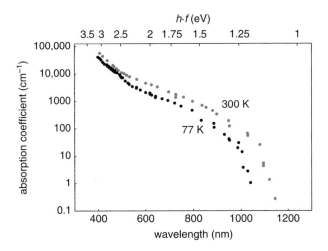

Fig. 10.6 Absorption coefficient for silicon at room temperature and $77\,\text{K}$.

from 1000 nm to 200 nm, silicon is the most common semiconductor material with its band gap energy of 1.25 eV. For longer wavelengths, e.g. the typical choice in optical fiber communication ($\lambda \approx 1300$ nm and ≈ 1550 nm) are III-V semiconductors like GaAs or InGaAs.

For some very fast photodetectors, the depletion region in a Schottky contact (i.e. a metalsemiconductor interface) is utilized instead of a pn junction.

Semiconductor photodiodes typically exhibit a very high *quantum efficiency* η. This quantity describes the probability that light is converted into photo-electrons or electron–hole pairs, and detected. With appropriate antireflective coatings of the semiconductor to avoid the surface reflection at the large refractive-index contrast interface, most of the light can be guided into the semiconductor. A proper dimension of the depletion region allows for a complete absorption of the light. It is not uncommon to find photodiodes with a quantum efficiency of $\eta = 95\%$ or higher.

For a monochromatic light field at an optical frequency $f = c/\lambda$, there is a simple relationship between the observed photocurrent I_{ph} and the optical power P of the incident light. The rate of electron–hole pair creation, r_e, is just given by the rate of elementary absorption processes:

$$r_a = P/(hf) = \frac{P\lambda}{hc}. \qquad (10.169)$$

The resulting photocurrent, considering a quantum efficiency η, is simply given by the rate r_e and charge per electron:

$$I_{ph} = \eta e r_e = \eta e r_a = \eta \frac{e\lambda}{hc} P =: SP. \qquad (10.170)$$

For the wavelength of a HeNe laser, $\lambda = 633$ nm, the *sensitivity* S of a photodiode with $\eta = 98\%$ is $S = 0.5$ A/W.

One of the practical advantages of semiconductor photodiodes in comparison with photomultipliers is their low cost, a typically very small size and the absence of high voltages. On the physics side, we will see that many measurements of quantum states of light will require a high quantum efficiency, which is currently unparalleled with any other photodetection techniques.

Avalanche photodetectors. One of the shortcomings of a simple photodiode in comparison with a photomultiplier is the difficulty to observe single absorption processes, as the charge of a single electron–hole pair is hard to distinguish from normal electronic noise in a system.

However, it is possible to find an analogon to the electron-multiplication process of a photomultiplier in a solid-state device. In so-called avalanche diodes, a region with a high electrical field allows a charge carrier to acquire enough energy to create additional electron–hole pairs in scattering processes, similar to the ionization processes in an electrical discharge through a gas.

Such a semiconductor device can be combined with a charge-depleted region, where electron–hole pairs are generated as a consequence of light absorption (see Figure 10.7). An avalanche photodiode with a built-in charge amplification mechanism can then also be used to detect a single absorption process.

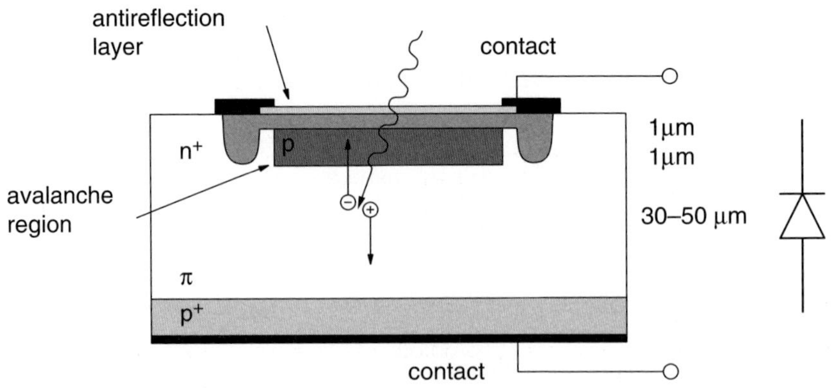

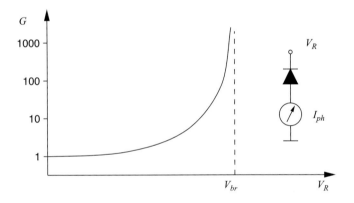

Fig. 10.7 Structure of a reach-through Silicon avalanche diode. A large region where electron–hole pairs are created due to absorption of light is combined with a region of high electric-field strength (p-n+ junction) where an avalanche of charge carriers is triggered.

Fig. 10.8 Gain G of an avalanche photodiode as a function of the reverse bias voltage V_R.

The gain G of such photodiodes increases with the applied reverse bias voltage V_R. These devices are often operated in a regime where one photo-electron creates a charge avalanche of about 100 electron–hole pairs. In this regime, the avalanche photodiode is used in a similar way as a normal pin-photodiode. The gain of the multiplication region diverges at the so-called *breakdown voltage* V_{br}, where the stationary operation of the device leads to a self-sustaining conduction in the reverse direction even without additional light from outside. Such a mode of operation is similar to an electrical discharge in a gas, where electrons and ions are accelerated in an electrical field and create more conduction carriers via impact ionization of the residual neutral gas. For semiconductor devices, such an operation over an extended time would deposit a destructive amount of heat into the device.

However, if the energy deposited in the device is limited, this operation regime can be used to identify individual photo-electrons, in a very similar way that a Geiger counter can be operated to observe the breakdown triggered by a few ions in a gas

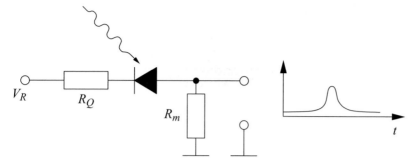

Fig. 10.9 Operation of an avalanche photodiode to observe single absorption processes. The device is reverse-biased above its breakdown voltage V_{br} such that a single photo-electron can switch the device into a conducting mode. The subsequent voltage drop across the quenching resistor R_Q restores the non-conducting mode again.

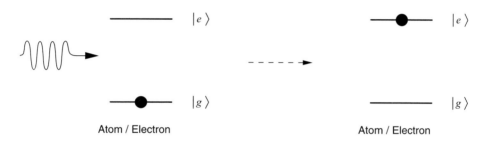

Fig. 10.10 An atom gets into an excited state after absorption of light.

cell created by an incident particle. In a very simple approach, the discharge current during a breakdown of the diode is limited by a large resistor, which leads to a voltage drop below the breakdown voltage V_{br}.

10.2.3.2 *How to describe photodetection in terms of quantum mechanics?*

We have seen several photodetector schemes so far – most of them rely on an elementary process of absorption of light, which is then detected by some other mechanism. Let us therefore consider the physical process of creating a photo-electron. We further simplify our detector to a two-level system instead of the more commonly found continuum of excitation states of usual photodetectors. The absorption process is then modelled by the following transition:

Energy conservation makes it necessary that the energy of the field and the detector system stays constant. Therefore, the operator associated with the transition in the photodetector model has not only to "destroy" an energy quantum of the field, but also create an excitation of the detector. Restricting to a single field mode, we could represent this process by an operator

$$\hat{h} = \hat{a}\hat{d}^{\dagger},\qquad(10.171)$$

where $\hat{a}$ is our usual lowering operator for the field, and

$$\hat{d}^\dagger = |e\rangle\langle g| \tag{10.172}$$

describes the rising operator for the atom or electron. If the excitation operator $\hat{h}$ should be the outcome of a physical interaction process, it has to be derived from an interaction Hamiltonian of the form

$$\hat{H}_I = \text{const} \cdot (\hat{a}\hat{d}^\dagger + \hat{a}^\dagger\hat{d}). \tag{10.173}$$

To ensure hermiticity, with an atomic lowering operator $\hat{d} = |g\rangle\langle e|$. We will see such a process later.

Single photo-electrons. Let's now consider the probability of observing a photo-electron. Assume the initial field before the creation of a photo-electron is in the state $|\psi_i\rangle$, and afterwards in the final state $|\psi_f\rangle$. The probability for such a transition is proportional to $|\langle\psi_f|\hat{a}|\psi_i\rangle|^2$.

If a photo-electron can be generated by a large bandwidth of frequencies from the electrical field, this probability can be expressed in terms of the sum over different modes of the electrical field operator $\hat{E}^{(+)}$ containing the lowering operators:

$$\tilde{w} \propto \left|\langle\psi_f|\hat{E}^{(+)}(\mathbf{x}, t)|\psi_i\rangle\right|^2. \tag{10.174}$$

This may be interpreted as the probability of creating a photo-electron at a position $\mathbf{x}$ and a time t – the usual model of light – matter interaction indeed justifies that, where we assign a well-defined location to the electron.

To use this probability to come up with some meaningful field measurement technique, let's keep in mind that for a given situation, we do not really do anything with the field state afterwards, and we consider only the photo-electron. We thus can obtain the probability of observing a photo-electron as a sum over all possible final field states,

$$w_1 = \sum_f |\langle\psi_f|\hat{E}^{(+)}(\mathbf{x}, t)|\psi_i\rangle|^2 \tag{10.175}$$

$$= \sum_f \langle\psi_i|\hat{E}^{(-)}|\psi_f\rangle\langle\psi_f|\hat{E}^{(+)}|\psi_i\rangle$$

$$= \langle\psi_i|\hat{E}^{(-)}\hat{E}^{(+)}|\psi_i\rangle,$$

where the completeness property has been used:

$$\sum_f |\psi_f\rangle\langle\psi_f| = \mathbb{I}. \tag{10.176}$$

We end up with the probability w_1 of observing a photo-electron to be proportional to $\langle\hat{E}^{(-)}\hat{E}^{(+)}\rangle$. For stationary fields, this photo-electron's creation probability may be used to calculate a photo counting rate r.

However, the fact that we observe a number of discrete photo-electrons is not really an outcome of the field quantization procedure – you can derive a probability distribution for creating photo-electrons equally well assuming assuming a classical field interacting with a quantized detector system. There, the interaction Hamiltonian between field and system has a contribution

$$|e\rangle\langle g|E_{cl}^{(+)}d^{(+)} + |g\rangle\langle e|E_{cl}^{(-)}d^{(-)}, \tag{10.177}$$

where $E_{cl}^{(+)}$ and $E_{cl}^{(-)}$ are components with positive and negative frequency, respectively, and the $d^{\pm}$ are the corresponding components of the electric dipole matrix elements of the considered transition. The photo-electron count rate then is proportional to

$$r = E_{cl}^{(-)}E_{cl}^{(+)} \propto I_{cl}, \tag{10.178}$$

where I_{cl} is the classical intensity of the light field, derived out of the Poynting vector:

$$I = \left\langle \frac{1}{\mu_0}\mathbf{E} \times \mathbf{B} \right\rangle \cdot \mathbf{n}. \tag{10.179}$$

Photo-electron pairs. In the last section, we made the transition from a single photo-electron probability to a rate rather silently, assuming we can consider all photo-electron creation processes independently and summing them up, e.g. by assuming we evaluate the single photo-electron detection probability for infinitesimal time intervals Δt.

This may not always be appropriate. Let us therefore construct an expression for the probability w_2 of observing a pair of photo-electrons at two locations $\mathbf{x}_1$ and $\mathbf{x}_2$ at two times t_1 and t_2 (in the time intervals $[t_1, t_1 + \Delta t_1]$ and $[t_2, t_2 + \Delta t_2]$) similar to the single-photo-electron case:

$$w_2(\mathbf{x}_1, \mathbf{x}_2, t_1, t_2) \propto |\langle \psi_f|\hat{E}^{(+)}(\mathbf{x}_2, t_2)\hat{E}^{(+)}(\mathbf{x}_1, t_1)|\psi_i\rangle|^2. \tag{10.180}$$

$\hat{E}^{(+)}(\mathbf{x}_1, t_1)$ and $\hat{E}^{(+)}(\mathbf{x}_2, t_2)$ refer to the creation of first and second photo-electrons, respectively. Again, this should be summed over all the final field states, since we are only interested in the observation of the photo-electrons. This leads to

$$w \propto \langle \psi_i|\hat{E}^{(-)}(\mathbf{x}_1, t_1)\hat{E}^{(-)}(\mathbf{x}_2, t_2)\hat{E}^{(+)}(\mathbf{x}_2, t_2)\hat{E}^{(+)}(\mathbf{x}_1, t_1)|\psi_i\rangle. \tag{10.181}$$

Again, this probability may be associated with a count rate – this time, the observed quantity would be a coincidence count rate of detectors at $\mathbf{x}_1$ and $\mathbf{x}_2$ at times t_1 and t_2.

10.2.3.3 Correlation functions: First and second order

The expression for the single photodetection probability w_1 looks like an expectation value of a product of fields,

$$\langle \hat{E}^{(-)}\hat{E}^{(+)}\rangle \tag{10.182}$$

at the same position $\mathbf{x}$ and time t.

This concept can be generalized to different points $\mathbf{x}_1$, $\mathbf{x}_2$, t_1 and t_2 to a quantity

$$G^{(1)}(\mathbf{x}_1, \mathbf{x}_2, t_1, t_2) := \langle \hat{E}^{(-)}(\mathbf{x}_1, t_1)\hat{E}^{(+)}(\mathbf{x}_2, t_2)\rangle, \qquad (10.183)$$

which is called the *first-order correlation function* of the field.

Similarly to the generalization of the simple photo-electron, this coincidence counting-rate expression can be generalized into a definition of a *second-order correlation function* for the electromagnetic field:

$$G^{(2)}(\mathbf{x}_1, \mathbf{x}_2, \mathbf{x}_3, \mathbf{x}_4, t_1, t_2, t_3, t_4) := \langle \hat{E}^{(-)}(\mathbf{x}_1, t_1)\hat{E}^{(-)}(\mathbf{x}_2, t_2)\hat{E}^{(+)}(\mathbf{x}_3, t_3)\hat{E}^{(+)}(\mathbf{x}_4, t_4)\rangle.$$
$$(10.184)$$

Let's consider the case of a stationary field, where the two times involved in the definition of $G^{(1)}(\mathbf{x}_1, \mathbf{x}_2, t_1, t_2)$ are reduced to a time difference:

$$G^{(1)}(\mathbf{x}_1, \mathbf{x}_2, t_1, t_2) \rightarrow G^{(1)}(\mathbf{x}_1, \mathbf{x}_2, \tau) \qquad \text{with } \tau = t_2 - t_1. \qquad (10.185)$$

The count rate (or intensity) is then given by $G^{(1)}(\mathbf{x}, \mathbf{x}, 0)$ for a given position $\mathbf{x}$. We can perform a normalization of the correlation function using the intensities:

$$g^{(1)}(\mathbf{x}_1, \mathbf{x}_2, \tau) = \frac{\langle \hat{E}^{(-)}(\mathbf{x}_1, t)\hat{E}^{(+)}(\mathbf{x}_2, t+\tau)\rangle}{\sqrt{\langle \hat{E}^{(-)}(\mathbf{x}_1, t)\hat{E}^{(+)}(\mathbf{x}_1, t)\rangle\langle \hat{E}^{(-)}(\mathbf{x}_2, t+\tau)\hat{E}^{(+)}(\mathbf{x}_2, t+\tau)\rangle}}.$$
$$(10.186)$$

Similarly for the second-order correlation function,

$$g^{(2)}(\mathbf{x}_1, \mathbf{x}_2, \tau) = \frac{\langle \hat{E}^{(-)}(\mathbf{x}_1, t)\hat{E}^{(-)}(\mathbf{x}_2, t+\tau)\hat{E}^{(+)}(\mathbf{x}_2, t+\tau)\hat{E}^{(+)}(\mathbf{x}_1, t)\rangle}{\langle \hat{E}^{(-)}(\mathbf{x}_1, t)\hat{E}^{(+)}(\mathbf{x}_1, t)\rangle\langle \hat{E}^{(-)}(\mathbf{x}_2, t+\tau)\hat{E}^{(+)}(\mathbf{x}_2, t+\tau)\rangle}. \qquad (10.187)$$

The normalization of this function is chosen such that the denominator contain two expressions $\langle \hat{E}^{(-)}(\mathbf{x}_i, t)\hat{E}^{(+)}(\mathbf{x}_i, t)\rangle$, which are intensities at the two locations, and independent of time for stationary fields.

These quantities are referred to as first-order and second-order *coherence functions*. They have a relatively simple interpretation for many optical experiments.

10.2.3.4   Double-slit experiment

To understand the first-order coherence function, we consider the double-slit experiment as shown in Figure 10.11.

Simple propagation of the field according to Huygens principle (while ignoring any fine emission structure due to diffraction, and attenuation at same distance from the openings in the screen) leads to an electrical field at the detector location P of

$$\hat{E}^{(\pm)}(\mathbf{r}, t) = \hat{E}^{(\pm)}\left(\mathbf{r}_1, t - \frac{s_1}{c}\right) + \hat{E}^{(\pm)}\left(\mathbf{r}_2, t - \frac{s_2}{c}\right). \qquad (10.188)$$

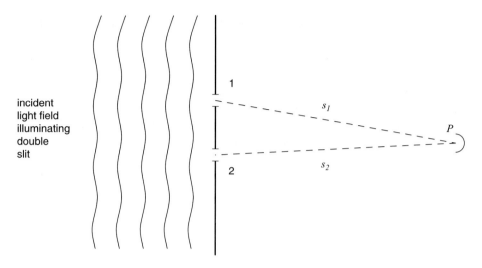

Fig. 10.11 Double-slit experiment. The field is detected at point P separated by distances s_1, s_2 from the two openings in the screen.

For the intensity at the point P of observation behind the slit, we have

$$
\begin{aligned}
I(\mathbf{r}, t) &= \langle \hat{E}^{(-)}(\mathbf{r}, t) \hat{E}^{(+)}(\mathbf{r}, t) \rangle \\
&= \langle \hat{E}^{(-)}\left(\mathbf{r}_1, t - \frac{s_1}{c}\right) \hat{E}^{(+)}\left(\mathbf{r}_1, t - \frac{s_1}{c}\right) \rangle \\
&\quad + \langle \hat{E}^{(-)}\left(\mathbf{r}_2, t - \frac{s_2}{c}\right) \hat{E}^{(+)}\left(\mathbf{r}_2, t - \frac{s_2}{c}\right) \rangle \\
&\quad + 2\mathrm{Re}[\langle \hat{E}^{(-)}\left(\mathbf{r}_1, t - \frac{s_1}{c}\right) \hat{E}^{(+)}\left(\mathbf{r}_2, t - \frac{s_2}{c}\right) \rangle] \\
&= I_1 + I_2 + 2\mathrm{Re}[\sqrt{I_1 I_2} g^{(1)}(\mathbf{r}_1, \mathbf{r}_2, \tau)],
\end{aligned}
\tag{10.189}
$$

with

$$
I_{1,2} = \left\langle \hat{E}^{(-)}\left(\mathbf{r}_{1,2}, t - \frac{s_{1,2}}{c}\right) \hat{E}^{(+)}\left(\mathbf{r}_{1,2}, t - \frac{s_{1,2}}{c}\right) \right\rangle,
\tag{10.190}
$$

and

$$
\tau = \frac{s_2 - s_1}{c}
\tag{10.191}
$$

the time difference of propagation from the slits/holes to the detector position. This can be rewritten using our new coherence function

$$
g^{(1)}(\mathbf{x}_1, \mathbf{x}_2, \tau) = \left| g^{(1)}(\mathbf{r}_1, \mathbf{r}_2, \tau) \right| e^{i\varphi(r_1, r_2, t)},
\tag{10.192}
$$

where the fastest change along the screen comes from the varying time difference. For a light field with a fixed frequency ω_0, the phase φ can be decomposed into

$$\varphi = \alpha(\mathbf{r}_1, \mathbf{r}_2, \tau) - \omega_0 \tau, \tag{10.193}$$

where α is a slowly varying function.[7]

With this, the intensity at the observation point P is given by

$$I(\mathbf{r}) = I_1 + I_2 + 2\sqrt{I_1 I_2} \cos\left(\alpha - \frac{s_1 - s_2}{c}\omega_0\right) |g^{(1)}(\mathbf{r}_1, \mathbf{r}_2, \tau)|, \tag{10.194}$$

leading to the well-known interference pattern of a double slit,[8] where the $|g^{(1)}(\mathbf{r}_1, \mathbf{r}_2, \tau)|$ term determines the "visibility" of the interference pattern. The quantitative definition of visibility of an interference pattern is given by

$$V := \frac{I_{\max} - I_{\min}}{I_{\max} + I_{\min}}, \tag{10.195}$$

which can be expressed in terms of the coherence function:

$$V = \frac{2\sqrt{I_1 I_2}}{I_1 + I_2} |g^{(1)}(\mathbf{r}_1, \mathbf{r}_2, \tau)|. \tag{10.196}$$

For $I_1 = I_2$, the visibility itself is equal to the modulus of the first-order coherence.

If light at two positions $\mathbf{r}_1$ and $\mathbf{r}_2$ is mutually incoherent, no interference pattern forms, or $V = 0$ and $g^{(1)}(\mathbf{r}_1, \mathbf{r}_2, \tau) = 0$. Maximal visibility of $V = 1$ occurs when $|g^{(1)}(\mathbf{r}_1, \mathbf{r}_2, \tau)| = 1$ or the fields are mutually coherent.

This is a result perfectly compatible with classical optics. In fact, the notion of a complex coherence function is very well established in classical optics, and can be used to describe incoherent or partially coherent light. For the first-order coherence describing field – field correlations, there are in fact no differences between the prediction of classical optics and the fact that we had to describe the field quantum mechanically.

10.2.3.5 *Power spectrum*

Another important coherence property is the connection between the power spectrum for different frequencies and the temporal coherence. One can show that the power density defined by

$$S(\mathbf{r}, \omega) = |\mathscr{E}(\omega)|^2, \tag{10.197}$$

where

$$\mathscr{E}(\omega) = \frac{1}{\sqrt{\pi}} \int_{-\infty}^{\infty} E(\mathbf{x}, t) e^{-i\omega t} dt \tag{10.198}$$

is just the Fourier component of the electrical field at a given (angular) frequency ω.

The spectral power density is also related to the first-order correlation function via

$$S(\mathbf{r}, \omega) = \frac{1}{\pi} \mathrm{Re} \int_{-\infty}^{\infty} G^{(1)}(\mathbf{r}, \mathbf{r}, \tau) e^{i\omega\tau} \mathrm{d}\tau \tag{10.199}$$

Therefore, there is a close connection between the form of the power spectrum and the coherence length.

As an example, consider the green-light component in common fluorescent lamps (resulting from mercury atoms emitting at around 546 nm). The atoms move with a velocity given by their thermal distribution, and thereby exhibit a Doppler effect for the emitted wavelength (which will dominate the spectral broadening). Assume the frequency distribution is Gaussian, with a center frequency ω_0 and a certain width σ:

$$S(\omega) = A e^{-\frac{(\omega - \omega_0)^2}{2\sigma^2}}. \tag{10.200}$$

The corresponding coherence function is a Gaussian distribution again, this time centered around $\tau = 0$:

$$G(\tau) \propto e^{-\frac{t^2\sigma^2}{2}} = e^{-\frac{t^2}{2\tau_c^2}} \qquad \text{with } \tau_c = \frac{1}{\sigma}. \tag{10.201}$$

τ_c may be considered as the coherence time of the light field. Such a definition always makes sense if the whole distribution $G^{(1)}(\tau)$ can be characterized by a single number. Using the complex degree of coherence $g^{(1)}$, we obtain a function that is normalized to 1 for $\tau = 0$.

10.2.3.6   Coherence functions of the various field states

To get an understanding of the second-order correlation function, we restrict ourselves to the case when only one mode is present and evaluate them at a fixed location X. For

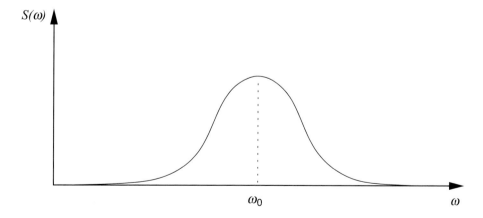

Fig. 10.12 Spectral density of a light, with a Gaussian distribution centered around ω_0.

comparison, we do the same thing also for the first-order correlations. These simplified correlation functions become

$$g^{(1)}(\tau) = \frac{\langle \hat{a}^\dagger(t)\hat{a}(t+\tau)\rangle}{\langle \hat{a}^\dagger \hat{a}\rangle} \tag{10.202}$$

$$g^{(2)}(\tau) = \frac{\langle \hat{a}^\dagger(t)\hat{a}^\dagger(t+\tau)\hat{a}(t+\tau)\hat{a}(t)\rangle}{\langle \hat{a}^\dagger \hat{a}\rangle^2}. \tag{10.203}$$

Let's now evaluate these functions for the three classes of states we have considered before. The first one that we consider is the number states:

$$g^{(1)}(\tau) = \frac{\langle \hat{a}^\dagger(t)\hat{a}(t+\tau)\rangle}{\langle \hat{a}^\dagger \hat{a}\rangle} = \frac{\langle n|\hat{a}^\dagger(t)\hat{a}(t)e^{-i\omega\tau}|n\rangle}{\langle n|\hat{a}^\dagger \hat{a}|n\rangle} = e^{-i\omega\tau} \tag{10.204}$$

$$g^{(2)}(\tau) = \frac{\langle \hat{a}^\dagger(t)\hat{a}^\dagger(t+\tau)\hat{a}(t+\tau)\hat{a}(t)\rangle}{\langle \hat{a}^\dagger \hat{a}\rangle^2} = \frac{\sqrt{n}\sqrt{n-1}\sqrt{n-1}\sqrt{n}}{n^2} = 1 - \frac{1}{n^2}, \tag{10.205}$$

where we have made use of the following:

$$\hat{a}(t) = \hat{a}(t=0)e^{-i\omega t} \tag{10.206}$$

$$\hat{a}^\dagger(t) = \hat{a}^\dagger(t=0)e^{i\omega t}. \tag{10.207}$$

Note that the first-order coherence function of number states has the property that $|g^{(1)}(\tau)| = 1$. The second-order coherence function has the property $g^{(2)}(\tau) < 1$, which implies that there is photon anti-bunching for number states. This means that given two detectors, if we detected a photon at the first detector, the probability of seeing another photon at the second detector at the same time is lowered.

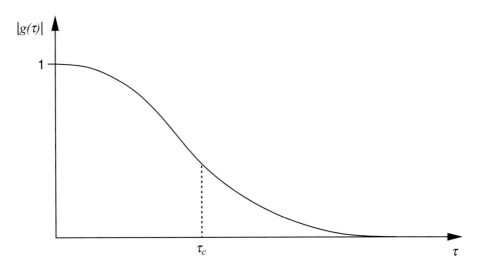

Fig. 10.13 The coherence function of a light with a Gaussian frequency spectrum.

Next, we proceed to coherent states where we have:

$$g^{(1)}(\tau) = \frac{\alpha^* \alpha e^{-i\omega\tau}}{\alpha\alpha^*} = e^{-i\omega\tau} \tag{10.208}$$

$$g^{(2)}(\tau) = \frac{\alpha^{*2}\alpha^2}{(\alpha^*\alpha)^2} = 1. \tag{10.209}$$

And finally, we have for thermal states

$$g^{(1)}(\tau) = \frac{\sum_n p_n n e^{-i\omega\tau}}{\sum_n p_n n} = e^{-i\omega\tau} \tag{10.210}$$

$$g^{(2)}(\tau) = \frac{\sum_n (1-A)A^n n(n-1)}{(\sum_n (1-A)A^n n)^2} = 2. \tag{10.211}$$

Here, in contrast to the number states, we observe photon bunching due to $g^{(2)}(\tau) = 2$. We also note that the different states of the light field do not reveal themselves in the first-order correlation function, but in the second-order correlation function.

10.2.3.7 *Interpretation of the $g^{(2)}(\tau)$ function*

An interpretation of the $g^{(2)}(\tau)$ may be obtained if we go back to the definition of w_2. There, we find the probability w_2 of finding a photo-electron in the time interval $[t, t + \Delta t]$, and another one at the time interval $[t + \tau, t + \tau + \Delta t]$.

$g^{(2)}(\tau)$ is then simply the probability of finding two photo-electrons separated by a duration τ (in a given time interval Δt), compared to the squared probability of finding one photo-electron. Therefore, cases with $g^{(2)}(\tau) = 1$ correspond to the case where the pair probability is just the squared probability of a single count. The quasi-classical state $|\alpha\rangle$ shows exactly such a behavior.

For $g^{(2)}(\tau) < 1$, as seen for photon number states, this probability is *reduced*. This fact is referred to as "photon anti-bunching", i.e. for this state it looks like photons prefer to be detected separately. States of the light field with $g^{(2)}(\tau) > 1$ are referred to as exhibiting photon bunching, with an increased probability of finding two photo-electrons at the same time.

We continue to discuss what happens to the $g^{(2)}(\tau)$ if light from two uncorrelated sources is detected. We use a statistical argument here, where we denote the probability of seeing one photon from A and B to be P_A and P_B, respectively, with

$$P_A + P_B = 1. \tag{10.212}$$

Now, we consider the case when two photons from source A are detected. This happens with a probability of P_A^2 with the corresponding $g_A^{(2)}(\tau)$ for source A. Similarly, we also have detection of two photons from source B with probability P_B^2 and $g_B^{(2)}(\tau)$. Finally, there is also a possibility that one photon is from source A, while the other one is from source B. The probability of this happening is $1 - P_A^2 - P_B^2$ and the $g_{AB}^{(2)}(\tau)$ associated is 1 for uncorrelated light source.

Now, the overall $g^{(2)}$ function for light from two uncorrelated sources is given by

$$g^{(2)}(\tau) = P_A^2 g_A^{(2)}(\tau) + P_B^2 g_B^{(2)}(\tau) + (1 - P_A^2 - P_B^2). \qquad (10.213)$$

If the two sources A and B are similar, that is their $g^{(2)}$ functions are the same $g_A^{(2)}(\tau) = g_B^{(2)}(\tau)$, and $P_A = P_B$, we have

$$g^{(2)}(\tau) = \frac{1}{2} g_A^{(2)}(\tau) + \frac{1}{2}. \qquad (10.214)$$

If the light is from two uncorrelated sources, the non-classical property of $g^{(2)} \neq 1$ is diluted.

10.2.3.8 *Experimental measurements of photon pair correlations*

With the second-order correlation function $g^{(2)}(\tau)$ we have seen a mathematical object or *measure* to distinguish between different quantum mechanical states of a light field, which classically is only characterized by an intensity and a frequency.

The differences between different states come about in the probability of observing photo-electron pairs at a given time difference. Such experiments have been pioneered by H. Hanburry-Brown and R. Twiss starting from 1956, and are now in widespread use.

The definition of $g^{(2)}(\tau)$ is already quite operational. The light of an optical mode under consideration is sent to two photodetectors using a beamsplitter, and a photo-electron pair creation is detected as a coincidence count between the two outputs:

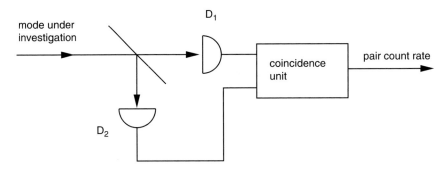

Fig. 10.14 The mode under investigation is sent through a beamsplitter where it is detected at the two detectors D_1 or D_2 before being sent through a coincidence unit.

By varying the time difference of the photodetection event, $g^{(2)}(\tau)$ can be measured. Furthermore, the observation of individual rates at detector D_1 and D_2 allows a proper normalization.

Such a setup is referred to as a Hanbury-Brown–Twiss configuration, according to the first experiments of this kind (Hanbury-Brown and Twiss, 1954; 1956; 1957)

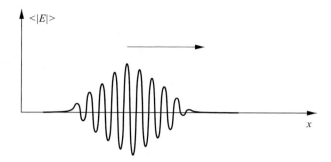

Fig. 10.15 Wave packet, made up by modes with a similar wavevector k_0. Such a packet can be localizable in space and time, and present a light field that can lead to a localizable detection event.

. In initial experiments, the coincidence was obtained in a way by multiplying the photocurrents of the two photomultipliers.

This setup was first used to investigate light from a spectral line emitted by mercury, where intensity fluctuations have been found. Subsequently, this technique was applied to investigate light from stars, and to measure the transverse coherence length, allowing people to determine the diameter of a light source with a large virtual aperture and without running into phase-stability problems.

Recently, this measurement technique was used to investigate and more or less *define* so-called single-photon sources as physical systems that exhibit a vanishing second-order correlation function for $\tau = 0$, meaning that the probability of creating more than one photo-electron at any single time vanishes. We come back to this in Section 10.2.5.2.

10.2.3.9 *Localized wave packets*

So far, we have considered photodetectors that generate a single photo-electron, and described their outcomes in statistical terms. Such a description does not imply the localization of a light field at all, but the observation of a single photo-electron or breakdown of an avalanche diode is a macroscopic signal, which one is tempted to give a localized light field as a physical reason. Later, we will see that it is indeed possible to create light fields that are very well localized in time or space, and one would like to think of a light field that can generate exactly one photodetection event as a particle-like object – that would be our localized photon.

This concept, however, seems to be at odds with the definition of photons we encountered earlier, namely some sort of Fock states in well-defined discrete modes, which exhibit no localization in space or time, but are energy eigenstates of the field and thus stationary.

So how can we have a localization of a photon in time in a way that seems compatible with the observation of a single photodetection event, or a well-defined pair of them? The answer is reasonably simple: We can use a wave packet or a linear combination of different modes, and populate each of these modes with a certain state.

A simple example would be Gaussian wave packets. Take $f(\mathbf{k})$ as an amplitude for a component $\mathbf{k}$ of the field decomposition in plane waves indexed by $\mathbf{k}$, with

$$f(\mathbf{k}) = \frac{1}{A} e^{-\frac{(\mathbf{k}-\mathbf{k}_0)^2}{2\sigma_k^2}} \qquad \text{for } t = 0. \tag{10.215}$$

If $f(\mathbf{k})$ is the amplitude of a particular Fourier component of a classical electromagnetic field, this would result in an electric field $E(\mathbf{r})$ of

$$E(\mathbf{r}) = \int f(\mathbf{k}) e^{i\mathbf{k}\cdot\mathbf{r}} d\mathbf{k}. \tag{10.216}$$

If the time evolution is included, we have

$$E(\mathbf{r}, t) = \int f(\mathbf{k}) e^{i(\mathbf{k}\cdot\mathbf{r} - \omega_k t)} d\mathbf{k}, \qquad \text{with } \omega_k = |\mathbf{k}|c. \tag{10.217}$$

This is a localized moving wave packet with a well-defined center (in space) moving with the speed of light, c, and a constant spatial extent with a variance of $\frac{1}{\sigma_k^2} = \sigma_r^2$.

In order to write a creation operator for such a field state, we can just use the idea introduced with the beamsplitter, where we expressed the lowering and raising operator at the output ports as linear combinations of the modes at the input:

$$\hat{c} = \sum_i \lambda_i \hat{a}_i. \tag{10.218}$$

This linear combination of modes can be generalized for wave packets to

$$\hat{c}_{k_0} = \sum_k f(\mathbf{k}) \hat{a}_k. \tag{10.219}$$

Such an object would be able to generate exactly one photodetection event (assuming a wide-band photodetector, i.e. that each of the contributing components $\hat{a}_k^\dagger |0\rangle$ would generate a detection event as well), and would exhibit a certain localization in time. If the distribution is reasonably restricted to a small number of mode indices k with a similar frequency, it still would appear to make sense to assign a center wavelength to the object $|\Psi\rangle = \hat{c}_{k_0} |0\rangle$ – which with some justification may be referred to as a localized photon.

10.2.4 Direct measurement of electric fields

Up to now, the optical measurements on quantized light fields we considered were related to detecting photo-electron rates. In order to more closely investigate the electrical field forming the light directly, we need to find a way to measure the electrical fields directly.

One approach to understanding how to measure fields connects to the basic setup in a Hanbury-Brown–Twiss measurement of photo-electron pairs, where two photodetectors are located behind a beamsplitter. We used that beamsplitter merely to divert a fraction of the light in a field mode of interest onto each detector.

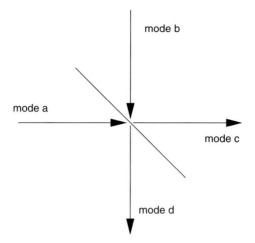

Fig. 10.16 The beamsplitter with two input modes a and b and two output modes c and d.

However, the beamsplitter has not only two output modes c and d receiving contributions from one input mode a, but also from a second input port b, with the associated field fluctuations even if no light is entering this port. We therefore should consider this element more carefully.

10.2.4.1 Beamsplitter

In classical optics, one can derive a relation between in and outgoing field amplitudes of a beamsplitter; the outgoing fields are a linear combination of the incoming fields of the form

$$\begin{pmatrix} E_C \\ E_D \end{pmatrix} = \begin{pmatrix} \sqrt{T} & i\sqrt{1-T} \\ i\sqrt{1-T} & \sqrt{T} \end{pmatrix}, \begin{pmatrix} E_A \\ E_B \end{pmatrix} = S \begin{pmatrix} E_A \\ E_B \end{pmatrix}, \tag{10.220}$$

where we define the "transfer matrix"

$$S := \begin{pmatrix} \sqrt{T} & i\sqrt{1-T} \\ i\sqrt{1-T} & \sqrt{T} \end{pmatrix}. \tag{10.221}$$

Here, we assume that we choose only one polarization, e.g. the polarization vector perpendicular to the plane of incidence. In this expression, T characterizes the fraction of transmitted *intensity*. For a balanced or symmetric beamsplitter, $T = 0.5$, i.e. half of the light is transmitted and the other half reflected.

A somewhat surprising aspect of that transfer matrix is the i appearing on the reflection entries – where should that asymmetry come from? It can be understood when looking at a particular physical implementation of a beamsplitter: One example would be a reflection from an airglass interface, where the reflection from the inside surface differs from the reflection on the outside surface by a sign change in the electrical field. The choice of the phases of reflection is somewhat arbitrary and depends on the specific implementation, but it is always 180 degrees between the two reflections.

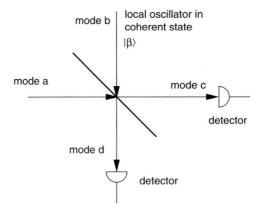

Fig. 10.17 Basic homodyne detection configuration. The field in mode a gets superimposed with the field of a local oscillator in mode b in a coherent state $|\beta\rangle$.

The choice i for a model beamsplitter is just a possibility that leads to a symmetric scattering matrix.

The linearity in fields carries simply over into a linear relationship between field operators, and the beamsplitter itself may be regarded as an element that transforms creation and annihilation operators:

$$\begin{pmatrix} \hat{c} \\ \hat{d} \end{pmatrix} = S \begin{pmatrix} \hat{a} \\ \hat{b} \end{pmatrix} = \begin{pmatrix} \sqrt{T}\hat{a} & i\sqrt{1-T}\hat{b} \\ i\sqrt{1-T}\hat{a} & \sqrt{T}\hat{b} \end{pmatrix}. \tag{10.222}$$

We now can express the photocurrents or count rates recorded at detectors C and D as expectation values $\langle \hat{n}_C \rangle$ and $\langle \hat{n}_D \rangle$, where $\hat{n}_{C,D} = \hat{a}^\dagger_{C,D}\hat{a}_{C,D}$:

$$\langle \hat{n}_C \rangle = \langle \hat{c}^\dagger \hat{c} \rangle = T\langle \hat{a}^\dagger \hat{a} \rangle + (1-T)\langle \hat{b}^\dagger \hat{b} \rangle + i\sqrt{T}\sqrt{1-T}\langle \hat{a}^\dagger \hat{b} - \hat{b}^\dagger \hat{a} \rangle, \tag{10.223}$$

$$\langle \hat{n}_D \rangle = \langle \hat{d}^\dagger \hat{d} \rangle = (1-T)\langle \hat{a}^\dagger \hat{a} \rangle + T\langle \hat{b}^\dagger \hat{b} \rangle - i\sqrt{T}\sqrt{1-T}\langle \hat{a}^\dagger \hat{b} - \hat{b}^\dagger \hat{a} \rangle. \tag{10.224}$$

Both terms consists of intensities contributions $\langle \hat{a}^\dagger \hat{a} \rangle$ and $\langle \hat{b}^\dagger \hat{b} \rangle$, and a mixed term with $\langle \hat{a}^\dagger \hat{b} \rangle$ and $\langle \hat{b}^\dagger \hat{a} \rangle$.

10.2.4.2 Homodyne detection

Let's now assume that we have a "local oscillator field" in mode $\hat{b}$, where field b is in a coherent state $|\beta\rangle$. Such a situation can be realized using a classical light source. It turns out that a laser light source far above threshold may be considered as such a light source.

Assuming that the combined field state in modes a and b separates in the form

$$|\psi\rangle = |\psi_A\rangle \otimes |\beta\rangle, \tag{10.225}$$

the expectation value of the photon number in the detection mode c is given by

$$\langle \hat{n}_C \rangle = T \langle \hat{n}_A \rangle + (1 - T)|\beta|^2 + i\sqrt{T}\sqrt{1 - T}\langle \hat{a}^\dagger \hat{b} - \hat{b}^\dagger \hat{a} \rangle. \tag{10.226}$$

If we interpret this as an expectation value of the number of detected photo-electrons, this vale refers to the expectation value of a discrete count rate (or photo current) at detector C. Using $\beta = |\beta|e^{i\varphi}$, this turns into

$$\langle \hat{n}_C \rangle = T \langle \hat{n}_A \rangle + (1 - T)|\beta|^2 + i\sqrt{T}\sqrt{1 - T}\,|\beta|\,\langle \hat{a}e^{i\varphi} - \hat{a}e^{-i\varphi} \rangle. \tag{10.227}$$

For $\varphi = 0$, the interference term at the end contains an expression for the expectation value for the field $\hat{E}$, so this method allows one to really measure the electric field.

We can define a generalized quadrature component along a phase angle φ by

$$\hat{a}_\varphi := \frac{1}{2}\left(\hat{a}^\dagger e^{i\varphi} + \hat{a}e^{-i\varphi}\right) = \hat{a}_Q \cos\varphi + \hat{a}_P \sin\varphi, \tag{10.228}$$

which can be used to simplify the expectation value in eq. (10.227):

$$\langle \hat{n}_C \rangle = T \langle \hat{n}_A \rangle + (1 - T)|\beta|^2 + 2\sqrt{T(1 - T)}\,|\beta|\,\langle \hat{a}_{(\varphi + \pi/2)} \rangle. \tag{10.229}$$

To compensate for the residual noise in the local oscillator, one often takes the difference of photocurrents in the two photodetectors, $i_C - i_D$, corresponding to an observable $\hat{n}_C - \hat{n}_D$. Then, only the interference term survives, and the noisy terms due to the local oscillator and the variance in the numbers of photons in the input state cancels out for $T = 0.5$:

$$\langle \hat{n}_C - \hat{n}_D \rangle = 2i\sqrt{T(1 - T)}|\beta|\langle \hat{a}^\dagger e^{i\varphi} - \hat{a}e^{-i\varphi} \rangle$$
$$= -|\beta|\langle \hat{a}_{(\varphi + \pi/2)} \rangle. \tag{10.230}$$

This technique can now be used to measure the quadrature amplitude $\hat{a}_\varphi$ directly and get information about the variances of the light field directly, and is referred to as a *balanced homodyne detection*. It became an important detection tool for detecting squeezed states of light, with a reduced noise level in one of the quadrature components, as well as for more complex interacting systems in quantum information processing using continuous variables.

10.2.4.3   *Heterodyne detection*

The idea in a homodyning setup is to remove all the contributions containing intensities by looking for the difference in photocurrents, and extract the information about the electrical field out of that difference. In practice, the difference will always be a small contribution to the total photocurrent. Furthermore, the small difference will be contaminated by significant noise in the photodetection signal at low frequencies due to sources other than the photocurrent, which tends to be far above the shot-noise limit corresponding to the photo-electron number fluctuation from the coherent-state contributions.

To overcome the low-frequency noise problem, another slightly modified configuration for field detection is typically used, referred to as *heterodyne detection*. The main idea behind this scheme is rather technical and takes the difference not at low frequencies, but moves it to a frequency where (a) the photodetetors exhibit a low intrinsic noise, and (b) the necessary amplifiers can be built with better noise properties. The shift is realized by using a difference $\Omega/2\pi$ in the frequencies for the mode a under investigation and the local oscillator mode in the coherent state $|\beta\rangle$. Then, the photocurrent difference component containing information about the electrical field in mode a is contained in a spectral component at $\Omega/2\pi$ in the photocurrent difference signal, which can be amplified without adding significant electronic noise. From there, the interesting spectral component can be brought back to a DC level with a second homodyning process, this time for classical electronic signals. The necessary element is just a mixer, which multiplies the amplified photocurrent difference with a sinusoidal signal at $\Omega/2\pi$. A phase shifter in the radio-frequency path allows an easy access to both quadrature components of the electromagnetic field. A sketch summarizing this scheme is shown in Figure 10.18.

In both homodyne and heterodyne detection schemes the identification of a single photon is not exactly trivial; even the assignment of a well-defined value of α in a homodyne measurement is difficult. In practice, there is a time interval necessary, or a corresponding frequency window, over which a photocurrent or its fluctuations are registered. Only with this notion do we arrive at well-localizable light states, which can be used as qubits or carriers of some information later on.

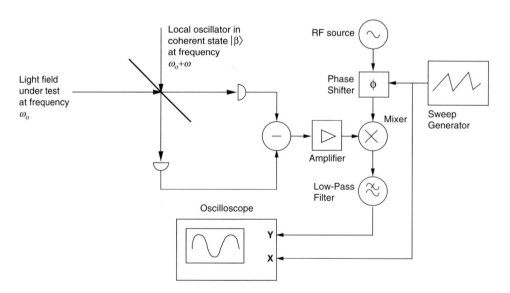

Fig. 10.18 A setup for heterodyne measurement of a light field at frequency ω_0.

10.2.5 Tying "the photon" to the generation process

So far, we have seen how a photon can be defined via the detection scheme in particular, the conversion into a single photo-electron seemed a very good candidate to define localizable photons. We are now looking for generation processes that can provide us with a state that makes a photodetector give a localizable signal, and would be rightly described by a wavepacket object generated by eq. (10.219).

10.2.5.1 *Spontaneous emission*

An explicit example for a compound "photon" is the light field emitted by spontaneous emission from an atom. Typically, the two levels can only be part of atomic levels, where there is a multiplicity in the levels. This allows interaction of the light field with the electronic states according to $\mathbf{E} \cdot \mathbf{p}$, where $\mathbf{E}$ is the electric field and $\mathbf{p}$ is the atomic electric polarization.

We also realize that there are a few possible decay paths in a typical atomic transitions, which are summarized in Figure 10.19.

Now, for the spontaneous emission, Wigner and Weisskopf have given a closed expression for the state of the system (Weisskopf and Wigner, 1930):

$$|\psi(t)\rangle = a(t)|e\rangle_A \otimes |0\rangle_{field} \tag{10.231}$$

$$+ \sum_{\rho} b_{\rho,-1}(t)|g_{-1}\rangle \otimes |n_{\rho} = 1, n_{\rho' \neq \rho} = 0\rangle$$

$$+ \sum_{\rho} b_{\rho,0}(t)|g_0\rangle \otimes |n_{\rho} = 1, n_{\rho' \neq \rho} = 0\rangle$$

$$+ \sum_{\rho} b_{\rho,+1}(t)|g_{+1}\rangle \otimes |n_{\rho} = 1, n_{\rho' \neq \rho} = 0\rangle,$$

with

$$a(t) = e^{-\gamma t/2}, \tag{10.232}$$

$$b_{\rho,m}(t) = \frac{w_{eg}}{\hbar} \frac{e^{-\gamma t/2} - e^{-i\Delta(\rho)t}}{i\gamma/2 - \Delta(\rho)} \, CG\,[0,0;1,m|1,m]. \tag{10.233}$$

Therein, ρ, m is a mode index corresponding to a spherical vector harmonic and an outgoing radial part. The details and derivation of this expression are part of atomic physics, so we just mention that w_{eg} is some form of reduced electric dipole matrix

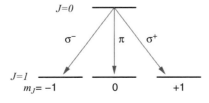

Fig. 10.19 Three possible decay paths from $J = 0$ to $J = 1$.

element between the two levels, $\gamma = 1/\tau$ corresponds to the natural linewidth of that transition, given by the lifetime τ of the excited state, $\Delta(\rho) = ck_\rho - \omega_0$ is the detuning of a particular radial mode from the atomic resonance frequency ω_0, CG is a Clebsch–Gordan coefficient corresponding to the angular momentum modulus of ground and excited level (here chosen to be 0 and 1, respectively), and m is one of -1, 0 or $+1$, describing the type of transition (σ^-, π or σ^+).

For the spherical symmetry of the problem it is adequate to formulate the electrical field operator

$$\hat{\mathbf{E}}(\mathbf{x}) = i \sum_{k,m} \mathscr{E}_{\omega_k} \left(\hat{a}_{k,m} \mathbf{g}_{k,m}(\mathbf{x}) - \hat{a}^\dagger_{k,m} \mathbf{g}^*_{k,m}(\mathbf{x}) \right), \qquad (10.234)$$

with two scalar mode indices k, m and a mode function $\mathbf{g}_{k,m}(\mathbf{x})$ expressing the position $\mathbf{x}$ in spherical coordinates r, θ, φ:

$$\mathbf{g}_{m,k}(r, \theta, \varphi) \approx \mathrm{Re}\left[\frac{e^{-ikr}}{kr} \right] \frac{\mathbf{r}}{|\mathbf{r}|} \times \mathbf{X}_{l,m}(\theta, \varphi), \qquad (10.235)$$

An exact expression needs to include the field at the atom more cleanly; the above expression, however, is a good approximation a few wavelengths away from the atom (Jackson, 1999).

The dominating part is a spherical wave propagating away from the atom at the coordinate origin, with a certain width due to the fact that the emission process takes only a finite time (see Figure 10.20). The vector spherical harmonics $\mathbf{X}_{l,m}(\theta, \varphi)$ basically contain information of the polarization in the various directions. For example, along the z-direction ($\theta = 0$, often referred to as quantization axis) the $m = \pm 1$ or $\sigma^\pm$ transitions correspond to left- and right-circular polarization. This function also contains the emission pattern of the different transitions, e.g. the fact that for the $m = 0$ or π transition, the field in the z-direction vanishes.

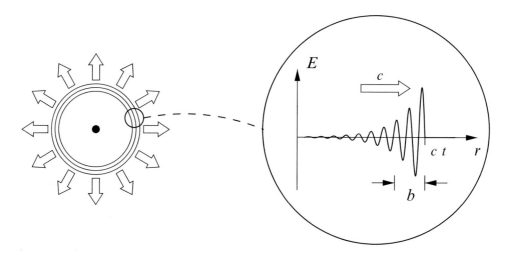

Fig. 10.20 Mode function of the field around an atom after the radiative decay from an excited state: one of the role model examples of a localizable photon.

We typically regard the outgoing state (a linear combination of single excitations in many modes) as a single photon, since the energy content of the field is limited by the initial atomic excitation. Moreover, we would typically expect only one photo-electron to be created in a detector. We also have a localized wave packet, centered around a frequency ω_0 corresponding to the initial energy difference $E_e - E_g = \hbar\omega_0$.

What we would like to do now is to combine the time dependencies of all the components $\{\rho, m\}$ in eq. (10.231) into simpler ones, corresponding to the three possible classes of transitions, $\sigma^{\pm}, \pi$ into a set of new mode functions $\tilde{\mathbf{g}}_m(\mathbf{r})$ in the very same way as we express the modes at the output of a beamsplitter as a linear combination of modes at its inputs. This will be done more formally in Section 10.4, where we make the connection between photons and qubits.

10.2.5.2 Single-photon sources

Apart from the definition of what single-photon states could be, we should consider specific physical implementations. While it is possible to prepare a single atom with a high probability into an excited state to establish the initial condition for a single photon emitted via subsequent spontaneous emission, it is not the simplest experimental approach. A much simpler approach is to excite an atomic system continuously, and observe the photodetection statistics of the scattered light of the microscopic system.

NV center single-photon source. One of the perhaps simplest experimental implementations of such a single photon source is shown in Figure 10.21, where a single charged nitrogen atom, which is embedded in a diamond lattice with an adjacent vacancy, is

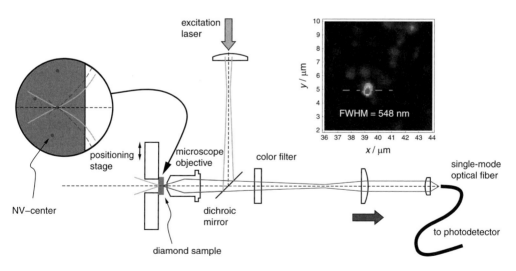

Fig. 10.21 Experimental setup to observe single-photon light from nitrogen-vacancy color centers in diamond. The excitation of the atom-like color center takes place with a short wavelength to an ensemble of states, which eventually leads to a spontaneous emission of a photon at a longer wavelength.

used as the single quantum system. This nitrogen atom has an electronic transition that can emit light in the red wavelength range, and can be driven into the excited state by optical excitation with green light, by passing through some higher excited states and a subsequent relaxation to something close to a two-level system. Being embedded in a material with a high band gap, the electronic excitation is released with a high probability via a radiative decay, under additional broadening due to a change in the vibrational state of the atom – host lattice combination.

In an experimental setup, excitation light is focused down onto a single such color center, which can be done because these centers can be prepared with a very low density in a diamond host, such that the average distance between these centers is larger than a few optical wavelengths. Then, excitation light can be focused on a single color center, and the emitted light from the very same center can be collected with a confocal microscope geometry. Since the spontaneous emitted light has a wavelength that is far enough away from the excitation light (usually, a frequency-doubled Nd : YVO_4 laser at a wavelength of 532 nm is used), color separation between the excitation light and the emitted light can be done with a high extinction ratio such that after spectral filtering, the photon statistics of the scattered light can be detected easily.

A typical photon-statistics experiment following the setup of Hanbury-Brown and Twiss is shown in Figure 10.22. The emitted light from the color center is directed

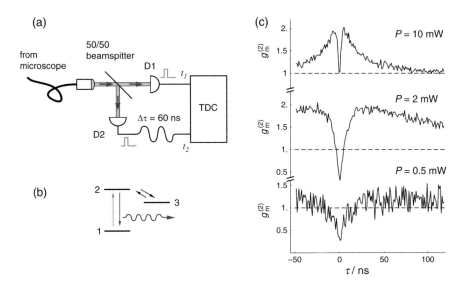

Fig. 10.22 Observation of photon anti-bunching behavior in a Hanburry-Brown–Twiss geometry (a), where the time difference between photoevent pairs is analyzed. The corresponding level model of the NV center consists of three levels (b), and at room temperature the transitions between them can be well described using simple rate equations. The uncorrected measurement results (c) for different excitation powers reveal a characteristic signature of a photon anti-bunching for $\tau=0$, thus indicating that there is a strong suppression of emission of more than one photon at the same time.

onto a beamsplitter, which distributes the field onto two photodetectors with a single photo-electron detection mechanism. The photo-electron detection signals are then histogrammed with respect to their arrival-time difference to experimentally obtain a second-order correlation function $g^{(2)}(\tau)$.

The experimental traces (c) in this figure shows a clear reduction of $g^{(2)}$ below 1 around $\tau = 0$ for low optical excitation powers, indicating that no two photo-electrons are generated at the same time. The interpretation of such an experimental signature is that the emitted light field is made up by isolated, or single photons. For larger excitation power levels, the "anti-bunching dip" gets narrower quickly, reflecting the fact that the NV center is transferred faster into the excited state again, ready for the emission of the next photon. The anti-correlation signature $g^{(2)}(\tau = 0) = 0$ gets more and more washed out, as the recovery time for the NV center excitation comes closer to the detector timing uncertainty.

The internal dynamics of the NV center guarantees that, once a photon has been emitted, the center is in the electronic ground state, and can only emit the next photon once the probability of being transferred into the excited state due to the presence of the excitation light has increased again. This particular experiment has been carried out at room temperature, where the presence of a huge phonon background in the diamond host leads to a very fast decoherence between ground and excited state. Thus, the internal dynamics of the NV center is adequately described by a set of rate equations for the populations in the participating internal levels. From the presence of two exponential components in $g^{(2)}(\tau)$, it can be inferred that there are at least three participating levels.

Single photons from single atoms. A measurement on a physical system that shows much less decoherence between the internal electronic states is shown in Figure 10.23. A single rubidium atom was trapped in an optical tweezer, while it was exposed to near-resonant excitation light. The atom occasionally undergoes a spontaneous emission, which is captured by a microscope objective onto a pair of photodetectors.

This time, the coherences between ground and excited state can not be neglected, and $g^{(2)}(\tau)$ shows a damped oscillatory behavior together with a clear anti-bunching signature at $\tau = 0$. The damping time constant is related to the radiative lifetime of the excited state (26 ns for this transition), and the Rabi oscillation reflects the coherent population transfer under the electric-field strength of the excitation light.

This atomic single-photon source, however, is far from being deterministic: While the probability of observing two photodetection events at the same time is very low, the probability of detecting a photon at a desired time is very small. This is due to the fact that the atom emits the photon in a spherical harmonics mode, but the collection optics manages to receive only a small fraction of the full solid angle. Furthermore, this particular excitation scheme is not deterministic at all, but driven by a continuous light source.

Deterministic single-photon sources. To overcome the statistical nature of the single photons emitted in the previous two examples, two aspects need to be considered:

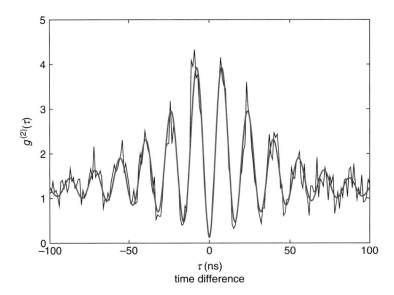

Fig. 10.23   second-order correlation function for light scattered by a single atom under the exposure of near-resonant optical radiation. The atom was held in an optical tweezer, and the excitation light was driving a Rabi oscillation between ground- and excited state. The photon anti-bunching is still present for time differences $\tau = 0$ between photodetection events.

1. The excitation process has to happen at a well-defined time, and with a high probability. This can be achieved with an optical excitation either by a short pulse performing half a Rabi oscillation, or via an adiabatic transfer scheme. Both methods require a light pulse much shorter than the radiative lifetime of the excited state.
2. The collection process into a particular target mode has to be efficient. If the target mode should be mapped into a propagating beam, or into an optical fiber, the matching of the spontaneous emission mode in free space is very limited. A method to circumvent this problem is to enhance the electrical field strength of a particular target mode with an optical cavity, such that the spontaneous emission preferably takes place into that specific mode. This requires usually optical cavities with a high finesse. Implementations of such sources have been demonstrated both with atoms in free space and with single emitters (quantum dots) in a solid-state matrix.

10.2.5.3   *Heralded photon sources*

Another method to prepare light fields into single-photon states makes use of the fact that photon pairs can be generated with a strong temporal correlation via parametric down-conversion. Then, one photon of a pair acts as a witness for the other photon, and an experiment can be triggered on a successful observation of the first photon. We will deal with this process in the next section.

10.3 Parametric down-conversion: a common workhorse

A key resource for many experiments in quantum information are sources of entangled photon pairs, based on parametric down-conversion. In this process, energy from a pump field gets transformed into correlated photon pairs. In this section, we will present the underlying physical process for parametric down-conversion.

We start with an extension of the electromagnetic field quantization into a regime of optical materials that respond to external fields. There is actually very little change in the field description to what we have seen so far. Then we will introduce a description of non-linear optical materials, which allow for an interaction between the electromagnetic field in different modes. We tie this together to present one of the typical down-conversion processes, and analyze quantitatively the generation rate of photon pairs.

10.3.1 Optics in media

Optical materials like glasses play an important role in manipulating electromagnetic fields: They allow us to change the mode structure. It should be of no surprise that they also play an important role in quantum optics. We quickly go through the ideas of how the interaction with such media is modelled, and how this affects the field quantization.

10.3.1.1 *Macroscopic Maxwell equations*

The response of an optical material to incoming fields is the combination of responses from a large number of atoms making up that material. In the optical regime, with wavelengths on the order of $10^{-7} - -10^{-6}$ m and a spacing of the atoms of $10^{-10} - -10^{-9}$ m can be well described by averaging over the material's response, and by introducing new mean-field quantities connected to the electric and magnetic response of the material. The resulting Maxwell equations for these macroscopic fields can be written as:

$$\nabla \cdot \mathbf{D}(\mathbf{x}, t) = \rho_{\text{free}}(\mathbf{x}, t), \tag{10.236}$$

$$\nabla \times \mathbf{E}(\mathbf{x}, t) = -\frac{\partial}{\partial t}\mathbf{B}(\mathbf{x}, t), \tag{10.237}$$

$$\nabla \cdot \mathbf{B}(\mathbf{x}, t) = 0, \tag{10.238}$$

$$\nabla \times \mathbf{H}(\mathbf{x}, t) = \frac{\partial}{\partial t}\mathbf{D}(\mathbf{x}, t) + \mathbf{j}(\mathbf{x}, t), \tag{10.239}$$

with new quantitates of the dielectric displacement $\mathbf{D}(\mathbf{x}, t)$ and a magnetic field strength $\mathbf{H}(\mathbf{x}, t)$. In the simplest case, these quantities are linearly related to the electrical field $\mathbf{E}$ and the induction flux density $\mathbf{B}$ by

$$\mathbf{D}(\mathbf{x}, t) = \epsilon_o \epsilon_r \mathbf{E}(\mathbf{x}, t), \tag{10.240}$$

$$\mathbf{B}(\mathbf{x}, t) = \mu_o \mu_r \mathbf{H}(\mathbf{x}, t). \tag{10.241}$$

Here, the permittivity ϵ_r and the relative permeability μ_r summarize the material's response to electromagnetic fields. Most optical materials do not connect to the magnetic-field component at optical frequencies, so we can set $\mu_r = 1$. They do, however, respond to the electrical field, so we will have a closer look at this part.

First, the relationship in eq. (10.240) can be broken down into

$$\mathbf{D}(\mathbf{x}, t) = \epsilon_\circ \mathbf{E}(\mathbf{x}, t) + \mathbf{P}(\mathbf{x}, t), \qquad (10.242)$$

with a contribution of the original electric field $\mathbf{E}$ and a polarization $\mathbf{P}$, which characterizes the averaged induced electrical dipole moments of the material. For linear materials, this macroscopic polarization can be expressed as

$$\mathbf{P} = \epsilon_\circ \chi \mathbf{E}, \qquad (10.243)$$

where χ is the susceptibility tensor characterizing the material, or simply a matrix connecting two vector quantities. For isotropic media like a gas or a glass, it is a scalar. We will later drop this linear connection between electric field and material response.

For now, we consider the simple case that we have an isotropic, homogenous medium with no free charges, current densities and any magnetic susceptibility, i.e. $\rho_{\text{free}} = 0$, $\mathbf{j} = 0$, $\mu_r = 1$ and χ is constant for all $\mathbf{x}$. The dielectric permittivity is given by $\epsilon_r = 1 + \chi$, and the solutions to the macroscopic Maxwell equations can be derived from a wave equation similarly as in the vacuum case, but this time with a reduced speed of light,

$$c = \frac{1}{\sqrt{\epsilon_0 \epsilon_r \mu_0 \mu_r}} = \frac{1}{\sqrt{\epsilon_0 \mu_0}} \frac{1}{\sqrt{\epsilon_r}} = \frac{c_0}{n}, \qquad (10.244)$$

with n being the refractive index of the media.

Note that the introduction of the new field quantities $\mathbf{D}$ and $\mathbf{H}$ did not introduce any new degrees of freedom – they just capture the presence of a material responding to the field in a way that averages over the response of a large number of atoms. Thus, the main strategy to quantize the electromagnetic field remains the same.

10.3.1.2 *Energy density in dielectric media*

The important quantity to carry out a field quantization in practice is an expression for the total energy in the electromagnetic field. With the quantities introduced to capture the material response, this quantity is given by

$$H = \frac{1}{2} \int d\mathbf{x} \left[\mathbf{E}(\mathbf{x}) \cdot \mathbf{D}(\mathbf{x}) + \mathbf{B}(\mathbf{x}) \cdot \mathbf{H}(\mathbf{x}) \right], \qquad (10.245)$$

which in media with dielectric response only can be reduced to

$$H = \frac{1}{2} \int d\mathbf{x} \left[\mathbf{E}(\mathbf{x}) \cdot \mathbf{D}(\mathbf{x}) + \frac{1}{\mu_0} \mathbf{B}^2(\mathbf{x}) \right] = \frac{\epsilon_0}{2} \int d\mathbf{x} \left[\mathbf{E}(\mathbf{x})(1 + \chi) \mathbf{E}(\mathbf{x}) + c^2 \mathbf{B}(\mathbf{x})^2 \right]. \qquad (10.246)$$

This equation still has the same structure as eq. 10.67 for the electromagnetic field in free space, i.e. the same mode-decomposition strategy applies, leading to degrees of freedom that are governed by a harmonic oscillator-type dynamics. For an isotropic medium, the presence of the medium just leads to a modification of the electric-field contribution to the total energy by the permittivity $\epsilon_r = 1 + \chi$. For birefringent media, where the linear susceptibility depends on the direction of the electric field, the field decomposition can still be carried out as long as a proper polarization basis is chosen; the technical details are well covered in books on light propagation in crystalline media (see, e.g., (Saleh and Teich, 2007)), but would exceed the scope of this section.

Similarly, the field operators can be written in the same way as for the vacuum case,

$$\hat{\mathbf{E}} = \hat{\mathbf{E}}(\mathbf{x}, t) = i \sum_j \mathscr{E}_{\omega_j} \left(\mathbf{g}_j(\mathbf{x}) \hat{a}_j(t) - \mathbf{g}_j^*(\mathbf{x}) \hat{a}_j^\dagger(t) \right). \tag{10.247}$$

The presence of the medium manifests now in a modified constant $\mathscr{E}$ capturing the physical constants, and the connection between $\hat{\mathbf{E}}$ and $\hat{\mathbf{B}}$ needs to take care of the modified speed of light in the medium.

10.3.1.3 *Frequency dependence of refractive index*

Typically, the susceptibility χ of an optical medium dependent on the frequency of the exciting field. This is due to the fact that the response of the medium in the optical domain may be considered as an off-resonant excitation of electrons bound in the material. Typical binding energies of electrons in transparent materials are on the order of a few electron volts, corresponding to a resonance frequency for light in the ultraviolet regime. This is compatible with the fact that most materials that are transparent in the optical regime absorb ultraviolet light. The response of the medium in the visible regime then corresponds to the low-frequency tail of a resonance in the ultraviolet.

A semi-heuristic description of this behavior used to characterize the dispersion property of transparent materials is based on such a model, assuming that the susceptibility is due to one or several resonances in the material. This model is referred to as a Sellmeier equation of a given material, typically formulated as a dependency of the refractive index n from the vacuum wavelength λ_0:

$$n^2 - 1 = \sum_i \frac{A_i}{1 - B_i/\lambda_0^2} + \ldots + C_i/\lambda_0^i. \tag{10.248}$$

Note that this is not a unique way of expressing the ultraviolet resonances, and serves mostly as an engineering tool where the coefficients A_i, B_i, C_i don't have an immediate physical interpretation, but are chosen to give an accurate (typically good to 10^{-6}) estimation of the refractive index in the visible regime, based on a few refractive-index measurements at a few wavelengths. When you encounter a set of Sellmeier equations characterizing a particular material, make sure you are using the corresponding model function.

The general structure of this dispersion relation is that the refractive index *increases* with the frequency. If you encounter a birefringent material, the Sellmeier equations are typically given for electrical fields polarized in particular directions and propagation directions, where plane waves are solutions the set of Maxwell equations. Depending on the symmetry of the material, two or three sets of Sellmeier equations are required to give a full description of the dispersion properties for the optical medium. The resulting description is reasonably messy.

10.3.1.4   *Non-linear response of a medium*

The linear response of a medium to an electric field has ensured that the complete electromagnetic field still can be decomposed into decoupled modes, and the field states in these individual modes evolve independently from each other.

We now consider a non-linear response of the medium to an exciting field. Such a non-linearity may be thought of as originating from a non-harmonic potential of the electrons in the media. As such a non-linear response is usually a small effect, it can be captured well by a Taylor expansion of the polarization $\mathbf{P}$ of the material in the exciting electrical field $\mathbf{E}$:

$$\mathbf{P} = \epsilon_0 \left(\chi \mathbf{E} + \chi^{(2)} \mathbf{E}^2 + \chi^{(3)} \mathbf{E}^3 \dots \right). \tag{10.249}$$

The newly introduced susceptibility tensors $\chi^{(n)}$ capture the higher-order terms of the material response, and are tensor objects of higher order; as an example, a contribution to the polarization vector due to the first higher-order term $\chi^{(2)}$ can be written in components as:

$$P_j^{(2)} = \epsilon_0 \sum_{k,l=x,y,z} \chi_{jkl}^{(2)} E_k E_l, \qquad \text{with} \quad i = x, y, z \tag{10.250}$$

The second-order non-linear susceptibility tensor $\chi^{(2)}$ must reflect the symmetry of the underlying material; this usually reduces the number of independent entries in this tensor substantially. One of the most important symmetry constraints of this type is that the material must lack an inversion symmetry in order to have $\chi^{(2)} \neq 0$ (try to prove this!). Thus, all gases, amorphous materials like glass or polymers, and a large number of crystalline materials do not exhibit this type of non-linear response to an external electrical field.

A typical signature of these higher-order processes in classical optics is higher harmonics generation: If you consider a monochromatic electrical field (e.g. in form the of a plane wave) at a frequency ω,

$$\mathbf{E}(\mathbf{x}, t) = \mathbf{E}_0 \, e^{i(\mathbf{k} \cdot \mathbf{x} - \omega t)}, \tag{10.251}$$

the second-order non-linear susceptibility will result in polarization components that oscillate at twice the original frequency,

$$\mathbf{P}^{(2)}(t) \propto \chi^{(2)} \mathbf{E}^2 \propto e^{-2i\omega t}. \tag{10.252}$$

This polarization component at a new frequency can be considered as a source term in the Maxwell equations, and will propagate through the medium according to the dispersion relation.

An example where such a process takes place are green laser pointers, which have a laser emitting light at a vacuum wavelength of 1064 nm, and a small piece of potassium titanylphosphate (KTP) as a material with a non-linear susceptibility to convert part of this light into radiation with a vacuum wavelength of 532 nm appearing as green. We will see later, however, that a non-linear susceptibility alone is not enough to observe this process, but the dispersion properties of the material must allow the fundamental and second-harmonic wave to propagate through the crystal with the same speed.

10.3.2 Non-linear optics: Three-wave mixing

To describe the energy transfer between different modes due to the non-linear susceptibility quantitatively, we consider the Hamiltonian with the higher terms in the susceptibility:

$$\hat{H} = \frac{\epsilon_0}{2} \int d\mathbf{x} \left[\mathbf{E}(\mathbf{x}) \cdot \left(\epsilon_r \mathbf{E}(\mathbf{x}) + \chi^{(2)} \mathbf{E}^2(\mathbf{x}) \right) + c^2 \mathbf{B}^2(\mathbf{x}) \right] \tag{10.253}$$

$$= \frac{\epsilon_0}{2} \int d\mathbf{x} \left[\epsilon_r \mathbf{E}^2(\mathbf{x}) + c^2 \mathbf{B}^2(\mathbf{x}) \right] + \frac{\epsilon_0}{2} \int d\mathbf{x}\, \mathbf{E}(\mathbf{x}) \cdot \chi^{(2)} \mathbf{E}^2(\mathbf{x}) \tag{10.254}$$

$$=: \hat{H}_0 + \hat{H}_I^{(2)}. \tag{10.255}$$

Here, we have split up the interaction with the medium into a part H_0 with decoupled harmonic oscillator modes, and an interaction term H_I containing the effects induced by the non-linear susceptibility. Before we have a closer look at this interaction Hamiltonian, we quickly mention that in cases where higher-order susceptibilities have to be considered, the interaction Hamiltonian takes a similar form. For example, a $\chi^{(3)}$ or Kerr non-linearity leads to:

$$\hat{H}_I^{(3)} := \frac{\epsilon_0}{2} \int d\mathbf{x}\, \chi^{(3)} \hat{\mathbf{E}}^4(\mathbf{x}) \tag{10.256}$$

Returning to the $H_I^{(2)}$ again, we can gain some insight if we carry out the spatial integration. Recalling the electrical field operators from eq. (10.111),

$$\hat{\mathbf{E}}(\mathbf{x}, t) = i \sum_j \mathscr{E}_{\omega_j} \left(\mathbf{g}_j(\mathbf{x}) \hat{a}_j(t) - \mathbf{g}_j^*(\mathbf{x}) \hat{a}_j^\dagger(t) \right), \tag{10.257}$$

the interaction Hamiltonian is an integral over space, and a generalized sum (i.e. sum and/or integral) over three mode indices:

$$\hat{H}_I^{(2)} = -i \int d\mathbf{x} \sum_j \sum_k \sum_l \{ \mathscr{E}_j \mathscr{E}_k \mathscr{E}_l \tag{10.258}$$

$$\chi^{(2)} \left[\mathbf{g}_j(\mathbf{x}) \hat{a}_j - \mathbf{g}_j(\mathbf{x}) \hat{a}_j^\dagger \right] \left[\mathbf{g}_k(\mathbf{x}) \hat{a}_k - \mathbf{g}_k(\mathbf{x}) \hat{a}_k^\dagger \right] \left[\mathbf{g}_l(\mathbf{x}) \hat{a}_l - \mathbf{g}_l(\mathbf{x}) \hat{a}_l^\dagger \right] \} \tag{10.259}$$

This operator thus splits up into eight terms of the form

$$\hat{H}_I^{(2)} = C_0 \, \hat{a}_j \hat{a}_k \hat{a}_l + C_1 \, \hat{a}_j^\dagger \hat{a}_k \hat{a}_l + C_2 \, \hat{a}_j \hat{a}_k^\dagger \hat{a}_l + C_3 \, \hat{a}_j \hat{a}_k \hat{a}_l^\dagger + h.c., \tag{10.260}$$

where the terms come in hermitian conjugated pairs. Each term has three ladder operators mediating transitions between population of modes – this gives the associated process the name *three-wave mixing*. The notion of waves is motivated by plane wave as mode functions $\mathbf{g}(\mathbf{x})$.

To illustrate the conditions under which the different terms contribute, we consider the prefactor C_1 for the second term in eq. (10.260) as an example, which mediates a process removing a photon in mode j and generates one in modes k and l each:

$$C_1 = i \frac{\epsilon_0}{2} \sum_j \sum_k \sum_l \mathscr{E}_j \mathscr{E}_k \mathscr{E}_l \int d\mathbf{x} \, \chi^{(2)} \mathbf{g}_j^*(\mathbf{x}) \mathbf{g}_k(\mathbf{x}) \mathbf{g}_l(\mathbf{x}). \tag{10.261}$$

Assuming plane waves as mode functions j, k, l of the form

$$\mathbf{g}(\mathbf{x}) = \hat{e}_j \, e^{i \mathbf{k}_j \cdot \mathbf{x}}, \tag{10.262}$$

where $\hat{e}$ is a polarization vector, the integral in the expression for C_1 becomes

$$C_1 = i \frac{\epsilon_0}{2} \sum_j \sum_k \sum_l \mathscr{E}_j \mathscr{E}_k \mathscr{E}_l (\hat{e}_j \chi^{(2)} \hat{e}_k \hat{e}_l) \int d\mathbf{x} \, e^{-i \mathbf{k}_j \cdot \mathbf{x}} \, e^{i \mathbf{k}_k \cdot \mathbf{x}} \, e^{i \mathbf{k}_l \cdot \mathbf{x}} \tag{10.263}$$

$$= i \frac{\epsilon_0}{2} \sum_j \sum_k \sum_l \mathscr{E}_j \mathscr{E}_k \mathscr{E}_l \chi^{(2)}_{\text{eff}} \int d\mathbf{x} \, e^{i (\mathbf{k}_k + \mathbf{k}_k - \mathbf{k}_j) \cdot \mathbf{x}} \tag{10.264}$$

With

$$\int d\mathbf{x} \, e^{i(\mathbf{k}_k + \mathbf{k}_l - \mathbf{k}_j) \cdot \mathbf{x}} = 2\pi \delta(\mathbf{k}_k + \mathbf{k}_l - \mathbf{k}_j) \text{ for a infinite integration volume,}$$
$$\approx 2\pi \delta(\mathbf{k}_k + \mathbf{k}_l - \mathbf{k}_j) \text{ for a *finite* integration volume,} \tag{10.265}$$

the coefficient C_1 only survives if the wavevectors of the removed photon matches the sum of the two created ones. This specific process is referred to as down-conversion, and the above condition is referred to as the *phase-matching condition* for the corresponding non-linear optical process. For geometries that don't show a full translational symmetry, e.g. for an interaction region of finite size, the momentum conservation is only approximately fulfilled.

Remember that the terms in the interaction Hamiltonian always come in pairs: Hence, when the coefficient for a down-conversion from mode j to modes k, l is non-vanishing, the reverse process is also possible. In this case, this would correspond to an up-conversion process.

The modes in the phase-matching conditions do not necessarily need to be distinct: The up-conversion process with $k = l$, for example, corresponds to the SHG process mentioned earlier.

Keep in mind that the phase-matching condition is only an overlap argument of various participating modes in a given interaction volume: It applies equally in a treatment of optical interactions in classical physics, and is not something that comes out of a quantum description of light. For many quantum optical experiments, it is nevertheless important to meet these phase-matching conditions. Thus, we will elaborate a little on them in the next section.

10.3.3 Phase matching

Apart from the phase-matching condition, which can be interpreted as momentum conservation in the limit of a large interaction region, the energy in conversion processes must be conserved. This puts another constraint on the processes taking place concurrently.

For the down-conversion process introduced above, the three participating modes are traditionally labelled as *pump* (that's the mode that has the lowering operator), and as *signal* and *idler* for the modes where photons are created in. The energy conservation can then be written as

$$\omega_p = \omega_s + \omega_i. \tag{10.266}$$

In the limit of large conversion regions considered for understanding the concept, the momentum conservation becomes:

$$\mathbf{k}_p = \mathbf{k}_s + \mathbf{k}_i. \tag{10.267}$$

Meeting both eqs. (10.266) and (10.267) at the same time for a range of modes can be accomplished by choosing proper polarizations and engineering the dispersion relation

$$\omega = c|\mathbf{k}|/n \tag{10.268}$$

and adjusting refractive index n with various methods. We will go through a few examples for parametric down-conversion from a pump frequency ω_p to a set of degenerate frequencies $\omega_s = \omega_i = \omega_p/2$ for the target modes. To meet eq. (10.267) in a collinear geometry (i.e. $\mathbf{k}_p \parallel \mathbf{k}_s \parallel \mathbf{k}_i$), the refractive indices for pump and target modes must be the same. As mentioned in Section 10.3.1.3, the refractive index for the pump (higher frequency) is larger than for the signal and idler frequencies. This problem is typically addressed by choosing different polarizations for the various modes since materials with non-vanishing $\chi^{(2)}$ are usually birefringent. Figure 10.24 shows this for the example of lithium iodate, $LiIO_3$, a negatively uniaxial birefringent material. A wave with a linear polarization parallel to the optical axis experiences the *extraordinary* refractive index n_e, while a plane wave with a polarization vector in a plane orthogonal to the optical axis experiences the larger *ordinary* refractive index.

The refractive index n_e for a pump wavelength of 295 nm (this refers to the vacuum wavelength, a commonly used proxy measure for the frequency) matches the ordinary index n_o at a degenerate target wavelength of 580 nm. Thus, for a pump polarized parallel to the optical axis, parametric conversion into a mode copropagating with the pump but with orthogonal polarization would be allowed.

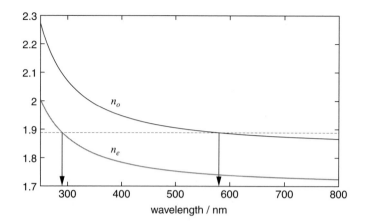

Fig. 10.24 Non-critical phase matching in LiIO$_3$ for SHG or degenerate parametric conversion. An ordinarily polarized plane wave at 580 nm and an extraordinarily polarized wave at half the (vacuum) wavelength have the same refractive index of $n = 1.88825$.

The combination of polarizations ($e - o - o$) for pump, signal and idler is referred to as type-I phase-matching. In this example, the propagation directions match the optical axes, which is referred to as non-critical phase matching.

The polarizations of the two target modes do not need to be the same; by choosing one of the target modes as e, and the other one as o-polarized, non-critical phase matching for a pump wavelength of 373.6 nm can be achieved in the LiIO$_3$, with the degenerate target modes at a wavelength of 747.3 nm. Such a polarization combination is referred to as type-II phase matching.

Obviously, the non-critical phase matching works only for very few wavelength combinations. There are, however, several efficient ways to extend the phase-matching range: Refractive indices of a given material can be altered by material engineering, temperature, the propagation direction and an artificially introduced periodic reorientation. We will briefly discuss the last three methods.

10.3.3.1 *Phase matching by temperature tuning*

A number of materials have a strong dependency of their refractive index with temperature; a prominent example is potassium niobate, KNbO$_3$, a biaxial birefringent material with a strong optical non-linearity and a strong temperature dependency of the refractive index (about 0.5% over 100 K, (Zysset, *et al.* 1992)).

10.3.3.2 *Phase matching by angle tuning*

Another common way to ensure phase matching is to manipulate the refractive index of one or more modes by choosing propagation directions through the conversion crystals where the electric field vector is not parallel to one of the principal axes of the susceptibility tensor. We refer the reader to any textbook on classical optics in

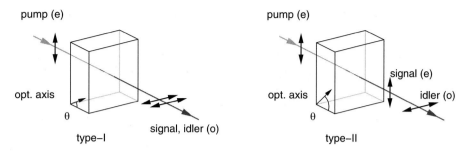

Fig. 10.25 Critical phase matching (type-I and type-II) for a uniaxial birefringent crystal in a collinear configuration. The refractive indices of extraordinary beams are tuned by choosing an angle θ between the wavevectors and the optical axis of the crystal.

birefringent crystalline materials for details, and show the basic idea only for a simple example for this so-called critical phase matching.

For birefringent materials and a given propagation direction with respect to the principal axes of the material, there are always two orthogonal polarization vectors where the solutions of Maxwell equations can be written as plane waves with a well-defined propagation speed c/n. For uniaxial birefringent materials, two of the principal linear susceptibilities are the same, and correspond to the so-called ordinary refractive index $n_o = \sqrt{1 + \chi_o}$, while the response to a dielectric displacement along the optical axis is governed by the extraordinary refractive index n_e. A wave described by a propagation vector $\mathbf{k}$ that forms an angle θ with the optical axis has then one polarization mode ("ordinary wave") with the corresponding refractive index n_o and an electrical field vector in the plane normal the optical axis, while the other polarization mode ("extraordinary wave") has an orthogonal polarization and is n'_e is propagating corresponding to a refractive index

$$n'_e(\theta) = \frac{n_e n_o}{\sqrt{n_o^2 + (n_e^2 - n_o^2)\cos^2\theta}}. \tag{10.269}$$

Thus, by choosing a proper orientation angle θ, the refractive index for the extraordinary wave can be adjusted between n_e and n_o. Keep in mind that the refractive indices n_e and n_o still depend on the frequency. As an example, we consider the negatively uniaxial crystal BBO, and wavelengths of $\lambda_p = 351$ nm for the pump, and $\lambda_s = \lambda_i = 702$ nm for frequency-degenerate target modes in a down-conversion configuration. The refractive indices at these wavelengths are shown in the following table:

	n_e	n_o
$\lambda = 351$ nm	1.5784	1.7069
$\lambda = 702$ nm	1.5484	1.6648

This combination of refractive indices allows both for type-I and type-II phase matching for copropagating target and pump modes. For type-I, the target modes have to be both ordinary waves with $n = 1.66$, which lies between n_e and n_o for the

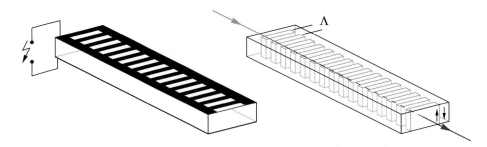

Fig. 10.26 Engineered phase-matching by electrically poling periodically various segments in a non-linear optical material like KTP or lithium niobate with a periode Λ. Additionally to the intrinsic wavevectors in the conversion material, an integer multiple of a quasi-phase matching wavevector $2\pi/\Lambda$ is added to the phase-matching condition.

pump wavelength. With an angle $\theta = 33.3°$, the extraordinary refractive index $n'_e(\theta)$ can be matched to the ordinary refractive index n_o of the target modes.

For type-II phase matching, one of the target modes is ordinarily, the other extraordinarily polarized. Since no combination of ordinary and extraordinary refractive index of the target modes can reach the ordinary refractive index at the pump wavelength, the pump wavelength is also extraordinarily polarized. Then with the dispersion relation $c|k| = n\omega$, the phase-matching condition for degenerate down-conversion $(\omega_p/2 = \omega_s = \omega_i)$ now reads

$$2n'_{e,351}(\theta) = n'_{e,702}(\theta) + n_{o,702}, \tag{10.270}$$

which has a solution for $\theta = 48.9°$. Such a configuration has been used for a large number of experiments with down-converted photon pairs. It should be mentioned, though, that the critical phase-matching approach has the disadvantage that the wavevector and Poynting vector, describing the energy flow in a beam, are no longer parallel for the extraordinary waves. For modes with a finite transverse extent, this leads to a "transverse walk-off" between ordinary and extraordinary modes, which may have a number of undesirable consequences, among them a limitation of the possible geometric overlap of different modes. On the other hand, it usually allows us to use popular laser wavelengths.

While the birefringent properties of a uniaxial crystal are completely determined by one angle θ, and thus the phase-matching conditions are fully met, it should not be forgotten that the orientation of the non-linear optical tensor elements may still depend on the other free orientation angle.

10.3.3.3   *Phase matching by periodic poling*

Another recent, very popular method to meet the phase-matching condition in non-linear optical materials is to engineer the optical properties of a material by breaking the translational symmetry of the conversion material, and modulate the optical property like the sign of the non-linear optical susceptibility periodically. This can be achieved by applying strong static electrical fields in small portions of a non-linear optical crystal under certain conditions. The field is applied to the material with the aid of microstructured electrodes, which later can be removed once the crystal is poled.

The new device still has a partial translational symmetry, since the optical properties repeat with the poling period Λ. The phase-matching condition is now replaced by a quasi-phase matching condition:

$$\mathbf{k}_p = \mathbf{k}_s + \mathbf{k}_i + \frac{2\pi\, m}{\Lambda}, \qquad m \in \mathbb{Z}_0 \tag{10.271}$$

By choosing the poling periode Λ and the quasiphasematching order m accordingly, wavelength combinations can be reached with materials with an intrinsic high non-linear optical susceptibility. Materials that have been used for this type of phase matching include the very common $LiNbO_3$ (and is in this modification referred to as PPLN, periodically poled lithium niobate), as well as potassium titanylphosphate (KTP, or PPKTP).

While with these materials, a tremendous increase of brightness for photon pair sources have been observed, one needs to keep in mind that the engineered periodicity is subject to manufacturing uncertainties; particularly the duty cycle of the poling structures can be noisy across the conversion material. As a consequence, the material can be considered as a region with a distribution of poling periods beyond the single value given by $2\pi/\Lambda$, and allow for a small contribution of phase matching at a large number of modes. This may contribute to noise in a variety of non-linear optical processes like upconverison or also parametric down conversion.

10.3.4 Calculating something useful: Absolute pair production rates

While many experiments with photon pairs from parametric down-conversion can be and have been carried out with considering the rather qualitative phase-matching criteria discussed in the previous section, it is helpful for developing these light sources to get a quantitative expression on how many pairs can be expected in a particular geometry (Ling, *et al.* 2008).

The example we consider in this chapter assumes that we generate photon pairs into modes which are defined by single mode optical fibers, which allows later on to manipulate the downconverted light conveniently in any interferometric arrangements - those typically require well-defined spatial modes, and an optical fiber acts as an efficient spatial mode filter for such purposes.

10.3.4.1 The model

To keep the calculation of an expected photon pair rate simple, we restrict ourselves to a geometry where the pump field is collinear with the target modes. Furthermore, we neglect all dispersion effects in the conversion crystal. The geometry of the model situation is shown in Figure 10.27.

The conversion is taking place in a crystal of thickness d in the main propagation direction of all modes, which we assume to have a Gaussian-mode profile. For simplicity, we also neglect a possible transverse beam walk-off in case we use critical phase-matching. Furthermore, we assume that we have chosen a type-II phase matching

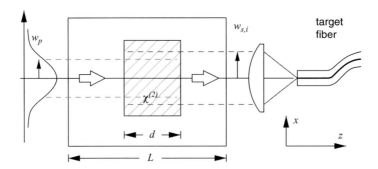

Fig. 10.27 Model geometry to calculate pair generation rates in PDC collected by a single-mode optical fiber. The pump and target modes are nearly collimated Gaussian beams with beam waists $w_{p,s,i}$, the conversion crystal should have a thickness d in the propagation direction z and a size much larger than the beam waists in the transverse directions. For convenience, the non-linear optical material is thought to be embedded in a material with the same refractive indices n_s, n_i, n_p as in the conversion region. We also introduce a quantization length L for the target modes.

condition, such that the two target modes are distinct by their polarization. By embedding the conversion crystal in a surrounding with the same linear susceptibilities (and thereby refractive indices) as in the conversion region, we can make use of the expression (eqns 10.118–10.120) in Section 10.1.5 for the electrical field operator $\hat{\mathbf{E}}(\mathbf{x})_{s,i}$. The corresponding beam waists for signal and idler modes are w_s, w_i, and the corresponding refractive indices are $n_s = \sqrt{\epsilon_s}, n_i = \sqrt{\epsilon_i}$. The quantization length L should not influence the result of the pair rate and is just kept for conceptual convenience. With this, and keeping in mind that we still have the longitudinal wavevector k_s and k_i as a one-dimensional mode index for the target modes, the target field operators become

$$\hat{\mathbf{E}}_{s,i}(\mathbf{x}) = i \sum_{k_{s,i}} \mathcal{E}_{k_{s,i}} \boldsymbol{\varepsilon} e^{ik_{s,i}z} e^{-\rho^2/w_0^2} \hat{a}_{k_{s,i}} + h.c., \quad \text{with} \quad \mathcal{E}_{k_{s,i}} = \sqrt{\frac{\hbar \omega_{k_{s,i}}}{\pi w_{s,i}^2 L \epsilon_0 n_{s,i}^2}}.$$

$$(10.272)$$

With the quantization length L and periodic boundary conditions, the mode indices $k_{s,i}$ are integer multiples of $2\pi/L$.

The pump field is assumed also to be monochromatic and in a Gaussian mode with waist w_p. The electrical field in the pump mode is treated as a classical amplitude, written in a similar form as the quantized field modes:

$$\mathbf{E}_p(\mathbf{x}, t) = iE_0 \left\{ \mathbf{g}_p(\mathbf{x}) e^{-i\omega_p t} - \mathbf{g}_p^*(\mathbf{x}) e^{i\omega t} \right\},$$

$$(10.273)$$

with the same Gaussian mode function $\mathbf{g}_p(\mathbf{x})$ from eq. 10.118 as the quantized fields, with a pump waist w_p. The optical power carried by this classical, real-valued electrical field is given by

$$P = E_0^2 n_p \pi w_p^2 \epsilon_0 c, \tag{10.274}$$

an expression that can be obtained by integrating the average Poynting vector, $\langle \mathbf{S} \rangle = \langle \mathbf{E}_p \times \mathbf{B}_p \rangle_t / \mu_0$ over the cross-section of the beam.

10.3.4.2 Interaction Hamiltonian

To evaluate the rate of photon pairs generated in the spatial modes collected into the optical fiber, we need to consider the interaction Hamiltonian $\hat{H}_I$ as outlined in eq. 10.254, which is assumed to be a small perturbation to the free field. For a classical driving field, this Hamiltonian becomes explicitly time dependent.

$$\hat{H}_I(t) = \frac{\epsilon_0}{2} \int d^3\mathbf{x}\, \mathbf{E}_p(\mathbf{x}, t) : \chi^{(2)} : \sum_{k_s} \hat{\mathbf{E}}_{k_s}(\mathbf{x}) : \sum_{k_i} \hat{\mathbf{E}}_{k_i}(\mathbf{x}) \tag{10.275}$$

Since we only want to consider processes where photon pairs are generated out of the vacuum of the target modes, it is sufficient to restrict the Hamiltonian to two terms for pair creation – annihilation, and we arrive at:

$$\hat{H}_I(t) = i\frac{\epsilon_0}{2}(\boldsymbol{\varepsilon}_p : \chi^{(2)}\boldsymbol{\varepsilon}_s : \boldsymbol{\varepsilon}_i)E_0 \int\limits_{-\infty}^{\infty} dxdy \int\limits_{-d/2}^{d/2} dz \sum_{k_s, k_i} \frac{\hbar\sqrt{\omega_s \omega_i}}{\pi\epsilon_0 w_s w_i L n_s n_i} \tag{10.276}$$

$$\times e^{i\Delta\omega t} e^{-i\Delta k z} e^{-(x^2+y^2)/w_s^2} e^{-(x^2+y^2)/w_i^2} \hat{a}_{k_s}(0)\hat{a}_{k_i}(0) + h.c. \tag{10.277}$$

We have introduced the detuning parameter $\Delta\omega = \omega_p - \omega_s - \omega_i$ and a wavevector mismatch $\Delta k = k_p - k_s - k_i$. With the effective non-linearity d_{eff} (capturing the contraction of the non-linear susceptibility for the polarizations used, $2d_{eff} = \boldsymbol{\varepsilon}_p : \chi^{(2)}\boldsymbol{\varepsilon}_s : \boldsymbol{\varepsilon}_i$), we can carry out the spatial integration of the mode-function overlap and simplify the above expression:

$$\hat{H}_I(t) = id_{eff}E_0 \sum_{k_s, k_i} \frac{\hbar\sqrt{\omega_s \omega_i}}{\pi w_s w_i L n_s n_i} \times e^{i\Delta\omega t}\Phi(\Delta k)\hat{a}_{k_s}(0)\hat{a}_{k_i}(0) + h.c., \tag{10.278}$$

with

$$\Phi(\Delta k) = \int\limits_{-\infty}^{\infty} dxdy \int\limits_{-d/2}^{d/2} dz\, e^{-i\Delta k z} e^{-(x^2+y^2)/w_s^2} e^{-(x^2+y^2)/w_i^2} e^{-(x^2+y^2)/w_p^2} \tag{10.279}$$

$$= \pi \left(\frac{1}{w_p^2} + \frac{1}{w_s^2} + \frac{1}{w_i^2} \right)^{-1} d\,\mathrm{sinc}\,(\Delta k d/2) \tag{10.280}$$

We observe that the interaction Hamiltonian does not vanish for a range of Δk due to the finite length d of the conversion crystal.

10.3.4.3   *Fermi's golden rule and spectral rates*

To make a quantitative statement on the number of photon pairs generated per unit time, we will use Fermi's golden rule for a transition rate $R(k_s)$ from a field in an initial vacuum state $|i\rangle = |0_{k_s}; 0_{k_i}\rangle$ into a final state $|f\rangle = |1_{k_s}; 1_{k_i}\rangle$ with one photon in each of the target modes k_s, k_i. Fermi's golden rule makes a statement about asymptotic scattering rates, thus energy must be conserved. We express this using the dispersion relation for signal and idler modes:

$$\Delta\omega = \omega_p - k_s \frac{c}{n_s} - k_i \frac{c}{n_i} = 0. \tag{10.281}$$

First, we consider now a transition rate $R(k_s)$ for photon pairs with a *fixed* target mode k_s. The density of states $\rho(\Delta E)$ per unit of energy $\Delta E = \hbar\Delta\omega$ is extracted out of the quasi-continuum of target modes k_i:

$$\rho(\Delta E) = \frac{\Delta m}{\Delta k_i} \cdot \frac{\partial k_i}{\partial(\hbar\Delta\omega)} = \frac{L}{2\pi} \cdot \frac{n_i}{\hbar c}. \tag{10.282}$$

With the approximation that the frequencies ω_s, ω_i vary only little over the range where $\Phi(\Delta k)$ contributes, the transition rate is then given by

$$R(k_s) = \frac{2\pi}{\hbar} \left| \langle f|\hat{H}_I|i\rangle \right|^2 \rho(\Delta E) \tag{10.283}$$

$$= \left| \frac{d_{eff}E_0}{\pi w_s w_i} \Phi(\Delta k) \right|^2 \frac{\omega_s \omega_i}{n_s^2 n_i cL}. \tag{10.284}$$

We would now like to map this transition rate into a fixed discrete mode k_s into a spectral rate density. For that, we just multiply the above expression with the number $Ln_s/2\pi c$ of modes k_s per frequency interval ω_s, and obtain

$$\frac{dR(\omega_s)}{d\omega_s} = \left[\frac{d_{eff}E_0 \Phi(\Delta k)}{\pi w_s w_i c} \right]^2 \frac{\omega_s \omega_i}{2\pi n_s n_i}. \tag{10.285}$$

At this point, the earlier introduced quantization length L has vanished.

We still need to express the wavevector mismatch Δk as a function of the frequency ω_s. For that, we use dispersion relations for the different modes and arrive at

$$\Delta k = \frac{1}{c} \left(\omega_p(n_p - n_i) - \omega_s(n_s - n_i) \right) = \frac{n_s - n_i}{c} (\omega_s^0 - \omega_s). \tag{10.286}$$

The sinc function leads to a spectral distribution of the down-converted light in the signal mode around a center frequency $\omega_s^0 = \omega_p(n_p - n_i)/(n_s - n_i)$ with a width of $\Delta\omega_s = 4\pi c/(n_s - n_i)/d$ between the first two zeros. The thicker the conversion crystal length d, the narrower the spectrum of the collected light.

To quote typical numbers, we consider degenerate type-II down-conversion in adequately aligned BBO from 351 nm to a center wavelength of 702 nm. There, $n_s - n_i \approx 0.111$. For a $d = 2$ mm thick crystal, the spectral width (between the zeros of the sinc function) would be $\Delta\omega = 2\pi \times 2.7$ THz corresponding to $\Delta\lambda \approx 4.4$ nm.

10.3.4.4   Connecting it together

The spectral conversion rate of photon pairs can be integrated over all frequencies ω_s. Assuming that only a small spectral region around ω_s^0 contributes to the total pairs generated, the integration over the spectral rate can be carried out over the mode overlap expression only; with $\delta\omega_s = \omega_s - \omega_s^0$ and

$$
\int_{-\infty}^{\infty} d\delta\omega_s\, \Phi^2\left(\frac{n_s - n_i}{c}\delta\omega_s\right) = \frac{2\pi^3 cd}{n_i - n_s}\left(w_p^{-2} + w_s^{-2} + w_i^{-2}\right)^{-2} \tag{10.287}
$$

we can finally write down the total pair rate R_T as

$$
R_T = \frac{d_{eff}^2 E_0^2}{w_s^2 w_i^2}\frac{\omega_s \omega_i cd}{n_s n_i (n_i - n_s)}\left(w_p^{-2} + w_s^{-2} + w_i^{-2}\right)^{-2} \tag{10.288}
$$

$$
= \frac{d_{eff}^2 P d\omega_s \omega_i}{n_p n_s n_i (n_i - n_s)\pi\epsilon_0 w_s^2 w_i^2 w_p^2 \left(w_p^{-2} + w_s^{-2} + w_i^{-2}\right)^2}. \tag{10.289}
$$

With the common choice of waists, $w_s = w_i = w_p/\alpha$, we get

$$
R_T = \frac{d_{eff}^2 P d\omega_s \omega_i}{n_p n_s n_i (n_i - n_s)\pi\epsilon_0 w_s^2 \left(\alpha^{-1} + 2\alpha\right)^2}, \tag{10.290}
$$

which for a fixed w_s maximizes for $\alpha = 1/\sqrt{2}$ to a value of

$$
R_T = \frac{d_{eff}^2 P d\omega_s \omega_i}{8\pi\epsilon_0 n_p n_s n_i (n_i - n_s)w_s^2}. \tag{10.291}
$$

A few observations on this expression should conclude this section:

- The rate of photon pairs grows linearly in the pump power and the crystal length.
- It also grows quadratically with the optical non-linearity; therefore, it is very advantageous to look for materials with a strong optical non-linearity; this makes it advantageous to consider waveguide structures in non-linear optical materials, since there the mode can be confined to an area of about λ^2 over a long distance d.
- The total rate grows inversely proportional to the pump waist area, and maximizes if the ratio between target and pump mode waists are $\sqrt{2}$.
- While the spectral pair density is independent of the refractive-index difference, the absolute rate grows inversely proportional to $n_i - n_s$, or the birefringence of the material.

10.3.5   Temporal correlations in photon pairs

One of the important aspects of light generated by parametric down-conversion is its strong temporal correlation between photodetection events observed between light in signal and idler modes. An experimental schematic for this observation is shown in Figure 10.28). First experimentally observed by Burnham and Weinberg

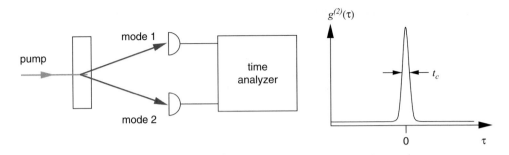

Fig. 10.28 Temporal correlations in photodetection events in parametric down-conversion. The second-order correlation function shows a prominent peak for a detection time difference $\tau = 0$. Its width is usually limited by the timing resolution of the detectors.

(Burnham and Weinberg, 1970), one usually finds a peaked pair correlation function for photodetection events where the width t_c is typically limited by the detector resolution. Thus, it appears that photo-electrons are preferably generated in coincidence.

This observation suggests that there are wave packets for signal and idler generated in a common "birth event"; it also leads to the view that each individual photon of a pair is a well-localized wave packet. More sophisticated measurements (see Hong–Ou–Mandel experiment in Section 10.4.2.3) indicate that the width t_c of the second-order correlation function is indeed given by the coherence time of the corresponding spectral distribution for each photon, and that Fourier-limited wave packets would be appropriate descriptions for the photons in each mode.

For practical purposes, this allows us to identify photon pairs by coincidence detection: A very powerful experimental method to take care of situations where one of the photons in a pair could not be detected, or was lost in either the collection or some optical elements between down conversion region and detectors. By looking for coincidences between two photodetectors, a selection of situations can be made where both of the photons made it through the entire experimental setup.

10.3.5.1 Heralded single-photon source

This strong correlation in time can be utilized to "prepare by detection" of one of the photons a situation where there will be, with a high probability, a photon in the other mode. Hence, a photon-pair source can in many cases be used to prepare single-photon states.

The limitation of the heralding efficiency is typically given by the detection efficiency of both photodetectors, and by the collection efficiency in a particular down-conversion setup. With many contemporary down-conversion sources with a peak emission in the sensitive region of silicon avalanche detectors, a ratio between pair events and single detection events of about 30% can be reached; This efficiency is a product of both detection and collection efficiency. With detection efficiencies around 50% for silicon avalanche photodiodes, this corresponds to collection efficiencies around

60%. However, sources have been reported with a very good overlap of pump and target modes that seem to achieve collection efficiencies around 80%.

In any case, the very well localized wave packets associated with photo-electron pair detections in very good timing definition gives rise to attempts to treat these photons as distinguishable particles. In the next section we will see how such individual particles, with an additional set of degrees of freedom, can be used to encode information.

10.4   Quantum information with photons

So far, we have treated photons – or more accurately, the electromagnetic field – with the aid of field operators and a variety of harmonic oscillator states. The electromagnetic field, similar, e.g., to a position of a particle, is a continuous variable, and quantum-mechanical properties in such a scenario are typically described and observed in terms of correlation functions between different continuous variables.

We have seen, however, that there is a set of physical phenomena where it makes sense to consider photons as localized electromagnetic fields, similar to how we perceive massive particles. We are used to the fact that particles can have "internal" degrees of freedom, which can be used to store or transport information. The simplest way of representing this concept in a quantum physics way is to think of such internal degrees as a qubit.

This section will review some of the practical considerations that arise when trying to select a physical system to perform quantum information tasks. Specifically, we will see how photons can be used as qubits, as they are virtually the universal choice when considering flying qubits used for quantum communication purposes. On the other hand, photons interact very weakly with each other and this limits the type of tasks that can be easily performed.

10.4.1   Single photons as qubits

Qubits are a convenient basic unit of storing or transporting information by a carrier governed by quantum physics. It is usually required that such a carrier is reasonably localized or otherwise distinguishable, such that it makes sense to talk about a system that is comprised by many qubits. In the previous section we have seen that in an attempt to generate and detect electromagnetic fields, photons can be often considered as fields that have this localizable property. We now need to find out how to make the transition from the field description to a qubit implementation with single photons.

10.4.1.1   What is a qubit?

In principle, any two-level system could be chosen as the physical basis for qubits. In practice, all qubits are not created equal, and they may vary strongly in the ease of preparation, complexity of performing generalized rotations (single-qubit gates), and their resistance to decoherence from interactions with the environment.

Whatever degrees of freedom we choose, a pure qubit state will be completely described by:

$$|\psi\rangle = \alpha|0\rangle + \beta|1\rangle \qquad \alpha, \beta \in \mathbb{C} \ , \ |\alpha|^2 + |\beta|^2 = 1. \qquad (10.292)$$

Any system that we choose to implement our qubits should not only support the full range of qubits states, but also provide a mechanism to transform one qubit state into another (single-qubit rotations).

10.4.1.2 *Preliminaries: How an electron spin becomes a separable degree of freedom*

In this section, we shall see how can we make connection between spin of an electron with internal, or much better, separable degree of freedom. As a first step, we should reconsider the way an electron spin is treated. This mechanism will help us understanding how to implement an equivalent with electromagnetic fields.

The spin of an electron is described by the spin wave function that arises naturally from the Dirac equation. In general, we can write the wave function of an electron as a vector quantity

$$|\Psi_e\rangle \equiv \begin{pmatrix} \phi_\uparrow(x) \\ \phi_\downarrow(x) \end{pmatrix}, \qquad (10.293)$$

where the two entries correspond to a spin-up component and a spin-down component. For completeness, it should be mentioned that in a relativistically correct description, this is a four-dimensional object, with the two additional entries corresponding to the two spin components of the positron mode, but we will not discuss this further.

We may write this spin wave now as a superposition of two components,

$$|\Psi_e\rangle \equiv \phi_\uparrow(x) \begin{pmatrix} 1 \\ 0 \end{pmatrix} + \phi_\downarrow(x) \begin{pmatrix} 0 \\ 1 \end{pmatrix}, \qquad (10.294)$$

with two terms written as a product of a spatial wave function and a vector with two components designating the spin state of the electron.

This rewriting of the spin wave function is an expression that, typically quoting observations of charge and mass conservation, we tend to consider the spatial aspects of the electron state as separable from its spin property. That process can be taken further by writing down a state in the form

$$\psi(x) \begin{pmatrix} \alpha \\ \beta \end{pmatrix}. \qquad (10.295)$$

At this point, we would consider $\psi(x)$ as the spatial wave function governing the center-of-mass motion (and refer to it as an external degree of freedom), whereas the column vector represents the spin part of the wave function (which, as it is not subject to any spatial variation, we can refer to as an internal degree of freedom). The whole object (here: electron) is described as a tensor product of the from

$$|\Psi_e\rangle = |\psi_{ext}\rangle \otimes |\psi_{int}\rangle, \tag{10.296}$$

which resides in a product space for the external and internal degree of freedom. We should keep in mind, however, that this is not a restriction of the electron state to a particular subclass of separable states, as we always can have superpositions of states of the form eq. (10.296) that don't separate, and finally compose the most general state eq. (10.293).

For electrons, this treatment is reasonably familiar: For the external degree of freedom, we continue to use the description of a massive particle with a scalar wave function describing its position, while we use a simple two-dimensional space to describe the spin properties, typically decoupled from the position degree of freedom.

In many physical systems, we can manipulate this internal degree of freedom independent of the external wave function, although we often have to go some length to suppress the coupling to the environment due to the charge, e.g. by using silver atoms with one unpaired electron spin to observe the Stern–Gerlach effect. Probably a better example is the nuclear spin of some atoms, where superposition states can last for hours without being affected by the position of the nucleus to the extent that spin-polarized atom can be prepared and then inhaled for medical imaging purposes without affecting the spin.

10.4.1.3 Generating "internal" degrees of freedom with photons

The concept we have seen in making the transition from a continuous Dirac state of an electron to a combination of internal and external degrees of freedom can easily be transferred to electromagnetic fields: First, we choose a light field that corresponds to a "single particle", as defined by the generation or detection process; in the previous section we have seen how this can be done for photons, and how they can be localized, e.g., by wave packets.

We now simply try to find a set of wave packet modes, and populate them in a fashion that can be described like an internal degree of freedom similar to the spin state of an electron. We will present a few common choices for appropriate degrees of freedom: polarization, which-way and time-bin are among the most widely used.

Polarization qubits. Let us first consider the polarization of a spontaneously emitted photon from an atom localized at the origin. Two possible decay options, that are $\Delta m = +1$ and $\Delta m = -1$, have similar spatial distribution for the light fields. We shall consider how we can describe the light field as a combination of internal and external degrees of freedom.

Besides this, we can also choose the basis for the photon's polarization as horizontal polarization ($|H\rangle$) and vertical polarization ($|V\rangle$). It is possible to use column vectors to represent the state:

$$|H\rangle \equiv \begin{pmatrix} 1 \\ 0 \end{pmatrix} \qquad |V\rangle \equiv \begin{pmatrix} 0 \\ 1. \end{pmatrix} \tag{10.297}$$

Similarly, we could equally well describe its polarization in terms of linear polarization along $+45°$ and $-45°$ directions or right-handed and left-handed circular polarization.

The states $|\pm45°\rangle$ and circular polarization $|L\rangle$ and $|R\rangle$ for left and right handed polarization can be written in terms of $|H\rangle$ and $|V\rangle$ as follows:

$$|+45°\rangle = \frac{1}{\sqrt{2}}\begin{pmatrix} 1 \\ 1 \end{pmatrix} = \frac{1}{\sqrt{2}}(|H\rangle + |V\rangle), \tag{10.298}$$

$$|-45°\rangle = \frac{1}{\sqrt{2}}\begin{pmatrix} 1 \\ -1 \end{pmatrix} = \frac{1}{\sqrt{2}}(|H\rangle - |V\rangle), \tag{10.299}$$

$$|L\rangle = \frac{1}{\sqrt{2}}\begin{pmatrix} 1 \\ i \end{pmatrix} = \frac{1}{\sqrt{2}}(|H\rangle + i|V\rangle), \tag{10.300}$$

$$|R\rangle = \frac{1}{\sqrt{2}}\begin{pmatrix} 1 \\ -i \end{pmatrix} = \frac{1}{\sqrt{2}}(|H\rangle - i|V\rangle). \tag{10.301}$$

So, we see that the polarization state of a single photon maps conveniently to a qubit. We now need to find out how to implement qubit rotations, i.e. arbitrary unitary transformations of this degree of freedom. For polarization, these rotations or singl-qubit gates can simply be implemented by the use of birefringent materials. In these materials the refractive index is dependent on the polarization direction.

The optical elements that implement polarization rotations are known as wave plates and are constructed such that two orthogonal linear polarizations acquire a definite phase difference. Generally, available wave plates implement a retardation of π ($\lambda/2$, half wave plate) or $\pi/2$ ($\lambda/4$, quarter-wave plate) for polarizations parallel to their principal axes (see Figure 10.29). For a half-wave plate aligned with the basis vectors (referred to sometimes as "computational basis" in quantum information), the action on the qubit can be represented by the matrix

$$\hat{U}_{\lambda/2} = \begin{pmatrix} 1 & 0 \\ 0 & -1 \end{pmatrix}. \tag{10.302}$$

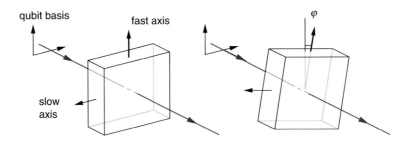

Fig. 10.29 Plates of birefringent materials exerting different phase shifts for polarizations along different crystal axes. By rotating such a wave plate by an angle φ, a number of single-qubit transformations can be implemented.

Similarly, it is straightforward to consider the effect of a $\lambda/2$ plate rotated with respect to the computational basis by an angle φ (Figure 10.29, right side) by concatenating rotation matrices and the retardation matrix:

$$\hat{U}_{\lambda/2}(\varphi) = \hat{R}(-\varphi) \cdot \hat{U}_{\lambda/2} \cdot \hat{R}(\varphi) \quad \text{with} \quad \hat{R}(\varphi) = \begin{pmatrix} \cos\varphi & \sin\varphi \\ -\sin\varphi & \cos\varphi \end{pmatrix}. \qquad (10.303)$$

One can show that a combination of $\lambda/2$, $\lambda/4$ and $\lambda/2$ plates that can be rotated individually is enough to transform any pure polarization into any other pure polarization.

An important component in qubit manipulation is also the detection or measurement process. We know how to detect a single photon already, but we do not yet have a way of measuring the qubit. Figure 10.30 shows how such a measurement is done for polarizations: the spatial mode carrying the photon with its two polarization components is sent on a polarization beamsplitter (PBS), which transmits one and reflects the other polarization component. Both outputs of he PBS are now covered with single-photon detectors. We will receive a binary answer (detector H or detector V) if there is a photon present, corresponding to measurement results seen in a Stern–Gerlach experiment.

Which-way qubits. Let's go back to the beamsplitter shown in Figure 10.31. There are two input ports, a and b, and two output ports, c and d. We can describe a photon passing through this beamsplitter from the input ports to output ports by artificially introducing an internal degree of freedom related to the ports of the beamsplitter that the photon passes, together with an external degree of freedom associated with the propagation of a wave packet describing the extent of the non-vanishing field in the main propagation direction z:

Similarly to what we did with polarizations, we can represent the state of "internal" degree of freedom by column vectors. For example, a photon that enters the beamsplitter through input port a can be represented by an internal state $|\Psi_i\rangle = \begin{pmatrix} 1 \\ 0 \end{pmatrix}$.

In this image, we view the action of a beamsplitter as a unitary operation acting on the states:

$$\hat{U}_{BS} = \frac{1}{\sqrt{2}} \begin{pmatrix} 1 & 1 \\ -1 & 1 \end{pmatrix}. \qquad (10.304)$$

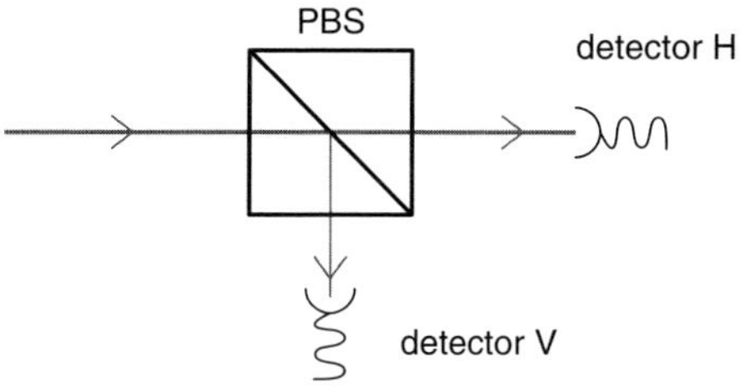

Fig. 10.30 Measurement scheme for a polarization qubit, based on a polarizing beamsplitter and two photodetectors.

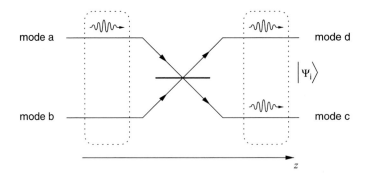

Fig. 10.31 Two spatial modes can be used to represent a qubit, and transformations can be established using beamsplitters.

for an ideal symmetric beamsplitter with a power transmission coefficient $T = 1/2$.

So, a photon in the initial state $|\Psi_i\rangle = \binom{1}{0}$ will pass through the beamsplitter and "evolve" into the state

$$|\Psi'_i\rangle = \hat{U}|\Psi_i\rangle = \frac{1}{\sqrt{2}}\begin{pmatrix} 1 \\ -1 \end{pmatrix}. \tag{10.305}$$

The internal degree of freedom in this picture is just a combination of different propagation-path possibilities, which could be implemented in a variety of ways. This method is not restricted to two-valued possibilities, equivalents to higher spin equivalents (or, in the language of quantum information, *qunits*, can be implemented using a larger base of possible propagation paths.

Typical elements to implement arbitrary unitary transformations in such an internal Hilbert space always involve beamsplitters (to combine two modes) and phase shifters; a concatenation of such elements is, e.g., a Mach–Zehnder interferometer. We can write the complete action of a Mach–Zehnder interferometer as:

$$|\psi\rangle_f = \hat{U}_{MZ}|\psi\rangle_i \tag{10.306}$$

$$\begin{pmatrix} c \\ d \end{pmatrix} = \frac{1}{\sqrt{2}}\begin{pmatrix} 1 & 1 \\ -1 & 1 \end{pmatrix}\begin{pmatrix} e^{i\phi} & 0 \\ 0 & 1 \end{pmatrix}\frac{1}{\sqrt{2}}\begin{pmatrix} 1 & 1 \\ -1 & 1 \end{pmatrix}\begin{pmatrix} a \\ b \end{pmatrix} \tag{10.307}$$

$$\begin{pmatrix} c \\ d \end{pmatrix} = \frac{1}{2}\begin{pmatrix} e^{i\phi} - 1 & e^{i\phi} + 1 \\ -e^{i\phi} - 1 & -e^{i\phi} + 1 \end{pmatrix} = \begin{pmatrix} a \\ b \end{pmatrix}. \tag{10.308}$$

For detailed explanations on how to efficiently manipulate polarization and which-way degrees of freedom, have a look at the classic textbook from Hecht, and Zajac (Hecht and Zajac, 1997), or the very detailed matrix-centric book of gerrand and Burch (Gerrard and Burch, 1994).

Time-bin qubits. Time bin qubits exploit the time-degree of freedom to implement a discrete basis (Brendel, *et al.* 1999). The state of the photon is distributed into several distinguishable time of arrival "bins". These can be labelled as "early" and "late"

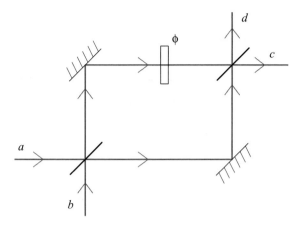

Fig. 10.32 A balanced Mach–Zehnder interferometer can be used to create an arbitrary qubit state by controlling the splitting ratio of the beamsplitters and the phase between the two possible paths.

or "long" and "short". The amplitude in each bin and the phase between the bins provide the full Hilbert space for a qubit, and the scheme can in principle be extended to higher-dimensional (qunits).

The simplest implementation uses all passive elements in an asymmetric Mach–Zehnder interferometer as shown in Figure 10.33. We can label the arms as *long* and *short*. If the beamsplitters are balanced, a photon entering through path a and after passing through the interferometer and exiting through c will be described as

$$\psi = \frac{1}{\sqrt{2}}\left(|s\rangle + e^{i\phi}|l\rangle\right),\tag{10.309}$$

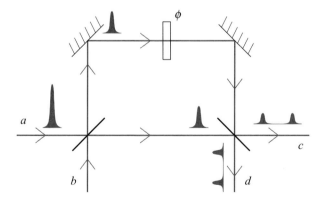

Fig. 10.33 An unbalanced Mach–Zehnder interferometer can be used to create an arbitrary qubit state as the superposition of two time bins corresponding to propagation along either the long or the short path.

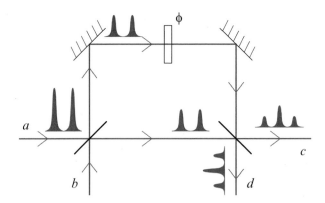

Fig. 10.34 An unbalanced Mach–Zehnder interferometer with path differences that correspond exactly to the time bin encoding can be used as an analyzer to project the state into superpositions of different time bins.

giving a photon that is in an equal superposition of arriving "early" or "late". To create a photon with different amplitudes in the two bins one can use a different splitting ratio in the first beamsplitter. To measure the state in the $|s\rangle$ and $|l\rangle$ bases it is enough to record their times of arrival. To measure in the superposition basis it is necessary to use another interferometer and "reverse" the operation (see Figure 10.34). On exit from the analysis interferometer, the photon will be distributed among three time bins. The first time bin clearly corresponds to a photon that has taken a short – short path. The last time bin equally clearly corresponds to a long – long path. However, the middle time bin has two contributions, that of a long – short path and that of a short – long path, and the two are indistinguishable, thus providing a projection onto the basis where the photon is in a superposition of the two time bins.

One disadvantage of this measurement scheme is that we only project onto the superposition basis with 50% probability. This could in principle be addressed by using active switches instead of the passive interferometers. In such a scenario, the early pulse would be directed towards the long path, and the late pulse through the short path. Both of them would arrive at the exit beamsplitter simultaneously, thus eliminating the satellite peaks.

10.4.1.4 *Qubit tomography and the good old Stokes parameters*

It has been known for a long time (Stokes, 1856) that the state of polarization of light is fully characterized by four quantities, the Stokes parameters. These are defined in relation to the measured quantities used to estimate them and are often grouped together into a vector form.

$$S_0 = I_t \tag{10.310}$$

$$S_1 = I_H - I_V \tag{10.311}$$

$$S_2 = I_{45°} - I_{-45°} \tag{10.312}$$

$$S_3 = I_{C+} - I_{C-}. \tag{10.313}$$

The first parameter is just the total intensity of the light. The others are, respectively, the difference in intensities in three orthogonal polarization basis. For example S_1 would be determined by the measuring the intensity of H and V polarized light and taking the difference. Equivalently for the S_2 and S_3 in the 45° and circular polarization basis. These four parameters form a complete basis of the space of all possible polarization states. It is also common to use a reduced normalized representation such that $\vec{S}_r = \frac{1}{S_0}(S_1, S_2, S_3)$. Using the reduced Stokes vector, any state of light can be visualized as a point in the Poincaré sphere.

The Stokes parameters are convenient because they take easily measurable quantities and reduce them to a minimal set that completely describes the light polarization; a corresponding experimental scheme is shown in Figure 10.35, where a fraction of the original light is sent onto various measurement branches, each corresponding to one of the three Stokes parameters. These parameters are useful and conveniently based on a physical measurement, however, there is nothing fundamental about them. It is possible to chose a different set of measurements and come up with an equally good equivalent representation.

Going back to qubits, consider now an arbitrary state of polarization written as a density matrix,

$$\rho = \begin{pmatrix} \rho_{HH} & \rho_{HV} \\ \rho_{VH} & \rho_{VV} \end{pmatrix} \quad \text{with} \quad \text{tr}\rho = 1. \tag{10.314}$$

Quantum tomography tries to completely determine all the entries in the matrix so as to completely describe the quantum state. A not too deep examination shows that the number of independent parameters in the density matrix is the same as for the Stokes representation. Not surprisingly, the two representations are equivalent and related simply by a linear transformation. In this sense, classical polarimetry is

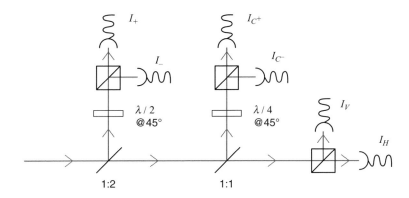

Fig. 10.35 A simple setup to measure the Stokes parameters.

equivalent to the state estimation of a single-qubit. In the case of qubits encoded in polarization of single photons, this equivalence is exact.

An uncomfortable point about the measurement scheme in figure 10.35 is that we record six quantities that get instantaneously reduced to four independent parameters; we should be able to do better than that (Řeháček, *et al.* 2004) and end up with a leaner alternative (Ling, *et al.* 2006).

When considering tomography schemes it is useful to keep some conditions in mind. The measurement should be universal and unbiased; that is, it should be able to characterize any possible input state, and do so with equal accuracy, or nearly so, for all of them. Additionally, when working with photons, we should remember that measurements are destructive, and that, given the efficiency limitations of detectors, we cannot use the failure of a photon to arrive as a projective measurement.

A measurement that fulfills these conditions is made up of four detector readings corresponding to the overlap of the unknown Stokes vector with four non-coplanar vectors b_j that define a tetrahedron in the Poincaré sphere (see figure 10.36). Each measurement operator B_j can be written as

$$B_j = \frac{1}{4}(\vec{b_j} \cdot \vec{\sigma}),$$ (10.315)

where $\vec{\sigma} = (\sigma_0, \sigma_1, \sigma_2, \sigma_3)$, σ_0 being the unit matrix and $\sigma_{1,2,3}$ the Pauli matrices. Figure 10.36 shows a possible experimental realization of these measurement operators. The average intensity falling on detector b_j is denoted as I_j. Expectation values of the tetrahedron operators are related to detected intensities as

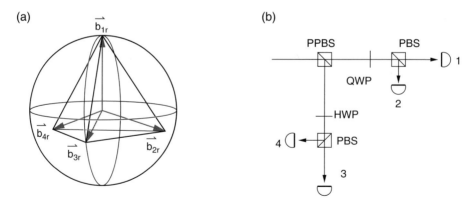

Fig. 10.36 Efficient polarimetry. (a) Polarization measurements are carried out corresponding to four measurement states $\vec{b_j}$, which form a tetrahedron on the Poincaré sphere. (b) Experimentally, four photon-counting detectors are used, each associated with the outcome of one of the measurements. The light is divided between the four detectors by a partially polarizing beamsplitter (PPBS), half- (HWP) and quarter- (QWP) wave plates, and polarizing beamsplitters (PBS).

$$\frac{I_j}{I_t} = \langle B_J \rangle = \frac{1}{4}(\vec{b_J} \cdot \vec{S}), \qquad \text{with} \qquad I_t = \sum_{j=1}^{4} I_j. \qquad (10.316)$$

A bit of manipulation allows us to relate the fractional intensities $\vec{I}$ directly to the components of the density matrix written in a vector form,

$$\vec{\rho} = \frac{1}{2}\Gamma_1 \Pi^{-1} \cdot \vec{I} = T \cdot \vec{I}. \qquad (10.317)$$

where $\vec{I} = \Pi \cdot \vec{S}$ and $\Gamma_1 = (\vec{\sigma_0}, \vec{\sigma_1}, \vec{\sigma_2}, \vec{\sigma_3})$. The matrix Π is sometimes referred to as the instrument matrix, as its exact values depend on the details of the experimental setup and will be adjusted during calibration. The relation between the fractional intensities and the entries in the density matrix is summarized in the so-called tomography matrix T.

10.4.2 Multiphoton stuff

Photons as single-qubits are only of moderate interest, the description is mostly a reformulation of phenomena that have been known in optics for more than 100 years into the language of quantum mechanics. The really fascinating effects present themselves when considering multiphoton states. In particular, those that arise from superpositions of multiphotons states, i.e. entangled states. Quantum physics allows the existence of states that cannot be described completely just in terms of their constituent parts. It was realized early on (Einstein, *et al.* 1935) that the existence of these states challenged the way we understood physical phenomena and the underlying physical theories accounting for them.

10.4.2.1 *Entangled photon pairs*

The theoretical implications of the existence of entangled states did not result in an experimental push to produce them until the formulation of Bell inequalities (Bell, 1964). Until that point, it was understood that entanglement brought up philosophical questions about the underlying properties of our theories, but there was no experimentally relevant measurement that could discriminate between the options presented. With the formulation of Bell's theorem this changed radically. There was now an experiment that could be used to resolve the issues in the EPR paradox. The first step in any experiment of this type was to prepare a pure entangled state of two particles – photons seemed very promising, since the measurements could be carried out independently, and even in space-like separated settings.

10.4.2.2 *Atomic cascades*

In this subsection, we illustrate how we can produce pairs of entangled photons using atomic cascades. First, let us consider the following three energy levels in the calcium atom used in the experiment by Aspect, et al. (Aspect, *et al.* 1981).

First, photon pairs were generated by a cascade decay from an excited atomic state as shown in Figure 10.37. Similarly to the single-photon sources discussed earlier, this

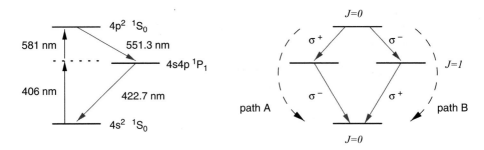

Fig. 10.37 The relevant energy levels of a calcium atom used to generate polarization entangled pairs of photons from an atomic cascade in Ca, and the two indistinguishable paths for a radiative decay of the upper excited state.

leads to photons that are well localized in time. The cascade decay ensures that there is a strong temporal correlation between the two photons as well.

The main idea to prepare an entangled state between this photon pair in a polarization degree of freedom is to consider the conservation of angular momentum for the complete decay process. As shown in Figure 10.37, there are two possible paths for the decay to proceed, each involving different magnetic orientation states in the intermediate level. Each partial emission process conserves the angular momentum, so decay via path A leads first to emission of a σ^--polarized photon for the first stage, and a σ^+-polarized one for the second decay. For path B, the polarizations change accordingly.

The atom is initially and after the cascade in a $J = 0$ level, which does not allow to storing of any angular momentum information about the decay path. Since it is therefore impossible in principle to know which path the decay process actually has taken, the two photons are in an entangled state

$$|\psi^-\rangle = \frac{1}{\sqrt{2}}(|\sigma^+\rangle_1|\sigma^-\rangle_2 - |\sigma^-\rangle_1|\sigma^+\rangle_2). \qquad (10.318)$$

In a practical experiment, a fraction of the emitted photons was captured into two opposite directions, and photon pairs were identified by a coincidence measurement with polarization analyzers under various angles.

After many hours of measurement time, the observed polarization correlations between the photon pairs violated the Bell inequality by 5 standard deviations.

10.4.2.3 *Hong–Ou–Mandel interference in parametric down-conversion*

The observation of correlated photon pairs by parametric down-conversion by (Burnham and Weinberg, 1970) suggested that this process is able to deliver correlated photon pairs as well. In order to prepare this photon pair in an entangled state, it was also necessary to have two completely indistinguishable photon-pair generation processes, similar to the different orientation paths for the atomic cascade decay.

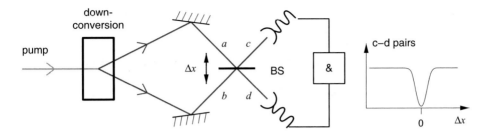

Fig. 10.38 Hong–Ou–Mandel experiment on the indistinguishability of photons generated in parametric down-conversion. Down-converted photons are superimposed on a beamsplitter BS with an adjustable delay Δx, and photon pair coincidences between the two output ports are recorded.

An important step to demonstrate this indistinguishability was presented in a paper in 1987 (Hong, *et al.* 1987) with a deceptive title: "Measurement of sub-picosecond time intervals between two photons by interference". The paper considered a scenario in which two photons enter the two input ports (a and b) of a 50:50 beamsplitter (see Figure 10.38). Classically, we would expect a binomial distribution of the possible outcomes. That is we expect that 25% of the time both photons exit through port c, 25% through port d and 50% of the time the photons are distributed to different ports. However, the quantum behavior is very different.

In terms of operators for generating removing photons from the vacuum, we can write the action of the beamsplitter as

$$\hat{a} = \frac{1}{\sqrt{2}}\left(\hat{c} + \hat{d}\right) \quad , \quad \hat{b} = \frac{1}{\sqrt{2}}\left(\hat{c} - \hat{d}\right). \tag{10.319}$$

This implies that the two photons will never exit through different ports. A very simple way to see this is to express a two-photon state in modes a, b in terms of creation operators acting on the vacuum, and then reformulate the state in terms of creation operators on modes c, d using the operator transformation rule eq. (10.319):

$$|1_a; 1_b\rangle = \hat{a}^\dagger \hat{b}^\dagger |0\rangle$$

$$= \frac{1}{2}\left(\hat{c}^{\dagger 2} - \hat{d}^{\dagger 2}\right)|0\rangle = \frac{1}{\sqrt{2}}\left(|2_c; 0_d\rangle - |0_c; 2_d\rangle\right). \tag{10.320}$$

This can be interpreted such that two photon interference effect gives complete cancellation of an outcome of a contribution $|1_c; 1_d\rangle$ leading to coincidence detection events. This simple description in terms of single modes needs to be completed to describe real systems: Each photon is a wave packet, e.g. described in the form of eq. (10.219). If the wave packets in the input modes do not perfectly overlap, the transfer relation eq. (10.319) for the input/output modes needs to be modified e.g. to include creation operators $\hat{a'}^\dagger$ that can take care of the part of wave packet mode a that does not overlap with the wave packet mode in b after superposition in the beamsplitter. The overlap of the two wave packets that determines the degree of cancellation.

The right part of Figure 10.38 a sketch and a corresponding experimental trace of this so-called Hong–Ou–Mandel dip. The H–O–M dips allow us to judge the degree of indistinguishability of two photon wave packets and has become a common tool in photonic quantum information.

10.4.3 Entangled photon pairs from spontaneous parametric down-conversion

The indistinguishability of the photons localized in time that originate from parametric down-conversion as demonstrated in the Hong–Ou–Mandel experiments opened the path to generate entangled photon pairs via this process. The additional element needed to arrive at a photon pair were two indistinguishable processes. In the following, we highlight a few approaches to this problem.

10.4.3.1 *"Energy–time" entanglement*

Apart from the polarization entangled photon pairs generated in cascade decays, one of the early suggestions was related to what is now referred to as time-bin qubits, and was proposed still with an atomic cascade as a photon pair source (Franson, 1989). The indistinguishability of two photon-pair generation processes there comes from the fact that it is not known when a particular pair-creation process takes place assuming the coherence length of the excitation light initiating the photon pair is long enough to not allow us to infer when it happened.

A SPDC-based version of the original proposal is shown in Figure 10.39. Each of two modes of target photons emanating from the non-linear optical crystal are sent into asymmetric Mach–Zehnder interferometers with a path-length difference

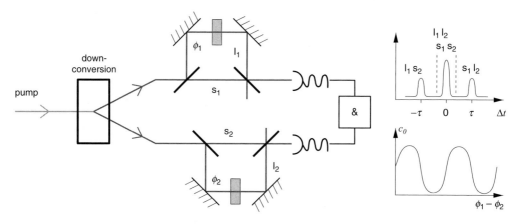

Fig. 10.39 Observation of photon pairs that are entangled in time bin qubits. The detection post-selects on coincidences between the two detectors for a time difference $\Delta t = 0$, where both short/long-path combinations become indistinguishable.

corresponding to a time delay that can be resolved electronically (typically a few ns) such a configuration is referred to as a "Franson interferometer". When looking for coincidences between the photodetectors at the output ports of the loops, the corresponding pair correlation function $g^{(2)}(\Delta t)$ has three distinct peaks, corresponding to the path combinations $s_1 l_2$, $s_1 s_2$ or $l_1 l_2$, and $l_1 s_2$. If the path-length differences of the two asymmetric MZI is the same within the coherence length of the down-converted light (typically a few $100\,\mu m$ for common down-conversion sources), and if there is no possible indication of the pair-generation time, then the possibilities $s_1 s_2$ and $l_1 l_2$ are indistinguishable. The necessary long coherence time for the pump mode implies a very narrow frequency distribution of the pump, or equivalently a well-defined energy of the pump photons – which probably is the reason for the otherwise not so obvious choice of the common term "energy – time entanglement" for this idea.

By restricting the observation to photon pairs to a time difference $\Delta T = 0$, simply by carrying out a coincidence selection with a narrow time window, the observed pairs can be thought of as having been in a time-bin entangled state. The entanglement, i.e. coherent superposition of the two early/late photon-generation processes, can be verified by modulating the phases in both Mach–Zehnder interferometers. The detected coincidence pair rate c_0 then varies sinusoidally only according to the difference of the two phase shifts ϕ_1 and ϕ_2, and could be used to test a Bell inequality.

An experiment with almost the same scheme has been published by the Geneva group (Brendel, *et al.* 1999), but with a pulsed source. This pulsed source has a very short coherence time; in order to ensure the indistinguishability of the two photon-pair generation processes – otherwise the two options $s_1 s_2$ and $l_1 l_2$ could be distinguished by looking at the detection time – the source path also had to include a Franson interferometer, preparing a coherent superposition of two pump pulses.

While preparing photons in a time-bin entangled state is not very demanding to the photon-pair generation process, the asymmetric Mach–Zehnder interferometers can be a substantial technical challenge, as they have to kept stable within a fraction of the optical wavelength for a path-length difference on the order of one meter to ensure proper electronic selection of the corresponding paths. Such a stability usually can only be maintained with relatively elaborate active stabilization schemes.

10.4.3.2 *Polarization entanglement from type-II non-collinear SPDC*

Technically much less demanding than time-bin qubits is the encoding into polarization states. An entangled photon-generation scheme utilizing this degree of freedom was suggested and experimentally demonstrated (Kwiat, *et al.* 1995). For a long time, this was probably one of the most widely used schemes for generation of polarization entangled photon pairs.

The basic idea there explores the angular dispersion properties in the birefringent conversion materials, and uses a non-collinear arrangement of pump and target modes in a type-II phase-matching configuration. The geometry of pump and target directions and polarizations is shown in Figure 10.40(a). For a pump mode with a fixed wavevector k_p, there is a wide spectrum wavevectors of signal and idler to meet

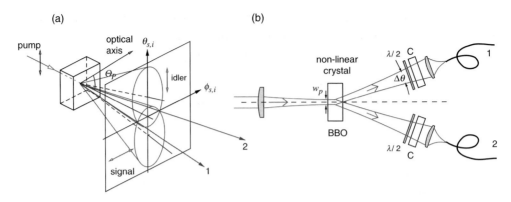

Fig. 10.40 Type-II non-collinear parametric down-conversion following (Kwiat, et al. 1995).

phase matching conditions and energy conservation. For a fixed target frequency, the possible combinations of signal- and idler modes form two cones, which – for type-II phase matching – have different main axes. For a proper choice of crystal orientations, these two cones intersect for two directions (labelled 1 and 2). Parametric down-conversion light collected into these two directions cannot be clearly identified as being either the signal (ordinary polarization) or idler (extraordinary polarization. Transverse momentum conservation now requires that if a e-polarized photon is present in direction 1, the o-polarized twin photon from the same conversion process must be present in direction 2, and vice versa.

To make sure that the two corresponding down-conversion possibilities leading to a photon pair in modes 1 and 2 are really indistinguishable, a residual distinguishability due to the birefringence in the conversion crystal needs to be removed, which would allow us to infer the ordinary/extraordinary information from the delay between the two photons. This "longitudinal walk-off" can be compensated by birefringent media with half of the total retardation as the conversion crystal; often crystals made from the same material are used, preceded by half-wave plates to rotate the polarization before the modes enter the compensators as shown in Figure 10.40 (b). This also partly reduces the effect of a transverse walk-off, which is present since the crystals are typically used in a critical phase-matching scenario.

Such a source geometry allows us to collect the down-converted light efficiently into single-mode optical fibers, which for a proper mode matching to the dispersion properties of the conversion crystal take care of the spectral filtering for the target modes (Kursiefer, *et al.* 2001). With such an arrangement, a reasonably high brightness (around 1000 observed pairs per mW pump power for a 3- mm thick BBO crystal) can be achieved. The pair-collection efficiency, quoted as observed pair events to single events (where one of the photons was not detected) for such a source is commonly around 30%, which would translate into a collection efficiency of 60% if a detector efficiency around 50% is assumed. For a cw- pumped source of this type, the desired singlet state $|\Psi^-\rangle = 1/\sqrt{2}(|H_1V_2\rangle - |V_1H_2\rangle)$ can be prepared with fidelities over 99.5%.

Such a geometry can also easily be used for generating pairs with ultrashort pump pulses, where the coherence length of the pump is on the order of the coherence length of the target photons. Such a pumping scheme is used for a large number of experiments where more than two photons needed to be generated.

10.4.3.3   *Polarization entanglement from type-I SPDC*

A somewhat simpler geometry to generate polarization-entangled photon pairs relies not on two different decay processes in one conversion crystal, but generates the two components with different crystals (Kwiat, *et al.* 1999). A schematic of this configuration is shown in Figure 10.41.

By choosing crystals with type-I phase matching, down-converted photon pairs are generated with the same polarization. The second decay process necessary to form an entangled state is provided by a second crystal located directly on top of the first one, but rotated by 90°. For collection into target modes 1,2 that are non-collinear with the pump beam, one needs to ensure that there is no distinction in any degrees of freedom from which of the crystals the pair emerged. Such a selection can either be done by spatial filtering, or by using single mode optical fibers, which efficiently remove any spatial information. This scheme is somewhat simpler than the type-II configuration discussed above, since it does not need any birefringence compensation as long as the coherence length of the pump field is larger than the physical extent of the two crystals. The photon pair is generated in a state

$$|\Psi\rangle = |H_1 H_2\rangle \cos\phi + |V_1 V_2\rangle \sin\phi, \qquad (10.321)$$

where ϕ indicates the orientation of the polarization of the pump with respect to the extraordinary polarization for a given crystal. This pair state different from a singlet state, but can easily be converted into one by applying local operations, i.e. polarization transformations on one of the modes only. This scheme has therefore the advantage to generate non-maximally entangled states, while the type-II source discussed before has by construction a fixed ratio between H and V components.

Another advantage of this geometry is that for type-I phase matching in BBO, the material commonly used for this process, a larger optical non-linearity can be used

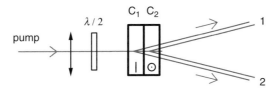

Fig. 10.41 Polarization-entangled photon-pair source based on two crystals C_1, C_2 cut for type-I phase-matching conditions, tilted by 90° with respect to each other. The indistinguishable decay paths correspond to a pair generation in the first or second conversion crystal. By adjusting the pump polarization with the hal-wave plate $\lambda/2$, the balance between the two components $|HH\rangle$ and $|VV\rangle$ can be adjusted.

than in type-II phase matching. Since this enters in the source brightness quadratically, these sources have the potential of being intrinsically brighter than sources based on type-II phase matching.

Recently, a very bright source based on the same idea, but using a collinear arrangement of pump- and target modes was reported (Trojek and Weinfurter, 2008). In this collinear configuration, the distinction between the two target modes is made by lifting the commonly used frequency degeneracy between the two target modes, and the photon pairs are separated with a wavelength-division multiplexer. The collinear conversion geometry allows also for a very good mode overlap, resulting not only in a high brightness, but also into a pair/single ratio of up to 39%; correction with the usual detector efficiency of around 50% suggests that this source has an extremely high pair collection or single-photon heralding efficiency, reaching up to 80%.

10.4.3.4   *Sagnac geometry*

Another approach to prepare polarization-entangled photon pairs in a collinear geometry makes use of a type-II phase matching process, but uses two different emission directions as the source for indistinguishable photon pairs (Kim, *et al.* 2006). A schematic of the experimental setup is shown in Figure 10.42.

The conversion crystal has to be pumped from two directions, and to make the two pair processes indistinguishable, the coherence length of the pump must be longer than the path-length difference; ideally it has a coherence length exceeding that of the crystal or a few crystal lengths.

Each of the two pair-generation processes leaves a H- and a V- polarized photon propagating in the same direction. In order to arrive at a polarization-entangled state in two target modes 1 and 2, the polarization of pair photons from one direction is rotated by 90°, and then light from both conversion directions is combined on a

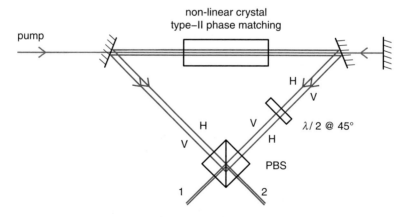

Fig. 10.42 Type-II parametric down-conversion in a Sagnac configuration. The two decay paths correspond to photon pairs generated in both directions in a collinear configuration; the pairs are combined via a beamsplitter such that they form a polarization-entangled state.

polarization beamsplitter. This distributes the photons from each conversion process onto the two output modes, resulting in a singled state for the photon pair.

Such sources have been implemented with periodically poled potassium titanyl phosphate (PPKTP), which leads to a very narrow spectral distribution of the photon pairs. An extremely high brightness of 28 000 detected pairs per mW pump power for a 25-mm long crystal has been reported for such a source (Fedrizzi, *et al.* 2007) after coupling into optical fibers, together with a fidelity of the targeted state comparable with the other bright sources described above. For this long crystal, a spectral bandwidth of 0.4 nm has been reported.

Such down-conversion sources have actually a high enough spectral brightness that interactions with atoms have been observed.

10.4.4 Multiphoton tomography

So far we have discussed the preparation of photonic qubit states – limited to single- and two-photon states, but using higher-order conversion processes, entangled states between up to 8 photons have been prepared.

Apart from techniques to prepare complex multiphoton states, we also need some tools to analyze such states. In this last section, we will visit a few common methods for state analysis if more than one photonic qubit is involved. While the techniques can be mapped onto most implementations of photonic qubits, here we focus on implementations with polarization.

10.4.4.1 Standard tomography

Following the same line of argument as in Section 10.4.1.4, we need to come up with a set of measurable quantities that can be used to determine all independent parameters necessary for a complete state description. A pure two-photon polarization state can be written as a linear combination of four product polarization states (we use H and V as our "computational basis"):

$$|\Psi_{12}\rangle = \alpha_{HH}|H_1 H_2\rangle + \alpha_{HV}|H_1 V_2\rangle$$

$$+\alpha_{VH}|V_1 H_2\rangle + \alpha_{VV}|V_1 V_2\rangle, \quad \text{with} \quad \sum_{i,j=H,V} \alpha_{ij}^2 = 1. \quad (10.322)$$

The density matrix of a two-photon polarization state can consequently be written as a 4×4 matrix in the same basis, and has 15 independent parameters (assuming normalization). It is clear from this that it is not enough to do polarimetry on the individual photons and just combine those results (3 free normalized Stokes parameters for each photon). This, in fact, is a reason why quantum information promises to be a powerful tool with not very complex systems.

To arrive at a sensible state description, we need to perform joint polarization measurements to fully account for the possible correlations present in the state. One way of doing this is to take the same basic measurements that are used for single-photon tomography and combine them. The traditional measurement scheme for the stokes parameters with a measurement device as shown in Figure 10.35 would require the recording of $6 \times 6 = 36$ detector correlations. A more sophisticated way of carrying out these measurements, and to reconstruct the photon pair state from them is given in (James, *et al.* 2001)

There are a few issues with this procedure. One problem with this is that it does not estimate all states equally well. Another is that the estimation sometimes produces unphysical states. Additionally, the measurement is overdetermined and therefore apparently inefficient.

10.4.4.2   *Efficient tomography*

Alternatively, we can use the minimal tomography setup based on tetrahedron POVMs. What we would like to do is extend the concept of the Stokes vector to a multiphoton state and extend the formalism presented in Section 10.4.1.4 to an arbitrary number of photons.

The simplest multi-photon system is a photon pair identified by the coincident time of arrival. In this scheme, each component photon is passed through a four-output polarimeter like the one described in Section 10.4.1.4. Given two polarimeters 1 and 2, each with four detectors b_{i_1} and b_{i_2}, respectively, $(i_1, i_2 = 0, 1, 2, 3)$, there are 16 possible coincidence combinations. Each coincidence rate is governed by an operator composed from the individual detectors' measurement operators. If we denote again the measurement operator of detectors b_{i_1} and b_{i_2} as B_{i_1} and B_{i_2}, and the coincidence count between them as c_{i_1, i_2}, we can express the coincidence rates as a linear function of a 2-photon polarization state vector S_2:

$$\frac{c_{i_1, i_2}}{c_t} = \langle B_{i_1} \otimes B_{i_2} \rangle = (\frac{1}{4}\vec{b_{i_1}} \otimes \frac{1}{4}\vec{b_{i_2}}) \cdot \vec{S_2}, \tag{10.323}$$

$$\text{with } c_t = \sum_{i_1, i_2=1}^{4} c_{i_1, i_2}.$$

Here, $\vec{S_2}$ is the Stokes vector equivalent for a 2-photon system and c_t is the total number of observed coincidences. We now have the set of measurement operators governing the coincidence pattern. The sixteen coincidences c_{i_1, i_2} can be written in column vector format $\vec{C_2} = (c_{1,1}, c_{1,2}, ..., c_{4,4})$. If we define the 2-polarimeter instrument matrix as Π_2, we obtain an instrument response:

$$\vec{C_2} = \Pi_2 \cdot \vec{S_2} \Leftrightarrow \vec{S_2} = \Pi_2^{-1} \cdot \vec{C_2}. \tag{10.324}$$

Thus, the 2-photon density matrix is given by:

$$\vec{\rho_2} = \frac{1}{2^2}\Gamma_2 \cdot \vec{S_2} = T_2 \cdot \vec{C_2} \tag{10.325}$$

Each column of Γ_2 is the product of two Pauli operators $\sigma_{i_1} \otimes \sigma_{i_2}$ $(i_1, i_2 = 0, 1, 2, 3)$ written in column vector format and T_2 is the complete tomography matrix for the 2-photon state.

Similarly, the procedure is generalized to states containing an arbitrary number of photons,

$$\vec{S_N} = \Pi_N^{-1} \cdot \vec{C_N}, \tag{10.326}$$

$$\vec{\rho_N} = \frac{1}{2^N} \Gamma_N \cdot \vec{S_N} = T_N \cdot \vec{C_N}. \tag{10.327}$$

Each row of the instrument matrix Π_N is given by $(\frac{1}{4}\vec{b_{i_1}} \otimes \frac{1}{4}\vec{b_{i_2}} ... \otimes \frac{1}{4}\vec{b_{i_N}})$ and each column of Γ_N is the product of N Pauli matrices $\sigma_{i_1} \otimes \sigma_{i_2} ... \otimes \sigma_{i_n}$ $(i_n = 0, 1, 2, 3$ and $n = 1, 2, ..., N)$.

A few more comments about multi-qubit tomography. The procedure just described uses the minimal number of measurements and thus is efficient in an experimental sense. It is also known to be optimal for single-qubits, but it is an open question whether this holds for multi-qubit states. We are only considering here methods that are "simple" and efficient from an experimental point of view. For example, we limit ourselves to individual measurements per photon rather than the in principle more powerful global POVMs for the simple reason that these are by themselves a challenge to implement. We have, however, extended our dictionary of available measurements beyond naive projective measurements, and expanded into simple POVMs. It is interesting to note that the classical optics community converged in very similar protocols for polarimetry (Azzam and De, 2003).

10.4.5 Bell-state analysis

The four Bell states are a complete orthogonal basis of the state of two qubits:

$$|\Psi^{\pm}\rangle = \frac{1}{\sqrt{2}} (|H_1 V_2\rangle \pm |V_1 H_2\rangle), \qquad |\Psi^{\pm}\rangle = \frac{1}{\sqrt{2}} (|H_1 H_2\rangle \pm |V_1 V_2\rangle). \tag{10.328}$$

They play a fundamental role in protocols such as teleportation, entanglement swapping and many others. None of these states can be identified by simply doing single-qubit measurements; each of them would result in a completely random result. A Bell-state measurement is now defined as a measurement, where an arbitrary two-photon state is projected onto the four Bell states, and the measurement result is one out of four values, indicating one of the four states in eq. 10.328.

Being orthogonal states, in principle it should be possible to distinguish the unambiguously. Unfortunately, it can be be shown that it is not possible to distinguish all four Bell-states using only linear optics. This problem is equivalent to the difficulty of implementing a universal two-qubit gate with linear optics (Sleator and Weinfurter, 1995), since a Bell state measurement can be implemented with a CNOT gate and a (cheap) one-qubit Hadamard gate.

10.4.5.1 Partial-Bell state analysis

The basic idea behind a partial Bell state analysis makes use of the fact that both photons of pair entering the input ports of a normal 50:50 beamsplitter will leave the beamsplitter at different output ports if the photon pair is in a $|\Psi^+\rangle$ state, as shown in Figure 10.43(a). This can be seen by extending the simple beamsplitter matrix eq. (10.304) to a 4×4 matrix for both polarization modes. One caveat for most beamsplitters is that they change the helicity of the polarization upon reflection;

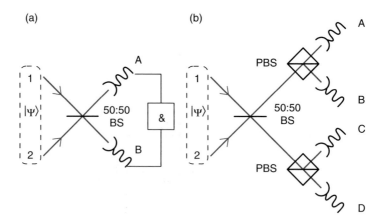

Fig. 10.43 Simple partial Bell-state analyzer, using the symmetry of photons entering a symmetric beamsplitter (a). For most physical beamsplitters, the photon pair gets distributed onto the two detectors A,B only for $|\Psi_{12}\rangle = |\Psi^+\rangle$. Another Bell state $|\Psi^-\rangle$ can be identified if each output port is further analyzed with a polarizing beamsplitter (b).

this means that the transfer matrix has the phase shift between the two reflections at the other ports. We leave it to the reader to generate the full transfer matrix, and to work out its action on various Bell states.

Most of the proof-of-principle experiments implementing a Bell-state analysis only perform this relatively simple part by looking for coincidences at detectors A and B behind beamsplitter to identify $|\Psi^+\rangle$ (Mattle, *et al.* 1996).

A simple extension of this method allows us to identify one more Bell state if the configuration shown in Figure 10.43 is used, where each of the output ports of the beamsplitter is followed by a polarizing beamsplitter, leading to a unique detector pattern in case both a H- and V- polarized photon was present. The identification of the Bell state can be obtained from the following lookup table:

detector coincidence	state
A–C or B–D	$|\Psi^+\rangle$
A–B or C–D	$|\Psi^-\rangle$

There have been several attempts to get around this limitation. Some invoke non-linear optical effects in detectors (Knill, *et al.* 2001), probabilistic identification (Zhao, *et al.* 2005) or heralded detection (Pittman, *et al.* 2002).

10.4.5.2 Complete Bell-state analysis

Surprisingly, there are ways to get around this limitation (Kwiat and Weinfurter 1998; Schuck, *et al.* 2006) if one takes advantage of additional entanglement that is present "for free" in the usual PDC sources used to produce Bell states.

When preparing entangled states via PDC (see Section 10.4.2.1), the photons are not only in a polarization entangled state but also in an energy – time entangled state,

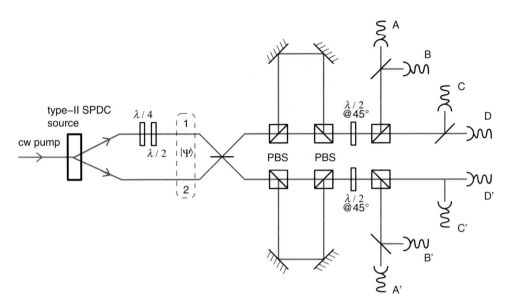

Fig. 10.44 Setup to perform a complete Bell-state measurement. This scheme relies on the fact that a polarization-entangled photon pair generated in a cw-pumped SPDC process exhibits also a time-bin entanglement. The larger Hilbert space this pair is embedded in then allows us to carry out a complete Bell-state analysis with linear optical elements.

i.e. they were generated at an unknown but identical time. This property, sometimes called hyper-entanglement, can be exploited to implement a full and deterministic Bell-state analyzer using only the tools of linear optics. Consider the setup in Figure 10.44. A PDC source produces pairs that are then combined into a beamsplitter. At zero delay, H–O–M interference will result in the state $|\psi^+\rangle$ split between the two output modes, thus can be uniquely identified. The other three states will send the two photons together either through a' or b'. Of these, $|\psi^-\rangle$ has its two photons in orthogonal polarizations. By sending this state into a delay line based on a pair of PBSs the two photons will be split deterministically with the V photon taking the long path (l) and the H photon taking the short path (s). This state can be uniquely identified by the different time of arrival given by the delay. Finally, the $|\phi\rangle$ states both have their photons in superpositions of identical polarizations with only the phase differentiating them. When going through the delay line the two photons will stick together and take either the long or the short path depending on their polarization, resulting in a state,

$$|\phi^\pm\rangle = \frac{1}{\sqrt{2}}(|H_s, H_s\rangle \pm e^{2i\phi}|V_l, V_l\rangle). \tag{10.329}$$

When $\phi = 0$, these two states can be easily distinguished by analyzing in the 45° basis.

$$|\phi^+\rangle = \frac{1}{\sqrt{2}}(|+45°, +45°\rangle + |-45°, -45°\rangle) \tag{10.330}$$

$$|\phi^-\rangle = \frac{1}{\sqrt{2}}(|+45°, -45°\rangle + |-45°, +45°\rangle). \tag{10.331}$$

Notes

1. This means that $\mathscr{B}(\mathbf{k}, t) = \mathscr{B}_\perp(\mathbf{k}, t)$.
2. Assume the region is far away from the source and we can neglect the contribution of the source, which is the longitudinal part of the energy.
3. The order of multiplication is retained for quantization purposes.
4. No quantization has been performed yet, even though the expression for $\mathscr{E}_\omega$ contains $\hbar$. This is a purely arbitrary choice of a normalization factor for the normal coordinates.
5. For simplicity, we consider this to be the same in all directions.
6. Keep in mind that this is not a complete set of modes, it just covers the ones with the lowest cutoff frequencies for $a > b$. There is also another field mode type, the TM modes. See, e.g., Jackson (Jackson, 1999) for a comprehensive list of modes.
7. For infinitesimally small pinholes, α is a constant.
8. This expression includes Fresnel and Fraunhofer diffraction.

References

Aspect, A, Grangier, P, and Roger, G (1981). Experimental tests of realistic local theories via Bell's theorem. *Phys. Rev. Lett.*, **47**, 460–463.

Azzam, R and De, A (2003). Optimal beam splitters for the division-of-amplitude photopolarimeter. *J. Opt. Soc. Am. A*, **20**, 955.

Bell, J S (1964). On the Einstein–Podolsky–Rosen paradox. *Physics*, **1**, 195–200.

Brendel, J, Gisin, N, Tittel, W, and Zbinden, H (1999). Pulsed energy-time entangled twin-photon source for quantum communication. *Phys. Rev. Lett.*, **82**, 2594.

Burnham, D C and Weinberg, D L (1970, july). Observation of simultaneity in parametric production of optical photon pairs. *Phys. Rev. Lett.*, **25**(2), 84.

Cohen-Tannoudji, C, Dupont-Roc, J, and Grynberg, G (1997). *Photons and Atoms: Introduction into quantum electrodynamics*. Wiley Interscience New York.

Einstein, A (1905). Über einen die Erzeugung und Verwandlung des Lichtes betreffenden heuristischen Gesichtspunkt. *Ann. d. Phys.*, **17**, 132.

Einstein, A, Podolsky, B, and Rosen, N (1935). Can quantum-mechanical description of physical reality be considered complete? *Phys. Rev.*, **47**, 777.

Fedrizzi, A, Herbst, T, Poppe, As, Jennewein, T, and Zeilinger, A (2007). A wavelength-tunable fiber-coupled source of narrowband entangled photons. *Opt. Exp.*, **15**, 15377.

Franson, J D (1989). Bell inequality for position and time. *Phys. Rev. Lett.*, **62**, 2205–2208.

Gerrard, A and Burch, J M (1994). *Introduction to matrix methods in optics.* Dover Mineola, N.

Glauber, R (1963). Coherent and incoherent states of the radiation field. *Phys. Rev.*, **131**, 2766–2788.

Hanbury-Brown, R and Twiss, R Q (1954). A new type of interferometer for use in radio astronomy. *Philo. Mag.*, **45**, 663.

Hanbury-Brown, R and Twiss, R Q (1956). A test of a new type of stellar interferometer on Sirius. *Nature*, **178**, 1046.

Hanbury-Brown, R and Twiss, R Q (1957). Interferometry of the intensity fluctuations in light. i. basic theory: the correlation between photons in coherent beams of radiation. *Proc. Roy. Soc.*, **A242**, 300.

Hecht, E and Zajac, A (1997). *Optics.* Addison Wesley Reading, MA.

Hertz, H R (1887). Ueber einen Einfluss des ultravioletten Lichtes auf die electrische Entladung. *Ann. Phys.*, **267**, 983–1000.

Hong, C. K., Ou, Z. Y., and Mandel, L. (1987). Measurement of subpicosecond time intervals between two photons by interference. *Phys. Rev. Lett.*, **59**, 2044.

Jackson, J D (1999). *Classical electrodynamics.* John Wiley & Sons New York.

James, D F V, Kwiat, P G, Munro, W J, and White, A G (2001). Measurement of qubits. *Phys. Rev. A*, **64**, 052312.

Kim, T, Fiorentino, M, and Wong, F N C (2006). Phase-stable source of polarization-entangled photons using a polarization sagnac interferometer. *Phys. Rev. A*, **73**, 012316.

Knill, E, Laflamme, R, and Milburn, G J (2001): A scheme for effecient quantum computation with linear optics. Nature **409**, 46.

Kursiefer, Cn, Oberparleiter, Ms, and Weinfurter, H (2001). High efficiency entangled photon pair collection in type ii parametric fluorescence. *Phys. Rev. A*, **64**, 023802.

Kwiat, P G., Mattle, Ks, Weinfurter, H, Zeilinger, An, Sergienko, A V., and Shih, Yanhua (1995). New high-intensity source of polarization – entangled photon pairs. *Phys. Rev. Lett.*, **75**, 4337.

Kwiat, P G, Waks, E, White, A G, Appelbaum, In, and Eberhard, P H (1999). Ultra-bright source of polarization-entangled photons. *Phys. Rev. A*, **60**, R773.

Kwiat, P G and Weinfurter, H (1998). Embedded Bell-state analysis. *Phys. Rev. A*, **58**, R2623.

Lamb, W E (1995). Anti-photon. *Appl. Phys. B*, **60**, 77–84.

Lenard, P (1902). Ueber die lichtelektrische Wirkung. *Ann. Physik*, **313**, 149–198.

Lewis, G N (1926). The conservation of photons. *Nature*, **118**, 874.

Ling, A, Lamas-Linares, A, and Kurtsiefer, C (2008). Absolute emission rates of spontaneous parametric down conversion into single transverse gaussian modes. *Phys. Rev. A*, **77**, 043834.

Ling, A, Pang, Soh Kee, Lamas-Linares, A, and Kurtsiefer, C (2006). Experimental polarization state tomography using optimal polarimeters. **74** 022309.

Mattle, K, Weinfurter, H, Kwiat, P G, and Zeilinger, A (1996). Dense coding in experimental quantum communication. *Phys. Rev. Lett.*, **76**, 4656–4659.

Millikan, R A (1916). Einstein's photoelectric equation and contact electromotive force. *Phys. Rev.*, **7**, 355–388.

Pittman, T B , Jacobs, B C, and Franson, J D (2002): Demonstration of Nondeterministic Quantum Logic Operations Using Linear Optical Elements. Phys. Rev. Lett. **88**, 257902.

Saleh, B E A and Teich, M C (2007). *Fundamentals of photonics*. John Wiley & Sons New York.

Schuck, C, Huber, G, Kurtsiefer, C, and Weinfurter, H (2006). Complete deterministic linear optics Bell state analysis. *Phys. Rev. Lett.*, **96**, 190501.

Sleator, T and Weinfurter, H (1995). Realizable univeral quantum logic gates. *Phys. Rev. Lett.*, **74**, 4087–4090.

Stokes, G G (1856). On the composition and resolution of streams of polarized light from diferent sources. *Trans. Cambr. Philos. Soc.*, **9**, 399.

Trojek, P and Weinfurter, H (2008). Collinear source of polarization-entangled photon pairs at non-degenerate wavelengths. *Appl. Phys. Lett.*, **92**, 211103.

Řeháček, J, Englert, Bd-G, and Kaszlikowski, D (2004). Minimal qubit tomography. *Phys. Rev. A*, **70**, 052321.

Weisskopf, V and Wigner, E (1930). Berechnung der natürlichen Linienbreite auf Grund der Diracschen Lichttheorie. *Z. Physik*, **63**, 54.

Wentzel, G (1926). Zur Theorie des photoelektrischen Effekts. *Zeitschr. Phys.*, **40**, 574.

Zhao, Z, Zhang, A N, Chen, Y A, Zhang, H, Du, J-F, Yang, T, Pan, J W (2005): Experimental Demonstration of a Nondestructive Controlled-NOT Quantum Gate for Two Independent Photon Qubits. Phys. Rev. Lett. **94**, 030501.

Zysset, B, Biaggio, I, and Guenter, P. (1992). Refractive indices of orthorhombic $KNbO_3$. I. Dispersion and temperature dependence. *JOSA B*, **9**, 380.